Uwe Mackenroth

Robust Control Systems

Springer

Berlin
Heidelberg
New York
Hong Kong
London
Milan
Paris
Tokyo

Uwe Mackenroth

Robust Control Systems

Theory and Case Studies

With 221 Figures

Springer

Professor Dr. Uwe Mackenroth
Fachhochschule Lübeck
University of Applied Sciences
FB Maschinenbau und Wirtschaftsingenieurwesen
Mönkhofer Weg 136-140
23562 Lübeck
Germany

ISBN 978-3-642-05891-2

Springer-Verlag is a part of Springer Science+Business Media
springeronline.com

Printed in Germany

Cover design: medio Technologies AG, Berlin
Printed on acid free paper 62/3020/M - 5 4 3 2 1 0

For my family:

Gabriela, Gisela and Julia

Preface

Control engineering is an exciting and challenging field within the engineering sciences, because by its very nature it is a multidisciplinary subject and because it plays a critical role in many practical technical applications. To be more specific: control theory brings together such different fields as electrical, mechanical and chemical engineering and applied mathematics. It plays a major role in engineering applications of all kinds of complexity. This may be a dc motor, a robot arm, an aircraft, a satellite, a power plant or a plant for the chemical industries such as a distillation column. On the other hand, the developments in the last two or three decades have shown that control engineering requires very solid skills in applied mathematics and therefore has become highly attractive for applied mathematicians, too.

The most fundamental idea of control theory is to change the dynamical behavior of a technical system (the "plant") by a device called the "controller" such that the dynamics gets certain desired properties. In almost all cases, the controller design is done by building a mathematical model, which can be emulated on a computer and serves as the basis for controller synthesis. It is obvious that the model represents the real system only up to a certain degree of accuracy. Despite this, the controller has to work for the real system as well as for the model; in other words, it has to be "robust" against errors in the mathematical model. Robustness has always played a key role in control theory, right from the beginning, but a considerable amount of theoretical research was necessary to find analysis and synthesis tools which work satisfactorily also for complex plants. In particular, this holds for plants with several inputs and outputs.

The intention of this textbook is to give a self-contained introduction to control theory which emphasizes the modern aspects concerning the design of controllers having prescribed performance and robustness properties. No prior knowledge about control theory is required; the most important facts from classical control are concisely presented in Chaps. 2–4. They are formulated in the light of the modern theory in order to get an early understanding of its most important ideas. The book is written at a graduate level for students of electrical or mechanical engineering sciences or of applied mathematics and may also serve as a reference for control practitioners.

During the last two decades, control theory has changed its character: although it is a well-established branch of the engineering sciences, it may also be viewed as a discipline of applied mathematics. This is a point which deserves some attention. Almost all control theory methods require some mathematics, and this is also true for the methods based on the Nyquist stability criterion presented in every undergraduate course, but for these the mathematical reasoning can be kept at a moderate level. Even for a complex plant, the role of the controller parameters is well understood and controller design can successfully be done step by step by an engineer who is experienced and has a good intuition.

For modern methods which use $\mathcal{H}_\infty$ optimization or the structured singular value μ the situation is more involved. Probably most people who see the conditions characterizing $\mathcal{H}_\infty$ suboptimal controllers for the first time will find them messy and hard to understand. In fact, there is no way to make these conditions plausible by intuitive engineering reasoning. The only way to accept them as natural and reasonable is to study the complete rigorous mathematical proofs. On the other hand, the application of the more advanced methods, where the weights for performance and the uncertainty structure are specified, requires again a solid engineering experience and intuition.

For the preparation of a textbook dealing with an advanced presentation of control theory, these considerations have some strong implications. In particular, there are only two possibilities concerning the proofs, namely to skip them all or to present them in full detail. Skipping them all would be an admissible way for students or practitioners who are only interested in applications, since the use of the methods, which means mainly to apply the commercial software, does not require a very detailed knowledge of the theory standing behind it. On the other hand, for readers who wish to get a deeper understanding of modern control theory, this possibility is unsatisfactory, and for this reason I have decided to present the modern theory completely as far as it is needed. A study of the theory requires a basic knowledge of linear algebra and analysis and in particular some familiarity with the mathematical way of thinking.

In the introduction it is shown how both kinds of readers can gain benefit from this book. In this context it must be emphasized that a large portion of the text is devoted to elementary examples and advanced case studies. The elementary examples answer questions like: How can a $\mathcal{H}_\infty$ or a μ controller for a simple second order SISO plant without or with uncertainty be compared with a conventional PID controller? The case studies are carried out in full detail and show how the modern methods can be applied to advanced problems. They make intensive use of MATLAB, in particular of the Control Systems Toolbox and the μ-Analysis and Synthesis Toolbox.

The main reason for writing this book was the following. There are many excellent textbooks which present control theory based on the Nyquist criterion and on elementary state space methods. On the other side, there are a lot of highly specialized books which report on recent developments described in the original literature. Books which present the modern theory in detail without being a research monograph or a research-oriented book are comparatively rare. It is in this sense that the present book should be understood. Its philosophy is to treat advanced theory only so far as commercial software is available in which the new algorithms are programmed.

The most fundamental theoretical parts of the book deal with $\mathcal{H}_\infty$ synthesis using two Riccati equations and robustness based on the structured singular value μ. They are written in the spirit of the famous monograph of Zhou, Doyle, and Glover [112]. A completely different approach to $\mathcal{H}_\infty$ synthesis is based on linear matrix inequalities (LMIs) and will also be presented here. A book treating $\mathcal{H}_\infty$ theory and robustness by LMIs in much greater detail is Dullerud and Paganini [37]. An (incomplete) list of further important books on advanced control theory in the sense discussed above, but with somewhat different intentions to that of this book, is Burl [17], Sanchez-Pena and Sznaier [80], Skogestad and Poslethwaite [92], and Trentelman, Stoorvogel, and Hautus [97].

Uwe Mackenroth
September 2003

Contents

Chapter 1

Introduction

1.1 Control Systems: Basic Definitions and Concepts

Suppose a physical system **G** with two inputs and two outputs is given. In the context of control systems, **G** is denoted as the **plant**. One of the inputs, the **control u**, can be prescribed, whereas the other, the **disturbance d**, is assumed to be unknown. Further, we suppose that one of the outputs, the variable **y**, can be measured (therefore **y** is denoted as a **measurement**). For the other output, the **variable to be controlled z**, a certain behavior is desired. In order to achieve this, a device called a **controller** is needed. The controller **K** has the measurement **y** and a **reference signal r** as inputs. The reference signal specifies the desired behavior of **z**. The output of the controller is the control **u**. The configuration of the described system is depicted in Fig. 1.1. Such a system is denoted as a **feedback system**.

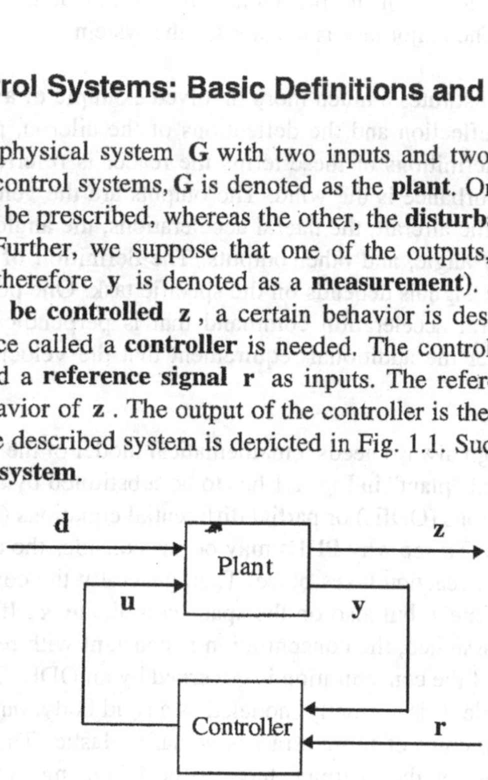

Fig. 1.1. Feedback system

In this context, the controller may be thought of as a mechanical or electrical device. If it is an electrical device, it may be analog or digital. In more complex applications, the controller is normally realized by a microprocessor.

Examples for plants.

1. DC motor. The input voltage is the control and a normally unknown moment acts as a disturbance. The variable to be controlled is the angular velocity ω of the shaft. The reference signal is the desired value for ω. The measurement may be ω or it may be ω together with the armature current. The controller can be realized as an analog or digital device.

2. Tank. Suppose a tank filled with a liquid is the plant. There are two valves, one at the inlet and one at the outlet of the tank. The setting of the inlet valve is the control and the setting of the outlet valve is the disturbance. The variable to be controlled is the tank level.

3. Inverted pendulum. In this example, a thin rod mounted on a motor-driven cart has to be balanced (cf. Sect. 11.1). The force exerted by the motor is the control and the angular deviation from the perpendicular position of the rod is the variable to be controlled. The major task is to stabilize the system.

4. Aircraft. An aircraft constitutes a much more involved example of a plant. The controls are the throttle deflection and the deflections of the aileron, rudder and elevator (for the precise definitions of these terms the reader is referred to Sect. 11.3.1). An important disturbance is the wind. The outputs are the velocity component in the direction of the aircraft, the lateral accelerations, the angle of attack, the sideslip angle, the roll angle, and other outputs. The definition of the measurements and the reference signals depends on the specific task. One possibility is to track a prescribed lateral acceleration command that is perpendicular to the planes of the aircraft under the additional requirement that the velocity remains constant.

Modern controller design always needs a mathematical model of the plant. This means that the block named "plant" in Fig. 1.1 has to be substituted by a system of ordinary differential equations (ODEs) or partial differential equations (PDEs) or a combination of both types. To see why PDEs may occur, consider the example of a tank in which a chemical reaction takes place. Then generally the concentration depends not only on the time t but also on the space coordinate $\mathbf{x}$. If a mixer is used and ideal mixing is assumed, the concentration is constant with respect to $\mathbf{x}$ and the time dependency of the concentration is governed by an ODE. Consider an aircraft as a second example. It is normally modeled as a rigid body, but this is not quite correct since the structure of the aircraft is actually elastic. There may be situations where vibrations of the airframe have to be taken into account and, again, PDEs come into play.

Since the methods described in this book are made for systems governed by ODEs, we cannot handle systems described by PDEs explicitly. This does not mean that such systems have to be excluded completely, because they often can be approximated by a system governed by ODEs with sufficient accuracy. Compare the missile case study (Sect. 14.3) as such an example.

Thus from now on the "plant" in Fig. 1.1 is a system of ODEs. In the overwhelming majority of cases this system is nonlinear. This is still too general, since our methods directly apply only to linear systems of ODEs with constant coefficients. At a first glance, this seems to be very restrictive, but, as will be explained now, in most cases nonlinear systems can be handled effectively by these methods. The reason for this is that the plant normally has only to be controlled in the vicinity of an equilibrium point. Such a point is a constant solution of the differential equation for constant inputs. As a very simple example consider the tank and suppose that the two valve positions are adjusted such that the tank level is constant. For an aircraft, trim positions are equilibrium points. Near an equilibrium point, the plant can be approximated by a linear system. It may be argued that the reference signal or a large disturbance could very well drive the state far away from the equilibrium point. For such a case it is possible to use a technique called gain scheduling or to interpret the nonlinearity as an uncertainty for the linear system. This will be explained later in the book. In particular we shall consider again the missile case study as an extreme of such a case.

With this discussion in mind, we may consider our "plant" as a linear system of ODEs with constant coefficients which is completed by two linear equations for the two output variables (all the quantities in capital letters are suitably dimensioned matrices):

$$\begin{aligned}\dot{\mathbf{x}} &= \mathbf{A}\,\mathbf{x} + \mathbf{B}_1\mathbf{d} + \mathbf{B}_2\mathbf{u}\\ \mathbf{z} &= \mathbf{C}_1\mathbf{x} + \mathbf{D}_{11}\mathbf{d} + \mathbf{D}_{12}\mathbf{u}\\ \mathbf{y} &= \mathbf{C}_2\mathbf{x} + \mathbf{D}_{21}\mathbf{d} + \mathbf{D}_{22}\mathbf{u}\,.\end{aligned}$$

This will be denoted as a **linear system**. Mathematically, the controller is also a linear system:

$$\begin{aligned}\dot{\mathbf{x}}_K &= \mathbf{A}_K\mathbf{x}_K + \mathbf{B}_{K1}\mathbf{r} + \mathbf{B}_{K2}\mathbf{y}\\ \mathbf{u} &= \mathbf{C}_K\mathbf{x} + \mathbf{D}_{K1}\mathbf{r} + \mathbf{D}_{K2}\mathbf{y}.\end{aligned}$$

Designing a controller means to find such a system which fulfills certain specifications. If the controller hardware is a microprocessor, these equations have to be programmed into the processor, but this step is not discussed in what follows.

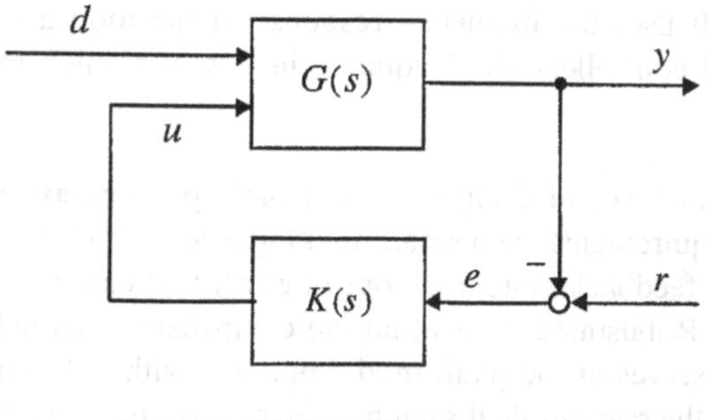

Fig. 1.2. SISO feedback system

We now discuss some basic ideas that will lead to a successful controller design and start with the DC motor as an example. Then $z = \omega$ and $y = \omega$, if we neglect that y is the measured ω which is corrupted by some measurement error. We choose the control error $e = r - y$ as the input of the controller. A similar concept may be applied to other **single input and single output** (**SISO**) problems, too. Then the feedback connection of Fig. 1.1 may be redrawn as depicted in Fig. 1.2. Herein, $G(s)$ and $K(s)$ are the transfer functions describing the (linear) plant and the controller.

Next, we need methods for determining $K(s)$. Therefore, it is necessary to formulate design objectives. Above all, the feedback system in Fig. 1.2 has to be **stable**. Stability implies that for constant inputs the output $y(t)$ has a limit for $t \to \infty$. Moreover, a minimum of performance is required: For constant inputs r, d, the control error $e(t)$ has to tend to 0 for $t \to \infty$.

Suppose that the plant $G(s)$ is stable and let the feedback law be purely proportional: $K(s) = k_P$ (with some constant k_P). If k_P is small enough, the feedback system will be stable, but the design objective concerning the control error will generally not be fulfilled. So we add integral action to the proportional controller by feeding back also the integrated control error. Then in a somewhat heuristic way we may argue that for a stable feedback system this integral becomes constant, which means that the control error tends to 0 for $t \to \infty$. It is plausible that more elaborate requirements for the feedback loop can be fulfilled if the control error is not only integrated but also differentiated. This leads to the so-called PIDT_1 controller, which can be written in the following form:

$$K(s) = k_P \frac{1 + T_N s + T_N T_V s^2}{T_N s(1 + T_0 s)}.$$

For SISO systems, this has been up to now a very popular conventional controller. This may be somewhat surprising since in Fig. 1.2 the controller is not restricted to be of order 2. In fact, any realizable (rational) transfer function of arbitrary order such that the feedback system is stable is a candidate for a controller. One of the reasons for the popularity of the PIDT_1 controller is that there are simple synthesis methods for it, which basically require only pencil and paper. Moreover, with some skill the role of the controller parameters can be very well explained. The theoretical tool is the Nyquist stability criterion, which is the heart of classical control. It uses the frequency response of the loop transfer function $L = KG$. For general controllers, the Nyquist criterion offers no general way for designing them.

There are two main types of design goals, namely **performance** and **robustness**. Performance requirements are a small tracking error, a high bandwidth and a good damping of the feedback system, a control effort that can be realized by the actuators, and so on. Robustness means that the controller works not only for the linear system which serves as the plant model but also with only minor performance degradation for the real physical system. The real system always differs from the linear model. The main kinds of uncertainty are **parameter uncertainty** and neglected **high-frequency dynamics**. Consider as an example again an aircraft. Then the inaccuracies of the aerodynamic coefficients are the parameter uncertainties and the neglected elastic behavior of the airframe leads to high-frequency model errors. Another source of uncertainty is neglected cross-couplings in **MIMO (multiple input and multiple output)** systems. Normally, performance and robustness are contradictory design goals. For the classical PIDT_1 controller some basic performance and robustness properties can be guaranteed by establishing a certain behavior of the loop transfer function L (in terms of gain crossover frequency and gain margin).

Next, we sketch a different way for designing a controller. Again, let the DC motor be the plant. Then the (linear) mathematical model for the plant consists of a system of two differential equations for the angular velocity ω and the armature current i_A. Instead of feeding back only ω, a different approach is to take ω and i_A as the measurements. These two quantities are the states of the system. For a general plant with state vector $\mathbf{x}$, this idea works in the same way, even if the plant is a MIMO system. Then the control law based on this concept is given by

$$\mathbf{u} = \mathbf{F}\mathbf{x} \tag{1.1}$$

with a matrix $\mathbf{F}$. The idea is now that with such an approach the poles of the feedback system can be very well manipulated so that stability is always achievable. More precisely, under reasonable assumptions it can be shown that $\mathbf{F}$ can be found such that the closed-loop system has arbitrarily assigned poles. Perfect tracking can always be fulfilled by adding the integrated control error as an additional state. This kind of controller design was developed in the early 1960s to the 1970s under the name "state space methods".

A problem arises if not all states of the system are measurable. Then, under some mild assumptions, it is possible to estimate the system state from the measurement $\mathbf{y}$. This leads to the Luenberger observer. The control law (1.1) is now applied to the estimate $\tilde{\mathbf{x}}$. Using (1.1), such an observer can be written in the form:

$$\dot{\tilde{\mathbf{x}}} = \mathbf{A}_K \tilde{\mathbf{x}} - \mathbf{L}\mathbf{y}\,. \tag{1.2}$$

The equations (1.1), (1.2) now describe the controller. The matrix $\mathbf{A}_K$ can be written in terms of the plant matrices and of $\mathbf{F}$ and $\mathbf{L}$, so that the controller design consists in finding these two matrices.

How can this be done? As already mentioned, one way is pole positioning, which works also for the observer, but there are several design objectives, which cannot be taken into account by this method. For example, concerning $\mathbf{F}$ this is the control effort and concerning $\mathbf{L}$ this is the measurement error. The idea is now to formulate the design goals in such a way that a certain functional is minimized. For the state feedback law (1.1) this is

$$\int_0^\infty (\mathbf{x}^T(t)\mathbf{Q}\mathbf{x}(t) + \mathbf{u}^T(t)\mathbf{R}\mathbf{u}(t))\,dt \tag{1.3}$$

with suitably chosen matrices $\mathbf{Q}, \mathbf{R}$. The control effort is specified indirectly by the matrix $\mathbf{R}$. The result is the LQ (linear quadratic) controller. If an observer is required, this concept has to be generalized and leads to the so-called LQG controller ("G" for Gaussian, since measurement error and plant disturbance have to be certain Gaussian stochastic processes). The LQG controller is a combination of an LQ controller and the famous Kalman filter. Both, the LQ and the LQG controller are special cases of $\mathcal{H}_2$ optimal controllers.

It should be noted that the matrices $\mathbf{Q}, \mathbf{R}$ express the performance properties of the feedback loop. Despite this fact, the LQ controller has some favorable robustness properties similar to that of the PIDT_1 controller designed with the Nyquist criterion. In general, these are lost if an observer is added. Another drawback of the LQ and LQG approach is the fact that it is not easy to translate the design goals into adequate choices for the matrices $\mathbf{Q}, \mathbf{R}$.

The LQ and the LQG controller fit into the general framework of the feedback loop depicted in Fig. 1.1. For the LQ controller we have $\mathbf{y} = \mathbf{x}$, and if no integrator for tracking is added, the controller order is zero. With a general observer, the controller order equals the order n of the plant (respectively $n+$ number of integrators).

The LQ and LQG controllers are designed in the state space, but the performance specifications are more naturally formulated in the frequency domain. Thus one may ask for a design method which can be directly applied to a performance specification in the frequency domain. To give an idea of how this can be achieved we look at the SISO system of Fig. 1.2 and suppose that the disturbance is such that it is simply added to the output of G. Then the transfer function from d to the control error e is the so-called sensitivity function S and the following equations hold:

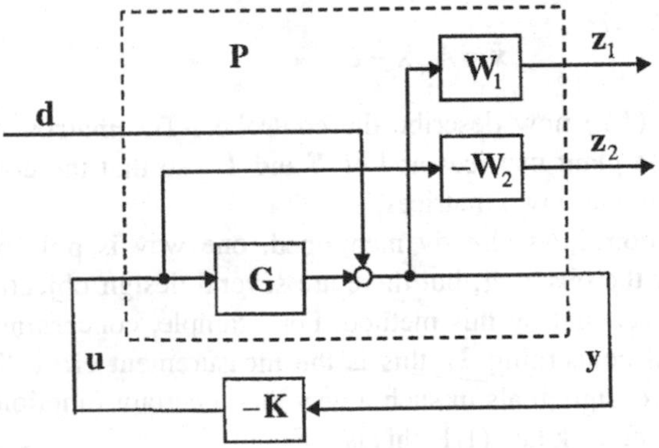

Fig. 1.3. Feedback system with frequency-dependent weightings

$$\hat{e} = -S\hat{d}, \quad \hat{u} = -KS\hat{d}.$$

The disturbance rejection properties of S can be specified by a frequency-dependent weight W_1 which has the property of having a large absolute value for small frequencies, whereas the absolute value tends to a limit for large frequencies. The control effort is expressed by the transfer function KS, which can be specified by a second weight W_2. Thus it is possible to formulate performance requirements by the following inequalities:

$$|S(i\omega)W_1(i\omega)| \leq 1, \qquad |K(i\omega)S(i\omega)W_2(i\omega)| \leq 1. \tag{1.4}$$

This concept works also for MIMO plants. It is helpful to define an extended plant $\mathbf{P}$, which contains the original plant $\mathbf{G}$ and the weights (cf. Fig. 1.3).

Denote now the closed-loop transfer function which maps $\mathbf{d}$ onto $\mathbf{z} = (\mathbf{z}_1, \mathbf{z}_2)$ by $\mathbf{F}_{zd}$. It is possible to define a norm for this operator, namely the so-called $\mathcal{H}_\infty$ norm $\|\cdot\|_\infty$, such that

$$\|\mathbf{F}_{zd}\|_\infty \leq 1 \tag{1.5}$$

expresses exactly the condition (1.4). This idea can be generalized with the result that a controller specification in the frequency domain leads to a requirement of the form (1.5) for the $\mathcal{H}_\infty$ norm of a certain operator which describes the I/O behavior of the feedback system. It should be emphasized that this operator is related to the extended plant $\mathbf{P}$. As for the LQG case, the controller can be written as a combination of state feedback and an observer, but if one looks at the details, there are some important differences. The controller order is the order of the extended plant.

The $\mathcal{H}_\infty$ controller developed on the basis of Fig. 1.3 has some robustness properties which are caused by the weighting of $\mathbf{S}$. For a **structured uncertainty**, which, for example, is given by a combination of a parameter uncertainty

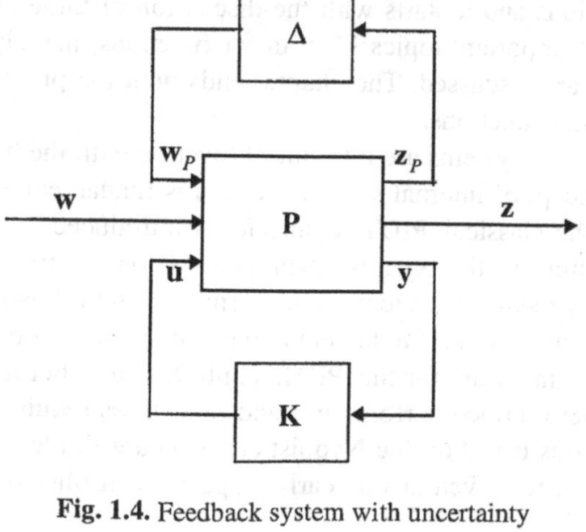

Fig. 1.4. Feedback system with uncertainty

and neglected dynamics at the plant input, things become more complicated. Then robust stability can be described by the so-called **structured singular value** μ. The uncertainty has to be explicitly modeled in the way shown in Fig. 1.4, which generalizes Fig. 1.3. Here, Δ denotes an unknown, but stable transfer function. The frequency dependency of the unmodeled dynamics or the inaccuracies of the plant parameters are incorporated into the plant by using suitable weights (in addition to the performance weights). The task is now to find a controller such that the feedback system is not only stable for all allowed perturbations Δ (**robust stability**), but also fulfills the performance specification for all these perturbations (**robust performance**). The maximum size of the allowed perturbations can be expressed by the structured singular value μ of the feedback system, consisting of $\mathbf{P}$ and $\mathbf{K}$. The mathematical tool for controller synthesis this is the so-called D-K iteration. The controller order for such a robust controller is typically high.

1.2 Outline of the Book

The book begins with a short introduction into classical controller design for SISO systems (Chaps. 2– 4) as found in many basic textbooks. Although the focus of the book is more on advanced topics such as $\mathcal{H}_2$, $\mathcal{H}_\infty$, and robust control, there are two reasons for starting in this way. The first one is that the book is intended to be a self-contained course on the design of control systems, which is accessible also for readers with no prior knowledge of control theory at all. The second reason is that $\mathcal{H}_\infty$ and robust control have become more and more standard techniques which have their roots in classical control, although the methods are much more elaborate and advanced. A careful study of the classical concepts leads to a well-founded motivation and a basic understanding of the modern approach. Vice versa, the modern approach brings some new insight into the classical results.

Chapter 2 is an elementary introduction to SISO systems, which are here described by transfer functions and it starts with the discussion of three elementary examples. Then the most important topics for transfer functions, namely stability and frequency response, are discussed. The chapter ends with the presentation of some basic types of transfer functions.

In Chap. 3, SISO feedback systems are introduced together with the basic transfer functions and the concept of internal stability, which is fundamental for every feedback system. Then the classical PIDT_1 controller is introduced and the most important tool for designing it, the Nyquist stability criterion, is presented. The basic design goals for any feedback system are performance and robustness. They are discussed next and thereby some fundamental ideas of $\mathcal{H}_\infty$ and robust control are introduced. These are important for the PIDT_1 controller, too, but its structure is too simple to fulfill more refined performance and robustness requirements. On the other hand, the methods based on the Nyquist criterion are limited essentially to the PIDT_1 controller. Thus, even at this early stage of controller design, it is seen that there is actually a need for more advanced methods.

In the next chapter, it is shown how conventional controller design based on the Nyquist stability criterion works. The plant is assumed to be stable with the possible exception of an integrator. The results may be compared with that of Sect. 10.5, where for a second order plant a $\mathcal{H}_\infty$ controller is designed. The result is that even for this simple plant the requirements are easier to fulfill with a $\mathcal{H}_\infty$ controller, but it should be noted that there is the following big difference in the design philosophy. The role of the parameters of the PIDT$_1$ controller are very well understood by an experienced control engineer and if one requirement is not quite met, the parameters often can be tuned to get a better result. In contrast to this, $\mathcal{H}_\infty$ controller synthesis is completely done by the software, which works like a black box. The tuning is then done by changing the weights. Chapter 4 ends with some remarks on the controller structure for cases where more information than only one single measurement is available.

Most of the modern methods are based on a state space description of the plant. For this reason, a rather long chapter on MIMO systems follows. It contains fundamental concepts such as linearization, frequency response, stability, controllability, observability, the Kalman decomposition and so on. Besides this, a short introduction to systems driven by random processes is given. Although this topic is somewhat outside the scope of the book, some basic knowledge is necessary for control engineering, since controller design often uses the properties of such processes. In particular, this holds for $\mathcal{H}_2$ optimal control.

Chapter 6 begins with two fundamental concepts of control theory, namely state feedback and observer-based controllers. If all states of the plant can be measured, a controller can be designed by pole placement. For SISO systems, the controller coefficients are uniquely determined by the closed-loop poles. In Sect. 6.5, we show how these poles have to be located such that the closed-loop system fulfills some basic robustness properties which are similar to those discussed in Sect. 3.5. For MIMO systems the situation is different, because the poles no longer uniquely determine the controller coefficients, so that there is additional freedom. We prove the existence of pole placement and sketch a method for how it can be done in a numerically stable manner. The chapter ends with a section on the structure of general MIMO feedback systems. It is fundamental for all that follows and generalizes the basic concepts for SISO systems introduced in Chap. 3. The main topics are internal stability and the generalized Nyquist criterion.

One of the most basic ideas of modern control is to extend the plant by weights which express the design goals and to view the feedback system with the extended plant as an operator between suitable function spaces so that a norm can be defined. The most important norms are the $\mathcal{H}_2$ and the $\mathcal{H}_\infty$ norm. They are introduced in Sects. 7.1 and 7.2. This follows a discussion of how the performance can be specified. Chapter 7 ends with some remarks on performance limitations and with a short introduction to coprime factorization.

Chapter 8 is devoted to the synthesis of $\mathcal{H}_2$ optimal controllers. It starts with the design of state feedback controllers by minimizing a certain quadratic functional. This leads to the LQ (linear quadratic) controller. The existence of such a controller strongly depends on the solution of a certain nonlinear algebraic matrix equation, the Riccati equation. Their basic properties are introduced in Sect. 8.1.2 and are also needed for $\mathcal{H}_\infty$ optimal control. It is possible to solve the general $\mathcal{H}_2$ control problem very elegantly by introducing a sequence of special problems, which can be sketched as follows. In the first step, the general $\mathcal{H}_2$ problem can be reduced to a so-called output estimation problem, which is the dual of a disturbance feedforward problem. This can be viewed as a special case of a full information problem, which is essentially the problem where the controller is of the form (1.1). The important special cases of $\mathcal{H}_2$ optimal controllers arise when the disturbance and the measurement are white noise processes. They are the Kalman filter as $\mathcal{H}_2$ state estimator and the LQG controller, which is a combination of LQ state feedback and Kalman filter. The chapter ends with an example and the description of some basic properties of such controllers, which can also be used for their synthesis.

In Sect. 8.5 we give a very short outlook to a related class of problems, namely optimal control problems. They arise when integral functionals like (1.3) are minimized, but in optimal control, much more general functionals are allowed, and the process dynamics may be nonlinear. We sketch also a few basic concepts for processes governed by a PDE.

The main part of Chap. 9 is concerned with the proof of the characterization theorem for $\mathcal{H}_\infty$ optimal controllers. As in the $\mathcal{H}_2$ case it is possible to use a sequence of subproblems to reduce the general problem to the full information problem, but now the things are much more complicated. In particular, this holds for the full information problem itself. The characterization of $\mathcal{H}_\infty$ optimal controllers based on this approach is given in terms of two Riccati equations. To make the problem tractable, it is necessary to start with a special case, which later can be generalized by a procedure called loop shifting (Sect. 9.5). In the last subsection of this chapter, weighting schemes for $\mathcal{H}_\infty$ synthesis are discussed. We also describe some basic properties of $\mathcal{H}_\infty$ suboptimal controllers resulting from this approach.

Another way of characterizing $\mathcal{H}_\infty$ suboptimal controllers is obtained when linear matrix inequalities (LMIs) are used. This approach is described in Sect. 10.1 and is, in some sense, more flexible than the one with the Riccati equations. After this, the connection between both approaches and the limiting behavior of $\mathcal{H}_\infty$ suboptimal controllers are discussed. With the LMI design method, it is possible to incorporate pole placement constraints. It also offers a way to apply gain scheduling in a theoretical well founded manner. This is shown in Sects. 10.3 and 10.4. In the last subsection, $\mathcal{H}_\infty$ controller design is applied to a SISO second-order plant. In this way, the reader can gain familiarity in a rather easy way with the more advanced techniques and can compare them with conventional methods.

Chapter 11 is the first big chapter with case studies. It begins with the very popular inverted pendulum. Although this example is somewhat academic, it offers a nice way to compare many of the design methods developed up to now and to get a feeling of how they work. The second case study is concerned with a continuously stirred tank, which is an unstable MIMO plant. We design a controller in a rather simple way by a combination of pole positioning and a reduced observer.

In the last example we design several controllers for the lateral motion of an aircraft. After a careful analysis of the plant dynamics, it is shown how the bad damping of the plant can be improved in a conventional way by using a so-called washout filter. Then more advanced controllers (LQ and $\mathcal{H}_\infty$) are developed. The next two chapters are both concerned with uncertainty. In Chap. 12 it is shown which kinds of uncertainties exist and how they can be implemented into the plant model. As a result, the extended plant now contains weights for performance and weights which describe uncertainty. It is important to notice that the uncertainty Δ is connected to the plant by feedback (cf. Fig. 1.4.). Thus, for the feedback system which consists of the plant and the controller, the uncertainty formally acts like a controller. This leads to an understanding how the small gain theorem can be applied to robustness analysis (Sect. 13.1). As a result, several tests for robust stability can be derived for unstructured uncertainties. If the uncertainty is structured (for example, because it consists of parameter uncertainty and neglected high frequency dynamics), it is necessary to introduce the structured singular value μ to perform the necessary analysis (Sect. 13.3). With this tool, refined criteria for robust stability can be given, and, what is of great importance, it is even possible to derive a necessary and sufficient condition for robust performance. A synthesis method based on these results is the D-K iteration, which can be used for the design of controllers having robust performance (Sects. 13.4 and 13.5). The order of such controllers is typically high, but can be reduced by suitable methods. As for the $\mathcal{H}_\infty$ synthesis, a simple second-order example is used to get familiar with these techniques.

Chapter 14 contains the second series of advanced case studies. It is completely devoted to the design of robust controllers. The first case study continues the controller design for aircraft of Sect. 11.3, but now the controller for lateral motion is required to have robust performance for a defined uncertainty structure of the plant. The second case study is concerned with the control of a ship. The controller shall not only have the property that the ship can track a prescribed course, it is also asked to improve the damping of the roll motion, again for an uncertain plant. In the third case study, the key part of the autopilot of a guided missile is developed. The problem is that the controller must be capable of handling large commands so that the region where the linearized model is valid will be left. Moreover, the limited actuator dynamics and elastic behavior of the airframe have to be taken into account. The final case study is concerned with high-purity distillation. This problem is challenging since the plant is extremely ill-conditioned.

1.3 The Main Steps of Controller Design

The book contains the full mathematical theory together with (almost) all the proofs. Readers who want to get a deeper understanding of $\mathcal{H}_2$ and $\mathcal{H}_\infty$ controller design will probably read these proofs, particularly, if they later want to carry out their own research in this field. On the other hand, it cannot be denied that the full study of the proofs requires some experience in the mathematical way of thinking and even for readers having these skills some endurance is required. For this reason, it may be asked which parts of the book are really necessary for readers who are not so much interested in the theory itself but who want to be able to apply successfully the design and analysis methods of modern control theory. In order to answer this question, the following very pragmatic approach is made. We define the major steps which are necessary for controller design, give some explanations of what has to be done in each single step and relate them to the corresponding sections of the book. Not all of the steps are actually needed in each case.

Step 1. *Building a mathematical plant model and a simulation.*

Step 2. *Definition and calculation of equilibrium points and calculation of the linearized model. Analysis of the linearized model.*

Step 3. *Synthesis of a "simple conventional" controller.*

Step 4. *Definition of performance weights. Synthesis of a* $\mathcal{H}_\infty$ *(or* $\mathcal{H}_2$ *) controller.*

Step 5. *Definition of the uncertainty structure and of the weights for uncertainty.*

Step 6. *Stability analysis of the "simple" controller.*

Step 7. *Synthesis of a* μ *controller. Reduction of the controller order.*

Step 8. *Simulation of the feedback system with the linearized and the full model.*

Step 9. *Test of the controller in a HIL (hardware in the loop) system and test of the controller in the real system.*

Step 1. Building a mathematical model of the plant means that physical laws have to be applied to find the differential equations which govern the dynamics of the system. These equations may also be found by identification or by a combination of the two methods. In any case, a decision has to be made on the level of detail to which the mathematical model has to be developed. This requires a good engineering judgment and some experience. For example, the body-fixed accelerations and angular velocities of an aircraft are measured by an IMU (inertial measurement unit), which for controller design may simply be modeled as a first-order system with white noise as input. In reality, the IMU is a complex technical system itself and the manufacturer of the IMU will have a full mathematical model. This may also be used in the detailed simulation of the aircraft, but for the design phase of the controller it is not required.

If the system of differential equations which governs the real plant is found, a computer simulation has to be built. This may be done by programming it in a high-level programming language such as FORTRAN, C++ or Ada or by using a simulation tool such as MATLAB/SIMULINK or MATRIXx. The latter way is much more comfortable and for the case studies in this book we always use such a tool.

There is no simple recipe for model building which is applicable to any thinkable situation. Thus concerning step 1, the reader is recommended to look at the examples and case studies given in this book (in particular, see Chap. 11 and Chap. 14). In all of them, the modeling process is thoroughly described step by step. In this way, a first insight into the basic properties of the plant is obtained, which is needed later for controller synthesis.

Step 2. In the next step, the equilibrium points have to be defined and calculated. Mathematically, they are solutions of a set of nonlinear algebraical equations. It may happen that not all equilibrium points which are mathematically possible are reasonable from an engineering point of view, so that they have to be selected with some care. For a given equilibrium point a linearized plant has to be calculated by applying the usual procedure, which is described in Sect. 5.1. This is more or less a routine task but may become difficult if the nonlinearity is nondifferentiable. Fortunately, this occurs only seldom, but there are some interesting examples of such cases.

If the linear model is established, its basic properties have to be exploited. The most important ones are stabilizability and detectability. They must be fulfilled to find a stabilizing controller at all. If not, the physical structure of the controllers and sensors has to be changed. In any case, it is extremely useful to calculate the poles of the plant because this shows whether or not it is stable. Even if the plant is stable it is important to calculate the damping of the complex pole pairs or to see whether there are far-away (stable) poles which do not contribute much to the dynamics. An interesting example of the latter case is the distillation column, which is a stable but very stiff system of high order. Despite this, for controller design it can be approximated by a system with a 2×2 state matrix. A difficulty for the controller design arises since the plant is extremely ill-conditioned. The latter fact is also a result of this preliminary analysis step. The necessary system theory background is presented in Chap. 5 and any reader should be familiar with most of the material presented there.

Step 3. The design of a "simple conventional" controller is optional, but in some cases it may be very helpful for several reasons. For example, in the distillation column example two simple PI controllers may be used to see what performance ideally could be achieved. Practically, they do not work since with them the feedback system is extremely sensitive with respect to certain perturbations. Another reason is to see what progress can be made with the new methods.

The necessary theory for this step can be found in a very concise form in Chaps. 2–4. For readers having no background in control systems at all, these introductory chapters should be read in detail. Readers with a background of a basic course on control systems will be familiar with most of the material presented there but they are recommended to look at the parts of these chapters where the classical concepts are related to the concepts of $\mathcal{H}_\infty$ and robust control. This mainly concerns Chap. 3.

Step 4. The specification of some performance properties is always necessary for controller design. It is one of the great benefits of $\mathcal{H}_\infty$ synthesis that it can be based on weights, which can be directly related to the performance properties of the closed-loop system such as tracking error, bandwidth, control effort and so on. On the other hand, there are the physical limitations of the system, which may be certain properties of the plant (for example, nonminmum-phase behavior), or the limitations of the actuators, sensor noise and so on. A successful $\mathcal{H}_\infty$ controller design ends with an optimal value near one. If it is greater then one, the performance requirements are too stringent and if it is smaller than one, it would be possible to achieve more with this plant and instrumentation. Normally, some (moderate) trial and error is necessary to find weightings leading to a successful design. For the design of $\mathcal{H}_2$ optimal controllers it is explained in Sects. 8.1.1, 8.3.1 and 8.4.1 how weighting works.

For the practical application of $\mathcal{H}_\infty$ control, the definition of the weights is perhaps the most important step. It requires the understanding of the $\mathcal{H}_\infty$ norm as an operator norm (Sect. 7.2) and familiarity with the weighting schemes (Sect. 9.6).

If the weights are defined, $\mathcal{H}_\infty$ synthesis is an easy step since it is fully automated within the corresponding software tools. In MATLAB, it is done by the call of a certain procedure that essentially consists of one line of code. The procedure must be fed with the extended plant, which has to be implemented first. Suitable procedures supporting this exist, so that the whole synthesis process can be achieved with a moderate effort. The result is a state space representation of the $\mathcal{H}_\infty$ controller.

Sometimes the software complains. Then the problem has to be reformulated to make it applicable and this is possible in almost any situation. Thus the conditions under which $\mathcal{H}_\infty$ synthesis works must be understood and the reader should know what has to be done if some of them are violated (see Sects. 9.5 and 9.6.2).

With this basic knowledge of $\mathcal{H}_\infty$ optimal control it is possible to switch directly to Sect. 10.5, where some basic examples are analyzed and compared with the classical PIDT_1 controller. It would be very helpful for readers to accompany this reading by some of their own numerical experiments.

Steps 5, 6. The $\mathcal{H}_\infty$ controller may have good robust stability properties but this must not necessarily be the case. For SISO problems, robust stability can be checked by looking at the loop transfer function, as is done for the classical PIDT_1 controller. Normally it is preferable to define an uncertainty structure, which may be set up by uncertain parameters and weights for uncertain dynamics. Then the extended plant consists of the original physical plant and weights for performance and uncertainty. How this can be done is explained in Chap. 12.

If a conventional controller is already given, it is possible to analyze its robust stability properties with respect to this uncertainty structure (Step 6).

Step 7. Controller synthesis is now carried out by the D-K iteration, which is based on the $\mathcal{H}_\infty$ synthesis (Sect. 13.5). If the extended plant is already implemented, this algorithm is started simply by writing "dkit" (if MATLAB is used). It may be run fully automatically and ends with a state space description of the μ controller. The design is successful if the iteration finishes with μ close to one. This means that the weights for performance and robustness are balanced. Of course, whether the weights are reasonable at all and really express what the designer wants, cannot be guaranteed by any software and is still left to engineering judgement.

The essential terms describing a μ controller are robust stability and robust performance. Both are measured by the structured singular value μ (Sects. 13.3 and 13.4). The μ optimal controller is normally of high order. It may be reduced by applying suitable algorithms (Sect. 13.6). As for $\mathcal{H}_\infty$ optimal control a simple numerical example is given which may help the reader to get rapidly familiar with these concepts and again the results can be compared with those for a conventional $PIDT_1$ controller (Sect. 13.7).

We conclude this section with some remarks concerning MATLAB. The most important toolbox concerning controller design is the "Control System Toolbox", [113]. Each of the examples makes use of it. Concerning $\mathcal{H}_\infty$ and robust control, there are three toolboxes: the "Robust Control TB", [24], the "LMI Control TB", [47], and the "μ-Analysis and Synthesis TB", [7]. Each of them is very powerful, but in the majority of cases our numerical analysis is based on the last-mentioned toolbox.

Chapter 2

Rational Transfer Functions

In this chapter we give a short overview of rational transfer functions with one input and one output. By deriving mathematical models for some selected elementary plants, we get typical examples for SISO transfer functions occurring in classical control. In many practical cases such a transfer function is stable, with the possible exception of one pole at the origin. Moreover, the complex poles are well damped.

With the background of these examples we then analyze rational transfer functions in greater generality. First, we define a normalized representation and derive a decomposition into a product of factors that are transfer functions of order one or two. Next the step response of a general rational transfer function will be calculated. This leads us easily to a definition and a characterization of stability. After this, a short introduction to the frequency response of a linear system is given.

The chapter concludes with an elementary discussion of basic transfer functions of order one or two. It is very helpful to get a good understanding of their behavior in the time and the frequency domains. They occur as models of simple linear systems, as elements in the product decomposition of a rational transfer function and some of them later as weights in $\mathcal{H}_\infty$ optimal control.

2.1 Introductory Examples

2.1.1 Plants for Pressure Control

A single container. We want to determine a mathematical model of the plant shown in Fig. 2.1. Here, the pressure p in the container is the controlled variable. The setting of the inlet valve is the control, whereas the setting of the outlet valve is influenced by an unknown consumer and acts as a disturbance. The aim of the pressure control is to bring the pressure to a desired value and to maintain this

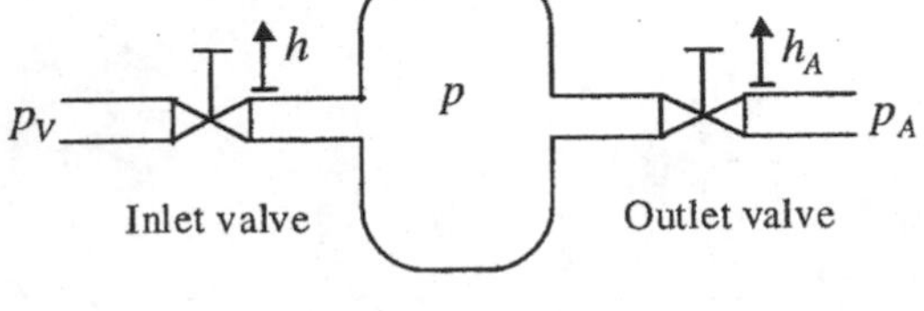

Fig. 2.1. Plant for pressure control with one container

value in the presence of disturbances. Here, we assume that the external pressures p_V and p_A are constant.

The gas equation for the pressure p is given by $pV = mR_G T$. Here, V denotes the volume of the container, T the temperature (which is assumed to be constant), m the mass of the gas in the container and R_G the gas constant. Differentiation of the gas equation leads to $\dot{p}V = \dot{m}R_G T$. At the inlet valve, a gas mass with flow rate q_V enters the container and at the outlet valve a gas mass with flow q_A leaves the container. Thus we have $\dot{m} = q_V - q_A$ and hence

$$K\dot{p} = q_V - q_A, \tag{2.1}$$

with $K = V/(R_G T)$. For a valve we can write $q = F(p_1, p_2, h)$. Here, p_1 denotes the pressure before the valve, p_2 denotes the pressure behind the valve, and h is the valve setting. At the moment, the precise structure of F is not so important. Let the outlet valve similarly be described by the function F_A. Then we get from (2.1) the nonlinear differential equation

$$K\dot{p} = F(p_V, p, h) - F_A(p, p_A, h_A). \tag{2.2}$$

In order to apply control theory methods, one has to linearize (2.2). First, an equilibrium state has to be calculated. Such a state is given when for constant external quantities p_V, p_A h, h_A the pressure p is also constant. In this case, the equation

$$F(p_{V0}, p_0, h_0) = F_A(p_0, p_{A0}, h_{A0})$$

holds. The index "0" indicates that the relevant quantity is constant. Let now

$$\frac{1}{W_1} = -\frac{\partial F}{\partial p}(p_{V0}, p_0, h_0), \qquad \frac{1}{W_2} = \frac{\partial F_A}{\partial p} F_A(p_0, p_{A0}, h_{A0}),$$

$$c_1 = \frac{\partial F}{\partial h}(p_{V0}, p_0, h_0), \qquad c_2 = \frac{\partial F}{\partial h_A}(p_{V0}, p_{A0}, h_{A0}),$$

and $\Delta p = p - p_0$, $\Delta h = h - h_0$, $\Delta h_A = h_A - h_{A0}$. Then we get the linear differential equation

$$T\Delta\dot{p} = -\Delta p + k_1 \Delta h - k_2 \Delta h_A,$$

where

$$T = \frac{K W_1 W_2}{W_1 + W_2}, \quad k_1 = \frac{W_1 W_2 c_1}{W_1 + W_2}, \quad k_2 = \frac{W_1 W_2 c_2}{W_1 + W_2}.$$

Letting

$$G(s) = \frac{k_1}{1+Ts}, \quad G_d(s) = \frac{k_2}{1+Ts},$$

we can write

$$\Delta\hat{p} = G(s)\Delta\hat{h} - G_d(s)\Delta\hat{h}_A. \tag{2.3}$$

Finally we give a short description of the function which describes the valve. Behind the valve there is complete turbulence and the mass flow is given by

$$q = \alpha A(h)\sqrt{p_1 - p_2} \quad (\text{if } p_1 > p_2).$$

Here, A denotes the cross-section of the valve and α is a constant. A depends on the valve position: $A = A(h)$.

A series of two containers. Fig. 2.2 shows a series of two containers with pressures p_1 and p_2, respectively. Let the pressure p_2 in the second container be the plant output. Between the containers there is an orifice with resistance W. Equation (2.2) has now to be replaced by the system:

$$K_1\dot{p}_1 = F(p_V, p_1, h) - \frac{1}{W}(p_1 - p_2),$$

$$K_2\dot{p}_2 = \frac{1}{W}(p_1 - p_2) - F_A(p_2, p_A, h_A).$$

It can be linearized as the single equation (2.2), and one obtains for the differences Δp_1, Δp_2 the linear system:

$$K_1\Delta\dot{p}_1 = -(\frac{1}{W_1} + \frac{1}{W})\Delta p_1 + \frac{1}{W}\Delta p_2 + c_1\Delta h$$

$$K_2\Delta\dot{p}_2 = \frac{1}{W}\Delta p_1 - (\frac{1}{W} + \frac{1}{W_2})\Delta p_2 - c_2\Delta h_A.$$

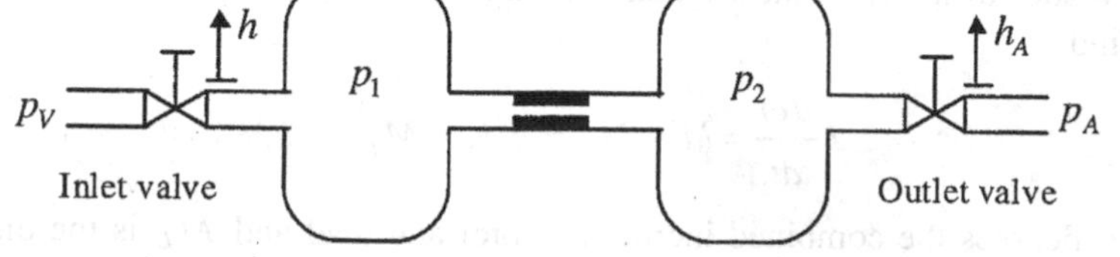

Fig. 2.2. Plant for pressure control with two containers

Using the Laplace transform and resolving the resulting equations for $\Delta\hat{p}_2$ one gets again (2.3), but now with

$$G(s) = \frac{k}{(1+T_1 s)(1+T_2 s)}, \qquad G_d(s) = \frac{k_1(1+T_0 s)}{(1+T_1 s)(1+T_2 s)}.$$

Here, all time constants are positive and we have $T_1 \neq T_2$. To see this, one has to show that the quadratic denominator of the transfer functions has two negative real zeros. This can easily be done by taking into account the sign of the coefficients.

2.1.2 Electromechanical Plants

Direct-current (dc) motor. In this section we consider the control of a dc motor. The angular velocity ω of the shaft or the position x of a tool mounted on the lead screw of the motor is assumed to be the plant output (cf. Fig. 2.3). The rotational motion results from the input voltage u_A. The motor shaft has to drive a load that exerts a torque on the shaft. The load moment is considered as a disturbance and the input voltage u_A is the control.

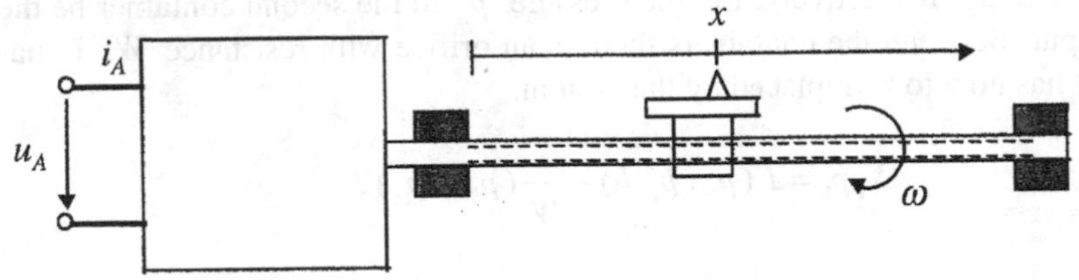

Fig. 2.3. Position control with a dc motor

We want to derive a mathematical model of the motor and start with the following basic equations:

$$M = c_A i_A, \quad u_i = c_A \omega. \tag{2.4}$$

Here, M is the torque generated by the armature current i_A, u_i is the voltage induced by the rotation of the armature and c_A denotes a constant. The equations follow easily from basic physical laws, but the fact that the constant in both equations is the same requires some additional reasoning. Concerning this point we refer the reader to the relevant literature. Using Newton's law for rotational motion we find

$$J\frac{d\omega}{dt} = M - M_L = c_A i_A - M_L.$$

Herein J denotes the combined inertia of motor and load and M_L is the disturbance torque. This differential equation has to be completed by one which describes the connection between the current i_A and the input voltage u_A. It will be obtained by taking into account all voltages appearing in the armature. These are the already mentioned induced voltage u_i, the voltage $R_A i_A$ caused by the resistance R_A of the armature and the self-induced voltage

$$u_L = L_A \frac{di_A}{dt} .$$

Here, L_A denotes the inductance of the armature winding. The sum of these three voltages is equal to the external input voltage. Hence, using (2.4) we get

$$L_A \frac{di_A}{dt} = -R_A i_A - c_A \omega + u_A .$$

An application of the Laplace transform gives

$$\hat{\omega} = G(s)\hat{u}_A - G_d(s)\hat{M}_L ,$$

with the transfer functions

$$G(s) = \frac{k}{T^2 s^2 + 2dTs + 1} ,$$

$$G_d(s) = \frac{R_A}{c_A^2} \frac{1 + T_e s}{T^2 s^2 + 2dTs + 1}$$

and the constants

$$k = \frac{1}{c_A}, \quad T = \frac{\sqrt{JL_A}}{c_A}, \quad d = \frac{R_A}{2c_A}\sqrt{\frac{J}{L_A}}, \quad T_e = \frac{L_A}{R_A} .$$

We now consider the position control for the machine tool. From Fig. 2.3 it is seen that the screw converts the rotational motion into a translational motion. Let n be the number of revolutions of the shaft and let the position x of the tool be changed by the value h_G for one rotation. Then we have $\omega = 2\pi n$ and $\dot{x} = h_G n$. Thus the transfer function which transforms u_A into x is given by (with $k_1 = k\, h_G / 2\pi$)

$$G_I(s) = \frac{k_1}{s} \frac{1}{T^2 s^2 + 2dTs + 1} .$$

Hovering ball. We consider now the position control for a ball containing iron, as shown in Fig. 2.4. It is assumed that an electromagnetic force F_M acts on the ball which can be manipulated by the input voltage u_A. The gravitational force is exerted precisely in the opposite direction. The task is to keep the center of gravity of the ball at a prescribed distance h from the coil so that the ball "hovers" under the coil in a fixed position. It is assumed that no disturbance occurs.

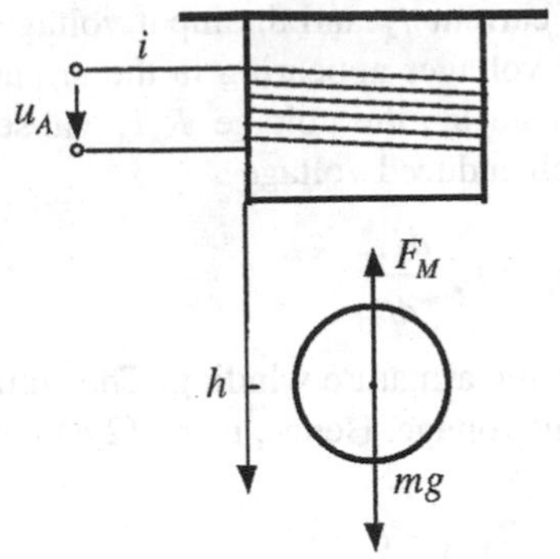

Fig. 2.4. Hovering ball

As in the preceding examples, we start our analysis by establishing a mathematical model for the plant. The equilibrium of the forces is given by the equation:

$$m\ddot{h} = mg - F_M(h, i)\,. \tag{2.5}$$

Here i denotes the current in the coil and m is the mass of the ball. It has to be noticed that the force caused by the electromagnet depends nonlinearly on the current i and the distance h. This dependency may be given by a closed formula or by a diagram. In any case, one has to linearize the nonlinear differential equation (2.5) at an equilibrium point. It is described by a constant current i_0 and a constant distance h_0, which are given such that the electromechanical force and the gravitational force compensate each other exactly: $F_M(h_0, i_0) = mg$.

Let now Δi and Δh be the corresponding deviations from the equilibrium point. Then the differential equation

$$m\frac{d^2\Delta h}{dt^2} = -\frac{\partial F_M}{\partial h}(h_0, i_0)\Delta h - \frac{\partial F_M}{\partial i}(h_0, i_0)\Delta i\,.$$

holds. Let us define

$$k_1 = -\frac{1}{m}\frac{\partial F_M}{\partial h}(h_0, i_0), \quad k_2 = \frac{1}{m}\frac{\partial F_M}{\partial i}(h_0, i_0)\Delta i\,.$$

Notice that in this definition k_1 and k_2 are positive, since F_M increases with increasing distance h. Thus we have

$$\frac{d^2\Delta h}{dt^2} = k_1\Delta h - k_2\Delta i\,.$$

This equation can be written in the form:

$$T^2\frac{d^2\Delta h}{dt^2} = \Delta h - k_3\Delta i \quad \text{with } T = \frac{1}{\sqrt{k_1}}, \quad k_3 = \frac{k_2}{k_1}\,. \tag{2.6}$$

A further part of the plant is the coil. Let L be the inductance and let R be the resistance of the coil. Then we can write

$$L\frac{d\Delta i}{dt} = -R\Delta i + u_A \,. \tag{2.7}$$

The equations (2.6) and (2.7) describe the dynamics of the plant for small deviations from the equilibrium state.

From (2.6) we get the transfer function which transforms the current into the position:

$$\Delta\hat{h} = \frac{K_3}{-T^2 s^2 + 1}\Delta\hat{i}_A \,.$$

The poles of this transfer function are $s_1 = 1/T$, $s_2 = -1/T$. Both of them are real, but one is positive, which means that the transfer function is unstable (cf. Sect. 2.2.2). From (2.7) we get

$$\Delta\hat{i} = \frac{k_4}{1+T_1 s}\Delta\hat{u}_A \qquad (\text{with } T_1 = \frac{L}{R},\; k_4 = \frac{1}{R}) \,.$$

Thus the transfer function for the complete plant is given by (with $k = k_4 k_3$)

$$G(s) = \frac{k}{(1+T_1 s)(1+Ts)(1-Ts)} \,.$$

2.2 Basic Properties

2.2.1 Definitions and Product Decomposition

We have seen in the examples of the preceding section that the connection of the Laplace transform $\hat{y}$ of the output variable and the Laplace transform $\hat{u}$ of the input variable is given by an equation of the form

$$\hat{y} = G(s)\hat{u} \,, \tag{2.8}$$

with a rational function

$$G(s) = \frac{b_m s^m + b_{m-1} s^{m-1} + \ldots + b_1 s + b_0}{a_n s^n + a_{n-1} s^{n-1} + \ldots + a_1 s + a_0} \tag{2.9}$$

(and real numbers a_i, b_i and $a_n \neq 0, b_m \neq 0$). Then, as in the examples, the function $G(s)$ is the **transfer function** of the linear system. For a realizable system the inequality $m \leq n$ holds. In this case, $G(s)$ is said to be **proper**. If the inequality is strict, i.e. $m < n$, then $G(s)$ is **strictly proper**. We assume that the numerator and denominator of $G(s)$ have no common zero. Then the possible common linear factors in the product decomposition of the numerator and denominator are already cancelled. Under this assumption, n is the order of the rational transfer function.

Let $g(t)$ be the function which satisfies $\mathcal{L}(g)(s) = G(s)$, where $\mathcal{L}$ denotes the Laplace transform. Then it follows from the convolution theorem that

$$y(t) = \int_0^t g(t-\tau)u(\tau)d\tau . \tag{2.10}$$

The function g is called the **impulse response** of the linear system. The impulse response describes the linear system in the **time domain**, whereas the transfer function describes it in the **frequency domain**. The first definition is simply explained by the time dependence of g. The second definition will become clear in Sect. 2.2.3, where the response of a linear system to a harmonic oscillation will be calculated.

The **step response** y_e is by definition obtained, if (2.8) is applied to the so called **unit step**:

$$u_e(t) = \begin{cases} 1, & \text{if} \quad t \geq 0 \\ 0, & \text{if} \quad t < 0. \end{cases}$$

Hence the equation

$$y_e(t) = \int_0^t g(\tau)d\tau$$

holds. Vice versa, if the step response is given, the impulse response can be calculated from $g = \dot{y}_e$. Moreover,

$$\hat{y}_e = \frac{1}{s} G(s) . \tag{2.11}$$

Nonrational transfer functions. The most important nonrational transfer function is the so called **delay**. It is defined by the equations

$$y(t) = k\, u(t-T) \quad \text{or} \quad \hat{y} = k e^{-Ts} \hat{u} .$$

Here, T is assumed to be positive. The delay can be approximated by rational transfer functions. For example, from the series development of the exponential function we get

$$e^{-Ts} \approx \frac{1}{1+Ts} .$$

A more accurate approximation is

$$e^{-Ts} \approx \frac{T^2s^2 - 6Ts + 12}{T^2s^2 + 6Ts + 12}.$$

By means of these approximations it is possible to stay in the domain of the rational transfer functions, but some care has to be taken that the rational substitute is only applied in a frequency region were the approximation actually holds.

Normalized representation. For later applications it is helpful to give a classification of $G(s)$. At first it may happen that $a_0 = a_1 = \ldots = a_\alpha = 0$ or $b_0 = b_1 = \ldots = b_\alpha = 0$. In this case we can write

$$G(s) = \frac{k}{s^\alpha} \frac{Z(s)}{N(s)}. \tag{2.12}$$

Here α is an integer, the so-called **type number**, and $Z(s)$ and $N(s)$ are polynomials with the property $Z(0) = N(0) = 1$. The factor k is the **gain** of the transfer function.

A classification can be made according to α. For $\alpha = 0$, the transfer function $G(s)$ has proportional behavior. For $\alpha = 1$ and $\alpha = 2$, $G(s)$ contains one or two pure integrators, respectively, as factors. For $\alpha = -1$, the transfer function contains a pure differentiator as factor. We shall return to this point in the next section, where step responses are discussed.

By decomposing $Z(s)$ and $N(s)$ into a product of linear factors, we get the following representation:

$$G(s) = \frac{k}{s^\alpha} \frac{a_0(s - s_{01})(s - s_{02})\ldots(s - s_{0k})}{a(s - s_1)(s - s_2)\ldots(s - s_l)}.$$

By combining the factors adequately, it is now possible to derive the basic types of transfer functions. To be more precise, let z be a zero of the numerator or denominator polynomial. If z is real, we can write

$$s - z = -z(1 + Ts),$$

where $T = -1/z$. If z is complex, then $\overline{z}$ is also a zero and the linear factors belonging to z and $\overline{z}$ can be multiplied to get a real quadratic factor:

$$\begin{aligned}(s - z)(s - \overline{z}) &= s^2 - 2(\operatorname{Re} z)s + |z|^2 \\ &= \omega_0^2(1 + 2d\,Ts + T^2s^2),\end{aligned}$$

where

$$\omega_0 = |z|, \quad d = -\frac{\operatorname{Re} z}{\omega_0}, \quad T = \frac{1}{\omega_0},$$

and $|d| < 1$. This leads us to the following result.

Each rational transfer function can be written as a product of transfer functions of the following kind:

Pure gains, integrators and differentiators: $k, \quad \frac{1}{s}, \quad s.$

Transfer functions of order 1 (with $T \neq 0$): $\frac{1+T_0 s}{1+Ts}.$

Transfer functions of order 2 (with $|d|<1,\ |d_1|<1, T \neq 0, T_3 \neq 0, T_4 \neq 0$):

$$\frac{1+T_0 s}{1+2dTs+T^2s^2}, \quad \frac{(1+T_1 s)(1+T_2 s)}{1+2dTs+T^2s^2}, \quad \frac{1+2d_1T_1s+T_1^2s^2}{1+2dTs+T^2s^2}, \quad \frac{1+2d_1T_1s+T_1^2s^2}{(1+T_3s)(1+T_4s)}.$$

The occurring time constants may be positive as well as negative.

Zeros. Let s_0 be a zero of $G(s)$, i.e. $G(s_0)=0$. Then

$$G_1(s) = G(s)\frac{1}{s-s_0} \tag{2.13}$$

is a transfer function which has the same denominator as $G(s)$. For the function $u(t)=e^{s_0 t}$ we get

$$\hat{y} = G(s)\hat{u} = G(s)\frac{1}{s-s_0} = G_1(s),$$

which implies $y = \dot{y}_{1e}$, where y_{1e} is the step response of G_1. Thus the following holds.

Let s_0 be a zero of $G(s)$ and let y_{1e} be the step response of the transfer function G_1 defined in (2.2.6). Then the input $u(t)=e^{s_0 t}$ for $G(s)$ leads to the output $\dot{y}_{1e}$.

In Sect. 2.2.3, stability will be defined and it will be seen that for a stable system the step response $y_{1e}(t)$ becomes constant for $t \to \infty$. Therefore, we get:

Let s_0 be a zero of $G(s)$ and assume that $G(s)$ is stable. Then the output y of $G(s)$ for the input $u(t)=e^{s_0 t}$ vanishes for $t \to \infty$.

The latter result is of particular interest for a zero on the imaginary axis: $s = i\omega_0$. Then we may conclude that the harmonic oscillation $u = \sin \omega_0 t$ as the input leads to an output that vanishes for $t \to \infty$.

Block diagrams. In order to represent a plant or a closed-loop system graphically, it is convenient to use a so-called **block diagram**. It consists of linear systems (the blocks) and arrows which represent the signals that are exchanged between the blocks. A single system with transfer function $G(s)$ is shown in Fig. 2.5.

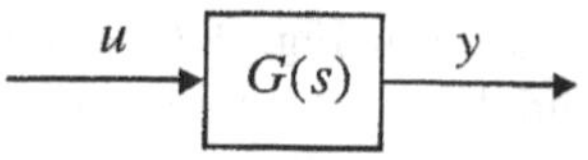

Fig. 2.5. Block diagram for a single transfer function

We discuss first which possibilities exist in creating block diagrams for one or two transfer functions. For the elementary block diagrams shown in Fig. 2.6, we get the following transfer functions:

$$G(s) = G_2(s)G_1(s) \qquad \text{(Series combination)},$$

$$G(s) = G_1(s) + G_2(s) \qquad \text{(Parallel combination)},$$

$$G(s) = \frac{G_0(s)}{1 \pm G_0(s)} \qquad \text{(Feedback)}.$$

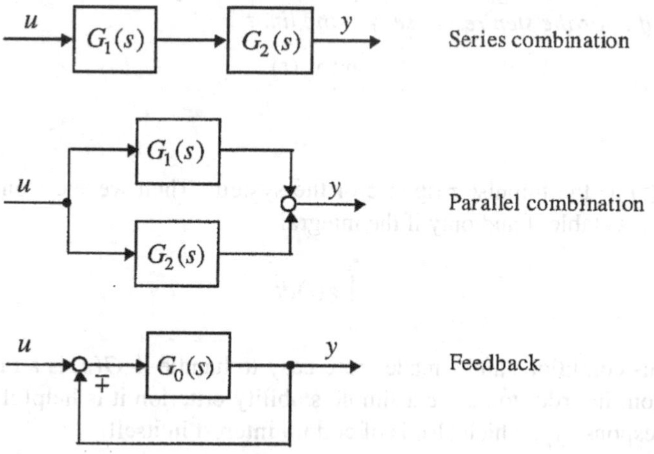

Fig. 2.6. Elementary block diagrams

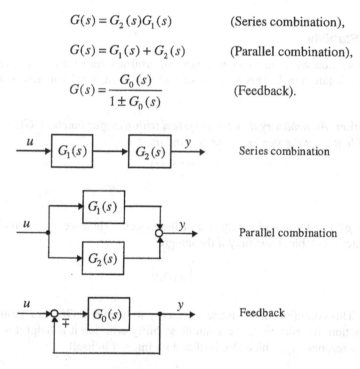

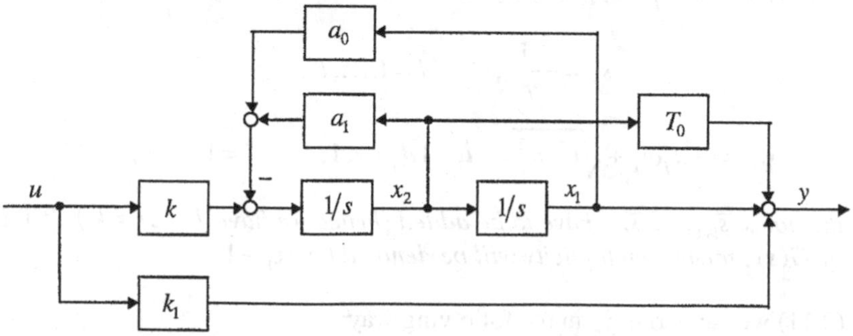

Fig. 2.7. Representation of a general transfer function of order 2 by pure gains and integrators

Example. We consider the block diagram of Fig. 2.7. In order to calculate the transfer function it is convenient to introduce two additional variables x_1 and x_2. It is easy to see that the transfer function is given by

$$G(s)=\frac{k_1 s^2+(kT_0+k_1 a_1)s+k+k_1 a_0}{s^2+a_1 s+a_0}.$$

2.2.2 Stability

In this section we want to analyze how an arbitrary transfer function reacts to a constant input signal. This leads to one of the most important terms of control theory.

Definition. *An arbitrary dynamical system with transfer function $G(s)$ is said to be* ***stable*** *if, for the step response y_e, the limit*

$$\lim_{t\to\infty} y_e(t)$$

exists.

Let $g(t)$ be the impulse response of the system. Then we see from (2.10) that the system is stable if and only if the integral

$$\int_0^\infty g(t)dt$$

exists. This condition can be made more easy to handle if $G(s)$ is a rational transfer function. In order to derive a simple stability criterion it is helpful to calculate the step response y_e, which also is of certain interest in itself.

Let $s_1,\ldots,s_k$ be the distinct poles of $G(s)$ which are different from 0 and let the multiplicity of s_j be m_j. Suppose they are written in the form

$$s_j=-\frac{1}{T_j},\qquad j=1,\ldots,l,$$

$$s_{l+j}=-d_j\omega_j+\sqrt{1-d_j^2}\,\omega_j i,\quad |d_j|<1,\qquad j=1,\ldots,r,$$

where the poles $\overline{s}_{l+1},\ldots,\overline{s}_{l+r}$ have to be added (hence we have $l+2r=k$). If 0 is a pole of $G(s)$, then its multiplicity will be denoted by m_0-1.

Using (2.11) we can write $\hat{y}_e$ in the following way:

$$\hat{y}_e = \frac{1}{s}G(s) = \sum_{\mu=1}^{m_0}\frac{c_{0\mu}}{s^\mu} + \sum_{j=1}^{k}\sum_{\mu=1}^{m_j}\frac{c_{j\mu}}{(s-s_j)^\mu}.$$

Here we have

$$c_{jm_j} \neq 0 \quad \text{for} \quad j = 0,\ldots,k \tag{2.14}$$

$$c_{j_1\mu} = \overline{c}_{j_2\mu}, \quad \text{if} \quad s_{j_1} = \overline{s}_{j_2}. \tag{2.15}$$

Moreover, $m_0 > 1$ holds if and only if 0 is a pole of $G(s)$. From this formula we get y_e by applying the inverse Laplace transform to each summand. This yields

$$y_e(t) = \sum_{\mu=1}^{m_0}\frac{c_{0\mu}}{(\mu-1)!}t^{\mu-1} + \sum_{j=1}^{k}\sum_{\mu=1}^{m_j}\frac{c_{j\mu}}{(\mu-1)!}t^{\mu-1}e^{s_j t}.$$

The complex coefficients $c_{l+j,\mu}$ can be written in the form $c_{l+j,\mu} = R_{j\mu}e^{i\varphi_{j\mu}}$. Hence we have for a complex pole pair:

$$\begin{aligned}
c_{l+j,\mu}e^{s_{l+j}t} + \overline{c}_{l+j,\mu}e^{\overline{s}_{l+j}t} &= 2\,\mathrm{Re}\,(c_{l+j,\mu}e^{s_{l+j}t})\\
&= 2\,\mathrm{Re}\,(R_{j\mu}e^{(\mathrm{Re}\,s_{l+j})t}e^{i((\mathrm{Im}\,s_{l+j})t+\varphi_{j\mu})})\\
&= 2R_{j\mu}e^{-d_j\omega_j t}\cos(\sqrt{1-d_j^2}\,\omega_j t + \varphi_{j\mu}).
\end{aligned}$$

By defining the functions

$$p_j(t) = \sum_{\mu=1}^{m_j}\frac{c_{j\mu}}{(\mu-1)!}t^{\mu-1}, \quad j = 0,\ldots,l, \tag{2.16}$$

$$q_j(t) = 2\sum_{\mu=1}^{m_{l+j}}\frac{R_{j\mu}}{(\mu-1)!}t^{\mu-1}\cos(\sqrt{1-d_j^2}\,\omega_j t + \varphi_{j\mu}), \quad j = 1,\ldots,r, \tag{2.17}$$

we get the following *representation of the step response*:

$$y_e(t) = p_0(t) + \sum_{j=1}^{l}p_j(t)e^{-\frac{t}{T_j}} + \sum_{j=1}^{r}q_j(t)e^{-d_j\omega_j t}. \tag{2.18}$$

From this equation the behavior of $y_e(t)$ for $t \to \infty$ can easily be seen. To this end, we assume that all poles of $G(s)$ have a negative real part. Then, in particular, 0 is not a pole of $G(s)$ and thus $m_0 = 1$ holds. From this we may conclude that $p_0(t) = c_{01}$. Moreover, the time constants T_j and the dampings are d_j positive. By using the rule of de L'Hospital it is easy to see that the corresponding summands tend to 0 for $t \to \infty$. Hence $y_e(t)$ tends to c_{01}, which shows the stability of $G(s)$. More precisely, the following theorem holds.

Theorem 2.2.1. *A rational transfer function is stable if and only if all its poles have a negative real part.*

Proof. The sufficiency of the above condition for stability has already been shown. In order to prove the necessity, we assume that there exists at least one pole that lies on the imaginary axis or in the open right half-plane. Let a be the maximum of the real parts of the poles. By our assumption, we have $a \geq 0$. From the poles with real part a we choose those with the largest multiplicity. Denote this multiplicity with respect to $s^{-1}G(s)$ by λ and let κ be the number of such poles. We may assume that these poles are the first κ ones. If one of these poles is real, let its index be κ. Dividing (2.18) by $t^{\lambda-1}e^{at}$, we obtain the equation

$$\frac{y_e(t)}{t^{\lambda-1}e^{at}} = h(t) + 2\sum_{j=1}^{\kappa-1} \frac{R_{j\lambda}}{(\lambda-1)!} \cos(\sqrt{1-d_j^2}\;\omega_j t + \varphi_{j\lambda}) + c_{\kappa\lambda}\,.$$

Here $h(t)$ denotes a function which tends to 0 for $t \to \infty$. It is possible that in this equation the sum of the trigonometric functions or the constant $c_{\kappa\lambda}$ is missing. We now assume that the limit of $y_e(t)$ for $t \to \infty$ exists. In this case, the sum of the trigonometric functions and the constant must also tend to 0 for $t \to \infty$ if $\lambda > 1$ or $a > 0$. This is only possible if these functions are identical to 0 for itself, i.e. if

$$R_{j\lambda} = 0 \quad j = 1, \ldots, \kappa-1, \qquad c_{\kappa\lambda} = 0\,.$$

This contradicts (2.14). If $\lambda = 1$ and $a = 0$, then the trigonometric sum actually occurs and we get a contradiction for $t \to \infty$ in a similar way. Hence $y_e(t)$ has no limit and $G(s)$ is not stable.

We now want to discuss the representation (2.18). To this end, it is useful to consider the following cases.

Case 1: $G(s)$ is stable.
Then $y_e(t)$ tends to a finite limit. An oscillating behavior occurs if and only if $G(s)$ has complex poles.

Case 2: $G(s)$ has $s = 0$ as a unique pole not lying in the open left half-plane.
In this case, $y_e(t)$ is dominated by the polynomial $p_0(t)$. The remaining parts of the step response tend to 0. Then in the sense of the definition of the preceding section, $G(s)$ has integrating behavior. If $s = 0$ is a simple pole (i.e. if $G(s)$ is of type 1), then $p_0(t)$ is the step response of an integrator: $p_0(t) = c_{01}t$.

Case 3: $G(s)$ has several poles on the imaginary axis.
If $s = 0$ does not belong to these poles, $y_e(t)$ is a permanent oscillation consisting of possibly several harmonic oscillations. If $s = 0$ is a pole of $G(s)$, this oscillation is superposed by the polynomial $p_0(t)$.

Case 4: $G(s)$ has poles in the right half-plane.

In this case, the absolute value of $y_e(t)$ exceeds all limits for $t \to \infty$.

The following application of the final-value theorem of the Laplace transform to stable systems is very useful.

Let $G(s)$ be a stable rational transfer function. Then

$$\lim_{t\to\infty} y_e(t) = G(0). \tag{2.19}$$

This can be seen as follows. Since $G(s)$ is stable, the limit for $t \to \infty$ exists. Thus, from the final-value theorem we get (using that $s = 0$ is not a pole of $G(s)$)

$$\begin{aligned}\lim_{t\to\infty} y_e(t) &= \lim_{s\to 0}\left[s\,\hat{y}_e(s)\right] = \lim_{s\to 0}\left[s\,G(s)\frac{1}{s}\right] \\ &= \lim_{s\to 0} G(s) = G(0).\end{aligned}$$

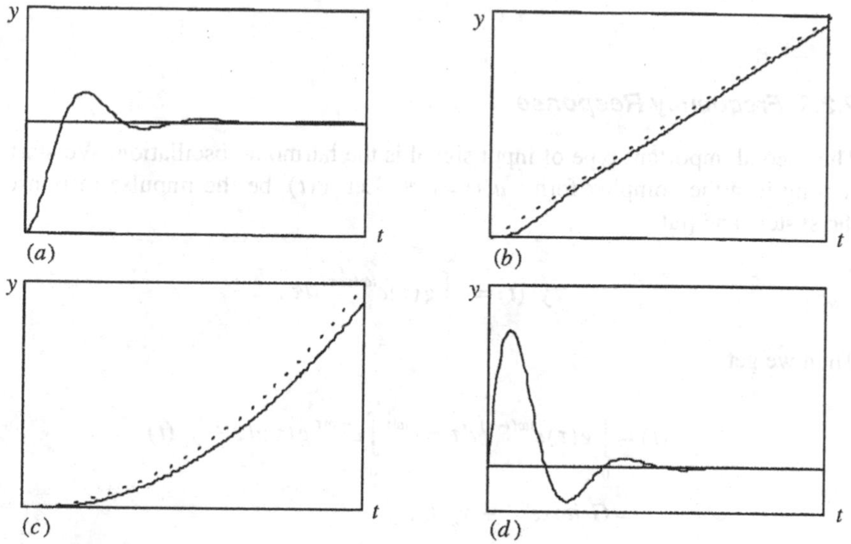

Fig. 2.8. Step responses for plants of type 0 (a), 1 (b), 2 (c), −1 (d)

We conclude this section by studying examples for the different types of transfer functions which were introduced in Sect. 2.2.1. Here, G is written in the normalized form (2.12) and Z/N is assumed to be stable.

Plants with of type 0: From the preceding result we get

$$\lim_{t\to\infty} y_e(t) = k.$$

This means that for large times the step response approximately coincides with the gain k. For small times the behavior of the step response may be quite different.

Plants of type $\alpha > 0$: In this case, $G(s)$ may be interpreted as a series combination of a stable system with proportional behavior and one or more pure integrators. Hence, the step response of the system with proportional behavior will be integrated one or two times. For large times, it is a straight line with slope k or a parabola.
Plants of type -1: In this case, the step response of the system with proportional behavior will be differentiated. Consequently, for large times the step response of the system tends to 0.

Fig. 2.8 shows the step responses for the system

$$G(s) = \frac{k}{s^{\alpha}} \frac{1}{1 + 2dTs + T^2 s^2}$$

and the cases $\alpha = 0, 1, 2, -1$ (parts (a), (b), (c), (d) of Fig. 2.8, respectively). The dotted line in Fig. 2.8(b) shows the step response of the pure integrator and the dotted line in Fig. 2.8(c) shows the step response of the double integrator.

2.2.3 Frequency Response

The second important type of input signal is the harmonic oscillation. We start by writing it in the complex form: $u(t) = e^{i\omega t}$. Let $g(t)$ be the impulse response of the system and put

$$y_a(t) = -\int_t^{\infty} g(\tau) e^{i\omega(t-\tau)} d\tau .$$

Then we get

$$\begin{aligned} y(t) &= \int_0^t g(\tau) e^{i\omega(t-\tau)} d\tau = e^{i\omega t} \int_0^{\infty} e^{-i\omega\tau} g(\tau) d\tau + y_a(t) \\ &= G(i\omega) e^{i\omega t} + y_a(t). \end{aligned}$$

The above calculation assumes the existence of the integral

$$\int_0^{\infty} e^{-i\omega\tau} g(\tau) d\tau ,$$

which is true if g is absolutely integrable. In this case we also have

$$| y_a(t) | = \left| \int_t^{\infty} g(\tau) e^{i\omega(t-\tau)} d\tau \right| \le \int_0^{\infty} | g(\tau) | \, d\tau - \int_0^t | g(\tau) | d\tau .$$

The right-hand side of this inequality tends to 0 for $t \to \infty$. For a rational transfer function, stability implies the absolute integrability of g.

For a real harmonic oscillation $\sin \omega t$ as the input we use the representation $\sin \omega t = \mathrm{Im}(e^{i\omega t})$. If we put

$$A = |G(i\omega)|, \quad \varphi = \arg G(i\omega), \tag{2.20}$$

then the above calculation implies

$$\begin{aligned} y(t) &= \mathrm{Im}(G(i\omega)e^{i\omega t}) + \mathrm{Im}\, y_a(t) \\ &= \mathrm{Im}(Ae^{i(\omega t+\varphi)}) + \mathrm{Im}\, y_a(t) \\ &= A\sin(\omega t + \varphi) + \mathrm{Im}\, y_a(t). \end{aligned}$$

Thus we have shown the following result.

Theorem 2.2.2. *Let $G(s)$ be a stable transfer function. Then the following assertions hold.*
(a) For a complex oscillation $u = e^{i\omega t}$ as the input signal, the output is

$$y(t) = G(i\omega)e^{i\omega t} + y_0(t).$$

(b) For a real harmonic oscillation $u(t) = \sin \omega t$ as the input signal, the output is

$$y(t) = A\sin(\omega t + \varphi) + y_0(t).$$

In both cases, $y_0(t)$ is a (complex or real, respectively) function which tends to 0 for $t \to \infty$. The amplitude and phase can be calculated from $G(s)$ by applying the equations (2.20).

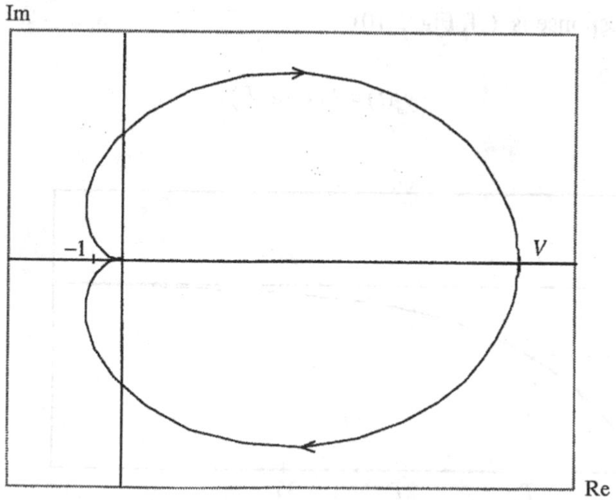

Fig. 2.9. Nyquist plot of a transfer function with 3 negative real poles and gain V.

The complex-valued function $G(i\omega)$ with $-\infty \le \omega < \infty$ is by definition the **frequency response** of the system. It plays an important role in the synthesis of controllers. Drawn as a curve in the complex plane, $G(i\omega)$ is the **Nyquist plot** of the system. The plot of $|G(i\omega)|$, expressed in dB as

$$|G(i\omega)|_{dB} = 20\log|G(i\omega)|,$$

is the **magnitude plot** of the system. The **phase plot** is obtained by drawing $\varphi = \arg G(i\omega)$. Here, for the frequency ω a logarithmic scale is used. The magnitude plot and phase plot together give the **Bode plot.**

2.3 Elementary Transfer Functions

2.3.1 Type-0 Systems of First or Second Order

Type-0 system of first order. A system of type 0 and order 1 without a zero can be written as a differential equation in the form

$$T\dot{y} + y = ku.$$

Here, we assume as the initial condition $y(0) = 0$. The so-called **time constant** T may be positive or negative (but $T \neq 0$). The transfer function of this system is

$$G(s) = \frac{k}{1+Ts}$$

and the step response is (cf. Fig. 2.10)

$$y_e(t) = k(1 - e^{-\frac{t}{T}}).$$

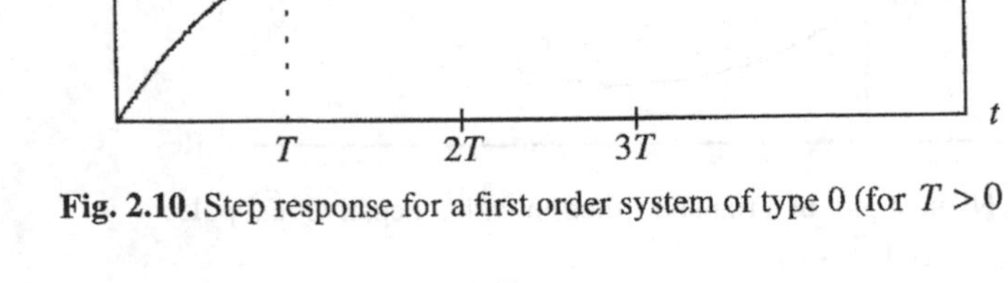

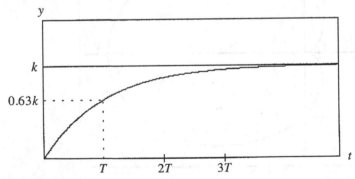

Fig. 2.10. Step response for a first order system of type 0 (for $T > 0$)

For $t=T$ the step response y_e reaches a value of 63% of the constant k (if $T>0$). The Bode plot (cf. Fig. 2.11) can be calculated from

$$|G(i\omega)| = \frac{k}{\sqrt{1+T^2\omega^2}}, \qquad \arg(G(i\omega)) = -\arctan T\omega .$$

Thus the magnitude in dB is

$$|G(i\omega)|_{dB} \approx \begin{cases} k_{dB}, & \text{if } \omega \ll \omega_0, \\ k_{dB} - 20\log T\omega, & \text{if } \omega \gg \omega_0. \end{cases}$$

This is a piecewise linear approximation of the magnitude plot. Here, $\omega_0 = 1/T$ is the **break frequency** of the system. It is easy to see that the true magnitude is at this frequency 3dB smaller then the approximation.

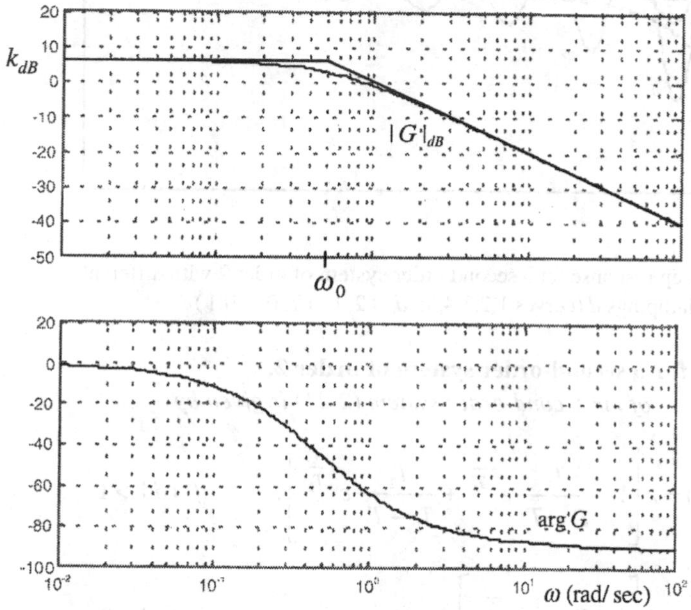

Fig. 2.11. Bode plot for a first order system of type 0 (for $T>0$)

Type-0 System of second order. An important kind of system of type 0 and order 2 is described by the transfer function

$$G(s) = \frac{k}{1+2dTs+T^2s^2}. \tag{2.21}$$

Here, T is the **time constant,** $\omega_0 = 1/T$ is **the break frequency** and d is the **damping**. The transfer function has the poles

$$s_{1,2} = -d\omega_0 \pm \sqrt{d^2 - 1}\,\omega_0 i\,.$$

They can be real or complex. For $|d| \geq 1$ they are real and in this case the system of second order can be written as a series combination of two first-order systems with time constants $T_1 = -1/s_1$, $T_2 = -1/s_2$.
The step response is easily obtained by evaluating (2.18), cf. Fig. 2.12.

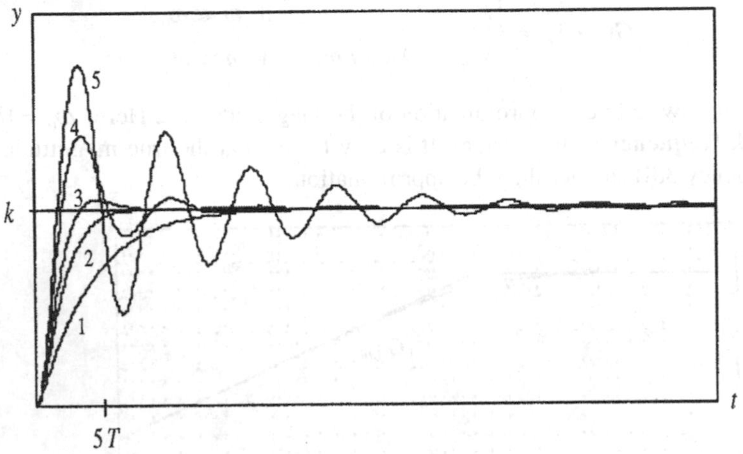

Fig. 2.12. Step response for a second order system of order 2 with different dampings d (curves 1,2,3,4,5: $d = 2, 1, 0.7, 0.3, 0.1$)

Step response for a second-order system of order 2.
The step response of the second-order system (2.21) is given by

$$y_e(t) = k\left[1 - \frac{T_1}{T_1 - T_2} e^{-\frac{t}{T_1}} + \frac{T_2}{T_1 - T_2} e^{-\frac{t}{T_2}}\right], \qquad \text{if } |d| > 1,$$

$$y_e(t) = k\left[1 - (1 + \frac{t}{T}) e^{-\frac{t}{T}}\right], \qquad \text{if } |d| = 1,$$

$$y_e(t) = k\left[1 - \frac{e^{-d\omega_0 t}}{\sqrt{1-d^2}} \sin(\sqrt{1-d^2}\,\omega_0\, t + \psi)\right], \qquad \text{if } |d| < 1,$$

where ψ is defined by $\cos\psi = d$.

From the above figure it is seen that for $0 < d < 1$ the step response oscillates with a decreasing amplitude. With decreasing damping these oscillations get larger.

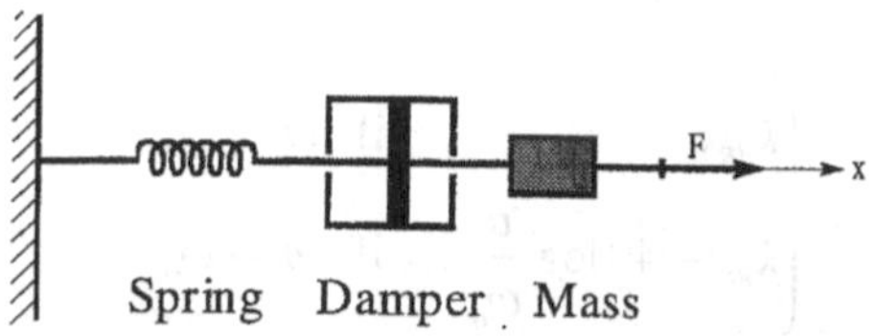

Fig. 2.13. Spring–mass–damper system

Example: *Spring–mass–damper system.*
For the mechanical system shown in Fig. 2.13 the acting forces are the force cx caused by the spring, the force $r\dot{x}$ caused by friction and the force $m\ddot{x}$. Besides these forces, it is assumed that an external force F is given. Then the equilibrium of the forces is

$$m\ddot{x} + r\dot{x} + cx = F .$$

This leads to a second-order system with

$$k = \frac{1}{c}, \quad T = \sqrt{\frac{m}{c}}, \quad d = \frac{r}{2\sqrt{cm}} .$$

We have $d < 1$ if and only if $r < \sqrt{cm}$. In this case, the system is able to oscillate.

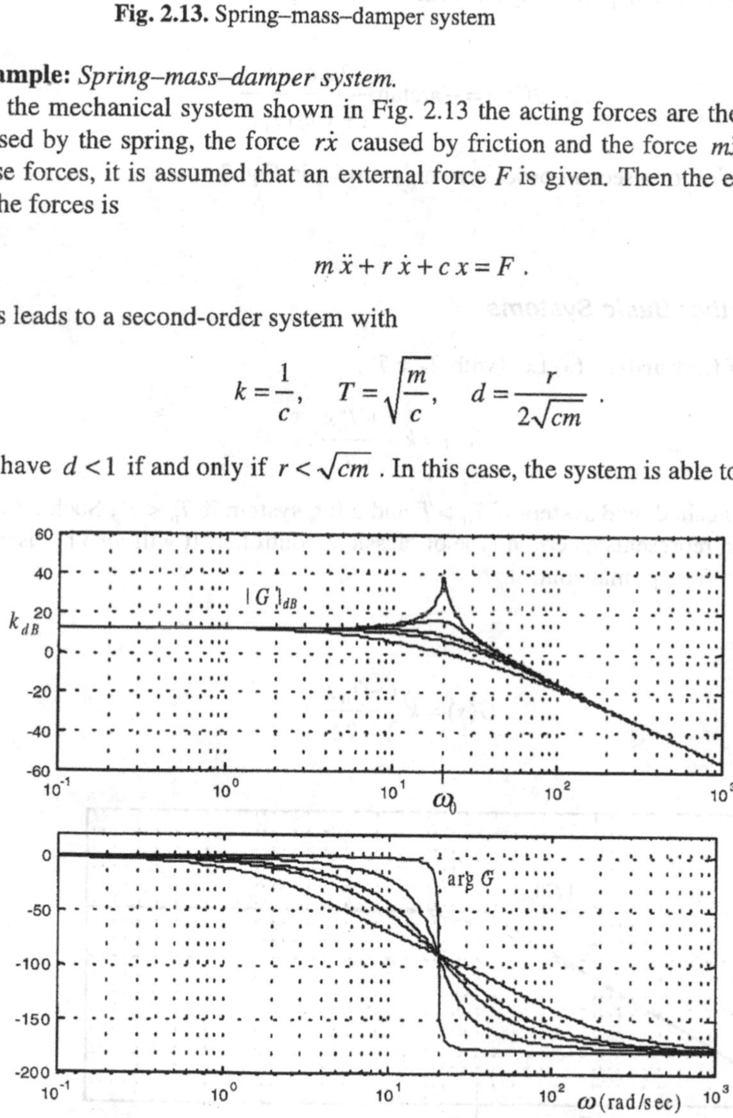

Fig. 2.14. Frequency response for a second-order system of order 2 for the dampings $d = 2, 1, 0.7, 0.3, 0.05$

A short calculation shows that a piecewise linear approximation of the second order system is

$$|G(i\omega)|_{dB} \approx \begin{cases} k_{dB}, & \text{if } \omega \ll \omega_0, \\ k_{dB} - 40\log\dfrac{\omega}{\omega_0}, & \text{if } \omega \gg \omega_0. \end{cases}$$

Hence the magnitude plot decays at a rate of 40dB per decade. The phase is given by

$$\arg G(i\omega) = -\arctan\frac{2d\,\omega/\omega_0}{1-(\omega/\omega_0)^2}.$$

The Bode plot for a second-order system is shown in Fig. 2.14.

2.3.2 Further Basic Systems

Systems of first order. (a) Let (with $T_0 \neq T$)

$$G(s) = k\frac{1+T_0 s}{1+Ts}.$$

This is a so called **lead system** if $T_0 > T$ and a **lag system** if $T_0 < T$. Such a transfer function represents a certain type of classical controller. It will also be used as a weight in $\mathcal{H}_\infty$ optimal control.

(b) Let now

$$G(s) = k\frac{1-T_0 s}{1+Ts}.$$

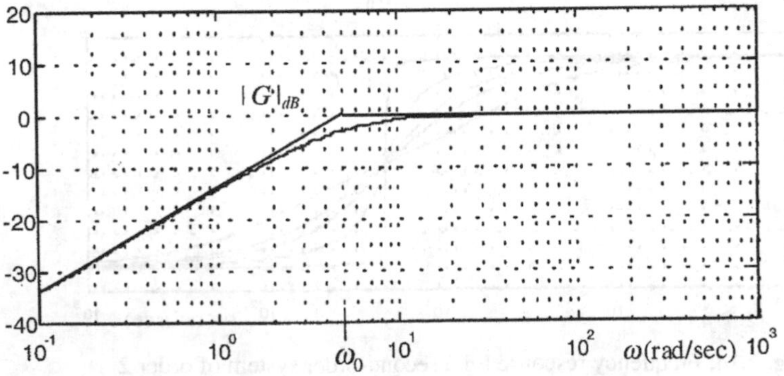

Fig. 2.15. Magnitude plot of a realizable differentiator

Then $G(s)$ has a zero in the right half-plane. It is typical for such a system that the step response at first tends into the "wrong" direction. A case of particular interest is given by $T_0 = T,\ k = 1$:

$$G(s) = \frac{1 - Ts}{1 + Ts}\ . \tag{2.22}$$

Therefore, we have $|G(i\omega)| = 1$. This transfer function describes an **all-pass** system. It leaves all harmonic oscillations unchanged with respect to their amplitude.

(c) Next we consider a realizable differentiator of the form

$$G(s) = \frac{kTs}{1 + Ts}\ .$$

Such a kind of transfer function occurs in control systems in connection with disturbances. Fig. 2.15 shows the Bode plot. It is seen that due to the differentiation the amplitude of oscillations with a low frequency is reduced (in particular, oscillations "with frequency zero" do not pass the system).

Systems of second order. Another kind of differentiator is given by

$$G(s) = \frac{Ts}{1 + 2dTs + T^2 s^2}\ .$$

Its step responses have been shown in Fig. 2.8 and for low frequencies, its Bode plot resembles the Bode plot of the differentiator in the previous example. For high frequencies, the magnitude decays at a rate of 20 dB/decade.

Next, for $d = 0.5\sqrt{2}$ let

$$G(s) = \frac{1 + \lambda T_0 s}{1 + 2dTs + T^2 s^2} \quad \text{for} \quad \lambda = -1, 0, 1\ . \tag{2.23}$$

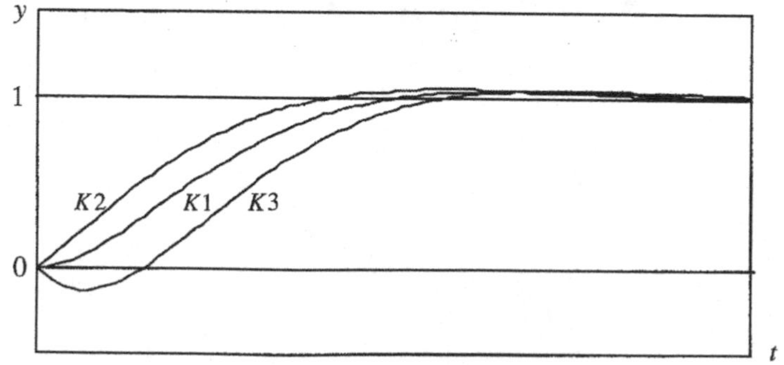

Fig. 2.16. Step responses for $G(s)$ defined in (2.23)
(K1: $\lambda = 1$; K2: $\lambda = 0$; K3: $\lambda = -1$)

The effect of the zero can be seen from Fig 2.16, where the step responses are plotted.

A second-order all-pass transfer function has the form (for $d \geq 0$)

$$G(s) = \frac{T^2 s^2 - 2dTs + 1}{T^2 s^2 + 2dTs + 1}. \tag{2.24}$$

Minimum-phase transfer functions. A proper rational transfer function $G(s)$ is said to be **minimum-phase** if it has no zeros in the open RHP. The reason for this name can be explained as follows. Let F be an all-pass transfer function. Then GF has the same magnitude as G, but a greater phase. Thus, of all the transfer functions having the magnitude of G, the one with minimum-phase is G.

Let now G be a proper rational transfer function and denote the zeros of G in the open RHP by $z_1, \ldots, z_l$. Then the transfer function

$$G_{ap}(s) = \prod_{k=1}^{l} \frac{s - z_k}{s + \overline{z}_k} \tag{2.25}$$

is all-pass and the transfer function

$$G_{mp} = \frac{G}{G_{ap}}$$

is minimum-phase. Thus G can be written as a series combination of a minimum-phase transfer function and an all-pass transfer function:

$$G = G_{mp} G_{ap}.$$

If s_k is real, we can write the corresponding factor in (2.25) in the form (2.22). If s_k is complex, then $\overline{s}_k$ is also a zero of G and the product of the factors of (2.25) belonging to $s_k, \overline{s}_k$ can be written in the form (2.24).

Chapter 3

SISO Feedback Systems

In this chapter, the closed-loop control system will be introduced for SISO plants. We start by calculating the transfer functions from the input signals (command r, disturbance d, and noise n) to the plant output y and to the internal quantities (the control error $r-y$ and the control u). The most basic requirement for a feedback loop is that all these transfer functions have to be stable. Then the control loop is said to be internally stable. Next, we consider the closed-loop system for constant inputs and require that in this case the steady-state control error vanishes, if no noise is present. The leads to the PID controller, which is the most important controller in classical control theory. The main tool for synthesizing such a controller uses the Nyquist stability theorem, which plays an important role in modern control theory, too. It is presented in Sect. 3.4.

In almost all cases controller design is a trade-off between performance and stability. In Sect. 3.5.1 it is shown how performance can be specified by using weighting functions and it is seen that controller synthesis leads to an $\mathcal{H}_\infty$ optimization problem. In Sect. 3.5.2 we take a first look at robust stability. Robust stability means that the closed-loop system is not only stable for the nominal system for which the controller has been designed, but also for the real physical system. For a special kind of perturbation, a necessary and sufficient condition for robust stability is given. It is important to notice that the ideas leading to this result can be generalized to the MIMO case and more general types of perturbations. In modern control one wants even more: the closed-loop system has not only to be robustly stable, it is also required to have robust performance. This means that the control loop for the real plant fulfills the same performance specification as for the nominal plant. A first analysis of this problem is given in Sect. 3.5.3. Again it is seen that $\mathcal{H}_\infty$ optimization comes into play at the very beginning of controller synthesis.

3.1 The Basic Transfer Functions

Structure of the closed-loop system, transfer functions. We assume that a plant is given which has two inputs: The **actuating signal** u_A and a **disturbance** d. Both of them influence the plant output z, which is the **variable to be controlled.** Then the plant can be described by two transfer functions, namely $G_1(s)$ for the actuating signal and $G_2(s)$ for the disturbance. The signal u_A will be generated by the actuator with transfer function $G_A(s)$. The input for the actuator is the **control** u. The plant output is measured by a **sensor** with transfer function $G_M(s)$ and output y. It is corrupted by an additive **measurement error** n. The

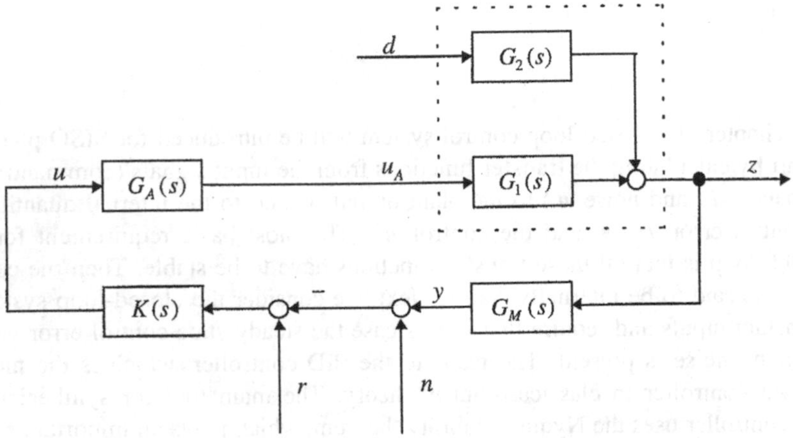

Fig. 3.1. Physical structure of a closed-loop system

result is the measured plant output $y+n$. The **command** (or **reference input**) will be denoted by r. Finally, the controller has the transfer function $K(s)$. The block diagram in Fig. 3.1 shows the connection of these systems and describes the structure of a SISO closed-loop system.

By defining transfer functions

$$G(s) = G_M(s)G_1(s)G_A(s)$$

$$G_d(s) = G_M(s)G_2(s),$$

the representation of the closed-loop system can be simplified. Here, $G(s)$ and $G_d(s)$ are the transfer functions of the plant extended by the actuator and sensor dynamics, and this system will again be denoted as the plant. Using these transfer functions, the closed-loop system can be drawn as shown in Fig. 3.2. It has the command r, the disturbance d and the measurement noise n as inputs. In this diagram, the plant output y is the output of the original plant after having passed the sensor dynamics. The "physical" variable to be controlled is z. The variables y and z differ by the sensor dynamics. It is a common practice to denote y as

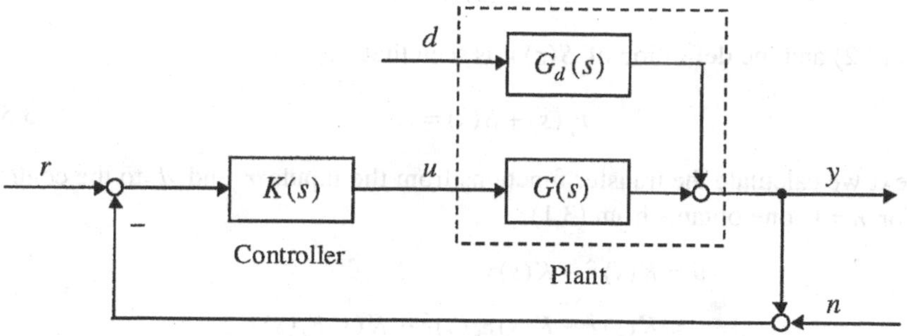

Fig. 3.2. Standard feedback configuration

the variable to be controlled or as the plant output. Having these ideas in mind, we may consider Fig. 3.2 as the standard closed-loop system for SISO plants.

Next, we want to derive the transfer functions for the different inputs. The controller and the plant are a series combination that leads to the so-called **loop transfer function**

$$L(s) = K(s)G(s) .$$

From Fig. 3.2 we obtain

$$\hat{y} = G_d(s)\hat{d} + L(s)(\hat{r} - \hat{y} - \hat{n}) .$$

Thus $\hat{y}$ is given by

$$\hat{y} = F_r(s)\hat{r} + F_d(s)\hat{d} - F_r(s)\hat{n} , \tag{3.1}$$

where

$$F_r(s) = \frac{L(s)}{1 + L(s)} \tag{3.2}$$

$$F_d(s) = \frac{G_d(s)}{1 + L(s)} . \tag{3.3}$$

The function $F_r(s)$ is called the **complementary sensitivity function** (the reason for this name will become clear immediately from (3.5) below). If the disturbance affects the output directly, i.e. if $G_d(s) = 1$, then

$$\hat{y} = S(s)\hat{d} ,$$

where

$$S(s) = \frac{1}{1 + L(s)}$$

is the so-called **sensitivity function.** Using this function, we can write

$$F_d(s) = S(s)G_d(s)\ . \tag{3.4}$$

From (3.2) and the definition of $S(s)$ it is seen that

$$F_r(s) + S(s) = 1\ . \tag{3.5}$$

Next we calculate the transfer functions from the inputs r and d to the control u. For $n = 0$ one obtains from (3.1)

$$\begin{aligned}\hat{u} &= K(s)\hat{r} - K(s)\hat{y}\\ &= K(s)\hat{r} - K(s)F_r(s)\hat{r} - K(s)F_d(s)\hat{d}\\ &= K(s)(1 - F_r(s))\hat{r} - K(s)F_d(s)\hat{d}\ .\end{aligned}$$

Thus (3.4) and (3.5) give

$$\hat{u} = K(s)S(s)\hat{r} - K(s)G_d(s)S(s)\hat{d}\ . \tag{3.6}$$

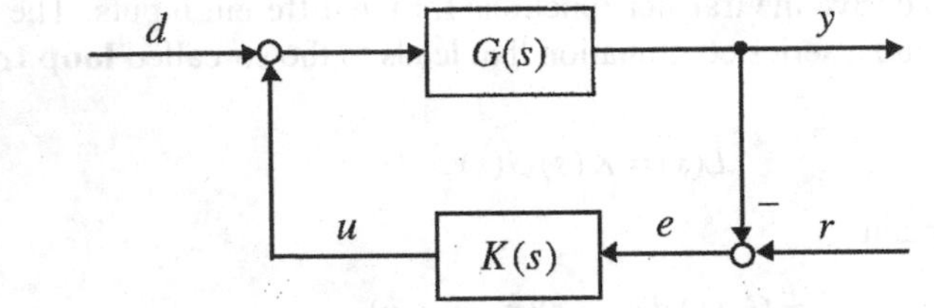

Fig. 3.3. Simplified standard feedback configuration

If the disturbance is directly added to the control and if no noise is present, the standard feedback configuration reduces to the block diagram of Fig. 3.3. Here, we have introduced the **control error** $e = r - y$. An application of (3.1) and (3.6) leads immediately to

$$\hat{e} = S(s)\hat{r} - S(s)G(s)\hat{d} \tag{3.7}$$

$$\hat{u} = K(s)S(s)\hat{r} - F_r(s)\hat{d}\ . \tag{3.8}$$

Thus for the simplified standard feedback configuration the complete input/output behavior is described by the four transfer functions S, SG, KS, F_r.

3.2 Internal Stability

The transfer function $G(s)$ of the plant is proper because it is the mathematical model of a real physical system. Since the controller is required to be realizable, it is a standing assumption that $K(s)$ is proper. The simplified feedback configuration of Fig. 3.3 is said to be **well-posed** if the closed-loop transfer functions S, SG, KS, F_r are proper. It is easy to see that the following holds:

The simplified standard feedback configuration is well-posed if and only if $L(\infty) \neq -1$.

In what follows, this will always be assumed. In particular, well-posedness is given if the plant is strictly proper.

The input/output behavior of the closed-loop system is described by the equation

$$\hat{y} = F_r \hat{r} + SG\hat{d}\,.$$

Hence, the closed loop is stable if and only if the transfer functions F_r and SG are stable. Actually, we require more, namely that the transfer functions from the external quantities r, d to the internal quantities e, u are also stable. This leads to the following definition.

Definition. *A feedback system is said to be* ***internally stable*** *if the four transfer functions* S, SG, KS, F_r *are stable.*

Internal stability is the most important requirement for a closed-loop system. It will always be posed. Therefore, we first ask for a criterion for internal stability. With coprime polynomials Z, N and Z_K, N_K let

$$G = \frac{Z}{N}\,, \qquad K = \frac{Z_K}{N_K}\,.$$

We then obtain the representations

$$S = \frac{NN_K}{ZZ_K + NN_K}\,, \quad F_r = \frac{ZZ_K}{ZZ_K + NN_K}\,, \tag{3.9}$$

$$KS = \frac{Z_K N}{ZZ_K + NN_K}\,, \quad GS = \frac{ZN_K}{ZZ_K + NN_K}\,. \tag{3.10}$$

Theorem 3.2.1. *A feedback system is internally stable if and only if one of the following conditions is fulfilled:*

(i) *The polynomial* $ZZ_K + NN_K$ *has no zeros in* $\overline{\mathbb{C}}_+$.

(ii) *The sensitivity function* S *is stable and there is no cancellation of a linear factor* $s - s_0$ *with* $s_0 \in \overline{\mathbb{C}}_+$ *when the loop transfer function* $L = KG$ *is formed.*

Proof. First, we show that internal stability is equivalent to (i). It is immediately seen from the (3.9) and (3.10) that the given condition is sufficient for internal stability. Necessity is not so obvious, since it could possibly happen that a linear factor $s - s_0$ with $s \in \overline{\mathbb{C}}_+$ will be cancelled in each of the four fractions. We assume this and show that a contradiction will arise. The assumption implies that $s - s_0$ is a linear factor of $ZZ_K + NN_K$ and that the following statements hold:

(a) $s - s_0$ is a linear factor of N or N_K;
(b) $s - s_0$ is a linear factor of Z or Z_K;
(c) $s - s_0$ is a linear factor of Z_K or N;
(d) $s - s_0$ is a linear factor of Z or N_K.

Case 1: $s - s_0$ is a linear factor of N. Then $s - s_0$ is not a linear factor of Z and, because of (d), $s - s_0$ is a linear factor of Z_K and because of (b) also a linear factor of Z_K, which is in contradiction to the coprimeness of Z_K and N_K.
Case 2: $s - s_0$ is a linear factor of N_K. Then $s - s_0$ is not a linear factor of Z_K and because of (b) $s - s_0$ is a linear factor of Z and, because of (c), also a linear factor of N, which is in contradiction to the coprimeness of Z and N.

We now show that internal stability is equivalent to (ii). Let the feedback system be internally stable. Then S is stable and $ZZ_K + NN_K$ has no zeros in $\overline{\mathbb{C}}_+$. Hence neither (Z, N_K) nor (Z_K, N) has a divisor $s - s_0$ with $s \in \overline{\mathbb{C}}_+$, and consequently (ii) is fulfilled. Let now (ii) be true. Assume that s_0 is a zero of $ZZ_K + NN_K$ with $s \in \overline{\mathbb{C}}_+$. At first suppose $(NN_K)(s_0) = 0$. Then we have $(ZZ_K)(s_0) = 0$ and $s - s_0$ is a linear factor of ZZ_K and NN_K, which is in contradiction to the assumption. Hence it can be supposed that $(NN_K)(s_0) \neq 0$. Thus no factor $s - s_0$ with $s \in \overline{\mathbb{C}}_+$ cancels when the sensitivity function

$$S = \frac{NN_K}{ZZ_K + NN_K}$$

is formed. Since S is stable, (i) must hold and the feedback loop is internally stable.

Condition (ii) may be concisely formulated as follows: when the loop transfer function $L = KG$ is formed, the controller is not allowed to cancel a pole or a zero of the plant that lies in the closed RHP.

Concerning the disturbance, we return to the general case. Then the input/output stability of the feedback loop is described by the stability of the transfer functions F_r and SG_d and internal stability is characterized by the stability of the transfer functions S, G_dS, KS, KG_dS. Let G_d be proper and write

$$G_d = \frac{Z_d}{N_d}.$$

Then again, the feedback loop is well-posed if and only if

$$L(\infty) \neq -1$$

and we have

$$G_d S = \frac{N}{N_d} \frac{Z_d N_K}{ZZ_K + NN_K}, \quad KG_d S = \frac{N}{N_d} \frac{Z_d Z_K}{ZZ_K + NN_K}.$$

Let N_d be a divisor of N. Then we can write $N = N_d N_0$ with a polynomial N_0. This assumption is in a natural way normally fulfilled, as we shall see when state-space representations are considered. The following formulas hold:

$$G_d S = \frac{Z_d N_K N_0}{ZZ_K + NN_K}, \quad KG_d S = \frac{Z_d Z_K N_0}{ZZ_K + NN_K}. \tag{3.11}$$

These transfer functions are stable if the feedback loop is internally stable.

Examples. 1. We first consider an example where the controller cancels an unstable pole of the plant. Let (with $T > 0, k_R > 0$)

$$G(s) = \frac{1}{1 - Ts}, \quad K(s) = k_R \frac{1 - Ts}{s}.$$

Then

$$S = \frac{s}{s + k_R}, \quad GS = \frac{s}{(s + k_R)(1 - Ts)}, \quad KS = k_R \frac{1 - Ts}{s + k_R}, \quad F_r = \frac{k_R}{s + k_R}.$$

and consequently, the feedback loop is not internally stable.

2. In the second example, the controller cancels a zero of the plant in the RHP. Let (with $T > 0$, $T_1 > 0$)

$$G(s) = \frac{1 - T_1 s}{1 + Ts}, \quad K(s) = k_R \frac{1 + Ts}{s(1 - T_1 s)}.$$

Then we get

$$S = \frac{s}{k_R + s}, \quad GS = \frac{s(1 - T_1 s)}{(1 + Ts)(k_R + s)},$$

$$KS = \frac{k_R(1 + Ts)}{(1 - T_1 s)(k_R + s)}, \quad F_r = \frac{k_R}{s + k_R}.$$

Again, the feedback loop is not internally stable. It should be noted that the instability of the controller does not cause this problem. In fact, there are internally stable control loops with an instable controller.

3.3 Stationary Behavior of the Feedback Loop

The main task of the feedback loop is to compensate disturbances and to track the command r. This means that, also in the case where a disturbance is present, the control error $e = r - y$ ideally has to be made to zero. We ask under which conditions this will be the case. From (3.1), (3.5) and (3.4) it follows that

$$\hat{e} = S(s)\hat{r} - (SG_d)(s)\hat{d} + (1 - S(s))\hat{n}. \tag{3.12}$$

Suppose now that the feedback loop is internally stable. Then the transfer functions occurring in (3.12) are stable. Moreover, let the command, disturbance and noise be **constant**: $r = r_0$, $d = d_0$ and $n = n_0$. Then (3.12) shows that the limit e_∞ of $e(t)$ for $t \to \infty$ is given by

$$e_\infty = S(0)r_0 - (SG_d)(0)d_0 + (1 - S(0))n_0. \tag{3.13}$$

Thus if there is no noise, the control error is zero if and only if

$$S(0) = 0, \quad (SG_d)(0) = 0. \tag{3.14}$$

If noise is present, then under the assumption of (3.14) the control error is $e_\infty = n_0$, i.e. the control error coincides with the measurement error. Then one has to make a constant part in the measurement error as small as possible by choosing a suitable sensor, and the effect of the command or the disturbance on the control error is minimized by fulfilling condition (3.14).

Case 1. Suppose that the loop transfer function is of type 0. Then we can define $V = L(0)$. If G_d has no pole at $s = 0$ (which is not a severe restriction), we obtain

$$e_\infty = \frac{1}{1+V} r_0 - \frac{G_d(0)}{1+V} d_0$$

(if $n_0 = 0$). Thus, a small stationary control error requires V to be large.

Case 2. Assume now that the loop transfer function contains at least one integrator:

$$L(s) = \frac{p(s)}{s^\alpha q(s)} \quad \text{with } \alpha > 0.$$

Then we get

$$S(s) = \frac{s^\alpha q(s)}{p(s) + s^\alpha q(s)}.$$

In this case we cannot simply assume that G_d has no pole at $s=0$. Let $s=0$ be a pole of G_d with multiplicity $l \geq 0$. It is now again necessary to consider different cases. At first, let the controller be of type 1. Then we have $\alpha > l$ and $s=0$ is a zero of SG_d. This leads to the following important result:

If the controller is integrating (more precisely, if it is of type 1), then the stationary control error vanishes with respect to the command as well as with respect to the disturbance.

Now let the controller not be integrating. Then, by our assumption concerning L, the plant must have a pole at $s=0$ with multiplicity α. In this case, there is no stationary control error with respect to the command. There is also no stationary control error with respect to the disturbance if and only if $\alpha > l$. This means that the plant contains at least one more integrator with respect to the control than with respect to the disturbance.

From these considerations, we conclude that the controller should be integrating. In its simplest form, such a controller feeds back the control difference $e = r - y$ together with its integral. Slightly more general is the so-called **PID controller**. It is of the form

$$K(s) = k_P\left(1 + \frac{1}{T_N s} + T_V s\right) \quad \text{with } T_N > 0,\ T_V > 0. \tag{3.15}$$

This controller is the most commonly used one in classical control (together with the special cases $T_V = 0$ (PI controller) and $T_N \to \infty$ (PD controller)). Often the PID controller is written in the form

$$K(s) = k_R \frac{(1 + T_I s)(1 + T_D s)}{T_I s} \quad \text{with } T_I \geq T_D > 0. \tag{3.16}$$

This representation is somewhat more special than that of (3.15). The controller of (3.15) can be written in the form (3.16) if and only if $T_N \geq 4T_V$.

In the above definition, the PID-controller is not proper. It can be made proper (and realizable) by modifying it as follows:

$$K(s) = k_P \frac{1 + T_N s + T_N T_V s^2}{T_N s(1 + T_0 s)}.$$

Note that the realizable PID-controller is not in a position to cancel the linear factors $s - s_0$ with $s \in \overline{\mathbb{C}}_+$ from $G(s)$ (if $s=0$ is not a zero of $G(s)$), neither from the denominator nor from the numerator. Hence, for a feedback loop with such a controller it is not necessary to distinguish between input/output stability and internal stability (cf. Theorem 3.2.1).

3.4 Nyquist Stability Criterion

The most important stability criterion for the feedback loop is the so-called Nyquist stability criterion. It allows us to determine from the properties of the Nyquist plot $L(i\omega)$ of the loop transfer function the stability of the feedback loop. Moreover, it can be used for analyzing and synthesizing controllers for SISO feedback systems. This is why it is of such great importance in classical control theory. We start with a well-known theorem from complex analysis.

Theorem 3.4.1. *Let $f(s)$ be a meromorphic non-constant function on an open set $\Omega \subset \mathbb{C}$ and let C be the piecewise regular boundary curve of a compact subset K of Ω, which has a positive orientation. Suppose that $f(s)$ has neither a pole nor a zero on C. Finally, denote by Z_C the number of zeros and by P_C the number of poles $f(s)$ in K (counted according to their multiplicities). Then*

$$\frac{1}{2\pi i}\int_C \frac{f'(s)}{f(s)}ds = Z_C - P_C .$$

It is possible to give a different representation of the above integral. To this end, let C_f be the image curve of C under the map $f(s)$ and denote by $s = h(t)$, $\alpha \le t \le \beta$, a continuously differentiable parameterization of C. Then a parameterization of C_f is given by $z = h_1(t) = f(h(t))$, $\alpha \le t \le \beta$, and we have

$$\begin{aligned}\frac{1}{2\pi i}\int_C \frac{f'(s)}{f(s)}ds &= \frac{1}{2\pi i}\int_\alpha^\beta \frac{f'(h(t))}{f(h(t))}\dot{h}(t)\,dt \\ &= \frac{1}{2\pi i}\int_\alpha^\beta \frac{1}{h_1(t)}\dot{h}_1(t)\,dt \\ &= \frac{1}{2\pi i}\int_{C_f} \frac{1}{z}dz.\end{aligned}$$

In this calculation it is assumed that the origin does not lie on C_f. Then the latter integral is nothing else but the index (or the winding number) $n(C_f, 0)$ of the image curve C_f with respect to the origin. Hence the following corollary is true, which is the argument principle in complex analysis.

Corollary 3.4.1. *In the situation of Theorem 3.4.1 and under the additional assumption that the origin does not lie on the image curve C_f, the number of encirclements $n(C_f, 0)$ of C_f with respect to the origin in the mathematical positive sense is given by*

$$n(C_f, 0) = Z_C - P_C .$$

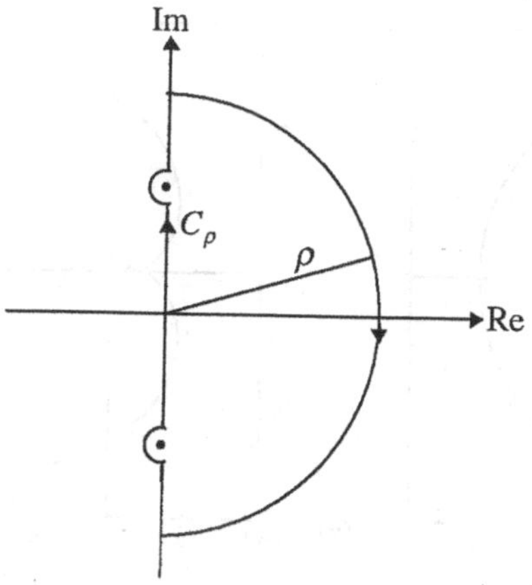

Fig. 3.4. Nyquist contour

Denote by $p(s)$ the denominator of the sensitivity function $S(s)$ and by $q(s)$ the denominator of the loop transfer function $L(s)$. Then we have

$$1+L(s)=\frac{p(s)}{q(s)} .$$

We now apply Corollary 3.4.1 to $1+L(s)$. To this end, it is necessary to fix the curve C. We denote it here by C_ρ and construct it as shown in Fig. 3.4, where ρ is the radius of the half-circle. It is assumed that ρ is chosen so large that C_ρ encircles all poles and zeros of $1+L(s)$, i.e. all poles of the open-loop and the closed-loop system. Moreover, all poles of $L(s)$ on the imaginary axis are encircled by the construction of C_ρ from the left-hand side by a small half-circle. The image curve $C_{\rho L}$ of C_ρ under the map $L(s)$ will be denoted as the **Nyquist plot** and C_ρ as the **Nyquist contour**. An application of Corollary 3.4.1 leads to the following theorem. Note that -1 lies on the Nyquist plot if and only if there exists a frequency ω such that $p(i\omega)=0$.

Theorem 3.4.2 (Nyquist Stability Criterion). *Let n be the number of poles of the open control loop (i.e. of $L(s)$) in the closed right half-plane. Then the sensitivity function is stable if and only if the critical point -1 does not lie on the Nyquist plot $C_{\rho L}$ and if the Nyquist plot does not encircle the critical point for $n=0$ and encircles it n times in the mathematical positive sense for $n>0$.*

Proof. In order to apply Corollary 3.4.1 to $f(s)=1+L(s)$, the curve C_ρ has to be replaced by the curve C_ρ^-, which is obtained from C_ρ by inverting the orientation. Let m be the number of closed-loop poles in the closed right half-plane. Then we get from Corollary 3.4.1 the equation $n(C_{\rho f}^-,0)=m-n$. Hence $S(s)$ is stable if and only if $n(C_{\rho f}^-,0)=-n$. This in turn is equivalent to $n(C_{\rho f},0)=n$ and to $n(C_{\rho L},-1)=n$, which proves the theorem.

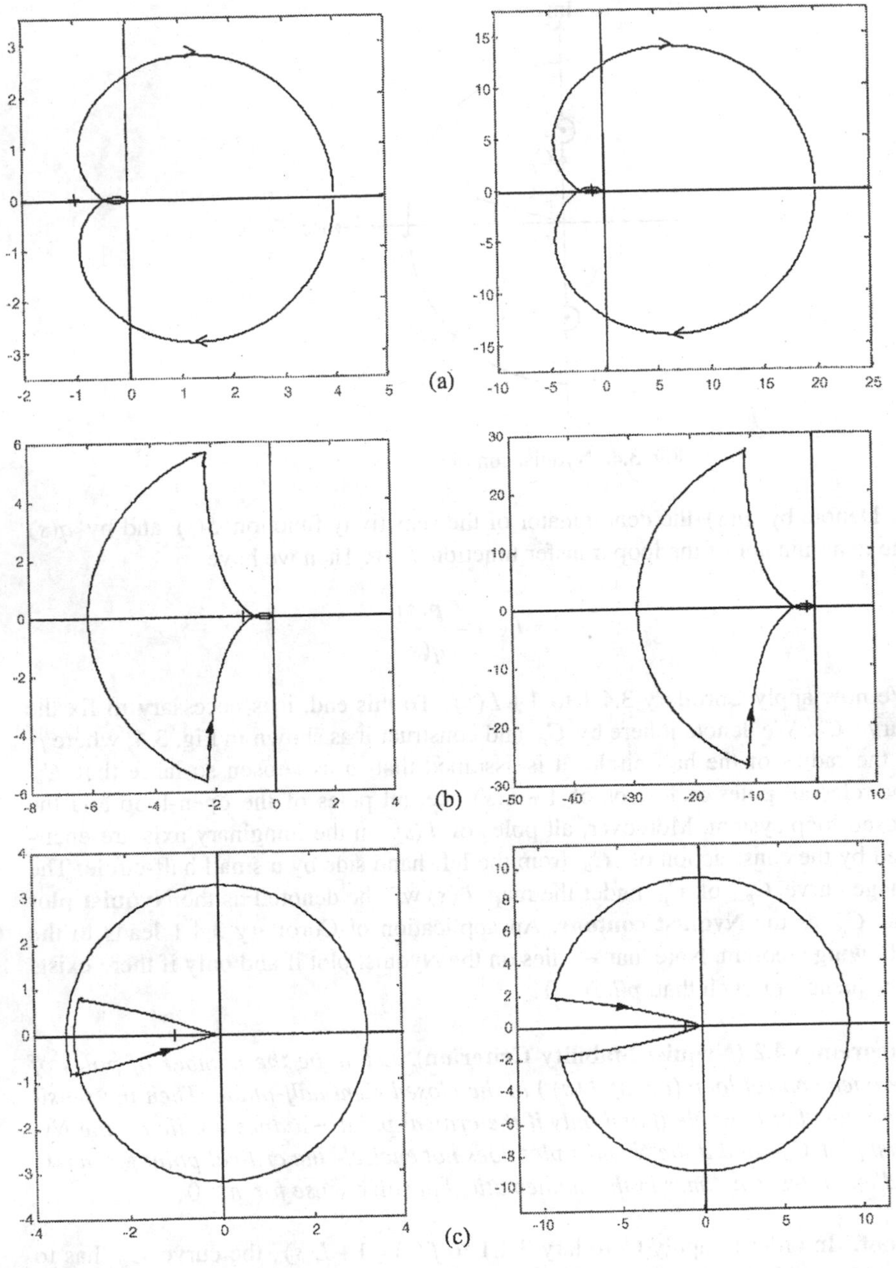

Fig. 3.5. Nyquist plots of the open loop for $L_1(s)$, (a); $L_2(s)$, (b); $L_3(s)$, (c).
Left-hand plots: closed loop stable;
right-hand plots: closed loop unstable.

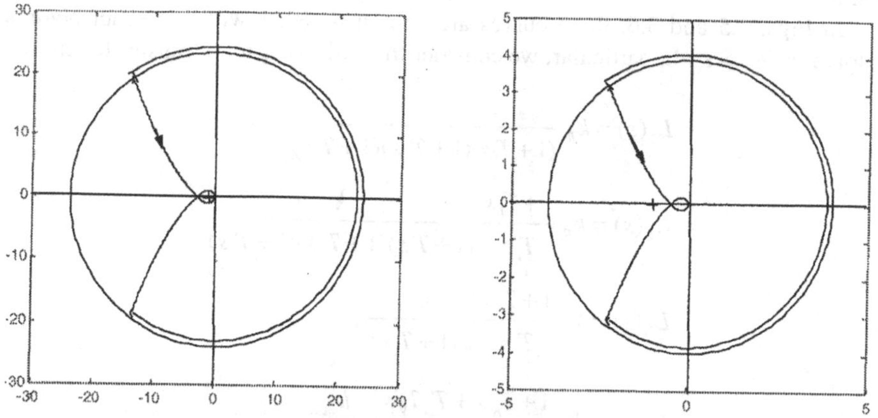

Fig. 3.6. Nyquist plot of the open loop for $L_4(s)$.
Left-hand plots: closed loop stable;
right-hand plots: closed loop unstable.

Remark. At the first sight, the direct application of Theorem 3.4.2 seems to be somewhat difficult, since a priori it is not known how large the radius ρ has to be chosen. Because $L(s)$ is proper, the set $(-\infty,-\rho)\cup(\rho,\infty)$ will be mapped under L on a small neighborhood of a point on the real axis. Since $S(s)$ is realizable, this point cannot be -1. Let C_∞ be the imaginary axis oriented in the sense of increasing frequency ω, where the poles of the open loop are encircled by half circles lying on the left-hand side of the imaginary axis and denote by $C_{\infty L}$, the image curve of C_∞. Then $C_{\infty L}$ has the same index with respect to -1 as $C_{\rho L}$, if ρ is sufficiently large. Therefore, in what follows we will always use $C_{\infty L}$ for stability investigations.

The Nyquist criterion will be mainly applied to plants which are stable, with the possible exception of a pole at the origin. Then $L(s)$ is of the form

$$L(s)=\frac{V}{s^\alpha}L_1(s) \quad \text{with } L_1(0)=1,\ \alpha\geq 0$$

and a stable transfer function $L_1(s)$. In this case, closed-loop stability is given if and only if $n(C_{\rho L},-1)=\alpha$. We now consider as curve a half-circle with radius $\varepsilon>0$ and origin 0, where 0 lies on the right-hand side of the half-circle. This curve will be mapped by $L(s)$ on a section of the Nyquist plot, which for small ε has the following parameterization:

$$w(\varphi)=\frac{V}{\varepsilon^\alpha}e^{i\alpha(\frac{\pi}{2}+\varphi)}, \quad 0\leq\varphi\leq\pi\,.$$

This is an arc of a circle with sections, which are passed round possibly several times.

In Figs. 3.5 and 3.6, these curves are shown together with the exact Nyquist plots for $|\omega| > \varepsilon$. In particular, we consider the following loop transfer functions:

$$L_1(s) = k_R \frac{k}{(1+T_1 s)(1+T_2 s)(1+T_3 s)},$$

$$L_2(s) = k_R \frac{1+T_I s}{T_I s} \frac{k}{(1+T_1 s)(1+T_2 s)(1+T_3 s)},$$

$$L_3(s) = k_R \frac{1+T_I s}{T_I s} \frac{k}{s(1+T_1 s)},$$

$$L_4(s) = k_P \frac{1+T_N s + T_N T_V s^2}{T_N s} \frac{k}{s^2}.$$

In the first example a purely proportional controller is used and in the second and third examples a PI controller is used. In both these examples, the plant is of type 0 and 1, respectively. In the last case, a double integrator is controlled by a PID controller, which gives a loop transfer function of type 3.

One observes from the plots that in the stable case the point -1 lies on the left-hand side of the Nyquist plot. In the first two examples, the influence of the controller gain k_R on the closed-loop stability can directly be seen: for increasing k_R the closed-loop system tends to be instable.

3.5 Requirements for Feedback Systems

3.5.1 Performance

In Sect. 3.3, we have already formulated the most elementary performance requirement, namely that the stationary control error has to be zero for constant inputs. In the frequency domain, this can be interpreted as follows: the control error has to vanish for a harmonic oscillation with zero frequency. For a harmonic oscillation with frequency $\omega > 0$, the amplitude of the output can normally not be expected to vanish. We consider the following simple, but illustrative, example. Let the plant be an integrator and the controller be a pure gain: $G(s) = 1/s$, $K(s) = \omega_1$. Then $L(s) = \omega_1 / s$ and

$$S(s) = \frac{s}{s+\omega_1}. \tag{3.17}$$

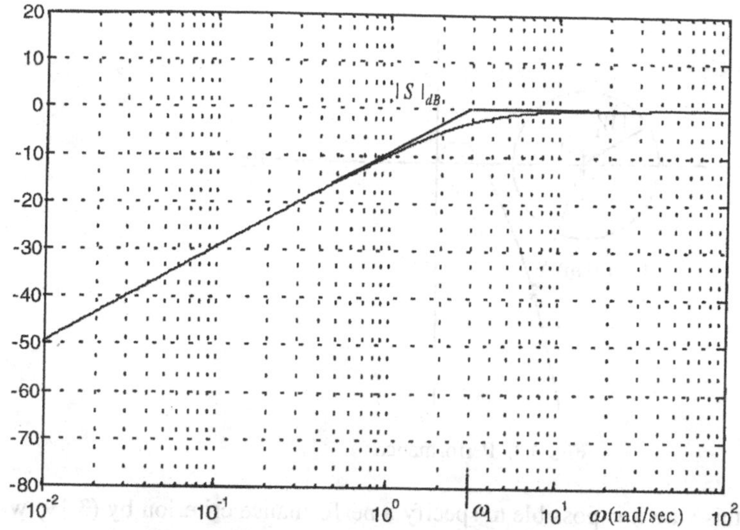

Fig. 3.7. Magnitude plot of S as defined in (3.17)

Figure 3.7 shows the magnitude plot of S. It is seen that the control error increases linearly up to the corner frequency ω_1 and then has the same amplitude as the input signal. This behavior is typical also of more complicated feedback loops.

It is now natural to introduce the requirement

$$|W_1(i\omega)S(i\omega)| \leq 1 \quad \text{for every } \omega \geq 0, \tag{3.18}$$

with a weighting function $W_1(s)$. If we define for a stable transfer function $G(s)$

$$\|G\|_\infty = \sup_{\omega \in \mathbb{R}} |G(i\omega)| ,$$

then (3.18) can be written in the form

$$\|W_1 S\|_\infty \leq 1. \tag{3.19}$$

Later we will see that $\|G\|_\infty$ is really a norm. It can be interpreted as the maximum of the magnitude plot of G.

In our example the requirement (3.18) is exactly fulfilled for the weight

$$W_1(s) = \frac{s + \omega_1}{s} .$$

As it will become clear later, only stable weights are feasible. In the example this can be achieved by modifying W_1 with a small $\varepsilon > 0$:

$$W_{1\varepsilon}(s) = \frac{s + \omega_1}{s + \varepsilon} .$$

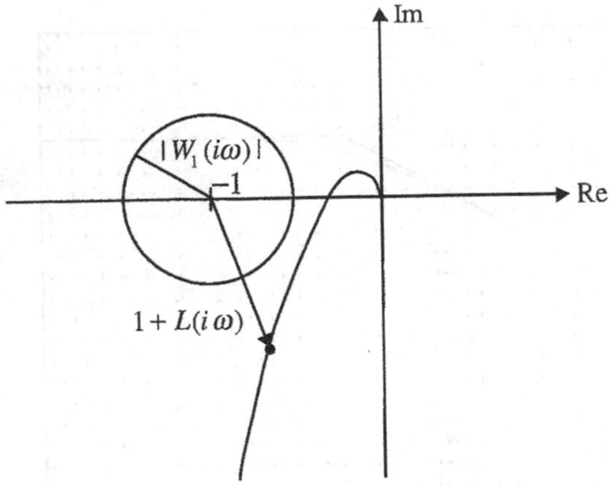

Fig. 3.8. Performance

Hence in this way it is possible to specify a performance criterion by (3.19) with a suitably chosen weight W_1.

The requirement (3.19) can be easily illustrated graphically. One has only to note that (3.18) is equivalent to

$$|W_1(i\omega)| \leq |1+L(i\omega)| \quad \text{for every } \omega \geq 0 .$$

In other words, the Nyquist plot $L(i\omega)$ has to avoid a circle with center -1 and radius $|W_1(i\omega)|$.

3.5.2 Robust Stability

A basic requirement for the feedback loop is robust stability. This means that it has not only to be stable for the nominal plant but also for the real physical plant. Suppose that the plant is modeled by a transfer function $G(s)$, whereas the true plant has the transfer function

$$G_\Delta(s) = (1+\Delta(s)W_2(s))G(s) . \tag{3.20}$$

Here $W_2(s)$ is assumed to be a stable weighting function which is used to model the expected deviation of the true plant from the nominal one. This transfer function will be specified by the magnitude plot. The transfer function $\Delta(s)$ is unknown and also assumed to be stable. It is only supposed that an inequality of the kind $\|\Delta\|_\infty < \gamma^{-1}$ holds, with $\gamma > 0$. The feedback loop is denoted as **robustly stable** if it is stable for all such transfer functions $\Delta(s)$, where $G(s)$ is replaced by $G_\Delta(s)$ as defined in (3.20).

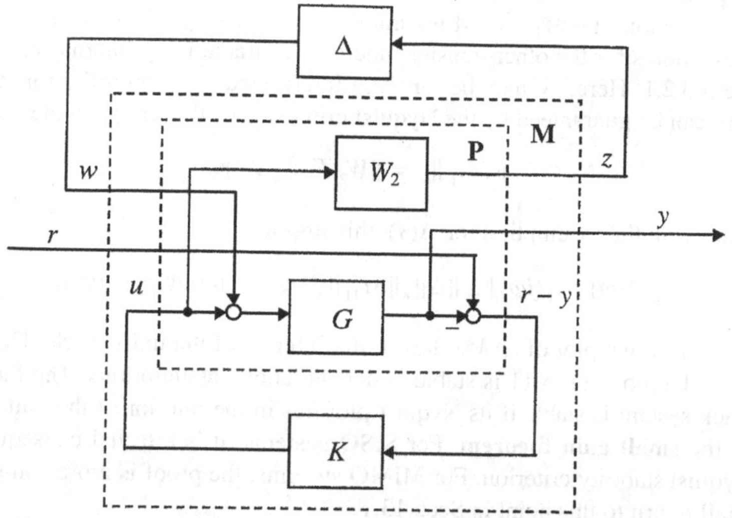

Fig. 3.9. SISO feedback system with a multiplicative perturbation at the plant input

It is now possible to represent the feedback loop as shown in Fig. 3.9. In this way one obtains a generalized plant **P**, which has three inputs: the disturbance w for modeling the uncertainty, the command r and the control u. The outputs are the weighted control $z = W_2 u$ (which is also used for modeling the uncertainty), the plant output y, and the control error $r - y$.

Denote the transfer function for the feedback system consisting of **P** and K by **M**. Then the diagram shown in Fig. 3.10 results and we get

$$\mathbf{M} = \begin{pmatrix} M_{11} & M_{12} \\ M_{21} & M_{22} \end{pmatrix} = \begin{pmatrix} -F_r W_2 & W_2 K S \\ SG & F_r \end{pmatrix}.$$

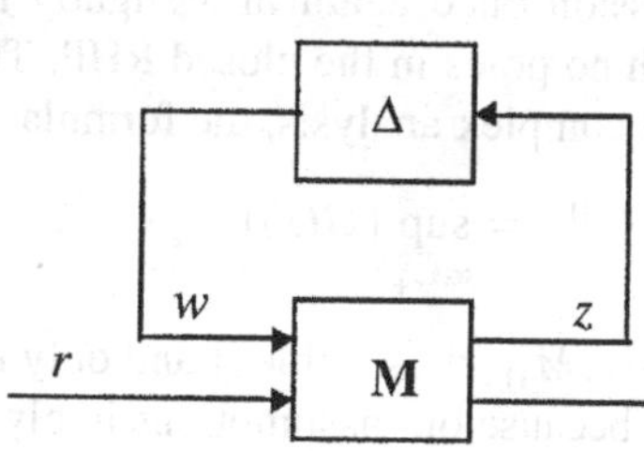

Fig. 3.10. Perturbed feedback loop

The transfer function from r to y of the perturbed feedback system (cf. Fig. 3.10) system is given by

$$F_{r\Delta}(s) = M_{22} + M_{21}\Delta \frac{1}{1 - M_{11}\Delta} M_{12}.$$

This equation shows that the stability of $F_{r\Delta}$ is determined by the stability of the transfer function $(1-M_{11}\Delta)^{-1}$ if the nominal feedback system is internally stable. The same holds for the other transfer functions characterizing internal stability, cf. Theorem 3.2.1. Here, Δ may be formally interpreted as a controller for M_{11} and stability can be guaranteed by the Nyquist criterion. To this end, assume

$$\| M_{11} \|_\infty = \| W_2 F_r \|_\infty \le \gamma .$$

Together with the assumption for $\Delta(s)$ this implies

$$| \Delta(i\omega) M_{11}(i\omega) | \le \|\Delta\|_\infty \| M_{11} \|_\infty < 1 \quad \text{for every } \omega \in \mathbb{R} .$$

Thus, the Nyquist plot of ΔM_{11} lies in the interior of the unity circle. Hence, the feedback loop of Fig. 3.11 is stable under the above assumptions. The fact that a feedback system is stable if its Nyquist plot lies in the interior of the unit circle is called the **small gain theorem**. For SISO systems, it is a trivial consequence of the Nyquist stability criterion. For MIMO systems, the proof is more complicated. We shall return to this point in Sect. 13.1.

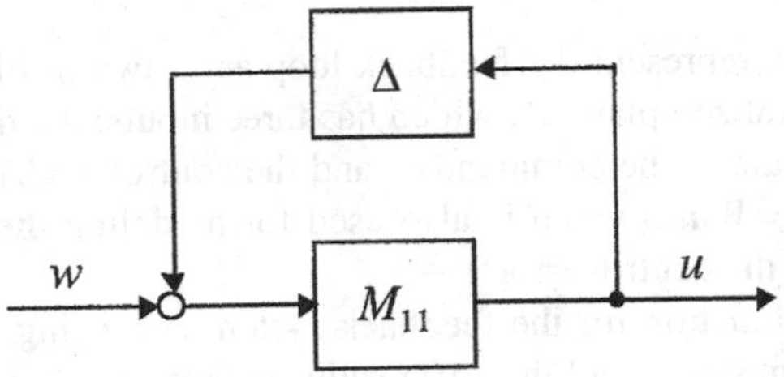

Fig. 3.11. Stability of the perturbed closed-loop system

In particular, with regard to later generalizations in Sects. 13.1 and 13.2 we want to repeat the above discussion once again in a slightly modified way. Let $G(s)$ be a transfer function with no poles in the closed RHP. Then, due to the the boundary maximum principle of complex analysis, the formula

$$\| G \|_\infty = \sup_{s \in \overline{\mathbb{C}}_+} | G(s) |$$

holds. The transfer function $(1-\Delta M_{11})^{-1}$ is stable if and only if $1-\Delta M_{11}$ has no zero in $\overline{\mathbb{C}}_+$. But this is the case, because our assumptions imply

$$1-\Delta(s) M_{11}(s) \ge 1 - \| \Delta M_{11} \|_\infty \ge 1 - \| M_{11} \|_\infty \| \Delta \|_\infty > 0 .$$

We now get the following theorem.

Theorem 3.5.1. *Let the unperturbed feedback system be internally stable. Moreover, suppose that* W_2 *is stable and let* $\gamma > 0$. *Then the following assertions are equivalent:*

(i) $\|W_2 F_r\|_\infty \leq \gamma$.

(ii) The system shown in Fig 3.11 is well-posed and internally stable for all stable perturbations Δ *with*

$$\|\Delta\|_\infty < \gamma^{-1}.$$

Proof. The above discussion shows that $(i) \Rightarrow (ii)$. We now prove the inverse direction. Without loss of generality it may be assumed that $\gamma = 1$ (otherwise, replace W_2 by $\gamma^{-1}W_2$). Suppose that *(i)* is not fulfilled. Then there exists an $\omega_0 \in \mathbb{R} \cup \{\infty\}$ with $|M_{11}(i\omega_0)| > 1$.

Case 1: $\omega_0 = 0$ or $\omega_0 = \infty$. Then $M_{11}(i\omega_0)$ is real. Let $\Delta = 1/M_{11}(i\omega_0)$. Hence we have $\|\Delta\|_\infty < 1$ and also

$$1 - M_{11}(i\omega_0)\Delta(i\omega_0) = 0.$$

Thus for $\omega_0 = 0$ the system of Fig. 3.11 is not internally stable and for $\omega_0 = \infty$ it is not well-posed.

Case 2: $0 < \omega_0 < \infty$. Then we may write with a real number v

$$M_{11}(i\omega_0) = v e^{i\varphi} \quad \text{with} \quad -\pi \leq \varphi < 0,$$

and define

$$\Delta(s) = \frac{1}{v}\frac{\alpha - s}{\alpha + s},$$

where $\alpha \geq 0$ is chosen such that

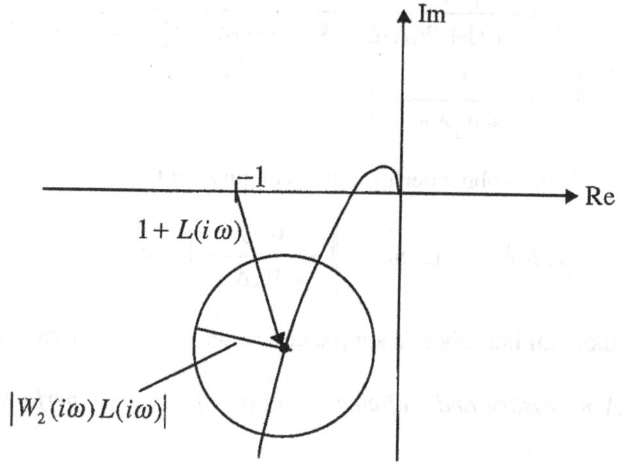

Fig. 3.12. Robust stability

$$\frac{\alpha - i\omega_0}{\alpha + i\omega_0} = e^{-i\varphi}.$$

Consequently, Δ is stable and $\|\Delta\|_\infty = 1/\nu < 1$. Moreover,

$$1 - \Delta(i\omega_0)M_{11}(i\omega_0) = 0.$$

Hence, in contradiction to our assumption, the system of Fig. 3.11 is not internally stable.

The condition $\|W_2 F_r\|_\infty < 1$ is equivalent to

$$|W_2(i\omega)L(i\omega)| < |1 + L(i\omega)| \quad \text{for all } \omega \geq 0.$$

This means that for every ω the point -1 lies outside a circle with center $L(i\omega)$ and radius $|W_2(i\omega)L(i\omega)|$ (cf. Fig. 3.12).

3.5.3 Robust Performance

Next we pose the question of under which assumptions the perturbed feedback loop additionally fulfills the condition for performance. This is the case if and only if (with γ normed to 1)

$$\|W_2 F_r\|_\infty < 1 \quad \text{and} \quad \|W_1 S_\Delta\|_\infty < 1.$$

Here S_Δ denotes the perturbed sensitivity function. We have

$$S_\Delta = \frac{1}{1 + (1 + W_2\Delta)L} = \frac{S}{S + (1 + W_2\Delta)F_r}$$
$$= \frac{S}{1 + W_2\Delta F_r}$$

(note that $F_r = LS$). Thus, robust performance is equivalent to

$$\|W_2 F_r\|_\infty < 1 \quad \text{and} \quad \left\| \frac{W_1 S}{1 + W_2 \Delta F_r} \right\|_\infty < 1. \tag{3.21}$$

The following theorem introduces a simple condition for robust performance.

Theorem 3.5.2. *A necessary and sufficient condition for robust performance is given by*

$$\|\,|W_1 S| + |W_2 F_r|\,\|_\infty < 1. \tag{3.22}$$

Proof. We only show that the condition is sufficient. First, we observe that it is equivalent to

$$\| W_2 F_r \|_\infty < 1 \quad \text{and} \quad \left\| \frac{W_1 S}{1 - |W_2 F_r|} \right\|_\infty < 1.$$

Now the inequality

$$1 = |1 + \Delta W_2 F_r - \Delta W_2 F_r| \le |1 + \Delta W_2 F_r| + |W_2 F_r| .$$

holds. This implies

$$0 < 1 - |W_2 F_r| \le |1 + \Delta W_2 F_r|.$$

Hence

$$1 > \left| \frac{W_1 S}{1 - |W_2 F_r|} \right| \ge \left| \frac{W_1 S}{1 + \Delta W_2 F_r} \right| ,$$

and (3.21) is shown.

It is also possible to visualize robust performance graphically. To this end, note that condition (3.22) is equivalent to

$$|W_1(i\omega)| + |W_2(i\omega) L(i\omega)| < |1 + L(i\omega)| \quad \text{for every } \omega \ge 0 .$$

This means that the circle with center -1 and radius $|W_1(i\omega)|$ and the circle with center $L(i\omega)$ and radius $|W_2(i\omega) L(i\omega)|$ have no point in common (cf. Fig. 3.13).

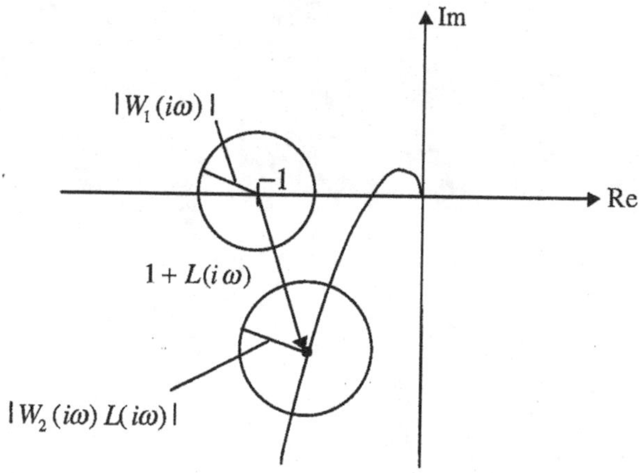

Fig. 3.13. Robust Performance

Chapter 4

Classical Design Methods

In this chapter, it will be shown how the Nyquist stability criterion can be used to synthesize a PID controller so that the feedback system is in a certain sense robustly stable and fulfills also some specified performance requirements. These properties are typically measured by the phase margin and the gain crossover frequency. The method will be most successfully applied if the plant is stable with the possible exception of a pole at the origin, but even in this case it has limitations if the performance requirements become too complex. We show that in such a situation an improvement can be achieved by generalizing the structure of the classical SISO feedback loop. This leads to a cascade of several controllers, to a two degrees-of-freedom control configuration or to feedforward of the disturbance.

4.1 Specification of the Closed-Loop System

4.1.1 Robust Stability and Phase Margin

We have shown in Sect. 3.4 that the position of the critical point -1 relative to the Nyquist plot of the loop transfer function determines whether the closed-loop system is stable. If this point lies on the Nyquist plot, then the closed loop system is marginally stable and has a pole on the imaginary axis. Thus it may be suspected that the stability properties of the feedback system get better if the distance of -1 from the Nyquist plot increases. To be more precise, assume that the Nyquist plot avoids the interior of a circle with radius R and center -1 (cf. Fig. 4.1). Then the inequality

$$|1+L(i\omega)| \geq R \tag{4.1}$$

holds. It may be written in the form $\| S \|_\infty \leq R^{-1}$ and we get

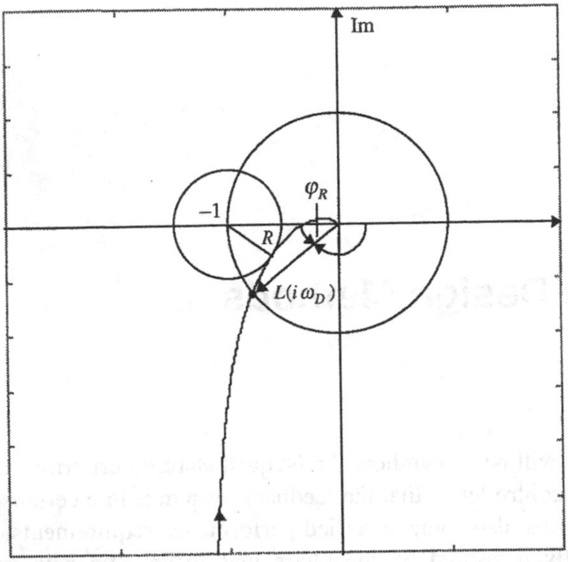

Fig. 4.1. Nyquist plot of the loop transfer function and phase margin

$$\| F_r \|_\infty = \| S - 1 \|_\infty \le \| S \|_\infty + 1 \le R^{-1} + 1 .$$

From Theorem 3.5.1 it is now seen that the feedback system is internally stable for all stable input multiplicative perturbations which fulfill

$$\| \Delta \|_\infty < \frac{R}{R+1}$$

(with the constant weight $W_2 = 1$). Consequently, R should be made as large as possible, where $R = 1$ is the optimum.

Let $L(s)$ be strictly proper. Then the Nyquist plot $L(i\omega)$ ends in the origin and starts from infinity if the loop transfer function is of type $\alpha \ge 1$. If it is of type zero, it starts outside the unit circle centered at the origin if the gain $V = L(0)$ is large enough. In both cases, $L(i\omega)$ enters the unit circle from the outside. In order to avoid unnecessary complications, we assume that $L(i\omega)$ remains within the unit circle. Then we can define the **gain crossover frequency** ω_D by

$$| L(i\omega_D) | = 1 .$$

The **phase margin** is given by (cf. Fig. 4.1)

$$\varphi_R = 180° + \arg L(i\omega_D) .$$

In the situation of this figure, it is possible to take φ_R instead of R as a measure of the distance of the Nyquist plot from the point -1. In classical control, this is a common practice. The reason is that φ_R is much easier to manage than R. One can imagine that there are situations where this distance is small and φ_R is large, and such cases can actually be constructed. In this sense, φ_R is not a fully adequate substitute for R. Nevertheless, in many practical examples of classical control it is reasonable to work with the phase margin.

4.1.2 Performance and Gain Crossover Frequency

Damping. Normally an important design objective is that the closed-loop system is sufficiently well damped. We ask whether it is possible to obtain a relationship between the damping of the feedback system and the phase margin.

In order to get an orientation, we start with an example and consider the following plant of order 3:

$$G(s)=\frac{k}{(1+T_0 s)(1+Ts)^2} \qquad \text{(with } T_0 \geq T\text{)}.$$

Suppose it has to be controlled with a PI controller

$$K(s)=k_R\,\frac{1+T_I s}{T_I s},$$

where T_I is chosen such that it compensates the largest time constant of the plant. Then we have $T_I = T_0$ and

$$L(s)=\frac{V}{Ts+2T^2s^2+T^3s^3} \qquad \text{(with } V=k_R k\frac{T}{T_0}\text{)}.$$

This leads immediately to

$$F_r(s)=\frac{1}{1+\frac{T}{V}s+2\frac{T^2}{V}s^2+\frac{T^3}{V}s^3}.$$

Let $F_r(s)$ be a series combination of a type 1 and a type 2 system:

$$F_r(s)=\frac{1}{(1+T_1 s)(1+2dT_2 s+T_2^2 s^2)}.$$

Comparison of the coefficients yields the equations

$$\frac{T}{V}=2dT_2+T_1, \quad 2\frac{T^2}{V}=2dT_1T_2+T_2^2, \quad \frac{T^3}{V}=T_1T_2^2.$$

The gain V will be determined such that a given value for d results. Then the

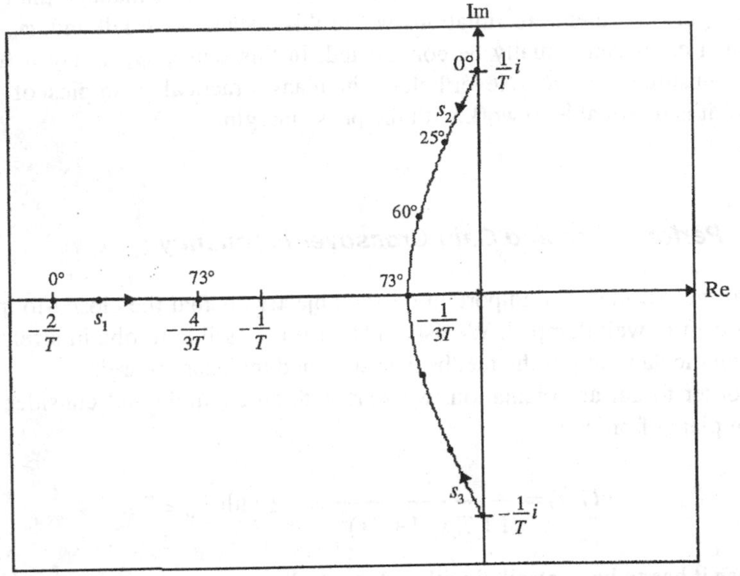

Fig. 4.2. Phase margin and closed-loop poles for a third-order system with a PI controller

above equations can be solved uniquely for V and T_1, T_2, and a short calculation gives (if $d \le 1$)

$$V = \frac{2d+2}{(2d+1)^3}, \quad T_1 = \frac{2d+1}{2d+2}T, \quad T_2 = (2d+1)T. \tag{4.2}$$

Moreover, it can be seen that the gain crossover frequency and phase margin fulfill

$$T\omega_D\,(1 + T^2\omega_D^2) = V\ , \quad \omega_D = \frac{1}{T}\tan\left(\frac{\pi}{4} - \frac{\varphi_R}{2}\right).$$

This implies

$$\tan\left(\frac{\pi}{4} - \frac{\varphi_R}{2}\right)\left(1 + \tan^2\left(\frac{\pi}{4} - \frac{\varphi_R}{2}\right)\right) = \frac{2d+2}{(2d+1)^3}.$$

Hence we have obtained a connection between the damping d of the conjugate-complex pole pair and the gain margin φ_R. It is now possible to plot the poles as a function of d and to indicate the corresponding phase margin. This leads to Fig. 4.2 and it is seen that good, but not too strongly, damped behavior results for $\varphi_R \approx 60°$. In this example, a sufficiently high phase margin leads to good damping as well as to robust stability.

Unfortunately, these considerations cannot be generalized. Let us choose as the plant

$$G(s) = \frac{k}{(1 + 2dTs + T^2 s^2)(1 + T_1 s)} \quad \text{with } 0 < d \ll 1$$

and as the controller

$$K(s) = k_R \frac{1 + 2dTs + T^2 s^2}{s}.$$

Then the loop transfer function is given by

$$L(s) = \frac{V}{s(1 + T_1 s)} \quad \text{with } V = k_R k.$$

In the sense of Sect. 3.5, the closed-loop system is robustly stable and has also good performance properties if V is suitably chosen. This means in particular that $F_r(s)$ is well damped. Now suppose that additionally a disturbance at the plant input occurs. The corresponding closed-loop transfer function for the disturbance is given by

$$F_d(s) = \frac{s}{V + s(1 + T_1 s)} \frac{k}{1 + 2dTs + T^2 s^2},$$

and is only badly damped.

For this reason, a PID controller would not be used to cancel a weakly damped pole pair. Even if one avoids the above situation, it is not possible to deduce the damping from the phase margin. In order to illustrate this, we consider a further example. Let

$$G_1(s) = \frac{1}{s(1 + s)^2}, \quad G_2(s) = \frac{s^2 + 0.09s + 0.81}{s^2 + 0.1s + 1},$$

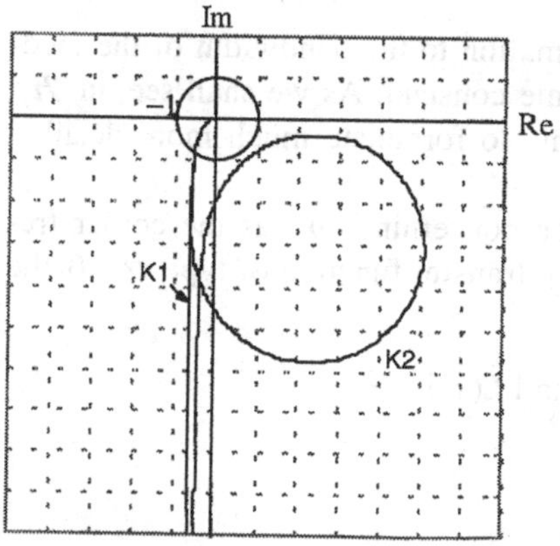

Fig. 4.3. Nyquist plot of $L(i\omega)$ (curve K1) and of $L_1(i\omega)$ (curve K2)

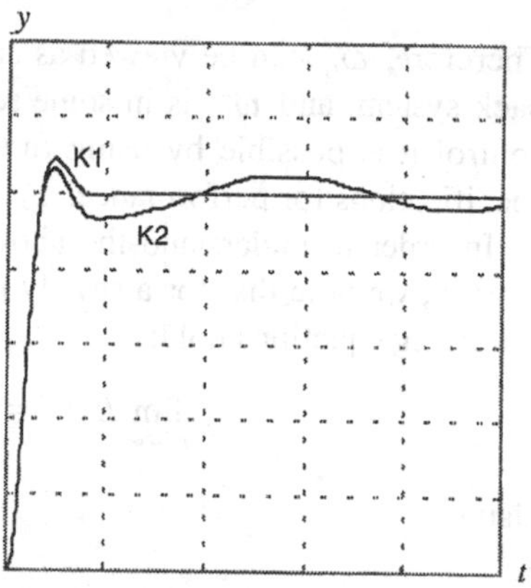

Fig. 4.4. Step response of $F_r(s)$ (curve K1) and of $F_{r1}(s)$ (curve K2)

$$L(s) = k_R G_1(s), \quad L_1(s) = k_R G_1(s) G_2(s).$$

The corresponding closed-loop transfer functions for the command are denoted by $F_r(s)$ and $F_{r1}(s)$. For a proportional controller with $k_R = 3$, the Nyquist plots of Fig. 4.3 result. It is seen that the phase margin in both cases is approximately 60°. In contrast to this, the step responses for the two systems differ considerably (Fig. 4.4). The reason for this behavior results from the factor $G_2(s)$, whose denominator has the damping $d = 0.05$. The quadratic numerator causes that for sufficiently large ω the approximation $|G_2(i\omega)| \approx 1$ holds. Then this factor contributes only slightly to the shape of the Nyquist plot for large ω, and in particular this holds for frequencies near the gain crossover frequency (cf. Fig. 4.3). In contrast to this, the denominator of $F_{r1}(s)$ is strongly influenced by this factor. It leads to a term with a very low damping ($d = 0.086$), which causes the oscillatory behavior of the step response of $F_{r1}(s)$.

Finally, it should be mentioned that the inverse case is also possible, namely a high damping in combination with a low phase margin. This situation is undesirable, too. It occurs typically if an observer-based controller with a "slow" observer is used. We return to this point in Sect. 8.4.3 and notice the fact that generally it is not possible to deduce the damping of the closed-loop system from the phase margin. In any case, in order to guarantee robust stability, it is necessary to require a sufficiently high phase margin (typically 45°– 60°).

Bandwidth. In classical control, the performance is often specified by damping and gain crossover frequency. The above discussion has given some insight about the damping. The gain crossover frequency is the dominating parameter in the weight W_1 for the sensitivity function. For example, W_1 may be written in the form

$$W_1^{-1}(s) = \frac{s + \varepsilon}{s + \omega_D}.$$

Therefore, ω_D can be viewed as an approximation to the bandwidth of the feedback system, and ω_D^{-1} is in some sense its time constant. As we shall see, in $\mathcal{H}_\infty$ control it is possible by using further weights to formulate much more detailed specifications for performance.

In order to understand the above assertion concerning ω_D as the corner frequency, we note that for a strictly proper loop transfer function of type $\alpha > 0$ the following equations hold:

$$\lim_{s \to \infty} L(s) = 0, \qquad \lim_{s \to 0} |L(s)| = \infty.$$

Using

$$|F_r(i\omega)| = \frac{|L(i\omega)|}{|1 + L(i\omega)|},$$

we get

$$|F_r(i\omega)| \approx \begin{cases} 1, & \text{if } |L(i\omega)| \gg 1, \\ |L(i\omega)|, & \text{if } |L(i\omega)| \ll 1. \end{cases}$$

Thus, because of $|L(i\omega_D)| = 1$, the following approximation holds:

$$|F_r(i\omega)| \approx \begin{cases} 1, & \text{if } \omega \ll \omega_D, \\ |L(i\omega)|, & \text{if } \omega \gg \omega_D. \end{cases}$$

Hence, the gain crossover frequency may be considered as an approximate measure of the bandwidth of $F_r(s)$. In other words, if a fast step response of a closed-loop system is required, the gain crossover frequency has to be high.

4.2 Controller Design with the Nyquist Criterion

4.2.1 Design Method

In this section, we want to show how the Nyquist stability criterion (Theorem 3.4.2) can be used for controller design. The original criterion uses the Nyquist plot of the loop transfer function, but since the Bode plot is more easy to handle, it is reasonable to seek a formulation of the stability criterion that uses the Bode plot instead of the Nyquist plot.

We start with an example and consider again the third-order plant of Sect. 4.1.2. Then the loop transfer function is given by

$$L(s) = \frac{V}{T_0 s\,(1+Ts)^2} \quad \text{with } V = k_R k. \tag{4.3}$$

Its Nyquist plot has a shape similar to that depicted in Fig. 4.1 (cf. also Fig. 3.5(b)) and one sees that in the stable case $\varphi_R > 0$ holds, whereas the unstable case is characterized by $\varphi_R \le 0$. Figure 4.5 shows the Bode plot for both cases.

In the Bode plot, the gain crossover frequency is characterized by the equation

$$|L(i\omega_D)|_{dB} = 0,$$

and the phase margin is the distance of the phase plot from the $-180°$ - line at $\omega = \omega_D$. The phase margin is positive if the phase plot lies above the $-180°$ - line and it is negative if the phase plot lies below the $-180°$ - line. One observes also that the gain crossover frequency increases if the loop gain V increases, whereas the phase margin decreases. If V is large enough, the phase margin becomes negative and the closed-loop system is unstable, compare curves K_1 and K_2 in

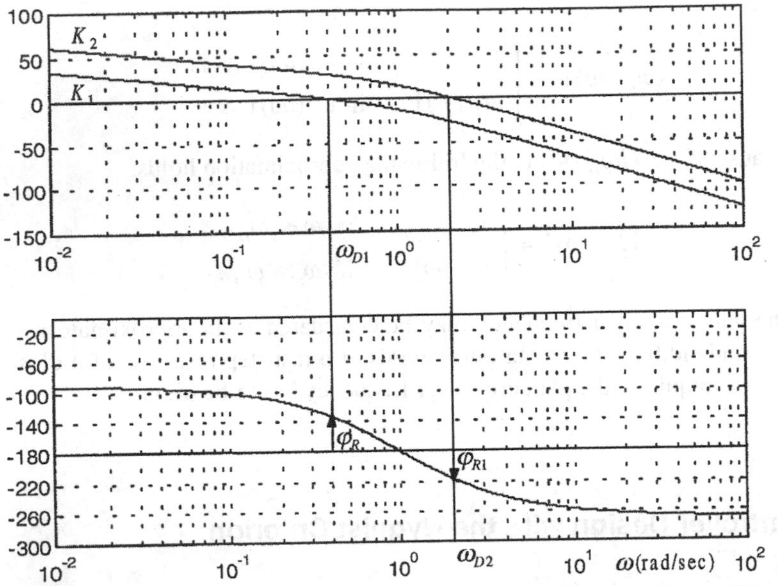

Fig. 4.5. Bode plot of $L(s)$ defined in (4.3) for different values of V

Fig. 4.5.

It is possible to generalize these considerations, which leads to the following theorem. The proof will not be presented here, it can be found in Föllinger [43].

Theorem 4.2.1. *Let the loop transfer function be given in the form*

$$L(s) = \frac{V}{s^{\alpha}} L_1(s) \quad \textit{with} \quad L_1(0) = 1,\ V > 0,\ \ \alpha \in \{0, 1, 2\}$$

where L is strictly proper and L_1 is stable. Moreover, assume that the magnitude plot $|L(i\omega)|$ intersects the 0 dB-line once and only once and let

$$-540° < \arg L(i\omega) < 180° \quad \textit{for any} \quad 0 < \omega \leq \omega_D .$$

Then the feedback system is stable if and only if $\varphi_R > 0$.

Note that Theorem 4.2.1 is only applicable to stable plants (with the possible exception of a pole at the origin with multiplicity 1 or 2), but the theorem can be generalized to instable plants (see again Föllinger [43]). We now show how this theorem can be applied in practical situations.

Plants of type 0. Let a stable plant of the form

$$G(s) = k \frac{Z(s)}{N(s)} \qquad \text{with} \quad Z(0) = N(0) = 1$$

be given and suppose that is has to be controlled by a PI controller:

$$K(s) = k_R \frac{1 + T_I\, s}{T_I\, s}.$$

Case 1: $G(s)$ *has a dominant real pole*. This means that there exists a real pole s_1 with $\operatorname{Re} s_k < \operatorname{Re} s_1$ for $k = 2, \ldots, n$, where $s_2, \ldots, s_n$ are the remaining poles of $G(s)$. Then we can write $s_1 = -1/T_1$, with $T_1 > 0$ and $N(s) = (1 + T_1 s)N_1(s)$ (where $N_1(s)$ is a polynomial). Choose now $T_I = T_1$ (compensation of the largest time constant of the plant). Then the loop transfer function is given by

$$L(s) = V \frac{Z(s)}{T_I\, s\, N_1(s)} \qquad \text{with } V = k_R k.$$

Case 2: $G(s)$ *has a dominant complex conjugate pole pair*. In this case there is a pole pair $s_1, \overline{s}_1$ such that $\operatorname{Re} s_k < \operatorname{Re} s_1$ for the remaining poles s_k. The dominant pole pair can be characterized by a time constant T_1 and damping d_1. If the damping is not too small (typically $d_1 \geq 0.5$ is required), the second-order system belonging to $s_1, \overline{s}_1$ can be approximated by a first-order system with time constant $T = 2d_1T_1$. Then it is reasonable to put $T_I = 2d_1T_1$. In this case, no cancellation occurs when $L(s)$ is calculated.

In both cases, T_I is prescribed in a quite reasonable way (note that this choice is made with respect to a good closed-loop performance concerning the reference signal r). It remains to determine the controller gain k_R (or the loop gain V).

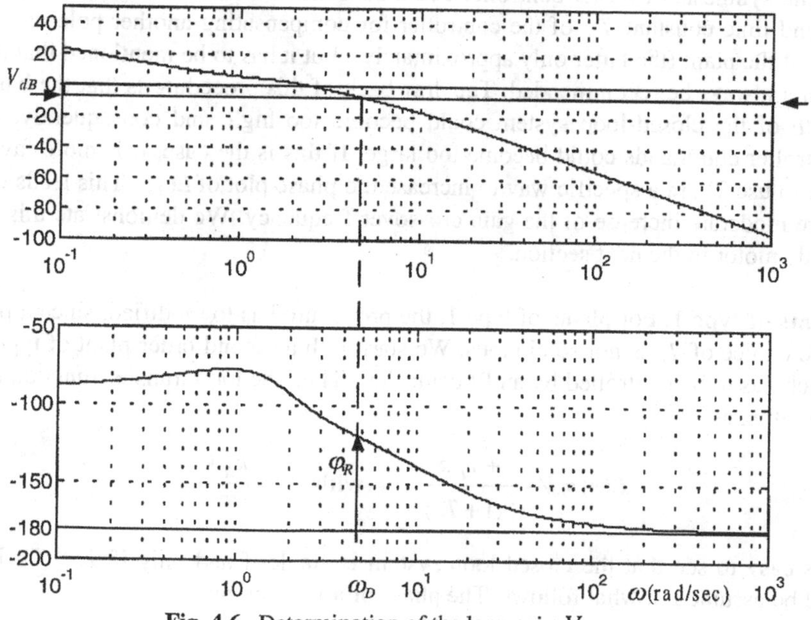

Fig. 4.6. Determination of the loop gain V

This will be achieved by requiring a given phase margin in order to fulfill the robust stability requirement.

We consider an example. The procedure is quite similar for different plants. Let

$$G(s)=\frac{2.4(1+0.5s)}{(1+0.7s+0.25s^2)(1+0.1s)} .$$

The time constant and the damping of the factor of order 2 are given by $T_1=0.5$ and $d_1=0.7$. Hence the plant has a dominant complex conjugate pole pair and we choose $T_I=0.7$. The loop transfer function is

$$L(s)=V\frac{1+0.7s}{0.7s}\frac{1+0.5s}{(1+0.7s+0.25s^2)(1+0.1s)} \quad \text{with } V=2.4k_R .$$

In order to obtain the loop gain *V*, first the Bode plot of $L(s)$ has to be drawn, cf. Fig. 4.6. The gain crossover frequency for $V=1$ is 2.3 s^{-1} with a phase margin of 81°. We now require a phase margin of 60°. Therefore, the loop gain *V* has to be altered such that the magnitude plot passes the 0dB - line at the same frequency at which the phase plot crosses the $-120°$ - line. In our example, the latter is true for $\omega_D=4.8\,s^{-1}$. Hence, as is seen from the figure, the magnitude plot has to be shifted upwards by an amount $V_{dB}=8$. Graphically, this can be done by introducing a new 0 dB - line, which intersects the magnitude plot at the frequency $\omega_D=4.8\,s^{-1}$. As a result, we get $V=2.51$ and $k_R=V/2.4=1.05$.

The synthesis of a PID-controller could be done in a similar way by using the second time constant T_D of the controller for compensating another pole or pole pair of the plant (the latter only approximately), but it has to be mentioned that this cannot always be recommended. The drawback of this procedure is that the bandwidth of the closed-loop system could become too high, and consequently, the controller commands could become too large. If this is the case, it is more favorable to use T_D in a specific way to increase the phase plot of $L(s)$. This leads to a more moderate increase of the gain crossover frequency. We demonstrate this for the dc motor in the next section.

Plants of type 1. For plants of type 1, the procedure has to modified, since a possible choice of T_I is not so obvious. We start with a second order plant of type 1, which has to be controlled by an PI controller. Then the loop transfer function can be written as

$$L(s)=V\frac{1+T_I s}{s^2(1+Ts)} \quad \text{with } V=\frac{k_R k}{T_I} .$$

It is easy to see that the closed-loop system is stable if and only if $T_I>T$. This will be assumed in what follows. The phase plot is given by

$$\arg L(i\omega) = -\pi + \arctan T_I\, \omega - \arctan T\omega\,.$$

This leads to the Bode plot shown in Fig. 4.7. The phase plot has a maximum, and therefore it is reasonable to choose the gain crossover ω_D such that it coincides with the frequency at which the phase plot reaches its maximum value. Then, as an elementary calculation shows, ω_D is given by

$$\omega_D = \frac{1}{\sqrt{T_I\, T}}\,.$$

On a logarithmic scale, ω_D is the arithmetic mean of the corner frequencies of the plant and the controller. For this reason, the design method under consideration is sometimes called the **method of the symmetrical optimum**.

Next, we have to calculate T_I. As an abbreviation, put $\mu = T_I/T$. Therefore it follows that $\mu > 1$ and

$$\varphi_R = \arctan\sqrt{\mu} - \arctan 1/\sqrt{\mu} = \arctan\left(1/2(\sqrt{\mu} - 1/\sqrt{\mu})\right).$$

This yields, for a given φ_R,

$$\mu = \mu(\varphi_R) = \left(\frac{1+\sin\varphi_R}{\cos\varphi_R}\right)^2.$$

The gain margin can be chosen arbitrarily in the interval $0 < \varphi_R < \pi/2$ and hence

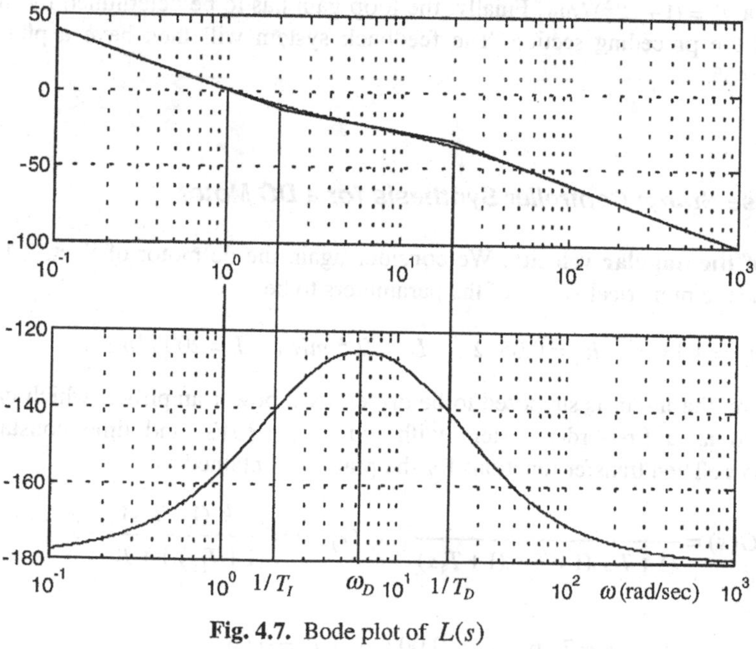

Fig. 4.7. Bode plot of $L(s)$

the gain crossover frequency and the controller time constant are

$$\omega_D = \frac{1}{\sqrt{\mu(\varphi_R)}\,T}\ , \qquad T_I = \mu(\varphi_R)T\ .$$

The next step consists in calculating the loop gain V. By using the equation $1 = |L(i\omega_D)|$ one gets $V = 1/(T_I(TT_I)^{1/2})$, and the controller is completely synthesized for a given phase margin φ_R. Moreover, the transfer function for the command is

$$F_r(s) = \frac{1+T_I s}{1+\sqrt{TT_I}\,s}\,\frac{1}{1+(T_I-\sqrt{TT_I})s+TT_I s^2}\ .$$

If the damping of the second-order factor is denoted by d, we get $\mu = (2d+1)^2$. Because of $T_I\omega_D = \mu^{1/2}$, the controller phase at ω_D results as

$$\arg K(i\omega_D) = -\frac{\pi}{2} + \arctan(2d+1)\ .$$

Hence, for the choice $d = 1/\sqrt{2}$ we obtain $\varphi_R = 45°$, $\arg K(i\omega_D) = -22.5°$ and $\arg G(i\omega_D) = -112.5°$.

The remaining question is how T_I is to be chosen for a general plant of type 1. In this case, one first calculates the frequency ω_D for which the phase of the plant is $-112.5°$. Then T_I is chosen such that the phase of the controller equals $-22.5°$ at ω_D (as in the previous example). This means that T_I has to be calculated from the formula $T_I = (1+2^{1/2})/\omega_D$. Finally, the loop gain has to be determined by the method of the preceding section. The feedback system will then have a phase margin of 45°.

4.2.2 Case Study: Controller Synthesis for a DC Motor

Control of the angular velocity. We consider again the dc motor of Sect. 2.1.2 and assume the numerical values of the parameters to be

$$c_A = 4.1\,Vs, \quad R_A = 0.05\,\Omega\ , \quad L_A = 2.5\,mH\ , \quad J = 200\,Nms^2\ .$$

Furthermore, the motor is supposed to be driven by a power amplifier, which can be modeled as a first-order system with gain $k_V = 30\,Vs$ and time constant $T_V = 0.005\ s$. Then transfer functions for the plant are obtained as

$$G(s) = \frac{k}{(1+T_1 s)(1+T_2 s)(1+T_3 s)}\ , \qquad G_d(s) = \frac{k_1(1+T_e s)}{(1+T_1 s)(1+T_2 s)}\ ,$$

where

$$k = 7.26, \quad k_1 = 0.00293, \quad T_e = 0.05\ ,$$

$$T_1 = 0.53, \quad T_2 = 0.055, \quad T_3 = T_V = 0.005.$$

Design of a PI controller. As described in the previous section, the time constant of the PI controller is used to compensate the largest time constant of the plant. This leads to the choice $T_I = 0.53\,s$ and the loop transfer function is given by

$$L(s) = V\frac{1}{0.53s(1+0.055s)(1+0.005s)}.$$

The corresponding Bode plot is depicted in Fig. 4.8 (curves K and K'). One gets $\omega_D = 9\,s^{-1}$ and $V_{dB} = 15$, thus $V = 5.62$ and $k_R = 0.77$.

Design of a realizable PID controller. A possible choice for T_D is $T_D = T_2$. In this case the controller compensates completely the dynamics of the dc motor. Then the loop transfer function is

$$L(s) = V\frac{1}{0.53s\,(1+0.005s)},$$

and the transfer function for the command is

$$F_r(s) = \frac{1}{1+\dfrac{0.53}{V}s+\dfrac{0.00265}{V}s^2},$$

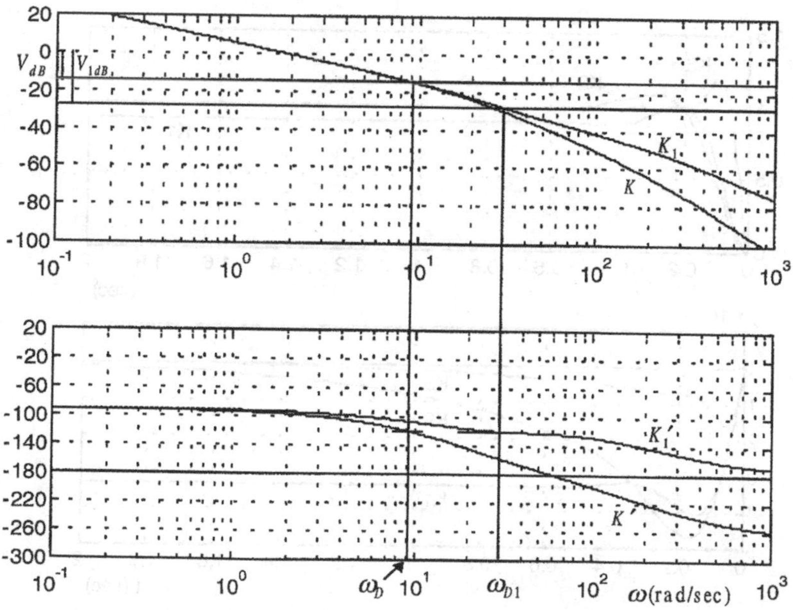

Fig. 4.8. Bode plot of L for a PI controller (curves K, K') and a PID controller (curves K_1, K_1')

with $V = 53$ (and $k_R = 7.3$) for $d = 0.701$. The gain crossover frequency is calculated as $\omega_D = 220\,s^{-1}$. Hence the bandwidth of the feedback loop has increased considerably. The price for this improvement of performance are larger control commands, which can lead to a saturation of the power amplifier.

For this reason, we try a different approach, where the increase of ω_D is not so high. We use the already-designed PI controller and determine T_D such that a prescribed value for ω_D results. In this example, $\omega_D = 30s^{-1}$ is a reasonable value. Then from the Bode plot in Fig. 4.8 it is seen that the phase plot has to be increased by about 37.5° by the PD part of the controller in order to get $\omega_{D1} = 30s^{-1}$ as the new gain crossover frequency (again for a phase margin of 60°). Thus for $\omega = 30s^{-1}$, the equation

$$\arg(1 + T_D\,\omega i) = \arctan T_D\,\omega = 37.5°$$

must hold. The resulting value is $T_D = \tan 37.5°/30 = 0.026$.

The next step consists in adding the magnitude and phase plot of the PD part of the controller to the Bode plot for the PI controller. This leads to the curves K_1, K_1' in Fig. 4.8. The open-loop gain is $V_{dB} = 28$, i.e. $V = 25.1$ and $k_R = 3.46$. For the $\mathrm{PIDT_1}$ controller we choose $T_0 = 0.25T_D$.

Fig. 4.9 shows the step responses for the PI controller, the second PID controller and additionally for a P controller (also with $\varphi_R = 60°$). Note that the performance of the PID controller with respect to the command is good, but somewhat worse with respect to the disturbance (it needs a comparatively long time before the plant output reaches zero).

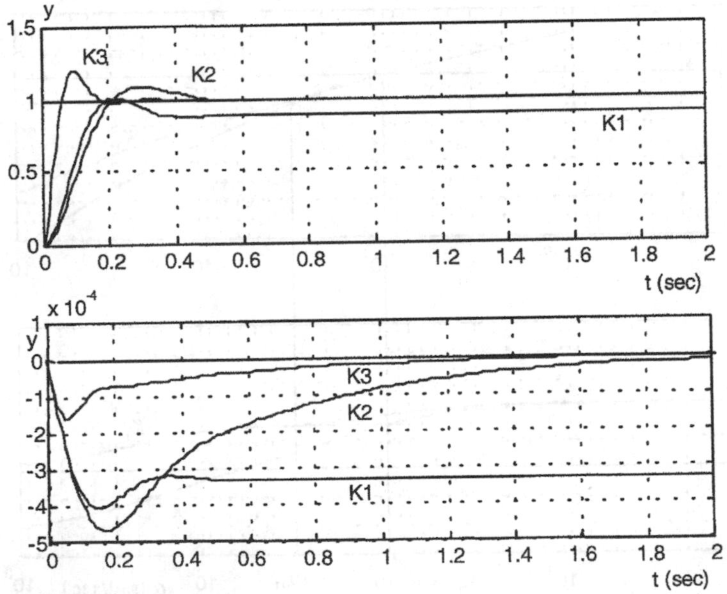

Fig. 4.9. Step responses for the command and the disturbance for the P controller (K1), PI controller (K2) and the PID controller (K3)

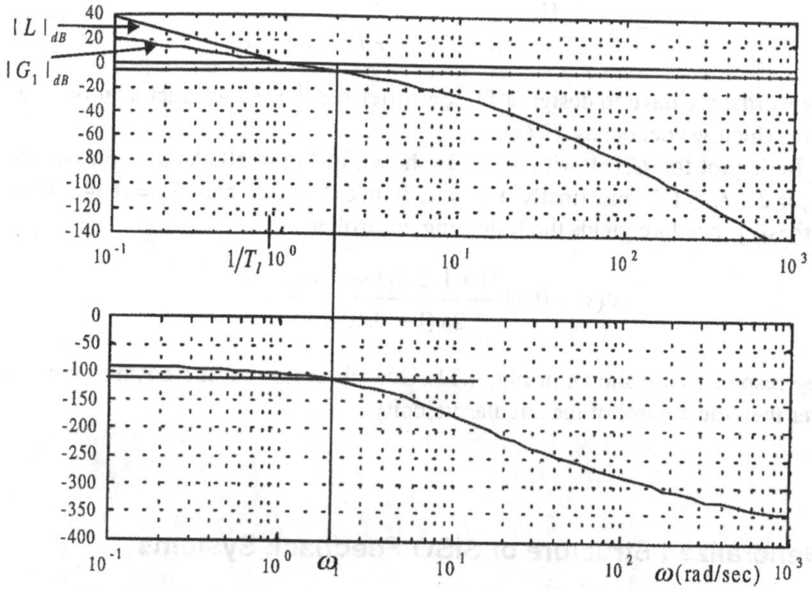

Fig. 4.10. Bode plot for position control with the symmetrical optimum

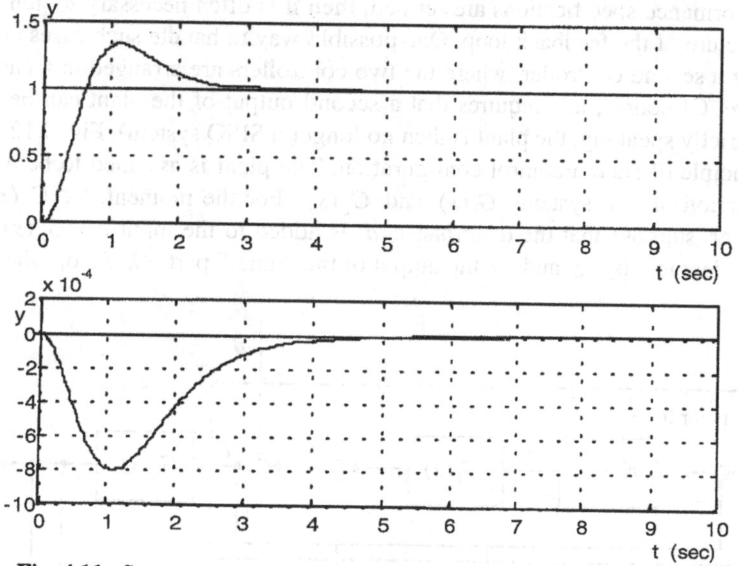

Fig. 4.11. Step responses for position control (command and disturbance)

Position control. We now consider the control of position for a dc motor (cf. Sect. 2.1.2) with a realizable PID controller. The controller is to be designed by the method of the symmetrical optimum. We define T_D such that the controller compensates the largest time constant: $T_D = T_1$. Moreover, as in the previous example, let $T_0 = 0.25T_D = 0.1325$. Then the loop transfer function is

$$L(s) = k_R \frac{1+T_I s}{T_I s} \frac{k}{s(1+T_0 s)(1+T_2 s)(1+T_3 s)}. \tag{4.4}$$

In other words, we have to design a PI controller for the second factor in (4.4) as the plant, which we denote by $G_1(s)$.

The Bode plot for $G_1(s)$ with $k=1$ is shown in Fig. 4.10. As a result we obtain $\omega_D = 2$, $T_I = 1.2$ and finally $V_{dB} = 8$, hence $V = 2.5$ and $k_R = 0.34$. Thus the synthesis procedure yields the following controller:

$$K(s) = 0.34 \frac{(1+1.2s)(1+0.53s)}{1.2s(1+0.1325s)}.$$

The step responses are shown in Fig. 4.11. One observes that the position control is slower than the control of the angular velocity.

4.3 Generalized Structure of SISO Feedback Systems

Cascaded feedback systems. If the plant has unfavorable dynamic properties or if the performance specifications are refined, then it is often necessary to generalize the structure of the feedback loop. One possible way to handle such cases consists in using a second controller, where the two controllers are arranged in a cascaded structure. Of course, this requires that a second output of the plant can be measured (strictly speaking, the plant is then no longer a SISO system). Fig. 4.12 shows the principle of such a control configuration. The plant is assumed to be a series combination of two systems $G_1(s)$ and $G_2(s)$. For the moment, let $G_3(s) = 0$. Moreover, suppose that the disturbance d is added to the input of $G_2(s)$. This sum is denoted by z and is the output of the "inner" part $G_1(s)$ of the plant.

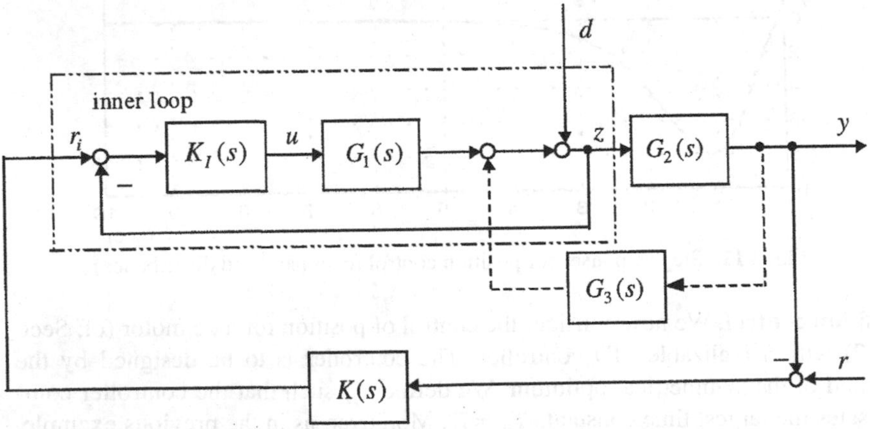

Fig 4.12. Cascaded feedback system

We assume that z can be measured. In this case, it is possible to control z by means of a controller $K_I(s)$. It is designed without referring to the "outer" part $G_2(s)$ of the plant. The inner feedback loop consisting of $G_1(s)$ and $K_I(s)$ is then the plant for the "outer" controller $K(s)$. In this procedure, both controllers are designed individually. Disturbance rejection is mainly done by the inner loop.

We consider now the case where the transfer $G_3(s)$ really occurs in the plant. This configuration differs from the previous one because there is feedback from the outer part of the plant to the inner loop. Then, in principle, the whole plant must be taken into account when the inner controller is designed. This can be avoided by making the bandwidth of the inner loop large in comparison to that of the outer loop, because under this assumption $G_3(s)\hat{y}$ may be considered as a nearly constant disturbance for the inner loop. Hence, again both controllers can be designed independently.

Direct current motor: control of the angular velocity. A typical application of a cascaded control configuration is the control of a dc motor. Then the inner loop has the task of controlling the current, cf. Fig. 4.13. The current may now be viewed as the control variable in the moment equation

$$J\dot{\omega} = c_A i_A - M_L$$

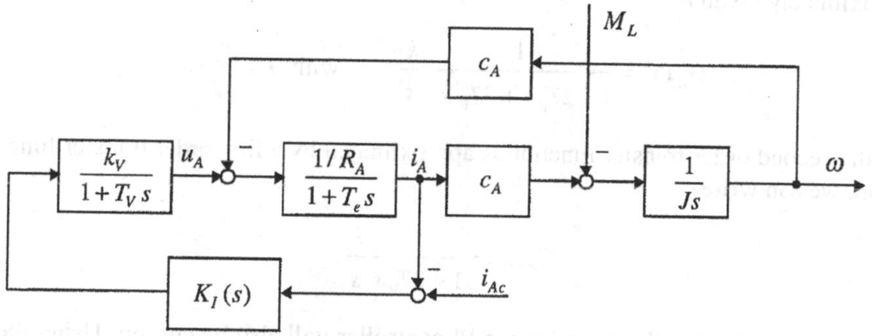

Fig. 4.13. Inner loop for the control of the current

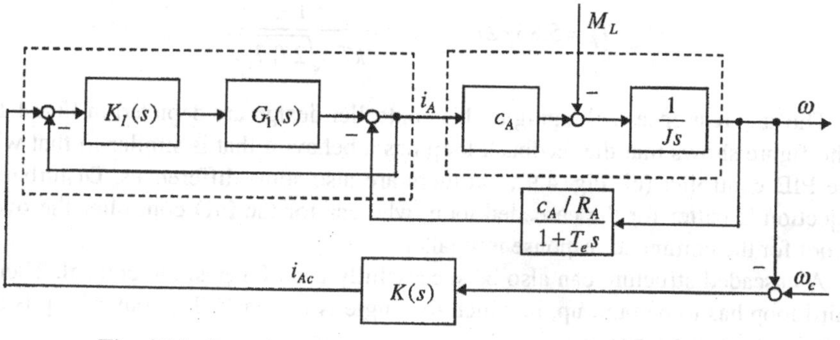

Fig. 4.14. Outer loop for the control of the angular velocity

(cf. Sect. 2.1.2). The block diagram shows that it is fed back into the inner loop. With the notation of Fig. 4.12 we have

$$G_1(s) = \frac{k_V / R_A}{(1+T_V s)(1+T_e s)}, \quad G_3(s) = -\frac{c_A / R_A}{1+T_e s}.$$

We now choose a PI controller for the inner loop. Then the loop is in a position to reject a constant perturbation. With the numerical values of Sect. 4.2.2 we have $T_e > T_V$. Hence $T_I = T_e$ is a reasonable choice for the inner loop. Denoting the gain of the inner loop by V_I, the transfer function for the command is

$$F_{rI}(s) = \frac{V_I}{V_I + T_e s + T_V T_e s^2}.$$

Defining V_I such that $F_{rI}(s)$ has the damping $d = 0.5\sqrt{2}$, we obtain as the loop gain $V_I = T_e /(2T_V)$ and as the transfer function for the command

$$F_{rI}(s) = \frac{1}{1 + 2T_V s + 2T_V^2 s^2}$$

The transfer function of the plant for the outer loop (cf. Fig. 4.14) is now approximately given by

$$G_A(s) = \frac{1}{1 + 2T_V s + 2T_V^2 s^2}\frac{k}{s} \quad \text{with } k = \frac{c_A}{J}.$$

If the second order transfer function is approximated by a first order transfer function, we can write

$$G_A(s) \approx \frac{1}{1 + 2T_V s}\frac{k}{s}.$$

As the controller for the outer loop, a PI controller will also be chosen. Using the formulas for the symmetrical optimum one gets (with a phase margin of 45°)

$$T_I = 5.83 \cdot 2T_V, \quad k_R = \frac{1}{kT_I\sqrt{2T_V T_I}}.$$

The step responses belonging to this controller design are depicted in Fig. 4.15. The figure shows that the feedback loop has a behavior that is similar to that with the PID controller (cf. Fig. 4.9), but there are also some differences. Disturbance rejection is better for the cascaded loop, whereas for the PID controller the overshoot for the command response is smaller.

A cascaded structure can also be successfully used for position control. Then a third loop has to be built up, in which the angle is fed back. This outer loop is decoupled from the inner ones.

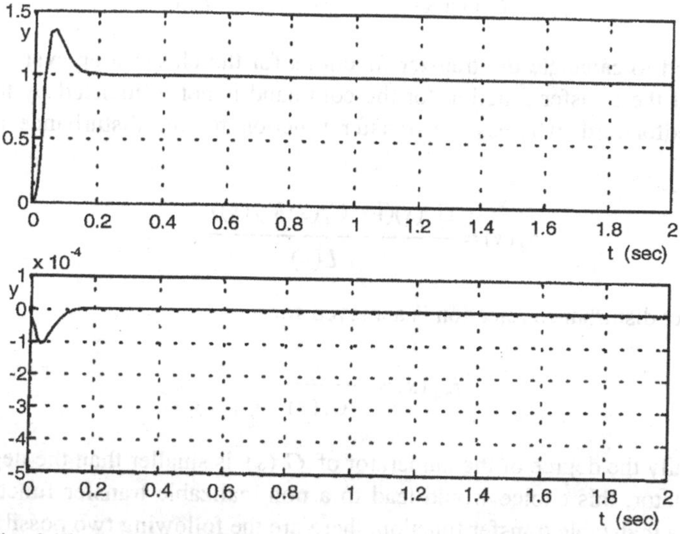

Fig. 4.15. Step responses for the cascaded control of the angular velocity

Disturbance feedforward. For some control problems it is possible to measure the disturbance. In such a case it may be expected that a better disturbance rejection will be obtained when the controller makes use of the measured disturbance. This can be done by using the structure shown in Fig. 4.16. Here the disturbance passes a transfer function $K_d(s)$, which may be viewed as a part of the extended controller. Then we can write

$$\hat{u} = K(s)(\hat{r} - \hat{y}) + K_d(s)\hat{d} .$$

The system $K_d(s)$ comprises also the sensor for the disturbance. We assume for simplicity that the sensor works purely proportionally. Then $K_d(s)$ is freely at our disposal.

From Fig. 4.16 follows immediately the equation

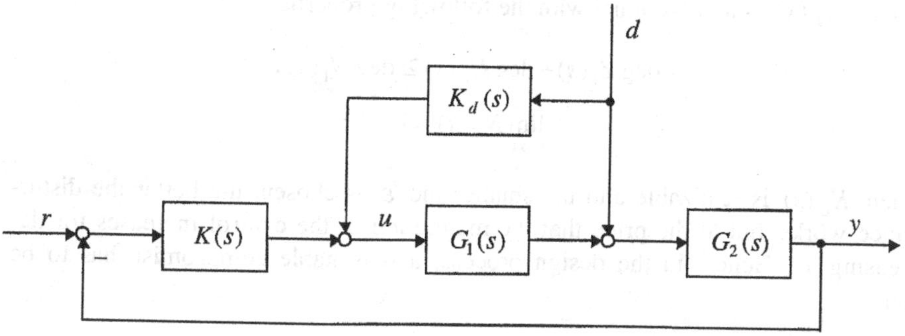

Fig. 4.16 Feedback loop with disturbance feedforward

$$\hat{y} = G_2(s)(G_1(s)(K(s)(\hat{r} - \hat{y}) + K_d(s)\hat{d}) + \hat{d}).$$

It can be used to calculate the transfer functions for the closed-loop system. One observes that the transfer function for the command is not influenced by the disturbance feedforward, whereas the transfer function for the disturbance is now given by

$$F_d(s) = \frac{G_2(s)(1 + G_1(s)K_d(s))}{1 + L(s)}.$$

Hence perfect disturbance rejection is achieved for

$$K_d(s) = -\frac{1}{G_1(s)}.$$

Since normally the degree of the numerator of $G_1(s)$ is smaller than the degree of the denominator, this choice would lead to a non-realizable transfer function. In order to get a realizable transfer function, there are the following two possibilities.

Approach 1. $K_d(s)$ is chosen as a constant: $K_d(s) = k_d$, and the constant is determined such that at least in the stationary case a constant disturbance will be rejected. This leads to the equation

$$k_d = -\frac{1}{G_1(0)}.$$

In this approach, $G_1(s)$ is assumed to be stable.

Approach 2. Let $G_1(s) = Z_1(s)/N_1(s)$ and choose

$$K_d(s) = -\frac{N_1(s)}{Z_1(s)N_\varepsilon(s)},$$

where $N_\varepsilon(s)$ is a polynomial with the following properties:

$$\deg Z_1(s) + \deg N_\varepsilon(s) \geq \deg N_1(s),$$

$$\lim_{\varepsilon \to 0} N_\varepsilon(s) = 1.$$

Then $K_d(s)$ is realizable and the smaller the ε is chosen, the better the disturbance works, but at the price that the magnitude of the control increases for decreasing ε. Hence, in the design process, a reasonable compromise has to be found.

Feedback loops with a prefilter. Since the performance requirements with re-

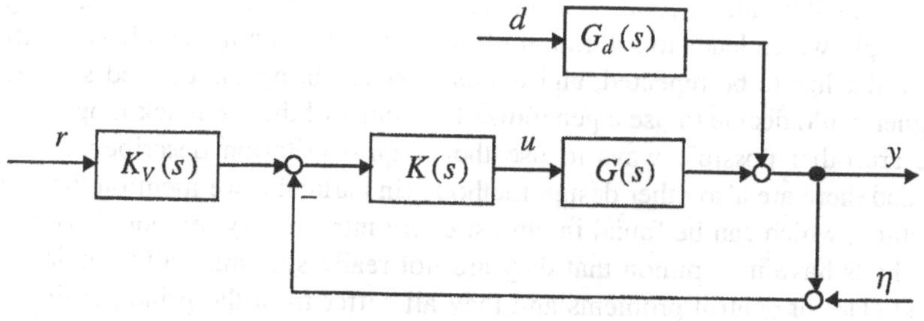

Fig. 4.17. Feedback loop with a prefilter

spect to the reference signal and the disturbance may lead to a conflict, it may sometimes be helpful to use a so-called prefilter. This is a dynamic system which the reference signal has to pass before it is fed into the closed-loop system (cf. Fig. 4.17). In this way, one gets a **two degrees-of-freedom control configuration.**

Let $K_V(s)$ be the transfer function of the prefilter. Then the transfer function for the command is

$$F_{re}(s) = K_V(s)F_r(s),$$

where $F_r(s)$ denotes the transfer function for the command for the loop without prefilter.

4.4 Summary and Outlook

We want to summarize some of the main ideas of the last three chapters. One of them was to introduce the PID controller in a quite natural way and we have seen that in many important cases with this controller the basic requirements for the closed-loop system can be fulfilled. On the other hand, it should have become clear that the PID controller is a rather specialized type of controller: its realizable version is only of order 2. Hence, it may assumed that with a controller of higher order a feedback system having much better properties with respect to stability and performance could be designed. A first step in this direction is to generalize the structure of the feedback loop as shown in the previous section, but this was not a real systematic approach.

The design method presented up to now is based on the Nyquist criterion and uses some ad hoc ideas. One of them is to design a PI controller which compensates the largest time constant of the plant (this is an attempt to fulfill a performance requirement which specifies the bandwidth of the closed-loop system). Then the controller gain is chosen such that the gain margin is large enough to give the feedback system some robustness properties. If the bandwidth with this design is not large enough, a PD controller may be added. Perhaps the design with respect to the reference signal is then satisfactory. But with this approach the disturbance rejection is possibly not as good as specified, or the magnitude of the control signal is too high, which leads to saturation of the actuator. Then the synthesis of the PID controller has to be repeated, and if this does not bring the desired success, the designer could decide to use a generalized structure of the feedback loop.

There are other possible ways to use the Nyquist criterion described in this chapter, and there are also other design methods (in particular we mention the root locus method, which can be found in almost every introductory textbook). But all these methods have in common that they are not really systematically applicable to a broad class of control problems and they all suffer from the principal limitations of the PID controller. These problems increase considerably when MIMO plants are considered. Despite these drawbacks, the methods have been shown to be of great practical value for many plants. Moreover, they need almost no tools and, in fact, they have been used in the past when computers still did not exist.

On the other hand, in Sect. 3.5 we have obtained an idea of how more complicated requirements concerning stability and performance could be specified. This has led us to generalized plants (which include weights for performance and stability) and to $\mathcal{H}_\infty$ optimization problems. Thus, a systematic design procedure should be based on this or a similar type of optimization. Later on in the book, it will be seen that $\mathcal{H}_\infty$ optimization is actually at the heart of modern control theory. It leads to an enormous improvement in controller design, for SISO plants as well as for MIMO plants, but, probably not surprisingly, the theoretical derivation of the results requires an effort that is considerably higher than for the classical methods. It will be done step by step and we start by analyzing state space realizations of general MIMO plants. Then in Chaps. 6 and 7 general feedback systems together with their basic properties are introduced and the ideas developed for SISO plants in Chap. 3 will be generalized in these chapters. In the remaining chapters, controller design by minimizing a certain norm is considered.

Notes and References

Most of the material presented in the Chaps. 1– 4 can be found in much greater detail in any introductory textbook on control. We mention here only Dorf and Bishop [30] and Föllinger [43], but there are many further excellent books on classical control. Some effort has been made to introduce the basic ideas of $\mathcal{H}_\infty$ and robust control for SISO systems and to relate them as far as possible to well-known classical concepts. In particular, this holds for Chap. 3. Concerning these questions, the reader is referred to Doyle, Francis and Tannenbaum [36], where $\mathcal{H}_\infty$ and robust control for SISO systems in the frequency domain is introduced in detail.

Chapter 5

Linear Dynamical Systems

In this chapter, we consider state space representations of linear systems of ordinary differential equations with several inputs and outputs (MIMO systems). As in the SISO case, a linear system arises when a nonlinear system is linearized at an equilibrium state. All control theory methods developed in this book refer to linear systems and the nonlinearity of the original system has to be taken into account indirectly. After having introduced the linearization procedure, we give solution formulas for linear MIMO systems, characterize the stability and calculate the frequency response. Then, by deriving a state space representation for a general SISO transfer function, we are led in a natural way to four very important properties of linear systems, namely controllability, observability, stabilizability and detectability. Several characterizations of these properties are given and the Kalman decomposition of the state space is presented. They are used in the formulation and the proofs of the theorems describing $\mathcal{H}_2$ and $\mathcal{H}_\infty$ optimal controllers.

Section 5.5 is concerned with state space realizations of transfer matrices and operations on such systems. After this, zeros of MIMO systems are defined and characterized. In Sect. 5.7, Lyapunov equations are considered. These are certain linear matrix equations and the positive definiteness of their solutions can be characterized in terms of observability. The last section of this chapter is concerned with linear systems having stochastic inputs. If it is possible to model the disturbance and the sensor noise of the feedback system as a stochastic process with certain known properties, it can be helpful to use this information in the synthesis procedure. In particular, this holds for the synthesis of $\mathcal{H}_2$ optimal controllers.

5.1 Linearization

Let a physical plant with m control variables $u_1(t),\ldots,u_m(t)$, r disturbance variables $d_1(t),\ldots,d_r(t)$ and l output variables $y_1(t),\ldots,y_l(t)$ be given. Suppose that the mathematical plant model consists of a system of n ordinary differential equations for certain functions $x_1(t),\ldots,x_n(t)$. They are denoted as **states** of the system. The output variables can be calculated from the states and the input variables by a system of l algebraically equations. Such a system is denoted as a **MIMO system** (multiple input and multiple output).

It is very helpful to use vector notation:

$$\mathbf{x}(t)=\begin{pmatrix} x_1(t) \\ \vdots \\ x_n(t) \end{pmatrix}, \quad \mathbf{u}(t)=\begin{pmatrix} u_1(t) \\ \vdots \\ u_m(t) \end{pmatrix}, \quad \mathbf{y}(t)=\begin{pmatrix} y_1(t) \\ \vdots \\ y_l(t) \end{pmatrix}, \quad \mathbf{d}(t)=\begin{pmatrix} d_1(t) \\ \vdots \\ d_r(t) \end{pmatrix}$$

Then the plant can be written in the form

$$\mathbf{f}(\dot{\mathbf{x}}(t),\mathbf{x}(t),\mathbf{d}(t),\mathbf{u}(t))=\mathbf{0} \tag{5.1}$$

$$\mathbf{y}(t)=\mathbf{g}(\mathbf{x}(t),\mathbf{d}(t),\mathbf{u}(t))\,. \tag{5.2}$$

Note that the independent variable t does not occur explicitly in $\mathbf{f}$ and $\mathbf{g}$. Moreover, the derivative depends implicitly on the state vector and the inputs. The components of $\mathbf{f}$ and $\mathbf{g}$ are denoted by f_i and g_i, respectively.

Normally the functions $\mathbf{f}$ and $\mathbf{g}$ are nonlinear. In order to apply the methods developed later in the book, the above system must be linearized. To this end, we define at first an **equilibrium point**. This is a constant solution $\mathbf{x}(t)=\mathbf{x}_0$ of (5.1), where it is assumed that the control and the disturbance are also constant (i.e. $\mathbf{d}(t)=\mathbf{d}_0$, $\mathbf{u}(t)=\mathbf{u}_0$). Then $\mathbf{x}_0,\mathbf{d}_0,\mathbf{u}_0$ fulfill the algebraic equation

$$\mathbf{f}(\mathbf{0},\mathbf{x}_0,\mathbf{d}_0,\mathbf{u}_0)=\mathbf{0} \tag{5.3}$$

In this case, the output signal is

$$\mathbf{y}_0=\mathbf{g}(\mathbf{x}_0,\mathbf{d}_0,\mathbf{u}_0)\,. \tag{5.4}$$

We define now deviations from the equilibrium point as follows:

$$\Delta\mathbf{x}=\mathbf{x}-\mathbf{x}_0, \quad \Delta\mathbf{d}=\mathbf{d}-\mathbf{d}_0, \quad \Delta\mathbf{u}=\mathbf{u}-\mathbf{u}_0, \quad \Delta\mathbf{y}=\mathbf{y}-\mathbf{y}_0\,.$$

The functions $\mathbf{f}$ and $\mathbf{g}$ are replaced by their linear approximation at the point $(\mathbf{0},\mathbf{x}_0,\mathbf{d}_0,\mathbf{u}_0)$. Here, the deviations are assumed to be small. Suppose that $\mathbf{f}$ is differentiable. Normally, this is the case, but compare Chen, Lee, and Venkataramanan [22] for an interesting example with a nondifferentiable function $\mathbf{f}$. Denote the Jacobi matrix of $\mathbf{f}$ at $(\mathbf{0},\mathbf{x}_0,\mathbf{d}_0,\mathbf{u}_0)$ by $\mathbf{J}_\mathbf{f}(\mathbf{0},\mathbf{x}_0,\mathbf{d}_0,\mathbf{u}_0)$. Then we have

$$\mathbf{0}=\mathbf{f}(\Delta\dot{\mathbf{x}},\mathbf{x}_0+\Delta\mathbf{x},\mathbf{d}_0+\Delta\mathbf{d},\mathbf{u}_0+\Delta\mathbf{u})$$

$$\approx \mathbf{f}(0,\mathbf{x}_0,\mathbf{d}_0,\mathbf{u}_0) + \mathbf{J}_\mathbf{f}(0,\mathbf{x}_0,\mathbf{d}_0,\mathbf{u}_0)\begin{pmatrix}\Delta\dot{\mathbf{x}}\\ \Delta\mathbf{x}\\ \Delta\mathbf{d}\\ \Delta\mathbf{u}\end{pmatrix}$$

$$= \mathbf{J}_\mathbf{f}(0,\mathbf{x}_0,\mathbf{d}_0,\mathbf{u}_0)\begin{pmatrix}\Delta\dot{\mathbf{x}}\\ \Delta\mathbf{x}\\ \Delta\mathbf{d}\\ \Delta\mathbf{u}\end{pmatrix}.$$

The Jacobi matrix can be written in the form

$$\mathbf{J}_\mathbf{f}(\mathbf{x}_0,\mathbf{d}_0,\mathbf{u}_0) = \begin{pmatrix}-\mathbf{E} & \mathbf{A}_0 & \mathbf{B}_{10} & \mathbf{B}_{20}\end{pmatrix},$$

where the partition is done in accordance with the variables. Then we have

$$\mathbf{A}_0 = \begin{pmatrix}\frac{\partial f_1}{\partial x_1}(0,\mathbf{x}_0,\mathbf{d}_0,\mathbf{u}_0) & \cdots & \frac{\partial f_1}{\partial x_n}(0,\mathbf{x}_0,\mathbf{d}_0,\mathbf{u}_0)\\ \vdots & & \vdots\\ \frac{\partial f_n}{\partial x_1}(0,\mathbf{x}_0,\mathbf{d}_0,\mathbf{u}_0) & \cdots & \frac{\partial f_n}{\partial x_n}(0,\mathbf{x}_0,\mathbf{d}_0,\mathbf{u}_0)\end{pmatrix},$$

$$\mathbf{B}_{20} = \begin{pmatrix}\frac{\partial f_1}{\partial u_1}(0,\mathbf{x}_0,\mathbf{u}_0,\mathbf{d}_0) & \cdots & \frac{\partial f_1}{\partial u_m}(0,\mathbf{x}_0,\mathbf{u}_0,\mathbf{d}_0)\\ \vdots & & \vdots\\ \frac{\partial f_n}{\partial u_1}(0,\mathbf{x}_0,\mathbf{u}_0,\mathbf{d}_0) & \cdots & \frac{\partial f_n}{\partial u_m}(0,\mathbf{x}_0,\mathbf{u}_0,\mathbf{d}_0)\end{pmatrix},$$

and the matrices $\mathbf{E}$ and $\mathbf{B}_{10}$ are defined analogously. Let now $\Delta\mathbf{x}$, $\Delta\mathbf{d}$, $\Delta\mathbf{u}$ be suitable *approximations* of the true deviations. Then we get

$$\mathbf{E}\,\Delta\dot{\mathbf{x}} = \mathbf{A}_0\,\Delta\mathbf{x} + \mathbf{B}_{10}\,\Delta\mathbf{d} + \mathbf{B}_{20}\,\Delta\mathbf{u}\,.$$

Under the assumption that $\mathbf{E}$ is nonsingular, the above system is equivalent to

$$\Delta\dot{\mathbf{x}} = \mathbf{A}\,\Delta\mathbf{x} + \mathbf{B}_1\,\Delta\mathbf{d} + \mathbf{B}_2\,\Delta\mathbf{u}\,, \tag{5.5}$$

with matrices

$$\mathbf{A} = \mathbf{E}^{-1}\mathbf{A}_0, \quad \mathbf{B}_1 = \mathbf{E}^{-1}\mathbf{B}_{10}, \quad \mathbf{B}_2 = \mathbf{E}^{-1}\mathbf{B}_{20}\,.$$

Equation (5.5) is a linear system of differential equations. With regard to the output, the linearization procedure is similar and one obtains

$$\begin{aligned}\Delta\mathbf{y} &= \mathbf{g}(\mathbf{x}_0+\Delta\mathbf{x},\mathbf{d}_0+\Delta\mathbf{d},\mathbf{u}_0+\Delta\mathbf{u})-\mathbf{g}(\mathbf{x}_0,\mathbf{d}_0,\mathbf{u}_0)\\ &\approx \mathbf{J}_g(\mathbf{x}_0,\mathbf{d}_0,\mathbf{u}_0)\begin{pmatrix}\Delta\mathbf{x}\\ \Delta\mathbf{d}\\ \Delta\mathbf{u}\end{pmatrix}.\end{aligned}$$

This can be written as

$$\Delta\mathbf{y} = \mathbf{C}\,\Delta\mathbf{x}+\mathbf{D}_1\,\Delta\mathbf{d}+\mathbf{D}_2\,\Delta\mathbf{u}\,. \tag{5.6}$$

Here, the definition of the matrices $\mathbf{C},\mathbf{D}_1,\mathbf{D}_2$ is obvious.

Equations (5.5) and (5.6) describe approximately the nonlinear system (5.1) and (5.2) near the equilibrium point. In the remaining parts of this chapter, we study the properties of linear systems which have the form (5.5) and (5.6).

5.2 General Properties of Linear Systems

5.2.1 Solution Formula and Transfer Matrices

We now want to consider linear systems of the form (5.5) and (5.6) in detail. In doing so, the symbol "Δ" will be omitted and it is sufficient to consider only one input variable. Then a general linear system with constant coefficients can be written in the form

$$\dot{\mathbf{x}} = \mathbf{A}\,\mathbf{x}+\mathbf{B}\mathbf{u} \tag{5.7}$$

$$\mathbf{y} = \mathbf{C}\mathbf{x}+\mathbf{D}\mathbf{u}\,. \tag{5.8}$$

Here, $\mathbf{x}(t)\in\mathbb{R}^n$ is the **state** of the system and $\mathbf{u}(t)\in\mathbb{R}^m$ denotes an input (a control or a disturbance). The output $\mathbf{y}(t)$ is assumed to be an element of $\mathbb{R}^l$. The matrices $\mathbf{A},\mathbf{B},\mathbf{C}$ and $\mathbf{D}$ are assumed to be appropriately dimensioned. The differential equation (5.7) has a unique solution if in addition an initial condition

$$\mathbf{x}(0) = \mathbf{x}_0 \tag{5.9}$$

is given (with a given vector $\mathbf{x}_0$).

It is not difficult to derive a solution formula for $\mathbf{x}(t)$. To this end, we first consider the homogenous system

$$\dot{\mathbf{x}} = \mathbf{A}\,\mathbf{x}\,.$$

For the initial condition (5.9), it has the solution

$$\mathbf{x}(t) = e^{\mathbf{A}\,t}\,\mathbf{x}_0\,. \tag{5.10}$$

Here, the matrix exponential function is defined by

$$e^{\mathbf{A}t} = \sum_{k=0}^{\infty} \frac{t^k}{k!} \mathbf{A}^k .$$

The solution property may be verified directly by differentiating the power series. It can be shown that all the necessary steps are feasible in a rigorous mathematical sense. The solution of the inhomogeneous system is formally completely analogous to the scalar case:

$$\mathbf{x}(t) = e^{\mathbf{A}t}\mathbf{x}_0 + \int_0^t e^{\mathbf{A}(t-\tau)}\mathbf{B}\mathbf{u}(\tau)\,d\tau . \tag{5.11}$$

This may be seen by inserting (5.11) into (5.7) or by variation of the constants as in the scalar case.

The output $\mathbf{y}$ is obtained by using (5.11) and (5.8):

$$\begin{aligned}\mathbf{y}(t) &= \mathbf{C}e^{\mathbf{A}t}\mathbf{x}_0 + \int_0^t \mathbf{C}e^{\mathbf{A}(t-\tau)}\mathbf{B}\mathbf{u}(\tau)\,d\tau + \mathbf{D}\mathbf{u}(t) \\ &= \mathbf{C}e^{\mathbf{A}t}\mathbf{x}_0 + \int_0^t (\mathbf{C}e^{\mathbf{A}(t-\tau)}\mathbf{B} + \mathbf{D}\delta(t-\tau))\mathbf{u}(\tau)\,d\tau .\end{aligned}$$

In this equation, $\delta(t)$ denotes the unit impulse. Let now

$$\mathbf{g}(t) = \mathbf{C}e^{\mathbf{A}t}\mathbf{B} + \mathbf{D}\delta(t) . \tag{5.12}$$

Then for a vanishing initial value ($\mathbf{x}_0 = \mathbf{0}$), the input/output (I/O) behavior of the system can be described by the convolution integral

$$\mathbf{y}(t) = \int_0^t \mathbf{g}(t-\tau)\mathbf{u}(\tau)\,d\tau . \tag{5.13}$$

As in the scalar case (c.f. (2.10)), the function $\mathbf{g}(t)$ is denoted as the **impulse response.**

It is also possible to describe the I/O relationship by a transfer matrix. To this end, the Laplace transform has to be applied to (5.7), which leads to

$$s\hat{\mathbf{x}} - \mathbf{x}(0) = \mathbf{A}\hat{\mathbf{x}} + \mathbf{B}\hat{\mathbf{u}} .$$

Hence, the output can be calculated as

$$\hat{\mathbf{y}} = \mathbf{C}(s\mathbf{I}-\mathbf{A})^{-1}\mathbf{x}_0 + (\mathbf{C}(s\mathbf{I}-\mathbf{A})^{-1}\mathbf{B} + \mathbf{D})\hat{\mathbf{u}} .$$

By defining

$$\mathbf{G}(s) = \mathbf{C}(s\mathbf{I} - \mathbf{A})^{-1}\mathbf{B} + \mathbf{D} \tag{5.14}$$

the input/output relationship is given by

$$\hat{\mathbf{y}} = \mathbf{G}(s)\hat{\mathbf{u}} \tag{5.15}$$

if the initial value is assumed to be the zero vector. This equation is the direct generalization of (2.8) to MIMO systems. $\mathbf{G}(s)$ is denoted as the **transfer matrix** of the linear system (5.7) and (5.8). It is convenient to represent $\mathbf{G}(s)$ in the form

$$\mathbf{G}(s) = \left(\begin{array}{c|c} \mathbf{A} & \mathbf{B} \\ \hline \mathbf{C} & \mathbf{D} \end{array}\right).$$

Moreover, it may easily be seen that the transfer matrix is the Laplace transform of the impulse response: $\mathbf{G}(s) = \hat{\mathbf{g}}(s)$.

The inverse matrix occurring in (5.14) can be represented in the form

$$(s\mathbf{I} - \mathbf{A})^{-1} = \frac{\mathbf{M}(s)}{q(s)}.$$

Here, $q(s)$ denotes the characteristic polynomial of $\mathbf{A}$,

$$q(s) = \det(s\mathbf{I} - \mathbf{A}),$$

and $\mathbf{M}(s)$ is a n x n-matrix, which has polynomials of maximum degree $n-1$ as entries. Thus, it is possible to write

$$\mathbf{G}(s) = \frac{1}{q(s)}\mathbf{C}\mathbf{M}(s)\mathbf{B} + \mathbf{D}. \tag{5.16}$$

Hence each component of $\mathbf{G}(s)$ is a proper rational function and the denominator of each component is a divisor of $q(s)$. Note that cancellations may occur.

As in the scalar case, a transfer matrix of the form (5.16) is said to be **proper**. For $\mathbf{D} = \mathbf{0}$, the transfer matrix is **strictly proper**.

Examples.

1. DC motor

The dynamics of a dc motor can be described by the following differential system (c.f. Sect. (2.1.2)):

$$\frac{d\omega}{dt} = \frac{c_A}{J} i_A - \frac{1}{J} M_L$$

$$\frac{di_A}{dt} = -\frac{R_A}{L_A} i_A - \frac{c_A}{L_A}\omega + \frac{1}{L_A} u_A.$$

Using matrices, the system can be written as

$$\begin{pmatrix} \dfrac{d\omega}{dt} \\ \dfrac{di_A}{dt} \end{pmatrix} = \begin{pmatrix} 0 & \dfrac{c_A}{J} \\ -\dfrac{c_A}{L_A} & -\dfrac{R_A}{L_A} \end{pmatrix} \begin{pmatrix} \omega \\ i_A \end{pmatrix} + \begin{pmatrix} 0 \\ \dfrac{1}{L_A} \end{pmatrix} u_A + \begin{pmatrix} -\dfrac{1}{J} \\ 0 \end{pmatrix} M_L ,$$

$$\omega = \begin{pmatrix} 1 & 0 \end{pmatrix} \begin{pmatrix} \omega \\ i_A \end{pmatrix} .$$

2. Crane positioning system.

Figure 5.1 shows a crane positioning system. The physical constants are the mass m_K of the carriage, the mass m_G of the bucket and the distance d of the center of gravity of the carriage to the center of gravity to the bucket. The dynamical quantities are the position x of the carriage, the angle θ of the rope and the force F exerted by the motor of the carriage.

We do not want to derive here the dynamics of the system (but compare Sect. 11.1, where the inverted pendulum is analyzed in detail). There it will be seen that for small angles θ the following equations hold:

$$\ddot{x} + d\ddot{\theta} + g\theta = 0,$$

$$(m_G + m_K)\ddot{x} + m_G d\ddot{\theta} = F .$$

It is convenient to work with the position $x_G = x + d\theta$ of the bucket instead of the angle θ and to use the abbreviations

$$\omega_0 = \sqrt{\frac{g}{d}}, \quad \omega_1 = \sqrt{1 + \frac{m_G}{m_K}}\,\omega_0 , \quad k = \frac{1}{m_K} .$$

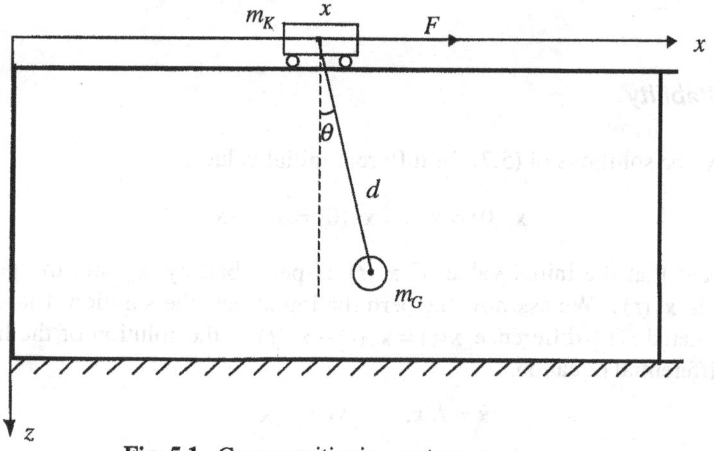

Fig. 5.1. Crane positioning system

Then we get

$$\ddot{x} = -(\omega_1^2 - \omega_0^2)x + (\omega_1^2 - \omega_0^2)x_G + kF,$$
$$\ddot{x}_G = \omega_0^2 x - \omega_0^2 x_G .$$

By defining states

$$x_1 = x, \quad x_2 = \dot{x}, \quad x_3 = x_G, \quad x_4 = \dot{x}_G,$$

the dynamics of the crane positioning system can be written in the form

$$\begin{pmatrix} \dot{x}_1 \\ \dot{x}_2 \\ \dot{x}_3 \\ \dot{x}_4 \end{pmatrix} = \begin{pmatrix} 0 & 1 & 0 & 0 \\ -(\omega_1^2 - \omega_0^2) & 0 & \omega_1^2 - \omega_0^2 & 0 \\ 0 & 0 & 0 & 1 \\ \omega_0^2 & 0 & -\omega_0^2 & 0 \end{pmatrix} \begin{pmatrix} x_1 \\ x_2 \\ x_3 \\ x_4 \end{pmatrix} + \begin{pmatrix} 0 \\ k \\ 0 \\ 0 \end{pmatrix} F .$$

From these equations, the following transfer functions can easily be calculated:

$$\hat{x} = k \frac{s^2 + \omega_0^2}{s^2 (s^2 + \omega_1^2)} \hat{F}, \quad \hat{x}_G = k \frac{\omega_0^2}{s^2 (s^2 + \omega_1^2)} \hat{F} .$$

The rope angle is given by

$$\theta = \frac{x_G - x}{d}.$$

It has the transfer function

$$\hat{\theta} = -\frac{k}{d} \frac{1}{s^2 + \omega_1^2} \hat{F} .$$

5.2.2 Stability

Let $\mathbf{x}_1, \mathbf{x}_2$ be solutions of (5.7) for different initial values:

$$\mathbf{x}_1(0) = \mathbf{x}_{10}, \quad \mathbf{x}_2(0) = \mathbf{x}_{10} + \mathbf{x}_0 .$$

This means that the initial value of $\mathbf{x}_1(t)$ is perturbed by $\mathbf{x}_0$ and the perturbed solution is $\mathbf{x}_2(t)$. We ask how the perturbation affects the solution if large times are considered. The difference $\mathbf{x}(t) = \mathbf{x}_2(t) - \mathbf{x}_1(t)$ is the solution of the homogeneous differential equation

$$\dot{\mathbf{x}} = \mathbf{A}\mathbf{x}, \qquad \mathbf{x}(0) = \mathbf{x}_0 . \tag{5.17}$$

Definition. *The system (5.7) is* ***stable,*** *if for every* $\mathbf{x}_0$ *the solution of the homogeneous system (5.17) tends to* $\mathbf{0}$ *for* $t \to \infty$ *:*

$$\lim_{t\to\infty} \mathbf{x}(t) = \mathbf{0}\,.$$

Thus, for large t the solution $\mathbf{x}_1$ defined above is only changed by small values if the initial value is perturbed.

Next, we want to derive a criterion for stability. As in the SISO case, this will be done by establishing a solution formula, which here applies to the homogeneous system and is also of interest in itself. To this end, let $\mathbf{W}$ be a nonsingular matrix and put $\mathbf{c} = \mathbf{W}^{-1}\mathbf{x}_0$. Then (5.10) implies

$$\mathbf{x}(t) = e^{\mathbf{A}t}\,\mathbf{x}_0 = e^{\mathbf{A}t}\,\mathbf{W}\mathbf{c}\,.$$

We now represent $\mathbf{W}$ by its columns and write $\mathbf{W} = (\mathbf{w}_1\ \dots\ \mathbf{w}_n)$. This yields

$$\mathbf{x}(t) = e^{\mathbf{A}t}\,(\mathbf{w}_1\ \dots\ \mathbf{w}_n)\mathbf{c} = (e^{\mathbf{A}t}\,\mathbf{w}_1\ \dots\ e^{\mathbf{A}t}\,\mathbf{w}_n)\mathbf{c}\,.$$

Denoting by c_k the components of $\mathbf{c}$, we get

$$\mathbf{x}(t) = \sum_{k=0}^{n} c_k e^{\mathbf{A}t}\,\mathbf{w}_k\,. \tag{5.18}$$

The usefulness of this formula is given by the fact that for certain vectors $\mathbf{v}$ it is possible to evaluate the term $e^{\mathbf{A}t}\,\mathbf{v}$ before having calculated the matrix exponential function in advance. For example, let $\mathbf{v}$ be an eigenvector of $\mathbf{A}$, which belongs to the eigenvalue λ. Then the equation $\mathbf{A}^k\mathbf{v} = \lambda^k\mathbf{v}$ holds and hence

$$e^{\mathbf{A}t}\,\mathbf{v} = \sum_{k=0}^{\infty} \frac{t^k}{k!}\lambda^k\,\mathbf{v} = e^{\lambda t}\,\mathbf{v}\,.$$

This leads directly to the following assertion:

Suppose the matrix $\mathbf{A}$ *has n eigenvalues* $\lambda_1,\dots,\lambda_n$ *with n linearly independent eigenvectors* $\mathbf{v}_1,\dots,\mathbf{v}_n$. *Let* $\mathbf{W} = (\mathbf{v}_1\ \dots\ \mathbf{v}_n)$ *be nonsingular and put* $\mathbf{c} = \mathbf{W}^{-1}\mathbf{x}_0$. *Then the solution of (5.17) is given by*

$$\mathbf{x}(t) = \sum_{k=0}^{n} c_k e^{\lambda_k t}\,\mathbf{v}_k\,. \tag{5.19}$$

The assumption of the previous assertion is fulfilled if and only if $\mathbf{A}$ is diagonizable. For a nondiagonizable matrix $\mathbf{A}$, the system of eigenvectors has to be completed by so-called principal vectors to a basis of $\mathbb{C}^n$. These are defined as follows.

Definition. *A vector* $\mathbf{v}$ *is called a* ***principal vector*** *corresponding to the eigenvector* λ *of the matrix* $\mathbf{A}$, *if there exists a number* $l \in \mathbb{N}$ *such that*

$$(\mathbf{A}-\lambda\mathbf{I})^l \mathbf{v} = \mathbf{0} \quad \textit{and} \quad (\mathbf{A}-\lambda\mathbf{I})^{l-1} \mathbf{v} \neq \mathbf{0}.$$

Each eigenvector is a principal vector with $l = 1$.

Let now $\lambda_1, \ldots, \lambda_k$ be the mutually different eigenvalues of $\mathbf{A}$ with multiplicities $n_1, \ldots, n_k$. The sum of these multiplicities is n. From linear algebra, the following facts are well known.

(a) *For each eigenvalue* λ_j $(1 \le j \le k)$, *there are* n_j *linearly independent principal vectors* $\mathbf{w}_{j\nu}$, $\nu = 1, \ldots, n_j$. *They form a basis of the* n_j*-dimensional solution space*

$$\{\mathbf{v} \in \mathbb{C}^n \mid \mathbf{A}-\lambda_j\mathbf{I})^{n_j} \mathbf{v} = \mathbf{0}\}.$$

(b) *The principal vectors* $\mathbf{w}_{j\nu}$, $j = 1, \ldots, k$, $\nu = 1, \ldots, n_j$, *are linearly independent and hence form a basis of* $\mathbb{R}^n$.

Theorem 5.2.1. *Let* $\mathbf{w}_1, \ldots, \mathbf{w}_n$ *be linearly independent principal vectors of* $\mathbf{A}$. *Define* $\mathbf{W} = (\mathbf{w}_1 \ldots \mathbf{w}_n)$ *and* $\mathbf{c} = \mathbf{W}^{-1}\mathbf{x}_0$. *Here,* $\mathbf{w}_k$ *is assumed to be a principal vector corresponding to the eigenvalue* λ_k *with* $l_k = l$ *as in the definition (of course, in this formulation the eigenvectors* $\lambda_1, \ldots, \lambda_n$ *are not assumed to be pairwise distinct). For* $k = 1, \ldots, n$ *define now the functions*

$$\mathbf{x}_k(t) = e^{\lambda_k t} \sum_{\nu=0}^{l_k-1} \frac{t^\nu}{\nu!} (\mathbf{A}-\lambda_k\mathbf{I})^\nu \mathbf{w}_k. \tag{5.20}$$

Then the solution of (5.17) is given by

$$\mathbf{x}(t) = \sum_{k=0}^{n} c_k \mathbf{x}_k(t). \tag{5.21}$$

Proof. The proof results from the above considerations by applying (5.18). We calculate $e^{\mathbf{A}t}\mathbf{w}_k$ and omit for the sake of notational simplicity the index k. First, note that

$$e^{\mathbf{A}t} = e^{\lambda t \mathbf{I} + t(\mathbf{A}-\lambda\mathbf{I})} = e^{\lambda t} e^{t(\mathbf{A}-\lambda\mathbf{I})}.$$

Because $(\mathbf{A}-\lambda\mathbf{I})^k \mathbf{w} = \mathbf{0}$ for $k \ge l$, the infinite series for $e^{t(\mathbf{A}-\lambda\mathbf{I})}$ reduces to the finite sum in (5.20) and everything is shown.

Note that the above theorem contains formula (5.19) as a special case. The formulas (5.19) and (5.20) represent the solution in a complex form. It is also possible to give a real version of these formulas. Let $\mathbf{w}_k$ be a complex principal vector.

Then the corresponding eigenvalue λ_k and the function $\mathbf{x}_k(t)$ of (5.20) are also complex. The real and imaginary parts of this function are solutions of the homogenous equation. Let $\mathbf{w}_k = \mathbf{a}_k + i\mathbf{b}_k$. Then we have $\operatorname{Re}\mathbf{x}_k(0) = \mathbf{a}_k$ and $\operatorname{Im}\mathbf{x}_k(0) = \mathbf{b}_k$. Now replace in the matrix $\mathbf{W}$ the principal vectors $\mathbf{w}_k, \overline{\mathbf{w}}_k$ by $\mathbf{a}_k, \mathbf{b}_k$ and in (5.21) the functions $\mathbf{x}_k(t), \overline{\mathbf{x}}_k(t)$ by $\operatorname{Re}\mathbf{x}_k(t), \operatorname{Im}\mathbf{x}_k(t)$. This must be done for all principal vectors. Thereby, the matrix $\mathbf{W}$ remains nonsingular and (5.21) becomes a linear combination of functions $\tilde{\mathbf{x}}_k(t)$, which are all of the form

$$\tilde{\mathbf{x}}(t) = e^{\alpha t}[(\cos\beta t)\mathbf{p}(t) + (\sin\beta t)\mathbf{q}(t)] \qquad \text{with } \lambda = \alpha + i\beta .$$

Here, $\mathbf{p}(t), \mathbf{q}(t)$ denote vectors whose coefficients are polynomials.

With the solution formula of Theorem 5.2.1 it is now easy to derive a criterion for stability.

Theorem 5.2.2. *The system (5.7) is stable if and only if all eigenvalues of* $\mathbf{A}$ *have a negative real part. In this case, the matrix* $\mathbf{A}$ *is said to be* ***stable***.

Proof. „$\Leftarrow$“. For $\lambda_k = \alpha_k + i\beta_k$ we can write $e^{\lambda_k t} = e^{\alpha_k t}(\cos\beta_k t + i\sin\beta_k t)$. Since $\alpha_k < 0$ by assumption, we now immediately see that $\mathbf{x}_k(t)$ tends to zero for $t \to \infty$.
„$\Rightarrow$“. Suppose $\operatorname{Re}\lambda_k \geq 0$ for a number k and choose $\mathbf{x}_0 = \mathbf{w}_k$. Then (5.20) implies (again with omitted index k)

$$\frac{1}{t^{l-1}e^{\alpha t}}\mathbf{x}(t) = \frac{1}{(l-1)!}(\cos\beta t + i\sin\beta t)(\mathbf{A} - \lambda\mathbf{I})^{l-1}\mathbf{w} + \mathbf{h}(t).$$

Here, $\mathbf{h}(t)$ denotes a function which tends to $\mathbf{0}$ for $t \to \infty$. Suppose now $\mathbf{x}(t) \to \mathbf{0}$ for $t \to \infty$. Because $(\mathbf{A} - \lambda\mathbf{I})^{l-1}\mathbf{w} \neq \mathbf{0}$, this implies that the complex harmonic oscillation $\cos\beta t + i\sin\beta t$ tends to 0 for $t \to \infty$, which is impossible.

Theorem 5.2.3. *Let* $\mathbf{A}$ *be stable and let* $\mathbf{u}(t) = \mathbf{u}_0$ *be constant. Then the output* $\mathbf{y}(t)$ *of (5.7) and (5.8) tends to a limit* $\mathbf{y}_\infty$ *for* $t \to \infty$. *This limit can be calculated from the formula*

$$\mathbf{y}_\infty = \mathbf{G}(0)\mathbf{u}_0 .$$

Proof. Since the system is stable, we may assume $\mathbf{x}_0 = \mathbf{0}$ without losing generality. The solution of (5.7) is given by

$$\mathbf{x} = \mathbf{A}^{-1}(e^{\mathbf{A}t} - \mathbf{I})\mathbf{B}\mathbf{u}_0 .$$

This may be verified by applying (5.11). Since $\mathbf{A}$ is stable, $\mathbf{x}(t)$ tends to $\mathbf{A}^{-1}\mathbf{B}\mathbf{u}_0$ as $t \to \infty$ and the theorem immediately follows.

Remark. The definition of stability given here also applies to SISO systems and may therefore also be compared with that introduced in Sect. 2.2.2. If each eigenvalue of $\mathbf{A}$ is a pole of $\mathbf{G}(s)$, then both definitions coincide, but it may happen that cancellations occur. In particular, a linear factor with an eigenvalue λ of $\mathbf{A}$ such that $\operatorname{Re}\lambda \geq 0$ could cancel. This fact will become very important when stability for feedback systems is considered, cf. Sect. 6.5.2.

5.2.3 Frequency Response

As in the SISO case, we are interested in the representation of the output if the input signal is a harmonic oscillation:

$$\mathbf{u} = e^{st}\mathbf{u}_0 \quad \text{with } s = i\omega .$$

We calculate the output for an arbitrary complex s. This leads to no additional difficulty but will be used in an application given later. The function $\mathbf{x} = e^{st}\mathbf{w}_0$ is a solution of (5.7), if s is not an eigenvalue of $\mathbf{A}$ and $\mathbf{w}_0$ is chosen as

$$\mathbf{w}_0 = (s\mathbf{I} - \mathbf{A})^{-1}\mathbf{B}\mathbf{u}_0 .$$

In order that the solution satisfies also the initial condition (5.9), a suitable solution $\mathbf{x}_h$ of the homogenous equation has to be added. It is given by

$$\mathbf{x}_h(t) = e^{\mathbf{A}t}\mathbf{x}_0 - e^{\mathbf{A}t}(s\mathbf{I} - \mathbf{A})^{-1}\mathbf{B}\mathbf{u}_0$$

and the output results as

$$\mathbf{y}(t) = \mathbf{C}\mathbf{x}_h(t) + e^{st}(\mathbf{C}(s\mathbf{I} - \mathbf{A})^{-1}\mathbf{B} + \mathbf{D})\mathbf{u}_0 .$$

If the initial value is appropriately chosen, the function $\mathbf{x}_h(t)$ vanishes. This is the case for

$$\mathbf{x}_0 = (s\mathbf{I} - \mathbf{A})^{-1}\mathbf{B}\mathbf{u}_0 .$$

Then we obtain

$$\mathbf{y}(t) = e^{st}(\mathbf{C}(s\mathbf{I} - \mathbf{A})^{-1}\mathbf{B} + \mathbf{D})\mathbf{u}_0 .$$

The next theorem follows directly from the above calculations and leads to an interpretation of the transfer matrix, which is the counterpart of Theorem 2.2.2 for MIMO systems.

Theorem 5.2.4. *Let $\mathbf{u}_0 \in \mathbb{C}^n$ be constant. Then the following assertions are true.*

(a) Suppose s is not an eigenvalue of $\mathbf{A}$. Then the solution $\mathbf{x}(t)$ of

$$\dot{\mathbf{x}} = \mathbf{A}\mathbf{x} + \mathbf{B}e^{st}\mathbf{u}_0, \quad \mathbf{x}(0) = (s\mathbf{I} - \mathbf{A})^{-1}\mathbf{B}\mathbf{u}_0$$

and the corresponding output signal $\mathbf{y}(t)$ *are given by*

$$\mathbf{x}(t) = e^{st}(s\mathbf{I} - \mathbf{A})^{-1}\mathbf{B}\mathbf{u}_0,$$
$$\mathbf{y}(t) = e^{st}\mathbf{G}(s)\mathbf{u}_0.$$

(b) Let $\mathbf{A}$ *be stable. Then the solution* $\mathbf{x}(t)$ *of*

$$\dot{\mathbf{x}} = \mathbf{A}\mathbf{x} + \mathbf{B}e^{i\omega t}\mathbf{u}_0, \quad \mathbf{x}(0) = \mathbf{x}_0$$

and the corresponding output signal $\mathbf{y}(t)$ *can be written as*

$$\mathbf{x}(t) = \mathbf{x}_h(t) + e^{i\omega t}(i\omega\mathbf{I} - \mathbf{A})^{-1}\mathbf{B}\mathbf{u}_0$$
$$\mathbf{y}(t) = \mathbf{y}_h(t) + e^{i\omega t}\mathbf{G}(i\omega)\mathbf{u}_0,$$

where $\mathbf{x}_h(t)$, $\mathbf{y}_h(t)$ *are functions which tend to* $\mathbf{0}$ *as* $t \to \infty$.

As in the SISO case, the function $\mathbf{G}(i\omega)$ is denoted as the **frequency response** of the system. The frequency response is the transfer matrix $\mathbf{G}(s)$ on the imaginary axis, i.e. the Fourier transform of the impulse response.

The response of the system for real harmonic oscillations is obtained as usual by passing to the imaginary part. Let

$$\mathbf{G}(s) = (G_{jk}(s)).$$

Here, the transfer functions $G_{jk}(s)$ are rational. We define now

$$\tilde{\mathbf{G}}(i\omega) = \left(|\,G_{jk}(i\omega)\,|\sin(\omega t + \arg G_{jk}(i\omega))\right).$$

Therefore, the input signal $\mathbf{u}(t) = (\sin\omega t)\mathbf{u}_0$ leads to the output signal

$$\mathbf{y}(t) = \tilde{\mathbf{G}}(i\omega)\mathbf{u}_0.$$

In this formula, the decaying part is omitted. Each component of $\mathbf{y}(t)$ is a superposition of several harmonic oscillations.

5.3 Controllability

5.3.1 Realization of Transfer Functions and Controller-Canonical Form

In Chap. 2 we have seen that the application of physical laws leads to a mathematical model of the plant, which consists of a system of differential equations. Then the transfer functions describing the plant are found by applying the Laplace transform to these equations. Sometimes it is necessary to proceed in the reverse direction. Then the transfer function (or transfer matrix) is given and a state space realization is sought. To be more precise, let a transfer matrix $\mathbf{G}(s)$ be given. The task is now to find a system of matrices $(\mathbf{A}, \mathbf{B}, \mathbf{C}, \mathbf{D})$ such that

$$\mathbf{G}(s) = \left(\begin{array}{c|c} \mathbf{A} & \mathbf{B} \\ \hline \mathbf{C} & \mathbf{D} \end{array}\right).$$

Then $(\mathbf{A}, \mathbf{B}, \mathbf{C}, \mathbf{D})$ is a **realization** of $\mathbf{G}(s)$.

It is easy to see that the solution of this problem is not unique. Let $(\mathbf{A}, \mathbf{B}, \mathbf{C}, \mathbf{D})$ be a realization of $\mathbf{G}(s)$ and let $\mathbf{T}$ be a nonsingular $n \times n$ -matrix. We now introduce a new state vector $\mathbf{z}$ by setting $\mathbf{z} = \mathbf{T}\mathbf{x}$. Then the equations

$$\dot{\mathbf{z}} = \mathbf{T}\mathbf{A}\mathbf{T}^{-1}\mathbf{z} + \mathbf{T}\mathbf{B}\mathbf{u}$$

$$\mathbf{y} = \mathbf{C}\mathbf{T}^{-1}\mathbf{z} + \mathbf{D}\mathbf{u}$$

hold and the system can be written in the form

$$\dot{\mathbf{z}} = \mathbf{A}_1\mathbf{z} + \mathbf{B}_1\mathbf{u} \tag{5.22}$$

$$\mathbf{y} = \mathbf{C}_1\mathbf{z} + \mathbf{D}_1\mathbf{u} \tag{5.23}$$

with

$$\mathbf{A}_1 = \mathbf{T}\mathbf{A}\mathbf{T}^{-1}, \quad \mathbf{B}_1 = \mathbf{T}\mathbf{B}, \quad \mathbf{C}_1 = \mathbf{C}\mathbf{T}^{-1}, \quad \mathbf{D}_1 = \mathbf{D}. \tag{5.24}$$

The two systems $(\mathbf{A}, \mathbf{B}, \mathbf{C}, \mathbf{D})$, $(\mathbf{A}_1, \mathbf{B}_1, \mathbf{C}_1, \mathbf{D}_1)$ are said to be **equivalent** if there exists a nonsingular matrix $\mathbf{T}$ such (5.24) holds. From (5.22) and (5.23) it follows that for equivalent systems a given input $\mathbf{u}$ leads to the same output $\mathbf{y}$. Hence, the following assertion holds:

If the realizations $(\mathbf{A}, \mathbf{B}, \mathbf{C}, \mathbf{D})$, $(\mathbf{A}_1, \mathbf{B}_1, \mathbf{C}_1, \mathbf{D}_1)$ *are equivalent, then the corresponding transfer matrices coincide:*

$$\left(\begin{array}{c|c} \mathbf{A} & \mathbf{B} \\ \hline \mathbf{C} & \mathbf{D} \end{array}\right) = \left(\begin{array}{c|c} \mathbf{A}_1 & \mathbf{B}_1 \\ \hline \mathbf{C}_1 & \mathbf{D}_1 \end{array}\right). \tag{5.25}$$

We now turn to the question of how such a realization can be constructed. In doing so, we limit ourselves in this section to SISO systems and assume the transfer function

$$G(s) = \frac{c_n s^n + \ldots + c_1 s + c_0}{s^n + a_{n-1}s^{n-1} + \ldots + a_1 s + a_0} \tag{5.26}$$

to be given. Let

$$\hat{z}_1 = \frac{1}{s^n + a_{n-1}s^{n-1} + \ldots + a_1 s + a_0}\hat{u}, \tag{5.27}$$

$$\hat{z}_i = s\,\hat{z}_{i-1}, \qquad i = 2,\dots,n. \tag{5.28}$$

Then $\hat{z}_i = s^{i-1}\hat{z}_1$ for $i = 2,\dots,n$. Equation (5.27) is equivalent to

$$a_0\hat{z}_1 + a_1 s\hat{z}_1 + \dots + a_{n-1}s^{n-1}\hat{z}_1 + s^n\hat{z}_1 = \hat{u},$$

and this can be written in the form

$$s\hat{z}_n = -a_0\hat{z}_1 - a_1\hat{z}_2 - \dots - a_{n-1}\hat{z}_n + \hat{u}. \tag{5.29}$$

Hence (5.27) and (5.28) are equivalent to the differential equations

$$\begin{aligned} &\dot{z}_1 = z_2,\ \dot{z}_2 = z_3,\ \dots,\ \dot{z}_{n-1} = z_n \\ &\dot{z}_n = -a_0 z_1 - a_1 z_2 - \dots - a_{n-1}z_n + u. \end{aligned}$$

Then (5.29) implies for $\hat{y} = G(s)\hat{u}$ that

$$\begin{aligned} \hat{y} &= c_n s^n\hat{z}_1 + \dots + c_1 s\hat{z}_1 + c_0\hat{z}_1 \\ &= c_n s\,\hat{z}_n + c_{n-1}\hat{z}_n + \dots + c_1\hat{z}_2 + c_0\hat{z}_1 \\ &= (c_{n-1} - c_n a_{n-1})\hat{z}_n + \dots + (c_1 - c_n a_1)\hat{z}_2 + (c_0 - c_n a_0)\hat{z}_1 + c_n\hat{u}. \end{aligned}$$

Consequently, we have obtained the following result.

Theorem 5.3.1 *The transfer function $G(s)$ defined in (5.26) has the realization*

$$\begin{aligned} \dot{\mathbf{z}} &= \mathbf{A}_1\mathbf{z} + \mathbf{b}_1 u \\ y &= \mathbf{C}_1\mathbf{z} + d_1 u, \end{aligned}$$

where

$$\mathbf{C}_1 = \mathbf{c}_1^T = \begin{pmatrix} c_0 - a_0 c_n & \dots & c_{n-1} - a_{n-1}c_n \end{pmatrix}, \qquad d_1 = c_n, \tag{5.30}$$

and

$$\mathbf{A}_1 = \begin{pmatrix} 0 & 1 & & & 0 \\ & 0 & 1 & & \\ & & \ddots & \ddots & \\ 0 & & & 0 & 1 \\ -a_0 & -a_1 & \cdots & -a_{n-2} & -a_{n-1} \end{pmatrix}, \qquad \mathbf{b}_1 = \begin{pmatrix} 0 \\ \vdots \\ \vdots \\ 0 \\ 1 \end{pmatrix}. \tag{5.31}$$

Hence, the following equation holds:

$$G(s) = \mathbf{c}_1^T(s\mathbf{I} - \mathbf{A}_1)^{-1}\mathbf{b}_1 + d_1.$$

The realization of the preceding theorem has a particularly simple form. This fact will be helpful in further investigations. If a different realization of $G(s)$ is given, it may be asked whether it can be brought into the form of Theorem 5.3.1 by a transformation described by the equations (5.24). More precisely, the following question arises: under which conditions does there exist a nonsingular matrix $\mathbf{T}$ such that for the equivalent system (with $\mathbf{z} = \mathbf{Tx}$)

$$\dot{\mathbf{z}} = \mathbf{A}_1\mathbf{z} + \mathbf{b}_1 u$$

the matrices $\mathbf{A}_1 = \mathbf{TAT}^{-1}$, $\mathbf{b}_1 = \mathbf{Tb}$ are of the form (5.31)? At first, it is easy to see that

$$\det(s\mathbf{I} - \mathbf{A}_1) = s^n + a_{n-1}s^{n-1} + ... + a_1 s + a_0 ,$$

i.e. the numbers $a_0, a_1, \ldots, a_{n-1}$ are the coefficients of the characteristic polynomial of $\mathbf{A}_1$ and therefore also of $\mathbf{A}$. We denote it by $p_A(s)$. We have

$$\mathbf{TA} = \mathbf{A}_1\mathbf{T}, \quad \mathbf{b}_1 = \mathbf{Tb} \tag{5.32}$$

and represent $\mathbf{T}$ by its rows in the form

$$\mathbf{T} = \begin{pmatrix} \mathbf{t}_1^T \\ \vdots \\ \mathbf{t}_n^T \end{pmatrix}.$$

Then (5.32) is equivalent to

$$\mathbf{t}_i^T \mathbf{A} = \mathbf{t}_{i+1}^T \qquad i = 1, \ldots, n-1, \tag{5.33}$$

$$\mathbf{t}_n^T \mathbf{A} = -\sum_{i=1}^{n} a_{i-1}\mathbf{t}_i^T , \tag{5.34}$$

$$\mathbf{t}_i^T \mathbf{b} = 0, \qquad i = 1, \ldots, n-1, \tag{5.35}$$

$$\mathbf{t}_n^T \mathbf{b} = 1 . \tag{5.36}$$

Equation (5.33) implies

$$\mathbf{t}_i^T = \mathbf{t}_1^T \mathbf{A}^{i-1} \qquad i = 1, \ldots, n$$

and consequently (5.35) and (5.36) yield

$$\mathbf{t}_1^T \mathbf{A}^{i-1}\mathbf{b} = 0 \qquad i = 1, \ldots, n-1, \text{ and } \mathbf{t}_1^T \mathbf{A}^{n-1}\mathbf{b} = 1.$$

Let $\mathbf{e}_n^T = (0, \ldots, 0, 1)$ and

$$\mathbf{C}_{Ab} = (\mathbf{b}\ \mathbf{Ab}\ \ldots\ \mathbf{A}^{n-1}\mathbf{b}) .$$

This implies $\mathbf{t}_1^T \mathbf{C}_{Ab} = \mathbf{e}_n^T$. If $\mathbf{C}_{Ab}$ is nonsingular, it follows that

$$\mathbf{t}_1^T = \mathbf{e}_n^T \mathbf{C}_{Ab}^{-1} .$$

Hence, in this case the equations (5.33), (5.35) and (5.36) determine the rows uniquely. It remains to prove that (5.34) is also fulfilled. An application of the Cayley–Hamilton theorem gives

$$\sum_{i=1}^{n} a_{i-1}\mathbf{t}_i^T + \mathbf{t}_n^T\mathbf{A} = \sum_{i=1}^{n} a_{i-1}\mathbf{t}_1^T\mathbf{A}^{i-1} + \mathbf{t}_1^T\mathbf{A}^n = \mathbf{t}_1^T p_A(\mathbf{A}) = \mathbf{0}.$$

Finally, it must be shown that $\mathbf{T}$ is nonsingular. Suppose

$$\sum_{i=1}^{n} c_i\mathbf{t}_i^T = \mathbf{0}. \tag{5.37}$$

This implies $c_n = 0$ (multiply (5.37) by $\mathbf{b}$), $c_{n-1} = 0$ (multiply (5.37) by $\mathbf{Ab}$) and so on. Finally, we have $c_1 = c_2 = \ldots = c_n = 0$ and the following theorem is shown.

Theorem 5.3.2. *Let the system* $(\mathbf{A}, \mathbf{b})$ *be given and suppose the matrix*

$$\mathbf{C}_{Ab} = (\mathbf{b}\ \ \mathbf{Ab}\ \ \ldots\ \ \mathbf{A}^{n-1}\mathbf{b}) \tag{5.38}$$

to be nonsingular. Denote the last row of $\mathbf{C}_{Ab}^{-1}$ *by* $\mathbf{t}_1^T$. *Then matrix*

$$\mathbf{T} = \begin{pmatrix} \mathbf{t}_1^T \\ \mathbf{t}_1^T\mathbf{A} \\ \vdots \\ \mathbf{t}_1^T\mathbf{A}^{n-1} \end{pmatrix}$$

is nonsingular, and by introducing a new state variable $\mathbf{z} = \mathbf{Tx}$, *the system* $(\mathbf{A}, \mathbf{b})$ *will be transformed into the equivalent system* $(\mathbf{A}_1, \mathbf{b}_1)$ *of (5.31). Herein, the numbers* $a_0, a_1, \ldots, a_{n-1}$ *are the coefficients of the characteristic polynomial of* $\mathbf{A}$.

The system $(\mathbf{A}_1, \mathbf{b}_1)$ is denoted as the **controller-canonical form** of $(\mathbf{A}, \mathbf{b})$. A system, for which the matrix $\mathbf{C}_{Ab}$ defined in (5.38) is nonsingular, is said to be **controllable**. Thus the preceding theorem implies that a controllable (SISO) system can be transformed into the controller-canonical form.

Examples. 1. Crane positioning system. It is easy to see that

$$\operatorname{rank}\mathbf{C}_{Ab} = \operatorname{rank}\begin{pmatrix} 0 & 1 & 0 & -(\omega_1^2-\omega_0^2) \\ 1 & 0 & -(\omega_1^2-\omega_0^2) & 0 \\ 0 & 0 & 0 & \omega_0^2 \\ 0 & 0 & \omega_0^2 & 0 \end{pmatrix} = 4.$$

Hence, the system is controllable.

2. Uncontrollable system. We consider the system

$$\dot{x}_1 = x_2$$
$$\dot{x}_2 = a\,x_2 + u$$
$$\dot{x}_3 = a_1 x_3 \,.$$

Here we have

$$\mathbf{A} = \begin{pmatrix} 0 & 1 & 0 \\ 0 & a & 0 \\ 0 & 0 & a_1 \end{pmatrix}, \quad \mathbf{b} = \begin{pmatrix} 0 \\ 1 \\ 0 \end{pmatrix}, \quad \text{rank}\,\mathbf{C}_{Ab} = \begin{pmatrix} 0 & 1 & a \\ 1 & a & a^2 \\ 0 & 0 & 0 \end{pmatrix} = 2 \,,$$

and the system is uncontrollable. The output $y = x_1$ leads to the transfer function

$$\hat{y} = G(s)\,\hat{u} \quad \text{with} \quad G(s) = \frac{1}{s(s-a)} \,.$$

A realization of $G(s)$ is given by

$$\mathbf{A}_0 = \begin{pmatrix} 0 & 1 \\ 0 & a \end{pmatrix}, \quad \mathbf{b}_0 = \begin{pmatrix} 0 \\ 1 \end{pmatrix}, \quad \mathbf{c}_0^T = \begin{pmatrix} 1 & 0 \end{pmatrix}, \quad d = 0 \,.$$

The system $(\mathbf{A}_0, \mathbf{b}_0)$ is controllable. The state x_3 occurs in the first state space representation, but plays no role for the I/O behavior of the system.

5.3.2 Conditions for Controllability

In the preceding section, we have seen that a system $(\mathbf{A}, \mathbf{b})$ is controllable if and only if it can be transformed into the controller-canonical form. There are other important characterizations of controllability. In order to get an idea of this, we consider again the uncontrollable system of Example 2, where, as it is directly seen, the state x_3 is not influenced by the control. This leads to the zero row in the controllability matrix and it may be supposed that controllability has something to do with the question of which points can be reached by the trajectories $\mathbf{x}(t)$ generated by the system. In fact, there are important results in this direction. They will be presented in this section for a general system $(\mathbf{A}, \mathbf{B})$.

We start our analysis by giving a precise version of the question posed above. To this end, let R_{t_1} be the set of all vectors of the form $\mathbf{x}_1 = \mathbf{x}(t_1)$ which result when $\mathbf{u}$ passes through all feasible controls, where $\mathbf{x}$ is the solution of

$$\dot{\mathbf{x}} = \mathbf{A}\mathbf{x} + \mathbf{B}\mathbf{u}, \qquad \mathbf{x}(0) = \mathbf{0} \tag{5.39}$$

(with an arbitrary time $t_1 > 0$). We ask now for a condition that guarantees $R_{t_1} = \mathbb{R}^n$. An answer can be found by applying the solution formula (5.11), where the exponential function is replaced by its series development. Then we get

$$\mathbf{x}(t_1) = \sum_{k=0}^{\infty} \mathbf{A}^k \mathbf{B} \int_0^{t_1} \frac{\tau^k}{k!} \mathbf{u}(t-\tau)\, d\tau .$$

Let now $a_{n-1}, \ldots, a_1, a_0$ be the coefficients of the characteristic polynomial of $\mathbf{A}$. Then the Cayley–Hamilton theorem gives

$$\mathbf{A}^n = -a_{n-1}\mathbf{A}^{n-1} - \ldots - a_1\mathbf{A} - a_0\mathbf{I} .$$

From this it is directly seen that all matrix powers $\mathbf{A}^k$ with $k \geq n$ can be expressed as a linear combination of the matrix powers $\mathbf{A}^{n-1}, \ldots, \mathbf{A}, \mathbf{I}$. This in turn implies that all matrices of the form $\mathbf{A}^k\mathbf{B}$ with $k \geq n$ are a linear combination of the matrices $\mathbf{A}^{n-1}\mathbf{B}, \ldots, \mathbf{A}\,\mathbf{B}, \mathbf{B}$. Hence $\mathbf{x}(t_1)$ can be written as a linear combination of the columns of the matrix

$$\mathbf{C}_{AB} = (\mathbf{B} \quad \mathbf{AB} \quad \ldots \quad \mathbf{A}^{n-1}\mathbf{B}) . \tag{5.40}$$

The matrix $\mathbf{C}_{AB}$ is a direct generalization of the controllability matrix (5.38) for systems with a single control.

Consequently, the validity of $R_{t_1} = \mathbb{R}^n$ is only possible if $\operatorname{rank} \mathbf{C}_{AB} = n$. Let this property now be given. Then we want to show that each $\mathbf{x}_1 \in \mathbb{R}^n$ lies in R_{t_1}. To this purpose, we construct a control which transfers $\mathbf{x}_0 = \mathbf{0}$ into $\mathbf{x}_1$. Define

$$\mathbf{X}_c(t) = \int_0^t e^{\mathbf{A}\tau}\mathbf{B}\mathbf{B}^T e^{\mathbf{A}^T\tau} d\tau . \tag{5.41}$$

This matrix is denoted as **controllability gramian.** The control

$$\mathbf{u}(\tau) = \mathbf{B}^T e^{\mathbf{A}^T(t_1-\tau)} \mathbf{X}_c(t_1)^{-1}\mathbf{x}_1$$

has the desired property (of course, it still has to be shown that $\mathbf{X}_c(t_1)$ is invertible). To see this, we insert the control into the solution formula (5.11) and get, as desired, the result

$$\mathbf{x}(t_1) = \int_0^{t_1} e^{\mathbf{A}(t_1-\tau)}\mathbf{B}\mathbf{B}^T e^{\mathbf{A}^T(t_1-\tau)} \mathbf{X}_c(t_1)^{-1}\mathbf{x}_1\, d\tau$$

$$= \int_0^{t_1} e^{\mathbf{A}(t_1-\tau)}\mathbf{B}\mathbf{B}^T e^{\mathbf{A}^T(t_1-\tau)} d\tau\, \mathbf{X}_c(t_1)^{-1}\mathbf{x}_1 = \mathbf{x}_1 .$$

The proof of the nonsingularity will be shown by contradiction. Suppose there exists a vector $\mathbf{v} \neq \mathbf{0}$ with $\mathbf{X}_c(t_1)\mathbf{v} = \mathbf{0}$. Then (5.41) implies

$$\int_0^{t_1} \mathbf{v}^T e^{\mathbf{A}\tau} \mathbf{B}\mathbf{B}^T e^{\mathbf{A}^T \tau} \mathbf{v} d\tau = 0 .$$

Since the integrand is nonnegative, we get

$$\mathbf{v}^T e^{\mathbf{A}\tau} \mathbf{B}\mathbf{B}^T e^{\mathbf{A}^T \tau} \mathbf{v} = 0 \quad \text{for every } \tau \in [0, t_1] .$$

Let $\mathbf{y}(\tau) = \mathbf{B}^T e^{\mathbf{A}^T \tau} \mathbf{v}$. Then the last equation implies $\mathbf{y}^T(\tau) = \mathbf{0}$ for $\tau \in [0, t_1]$. Hence, the derivatives of order $k \geq 0$ at zero satisfy

$$\frac{d^k \mathbf{y}^T}{d\tau^k}(0+) = \mathbf{0} .$$

On the other hand, the series for the matrix exponential function shows

$$\frac{d^k \mathbf{y}^T}{d\tau^k}(0+) = \mathbf{v}^T \mathbf{A}^k \mathbf{B} .$$

This leads to $\mathbf{v}^T \mathbf{A}^k \mathbf{B} = \mathbf{0}$ for $k = 0, 1, \ldots, n$. Since the controllability matrix is nonsingular by assumption, we get $\mathbf{v} = \mathbf{0}$, which is a contradiction.

These considerations show that it is reasonable to generalize the definition of controllability to MIMO systems in the following way.

Definition. *A system* $(\mathbf{A}, \mathbf{B})$ *is said to be* ***controllable*** *if the controllability matrix* $\mathbf{C}_{AB}$ *defined in (5.40) has rank* n.

The next theorem gives several characterizations of controllability.

Theorem 5.3.3. *The following assertions are equivalent.*

(a) The system $(\mathbf{A}, \mathbf{B})$ *is controllable.*

(b) For every $\mathbf{x}_1 \in \mathbb{R}^n$ *there is a control* $\mathbf{u}$ *such that (5.39) and* $\mathbf{x}(t_1) = \mathbf{x}_1$ *hold.*

(c) The controllability Gramian (5.41) is positive definite for every $t \geq 0$.

Proof. The following assertions have already been shown: (a)$\Rightarrow$(b), (a)$\Rightarrow$(c), (b)$\Rightarrow$(a).
(c) $\Rightarrow$(a): Let (c) be true, but suppose that $\mathbf{C}_{AB}$ is not of full rank. Then there is a vector $\mathbf{v} \neq \mathbf{0}$ with

$$\mathbf{v}^T \mathbf{A}^k \mathbf{B} = \mathbf{0} \quad \text{for } k = 0, 1, \ldots, n .$$

With the theorem of Cayley–Hamilton it is seen that this equation holds even for all $k \in \mathbb{N}$. Hence

$$\mathbf{v}^T e^{\mathbf{A}\tau}\mathbf{B} = \mathbf{0} \quad \text{for every } \tau \geq 0.$$

and therefore $\mathbf{v}^T \mathbf{X}_c(t) = \mathbf{0}$, but this is a contradiction to (c).

The next theorem describes the structure of an arbitrary linear system concerning controllability.

Theorem 5.3.4. *Suppose the controllability matrix* $\mathbf{C}_{AB}$ *of the system* $(\mathbf{A}, \mathbf{B})$ *is of rank* $r < n$. *Then there is a similarity transform* $\mathbf{T}$ *such that the following assertions are true.*

(a) The transformed pair has the form

$$\tilde{\mathbf{A}} = \mathbf{T}\mathbf{A}\mathbf{T}^{-1} = \begin{pmatrix} \tilde{\mathbf{A}}_{11} & \tilde{\mathbf{A}}_{12} \\ \mathbf{0} & \tilde{\mathbf{A}}_{22} \end{pmatrix}, \qquad \tilde{\mathbf{B}} = \mathbf{T}\mathbf{B} = \begin{pmatrix} \tilde{\mathbf{B}}_1 \\ \mathbf{0} \end{pmatrix}, \tag{5.42}$$

where $\tilde{\mathbf{A}}_{11} \in \mathbb{R}^{r \times r}$ *and* $\tilde{\mathbf{B}}_1 \in \mathbb{R}^{r \times m}$.

(b) The system $(\tilde{\mathbf{A}}_{11}, \tilde{\mathbf{B}}_1)$ *is controllable.*

Proof. (a) Let $\mathbf{v}_1, \mathbf{v}_2, ..., \mathbf{v}_r$ be linearly independent columns of the controllability matrix. We complete them by $n - r$ linearly independent vectors $\mathbf{v}_{r+1}, \mathbf{v}_{r+2}, ..., \mathbf{v}_n$, such that the matrix

$$\mathbf{Q} = (\mathbf{v}_1 \ \cdots \ \mathbf{v}_r \ \mathbf{v}_{r+1} \ \cdots \ \mathbf{v}_n)$$

is nonsingular and show that $\mathbf{T} = \mathbf{Q}^{-1}$ has the desired property. Because of the Cayley–Hamilton theorem, each vector $\mathbf{A}\mathbf{v}_1, ..., \mathbf{A}\mathbf{v}_r$ can be written as a linear combination of the columns of $\mathbf{C}_{AB}$. Hence there is an $r \times r$ - matrix $\tilde{\mathbf{A}}_{11}$ such that

$$(\mathbf{A}\mathbf{v}_1 \ ... \ \mathbf{A}\mathbf{v}_r) = (\mathbf{v}_1 \ ... \ \mathbf{v}_r)\tilde{\mathbf{A}}_{11},$$

and we can write with certain matrices $\tilde{\mathbf{A}}_{12}$, $\tilde{\mathbf{A}}_{22}$,

$$\begin{aligned} \mathbf{A}\mathbf{T}^{-1} &= (\mathbf{A}\mathbf{v}_1 \ ... \ \mathbf{A}\mathbf{v}_r \ \mathbf{A}\mathbf{v}_{r+1} \ ... \ \mathbf{A}\mathbf{v}_n) \\ &= (\mathbf{v}_1 \ ... \ \mathbf{v}_r \ \mathbf{v}_{r+1} \ ... \ \mathbf{v}_n)\begin{pmatrix} \tilde{\mathbf{A}}_{11} & \tilde{\mathbf{A}}_{12} \\ \mathbf{0} & \tilde{\mathbf{A}}_{22} \end{pmatrix} \\ &= \mathbf{T}^{-1}\begin{pmatrix} \tilde{\mathbf{A}}_{11} & \tilde{\mathbf{A}}_{12} \\ \mathbf{0} & \tilde{\mathbf{A}}_{22} \end{pmatrix}. \end{aligned}$$

Analogously it is possible to represent each column of $\mathbf{B}$ as a linear combination of the vectors $\mathbf{v}_1, \mathbf{v}_2, ..., \mathbf{v}_r$. Consequently there is a matrix $\tilde{\mathbf{B}}_1$ such that

$$\mathbf{B} = (\mathbf{v}_1 \; ... \; \mathbf{v}_r)\tilde{\mathbf{B}}_1 = \mathbf{Q}\begin{pmatrix} \tilde{\mathbf{B}}_1 \\ \mathbf{0} \end{pmatrix}.$$

Thus, (a) is demonstrated.
(b) We have

$$\mathbf{T}(\mathbf{B} \;\; \mathbf{AB} \; \dots \; \mathbf{A}^{n-1}\mathbf{B}) = (\mathbf{TB} \;\; \mathbf{TAT}^{-1}\mathbf{TB} \; \dots \; (\mathbf{TAT}^{-1})^{n-1}\mathbf{TB}).$$

Using (a), this implies

$$\mathbf{T}\mathbf{C}_{AB} = \begin{pmatrix} \tilde{\mathbf{B}}_1 & \tilde{\mathbf{A}}_{11}\tilde{\mathbf{B}}_1 & \cdots & \tilde{\mathbf{A}}_{11}^{r-1}\tilde{\mathbf{B}}_1 & \cdots & \tilde{\mathbf{A}}_{11}^{n-1}\tilde{\mathbf{B}}_1 \\ \mathbf{0} & \mathbf{0} & \cdots & \mathbf{0} & \cdots & \mathbf{0} \end{pmatrix}.$$

Since according to the Cayley–Hamilton theorem for every $k \geq r$ the matrix $\tilde{\mathbf{A}}_{11}^k$ is a linear combination of the matrices $\tilde{\mathbf{A}}_{11}^{r-1}, ..., \tilde{\mathbf{A}}_{11}, \mathbf{I}$, the rank condition

$$\text{rank}\,(\tilde{\mathbf{B}}_1 \;\; \tilde{\mathbf{A}}_{11}\tilde{\mathbf{B}}_1 \;\; \cdots \;\; \tilde{\mathbf{A}}_{11}^{r-1}\tilde{\mathbf{B}}_1) = \text{rank}\,\mathbf{C}_{AB} = r$$

holds, i.e. the system $(\tilde{\mathbf{A}}_{11}, \tilde{\mathbf{B}}_1)$ is controllable.

Conclusion. Let $\tilde{\mathbf{C}} = \mathbf{C}\mathbf{T}^{-1}$ and

$$\tilde{\mathbf{x}} = \begin{pmatrix} \tilde{\mathbf{x}}_1 \\ \tilde{\mathbf{x}}_2 \end{pmatrix} = \mathbf{T}\mathbf{x}$$

be the transformed state. Then the transformed system can be written as

$$\begin{pmatrix} \dot{\tilde{\mathbf{x}}}_1 \\ \dot{\tilde{\mathbf{x}}}_2 \end{pmatrix} = \begin{pmatrix} \tilde{\mathbf{A}}_{11} & \tilde{\mathbf{A}}_{12} \\ \mathbf{0} & \tilde{\mathbf{A}}_{22} \end{pmatrix}\begin{pmatrix} \tilde{\mathbf{x}}_1 \\ \tilde{\mathbf{x}}_2 \end{pmatrix} + \begin{pmatrix} \tilde{\mathbf{B}}_1 \\ \mathbf{0} \end{pmatrix}\mathbf{u} \tag{5.43}$$

$$\mathbf{y} = \begin{pmatrix} \tilde{\mathbf{C}}_1 & \tilde{\mathbf{C}}_2 \end{pmatrix}\begin{pmatrix} \tilde{\mathbf{x}}_1 \\ \tilde{\mathbf{x}}_2 \end{pmatrix} + \mathbf{Du}. \tag{5.44}$$

The corresponding transfer function has the representations

$$\mathbf{G}(s) = \mathbf{C}(s\mathbf{I} - \mathbf{A})^{-1}\mathbf{B} + \mathbf{D} = \tilde{\mathbf{C}}_1(s\mathbf{I} - \tilde{\mathbf{A}}_{11})^{-1}\tilde{\mathbf{B}}_1 + \mathbf{D}.$$

In (5.43), the part of the system which can be influenced by the control is separated from the part which cannot be influenced by the control. As we will see later, the controllable part can always be stabilized by an adequate controller, whereas the remaining part cannot be affected by a control at all. Hence the system (5.43) can only be stabilized if the system, which cannot be affected by a control, is stable. This leads to the following definition.

Definition. *The system* $(\mathbf{A},\mathbf{B})$ *is denoted as* ***stabilizable****, if the matrix* $\tilde{\mathbf{A}}_{22}$ *in the normal form (5.42) is stable.*

The following theorem gives an additional characterization of controllability and a characterization of stabilizability.

Theorem 5.3.5 (Popov–Belevitch–Hautus (PHB) test). *(a) The system* $(\mathbf{A},\mathbf{B})$ *is controllable if and only if*

$$\operatorname{rank}(\mathbf{A}-\lambda\mathbf{I} \quad \mathbf{B}) = n \quad \textit{for every } \lambda\in\mathbb{C}.$$

(b) The system $(\mathbf{A},\mathbf{B})$ *is stabilizable if and only if*

$$\operatorname{rank}(\mathbf{A}-\lambda\mathbf{I} \quad \mathbf{B}) = n \quad \textit{for every } \lambda\in\mathbb{C} \textit{ with } \operatorname{Re}\lambda \geq 0.$$

Proof. "$\Rightarrow$": Let $(\mathbf{A},\mathbf{B})$ be controllable and suppose

$$\operatorname{rank}(\mathbf{A}-\lambda\mathbf{I} \quad \mathbf{B}) < n.$$

Then there exists a nonzero vector $\mathbf{x}$ such that

$$\mathbf{x}^*\mathbf{A} = \lambda\mathbf{x}^* \quad \text{and} \quad \mathbf{x}^*\mathbf{B} = \mathbf{0}.$$

This implies in particular that

$$\mathbf{x}^*\mathbf{A}^k = \lambda^k\mathbf{x}^* \quad \text{for every } k \geq 1.$$

Hence

$$\mathbf{x}^*(\mathbf{B} \quad \mathbf{AB} \quad \cdots \quad \mathbf{A}^{n-1}\mathbf{B}) = \mathbf{0},$$

which is a contradiction to the assumed controllability.
"$\Leftarrow$": We use the decomposition of Theorem 5.3.4. Let the rank condition of (a) be fulfilled and suppose that $(\mathbf{A},\mathbf{B})$ is not controllable. For every eigenvalue λ of $\tilde{\mathbf{A}}_{22}$ we see that

$$\operatorname{rank}(\tilde{\mathbf{A}}-\lambda\mathbf{I} \quad \tilde{\mathbf{B}}) < n.$$

This implies

$$\operatorname{rank}(\mathbf{A}-\lambda\mathbf{I} \quad \mathbf{B}) = \operatorname{rank}\left(\mathbf{T}^{-1}(\tilde{\mathbf{A}}-\lambda\mathbf{I} \quad \tilde{\mathbf{B}})\begin{pmatrix}\mathbf{T} & \mathbf{0}\\ \mathbf{0} & \mathbf{I}\end{pmatrix}\right) < n,$$

which is in contradiction to our assumption.
Assertion (b) is shown in a similar manner.

The rank condition of (a) can be formulated as follows:

$$\mathbf{x}^*\mathbf{A} = \lambda\mathbf{x}^* \textit{ and } \mathbf{x}^*\mathbf{B} = \mathbf{0} \quad \Rightarrow \quad \mathbf{x} = \mathbf{0} \quad \textit{for every } \lambda,\ \mathbf{x}.$$

An eigenvalue for which the equation $\mathbf{x}^*\mathbf{A} = \lambda\mathbf{x}^*$ always implies $\mathbf{x}^*\mathbf{B} \neq \mathbf{0}$ is denoted as **controllable,** otherwise it is called **uncontrollable**. Condition (b) can be rewritten in a similar way. Using this definition, (a) says: *the system* $(\mathbf{A}, \mathbf{B})$ *is controllable if and only if each eigenvalue is controllable.*

Remark. The uncontrollable eigenvalues of $(\mathbf{A}, \mathbf{B})$ are precisely the eigenvalues of $\tilde{\mathbf{A}}_{22}$ in the decomposition of Theorem 5.3.4.

To show this, let λ be an eigenvalue of $\tilde{\mathbf{A}}_{22}$. Then λ is uncontrollable, as we have seen in part (b) of the proof of the preceding theorem. Vice versa, let λ be uncontrollable. Then there exist vectors $\mathbf{x}_1, \mathbf{x}_2$, not both vanishing , such that

$$\mathbf{x}_1^* \tilde{\mathbf{A}}_{11} = \lambda \mathbf{x}_1^* , \qquad \mathbf{x}_1^* \tilde{\mathbf{B}}_1 = \mathbf{0}$$
$$\mathbf{x}_1^* \tilde{\mathbf{A}}_{12} + \mathbf{x}_2^* \tilde{\mathbf{A}}_{22} = \lambda \mathbf{x}_2^* .$$

Since $(\tilde{\mathbf{A}}_{11}, \tilde{\mathbf{B}}_1)$ is controllable, we get $\mathbf{x}_1 = \mathbf{0}$. Hence λ is an eigenvalue of $\tilde{\mathbf{A}}_{22}$.

5.4 Observability

We consider again a transfer function of the form (5.26). Beside a realization in controllable canonical form there is another state space realization arising in a natural way. To see this, $G(s)$ has to be written in the form (with the notation defined in Theorem 5.3.1)

$$\begin{aligned} G(s) &= \mathbf{c}_1^T (s\mathbf{I} - \mathbf{A}_1)^{-1} \mathbf{b}_1 + d_1 \\ &= \mathbf{b}_1^T (s\mathbf{I} - \mathbf{A}_1^T)^{-1} \mathbf{c}_1 + d_1 . \end{aligned}$$

Thus by introducing matrices

$$\mathbf{A}_2 = \begin{pmatrix} 0 & & & 0 & -a_0 \\ 1 & 0 & & & -a_1 \\ & 1 & \ddots & & \vdots \\ & & \ddots & 0 & -a_{n-2} \\ 0 & & & 1 & -a_{n-1} \end{pmatrix}, \qquad \mathbf{b}_2 = \begin{pmatrix} c_0 - a_0 c_n \\ \\ \vdots \\ \\ c_{n-1} - a_{n-1} c_n \end{pmatrix}, \tag{5.45}$$

$$\mathbf{c}_2^T = (0 \quad \dots \quad 0 \quad 1), \tag{5.46}$$

we obtain the state space realization

$$G(s) = \mathbf{c}_2^T (s\mathbf{I} - \mathbf{A}_2)^{-1} \mathbf{b}_2 + d_1 .$$

As in Sect. 5.3.1 it may be asked under which assumptions the system $(\mathbf{c}^T, \mathbf{A})$ can be transformed by a similarity transform into the form $(\mathbf{c}_2^T, \mathbf{A}_2)$. Define

$$\mathbf{c}^T = (c_0 \quad \dots \quad c_{n-1}) .$$

An appropriate application of Theorem 5.3.2 shows that this is the case if and only if the matrix

$$\mathbf{O}_{cA} = \begin{pmatrix} \mathbf{c}^T \\ \mathbf{c}^T\mathbf{A} \\ \vdots \\ \mathbf{c}^T\mathbf{A}^{n-1} \end{pmatrix} \tag{5.47}$$

is nonsingular. The system $(\mathbf{c}_2^T, \mathbf{A}_2)$ defined in (5.45) and (5.46) is said to be the **observer-canonical form** of $(\mathbf{c}^T, \mathbf{A})$. Moreover, a system for which the matrix defined in (5.47) is nonsingular will be denoted as **observable**. Hence, a (SISO) system is observable if and only if it can be transformed by a similarity transform into the observer-canonical form.

There is also a correspondence to the assertions of Sect. 5.3.2, which are valid for MIMO systems. They are obtained by applying the results of this subsection to the so-called dual system $(\mathbf{B}^T, \mathbf{A}^T, \mathbf{C}^T)$. At first, we define the **observability matrix** for MIMO systems, which generalizes the matrix (5.47) in a natural way:

$$\mathbf{O}_{CA} = \begin{pmatrix} \mathbf{C} \\ \mathbf{CA} \\ \vdots \\ \mathbf{CA}^{n-1} \end{pmatrix} . \tag{5.48}$$

Definition. *The system* $(\mathbf{C}, \mathbf{A})$ *is said to be* ***observable*** *if the observability matrix (5.48) has rank* n.

The controllability gramian has to be replaced by the **observability gramian,** which is defined by

$$\mathbf{X}_o(t) = \int_0^t e^{\mathbf{A}^T\tau}\mathbf{C}^T\mathbf{C}e^{\mathbf{A}\tau}d\tau . \tag{5.49}$$

Theorem 5.4.1. *The following assertions are equivalent.*

(a) The system $(\mathbf{C}, \mathbf{A})$ *is observable.*

(b) For an arbitrary time $t_1 > 0$ *the initial state* $\mathbf{x}_0$ *is uniquely determined by the output* $\mathbf{y}(t)$ *and the control* $\mathbf{u}(t)$ *on the interval* $[0, t_1]$.

(c) The observability gramian (5.49) is positive definite for each $t \geq 0$.

Property (b) is of particular importance. It says that it is possible to proceed from the knowledge of the output to the whole state of the system. This will be used later, when observers are constructed. Assertion (b) can be reformulated as follows: if $\mathbf{y}(t)$ vanishes identically on an interval $[0, t_1]$ (and if $\mathbf{u} = \mathbf{0}$), then necessarily $\mathbf{x}_0 = \mathbf{0}$ holds.

The normal form of Theorem 5.3.4 has also an equivalence for observable systems:

Theorem 5.4.2. *Suppose the observability matrix $\mathbf{O}_{CA}$ of the system $(\mathbf{C}, \mathbf{A})$ has rank $r < n$. Then there is a similarity transform $\mathbf{T}$ such that the following assertions are true.*

(a) The transformed pair has the form

$$\mathbf{TAT}^{-1} = \begin{pmatrix} \tilde{\mathbf{A}}_{11} & \mathbf{0} \\ \tilde{\mathbf{A}}_{21} & \tilde{\mathbf{A}}_{22} \end{pmatrix}, \qquad \mathbf{CT}^{-1} = (\tilde{\mathbf{C}}_1 \;\; \mathbf{0}) \tag{5.50}$$

where $\tilde{\mathbf{A}}_{11} \in \mathbb{R}^{r \times r}$ and $\tilde{\mathbf{C}}_1 \in \mathbb{R}^{l \times r}$.

(b) The system $(\tilde{\mathbf{C}}_1, \tilde{\mathbf{A}}_{11})$ is observable.

Conclusion. Let

$$\tilde{\mathbf{x}} = \begin{pmatrix} \tilde{\mathbf{x}}_1 \\ \tilde{\mathbf{x}}_2 \end{pmatrix} = \mathbf{Tx}$$

be the transformed state. Then the transformed system can be written as

$$\begin{pmatrix} \dot{\tilde{\mathbf{x}}}_1 \\ \dot{\tilde{\mathbf{x}}}_2 \end{pmatrix} = \begin{pmatrix} \tilde{\mathbf{A}}_{11} & \mathbf{0} \\ \tilde{\mathbf{A}}_{21} & \tilde{\mathbf{A}}_{22} \end{pmatrix} \begin{pmatrix} \tilde{\mathbf{x}}_1 \\ \tilde{\mathbf{x}}_2 \end{pmatrix} + \begin{pmatrix} \tilde{\mathbf{B}}_1 \\ \tilde{\mathbf{B}}_2 \end{pmatrix} \mathbf{u} \tag{5.51}$$

$$\mathbf{y} = (\tilde{\mathbf{C}}_1 \;\; \mathbf{0}) \begin{pmatrix} \tilde{\mathbf{x}}_1 \\ \tilde{\mathbf{x}}_2 \end{pmatrix} + \mathbf{Du} \tag{5.52}$$

and the transfer function has the representations

$$\mathbf{G}(s) = \mathbf{C}(s\mathbf{I} - \mathbf{A})^{-1}\mathbf{B} + \mathbf{D} = \tilde{\mathbf{C}}_1 (s\mathbf{I} - \tilde{\mathbf{A}}_{11})^{-1} \tilde{\mathbf{B}}_1 + \mathbf{D}.$$

In equation (5.51), the observable and the unobservable part are separated. The state of the unobservable part cannot be reconstructed from the output, whereas the state of the observable part can be reconstructed, as we shall see later in detail. Therefore, an estimate of the whole state is only possible if the nonobservable part of the state tends to $\mathbf{0}$ as $t \to \infty$ (if $\mathbf{u} = \mathbf{0}$). This leads to the following definition.

Definition. *The system is said to be* $(\mathbf{C}, \mathbf{A})$ ***detectable*** *if the matrix* $\tilde{\mathbf{A}}_{22}$ *in the normal form* (5.50) *is stable.*

The next theorem corresponds to Theorem 5.3.5.

Theorem 5.4.3 (Popov–Belevitch–Hautus (PHB) test). *(a) The system* $(\mathbf{C}, \mathbf{A})$ *is observable if and only if*

$$\operatorname{rank}\begin{pmatrix} \mathbf{A} - \lambda\mathbf{I} \\ \mathbf{C} \end{pmatrix} = n \qquad \textit{for every } \lambda \in \mathbb{C}.$$

(b) The system $(\mathbf{C}, \mathbf{A})$ *is detectable if and only if*

$$\operatorname{rank}\begin{pmatrix} \mathbf{A} - \lambda\mathbf{I} \\ \mathbf{C} \end{pmatrix} = n \qquad \textit{for every } \lambda \textit{ with } \operatorname{Re}\lambda \geq 0.$$

The rank condition of (a) can be formulated as follows:

$$\mathbf{A}\mathbf{x} = \lambda\mathbf{x} \ \textit{ and } \ \mathbf{C}\mathbf{x} = \mathbf{0} \ \Rightarrow \ \mathbf{x} = \mathbf{0} \quad \textit{for every } \lambda,\ \mathbf{x}.$$

An eigenvalue that has the property that $\mathbf{A}\mathbf{x} = \lambda\mathbf{x}$ always implies $\mathbf{C}\mathbf{x} \neq \mathbf{0}$ is said to be **observable,** otherwise it is called **unobservable**. *Then the system* $(\mathbf{A}, \mathbf{C})$ *is observable if and only if each eigenvalue is observable.*

Remark. The unobservable eigenvalues of $(\mathbf{C}, \mathbf{A})$ are precisely the eigenvalues of $\tilde{\mathbf{A}}_{22}$ in the decomposition of Theorem 5.4.2.

Example. Crane positioning system.
We consider again the crane positioning system and check observability for several scalar outputs y.

1. *Measurement of the bucket position.* Then we have

$$y = x_G, \quad \mathbf{C} = (0 \ \ 0 \ \ 1 \ \ 0),$$

and

$$\operatorname{rank}\mathbf{O}_{CA} = \operatorname{rank}\begin{pmatrix} 0 & 0 & 1 & 0 \\ 0 & 0 & 0 & 1 \\ \omega_0^2 & 0 & -\omega_0^2 & 0 \\ 0 & \omega_0^2 & 0 & \omega_0^2 \end{pmatrix} = 4.$$

2. *Measurement of the carriage position.* Then we have

$$y = x, \quad \mathbf{C} = (1 \ \ 0 \ \ 0 \ \ 0),$$

and it is easily seen that in this case also $\operatorname{rank} \mathbf{O}_{CA} = 4$

2. *Measurement of the rope angle.* Then,

$$y = \theta = \frac{x_G - x}{d}, \quad \mathbf{C} = \frac{1}{d}(-1 \;\; 0 \;\; 1 \;\; 0),$$

and

$$\operatorname{rank} \mathbf{Q}_{CA} = \operatorname{rank} \frac{1}{\mathrm{d}} \begin{pmatrix} -1 & 0 & 1 & 0 \\ 0 & -1 & 0 & 1 \\ \omega_1^2 & 0 & -\omega_1^2 & 0 \\ 0 & 2\omega_0^2 - \omega_1^2 & 0 & -2\omega_0^2 + \omega_1^2) \end{pmatrix} = 2.$$

Hence, the crane positioning system is observable if the carriage position or the bucket position are measured. If the rope angle is measured, it is not observable.

The system matrix $\mathbf{A}$ has $s = 0$ as an eigenvalue and

$$\mathbf{x}^* = (1 \;\; 0 \;\; 1 \;\; 0)$$

as a corresponding eigenvector (and the dimension of the corresponding eigenspace is one). We have $\mathbf{C}\mathbf{x} \neq 0$ if the bucket position or the carriage position is measured and $\mathbf{C}\mathbf{x} = 0$ if the rope angle is measured. Thus with respect to the bucket position or the carriage position, $s = 0$ is observable and with respect to the rope angle, $s = 0$ is unobservable. The latter fact may equivalently be seen from the transfer function from F to θ: when it is calculated from the state space representation, the linear factor s cancels (cf. Sect. 5.2.1).

Finally, we introduce the famous **Kalman decomposition** of the state space. It is based on a combination of Theorem 5.3.4 and Theorem 5.4.2. The details of the reasoning will not be presented here.

Theorem 5.4.4. *Let a linear system of the form (5.7) and (5.8) be given. Then there exists a transformation* $\tilde{\mathbf{x}} = \mathbf{T}\mathbf{x}$ *of the coordinates such that*

$$\begin{pmatrix} \dot{\tilde{\mathbf{x}}}_1 \\ \dot{\tilde{\mathbf{x}}}_2 \\ \dot{\tilde{\mathbf{x}}}_3 \\ \dot{\tilde{\mathbf{x}}}_4 \end{pmatrix} = \begin{pmatrix} \tilde{\mathbf{A}}_{11} & \mathbf{0} & \tilde{\mathbf{A}}_{12} & \mathbf{0} \\ \tilde{\mathbf{A}}_{21} & \tilde{\mathbf{A}}_{22} & \tilde{\mathbf{A}}_{23} & \tilde{\mathbf{A}}_{24} \\ \mathbf{0} & \mathbf{0} & \tilde{\mathbf{A}}_{33} & \mathbf{0} \\ \mathbf{0} & \mathbf{0} & \tilde{\mathbf{A}}_{43} & \tilde{\mathbf{A}}_{44} \end{pmatrix} \begin{pmatrix} \tilde{\mathbf{x}}_1 \\ \tilde{\mathbf{x}}_2 \\ \tilde{\mathbf{x}}_3 \\ \tilde{\mathbf{x}}_4 \end{pmatrix} + \begin{pmatrix} \tilde{\mathbf{B}}_1 \\ \tilde{\mathbf{B}}_2 \\ \mathbf{0} \\ \mathbf{0} \end{pmatrix} \mathbf{u}$$

$$\mathbf{y} = (\tilde{\mathbf{C}}_1 \;\; \mathbf{0} \;\; \tilde{\mathbf{C}}_3 \;\; \mathbf{0}).$$

Here, the subsystems have the following properties:

(i) *The system* $(\tilde{\mathbf{A}}_{11}, \tilde{\mathbf{B}}_1)$ *is controllable and the system* $(\tilde{\mathbf{C}}_1, \tilde{\mathbf{A}}_{11})$ *is observable.*

(ii) *The following system is controllable:*

$$\left(\begin{pmatrix} \tilde{\mathbf{A}}_{11} & \mathbf{0} \\ \tilde{\mathbf{A}}_{21} & \tilde{\mathbf{A}}_{22} \end{pmatrix}, \begin{pmatrix} \tilde{\mathbf{B}}_1 \\ \tilde{\mathbf{B}}_2 \end{pmatrix} \right).$$

(iii) *The following system is observable:*

$$\left(\begin{pmatrix} \tilde{\mathbf{C}}_1 & \tilde{\mathbf{C}}_2 \end{pmatrix}, \begin{pmatrix} \tilde{\mathbf{A}}_{11} & \tilde{\mathbf{A}}_{13} \\ \mathbf{0} & \tilde{\mathbf{A}}_{33} \end{pmatrix} \right).$$

Conclusion. The corresponding transfer function has the representations

$$\mathbf{G}(s) = \mathbf{C}(s\mathbf{I} - \mathbf{A})^{-1}\mathbf{B} + \mathbf{D} = \tilde{\mathbf{C}}_1 (s\mathbf{I} - \tilde{\mathbf{A}}_{11})^{-1} \tilde{\mathbf{B}}_1 + \mathbf{D}.$$

Remark. If an eigenvalue of $\mathbf{A}$ is controllable and observable, it is an eigenvalue of $\tilde{\mathbf{A}}_{11}$.

5.5 State Space Realizations for Transfer Matrices

The realization for SISO transfer functions has led us in a natural way to the controller- and observer-canonical form. In this section, we consider state space realizations for MIMO systems.

Suppose that two SISO transfer functions

$$G_1(s) = \left(\begin{array}{c|c} \mathbf{A}_1 & \mathbf{b}_1 \\ \hline \mathbf{c}_1^T & d_1 \end{array} \right), \quad G_2(s) = \left(\begin{array}{c|c} \mathbf{A}_2 & \mathbf{b}_2 \\ \hline \mathbf{c}_2^T & d_2 \end{array} \right)$$

are given. Then

$$\begin{pmatrix} G_1(s) \\ G_2(s) \end{pmatrix} = \begin{pmatrix} \mathbf{c}_1^T (s\mathbf{I} - \mathbf{A}_1)\mathbf{b}_1 + d_1 \\ \mathbf{c}_2^T (s\mathbf{I} - \mathbf{A}_2)\mathbf{b}_2 + d_2 \end{pmatrix} = \left(\begin{array}{cc|c} \mathbf{A}_1 & \mathbf{0} & \mathbf{b}_1 \\ \mathbf{0} & \mathbf{A}_2 & \mathbf{b}_2 \\ \hline \mathbf{c}_1^T & \mathbf{0} & d_1 \\ \mathbf{0} & \mathbf{c}_2^T & d_2 \end{array} \right)$$

and

$$(G_1(s) \quad G_2(s)) = \left(\begin{array}{cc|cc} \mathbf{A}_1 & \mathbf{0} & \mathbf{b}_1 & \mathbf{0} \\ \mathbf{0} & \mathbf{A}_2 & \mathbf{0} & \mathbf{b}_2 \\ \hline \mathbf{c}_1^T & \mathbf{c}_2^T & d_1 & d_2 \end{array} \right).$$

Let now an arbitrary transfer matrix be given. Then a state space realization can be found by constructing as a first step a realization of each component (which may, for example, have controller-canonical form) and then in a second step, the realizations are composed according to the rules given above.

Example. We apply these results to the following system with two inputs and two outputs:

$$\mathbf{G}(s) = \begin{pmatrix} \frac{1}{s} & 0 \\ \frac{1}{(s+1)(s+2)} & \frac{2}{s+1} \end{pmatrix}.$$

It has the realization

$$\mathbf{G}(s) = \left(\begin{array}{cccc|cc} 0 & 0 & 0 & 0 & 1 & 0 \\ 0 & 0 & 1 & 0 & 0 & 0 \\ 0 & -2 & -3 & 0 & 1 & 0 \\ 0 & 0 & 0 & -1 & 0 & 2 \\ \hline 1 & 0 & 0 & 0 & 0 & 0 \\ 0 & 1 & 0 & 1 & 0 & 0 \end{array}\right). \tag{5.53}$$

Definition. *A state space realization* $(\mathbf{A}, \mathbf{B}, \mathbf{C}, \mathbf{D})$ *is said to be a* ***minimal realization*** *of* $\mathbf{G}(s)$ *if* $\mathbf{A}$ *has the smallest possible dimension.*

The construction described above leads normally to a nonminimal realization. It is possible to construct for this realization the Kalman decomposition and then cancel the noncontrollable and nonobservable states. If there are such states, the order will be reduced by this procedure. The next theorem characterizes minimal realizations (cf. Zhou, Doyle and Glover [112], Theorem 3.16 and Theorem 3.17).

Theorem 5.5.1. *(a) A state space realization* $(\mathbf{A}, \mathbf{B}, \mathbf{C}, \mathbf{D})$ *of a transfer matrix* $\mathbf{G}(s)$ *is minimal if and only if* $(\mathbf{A}, \mathbf{B})$ *is controllable and if* $(\mathbf{C}, \mathbf{A})$ *is observable.*

(b) Two minimal state space realizations of a transfer matrix $\mathbf{G}(s)$ *are equivalent.*

It is natural to ask whether it is possible to construct directly a minimal realization. Such a way really exists and is given by the so-called **Gilbert's realization,** which will be described in the following under some restricting assumptions.

Let $\mathbf{G}(s)$ be a $l \times r$ - transfer matrix. We write it in the form

$$\mathbf{G}(s) = \frac{\mathbf{Z}(s)}{p(s)},$$

where $p(s)$ is a scalar polynomial and where the matrix $\mathbf{Z}(s)$ is a matrix whose components are also polynomials. We denote the zeroes of $p(s)$ by $\lambda_1,\ldots,\lambda_r$ and assume for the sake of simplicity that they are pairwise distinct and real. Then

$$p(s)=(s-\lambda_1)\cdot\ldots\cdot(s-\lambda_r)$$

and $\mathbf{G}(s)$ has the following partial fractional expansion (with $r\times r$ - matrices $\mathbf{W}_k$):

$$\mathbf{G}(s)=\mathbf{D}+\sum_{k=1}^{r}\frac{1}{s-\lambda_k}\mathbf{W}_k\,.$$

Suppose

$$\operatorname{rank}\mathbf{W}_k=\rho_k\,.$$

Let $\mathbf{B}_k\in\mathbb{R}^{\rho_k\times m}$ and $\mathbf{C}_k\in\mathbb{R}^{l\times\rho_k}$ be matrices with the property

$$\mathbf{W}_k=\mathbf{C}_k\mathbf{B}_k\,.$$

Then a realization of $\mathbf{G}(s)$ is given by

$$\mathbf{G}(s)=\left(\begin{array}{ccc|c}\lambda_1\mathbf{I}_{\rho_1} & \cdots & \mathbf{0} & \mathbf{B}_1\\ \vdots & \ddots & \vdots & \vdots\\ \mathbf{0} & \cdots & \lambda_r\mathbf{I}_{\rho_r} & \mathbf{B}_r\\ \hline \mathbf{C}_1 & \cdots & \mathbf{C}_r & \mathbf{D}\end{array}\right).$$

Using the PHB tests it is easy to see that this realization is controllable and observable and hence minimal by Theorem 5.5.1. Note that each pole λ_k is of multiplicity 1 by assumption, whereas in the state space realization the eigenvalue λ_k is of multiplicity ρ_k, which may be larger then 1. The method can be extended to cases with complex or multiple roots of $p(s)$. In any case, each eigenvalue of the $\mathbf{A}$ - matrix of the Gilbert realization is a pole of the transfer matrix.

Example. We want to construct a Gilbert realization for

$$\mathbf{G}(s)=\begin{pmatrix}\dfrac{2s+1}{s(s+1)} & \dfrac{1}{s+1}\\[2ex] \dfrac{1}{(s+1)(s+2)} & -\dfrac{2}{s+2}\end{pmatrix}.$$

In this case, one easily obtains

$$\mathbf{G}(s)=\frac{1}{s}\mathbf{W}_1+\frac{1}{s+1}\mathbf{W}_2+\frac{1}{s+2}\mathbf{W}_3$$

where

$$\mathbf{W}_1 = \begin{pmatrix} 1 & 0 \\ 0 & 0 \end{pmatrix} = \begin{pmatrix} 1 \\ 0 \end{pmatrix} \begin{pmatrix} 1 & 0 \end{pmatrix} = \mathbf{C}_1 \mathbf{B}_1 ,$$

$$\mathbf{W}_2 = \begin{pmatrix} 1 & 1 \\ 1 & 0 \end{pmatrix} = \begin{pmatrix} 1 & 1 \\ 1 & 0 \end{pmatrix} \begin{pmatrix} 1 & 0 \\ 0 & 1 \end{pmatrix} = \mathbf{C}_2 \mathbf{B}_2 ,$$

$$\mathbf{W}_3 = \begin{pmatrix} 0 & 0 \\ -1 & 2 \end{pmatrix} = \begin{pmatrix} 0 \\ 1 \end{pmatrix} \begin{pmatrix} -1 & 2 \end{pmatrix} = \mathbf{C}_3 \mathbf{B}_3 ,$$

$$\rho_1 = \rho_3 = 1, \ \rho_2 = 2 .$$

Hence, we get the following realization of order 4:

$$\mathbf{G}(s) = \left(\begin{array}{cccc|cc} 0 & 0 & 0 & 0 & 1 & 0 \\ 0 & -1 & 0 & 0 & 1 & 0 \\ 0 & 0 & -1 & 0 & 0 & 1 \\ 0 & 0 & 0 & -2 & -1 & 2 \\ \hline 1 & 1 & 1 & 0 & 0 & 0 \\ 0 & 1 & 0 & 1 & 0 & 0 \end{array} \right) .$$

Next, we consider the state space realization of a system which results if an operation on two systems

$$\mathbf{G}_1(s) = \left(\begin{array}{c|c} \mathbf{A}_1 & \mathbf{B}_1 \\ \hline \mathbf{C}_1 & \mathbf{D}_1 \end{array} \right), \quad \mathbf{G}_2(s) = \left(\begin{array}{c|c} \mathbf{A}_2 & \mathbf{B}_2 \\ \hline \mathbf{C}_2 & \mathbf{D}_2 \end{array} \right)$$

is performed. We start with a series connection that corresponds to a multiplication of the transfer functions. A simple calculation shows

$$\mathbf{G}_1 \mathbf{G}_2 = \left(\begin{array}{cc|c} \mathbf{A}_1 & \mathbf{B}_1 \mathbf{C}_2 & \mathbf{B}_1 \mathbf{D}_2 \\ \mathbf{0} & \mathbf{A}_2 & \mathbf{B}_2 \\ \hline \mathbf{C}_1 & \mathbf{D}_1 \mathbf{C}_2 & \mathbf{D}_1 \mathbf{D}_2 \end{array} \right) = \left(\begin{array}{cc|c} \mathbf{A}_2 & \mathbf{0} & \mathbf{B}_2 \\ \mathbf{B}_1 \mathbf{C}_2 & \mathbf{A}_1 & \mathbf{B}_1 \mathbf{D}_2 \\ \hline \mathbf{D}_1 \mathbf{C}_2 & \mathbf{C}_1 & \mathbf{D}_1 \mathbf{D}_2 \end{array} \right) . \tag{5.54}$$

Similarly, a parallel connection of the two transfer functions corresponds to their sum and has the state space realization

$$\mathbf{G}_1 + \mathbf{G}_2 = \left(\begin{array}{cc|c} \mathbf{A}_1 & \mathbf{0} & \mathbf{B}_1 \\ \mathbf{0} & \mathbf{A}_2 & \mathbf{B}_2 \\ \hline \mathbf{C}_1 & \mathbf{C}_2 & \mathbf{D}_1 + \mathbf{D}_2 \end{array} \right) . \tag{5.55}$$

Definition. *The **dual system** (or **transpose**) of a system* $\mathbf{G}(s)$ *with state space realization* $(\mathbf{A}, \mathbf{B}, \mathbf{C}, \mathbf{D})$ *is defined as*

$$\mathbf{G}^T(s) = \mathbf{B}^T(s\mathbf{I} - \mathbf{A}^T)^{-1}\mathbf{C}^T + \mathbf{D}^T.$$

The dual system has the state space realization

$$\mathbf{G}(s) = \left(\begin{array}{c|c} \mathbf{A}^T & \mathbf{C}^T \\ \hline \mathbf{B}^T & \mathbf{D}^T \end{array}\right).$$

Note that for SISO systems the transfer functions of the system and its dual coincide, whereas the state space realizations are different. The state space realization for the dual system has led us from the controller-canonical form to the observer-canonical form.

Definition. *The conjugate system of a system* $\mathbf{G}(s)$ *with state space realization* $(\mathbf{A}, \mathbf{B}, \mathbf{C}, \mathbf{D})$ *is defined as*

$$\mathbf{G}^{\sim}(s) = \mathbf{G}^T(-s) = \mathbf{B}^T(-s\mathbf{I} - \mathbf{A}^T)^{-1}\mathbf{C}^T + \mathbf{D}^T.$$

It has the state space realization

$$\mathbf{G}^{\sim}(s) = \left(\begin{array}{c|c} -\mathbf{A}^T & -\mathbf{C}^T \\ \hline \mathbf{B}^T & \mathbf{D}^T \end{array}\right).$$

Defining $\mathbf{G}^*(s) = \mathbf{G}(s)^*$, we have in particular $\mathbf{G}^*(i\omega) = \mathbf{G}^{\sim}(i\omega)$.

Suppose $\mathbf{D}$ is invertible. Then there exists a transfer function $\mathbf{G}^{-1}(s)$ such that

$$\mathbf{G}(s)\mathbf{G}^{-1}(s) = \mathbf{G}^{-1}(s)\mathbf{G}(s) = \mathbf{I}$$

It called the **inverse** of $\mathbf{G}(s)$ and has the state space representation

$$\mathbf{G}^{-1}(s) = \left(\begin{array}{c|c} \mathbf{A} - \mathbf{B}\mathbf{D}^{-1}\mathbf{C} & -\mathbf{B}\mathbf{D}^{-1} \\ \hline \mathbf{D}^{-1}\mathbf{C} & \mathbf{D}^{-1} \end{array}\right).$$

This can be seen by applying (5.54) and performing the similarity transform

$$\mathbf{T} = \begin{pmatrix} \mathbf{I} & \mathbf{I} \\ \mathbf{0} & \mathbf{I} \end{pmatrix}.$$

5.6 Poles and Zeros of Multivariable Systems

In this section, we consider transfer matrices for MIMO systems of the form

$$\mathbf{G}(s) = \begin{pmatrix} G_{11}(s) & \cdots & G_{1m}(s) \\ \vdots & & \vdots \\ G_{l1}(s) & \cdots & G_{lm}(s) \end{pmatrix}.$$

Definition. *A number* $s \in \mathbb{C}$ *is denoted as a* ***pole*** *of* $\mathbf{G}(s)$, *if it is a pole of at least one of the component polynomials* $G_{jk}(s)$.

Let now $(\mathbf{A}, \mathbf{B}, \mathbf{C}, \mathbf{D})$ be a realization of $\mathbf{G}(s)$. Then each pole of $\mathbf{G}(s)$ is an eigenvalue of $\mathbf{A}$. The reverse is not necessarily true, since cancellations may occur when $\mathbf{G}(s)$ is calculated. The following theorem describes the connection between the poles of $\mathbf{G}(s)$ and the eigenvalues of $\mathbf{A}$.

Theorem 5.6.1. *Let* $(\mathbf{A}, \mathbf{B}, \mathbf{C}, \mathbf{D})$ *be a realization of the transfer matrix* $\mathbf{G}(s)$. *Then the following assertions hold.*

(a) If $(\mathbf{A}, \mathbf{B})$ *is controllable and* $(\mathbf{C}, \mathbf{A})$ *is observable, then each eigenvalue* λ *of* $\mathbf{A}$ *is a pole of* $\mathbf{G}(s)$.

(b) If $(\mathbf{A}, \mathbf{B})$ *is stabilizable and* $(\mathbf{C}, \mathbf{A})$ *is detectable, then each eigenvalue* λ *of* $\mathbf{A}$ *with* $\operatorname{Re}\lambda \geq 0$ *is a pole of* $\mathbf{G}(s)$.

(c) Every eigenvalue λ *of* $\mathbf{A}$ *is a pole of* $\mathbf{G}(s)$ *or it is uncontrollable or unobservable.*

Proof. (a) If $(\mathbf{A}, \mathbf{B})$ is controllable and $(\mathbf{C}, \mathbf{A})$ is observable, then by Theorem 5.5.1 (a) $(\mathbf{A}, \mathbf{B}, \mathbf{C}, \mathbf{D})$ is a minimal realization of $\mathbf{G}(s)$ and by Theorem 5.5.1 (b) this realization is equivalent to the Gilbert realization of $\mathbf{A}$. Hence each eigenvalue of $\mathbf{A}$ is also an eigenvalue of the corresponding system matrix of the Gilbert realization. Since each eigenvalue of this matrix is a pole of $\mathbf{G}(s)$, the assertion is shown.

(b) We use the Kalman decomposition of the system $(\mathbf{A}, \mathbf{B}, \mathbf{C})$ as described in Theorem 5.4.4. The assumption implies that the matrices $\tilde{\mathbf{A}}_{22}, \tilde{\mathbf{A}}_{33}, \tilde{\mathbf{A}}_{44}$ are stable and consequently each eigenvalue λ of $\mathbf{A}$ with $\operatorname{Re}\lambda \geq 0$ must be an eigenvalue of $\tilde{\mathbf{A}}_{11}$. Since $(\tilde{\mathbf{C}}_1, \tilde{\mathbf{A}}_{11}, \tilde{\mathbf{B}}_1, \mathbf{D})$ is a realization of $\mathbf{G}(s)$ such that $(\tilde{\mathbf{A}}_{11}, \tilde{\mathbf{B}}_1)$ is controllable and $(\tilde{\mathbf{C}}_1, \tilde{\mathbf{A}}_{11})$ is observable, we conclude from (a) that λ is a pole of $\mathbf{G}(s)$.

(c) This is true since λ is an eigenvalue of $\tilde{\mathbf{A}}_{11}$ or an eigenvalue of one the matrices $\tilde{\mathbf{A}}_{22}, \tilde{\mathbf{A}}_{33}, \tilde{\mathbf{A}}_{44}$. In the first case, λ is a pole of $\mathbf{G}(s)$ and in the second case λ is uncontrollable or unobservable.

Remark. It may happen that λ is a pole of $\mathbf{G}(s)$ and uncontrollable (or unobservable). Take as an example the system with the matrices

$$\mathbf{A} = \begin{pmatrix} a & 0 \\ 0 & a \end{pmatrix}, \quad \mathbf{B} = \begin{pmatrix} 1 \\ 0 \end{pmatrix}, \quad \mathbf{C} = \begin{pmatrix} 1 & 0 \end{pmatrix}.$$

It has $\lambda = a$ as an uncontrollable pole.

A meaningful definition of zeros for MIMO systems is not so obvious. Simply requiring $\mathbf{G}(s_0) = \mathbf{0}$ for a zero s_0 would not be adequate. The point is that the rank of the transfer matrix reduces at the zero. Therefore, the following definition is needed.

Definition. *The largest possible rank r for* $\operatorname{rank} \mathbf{G}(s)$, *where s is an arbitrary complex number, is said to be the* ***normal rank*** *of $\mathbf{G}(s)$ and is denoted as* $r = \operatorname{normalrank} \mathbf{G}(s)$.

For example, let

$$\mathbf{G}(s) = \frac{1}{(s+1)^2} \begin{pmatrix} 1 & 1 \\ s+2 & 2 \end{pmatrix}.$$

Then we have $\operatorname{normalrank} \mathbf{G}(s) = 2$ and $\operatorname{rank} \mathbf{G}(0) = 1$. Hence the normal rank is the rank of $\mathbf{G}(s)$ except at certain single points.

Definition. *Let $\mathbf{G}(s)$ be an $l \times m$ proper transfer matrix with full column normal rank or full row normal rank. If $\mathbf{G}(s)$ has full column normal rank, then a complex number s_0 is called a* ***transmission zero*** *of $\mathbf{G}(s)$ if there exists a $\mathbf{u}_0 \in \mathbb{C}^m$, $\mathbf{u}_0 \neq \mathbf{0}$ such that $\mathbf{G}(s_0)\mathbf{u}_0 = \mathbf{0}$. Similarly, if $\mathbf{G}(s)$ has full row column rank, s_0 is called a transmission zero of $\mathbf{G}(s)$ if there exists a $\mathbf{w}_0 \in \mathbb{C}^l$, $\mathbf{w}_0 \neq \mathbf{0}$ such that $\mathbf{w}_0^* \mathbf{G}(s_0) = \mathbf{0}$.*

It is possible to define transmission zeros also for transfer matrices which do not fulfill any rank condition, cf. Zhou, Doyle, and Glover [112], Sect. 3.11.

For a MIMO system it may happen that s_0 is a pole and a transmission zero. For example, the system

$$\mathbf{G}(s) = \begin{pmatrix} \dfrac{s+2}{s-1} & 0 \\ 0 & \dfrac{s+1}{s+2} \end{pmatrix}$$

has the zeros $s_0 = -2$ and $s_0 = -1$. The zero $s_0 = -2$ is also pole of the system.

For zeros, which are not a pole, the following equivalent descriptions easily follow from the definition.

Suppose s_0 *is not a pole of* $\mathbf{G}(s)$. *Then* s_0 *is a transmission zero if and only if*

$$\operatorname{rank} \mathbf{G}(s_0) < \operatorname{normalrank} \mathbf{G}(s).$$

If additionally $\mathbf{G}(s)$ *is square and* $\det \mathbf{G}(s)$ *does not vanish identically, then* s_0 *is a transmission zero if and only if*

$$\det \mathbf{G}(s_0) = 0.$$

Note that in the previous example the zero $s_0 = -2$ is not a zero of $\det \mathbf{G}(s)$.

The interpretation of transmission zeros is quite similar to that of zeros for SISO systems. Let $(\mathbf{A}, \mathbf{B}, \mathbf{C}, \mathbf{D})$ be a minimal realization of $\mathbf{G}(s)$. Suppose that s_0 is not a pole of $\mathbf{G}(s)$ and put $\mathbf{u}(t) = e^{s_0 t}\mathbf{u}_0$ with a vector $\mathbf{u}_0$ which still has to be defined. Then by Theorem 5.2.4(a)

$$\mathbf{y}(t) = \mathbf{G}(s_0)\mathbf{u}_0 e^{s_0 t},$$

if $\mathbf{x}(0)$ is adequately chosen. Let now s_0 be a transmission zero . Then there is a vector $\mathbf{u}_0 \neq \mathbf{0}$ such that $\mathbf{G}(s_0)\mathbf{u}_0 = \mathbf{0}$. Consequently, the corresponding output signal $\mathbf{y}(t)$ vanishes identically.

This result has a particularly interesting interpretation for transmission zeros on the imaginary axis: $s_0 = i\omega_0$. Suppose again that s_0 is not a pole of $\mathbf{G}(s)$. Then $\mathbf{u}(t) = e^{i\omega_0 t}\mathbf{u}_0$ is for each component a harmonic oscillation, which (for an appropriate initial value) is completely suppressed by the system. If the system is stable, the output $\mathbf{y}(t)$ tends to $\mathbf{0}$ as $t \to \infty$ for an arbitrary initial value.

Beside the transmission zeros, there is a different type of zero which is of importance. They refer to the state space description of the system and take into account its internal behavior. Let $(\mathbf{A}, \mathbf{B}, \mathbf{C}, \mathbf{D})$ be an arbitrary realization of $\mathbf{G}(s)$. Define for an arbitrary complex number s_0 and vectors $\mathbf{x}_0$, $\mathbf{u}_0$ the functions $\mathbf{x}(t) = e^{s_0 t}\mathbf{x}_0$, $\mathbf{u}(t) = e^{s_0 t}\mathbf{u}_0$. Then $\mathbf{x}(t)$ satisfies (5.7) if and only if

$$(s_0\mathbf{I} - \mathbf{A})\mathbf{x}_0 - \mathbf{B}\mathbf{u}_0 = \mathbf{0}.$$

The output vanishes identically if and only if

$$\mathbf{C}\mathbf{x}_0 + \mathbf{D}\mathbf{u}_0 = \mathbf{0}.$$

In this case, it is reasonable to say that s_0 is a zero of the system. The last two equations can be compactly be written as

$$\mathbf{R}(s_0)\begin{pmatrix} \mathbf{x}_0 \\ \mathbf{u}_0 \end{pmatrix} = \mathbf{0},$$

where

$$\mathbf{R}(s) = \begin{pmatrix} s\mathbf{I} - \mathbf{A} & -\mathbf{B} \\ \mathbf{C} & \mathbf{D} \end{pmatrix}$$

is the so-called **Rosenbrock matrix** of the system. By this way, we are led to the following definition.

Definition. *A complex number* s_0 *is denoted as an* ***invariant zero*** *of the system realization* $(\mathbf{A}, \mathbf{B}, \mathbf{C}, \mathbf{D})$ *if*

$$\operatorname{rank} \mathbf{R}(s_0) < \operatorname{normalrank} \mathbf{R}(s).$$

Suppose that the system has an unobservable eigenvalue s_0. Then there exist a vector $\mathbf{x}_0 \neq \mathbf{0}$ such that $(s_0\mathbf{I} - \mathbf{A})\mathbf{x}_0 = \mathbf{0}$ and $\mathbf{C}\mathbf{x}_0 = \mathbf{0}$. This implies

$$\mathbf{R}(s_0)\begin{pmatrix} \mathbf{x}_0 \\ \mathbf{0} \end{pmatrix} = \mathbf{0}.$$

Similarly, if s_0 is an uncontrollable eigenvalue, then there exists vector $\mathbf{x}_0 \neq \mathbf{0}$ such that $\mathbf{x}_0^* (s_0\mathbf{I} - \mathbf{A}) = \mathbf{0}$ and $\mathbf{x}_0^*\mathbf{B} = \mathbf{0}$. This yields

$$(\mathbf{x}_0^* \quad \mathbf{0})\,\mathbf{R}(s_0) = \mathbf{0}$$

and the following conclusion can be made.

The unobservable and the uncontrollable eigenvalues of the system matrix $\mathbf{A}$ *of a realization* $(\mathbf{A}, \mathbf{B}, \mathbf{C}, \mathbf{D})$ *of a transfer matrix* $\mathbf{G}(s)$ *are invariant zeros of* $\mathbf{G}(s)$.

It is reasonable to presume that for a minimal realization the invariant zeros and the transmission zeros coincide. In fact, this can be shown (cf. Zhou, Doyle, and Glover [112], Theorem 3.34). If the transmission zero is not a pole of the transfer matrix, the proof is rather simple and based on the decomposition

$$\begin{pmatrix} s\mathbf{I} - \mathbf{A} & -\mathbf{B} \\ \mathbf{C} & \mathbf{D} \end{pmatrix} = \begin{pmatrix} \mathbf{I} & \mathbf{0} \\ \mathbf{C}(s\mathbf{I} - \mathbf{A})^{-1} & \mathbf{I} \end{pmatrix} \begin{pmatrix} s\mathbf{I} - \mathbf{A} & -\mathbf{B} \\ \mathbf{0} & \mathbf{G}(s) \end{pmatrix}.$$

Theorem 5.6.2. *Let* $(\mathbf{A}, \mathbf{B}, \mathbf{C}, \mathbf{D})$ *be a minimal realization of a transfer matrix* $\mathbf{G}(s)$. *Then* s_0 *is a transmission zero of* $\mathbf{G}(s)$ *if and only if* s_0 *is an invariant zero of the realization* $(\mathbf{A}, \mathbf{B}, \mathbf{C}, \mathbf{D})$.

We finish this section by considering plants that are invertible. Let

$$\mathbf{G}(s) = \left(\begin{array}{c|c} \mathbf{A} & \mathbf{B} \\ \hline \mathbf{C} & \mathbf{D} \end{array}\right)$$

be square with invertible $\mathbf{D}$. The characteristic polynomial of $\mathbf{G}^{-1}(s)$ is given by

(cf. Sect. 5.5)

$$p(s) = \det(s\mathbf{I} - \mathbf{A} + \mathbf{B}\mathbf{D}^{-1}\mathbf{C}). \tag{5.56}$$

Note that the realization of $\mathbf{G}^{-1}(s)$ given in Sect. 5.5 is minimal if and only if the realization of $\mathbf{G}(s)$ is minimal. Using the determinant formulas of Appendix A.1.4, the determinant of the Rosenbrock matrix can be evaluated in two ways, namely

$$\det \mathbf{R}(s) = \det(\mathbf{D})\, p(s) \quad \text{and} \quad \det \mathbf{R}(s) = \det(s\mathbf{I} - \mathbf{A}) \det \mathbf{G}(s). \tag{5.57}$$

The second formula requires that s is not an eigenvalue of $\mathbf{A}$. Thus in this case the following relationship exists between the characteristic polynomial of $\mathbf{G}^{-1}(s)$ and $\det \mathbf{G}(s)$:

$$\det(\mathbf{D})\, p(s) = \det(s\mathbf{I} - \mathbf{A}) \det \mathbf{G}(s). \tag{5.58}$$

Corollary 5.6.1. *Let* $\mathbf{G}(s)$ *be an invertible transfer matrix. Then the following assertions hold.*

(a) A number s_0 *is an invariant zero of the realization of* $\mathbf{G}(s)$ *if and only if it is an eigenvalue of* $\mathbf{A} - \mathbf{B}\mathbf{D}^{-1}\mathbf{C}$.

(b) A number s_0 *is a transmission zero of* $\mathbf{G}(s)$ *if and only if it is a pole of* $\mathbf{G}^{-1}(s)$.

Proof. (a) A number s_0 is an invariant zero of $\mathbf{G}(s)$ if and only if $\det \mathbf{R}(s_0) = 0$. Due to (5.57), this is equivalent to $p(s) = 0$. This proves (a).

(b) Let $(\mathbf{A}, \mathbf{B}, \mathbf{C}, \mathbf{D})$ be a minimal realization of $\mathbf{G}(s)$. Using Theorem 5.6.2, a number s_0 is a transmission zero if and only if it is an invariant zero of $\mathbf{G}(s)$. Using (a), this is the case if and only if s_0 is an eigenvalue of $\mathbf{A} - \mathbf{B}\mathbf{D}^{-1}\mathbf{C}$. Since the realization of $\mathbf{G}^{-1}(s)$ is minimal, the latter property holds if and only if s_0 is a pole of $\mathbf{G}^{-1}(s)$.

5.7 Lyapunov Equations

For a later application in the proofs of the $\mathcal{H}_2$ and $\mathcal{H}_\infty$ optimization problems we need some results concerning the so-called **Lyapunov equation.** This is a matrix equation of the form

$$\mathbf{A}^T\mathbf{X} + \mathbf{X}\mathbf{A} + \mathbf{Q} = \mathbf{0} \tag{5.59}$$

with $n \times n$ matrices $\mathbf{A}, \mathbf{Q}$.

Theorem 5.7.1. *Let* $\mathbf{A}$ *be stable. Then*

$$\mathbf{X} = \int_0^\infty e^{\mathbf{A}^T t} \mathbf{Q} e^{\mathbf{A}t} \, dt$$

is the unique solution of the Lyapunov equation (5.59).

Proof. We have

$$\frac{d}{dt} e^{\mathbf{A}^T t} \mathbf{Q} e^{\mathbf{A}t} = \mathbf{A}^T e^{\mathbf{A}^T t} \mathbf{Q} e^{\mathbf{A}t} + e^{\mathbf{A}^T t} \mathbf{Q} e^{\mathbf{A}t} \mathbf{A}.$$

Integration gives

$$e^{\mathbf{A}^T T} \mathbf{Q} e^{\mathbf{A}T} - \mathbf{Q} = \mathbf{A}^T \int_0^T e^{\mathbf{A}^T t} \mathbf{Q} e^{\mathbf{A}t} \, dt + \int_0^T e^{\mathbf{A}^T t} \mathbf{Q} e^{\mathbf{A}t} \, dt \, \mathbf{A}.$$

Since $\mathbf{A}$ is assumed to be stable we may pass to the limit as $t \to \infty$. Then the first term on the left-hand side of this equation tends to the zero matrix and the solution property follows.

The uniqueness is easily seen by a dimension argument. Define a linear mapping $L: \mathbb{R}^{n \times n} \to \mathbb{R}^{n \times n}$ by

$$L(\mathbf{X}) = \mathbf{A}^T \mathbf{X} + \mathbf{X}\mathbf{A}.$$

Since, as we have already shown, the equation $L(\mathbf{X}) = -\mathbf{Q}$ has a solution for every $\mathbf{Q} \in \mathbb{R}^{n \times n}$, the dimension of the image of L is n^2. Because the domain of L has also the dimension n^2, the kernel of L consists only of the zero matrix and hence L is injective.

Remark 1. From the Lyapunov equation, we directly see that the following similarity transformation holds

$$\begin{pmatrix} \mathbf{I} & \mathbf{0} \\ \mathbf{X} & \mathbf{I} \end{pmatrix} \begin{pmatrix} \mathbf{A} & \mathbf{0} \\ \mathbf{Q} & -\mathbf{A}^T \end{pmatrix} \begin{pmatrix} \mathbf{I} & \mathbf{0} \\ -\mathbf{X} & \mathbf{I} \end{pmatrix} = \begin{pmatrix} \mathbf{A} & \mathbf{0} \\ \mathbf{0} & -\mathbf{A}^T \end{pmatrix}.$$

Remark 2. Theorem 5.7.1 implies the following statement (if $\mathbf{A}$ is stable):

$\mathbf{Q} \geq \mathbf{0}$ *implies* $\mathbf{X} \geq \mathbf{0}$, *and* $\mathbf{Q} > \mathbf{0}$ *implies* $\mathbf{X} > \mathbf{0}$.

The next lemma states the conditions under which the reverse directions of the implications of Remark 2 hold. Consequently, a characterization of the stability of $\mathbf{A}$ by means of the Lyapunov equation is obtained.

Lemma 5.7.1. *Let* $\mathbf{X}$ *be a solution of the Lyapunov equation (5.59). Then:*

(a) $\mathbf{A}$ *is stable if* $\mathbf{X} > \mathbf{0}$ *and* $\mathbf{Q} > \mathbf{0}$.

(b) $\mathbf{A}$ *is stable if* $\mathbf{X} \geq \mathbf{0}, \mathbf{Q} \geq \mathbf{0}$ *and* $(\mathbf{Q}, \mathbf{A})$ *is detectable.*

Proof. Let λ be an eigenvalue of $\mathbf{A}$ and let $\mathbf{v} \neq \mathbf{0}$ be a corresponding eigenvector. Premultiply (5.59) by $\mathbf{v}^*$ and postmultiply (5.59) by $\mathbf{v}$ to get

$$2\,\mathrm{Re}(\lambda)\mathbf{v}^*\mathbf{X}\mathbf{v} + \mathbf{v}^*\mathbf{Q}\mathbf{v} = 0. \tag{5.60}$$

Now if $\mathbf{X} > \mathbf{0}$ and $\mathbf{Q} > \mathbf{0}$, this equation implies $\mathrm{Re}(\lambda) < 0$, and hence $\mathbf{A}$ is stable. Let now the assumption of (b) be fulfilled and suppose $\mathrm{Re}(\lambda) \geq 0$. Then (5.60) implies $\mathbf{v}^*\mathbf{Q}\mathbf{v} = 0$ and therefore $\mathbf{Q}\mathbf{v} = \mathbf{0}$. This is according to Theorem 5.4.3(b) in contradiction to the detectability of $(\mathbf{Q}, \mathbf{A})$.

The next lemma characterizes the observability by means of the solution $\mathbf{X}$.

Lemma 5.7.2. *Let* $\mathbf{A}$ *be stable and suppose* $\mathbf{Q} \geq \mathbf{0}$. *Then:*

$(\mathbf{Q}, \mathbf{A})$ *is observable if and only if* $\mathbf{X} > \mathbf{0}$.

Proof. Decompose $\mathbf{Q}$ in the form $\mathbf{Q} = \mathbf{C}^T\mathbf{C}$ and define for a given initial value $\mathbf{x}_0$ functions $\mathbf{x}, \mathbf{y}$ by

$$\dot{\mathbf{x}} = \mathbf{A}\mathbf{x}, \quad \mathbf{x}(0) = \mathbf{x}_0$$

$$\mathbf{y} = \mathbf{C}\mathbf{x}.$$

Then, by Theorem 5.7.1 it follows that

$$\mathbf{x}_0^T\mathbf{X}\mathbf{x}_0 = \int_0^\infty \mathbf{x}_0^T e^{\mathbf{A}^T t}\mathbf{Q}\, e^{\mathbf{A}t}\mathbf{x}_0\, dt = \int_0^\infty \mathbf{y}^T(t)\mathbf{y}(t)\, dt. \tag{5.61}$$

"$\Rightarrow$": Let $\mathbf{x}_0^T\mathbf{X}\mathbf{x}_0 = 0$. Thus, by (5.61), $\mathbf{y}(t) = \mathbf{0}$ and hence $\mathbf{x}_0 = 0$, since $(\mathbf{Q}, \mathbf{A})$ is observable by assumption.
"$\Leftarrow$": Let $\mathbf{y}(t) = \mathbf{0}$. Then, by (5.61), $\mathbf{x}_0^T\mathbf{X}\mathbf{x}_0 = 0$ and therefore $\mathbf{x}_0 = 0$, since $\mathbf{X} > \mathbf{0}$ by assumption. Hence, $(\mathbf{Q}, \mathbf{A})$ is observable.

We define the **controllability gramian** $\mathbf{X}_c$ and the **observability gramian** $\mathbf{X}_o$ by the following equations (cf. (5.41) and (5.49)):

$$\mathbf{X}_c = \int_0^\infty e^{\mathbf{A}t}\mathbf{B}\mathbf{B}^T e^{\mathbf{A}^T t}dt, \quad \mathbf{X}_o = \int_0^\infty e^{\mathbf{A}^T t}\mathbf{C}^T\mathbf{C}\, e^{\mathbf{A}t}dt.$$

They are solutions of the following Lyapunov equations:

$$\mathbf{A}\mathbf{X}_c + \mathbf{X}_c\mathbf{A}^T + \mathbf{B}\mathbf{B}^T = \mathbf{0}, \tag{5.62}$$

$$\mathbf{A}^T\mathbf{X}_o + \mathbf{X}_o\mathbf{A} + \mathbf{C}^T\mathbf{C} = \mathbf{0}. \tag{5.63}$$

From Lemma 5.7.2 we see that for a stable $\mathbf{A}$ the pair $(\mathbf{C}, \mathbf{A})$ is observable if and only if $\mathbf{X}_o > \mathbf{0}$. Similarly, $(\mathbf{A}, \mathbf{B})$ is controllable if and only if $\mathbf{X}_c > \mathbf{0}$.

The following corollary is a direct consequence of Theorem 5.7.1 and Lemma 5.7.1.

Corollary 5.7.1. *The matrix* $\mathbf{A}$ *is stable if and only if there exists* $\mathbf{X} > \mathbf{0}$ *such that*

$$\mathbf{A}^T\mathbf{X} + \mathbf{X}\mathbf{A} < \mathbf{0}.$$

5.8 Linear Dynamical Systems with Stochastic Inputs

5.8.1 Gaussian Random Processes

We consider in this section the situation where the system state $\mathbf{x}(t)$ and the inputs (the disturbance and the measurement error) are random vectors. Then the map from the time t to the random vector $\mathbf{x}(t)$ is a **stochastic process.** If $\xi(t)$ is the result of an experiment which is performed according to the stochastic properties of $\mathbf{x}(t)$ at time t, then the time function $\xi(t)$ is called the **realization** of the stochastic process $\mathbf{x}(t)$. For a control theory analysis it is often useful to model the disturbance and the measurement error (and then inevitably also the state vector $\mathbf{x}(t)$) as stochastic processes. In the controller design process, it is often possible to take advantage of their properties.

In this section, we shortly summarize the basic properties of random processes that play a role in connection with control theory. The **mean** of a random process is obtained by forming at each time t the mean of the random vector:

$$\mathbf{m}_\mathbf{x}(t) = E(\mathbf{x}(t)).$$

The expected value of the product $\mathbf{x}(t_1)\mathbf{x}^T(t_2)$ is the **correlation function** (or **autocorrelation function**) and is denoted as follows:

$$\mathbf{R}_\mathbf{x}(t_1, t_2) = E(\mathbf{x}(t_1)\mathbf{x}^T(t_2)).$$

The quantity $E(\mathbf{x}^T(t)\mathbf{x}(t))$ is the so-called **mean square value** of the random process. It can be calculated from the correlation function by

$$E(\mathbf{x}^T(t)\mathbf{x}(t)) = \mathrm{trace}(\mathbf{R}_\mathbf{x}(t, t)).$$

If the time arguments in the correlation function are equal, one obtains the **correlation matrix:**

$$\Sigma_\mathbf{x}(t) = E(\mathbf{x}(t)\mathbf{x}^T(t)) = \mathbf{R}_\mathbf{x}(t, t).$$

The **covariance function** of the random process is defioned as follows:

$$\mathbf{C_x}(t_1,t_2)=E((\mathbf{x}(t_1)-\mathbf{m_x}(t_1))\,(\mathbf{x}(t_2)-\mathbf{m_x}(t_2))^T).$$

For a process with mean $\mathbf{0}$, the correlation function and the covariance function coincide. In any case, we have

$$\mathbf{C_x}(t_1,t_2)=\mathbf{R_x}(t_1,t_2)-\mathbf{m_x}(t_1)\,\mathbf{m_x}(t_2)^T.$$

The **covariance matrix** results if the time arguments are equal:

$$\Xi_\mathbf{x}(t)=E((\mathbf{x}(t)-\mathbf{m_x}(t))\,(\mathbf{x}(t)-\mathbf{m_x}(t)^T)).$$

A stochastic process is **wide-sense stationary** if the mean is independent of time,

$$E(\mathbf{x}(t))=\mathbf{m_x},$$

with constant $\mathbf{m_x}$ and if additionally the correlation function depends only on the time difference $\tau=t_2-t_1$. Then one can define.

$$\mathbf{R_x}(\tau)=\mathbf{R_x}(t+\tau,t),\quad t \text{ arbitrary}.$$

In this case, the covariance function depends also only on the time difference. For stationary random processes, the **spectral density** (also denoted as the **power spectral density**) is defined as the Fourier transform of the correlation function:

$$\mathbf{S_x}(\omega)=\int_{-\infty}^{\infty}\mathbf{R_x}(\tau)e^{-i\omega\tau}\,d\tau.$$

It describes the distribution of the frequencies, which build up the stochastic process.

For control theory applications the **Gaussian random processes** play the major role. These are processes with the property that for all times $t_1,t_2,\dots,t_m$ (arbitrary) the random vectors $\mathbf{x}(t_1),\mathbf{x}(t_2),\dots,\mathbf{x}(t_m)$ have a common Gaussian probability distribution. For a single random vector $\mathbf{x}$, the probability density is given by

$$f(\mathbf{x})=\frac{1}{((2\pi)^n\det\Xi_\mathbf{x})^{1/2}}\exp\left\{-\frac{1}{2}(\mathbf{x}-\mathbf{m_x})^T\,\Xi_\mathbf{x}^{-1}(\mathbf{x}-\mathbf{m_x})\right\}.$$

A Gaussian process is completely determined by its mean and its covariance function.

A process $\mathbf{w}(t)$ which has mean $\mathbf{0}$ and an autocorrelation function of the form

$$\mathbf{R_x}(t_1,t_2)=\mathbf{W}(t_1)\delta(t_1-t_2)$$

is denoted as **white noise**. For such a process the random vectors $\mathbf{x}(t_1)$, $\mathbf{x}(t_2)$ are uncorrelated for all different times t_1,t_2. For stationary white noise the autocorrelation function has the representation

$$\mathbf{R}_x(\tau) = \mathbf{W}\delta(\tau).$$

Hence, the spectral density is given by

$$\mathbf{S}_w(\omega) = \mathbf{W}$$

and is consequently constant. Thus, a white noise process is composed of all frequencies such that they contribute in an equal manner. Although no real process has this property, it is meaningful to work with this idealization.

For the design of the LQG controller (cf. Sect. 8.4) we will assume that all disturbances and measurement errors are white noise processes or can be modelled with the help of white noise processes. Then the question arises of how the solutions of the state equation can be characterized. This will be analyzed in the next two sections.

5.8.2 Solutions of Linear Differential Equations with White Noise Inputs

We consider the following linear system:

$$\dot{\mathbf{x}} = \mathbf{A}\,\mathbf{x} + \mathbf{B}\mathbf{w}, \qquad \mathbf{x}(0) = \mathbf{x}_0$$

$$\mathbf{y} = \mathbf{C}\,\mathbf{x}.$$

Here, $\mathbf{w}(t)$ denotes white noise with mean $\mathbf{0}$ and spectral density $\mathbf{W}$. The initial state is assumed to be a random vector with mean $\mathbf{0}$ and correlation matrix $\mathbf{R}_0$. Finally, we assume the white noise and the initial value to be uncorrelated:

$$E(\mathbf{x}_0\mathbf{w}^T(t)) = \mathbf{0}.$$

The mean of $\mathbf{x}(t)$ is obtained by forming the expected values in the solution formula:

$$\mathbf{m}_x(t) = E\left\{e^{\mathbf{A}t}\mathbf{x}_0 + \int_0^t e^{\mathbf{A}(t-\tau)}\mathbf{B}\mathbf{w}(\tau)\,d\tau\right\}$$

$$= e^{\mathbf{A}t}E(\mathbf{x}_0) + \int_0^t e^{\mathbf{A}(t-\tau)}\mathbf{B}E(\mathbf{w}(\tau))\,d\tau = \mathbf{0}.$$

Hence the mean of the output $\mathbf{y}$ vanishes also: $\mathbf{m}_y(t) = \mathbf{0}$. The assumption, that the initial value $\mathbf{x}_0$ has mean $\mathbf{0}$ is no restriction of generality. If this is not the case, one can separate the mean as a deterministic part and add the solution for this deterministic initial value to the stochastic solution for mean $\mathbf{0}$.

The correlation function for the state vector $\mathbf{x}$ can also be calculated by applying the solution formula for the state equation. We get

$$\mathbf{R_x}(t_1,t_2) = E(\mathbf{x}(t_1)\mathbf{x}^T(t_2))$$

$$= E\left\{\left(e^{\mathbf{A}t_1}\mathbf{x}_0 + \int_0^{t_1} e^{\mathbf{A}(t_1-\tau)}\mathbf{B}\mathbf{w}(\tau)\,d\tau\right)\left(e^{\mathbf{A}t_2}\mathbf{x}_0 + \int_0^{t_2} e^{\mathbf{A}(t_2-\sigma)}\mathbf{B}\mathbf{w}(\sigma)\,d\sigma\right)^T\right\}$$

$$= e^{\mathbf{A}t_1}E(\mathbf{x}_0\mathbf{x}_0^T)e^{\mathbf{A}^T t_2} + \int_0^{t_2} e^{\mathbf{A}t_1}E(\mathbf{x}_0\mathbf{w}^T(\sigma))\mathbf{B}^T e^{\mathbf{A}^T(t_2-\sigma)}\,d\sigma$$

$$+ \int_0^{t_1} e^{\mathbf{A}(t_1-\tau)}\mathbf{B}E(\mathbf{w}(\tau)\mathbf{x}_0^T)e^{\mathbf{A}^T t_2}\,d\tau$$

$$+ \int_0^{t_1}\int_0^{t_2} e^{\mathbf{A}(t_1-\tau)}\mathbf{B}\,E(\mathbf{w}(\tau)\mathbf{w}(\sigma)^T)\,\mathbf{B}^T e^{\mathbf{A}^T(t_2-\sigma)}\,d\sigma d\tau$$

$$= e^{\mathbf{A}t_1}\mathbf{R}_0 e^{\mathbf{A}^T t_2} + \int_0^{t_1}\int_0^{t_2} e^{\mathbf{A}(t_1-\tau)}\mathbf{B}\mathbf{W}\delta(\tau-\sigma)\,\mathbf{B}^T e^{\mathbf{A}^T(t_2-\sigma)}\,d\sigma d\tau$$

with $\mathbf{R}_0 = E(\mathbf{x}_0\mathbf{x}_0^T)$. By defining $s = \min(t_1, t_2)$, the desired formula follows:

$$\mathbf{R_x}(t_1,t_2) = e^{\mathbf{A}t_1}\mathbf{R}_0 e^{\mathbf{A}^T t_2} + \int_0^{s} e^{\mathbf{A}(t_1-\tau)}\mathbf{B}\mathbf{W}\mathbf{B}^T e^{\mathbf{A}^T(t_2-\tau)}\,d\tau\,. \tag{5.64}$$

The correlation function of the output variable is also given by this calculation:

$$\mathbf{R_y}(t_1,t_2) = \mathbf{C}\mathbf{R_x}(t_1,t_2)\mathbf{C}^T\,. \tag{5.65}$$

Hence the correlation matrix is

$$\Sigma_{\mathbf{x}}(t) = e^{\mathbf{A}t}\mathbf{R}_0 e^{\mathbf{A}^T t} + \int_0^{t} e^{\mathbf{A}(t-\tau)}\mathbf{B}\mathbf{W}\mathbf{B}^T e^{\mathbf{A}^T(t-\tau)}\,d\tau\,.$$

By substituting the difference $t-\tau$, we get

$$\Sigma_{\mathbf{x}}(t) = e^{\mathbf{A}t}\mathbf{R}_0 e^{\mathbf{A}^T t} + \int_0^{t} e^{\mathbf{A}\tau}\mathbf{B}\mathbf{W}\mathbf{B}^T e^{\mathbf{A}^T\tau}\,d\tau\,. \tag{5.66}$$

Thus, the correlation matrix of the output is computed as

$$\Sigma_{\mathbf{y}}(t) = \mathbf{C}\Sigma_{\mathbf{x}}(t)\mathbf{C}^T\,. \tag{5.67}$$

Instead of calculating $\Sigma_{\mathbf{x}}$ by the integral (5.66), it is also possible to describe this matrix as the solution of a certain matrix differential equation. Define

$$\mathbf{M}(t) = e^{\mathbf{A}t}\mathbf{B}\mathbf{W}\mathbf{B}^T e^{\mathbf{A}^T t}\,.$$

Then the equations

$$\Sigma_{\mathbf{x}}(t) = e^{\mathbf{A}t}\mathbf{R}_0 e^{\mathbf{A}^T t} + \int_0^t \mathbf{M}(\tau)\,d\tau \ ,$$

$$\dot{\mathbf{M}}(t) = \mathbf{A}\mathbf{M}(t) + \mathbf{M}(t)\mathbf{A}^T ,$$

hold and differentiation of (5.66) leads to

$$\dot{\Sigma}_{\mathbf{x}}(t) = \mathbf{A}e^{\mathbf{A}t}\mathbf{R}_0 e^{\mathbf{A}^T t} + e^{\mathbf{A}t}\mathbf{R}_0 e^{\mathbf{A}^T t}\mathbf{A}^T + \mathbf{M}(t)$$

$$= \mathbf{A}e^{\mathbf{A}t}\mathbf{R}_0 e^{\mathbf{A}^T t} + e^{\mathbf{A}t}\mathbf{R}_0 e^{\mathbf{A}^T t}\mathbf{A}^T + \int_0^t (\mathbf{A}\mathbf{M}(\tau) + \mathbf{M}(\tau)\mathbf{A}^T)\,d\tau + \mathbf{M}(0)$$

$$= \mathbf{A}\left\{ e^{\mathbf{A}t}\mathbf{R}_0 e^{\mathbf{A}^T t} + \int_0^t \mathbf{M}(\tau)\,d\tau \right\} + \left\{ e^{\mathbf{A}t}\mathbf{R}_0 e^{\mathbf{A}^T t} + \int_0^t \mathbf{M}(\tau)\,d\tau \right\}\mathbf{A}^T + \mathbf{B}\mathbf{W}\mathbf{B}^T .$$

Hence, we have proven the following theorem.

Theorem 5.8.1. *The correlation matrix $\Sigma_{\mathbf{x}}$ fulfills the matrix differential equation*

$$\dot{\Sigma}_{\mathbf{x}}(t) = \mathbf{A}\Sigma_{\mathbf{x}}(t) + \Sigma_{\mathbf{x}}(t)\mathbf{A}^T + \mathbf{B}\mathbf{W}\mathbf{B}^T . \tag{5.68}$$

For stable systems it is possible to pass in the representations (5.66), (5.67) and (5.68) to the limit. The next theorem summarizes the relevant results and is an immediate consequence of the mentioned formulas.

Theorem 5.8.2. *Let $\mathbf{A}$ be stable. Then the limits*

$$\Sigma_{\mathbf{x}}^{\infty} = \lim_{t\to\infty} \Sigma_{\mathbf{x}}(t), \quad \Sigma_{\mathbf{y}}^{\infty} = \lim_{t\to\infty} \Sigma_{\mathbf{y}}(t)$$

exist and have the representations

$$\Sigma_{\mathbf{x}}^{\infty} = \int_0^{\infty} e^{\mathbf{A}t}\mathbf{B}\mathbf{W}\mathbf{B}^T e^{\mathbf{A}^T t}\,dt , \quad \Sigma_{\mathbf{y}}^{\infty} = \int_0^{\infty} \mathbf{g}(t)\mathbf{W}\mathbf{g}^T(t)\,dt .$$

The stationary correlation matrix $\Sigma_{\mathbf{x}}^{\infty}$ fulfills the Lyapunov equation

$$\mathbf{A}\Sigma_{\mathbf{x}}^{\infty} + \Sigma_{\mathbf{x}}^{\infty}\mathbf{A}^T + \mathbf{B}\mathbf{W}\mathbf{B}^T = \mathbf{0} .$$

In a similar manner, it is possible by using (5.64) to get a representation of the stationary correlation function of the state vector and the output vector. From (5.64), a short calculation yields

$$\mathbf{R}_{\mathbf{x}}(t_1, t_1 + \tau) = e^{\mathbf{A}t_1}\mathbf{R}_0 e^{\mathbf{A}^T(t_1+\tau)} + \int_{\max(0,-\tau)}^{t_1} e^{\mathbf{A}\sigma}\mathbf{B}\mathbf{W}\mathbf{B}^T e^{\mathbf{A}^T(\tau+\sigma)}\,d\sigma . \tag{5.69}$$

For a stable matrix **A** it is possible to define

$$\mathbf{R}_\mathbf{x}^\infty(\tau) = \lim_{t_1\to\infty} \mathbf{R}_\mathbf{x}(t_1, t_1+\tau), \quad \mathbf{R}_\mathbf{y}^\infty(\tau) = \lim_{t_1\to\infty} \mathbf{R}_\mathbf{y}(t_1, t_1+\tau).$$

These matrices describe the correlation matrices for the corresponding processes, if the initial value is zero or if its contribution to the integral has faded away. Then the correlation matrices depend only on the time distance τ of the state vectors. Hence, the processes are stationary after the initial phase. Then the spectral densities can be formed. The next theorem collects the relevant formulas.

Theorem 5.8.3. *(a) The stationary correlation functions have the representations*

$$\mathbf{R}_\mathbf{x}^\infty(\tau) = \int_{\max(0,-\tau)}^{\infty} e^{\mathbf{A}\sigma}\mathbf{B}\mathbf{W}\mathbf{B}^T e^{\mathbf{A}^T(\tau+\sigma)} d\sigma \tag{5.70}$$

$$\mathbf{R}_\mathbf{y}^\infty(\tau) = \int_{-\infty}^{\infty} \mathbf{g}(\sigma)\mathbf{W}\mathbf{g}^T(\sigma+\tau)\, d\sigma\ . \tag{5.71}$$

(b) The following equation holds for the spectral density of the output variable:

$$\mathbf{S}_\mathbf{y}(\omega) = \mathbf{G}(-i\omega)\mathbf{W}\mathbf{G}^T(i\omega)\,. \tag{5.72}$$

Proof. Equation (5.70) follows directly from (5.69) by passing to the limit and (5.71) is a direct consequence of (5.70). From the definition of the spectral density and by interchanging the order of integration, we get

$$\mathbf{S}_\mathbf{y}(\omega) = \int_{-\infty}^{\infty}\int_{-\infty}^{\infty} \mathbf{g}(\sigma)\mathbf{W}\mathbf{g}^T(\sigma+\tau)\, d\sigma\, e^{-i\omega\tau} d\tau$$

$$= \int_{-\infty}^{\infty} \mathbf{g}(\sigma)\mathbf{W}\int_{-\infty}^{\infty} \mathbf{g}^T(\sigma+\tau)\, e^{-i\omega\tau} d\tau\, d\sigma\,.$$

The inner integral is calculated as

$$\int_{-\infty}^{\infty} \mathbf{g}^T(\sigma+\tau)\, e^{-i\omega\tau} d\tau = \int_{-\infty}^{\infty} \mathbf{g}^T(t)\, e^{-i\omega(t-\sigma)} dt = \mathbf{G}^T(i\omega)e^{i\omega\sigma},$$

and the assertion follows:

$$\mathbf{S}_\mathbf{y}(\omega) = \int_{-\infty}^{\infty} \mathbf{g}(\sigma)\mathbf{W}\mathbf{G}^T(i\omega)e^{i\omega\sigma} d\sigma = \int_{-\infty}^{\infty} \mathbf{g}(\sigma)e^{i\omega\sigma} d\sigma\,\mathbf{W}\mathbf{G}^T(i\omega)\,.$$

Remark. If the spectral density is given, the autocorrelation function can be calculated from the spectral density by applying the inverse Fourier transform. Thus

$$\mathbf{R_x}(\tau) = \frac{1}{2\pi}\int_{-\infty}^{\infty} \mathbf{S_x}(\omega) e^{i\omega\tau} d\omega .$$

5.8.3 Colored Noise and Shaping Filters

In many situations, it is not correct to model the disturbances or the measurement error directly as white noise. In such cases, it is often possible to model these quantities as a stochastic process that results if white noise passes through a certain linear system, which in this context is denoted as a **shaping filter**. The output is so-called **colored noise**. Here it is assumed that the white noise has $\mathbf{W} = \mathbf{I}$ as the spectral density. We consider two examples.

Shaping filter of order 1. A shaping filter of order 1 is nothing else but a type 0 system of order 1 without zeros, i.e.

$$G(s) = \frac{k}{1+Ts} .$$

The spectral density is according to Theorem 5.8.3 given by

$$S_y(\omega) = \frac{k^2}{1+T^2\omega^2}$$

and an application of the inverse Fourier transformation yields

$$R_y(t) = \frac{k^2}{2T} e^{-\frac{|t|}{T}} .$$

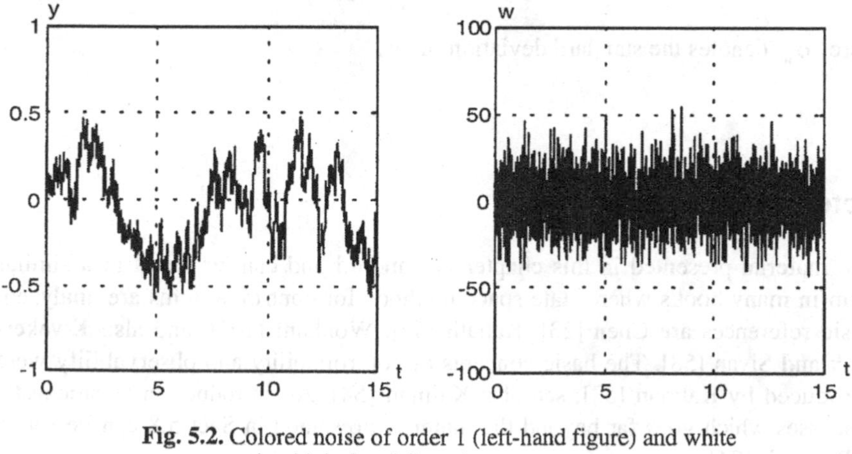

Fig. 5.2. Colored noise of order 1 (left-hand figure) and white noise (right-hand figure)

Thus the (stationary) standard deviation of the output signal y is

$$\sigma_y = \sqrt{R_y(0)} = k / \sqrt{2T} .$$

The filter will be specified by σ_y and the time constant T. Figure 5.2 shows the white noise together with the colored noise y which results, when the white noise passes through the shaping filter. The parameter values are $T = 1$, $\sigma_y = 0.3$. It should be noted that the standard deviation of the approximating discrete white noise has to be calculated according to the formula $\sigma_{w\Delta} = 1/\Delta^{1/2}$. The quantity Δ denotes the simulation step size. We shall not give here the arguments that lead to this formula.

Dryden model for vertical gusts. For vertical gusts, there is a stochastic model which is based on the transfer function

$$G(s) = k \frac{s + \omega_0}{(s + \omega_1)^2} .$$

The input is white noise with spectral density 1 and the output signal is the velocity w of the gusts in the vertical direction. The parameters can be calculated from the formulas

$$k = (3V / \pi L_2)^{1/2}, \quad \omega_0 = V / (\sqrt{3} L_2), \quad \omega_1 = V / L_2 .$$

Here, V denotes the velocity of the aircraft and L_2 is the so-called vertical turbulence scale. Then $T = L_2 / V$ is a measure of the correlation of the gusts with respect to time, similar to T in the last example. It can be shown that the correlation function is given by

$$R_w(t) = \sigma_w^2 e^{-\frac{t}{T}} \left(1 - \frac{t}{2T} \right) .$$

Here, σ_w denotes the standard deviation of w.

Notes and References

The material presented in this chapter is standard and can be found in a similar form in many books where state space methods for control systems are analyzed. Basic references are Chen [23], Kailath [52], Wonham [105], and also Kwakernaak and Sivan [58]. The basic concepts of controllability and observability were introduced by Kalman [53]; see also Kalman [54]. An introduction to stochastic processes which goes far beyond the material presented in Sect 5.8 can be found in Papoulis [74].

Chapter 6

Basic Properties of Multivariable Feedback Systems

As we have seen in Chap. 2, the linearized model of the plant consists of a system of linear differential equations for its physical states. The inputs of the plant are the control u and the disturbance d and output is the variable to be controlled y, which is also the measurement. In classical control theory, the differential equations are used to calculate the transfer functions and the controller makes use only of the output variable. The states are no longer present and consequently they are not fed back.

Thus the classical approach does not make use of the information given in the states and for this reason it is quite natural to introduce a different kind of controller which is based on state feedback. This will be done in Sect. 6.1 for general multivariable control problems, because the basic ideas are the same for SISO and MIMO problems. We start with pure state feedback for systems without input signals, which corresponds to a stabilization problem. If a reference variable has to be tracked, the concept of state feedback has to be slightly generalized.

Of course, state feedback requires that all states can be measured. If this is not the case, the state vector has to be estimated. This can be done by the Luenberger observer, which is introduced in Sect. 6.2. The combination of state feedback and observer leads to a so-called observer-based controller.

In Sect. 6.3 we consider the concept of state feedback for SISO problems. The intuitive idea is that state feedback makes it possible to influence the dynamics of the feedback system to a rather large degree. As we will see, this is actually true: if the plant is controllable, the closed-loop poles can be prescribed. Moreover, a formula for calculating the controller gains from the closed-loop poles can be given. Thus, it only remains to position the poles such that the feedback system has certain desired performance and robust stability properties. It is quite interesting to see that the Nyquist criterion combined with some modern, but at this stage elementary, arguments leads to rules for how the poles have to be positioned. Some of the arguments are quite similar to those given in Sect. 3.5. In Sect. 6.4, pole placement for MIMO systems is considered.

We are now ready to introduce in Sect. 6.5 the general framework for MIMO feedback systems. The plant **P** has again two inputs, namely a vector **w**, which consists of the disturbances and a reference signal and the control **u**. The outputs are the variable to be controlled **z** and the measurement **y**, which is input of the controller. The controller **K** is normally a dynamic system (but may be constant, as a special case) and it is easy to see that the classical PID controller together with its generalizations as well as state feedback (with or without observer) are special cases of this approach. In later applications, the plant may be even more general because it is allowed to contain weights for performance and robustness. Conceptually, the analysis follows exactly the lines introduced in Chap. 3 for SISO systems. This means in particular that well-posedness and internal stability have to be adequately defined and characterized for MIMO systems. The chapter ends with a version of the Nyquist criterion that is applicable to the general feedback loop.

6.1 Controllers with State Feedback

We start with the following state space representation for the dynamics of a MIMO plant:

$$\dot{\mathbf{x}} = \mathbf{A}\mathbf{x} + \mathbf{B}_1\mathbf{d} + \mathbf{B}_2\mathbf{u}\,. \tag{6.1}$$

The inputs are the disturbance $\mathbf{d}$ and the control $\mathbf{u}$. Internally, the plant is described by the state vector $\mathbf{x}$. As mentioned in the introduction to this chapter, we want to consider controllers which use the whole state vector $\mathbf{x}$. Thus the measurement is described by the equation $\mathbf{y} = \mathbf{x}$.

We now introduce a controller of the form

$$\mathbf{u} = \mathbf{F}\,\mathbf{x} \tag{6.2}$$

with a suitably dimensioned matrix $\mathbf{F} = (f_{ij})$. In the special case of a single control, $\mathbf{F}$ is a row and we can write

$$u = f_1 x_1 + \ldots + f_n x_n\,.$$

In the MIMO case, such an equation holds for each control. This kind of control is said to be **control by state feedback**. Of course, state feedback only makes sense when all states can be measured, and therefore we assume here that this is actually the case. If a measurement error (or noise) $\mathbf{n}$ is taken into account, the measurement equation is

$$\mathbf{y} = \mathbf{x}_m = \mathbf{x} + \mathbf{n}\,.$$

Figure 6.1 shows a closed-loop system with state feedback. For this system, no reference signal is injected and, moreover, no output is considered. Thus, the con-

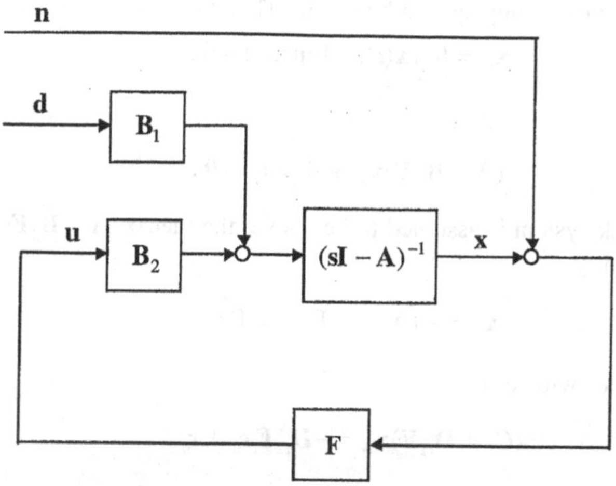

Fig. 6.1. State feedback

troller is only used to improve the dynamics of the plant, which is of particular importance if the plant is unstable. The question of under which conditions an instable plant can be stabilized by the feedback law (6.2) and how in this case $\mathbf{F}$ has to be chosen will be discussed in Sects. 6.3 and 6.4. Note that Fig. 6.1 even makes sense when no disturbance and no noise are present. Then the only input to the system is the initial value $\mathbf{x}(0) = \mathbf{x}_0$ (which is not shown in the diagram).

Tracking a reference signal. We do not take into account the noise and suppose that the variable to be controlled is defined by the output equation

$$\mathbf{z} = \mathbf{C}_1\mathbf{x} + \mathbf{D}_{11}\mathbf{d} + \mathbf{D}_{12}\mathbf{u}\,. \tag{6.3}$$

Let an additional task of the controller be that $\mathbf{z}$ tracks a given reference signal $\mathbf{r}$. We try to achieve this by the following approach:

$$\mathbf{u} = \mathbf{F}\mathbf{x} + \mathbf{F}_1\mathbf{r}\,. \tag{6.4}$$

Here, the matrix $\mathbf{F}_1$ has to be chosen such that at least for constant inputs in the stationary case the output and reference signal coincide. We suppose that there are as many reference variables as control variables. Then the matrix $\mathbf{F}_1$ is quadratic.

The state space representation of the closed-loop system is here given by

$$\dot{\mathbf{x}} = (\mathbf{A} + \mathbf{B}_2\mathbf{F})\mathbf{x} + \mathbf{B}_2\mathbf{F}_1\mathbf{r} + \mathbf{B}_1\mathbf{d}$$
$$\mathbf{z} = (\mathbf{C}_1 + \mathbf{D}_{12}\mathbf{F})\mathbf{x} + \mathbf{D}_{12}\,\mathbf{F}_1\mathbf{r} + \mathbf{D}_{11}\mathbf{d}\,.$$

Suppose now that no disturbance occurs ($\mathbf{d} = \mathbf{0}$) and that the reference signal is constant: $\mathbf{r} = \mathbf{r}_0$. The matrix $\mathbf{F}_1$ has to be chosen such that

$$\mathbf{z}_\infty = \lim_{t\to\infty} \mathbf{z}(t) = \mathbf{r}_0\,. \tag{6.5}$$

Moreover, let the closed-loop system be stable. Then there is a vector $\mathbf{x}_\infty$ with

$$\mathbf{x}_\infty = \lim_{t\to\infty} \mathbf{x}(t), \quad \lim_{t\to\infty} \dot{\mathbf{x}}(t) = \mathbf{0},$$

and we get

$$(\mathbf{A} + \mathbf{B}_2\mathbf{F})\mathbf{x}_\infty + \mathbf{B}_2\mathbf{F}_1\mathbf{r}_0 = \mathbf{0}.$$

Since the feedback system is assumed to be stable, the matrix $\mathbf{A} + \mathbf{B}_2\mathbf{F}$ is invertible and consequently

$$\mathbf{x}_\infty = -(\mathbf{A} + \mathbf{B}_2\mathbf{F})^{-1}\mathbf{B}_2\mathbf{F}_1\mathbf{r}_0.$$

Hence $\mathbf{z}_\infty$ coincides with $\mathbf{r}_0$ if

$$(\mathbf{C}_1 + \mathbf{D}_{12}\mathbf{F})\mathbf{x}_\infty = -\mathbf{D}_{12}\mathbf{F}_1\mathbf{r}_0 + \mathbf{r}_0.$$

Thus, we must have

$$(-(\mathbf{C}_1 + \mathbf{D}_{12}\mathbf{F})(\mathbf{A} + \mathbf{B}_2\mathbf{F})^{-1}\mathbf{B}_2 + \mathbf{D}_{12})\mathbf{F}_1\mathbf{r}_0 = \mathbf{r}_0.$$

This equation is satisfied if $\mathbf{F}_1$ is chosen as

$$\mathbf{F}_1 = (-(\mathbf{C}_1 + \mathbf{D}_{12}\mathbf{F})(\mathbf{A} + \mathbf{B}_2\mathbf{F})^{-1}\mathbf{B}_2 + \mathbf{D}_{12})^{-1}, \tag{6.6}$$

where the invertibility of the corresponding matrix will be assumed. Thus we have shown the following result.

Let the feedback law be given by (6.4) with a gain matrix $\mathbf{F}$ *which stabilizes the feedback system. Moreover, suppose* $\mathbf{F}_1$ *is defined by (6.6). Then the stationary control error* $\mathbf{r}_0 - \mathbf{z}_\infty$ *vanishes for a constant reference signal* $\mathbf{r}_0$.

It has to be noted that even a constant disturbance may cause a stationary control error. Inaccurate plant parameters also contribute to the control error. For these reasons, the above approach can only be recommended in special cases.

Integral control. The disadvantages of the controller (6.4) concerning the control error can be avoided by introducing an integrator into the feedback law. To this end, we define an additional output variable $\mathbf{h}_p$ by

$$\mathbf{h}_p = \mathbf{r} - \mathbf{z}$$

and use a controller of the form

$$\mathbf{u} = \mathbf{F}\mathbf{x} + \mathbf{F}_1\mathbf{h}, \quad \mathbf{h}(t) = \int_0^t \mathbf{h}_p(\tau)d\tau \tag{6.7}$$

with a suitably dimensioned matrix $\mathbf{F}_1$. The basic idea is now quite simple: Since

$\mathbf{h}$ is a state of the feedback system, the function $\mathbf{h}(t)$ becomes constant for $t \to \infty$, if the feedback system is stable and if the reference signal and the disturbance are constant. Consequently, $\dot{\mathbf{h}}(t)$ tends to $\mathbf{0}$ as $t \to \infty$ and (6.5) holds.

The output $\mathbf{z}$ now appears as an additional measurement with noise denoted by $\mathbf{n}_1$. For $\mathbf{D}_{11} = \mathbf{0}, \mathbf{D}_{12} = \mathbf{0}$, no extra sensor is required since $\mathbf{z}$ can be calculated from the measurement of $\mathbf{x}$: $\mathbf{z} = \mathbf{C}_1\mathbf{x}$, and the noise is given by $\mathbf{n}_1 = \mathbf{C}_1\mathbf{n}$. With noise, the new measurement equation is

$$\mathbf{h}_{pm} = \mathbf{r} - \mathbf{z} + \mathbf{n}_1 = \mathbf{r} - \mathbf{C}_1\mathbf{x} - \mathbf{D}_{11}\mathbf{d} - \mathbf{D}_{12}\mathbf{u} + \mathbf{n}_1, \tag{6.8}$$

and in (6.7) $\mathbf{x}, \mathbf{h}_p$ have to be replaced by $\mathbf{x}_m, \mathbf{h}_{pm}$ if noise is taken into account. The integral in (6.7) is then denoted by $\mathbf{h}_m$.

Although the integration is clearly a part of the controller, for design purposes it may by helpful to view $\mathbf{h}_m$ as a state of the plant, i.e. to extend the state vector $\mathbf{x}$ by $\mathbf{h}_m$. Then again, we get feedback of the full state vector with constant gains:

$$\mathbf{u} = \mathbf{F}(\mathbf{x} + \mathbf{n}) + \mathbf{F}_1\mathbf{h}_m. \tag{6.9}$$

We want to derive a state space representation of the feedback loop in this case. To this end, let

$$\mathbf{x}_e = \begin{pmatrix} \mathbf{x} \\ \mathbf{h}_m \end{pmatrix}, \quad \mathbf{A}_e = \begin{pmatrix} \mathbf{A} & \mathbf{0} \\ -\mathbf{C}_1 & \mathbf{0} \end{pmatrix}, \quad \mathbf{C}_{1e} = \begin{pmatrix} \mathbf{C}_1 & \mathbf{0} \end{pmatrix},$$

$$\mathbf{B}_{2e} = \begin{pmatrix} \mathbf{B}_2 \\ -\mathbf{D}_{12} \end{pmatrix}, \quad \mathbf{B}_{1e} = \begin{pmatrix} \mathbf{B}_1 \\ -\mathbf{D}_{11} \end{pmatrix}, \quad \mathbf{B}_s = \begin{pmatrix} \mathbf{0} \\ \mathbf{I} \end{pmatrix}.$$

Then the dynamics of the extended plant is given by

$$\begin{aligned} \dot{\mathbf{x}}_e &= \mathbf{A}_e\mathbf{x}_e + \mathbf{B}_{1e}\mathbf{d} + \mathbf{B}_s\mathbf{n}_1 + \mathbf{B}_s\mathbf{r} + \mathbf{B}_{2e}\mathbf{u} \\ \mathbf{z} &= \mathbf{C}_{1e}\mathbf{x}_e + \mathbf{D}_{11}\mathbf{d} + \mathbf{D}_{12}\mathbf{u}. \end{aligned} \tag{6.10}$$

Equation (6.9) can be written in the form

$$\mathbf{u} = \mathbf{F}_e\mathbf{x}_e + \mathbf{F}\mathbf{n} \quad \text{with} \quad \mathbf{F}_e = (\mathbf{F} \quad \mathbf{F}_1),$$

and the closed-loop system is immediately seen to be

$$\begin{aligned} \dot{\mathbf{x}}_e &= (\mathbf{A}_e + \mathbf{B}_{2e}\mathbf{F}_e)\mathbf{x}_e + \mathbf{B}_{1e}\mathbf{d} + \mathbf{B}_{2e}\mathbf{F}\mathbf{n} + \mathbf{B}_s\mathbf{n}_1 + \mathbf{B}_s\mathbf{r} \\ \mathbf{z} &= (\mathbf{C}_{1e} + \mathbf{D}_{12}\mathbf{F}_e)\mathbf{x}_e + \mathbf{D}_{12}\mathbf{F}\mathbf{n} + \mathbf{D}_{11}\mathbf{d}. \end{aligned} \tag{6.11}$$

If it is stable and if all inputs are constant, $\dot{\mathbf{x}}_e(t)$ and, in particular, $\mathbf{h}_{pm}(t)$ tend to $\mathbf{0}$ for $t \to \infty$. Hence the following statement is true.

If state feedback is extended by integral control as given in (6.9), the stationary control error is given by

$$\lim_{t\to\infty} \mathbf{z}(t) - \mathbf{r}_0 = \mathbf{n}_{10}$$

if the feedback system is stable and if the disturbance $\mathbf{d}_0$*, the reference signal* $\mathbf{r}_0$ *and the measurement error* $\mathbf{n}_{10}$ *are constant. In particular, this error vanishes for a vanishing error* $\mathbf{n}_{10}$.

Invariant zeros. The normal rank of the Rosenbrock matrix for the closed-loop system (6.10) (with $\mathbf{r}$ as input and $\mathbf{z}$ as output) can be calculated as follows:

$$\begin{aligned}
\text{normalrank}\,\mathbf{R}(s) &= \text{normalrank}\begin{pmatrix} s\mathbf{I}-\mathbf{A}_e-\mathbf{B}_{2e}\mathbf{F}_e & -\mathbf{B}_s \\ \mathbf{C}_{1e}+\mathbf{D}_{12}\mathbf{F}_e & \mathbf{0} \end{pmatrix} \\
&= \text{normalrank}\begin{pmatrix} s\mathbf{I}_1-\mathbf{A}-\mathbf{B}_2\mathbf{F} & -\mathbf{B}_2\mathbf{F}_1 & \mathbf{0} \\ \mathbf{C}_1+\mathbf{D}_{12}\mathbf{F} & s\mathbf{I}_2+\mathbf{D}_{12}\mathbf{F}_1 & -\mathbf{I}_2 \\ \mathbf{C}_1+\mathbf{D}_{12}\mathbf{F} & \mathbf{D}_{12}\mathbf{F}_1 & \mathbf{0} \end{pmatrix} \\
&= \text{rank}\,(\mathbf{I}_2)+\text{normalrank}\begin{pmatrix} s\mathbf{I}_1-\mathbf{A}-\mathbf{B}_2\mathbf{F} & -\mathbf{B}_2\mathbf{F}_1 \\ \mathbf{C}_1+\mathbf{D}_{12}\mathbf{F} & \mathbf{D}_{12}\mathbf{F}_1 \end{pmatrix} \\
&= \text{rank}\,(\mathbf{I}_2)+\text{normalrank}\begin{pmatrix} s\mathbf{I}_1-\mathbf{A} & -\mathbf{B}_2 \\ \mathbf{C}_1 & \mathbf{D}_{12} \end{pmatrix}\begin{pmatrix} \mathbf{I}_1 & \mathbf{0} \\ \mathbf{F} & \mathbf{F}_1 \end{pmatrix},
\end{aligned}$$

where $\mathbf{I}_1$ and $\mathbf{I}_2$ are suitably dimensioned identity matrices. If $\mathbf{F}_1$ is invertible, the following holds.

The plant (6.1) and (6.3) (with $\mathbf{u}$ *as input) has the same invariant zeros as the corresponding closed-loop system (with* $\mathbf{r}$ *as input) if state feedback combined with integral control of the form (6.9) is used.*

Remark. The equations for the extended plant can be written in a more compact form by defining

$$\tilde{\mathbf{B}}_1 = (\mathbf{B}_{1e} \quad \mathbf{0} \quad \mathbf{B}_s \quad \mathbf{B}_s),\quad \tilde{\mathbf{D}}_{11} = (\mathbf{D}_{11} \quad \mathbf{0} \quad \mathbf{0} \quad \mathbf{0}),\quad \tilde{\mathbf{D}}_{21} = \begin{pmatrix} \mathbf{0} & \mathbf{I} & \mathbf{0} & \mathbf{0} \\ \mathbf{0} & \mathbf{0} & \mathbf{0} & \mathbf{0} \end{pmatrix}.$$

The inputs $\mathbf{d}, \mathbf{n}, \mathbf{n}_1, \mathbf{r}$ are combined into one single vector $\mathbf{w}$. The plant inputs are then $\mathbf{u}$ and $\mathbf{w}$ and the first equation of (6.10) can be written as

$$\dot{\mathbf{x}}_e = \mathbf{A}_e\mathbf{x}_e + \tilde{\mathbf{B}}_1\mathbf{w} + \mathbf{B}_{2e}\mathbf{u}\,.$$

The plant output consists of the two quantities

$$\begin{aligned}
\mathbf{z} &= \mathbf{C}_{1e}\mathbf{x}_e + \tilde{\mathbf{D}}_{11}\mathbf{w} + \mathbf{D}_{12}\mathbf{u} \\
\mathbf{y} &= \mathbf{x}_e + \tilde{\mathbf{D}}_{21}\mathbf{w}.
\end{aligned}$$

We will see in Sect. 6.4 that the structure of the feedback system can now be viewed as a special case of the general feedback system introduced there.

6.2 Estimation of the State Vector

6.2.1 Luenberger Observer and Observer-Based Controllers

As we have seen, state feedback requires the measurement of all state variables. If it is not possible or not desirable for technical reasons to measure the whole state vector, the question arises whether the state vector $\mathbf{x}$ can be estimated from a certain measured output vector $\mathbf{y}$. To be more specific, consider the system

$$\begin{aligned}\dot{\mathbf{x}} &= \mathbf{Ax} + \mathbf{Bu}\\ \mathbf{y} &= \mathbf{Cx} + \mathbf{Du}.\end{aligned} \tag{6.12}$$

Disturbances and noise will be neglected for the moment. The initial value $\mathbf{x}(0) = \mathbf{x}_0$ is assumed to be unknown. We try to get an estimate $\tilde{\mathbf{x}}$ of the state vector $\mathbf{x}$ by the following approach:

$$\dot{\tilde{\mathbf{x}}} = \mathbf{A}\tilde{\mathbf{x}} + \mathbf{L}(\mathbf{C}\tilde{\mathbf{x}} + \mathbf{Du} - \mathbf{y}) + \mathbf{Bu}\,. \tag{6.13}$$

As it is seen, an additional term of the form "gain multiplied by the output error" is introduced. Here, $\mathbf{L}$ is an $n \times l$ - matrix which has still to be specified. The task is to determine $\mathbf{L}$ such that for all initial values $\mathbf{x}_0$ and all controls $\mathbf{u}$ the relationship

$$\lim_{t\to\infty}(\tilde{\mathbf{x}}(t) - \mathbf{x}(t)) = \mathbf{0} \tag{6.14}$$

holds. A system of the form (6.13) with this property is denoted as a **Luenberger observer**. It can equivalently be written in the form

$$\dot{\tilde{\mathbf{x}}} = (\mathbf{A} + \mathbf{L}\,\mathbf{C})\tilde{\mathbf{x}} + (\mathbf{B} + \mathbf{LD})\mathbf{u} - \mathbf{Ly}\,. \tag{6.15}$$

Theorem 6.2.1. *Let $\mathbf{L}$ be chosen such that the matrix $\mathbf{A} + \mathbf{LC}$ is stable. Then the system (6.15) is a Luenberger observer. A matrix $\mathbf{L}$ with this property exists if and only if the system $(\mathbf{C}, \mathbf{A})$ is detectable.*

Proof. Let $(\mathbf{C}, \mathbf{A})$ be detectable. The estimation error is given by $\mathbf{e} = \mathbf{x} - \tilde{\mathbf{x}}$. It fulfills the equation

$$\begin{aligned}\dot{\mathbf{e}} &= \mathbf{Ax} - \mathbf{A}\tilde{\mathbf{x}} - \mathbf{L}(\mathbf{C}\tilde{\mathbf{x}} + \mathbf{Du} - \mathbf{y})\\ &= \mathbf{Ax} - \mathbf{A}\tilde{\mathbf{x}} + \mathbf{L}(\mathbf{Cx} - \mathbf{C}\tilde{\mathbf{x}}),\end{aligned}$$

and therefore we also have

$$\dot{\mathbf{e}} = (\mathbf{A} + \mathbf{L}\,\mathbf{C})\mathbf{e}. \tag{6.16}$$

If $\mathbf{L}$ is chosen such that the matrix $\mathbf{A} + \mathbf{LC}$ is stable, then (6.14) holds. The problem is now to show that under the given assumption a matrix $\mathbf{L}$ with this property exists. We return to this point in Sect. 6.4, where the general pole placement problem is discussed.

If $(\mathbf{C}, \mathbf{A})$ is not detectable, then it is easily seen from the Theorem 5.4.2 that (6.14) cannot hold for all initial values. The details are left to the reader.

If state feedback requires an observer, the state feedback controller has to use the estimate $\tilde{\mathbf{x}}$ instead of the actual state $\mathbf{x}$:

$$\mathbf{u} = \mathbf{F}\tilde{\mathbf{x}}. \tag{6.17}$$

If $\mathbf{u}$ is eliminated from the observer equation (6.15), the relationship

$$\dot{\tilde{\mathbf{x}}} = (\mathbf{A} + \mathbf{BF} + \mathbf{LC} + \mathbf{LDF})\tilde{\mathbf{x}} - \mathbf{Ly}. \tag{6.18}$$

follows. The equations (6.17) and (6.18) are a state space representation of the controller, which can now be written as:

$$\hat{\mathbf{u}} = \mathbf{K}(s)\hat{\mathbf{y}},$$

$$\mathbf{K}(s) = \left(\begin{array}{c|c} \mathbf{A} + \mathbf{BF} + \mathbf{LC} + \mathbf{LDF} & -\mathbf{L} \\ \hline \mathbf{F} & \mathbf{0} \end{array}\right). \tag{6.19}$$

A controller of this kind is said to be **observer-based**.

Next we want to take a closer look at the dynamics of a feedback system with an observer-based controller. It is possible to eliminate the control from the equations (6.17) and (6.12). This is most easily done by replacing the estimate $\tilde{\mathbf{x}}$ by the estimation error $\mathbf{e}$. This yields

$$\begin{aligned} \dot{\mathbf{x}} &= \mathbf{Ax} + \mathbf{BF}(\mathbf{x} - \mathbf{e}) \\ &= (\mathbf{A} + \mathbf{BF})\mathbf{x} - \mathbf{BFe}. \end{aligned}$$

The equation for the error is again (6.16). Thus, the following result is shown.

Theorem 6.2.2 (Separation Theorem). *A state space representation of the feedback system which results for the plant (6.12) and the controller (6.19) is given by*

$$\begin{pmatrix} \dot{\mathbf{x}} \\ \dot{\mathbf{e}} \end{pmatrix} = \begin{pmatrix} \mathbf{A} + \mathbf{BF} & -\mathbf{BF} \\ \mathbf{0} & \mathbf{A} + \mathbf{LC} \end{pmatrix} \begin{pmatrix} \mathbf{x} \\ \mathbf{e} \end{pmatrix}. \tag{6.20}$$

Thus the eigenvalues of a closed-loop system with an observer-based controller are exactly the eigenvalues of the matrices $\mathbf{A} + \mathbf{BF}$ *and* $\mathbf{A} + \mathbf{LC}$, *i.e. the eigenvalues of the feedback system with state feedback and the eigenvalues of the observer.*

Hence, the state-feedback controller and the observer may be designed independently. The name of the theorem is based on this result.

Note that the feedback systems considered in this section have no external inputs and the controller has the sole task of improving the plant dynamics. In order to introduce the idea of the Luenberger observer this is sufficient. If a reference signal or disturbances are present, the same observer-based controller may be used and the system matrix for the closed-loop system remains the same.

6.2.2 Reduced Observer

We pose the question of whether it is possible to construct an observer whose order is lower then the order of the system if $l<n$. For the sake of simplicity, let $\mathbf{D}=\mathbf{0}$ and $\operatorname{rank}\mathbf{C}=l$. The idea is now as follows. The relationship $\mathbf{y}=\mathbf{C}\mathbf{x}$ for the output is a linear system of l equations for the n components of the state vector $\mathbf{x}$ which can be rearranged in such a way that l components of $\mathbf{x}$ can be uniquely expressed by the remaining $n-l$ ones. Then it is sufficient to design an observer for the state vector consisting of these $n-l$ components.

We want to describe this concept in detail. Since the rank of $\mathbf{C}$ is l by assumption, there are l linearly independent columns of $\mathbf{C}$. Suppose that the columns of $\mathbf{C}$ are interchanged such that the first l columns of the resulting matrix are linearly independent. Hereby, the column with number i of the original matrix becomes the column with number j_i of the new matrix. We define now a permutation matrix $\mathbf{P}=(p_{ik})$ by

$$p_{ik}=\begin{cases}1, & \text{if } j_i=k,\\ 0, & \text{if } j_i\neq k.\end{cases}$$

Then it is possible to write

$$\mathbf{C}\mathbf{P}=(\mathbf{C}_1 \quad \mathbf{C}_2),$$

where $\mathbf{C}_1$ is an invertible $l\times l$ - matrix. Since $\mathbf{P}$ is a permutation matrix, the equation $\mathbf{P}\mathbf{P}^T=\mathbf{I}$ holds and consequently the inverse matrix is given by $\mathbf{P}^{-1}=\mathbf{P}^T$. Let now

$$\begin{pmatrix}\mathbf{x}_1\\ \mathbf{x}_2\end{pmatrix}=\mathbf{P}^T\mathbf{x}$$

and $\tilde{\mathbf{A}}=\mathbf{P}^T\mathbf{A}\mathbf{P}$, $\tilde{\mathbf{B}}=\mathbf{P}^T\mathbf{B}$. Then it follows that (cf. (5.24))

$$\begin{pmatrix}\dot{\mathbf{x}}_1\\ \dot{\mathbf{x}}_2\end{pmatrix}=\tilde{\mathbf{A}}\begin{pmatrix}\mathbf{x}_1\\ \mathbf{x}_2\end{pmatrix}+\tilde{\mathbf{B}}\mathbf{u}$$

$$\mathbf{y}=\mathbf{C}_1\mathbf{x}_1+\mathbf{C}_2\mathbf{x}_2.$$

The vector $\mathbf{x}_1$ can be calculated from the last equation:

$$\mathbf{x}_1 = \mathbf{C}_1^{-1}(\mathbf{y} - \mathbf{C}_2\mathbf{x}_2). \tag{6.21}$$

This implies

$$\begin{pmatrix} \mathbf{x}_1 \\ \mathbf{x}_2 \end{pmatrix} = \begin{pmatrix} \mathbf{C}_1^{-1} & -\mathbf{C}_1^{-1}\mathbf{C}_2 \\ \mathbf{0} & \mathbf{I} \end{pmatrix} \begin{pmatrix} \mathbf{y} \\ \mathbf{x}_2 \end{pmatrix}.$$

With the matrix

$$\mathbf{Q} = \begin{pmatrix} \mathbf{C}_1^{-1} & -\mathbf{C}_1^{-1}\mathbf{C}_2 \\ \mathbf{0} & \mathbf{I} \end{pmatrix}^{-1}$$

it is possible to write

$$\begin{pmatrix} \mathbf{A}_{11} & \mathbf{A}_{12} \\ \mathbf{A}_{21} & \mathbf{A}_{22} \end{pmatrix} = \mathbf{Q}\tilde{\mathbf{A}}\mathbf{Q}^{-1}, \qquad \begin{pmatrix} \mathbf{B}_1 \\ \mathbf{B}_2 \end{pmatrix} = \mathbf{Q}\tilde{\mathbf{B}}. \tag{6.22}$$

Thus, we get

$$\dot{\mathbf{y}} = \mathbf{A}_{11}\mathbf{y} + \mathbf{A}_{12}\mathbf{x}_2 + \mathbf{B}_1\mathbf{u} \tag{6.23}$$

$$\dot{\mathbf{x}}_2 = \mathbf{A}_{21}\mathbf{y} + \mathbf{A}_{22}\mathbf{x}_2 + \mathbf{B}_2\mathbf{u}. \tag{6.24}$$

Hence, the original system is equivalently reformulated such that the output $\mathbf{y}$ is part of the state vector of the new system. The remaining task is to construct an observer for the vector $\mathbf{x}_2$ of (6.24). Because of (6.23), the vector

$$\bar{\mathbf{y}} = \dot{\mathbf{y}} - \mathbf{A}_{11}\mathbf{y} - \mathbf{B}_1\mathbf{u}$$

satisfies

$$\bar{\mathbf{y}} = \mathbf{A}_{12}\mathbf{x}_2.$$

This can be interpreted as the output equation for $\mathbf{x}_2$. Therefore the state equation for the observer is

$$\begin{aligned} \dot{\tilde{\mathbf{x}}}_2 &= \mathbf{A}_{22}\tilde{\mathbf{x}}_2 + \mathbf{A}_{21}\mathbf{y} + \mathbf{B}_2\mathbf{u} + \mathbf{L}(\mathbf{A}_{12}\tilde{\mathbf{x}}_2 - \bar{\mathbf{y}}) \\ &= (\mathbf{A}_{22} + \mathbf{L}\mathbf{A}_{12})\tilde{\mathbf{x}}_2 + \mathbf{A}_{21}\mathbf{y} + \mathbf{B}_2\mathbf{u} + \mathbf{L}(\mathbf{A}_{11}\mathbf{y} + \mathbf{B}_1\mathbf{u} - \dot{\mathbf{y}}). \end{aligned}$$

The derivative $\dot{\mathbf{y}}$ can be eliminated by the following transformation of variables:

$$\bar{\mathbf{x}}_2 = \tilde{\mathbf{x}}_2 + \mathbf{L}\mathbf{y}.$$

Hence, the equation for the observer can be rewritten as

$$\begin{aligned} \dot{\bar{\mathbf{x}}}_2 &= (\mathbf{A}_{22} + \mathbf{L}\mathbf{A}_{12})(\bar{\mathbf{x}}_2 - \mathbf{L}\mathbf{y}) + \mathbf{A}_{21}\mathbf{y} + \mathbf{B}_2\mathbf{u} + \mathbf{L}(\mathbf{A}_{11}\mathbf{y} + \mathbf{B}_1\mathbf{u}) \\ &= (\mathbf{A}_{22} + \mathbf{L}\mathbf{A}_{12})\bar{\mathbf{x}}_2 + (-(\mathbf{A}_{22} + \mathbf{L}\mathbf{A}_{12})\mathbf{L} + \mathbf{A}_{21} + \mathbf{L}\mathbf{A}_{11})\mathbf{y} + (\mathbf{B}_2 + \mathbf{L}\mathbf{B}_1)\mathbf{u}. \end{aligned}$$

We introduce as an abbreviation the matrices

$$\mathbf{L}_r = -(\mathbf{A}_{22} + \mathbf{L}\mathbf{A}_{12})\mathbf{L} + \mathbf{A}_{21} + \mathbf{L}\mathbf{A}_{11} \tag{6.25}$$

$$\mathbf{B}_r = \mathbf{B}_2 + \mathbf{L}\mathbf{B}_1 . \tag{6.26}$$

Then the desired equations for the observer are given by

$$\dot{\overline{\mathbf{x}}}_2 = (\mathbf{A}_{22} + \mathbf{L}\mathbf{A}_{12})\overline{\mathbf{x}}_2 + \mathbf{L}_r\mathbf{y} + \mathbf{B}_r\mathbf{u} \tag{6.27}$$

$$\tilde{\mathbf{x}}_2 = \overline{\mathbf{x}}_2 - \mathbf{L}\mathbf{y} . \tag{6.28}$$

This system is called a **reduced observer**.

Theorem 6.2.3. *Let the system* $(\mathbf{C}, \mathbf{A})$ *be detectable. Then there is a matrix* $\mathbf{L}$ *such that* $\mathbf{A}_{22} + \mathbf{L}\mathbf{A}_{12}$ *is stable (with the partition introduced in (6.22)). Let* $\tilde{\mathbf{x}}_2$ *be defined as in (6.25) – (6.28) and put*

$$\tilde{\mathbf{x}}_1 = \mathbf{C}_1^{-1}(\mathbf{y} - \mathbf{C}_2\tilde{\mathbf{x}}_2) .$$

Then the estimation errors $\mathbf{e}_1 = \mathbf{x}_1 - \tilde{\mathbf{x}}_1$ *and* $\mathbf{e}_2 = \mathbf{x}_2 - \tilde{\mathbf{x}}_2$ *satisfy*

$$\lim_{t\to\infty} \mathbf{e}_i(t) = \mathbf{0} \quad \textit{for } i = 1, 2 .$$

Here, $\mathbf{x}_1$, $\mathbf{x}_2$ *are defined by (6.21).*

Proof. Using the PHB test, it is easy to see that the detectability of $(\mathbf{C}, \mathbf{A})$ implies the detectability of $(\mathbf{A}_{22}, \mathbf{A}_{12})$. Thus, if the pair $(\mathbf{C}, \mathbf{A})$ is detectable, there exists a matrix $\mathbf{L}$ such that $\mathbf{A}_{22} + \mathbf{L}\mathbf{A}_{12}$ is stable (compare the proof of Theorem 6.2.1).

Taking into account (6.24) and (6.25), the estimation error $\mathbf{e}_2$ is seen to satisfy

$$\begin{aligned}
\dot{\mathbf{e}}_2 &= \dot{\mathbf{x}}_2 - \dot{\overline{\mathbf{x}}}_2 + \mathbf{L}\dot{\mathbf{y}} \\
&= \mathbf{A}_{21}\mathbf{y} + \mathbf{A}_{22}\mathbf{x}_2 + \mathbf{B}_2\mathbf{u} - (\mathbf{A}_{22} + \mathbf{L}\mathbf{A}_{12})\overline{\mathbf{x}}_2 - \mathbf{L}_r\mathbf{y} - \mathbf{B}_r\mathbf{u} \\
&\quad + \mathbf{L}(\mathbf{A}_{11}\mathbf{y} + \mathbf{A}_{12}\mathbf{x}_2 + \mathbf{B}_1\mathbf{u}) \\
&= (\mathbf{A}_{22} + \mathbf{L}\mathbf{A}_{12})\mathbf{x}_2 - (\mathbf{A}_{22} + \mathbf{L}\mathbf{A}_{12})\overline{\mathbf{x}}_2 + (\mathbf{A}_{22} + \mathbf{L}\mathbf{A}_{12})\mathbf{L}\mathbf{y} \\
&= (\mathbf{A}_{22} + \mathbf{L}\mathbf{A}_{12})\mathbf{e}_2 .
\end{aligned}$$

This implies that $\mathbf{e}_2(t) \to \mathbf{0}$ as $t \to \infty$. This fact and (6.21 lead immediately to the second one of the asserted relationships for the limits.

The controller can be written in the form

$$\mathbf{u} = \mathbf{F}\mathbf{x} = \mathbf{F}\,\mathbf{P}\begin{pmatrix}\tilde{\mathbf{x}}_1 \\ \tilde{\mathbf{x}}_2\end{pmatrix} = (\mathbf{F}_1 \quad \mathbf{F}_2)\begin{pmatrix}\tilde{\mathbf{x}}_1 \\ \tilde{\mathbf{x}}_2\end{pmatrix}$$

with a suitable partitioning of $\mathbf{F}\,\mathbf{P}$. The simplest way is to feed back the variable $\tilde{\mathbf{x}}_1$ instead of the output $\mathbf{y}$. Defining

$$\tilde{\mathbf{F}}_1 = \mathbf{F}_1\mathbf{C}_1^{-1}, \quad \tilde{\mathbf{F}}_2 = \mathbf{F}_2 - \mathbf{F}_1\mathbf{C}_1^{-1}\mathbf{C}_2,$$

it is possible to write

$$\mathbf{u} = \tilde{\mathbf{F}}_1\mathbf{y} + \tilde{\mathbf{F}}_2\tilde{\mathbf{x}}_2\,. \tag{6.29}$$

It is easy to verify that a state space representation of the closed-loop system is given by

$$\begin{pmatrix}\dot{\mathbf{y}}\\ \dot{\mathbf{x}}_2\\ \dot{\mathbf{e}}_2\end{pmatrix} = \begin{pmatrix}\mathbf{A}_{11}+\mathbf{B}_1\tilde{\mathbf{F}}_1 & \mathbf{A}_{12}+\mathbf{B}_1\tilde{\mathbf{F}}_2 & -\mathbf{B}_1\tilde{\mathbf{F}}_2\\ \mathbf{A}_{21}+\mathbf{B}_2\tilde{\mathbf{F}}_1 & \mathbf{A}_{22}+\mathbf{B}_2\tilde{\mathbf{F}}_2 & -\mathbf{B}_2\tilde{\mathbf{F}}_2\\ \mathbf{0} & \mathbf{0} & \mathbf{A}_{22}+\mathbf{L}\mathbf{A}_{12}\end{pmatrix}\begin{pmatrix}\mathbf{y}\\ \mathbf{x}_2\\ \mathbf{e}_2\end{pmatrix}.$$

Hence, the following holds.

The separation principle holds (in an obvious reformulation) also for the reduced observer.

6.3 Pole Placement for SISO Systems

6.3.1 Formula for the Feedback Gains

We start with a state space representation $(\mathbf{C},\mathbf{A},\mathbf{b})$ of a SISO plant which has to be controlled by state feedback and ask, whether it is possible to find a gain matrix $\mathbf{F}$ such that the closed-loop system has its poles at prescribed positions. Here, the system may be the original system or the extended system as described in Sect. 6.1. Since the gain matrix has n coefficients and since the feedback system has n poles, it may be supposed that the problem has a unique solution, at least under some reasonable assumptions.

The closed-loop poles determine the coefficients $\alpha_0,\alpha_1,\ldots,\alpha_{n-1}$ of the characteristic polynomial

$$p(s) = \det(s\mathbf{I} - (\mathbf{A}+\mathbf{b}\mathbf{F})) = s^n + \alpha_{n-1}s^{n-1} + \ldots + \alpha_1 s + \alpha_0 \tag{6.30}$$

uniquely. Hence we have to find a matrix $\mathbf{F}$ such that (6.30) holds, where the coefficients $\alpha_0,\alpha_1,\ldots,\alpha_{n-1}$ are given. This problem is most easily solved if the system $(\mathbf{A},\mathbf{b})$ has a controller-canonical form, i.e. $(\mathbf{A},\mathbf{b}) = (\mathbf{A}_1,\mathbf{b}_1)$ with the notation of Sect. 5.3.1. We first solve the problem for this case. Denote the gain matrix by

$$\mathbf{F}_1 = (f_1 \quad \cdots \quad f_n)\,.$$

Then it is easy to see that

$$\det(s\mathbf{I}-(\mathbf{A}_1+\mathbf{b}_1\mathbf{F}_1))=\begin{pmatrix} s & -1 & & \\ 0 & \ddots & \ddots & \\ \vdots & & s & -1 \\ a_0-f_1 & \dots & a_{n-2}-f_{n-1} & s+a_{n-1}-f_n \end{pmatrix}$$
$$=s^n+(a_{n-1}-f_n)s^{n-1}+\ldots+(a_1-f_2)s+a_0-f_1 .$$

Thus, problem is solved for

$$f_i=a_{i-1}-\alpha_{i-1} \qquad i=1,\ldots,n .$$

We have (with $\mathbf{T}$ as in Sect. 5.3.1)

$$u=\mathbf{F}_1\mathbf{z}=\mathbf{F}_1\mathbf{T}\mathbf{x} .$$

Hence, the matrix $\mathbf{F}=\mathbf{F}_1\mathbf{T}$ satisfies

$$\det(s\mathbf{I}-(\mathbf{A}+\mathbf{bF}))=\det(s\mathbf{I}-(\mathbf{T}^{-1}\mathbf{A}_1\mathbf{T}+\mathbf{T}^{-1}\mathbf{b}\mathbf{F}_1\mathbf{T}))$$
$$=\det(s\mathbf{I}-(\mathbf{A}_1+\mathbf{b}\mathbf{F}_1)) ,$$

and with this choice for $\mathbf{F}$, the feedback loop with $(\mathbf{A},\mathbf{b})$ as a state space realization of the plant has, as desired, $q(s)$ as the characteristic polynomial.

The remaining task is to find a representation of $\mathbf{F}$ which can easily be computed if the system is not of controller-canonical form. Using the equation

$$(\mathbf{A}_1+\mathbf{b}_1\mathbf{F}_1)\mathbf{T}=\mathbf{TA}+\mathbf{b}_1\mathbf{F}$$

and noting that $\mathbf{F}$ is the last row of $\mathbf{b}_1\mathbf{F}$, we obtain (cf. (5.33)):

$$\mathbf{F}=-\mathbf{t}_n^T\mathbf{A}-\sum_{i=0}^{n-1}\alpha_i\mathbf{t}_{i+1}^T$$
$$=-\mathbf{t}_1^T\mathbf{A}^{n-1}\mathbf{A}-\sum_{i=0}^{n-1}\alpha_i\mathbf{t}_1^T\mathbf{A}^i$$
$$=-\mathbf{t}_1^T p(\mathbf{A}).$$

Thus, the following theorem is shown (cf. Theorem 5.3.2).

Theorem 6.3.1. *Suppose that the system* $(\mathbf{A},\mathbf{b})$ *is controllable and denote by* $\mathbf{t}_1^T$ *the last row of the inverse controllability matrix of* $(\mathbf{A},\mathbf{b})$. *Let the polynomial*

$$p(s)=s^n+\alpha_{n-1}s^{n-1}+\ldots+\alpha_1 s+\alpha_0$$

be arbitrarily chosen. Then the equation

$$\det(s\mathbf{I} - (\mathbf{A} + \mathbf{bF})) = p(s),$$

holds if the matrix $\mathbf{F}$ *is defined by*

$$\mathbf{F} = -\mathbf{t}_1^T p(\mathbf{A}).$$

The representation of $\mathbf{F}$ given in this theorem is denoted as the **Ackermann-formula.**

6.3.2 Robust Pole Placement

In this section we investigate for a controllable SISO system

$$\begin{aligned} \dot{\mathbf{x}} &= \mathbf{A}\mathbf{x} + \mathbf{b}_2 u \\ z &= \mathbf{C}_1 \mathbf{x} \end{aligned} \tag{6.31}$$

how the closed-loop poles should be prescribed if certain performance and robustness requirements are given. It is assumed that a reference signal has to be tracked. Then the extended plant

$$\begin{aligned} \dot{\mathbf{x}}_e &= \mathbf{A}_e \mathbf{x} + \mathbf{b}_s r + \mathbf{b}_{2e} u \\ z &= \mathbf{C}_{1e} \mathbf{x}_e \end{aligned}$$

has to be considered (cf. Sect. 6.1). We denote the transfer function of the extended plant by $\mathbf{G}(s)$. It has r, u as inputs and $z, \mathbf{x}_e$ as outputs. Moreover, the plant (6.31) can be written in the form

$$\hat{z} = \frac{Z_1(s)}{q(s)} \hat{u}.$$

Using the PHB test, it is easy to see that the controllability of $(\mathbf{A}, \mathbf{b}_2)$ implies the controllability of the extended system $(\mathbf{A}_e, \mathbf{b}_{2e})$ if $Z_1(0) \neq 0$.

Figure 6.2 shows the state-vector feedback system with an additional multiplicative perturbation at the plant input. The situation is similar to that given in Fig. 3.9. We want to analyze under which conditions the feedback system is robustly stable. In doing so, it is possible repeat the discussion of Sect. 3.5 with some modifications. Again, the feedback system consisting of the plant $\mathbf{G}$ and the controller $\mathbf{F}_e$ will be denoted as $\mathbf{M}$. Then also Fig. 3.10 can be redrawn. As in Sect. 3.5, it may be interpreted as an equivalent representation of Fig. 6.2. The weight W_2 is given by $W_2 = 1$. Suppose that $\mathbf{F}_e$ is chosen such that $\mathbf{A}_e + \mathbf{b}_{2e}\mathbf{F}_e$ is stable.

Let $u_1 = u + w$ and denote by S_i the transfer function from w to u_1 and by L_i the transfer function from u_1 to $-u$ (cf. Fig. 6.2). Then we have

$$M_{11} = -L_i S_i \quad \text{and} \quad S_i = (1 + L_i)^{-1}.$$

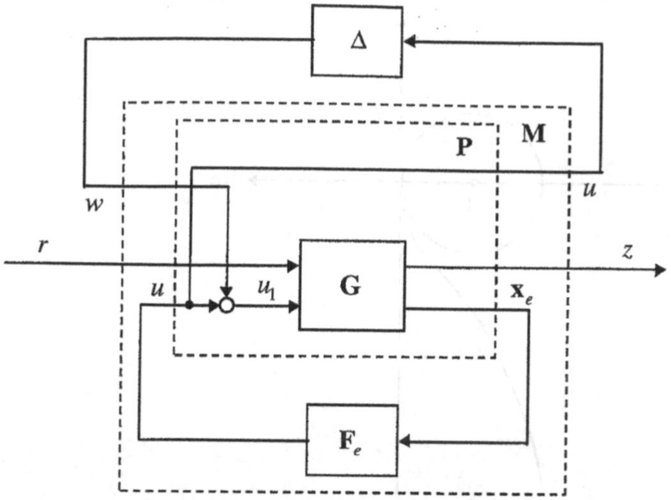

Fig. 6.2. Feedback system with state-vector feedback and multiplicative perturbation at the plant input

Moreover, the definition of L_i implies

$$L_i(s) = -\mathbf{F}_e\,(s\,\mathbf{I} - \mathbf{A}_e)^{-1}\,\mathbf{b}_{2e}\,.$$

Suppose that

$$\| M_{11} \|_\infty \le \gamma\,. \tag{6.32}$$

Then using the Nyquist criterion (Theorem 3.4.2) it may be seen exactly as in Sect. 3.5 that the following holds:

Under the assumption (6.32) the system in Fig. 3.11 (adapted to the situation given here) is stable for all stable perturbations with

$$\| \Delta \|_\infty < \gamma^{-1}.$$

Hence we have to find an upper bound of $\| M_{11} \|_\infty$. Observing $M_{11} = S_i - 1$, it follows that

$$\| M_{11} \|_\infty \le \| S_i \|_\infty + 1. \tag{6.33}$$

If it is possible to find an estimate of the form

$$| 1 + L_i(i\omega) | \ge R \quad \text{for every } \omega\,, \tag{6.34}$$

then $\gamma = (R+1)/R$ is a feasible choice. We have

$$\det(s\mathbf{I} - \mathbf{A}_e - \mathbf{b}_{2e}\mathbf{F}_e) = \det\{(s\mathbf{I} - \mathbf{A}_e)(\mathbf{I} - (s\mathbf{I} - \mathbf{A}_e)^{-1}\mathbf{b}_{2e}\mathbf{F}_e)\}$$

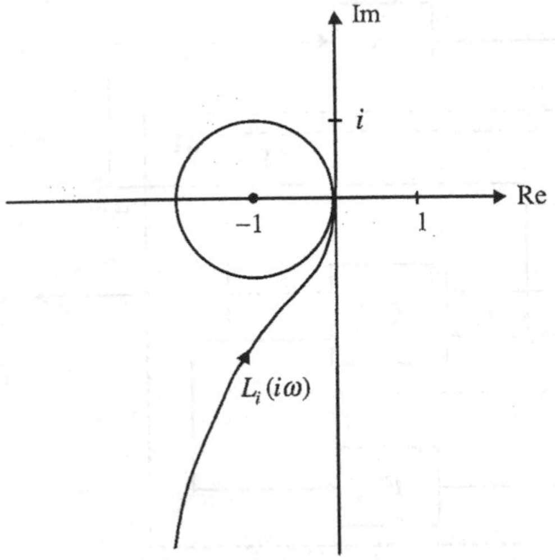

Fig. 6.3. Loop transfer function in the case of pole placement

$$= \det(s\mathbf{I} - \mathbf{A}_e)\det(\mathbf{I} - (s\mathbf{I} - \mathbf{A}_e)^{-1}\mathbf{b}_{2e}\mathbf{F}_e)$$
$$= \det(s\mathbf{I} - \mathbf{A}_e)(1 - \mathbf{F}_e(s\mathbf{I} - \mathbf{A}_e)^{-1}\mathbf{b}_{2e})$$
$$= s\det(s\mathbf{I} - \mathbf{A})(1 + L_i(s))$$

and consequently

$$S_i(s) = \frac{s\,q(s)}{p_e(s)} \tag{6.35}$$

with

$$p_e(s) = \det(s\mathbf{I} - \mathbf{A}_e + \mathbf{b}_e\mathbf{F}_e), \qquad q(s) = \det(s\mathbf{I} - \mathbf{A}).$$

The following lemma gives sufficient conditions for $R = 1$, which is the optimal case (cf. Fig. 6.3). Then the feedback system has a phase margin of at least 60°.

Lemma 6.3.1. *Let* $-s_{01}, -s_{02}, \ldots, -s_{0k}$ *be the real poles and let*

$$s_{\nu 1,2}^0 = -d_{0\nu}\omega_{0\nu} \pm \sqrt{1 - d_{0\nu}^2}\,\omega_{0\nu} i \quad \nu = 1,\ldots,l \quad \text{with} \quad |d_{0\nu}| < 1$$

be the complex poles of the loop transfer function ($k + l = n + 1$). *Similarly, let* $-s_1, -s_2, \ldots, -s_k$ *be the real poles and let*

$$s_{\nu 1,2} = -d_\nu\omega_\nu \pm \sqrt{1 - d_\nu^2}\,\omega_\nu i \qquad \nu = 1,\ldots,l$$

be the complex poles of the closed-loop system, and suppose

$$s_\nu \geq |s_{0\nu}| \qquad \mu = 1,\dots,k$$

$$\omega_\nu \geq \omega_{0\nu}, \quad d_\nu \geq |d_{0\nu}| \qquad \nu = 1,\dots,l.$$

Then the following estimate holds:

$$|1 + L_i(i\omega)| \geq 1 \qquad \textit{for every } \omega. \tag{6.36}$$

Proof. Because of (6.35), the asserted inequality (6.36) is equivalent to

$$\frac{|p_e(i\omega)|}{|(i\omega)q(i\omega)|} \geq 1 \qquad \text{for every } \omega. \tag{6.37}$$

By our assumptions the polynomials $s\,q(s)$, $p_e(s)$ can be written in the form

$$s\,q(s) = \prod_{\mu=1}^{k}(s + s_{0\mu}) \prod_{\nu=1}^{l}(s^2 + 2d_{0\nu}\omega_{0\nu}s + \omega_{0\nu}^2),$$

$$p_e(s) = \prod_{\mu=1}^{k}(s + s_\mu) \prod_{\nu=1}^{l}(s^2 + 2d_\nu\omega_\nu s + \omega_\nu^2)$$

and hence (6.37) is certainly true if the inequalities

$$|i\omega + s_\mu| \geq |i\omega + s_{0\mu}| \qquad \mu = 1,\dots,k, \tag{6.38}$$

$$|(i\omega)^2 + 2d_\nu\omega_\nu i\omega + \omega_\nu^2| \geq |(i\omega)^2 + 2d_{0\nu}\omega_{0\nu}i\omega + \omega_{0\nu}^2| \qquad \nu = 1,\dots,l, \tag{6.39}$$

hold for every ω. The validity of (6.38) is obvious under the given assumptions. Because of

$$|(i\omega)^2 + 2d_\nu\omega_\nu i\omega + \omega_\nu^2|^2 = \omega^4 + \omega_\nu^4 + 2(2d_\nu^2 - 1)\omega_\nu^2\omega^2,$$

(6.39) is equivalent to

$$\omega_\nu^4 + 2(2d_\nu^2 - 1)\omega_\nu^2\omega^2 \geq \omega_{0\nu}^4 + 2(2d_{0\nu}^2 - 1)\omega_{0\nu}^2\omega^2.$$

This inequality holds also under the assumptions of the lemma.

As an example, we consider an unstable plant of order 3 with a conjugate-complex pole pair in the RHP (cf. Fig. 6.4). At first, the unstable pole pair will be reflected on the imaginary axis and then rotated until it lies on the straight lines which form an angle of 45° with the negative real axis. The integrator pole has to be shifted to the left. For example, it may be placed so that all poles lie on the boundary of a circle. The closed-loop poles s_1, s_2, s_3 now satisfy the assumptions of the lemma. It is possible to increase the bandwidth of the feedback system by increasing the distance of the closed-loop poles from the origin, for example by increasing the radius of the circle.

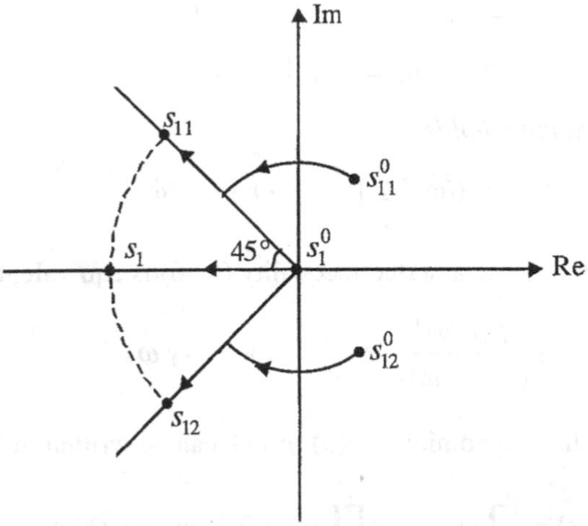

Fig 6.4. Pole placement for an unstable system of order 3

Pole positioning for systems with non-neglectable dynamics of the actuator and the sensors. The poles of the sensors and the actuator cannot be influenced by feedback. Their dynamics may limit the bandwidth of the feedback system. We model the dynamics of the sensors and the actuator in a very rough way together by a type 1 system which has no zero and whose time constant is T. Then the loop transfer function is given by

$$\tilde{L}_i(s) = \frac{1}{1+Ts} L_i(s) .$$

This implies

$$1 + \tilde{L}_i(s) = 1 + \frac{1}{1+Ts} \frac{p_e(s) - s\,q(s)}{s\,q(s)}$$

$$= \frac{Ts^2 q(s) + p_e(s)}{s(1+Ts)q(s)} .$$

The characteristic polynomial of the feedback system now has the degree $n+2$ due to the additive term $Ts^2q(s)$ and thus has one more zero. The following theorem gives an approximate decomposition into linear factors of this polynomial. Then it is possible to find an estimate similar to (6.34) for this more realistic plant model.

Theorem 6.3.2. *Let polynomials*

$$p(x) = x^n + a_{n-1}x^{n-1} + \ldots + a_1 x + a_0$$

$$q(x) = x^{n+1} + b_n x^n + \ldots + b_1 x + b_0$$

be given and put with an arbitrary $\varepsilon > 0$

$$p_\varepsilon(x) = \varepsilon\, q(x) + p(x).$$

Let $x_1, \ldots, x_n$ *be the zeros of* $p(x)$ *and denote by* $x_1(\varepsilon), \ldots, x_n(\varepsilon), x_{n+1}(\varepsilon)$ *the zeros of* $p_\varepsilon(x)$. *Finally, define*

$$\tilde{p}_\varepsilon(x) = (x - x_1(\varepsilon)) \cdots (x - x_n(\varepsilon)).$$

Then there are constants c, c_x *and a number* $\varepsilon_0 > 0$ *such that (with adequately chosen indices)*

$$\left| x_{n+1}(\varepsilon) + \frac{1}{\varepsilon} + b_n - a_{n-1} \right| \le c\,\varepsilon \tag{6.40}$$

and

$$| \tilde{p}_\varepsilon(x) - p(x) | \le c_x \varepsilon \quad \textit{for every } \varepsilon < \varepsilon_0. \tag{6.41}$$

The proof of this theorem will not be given here. In the sense of the approximations of the preceding theorem, it is possible to write

$$p_\varepsilon(x) \approx \varepsilon\Big(x + \frac{1}{\varepsilon} + b_n - a_{n-1}\Big)\tilde{p}_\varepsilon(x) \approx \varepsilon\Big(x + \frac{1}{\varepsilon} + b_n - a_{n-1}\Big)p(x).$$

The application of this approximation to the numerator of $1 + \tilde{L}_e(s)$ yields

$$Ts^2 q(s) + p_e(s) \approx T\Big(s + \frac{1}{T} + a_{n-1} - \alpha_n\Big)p_e(s).$$

We have the representation

$$L_i(s) = \frac{1}{s}\,\frac{(\alpha_n - a_{n-1})s^n + \ldots + (\alpha_1 - a_0)s + \alpha_0}{s^n + a_{n-1}s^{n-1} + a_{n-2}s^{n-2} + \ldots + a_0}.$$

Hence for complex numbers s with a large absolute value the approximation

$$L_i(s) \approx \frac{1}{s}(\alpha_n - a_{n-1}) \approx \frac{\omega_D}{s}$$

holds, where ω_D is the gain crossover frequency. Thus, $\omega_D \approx \alpha_n - a_{n-1}$ and

$$Ts^2 q(s) + p_e(s) \approx (Ts + 1 - T\omega_D)p_e(s).$$

This implies

$$1 + \tilde{L}_i(s) \approx \frac{(Ts + 1 - T\omega_D)p_e(s)}{s(1 + Ts)q(s)} = \frac{1 - T\omega_D + Ts}{1 + Ts}(1 + L_i(s)). \tag{6.42}$$

Define

$$R = 1 - T\omega_D \tag{6.43}$$

and suppose T and ω_D to be given such that $R > 0$. Then from (6.42) it may be concluded that the estimate

$$|1 + \tilde{L}_i(s)| \geq R|1 + L_i(s)|$$

approximately holds. It allows us to estimate the degradation of the robustness by the additional first order dynamics. If (6.36) is valid, we get

$$|1 + \tilde{L}_i(s)| \geq R\,.$$

In other words, the radius of the circle with center -1 which is avoided by the curve $L_i(i\omega)$ has decreased from 1 to $R < 1$. Hence, when the controller is synthesized, one has to take care that the product $T\omega_D$ remains sufficiently small.

The latter point has to be discussed in more detail and this will lead us to an additional restriction for the choice of the closed-loop poles $s_1, \ldots, s_{n+1}$. The equation

$$p_e(s) = (s - s_1)\cdots(s - s_{n+1}) = s^{n+1} + \alpha_n s^n + \ldots + \alpha_0$$

implies $\alpha_n = -(s_1 + \cdots + s_{n+1})$. Thus we conclude that

$$\sum_{\mu=1}^{n+1} s_\mu = -\omega_D - a_{n-1}\,. \tag{6.44}$$

A meaningful choice of ω_D follows from (6.43) (e.g. with $R = 0.75$). Then the closed-loop poles have to be chosen such that (6.44) holds and additionally the conditions of Lemma 6.3.1 are fulfilled.

6.3.3 Example: Pole Placement for a Second-Order System

As an example, we consider the following second-order plant:

$$G(s) = k\omega_0^2 \frac{T_1 s + 1}{s^2 + 2d\omega_0 s + \omega_0^2}\,.$$

The characteristic polynomial of the closed-loop system is given by

$$p_e(s) = s^3 + \alpha_2 s^2 + \alpha_1 s + \alpha_0$$

and the feedback gains are easily calculated as

$$k = \alpha_0, \quad k_1 = \alpha_1 - \omega_0^2 - \alpha_0 T_1, \quad k_2 = \alpha_2 - 2d\omega_0\,.$$

The gain crossover frequency is approximately $\omega_D \approx k_2$. We choose $R = 0.75$ and $T = 0.25/\omega_D$.

1. Suppose the plant has no zero and a pole with multiplicity 2 in the left open half plane (i.e. let $T_1 = 0$ and $d = 1$). We choose the closed-loop poles as $s_1 = s_2 = s_3 = -\mu\omega_0$, where $\mu \geq 1$. This leads to

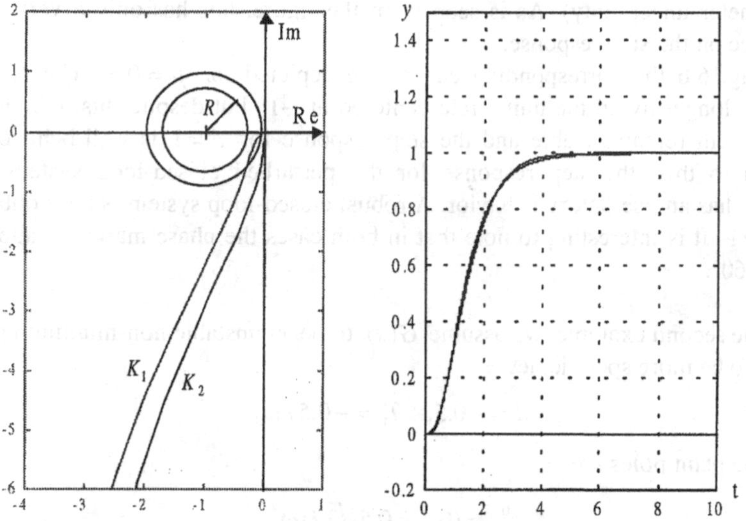

Fig. 6.5. Plots of $L_i(i\omega)$ and $\tilde{L}_i(i\omega)$ for $\mu = 2$ and the corresponding step responses (stable plant)

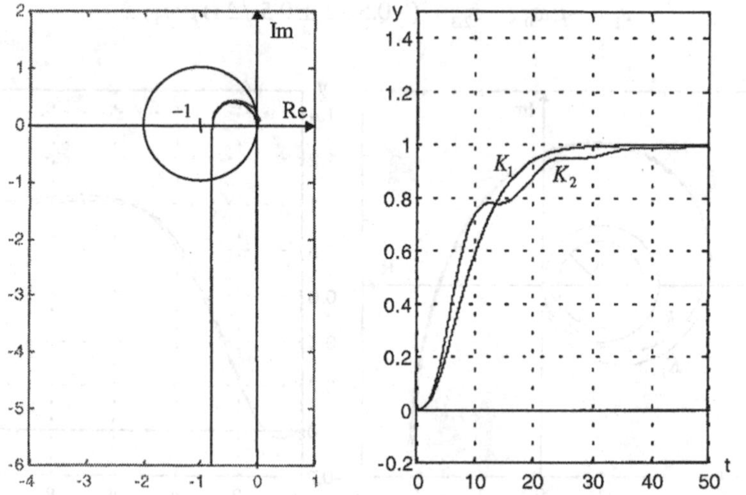

Fig. 6.6. Plots of $L_i(i\omega)$ and $\tilde{L}_i(i\omega)$ for $\mu = 0.3$ and the corresponding step responses (stable plant)

$$p_e(s) = s^3 + 3\mu\omega_0 s^2 + 3\mu^2\omega_0^2 s + \omega_0^3$$

and $\omega_D \approx (3\mu - 2)\omega_0$. Figure 6.5 (left-hand plot) shows $L_i(i\omega)$ for $\mu = 2$ without (curve K_1) and with (curve K_2) an additional first-order system. Its time constant results for $\mu = 2$ as $T = 0.063/\omega_0$ ($\mu = 1$ yields $T = 0.25/\omega_0$). The right-hand figure contains the corresponding step responses. For the system with the additional first-order system, the damping has been reduced to 0.7 (in the sense of a parameter uncertainty). As is seen from the figure, this has only a very small influence on the step response.

In Fig. 6.6 the corresponding curves are depicted for $\mu = 0.3$. The Nyquist plots no longer avoid the unit circle centered at -1, but despite this, the closed-loop system remains stable and the step response for $d = 1$ is well-behaved. In contrast to this, the step response for the perturbed closed-loop system with $d = 0.7$ has an oscillatory behavior. A robust closed-loop system is only obtained for $\mu \geq 1$. It is interesting to note that in both cases the phase margin is approximately 60°.

2. In the second example, we assume $G(s)$ to be an unstable non-minimum phase plant. To be more specific, let

$$d = -0.5, \quad T_1 = -0.5/\omega_0 .$$

Then the plant poles are

$$s_{1,2}^0 = (0.5 \pm 0.5\sqrt{3}\, i)\,\omega_0$$

and the closed-loop poles are chosen as

$$s_1 = -\mu\omega_0 , \quad s_{2,3} = (-0.5\sqrt{2} \pm 0.5\sqrt{2}\, i)\,\mu\omega_0 .$$

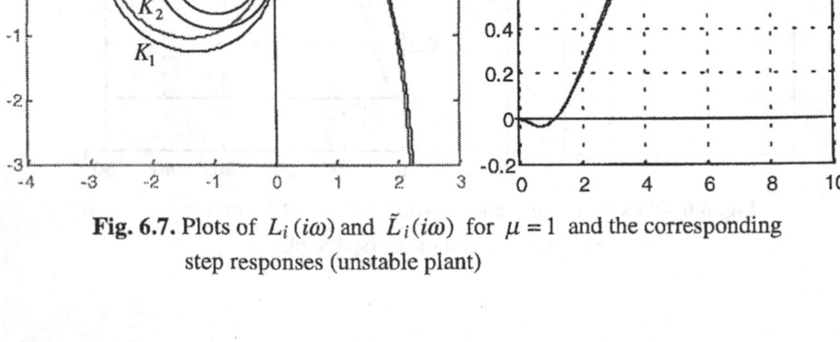

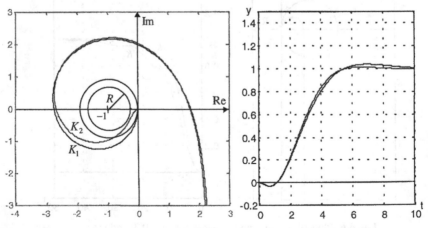

Fig. 6.7. Plots of $L_i(i\omega)$ and $\tilde{L}_i(i\omega)$ for $\mu = 1$ and the corresponding step responses (unstable plant)

This implies

$$p_e(s) = s^3 + (1+\sqrt{2})\mu\omega_0 s^2 + (1+\sqrt{2})\mu^2\omega_0^2 s + \mu^3\omega_0^3$$

and

$$\omega_D \approx (1+(1+\sqrt{2})\mu)\omega_0 .$$

Let now $\mu = 1$. Then the time constant of the additional first-order system has to be smaller or equal to $T = 0.073/\omega_0$. Hence, for this plant the bandwidth of the actuator and the sensors has to be considerably higher then for the plant in the first example. The Nyquist curves for the unperturbed and the perturbed system are shown in Fig. 6.7. The Nyquist plots are as expected and the degradation of the step response due to the additional dynamics and the changed damping is minimal.

6.4 Pole Placement for MIMO Systems

6.4.1 Pole Placement and Controllability

We consider now the pole placement problem for general MIMO plants. In this case, the situation is different, because the gain matrix $\mathbf{F}$ is no longer uniquely determined by the closed-loop poles.

Lemma 6.4.1. *Suppose* $(\mathbf{A}, \mathbf{B})$ *is controllable. If* $\mathbf{b}_1$ *is any nonzero column of* $\mathbf{B}$, *then there exists a matrix* $\mathbf{F}_1$ *such that* $(\mathbf{A}+\mathbf{B}\mathbf{F}_1, \mathbf{b}_1)$ *is controllable.*

Proof. For given input vectors $\mathbf{u}_k$ we define the sequence

$$\mathbf{x}_{k+1} = \mathbf{A}\mathbf{x}_k + \mathbf{B}\mathbf{u}_k, \quad \mathbf{x}_0 = \mathbf{b}_1 . \tag{6.45}$$

Suppose we can prove that there exists an input sequence $\{\mathbf{u}_0, \ldots, \mathbf{u}_{n-2}\}$ such that for the resulting state vectors

$$\text{span}\,\{\mathbf{x}_0, \ldots, \mathbf{x}_{n-1}\} = \mathbb{R}^n .$$

Then for any $\mathbf{u}_{n-1}$ there exists a matrix $\mathbf{F}_1$ such that

$$\mathbf{F}_1 (\mathbf{x}_0 \quad \mathbf{x}_1 \quad \ldots \quad \mathbf{x}_{n-1}) = (\mathbf{u}_0 \quad \mathbf{u}_1 \quad \cdots \quad \mathbf{u}_{n-1}) . \tag{6.46}$$

We show that $(\mathbf{A}+\mathbf{B}\mathbf{F}_1, \mathbf{b}_1)$ is controllable. Equations (6.45) and (6.46) imply

$$\mathbf{x}_{k+1} = (\mathbf{A}+\mathbf{B}\mathbf{F}_1)\mathbf{x}_k, \quad \mathbf{x}_0 = \mathbf{b}_1$$

and therefore

$$(\mathbf{x}_0 \quad \ldots \quad \mathbf{x}_{n-1}) = (\mathbf{b}_1 \quad (\mathbf{A}+\mathbf{B}\mathbf{F}_1)\mathbf{b}_1 \quad \ldots \quad (\mathbf{A}+\mathbf{B}\mathbf{F}_1)^{n-1}\mathbf{b}_1) .$$

Since the matrix on the left-hand side has full rank, the assertion follows.

Thus, it is sufficient to show there exists an input sequence $\{\mathbf{u}_0, \ldots, \mathbf{u}_{n-2}\}$ such that $\{\mathbf{x}_0, \ldots, \mathbf{x}_{n-1}\}$ are linearly independent. We use induction. Assume $k < n$ and let $\{\mathbf{u}_0, \ldots, \mathbf{u}_{k-1}\}$ be given such that the vectors $\{\mathbf{x}_0, \ldots, \mathbf{x}_k\}$ are linearly independent. Define the subspace

$$W = \operatorname{span}\{\mathbf{x}_0, \ldots, \mathbf{x}_k\}.$$

We have to show that there exists a vector $\mathbf{u}_k$ with the property

$$\mathbf{A}\mathbf{x}_k + \mathbf{B}\mathbf{u}_k \notin W. \tag{6.47}$$

Suppose now that such a vector does not exist. This means that for any $\mathbf{u}_k \in \mathbb{R}^n$ we have

$$\mathbf{A}\mathbf{x}_k + \mathbf{B}\mathbf{u}_k \in W.$$

Letting $\mathbf{u}_k = \mathbf{0}$, this implies $\mathbf{A}\mathbf{x}_k \in W$, and since W is a linear subspace, we get $\mathbf{B}\mathbf{u}_k \in W$. Since $\mathbf{u}_k$ was arbitrary, the inclusion $\operatorname{Im}(\mathbf{B}) \subset \mathrm{W}$ follows.

Next, we show that the inclusion $\mathbf{A}(\mathrm{W}) \subset \mathrm{W}$ also holds. The latter condition can be equivalently formulated as

$$\mathbf{A}\mathbf{x}_i \in W \text{ for } i = 0, \ldots, k.$$

As we have already seen, this holds for $i = k$. For $i < k$ we have

$$\mathbf{A}\mathbf{x}_i = \mathbf{x}_{i+1} - \mathbf{B}\mathbf{u}_i \in W$$

since $\mathbf{x}_{i+1} \in W$ and $\mathbf{B}\mathbf{u}_i \in W$. This implies $\mathbf{A}(\mathrm{W}) \subset \mathrm{W}$ and

$$\operatorname{Im}(\mathbf{AB}) \subset \mathbf{A}(\mathrm{W}) \subset \mathrm{W}.$$

The latter inclusion yields

$$\operatorname{Im}(\mathbf{A}^i\mathbf{B}) \subset \mathrm{W} \text{ for } i = 0, \ldots, n-1$$

and consequently

$$\operatorname{Im}(\mathbf{B} \quad \mathbf{AB} \quad \ldots \quad \mathbf{A}^{n-1}\mathbf{B}) \subset W.$$

Since W is of dimension less than n, we have a contradiction because $(\mathbf{A}, \mathbf{B})$ is controllable by hypothesis.

Theorem 6.4.1. *A matrix* $\mathbf{F}$ *such that* $\mathbf{A} + \mathbf{BF}$ *has arbitrarily prescribed eigenvalues exists if and only if the pair* $(\mathbf{A}, \mathbf{B})$ *is controllable.*

Proof. "$\Rightarrow$": Suppose that $(\mathbf{A}, \mathbf{B})$ is not controllable and that with a suitable $\mathbf{F}$ the matrix $\mathbf{A} + \mathbf{BF}$ has arbitrarily prescribed eigenvalues. Then the PHB test shows the existence of an eigenvalue λ of $\mathbf{A}$ and a vector $\mathbf{x} \neq \mathbf{0}$ with

$$\mathbf{x}^*\mathbf{A} = \lambda\mathbf{x}^*, \quad \mathbf{x}^*\mathbf{B} = \mathbf{0}.$$

This implies

$$\mathbf{x}^*(\mathbf{A}+\mathbf{BF}) = \mathbf{x}^*\mathbf{A} = \lambda\mathbf{x}^*.$$

Hence λ is a common eigenvalue of $\mathbf{A}$ and $\mathbf{A}+\mathbf{BF}$ and is independent of $\mathbf{F}$, which is a contradiction.

"$\Leftarrow$": Let $(\mathbf{A},\mathbf{B})$ be controllable. Then, due to Lemma 6.4.1, there exists a matrix $\mathbf{F}_1$ such that $(\mathbf{A}+\mathbf{BF}_1, \mathbf{b}_1)$ is controllable. This system has only a single input and we can apply Theorem 6.3.1. Hence there exists a matrix $\mathbf{F}_2$ such that

$$(\mathbf{A}+\mathbf{BF}_1) + \mathbf{b}_1\mathbf{F}_2$$

has the desired eigenvalues. Since $\mathbf{b}_1$ is a column of $\mathbf{B}$, there exists a matrix $\mathbf{F}$ such that $\mathbf{BF}_1 + \mathbf{b}_1\mathbf{F}_2 = \mathbf{BF}$ and the theorem is proven.

6.4.2 Robust Eigenstructure Assignment

It would be possible to construct an algorithm from the proof of Theorem 6.4.1 which solves the pole placement problem, but with respect to robustness and numerical reliability better methods exist. As already mentioned, except for the case of a single control input ($m = 1$), the gain matrix $\mathbf{F}$ is not uniquely determined by the closed-loop eigenvalues. Hence there is some degree of freedom when the eigensystem is chosen which can be used to improve the properties of the feedback system. To formulate and prove the results, we need some basic results from linear algebra concerning the singular value decomposition of a matrix and matrix norms. Readers not familiar with these topics are referred to Sect. 7.1.1, where a short introduction is given.

Denote the prescribed eigenvalues by $\lambda_1,\ldots,\lambda_n$ (which of course are supposed to lie in the open LHP) and assume that they are mutually distinct: $\lambda_k \neq \lambda_j$ for $k \neq j$. Let $\mathbf{x}_j$ be an eigenvector of $\mathbf{A}+\mathbf{BF}$. Then the vectors $\mathbf{x}_1,\ldots,\mathbf{x}_n$ are linearly independent. We define

$$\mathbf{D} = \mathrm{diag}(\lambda_1,\ldots,\lambda_n), \quad \mathbf{X} = (\mathbf{x}_1 \;\; \ldots \;\; \mathbf{x}_n).$$

Then we have

$$\mathbf{A}+\mathbf{BF} = \mathbf{XDX}^{-1}. \tag{6.48}$$

Since $\mathbf{X}$ is nonsingular, the condition number of $\mathbf{X}$ is defined by (compare Sect. 7.1.1)

$$\kappa(\mathbf{X}) = \|\mathbf{X}\|\|\mathbf{X}^{-1}\|.$$

For a given gain matrix $\mathbf{F}$ we put

$$\delta(\mathbf{F}) = \min_{\omega \in \mathbb{R}} \underline{\sigma}(i\omega\mathbf{I} - \mathbf{A} + \mathbf{BF}).$$

The following theorem gives an upper bound of the norm for perturbations Δ, which leave the feedback system stable.

Theorem 6.4.2. *The matrix* $\mathbf{A} + \Delta + \mathbf{BF}$ *is stable for all perturbations* Δ *with*

$$\|\Delta\| < \delta(\mathbf{F}).$$

A lower bound of $\delta(\mathbf{F})$ *is given by*

$$\delta(\mathbf{F}) \geq \min_j (\mathrm{Re}(-\lambda_j)) / \kappa(\mathbf{X}). \tag{6.49}$$

Proof. Let $\mathbf{M}$ be nonsingular and suppose $\|\Delta\| < \underline{\sigma}(\mathbf{M})$. We show that then the perturbed matrix $\mathbf{M} + \Delta$ must be nonsingular. Because of

$$\mathbf{M} + \Delta = \mathbf{M}(\mathbf{I} + \mathbf{M}^{-1}\Delta),$$

it suffices to show that $\mathbf{I} + \mathbf{M}^{-1}\Delta$ is nonsingular. Suppose that this matrix is singular. Then there is a vector $\mathbf{x} \neq \mathbf{0}$ such that $\mathbf{x} = -\mathbf{M}^{-1}\Delta\mathbf{x}$. This implies $\|\mathbf{M}^{-1}\Delta\mathbf{x}\| \geq 1$. On the other hand, we also have

$$\|\mathbf{M}^{-1}\Delta\| \leq \|\mathbf{M}^{-1}\| \|\Delta\| < \|\mathbf{M}^{-1}\| \underline{\sigma}(\mathbf{M}) = 1,$$

which is impossible.

We apply this result to $\mathbf{M} = i\omega\mathbf{I} - (\mathbf{A} + \mathbf{BF})$ and see in this way that $i\omega$ is not an eigenvalue of $\mathbf{A} + \mathbf{BF} + \Delta$. Suppose that there is an eigenvalue λ of this matrix with $\mathrm{Re}\,\lambda > 0$. Then there is a number $0 < \alpha < 1$ and an ω such that $i\omega$ is an eigenvalue of $\mathbf{A} + \mathbf{BF} + \alpha\Delta$. Since this cannot be true, $\mathbf{A} + \mathbf{BF} + \Delta$ must be stable.

Using (6.48), the second inequality follows from

$$\begin{aligned} \delta(\mathbf{F}) &= \min_{\omega \in \mathbb{R}} \underline{\sigma}(i\omega\mathbf{I} - \mathbf{XDX}^{-1})) \\ &\geq \min_{\omega \in \mathbb{R}} (\underline{\sigma}(\mathbf{X}) \underline{\sigma}(i\omega\mathbf{I} - \mathbf{D}) \underline{\sigma}(\mathbf{X}^{-1})) \\ &\geq \min_j (\mathrm{Re}(-\lambda_j)) / (\|\mathbf{X}^{-1}\| \|\mathbf{X}\|). \end{aligned}$$

From this theorem, we see that it is desirable to construct the matrix $\mathbf{X}$ such that its condition number is as small as possible. In the remaining part of this section we will sketch how this can be achieved.

Suppose $\mathbf{B}$ has full column rank: $\mathrm{rank}(\mathbf{B}) = m$. Then there is a nonsingular matrix $\mathbf{Z}$ and an orthogonal matrix $\mathbf{U} = (\mathbf{U}_0 \quad \mathbf{U}_1)$ with

$$(\mathbf{U}_0 \quad \mathbf{U}_1) \begin{pmatrix} \mathbf{Z} \\ \mathbf{0} \end{pmatrix} = \mathbf{B}.$$

One possible way to derive this decomposition is to use the singular value decomposition of $\mathbf{B}$,

$$\mathbf{U}\begin{pmatrix}\mathbf{\Sigma}_1\\ \mathbf{0}\end{pmatrix}\mathbf{V}^T=\mathbf{B},$$

and to define $\mathbf{Z}=\mathbf{\Sigma}_1\mathbf{V}^T$ and with a suitable partitioning $\mathbf{U}=(\mathbf{U}_0 \quad \mathbf{U}_1)$.

Lemma 6.4.2. *A matrix* $\mathbf{F}$ *and a nonsingular matrix* $\mathbf{X}$ *with (6.48) exist if and only if the equation*

$$\mathbf{U}_1^T(\mathbf{AX}-\mathbf{XD})=\mathbf{0} \tag{6.50}$$

has a nonsingular solution $\mathbf{X}$. *Then* $\mathbf{F}$ *is given by*

$$\mathbf{F}=\mathbf{Z}^{-1}\mathbf{U}_0^T(\mathbf{XDX}^{-1}-\mathbf{A}).$$

Proof. The proof consists in rewriting (6.48) by using the decomposition of $\mathbf{B}$. Pre-multiplication of (6.48) by $\mathbf{U}^T$ shows that (6.48) is equivalent to

$$\begin{pmatrix}\mathbf{ZF}\\ \mathbf{0}\end{pmatrix}=\begin{pmatrix}\mathbf{U}_0^T(\mathbf{XDX}^{-1}-\mathbf{A})\\ \mathbf{U}_1^T(\mathbf{XDX}^{-1}-\mathbf{A})\end{pmatrix},$$

from which the assertion directly follows.

For $j=1,\ldots,n$ define now subspaces

$$S_j=\operatorname{Ker}(\mathbf{U}_1^T(\mathbf{A}-\lambda_j\mathbf{I})).$$

From (6.50) we see directly that $\mathbf{x}_j\in S_j$. Moreover, the following equation holds:

$$\mathbf{U}^T(\mathbf{B}\quad \mathbf{A}-\lambda_j\mathbf{I})=\begin{pmatrix}\mathbf{Z} & \mathbf{U}_0^T(\mathbf{A}-\lambda_j\mathbf{I})\\ \mathbf{0} & \mathbf{U}_1^T(\mathbf{A}-\lambda_j\mathbf{I})\end{pmatrix}.$$

If $(\mathbf{A},\mathbf{B})$ is controllable, the dimension of S_j can be calculated as

$$\begin{aligned}\dim(S_j)&=\dim(\operatorname{Ker}(\mathbf{B}\quad \mathbf{A}-\lambda_j\mathbf{I})\\ &=(m+n)-\dim(\operatorname{Im}(\mathbf{B}\quad \mathbf{A}-\lambda_j\mathbf{I}))\\ &=(m+n)-n=m.\end{aligned}$$

The next corollary summarizes our results.

Corollary 6.4.1. *(a) Let* $\mathbf{X}=(\mathbf{x}_1 \quad \ldots \quad \mathbf{x}_n)$ *be a nonsingular matrix with* $\mathbf{x}_j\in S_j$. *Then the matrix*

$$\mathbf{F}=\mathbf{Z}^{-1}\mathbf{U}_0^T(\mathbf{XDX}^{-1}-\mathbf{A})$$

solves the pole placement problem.

(b) If the pair $(\mathbf{A}, \mathbf{B})$ *is controllable, there exists a matrix* $\mathbf{X}$ *with the properties described in (a). In this case,* $\dim(\text{Ker } S_j) = m$.

The problem is now to find a matrix $\mathbf{X}$ with $\mathbf{x}_j \in S_j$ such that the condition number of $\mathbf{X}$ is minimal. We note that there are two extreme cases. The first one is $m = 1$, where there is no freedom in choosing $\mathbf{X}$. The second one is $m = n$. Then $\mathbf{X} = \mathbf{I}$ is a feasible choice and the condition number is minimal: $\kappa(\mathbf{I}) = 1$. There are several possibilities for constructing $\mathbf{X}$ such that the condition number or a similar robustness measure is minimal. One of them is to construct an orthonormal set of n vectors $\tilde{\mathbf{x}}_j$ which have the property that an adequately defined distance between $\tilde{\mathbf{x}}_j$ and the subspace S_j is minimized. Then the eigenvectors $\mathbf{x}_j$ are taken as the normalized projections of the vectors $\tilde{\mathbf{x}}_j$ onto S_j. The resulting vectors $\mathbf{x}_j$ are "as orthogonal as possible". We do not describe this in detail and refer the interested reader to Kautsky, Nichols, and Van Dooren [57].

Finally, we mention that there is a method for parametrizing $\mathbf{F}$ which works if no eigenvalue of the feedback system is an eigenvalue of the plant. Put

$$\mathbf{P} = (\mathbf{p}_1 \ \cdots \ \mathbf{p}_n) = \mathbf{F}\mathbf{X}.$$

Then (6.48) is equivalent to

$$(\mathbf{A} - \lambda_j \mathbf{I})\mathbf{x}_j = -\mathbf{B}\mathbf{p}_j \qquad j = 1, \ldots, n. \tag{6.51}$$

If we assume that no λ_j is an eigenvalue of the state matrix $\mathbf{A}$, these equations have a unique solution $\mathbf{x}_j$. Now let the parameter matrix $\mathbf{P}$ be given. Then $\mathbf{X}$ can be uniquely determined from (6.51) and the gain matrix is given by $\mathbf{F} = \mathbf{P}\mathbf{X}^{-1}$.

6.5 General Feedback Systems

6.5.1 Structure of General Feedback Systems

In this section, we want to introduce a very general representation of feedback loops, which covers practically all thinkable cases. The corresponding block diagram is depicted in Fig. 6.8. The plant $\mathbf{P}$ has a disturbance $\mathbf{w}$ and the control $\mathbf{u}$ as inputs and the quantities $\mathbf{z}$ and $\mathbf{y}$ as outputs. The latter is the measurement and, as usual, the input to the controller $\mathbf{K}$, which has $\mathbf{u}$ as its output. The vector of disturbances contains all possible inputs, and, in particular, a possible reference variable. The plant output $\mathbf{z}$ is also an output of the feedback loop and is built up by the variables which have to be influenced by the controller. Some of them typically are errors that have to be small. It has to be emphasized that the plant

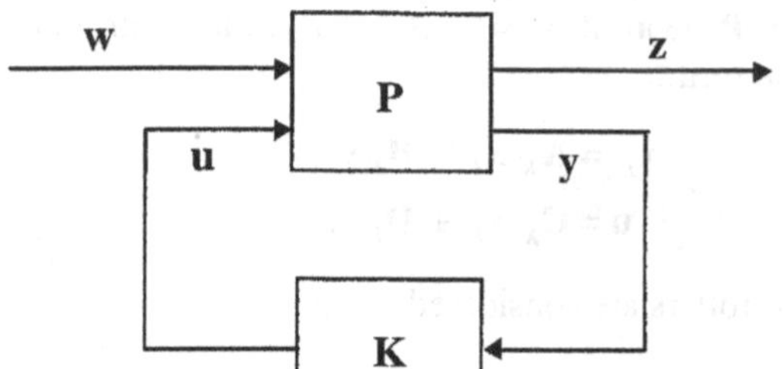

Fig. 6.8. General Feedback Loop

P contains not only the "physical" plant (which we normally denote by **G**), but also some weights. As a consequence, not all the inputs or outputs of the feedback loop are of a physical nature. We will return to this point later when $\mathcal{H}_2$ or $\mathcal{H}_\infty$ optimal controllers are considered. The controller is also a dynamic system.

We denote this controlled system as a **general feedback system**. The plant will always be assumed to be rational and proper. Then it has a state space representation of the form

$$\begin{aligned}\dot{\mathbf{x}} &= \mathbf{A}\,\mathbf{x} + \mathbf{B}_1\mathbf{w} + \mathbf{B}_2\mathbf{u}\\ \mathbf{z} &= \mathbf{C}_1\,\mathbf{x} + \mathbf{D}_{11}\mathbf{w} + \mathbf{D}_{12}\mathbf{u}\\ \mathbf{y} &= \mathbf{C}_2\,\mathbf{x} + \mathbf{D}_{21}\mathbf{w} + \mathbf{D}_{22}\mathbf{u}\,.\end{aligned} \tag{6.52}$$

This can be written in a more compact way as

$$\mathbf{P}(s) = \left(\begin{array}{c|cc} \mathbf{A} & \mathbf{B}_1 & \mathbf{B}_2 \\ \hline \mathbf{C}_1 & \mathbf{D}_{11} & \mathbf{D}_{12} \\ \mathbf{C}_2 & \mathbf{D}_{21} & \mathbf{D}_{22} \end{array}\right) = \begin{pmatrix} \mathbf{P}_{11}(s) & \mathbf{P}_{12}(s) \\ \mathbf{P}_{21}(s) & \mathbf{P}_{22}(s) \end{pmatrix},$$

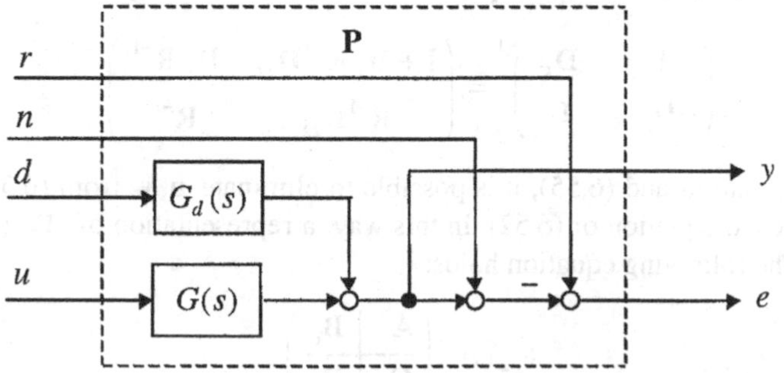

Fig. 6.9. Representation of the standard SISO feedback system in the generalized form

where the partitioning of the plant $\mathbf{P}(s)$ is given in an obvious manner. Thus we have in particular $\hat{\mathbf{y}} = \mathbf{P}_{22}(s)\hat{\mathbf{u}}$ if $\mathbf{w} = \mathbf{0}$. Similarly, the controller is supposed to be representable in the form:

$$\begin{aligned} \dot{\mathbf{x}}_K &= \mathbf{A}_K \mathbf{x}_K + \mathbf{B}_K \mathbf{y}, \\ \mathbf{u} &= \mathbf{C}_K \mathbf{x}_K + \mathbf{D}_K \mathbf{y}. \end{aligned} \tag{6.53}$$

Thus, only proper controllers are considered.

Examples. 1. Standard SISO feedback loop. Figure 6.9 shows how the standard SISO feedback system can be written in the generalized form (cf. Fig. 3.2). Here the vector $\mathbf{w}$ consists of the components r, n, d. The following equation is immediately seen from the block diagram:

$$\begin{pmatrix} \hat{y} \\ \hat{e} \end{pmatrix} = \left(\begin{array}{ccc|c} 0 & 0 & G_d(s) & G(s) \\ \hline 1 & -1 & -G_d(s) & -G(s) \end{array}\right) \begin{pmatrix} \hat{r} \\ \hat{n} \\ \hat{d} \\ \hat{u} \end{pmatrix}.$$

2. State feedback with integral control. These problems fit directly into the general framework (compare the discussion at the end of Sect. 6.1).

We start our analysis by calculating a state space representation of the transfer function $\mathbf{F}_{zw}(s)$ of the closed-loop system. Equations (6.52) and (6.53) imply

$$\begin{pmatrix} \dot{\mathbf{x}} \\ \dot{\mathbf{x}}_K \end{pmatrix} = \begin{pmatrix} \mathbf{A} & \mathbf{0} \\ \mathbf{0} & \mathbf{A}_K \end{pmatrix} \begin{pmatrix} \mathbf{x} \\ \mathbf{x}_K \end{pmatrix} + \begin{pmatrix} \mathbf{B}_1 \\ \mathbf{0} \end{pmatrix} \mathbf{w} + \begin{pmatrix} \mathbf{B}_2 & \mathbf{0} \\ \mathbf{0} & \mathbf{B}_K \end{pmatrix} \begin{pmatrix} \mathbf{u} \\ \mathbf{y} \end{pmatrix} \tag{6.54}$$

$$\begin{pmatrix} \mathbf{I} & -\mathbf{D}_K \\ -\mathbf{D}_{22} & \mathbf{I} \end{pmatrix} \begin{pmatrix} \mathbf{u} \\ \mathbf{y} \end{pmatrix} = \begin{pmatrix} \mathbf{0} & \mathbf{C}_K \\ \mathbf{C}_2 & \mathbf{0} \end{pmatrix} \begin{pmatrix} \mathbf{x} \\ \mathbf{x}_K \end{pmatrix} + \begin{pmatrix} \mathbf{0} \\ \mathbf{D}_{21} \end{pmatrix} \mathbf{w}. \tag{6.55}$$

Suppose the matrix $\mathbf{R} = \mathbf{I} - \mathbf{D}_{22}\mathbf{D}_K$ is invertible. Then it is easy to show that

$$\begin{pmatrix} \mathbf{I} & -\mathbf{D}_K \\ -\mathbf{D}_{22} & \mathbf{I} \end{pmatrix}^{-1} = \begin{pmatrix} \mathbf{I} + \mathbf{D}_K \mathbf{R}^{-1} \mathbf{D}_{22} & \mathbf{D}_K \mathbf{R}^{-1} \\ \mathbf{R}^{-1} \mathbf{D}_{22} & \mathbf{R}^{-1} \end{pmatrix}.$$

Using this equation and (6.55), it is possible to eliminate $\mathbf{u}, \mathbf{y}$ from (6.54) and $\mathbf{u}$ from the second equation of (6.52). In this way, a representation of $\mathbf{F}_{zw}(s)$ can be obtained. The following equation holds:

$$\mathbf{F}_{zw}(s) = \left(\begin{array}{c|c} \mathbf{A}_c & \mathbf{B}_c \\ \hline \mathbf{C}_c & \mathbf{D}_c \end{array}\right),$$

where

$$\mathbf{A}_c = \begin{pmatrix} \mathbf{A} & \mathbf{B}_2\mathbf{C}_K \\ \mathbf{0} & \mathbf{A}_K \end{pmatrix} + \begin{pmatrix} \mathbf{B}_2\mathbf{D}_K \\ \mathbf{B}_K \end{pmatrix} \mathbf{R}^{-1} (\mathbf{C}_2 \quad \mathbf{D}_{22}\mathbf{C}_K) \tag{6.56}$$

$$\mathbf{B}_c = \begin{pmatrix} \mathbf{B}_1 \\ \mathbf{0} \end{pmatrix} + \begin{pmatrix} \mathbf{B}_2\mathbf{D}_K \\ \mathbf{B}_K \end{pmatrix} \mathbf{R}^{-1}\mathbf{D}_{21}$$

$$\mathbf{C}_c = (\mathbf{C}_1 \quad \mathbf{D}_{12}\mathbf{C}_K) + \mathbf{D}_{12}\mathbf{D}_K\mathbf{R}^{-1}(\mathbf{C}_2 \quad \mathbf{D}_{22}\mathbf{C}_K)$$

$$\mathbf{D}_c = \mathbf{D}_{11} + \mathbf{D}_{12}\mathbf{D}_K\mathbf{R}^{-1}\mathbf{D}_{21}.$$

As in the SISO case, the feedback system is said to be **well-posed** if each input $\mathbf{w}$ is uniquely mapped on an output $\mathbf{z}$. The preceding calculation implies the following lemma.

Lemma 6.5.1. *The feedback system in Fig. 6.8 is well-posed if and only if the matrix* $\mathbf{R} = \mathbf{I} - \mathbf{D}_{22}\mathbf{D}_K$ *is nonsingular.*

As in the SISO case we now define internal stability for a well-posed feedback system.

Definition. *The general feedback system with state space representation (6.52) and (6.53) is said to be* ***internally stable*** *if the closed-loop system matrix* $\mathbf{A}_c$ *is stable.*

If the feedback system is internally stable, the closed-loop transfer function $\mathbf{F}_{zw}(s)$ is stable. The converse is not generally true, since cancellations may occur. Note that this definition is different from that given in Sect. 3.2 for SISO systems, but, as will be shown in the next section, both definitions are equivalent for such systems.

Definition. *A system* $\mathbf{P}$ *is said to be* ***stabilizable*** *if there is a proper controller* $\mathbf{K}$ *such that the feedback system is well-posed and internally stable. The controller is then said to be* ***admissible***.

The next theorem characterizes the existence of admissible controllers.

Theorem. 6.5.1. *(a) An admissible controller* $\mathbf{K}$ *exists if and only if* $(\mathbf{A}, \mathbf{B}_2)$ *is stabilizable and* $(\mathbf{C}_2, \mathbf{A})$ *is detectable.*

(b) Suppose $(\mathbf{A}, \mathbf{B}_2)$ *is stabilizable and* $(\mathbf{C}_2, \mathbf{A})$ *is detectable and let* $\mathbf{F}$ *and* $\mathbf{L}$ *be such that* $\mathbf{A} + \mathbf{B}_2\mathbf{F}$ *and* $\mathbf{A} + \mathbf{L}\mathbf{C}_2$ *are stable. Then an admissible controller is given by*

$$\mathbf{K}(s) = \left(\begin{array}{c|c} \mathbf{A} + \mathbf{B}_2\mathbf{F} + \mathbf{L}\mathbf{C}_2 + \mathbf{L}\mathbf{D}_{22}\mathbf{F} & -\mathbf{L} \\ \hline \mathbf{F} & \mathbf{0} \end{array}\right). \tag{6.57}$$

Proof. (a) ($\Leftarrow$) Since $(\mathbf{A}, \mathbf{B}_2)$ is stabilizable and $(\mathbf{C}_2, \mathbf{A})$ is detectable, an application of Theorem 6.4.1 shows that there are matrices $\mathbf{F}$ and $\mathbf{L}$ such that $\mathbf{A}+\mathbf{B}_2\mathbf{F}$ and $\mathbf{A}+\mathbf{L}\mathbf{C}_2$ are stable. The idea is now to use the observer-based controller constructed in Sect. 6.2, namely the controller given in (6.57) (cf. (6.19)). Because $\mathbf{D}_K = \mathbf{0}$, the feedback loop is well-posed. Then the closed-loop system matrix (6.56) is given by

$$\mathbf{A}_c = \begin{pmatrix} \mathbf{A} & \mathbf{B}_2\mathbf{F} \\ -\mathbf{L}\mathbf{C}_2 & \mathbf{A}+\mathbf{B}_2\mathbf{F}+\mathbf{L}\mathbf{C}_2 \end{pmatrix}.$$

Now define

$$\mathbf{T} = \begin{pmatrix} \mathbf{I} & \mathbf{0} \\ -\mathbf{I} & \mathbf{I} \end{pmatrix}.$$

With this matrix, the following similarity transform of $\mathbf{A}_c$ may be performed:

$$\mathbf{T}\mathbf{A}_c\mathbf{T}^{-1} = \begin{pmatrix} \mathbf{A}+\mathbf{B}_2\mathbf{F} & \mathbf{B}_2\mathbf{F} \\ \mathbf{0} & \mathbf{A}+\mathbf{L}\mathbf{C}_2 \end{pmatrix}.$$

(This corresponds to the introduction of the new variable $\mathbf{e}$ in Sect. 6.2.1). Hence the eigenvalues of $\mathbf{A}_c$ are precisely those of $\mathbf{A}+\mathbf{B}_2\mathbf{F}$ and $\mathbf{A}+\mathbf{L}\mathbf{C}_2$. Thus, $\mathbf{A}_c$ is stable.
($\Rightarrow$) Suppose $(\mathbf{A}, \mathbf{B}_2)$ is not stabilizable or $(\mathbf{C}_2, \mathbf{A})$ is not detectable. Then there are eigenvalues of $\mathbf{A}_c$ with a nonnegative real part which cannot be changed by the controller. The further details of the proof are left to the reader.
(b) The assertion of (b) is already shown in the proof of (a).

The next lemma gives a sufficient condition for the equivalence of the stability of $\mathbf{F}_{zw}$ and the internal stability of the closed-loop system.

Lemma 6.5.2. *Let the realizations for* $\mathbf{P}$ *and* $\mathbf{K}$ *both be stabilizable and detectable and assume that the following assumptions hold:*

(A1) The following matrix has full column rank for all λ *with* $\operatorname{Re}\lambda \geq 0$:

$$\left(\begin{array}{c|c} \mathbf{A}-\lambda\mathbf{I} & \mathbf{B}_2 \\ \hline \mathbf{C}_1 & \mathbf{D}_{12} \end{array}\right).$$

(A2) The following matrix has full row rank for all λ *with* $\operatorname{Re}\lambda \geq 0$:

$$\left(\begin{array}{c|c} \mathbf{A}-\lambda\mathbf{I} & \mathbf{B}_1 \\ \hline \mathbf{C}_2 & \mathbf{D}_{21} \end{array}\right).$$

Then $\mathbf{K}$ *is an internally stabilizing controller if and only if* $\mathbf{F}_{zw} \in \mathcal{RH}_\infty$.

Proof. We show that under the given conditions the feedback system is stabilizable and detectable. Then the assertion follows from Theorem 5.6.1. Let $\operatorname{Re}\lambda \geq 0$ and

$$\mathbf{A}_c\mathbf{x} = \lambda\mathbf{x}, \quad \mathbf{C}_c\mathbf{x} = \mathbf{0}.$$

If $\mathbf{x}$ is adequately partitioned, these equations can be written in the form

$$\begin{aligned}
&\mathbf{A}\mathbf{x}_1 - \lambda\mathbf{x}_1 + \mathbf{B}_2\mathbf{v} = \mathbf{0} \\
&\mathbf{B}_K\mathbf{R}^{-1}(\mathbf{C}_2\mathbf{x}_1 + \mathbf{D}_{22}\mathbf{C}_K\mathbf{x}_2) + \mathbf{A}_K\mathbf{x}_2 - \lambda\mathbf{x}_2 = \mathbf{0} \\
&\mathbf{C}_1\mathbf{x}_1 + \mathbf{D}_{12}\mathbf{v} = \mathbf{0},
\end{aligned}$$

where we have set

$$\mathbf{v} = \mathbf{D}_K\mathbf{R}^{-1}\mathbf{C}_2\mathbf{x}_1 + \mathbf{C}_K\mathbf{x}_2 + \mathbf{D}_K\mathbf{R}^{-1}\mathbf{D}_{22}\mathbf{C}_K\mathbf{x}_2.$$

Assumption (A1) now gives $\mathbf{x}_1 = \mathbf{0}$, $\mathbf{v} = \mathbf{0}$. This implies

$$(\mathbf{I} + \mathbf{D}_K\mathbf{R}^{-1}\mathbf{D}_{22})\mathbf{C}_K\mathbf{x}_2 = \mathbf{0}.$$

Multiplication of this equation with $\mathbf{I} - \mathbf{D}_K\mathbf{D}_{22}$ yields $\mathbf{C}_K\mathbf{x}_2 = \mathbf{0}$ and therefore $\mathbf{A}_K\mathbf{x}_2 = \lambda\mathbf{x}_2$. Since the realization of $\mathbf{K}$ is detectable, $\mathbf{x}_2 = \mathbf{0}$ follows, and thus the closed-loop system is detectable. The stabilizability is shown in a similar way.

6.5.2 Internal Stability and Generalized Nyquist Criterion

In this section, we want to take a closer look at internal stability and conceptually we proceed in quite a similar way to that used in Sect. 3.2. Our first step consists in noting that internal stability can be completely characterized by the part $\mathbf{P}_{22}$ of the plant and the controller $\mathbf{K}$. To this end, consider Fig. 6.10. Here, two disturbances $\mathbf{d}_1$ and $\mathbf{d}_2$ act as the inputs, and the outputs are the quantities $\mathbf{v}_1$ and $\mathbf{v}_2$. Whether the disturbances are actually present in the physical system does not matter. The next equations follow directly from this diagram:

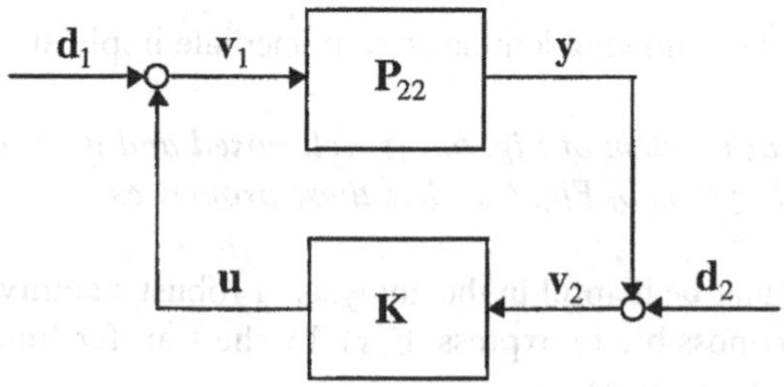

Fig. 6.10. Diagram for internal stability

$$\begin{pmatrix} \dot{\mathbf{x}} \\ \dot{\mathbf{x}}_K \end{pmatrix} = \begin{pmatrix} \mathbf{A} & \mathbf{0} \\ \mathbf{0} & \mathbf{A}_K \end{pmatrix}\begin{pmatrix} \mathbf{x} \\ \mathbf{x}_K \end{pmatrix} + \begin{pmatrix} \mathbf{B}_2 & \mathbf{0} \\ \mathbf{0} & \mathbf{B}_K \end{pmatrix}\begin{pmatrix} \mathbf{d}_1 \\ \mathbf{d}_2 \end{pmatrix} + \begin{pmatrix} \mathbf{B}_2 & \mathbf{0} \\ \mathbf{0} & \mathbf{B}_K \end{pmatrix}\begin{pmatrix} \mathbf{u} \\ \mathbf{y} \end{pmatrix}$$

$$\begin{pmatrix} \mathbf{y} \\ \mathbf{u} \end{pmatrix} = \begin{pmatrix} \mathbf{C}_2 & \mathbf{0} \\ \mathbf{0} & \mathbf{C}_K \end{pmatrix}\begin{pmatrix} \mathbf{x} \\ \mathbf{x}_K \end{pmatrix} + \begin{pmatrix} \mathbf{D}_{22} & \mathbf{0} \\ \mathbf{0} & \mathbf{D}_K \end{pmatrix}\begin{pmatrix} \mathbf{v}_1 \\ \mathbf{v}_2 \end{pmatrix}$$

$$\begin{pmatrix} \mathbf{v}_1 \\ \mathbf{v}_2 \end{pmatrix} = \begin{pmatrix} \mathbf{d}_1 \\ \mathbf{d}_2 \end{pmatrix} + \begin{pmatrix} \mathbf{0} & \mathbf{I} \\ \mathbf{I} & \mathbf{0} \end{pmatrix}\begin{pmatrix} \mathbf{y} \\ \mathbf{u} \end{pmatrix}.$$

The last two equations imply

$$\begin{pmatrix} \mathbf{I} & -\mathbf{D}_K \\ -\mathbf{D}_{22} & \mathbf{I} \end{pmatrix}\begin{pmatrix} \mathbf{v}_1 \\ \mathbf{v}_2 \end{pmatrix} = \begin{pmatrix} \mathbf{d}_1 \\ \mathbf{d}_2 \end{pmatrix} + \begin{pmatrix} \mathbf{0} & \mathbf{C}_K \\ \mathbf{C}_2 & \mathbf{0} \end{pmatrix}\begin{pmatrix} \mathbf{x} \\ \mathbf{x}_K \end{pmatrix}.$$

We define now a transfer function $\mathbf{F}(s)$ by

$$\begin{pmatrix} \mathbf{v}_1 \\ \mathbf{v}_2 \end{pmatrix} = \mathbf{F}(s)\begin{pmatrix} \mathbf{d}_1 \\ \mathbf{d}_2 \end{pmatrix} = \left(\begin{array}{c|c} \mathbf{A}_c & \mathbf{B}_c \\ \hline \mathbf{C}_c & \mathbf{D}_c \end{array}\right)\begin{pmatrix} \mathbf{d}_1 \\ \mathbf{d}_2 \end{pmatrix}. \tag{6.58}$$

It is now easy to show that $\mathbf{F}(s)$ has the following state space representation (if $\mathbf{R} = \mathbf{I} - \mathbf{D}_{22}\mathbf{D}_K$ is nonsingular):

$$\mathbf{D}_c = \begin{pmatrix} \mathbf{I} & -\mathbf{D}_K \\ -\mathbf{D}_{22} & \mathbf{I} \end{pmatrix}^{-1} = \begin{pmatrix} \mathbf{I} + \mathbf{D}_K\mathbf{R}^{-1}\mathbf{D}_{22} & \mathbf{D}_K\mathbf{R}^{-1} \\ \mathbf{R}^{-1}\mathbf{D}_{22} & \mathbf{R}^{-1} \end{pmatrix} \tag{6.59}$$

$$\begin{aligned} \mathbf{A}_c &= \begin{pmatrix} \mathbf{A} & \mathbf{0} \\ \mathbf{0} & \mathbf{A}_K \end{pmatrix} + \begin{pmatrix} \mathbf{B}_2 & \mathbf{0} \\ \mathbf{0} & \mathbf{B}_K \end{pmatrix}\mathbf{D}_c\begin{pmatrix} \mathbf{0} & \mathbf{C}_K \\ \mathbf{C}_2 & \mathbf{0} \end{pmatrix} \\ &= \begin{pmatrix} \mathbf{A} & \mathbf{B}_2\mathbf{C}_K \\ \mathbf{0} & \mathbf{A}_K \end{pmatrix} + \begin{pmatrix} \mathbf{B}_2\mathbf{D}_K \\ \mathbf{B}_K \end{pmatrix}\mathbf{R}^{-1}\begin{pmatrix} \mathbf{C}_2 & \mathbf{D}_{22}\mathbf{C}_K \end{pmatrix} \end{aligned} \tag{6.60}$$

$$\mathbf{B}_c = \begin{pmatrix} \mathbf{B}_2 & \mathbf{0} \\ \mathbf{0} & \mathbf{B}_K \end{pmatrix}\mathbf{D}_c, \quad \mathbf{C}_c = \mathbf{D}_c\begin{pmatrix} \mathbf{0} & \mathbf{C}_K \\ \mathbf{C}_2 & \mathbf{0} \end{pmatrix}. \tag{6.61}$$

From this representation the following lemma is an immediate implication.

Lemma 6.5.3. *The feedback system of Fig. 6.8 is well-posed and internally stable if and only if the feedback system of Fig. 6.10 has these properties.*

This lemma will in particular be helpful in the analysis of robust stability.

On the other hand it is possible to express $\mathbf{F}(s)$ by the transfer functions $\mathbf{P}_{22}$ and $\mathbf{K}$. Figure 6.10 directly shows that

$$\hat{\mathbf{v}}_1 = \hat{\mathbf{d}}_1 + \mathbf{K}\hat{\mathbf{v}}_2$$
$$\hat{\mathbf{v}}_2 = \hat{\mathbf{d}}_2 + \mathbf{P}_{22}\hat{\mathbf{v}}_1 .$$

This can be rewritten with matrices as

$$\begin{pmatrix} \mathbf{I} & -\mathbf{K} \\ -\mathbf{P}_{22} & \mathbf{I} \end{pmatrix}\begin{pmatrix} \mathbf{v}_1 \\ \mathbf{v}_2 \end{pmatrix} = \begin{pmatrix} \mathbf{d}_1 \\ \mathbf{d}_2 \end{pmatrix}.$$

The inverse of the matrix on the left-hand side of this equation is given by (if the transfer functions $\mathbf{I}-\mathbf{P}_{22}\mathbf{K}$ and $\mathbf{I}-\mathbf{K}\mathbf{P}_{22}$ are invertible)

$$\begin{pmatrix} \mathbf{I} & -\mathbf{K} \\ -\mathbf{P}_{22} & \mathbf{I} \end{pmatrix}^{-1} = \begin{pmatrix} (\mathbf{I}-\mathbf{K}\mathbf{P}_{22})^{-1} & \mathbf{K}(\mathbf{I}-\mathbf{P}_{22}\mathbf{K})^{-1} \\ \mathbf{P}_{22}(\mathbf{I}-\mathbf{K}\mathbf{P}_{22})^{-1} & (\mathbf{I}-\mathbf{P}_{22}\mathbf{K})^{-1} \end{pmatrix}. \tag{6.62}$$

The following formulas are easily verified:

$$\mathbf{P}_{22}(\mathbf{I}-\mathbf{K}\mathbf{P}_{22})^{-1} = (\mathbf{I}-\mathbf{P}_{22}\mathbf{K})^{-1}\mathbf{P}_{22} \tag{6.63}$$

$$(\mathbf{I}-\mathbf{K}\mathbf{P}_{22})^{-1} = \mathbf{I} + \mathbf{K}(\mathbf{I}-\mathbf{P}_{22}\mathbf{K})^{-1}\mathbf{P}_{22}. \tag{6.64}$$

Hence the invertibility of the transfer matrix in (6.62) is given if $\mathbf{I}-\mathbf{P}_{22}\mathbf{K}$ is invertible. This is the case if and only if

$$\mathbf{I}-\mathbf{P}_{22\infty}\mathbf{K}_\infty = \mathbf{I}-\mathbf{D}_{22}\mathbf{D}_K$$

is invertible.

As in the SISO case, it is helpful to define the following transfer functions. The transfer matrices

$$\mathbf{L}_o = -\mathbf{P}_{22}\mathbf{K}, \quad \mathbf{L}_i = -\mathbf{K}\mathbf{P}_{22}$$

are the **output loop transfer matrix** and **the input loop transfer matrix**, respectively. The transfer matrices

$$\mathbf{S}_o = (\mathbf{I}+\mathbf{L}_o)^{-1}, \quad \mathbf{S}_i = (\mathbf{I}+\mathbf{L}_i)^{-1}$$

are the **output sensitivity function** and **the input sensitivity function**, respectively. With these definitions we can write

$$\mathbf{F}(s) = \begin{pmatrix} \mathbf{I} & -\mathbf{K} \\ -\mathbf{P}_{22} & \mathbf{I} \end{pmatrix}^{-1} = \begin{pmatrix} \mathbf{S}_i & \mathbf{K}\mathbf{S}_o \\ \mathbf{P}_{22}\mathbf{S}_i & \mathbf{S}_o \end{pmatrix} \tag{6.65}$$

and

$$\mathbf{v}_1 = \mathbf{S}_i\mathbf{d}_1 + \mathbf{K}\mathbf{S}_o\mathbf{d}_2$$
$$\mathbf{v}_2 = \mathbf{P}_{22}\mathbf{S}_i\mathbf{d}_1 + \mathbf{S}_o\mathbf{d}_2 .$$

The next theorem states that (under some mild conditions) the internal stability and the stability of the closed-loop transfer function are equivalent for the feedback system of Fig. 6.10. For SISO systems, this was the definition of internal stability in Sect. 3.2.

Theorem 6.5.2. *(a) Let* $\mathbf{R} = \mathbf{I} - \mathbf{D}_{22}\mathbf{D}_K$ *be nonsingular. Then the transfer matrix* $\mathbf{F}(s)$ *defined in (6.58) exists and has the state space representation (6.59)–(6.61). Moreover,* $\mathbf{F}(s)$ *can be written in the form (6.65).*

(b) Suppose that $(\mathbf{A}, \mathbf{B}_2)$ *and* $(\mathbf{A}_K, \mathbf{B}_K)$ *are stabilizable and that* $(\mathbf{C}_2, \mathbf{A})$ *and* $(\mathbf{C}_K, \mathbf{A}_K)$ *are detectable. Then the feedback system of Fig. 6.10 is internally stable if and only if the transfer matrix (6.65) or the transfer matrices* $\mathbf{S}_i, \mathbf{S}_o, \mathbf{K}\mathbf{S}_o, \mathbf{P}_{22}\mathbf{S}_i$ *are stable.*

Proof. (a) This is already shown by our calculations above.
(b) "$\Rightarrow$": This is true since each pole of $\mathbf{F}(s)$ is an eigenvalue of $\mathbf{A}_c$.
"$\Leftarrow$": An application of the PHB test shows that since $(\mathbf{A}, \mathbf{B}_2, \mathbf{C}_2)$ and $(\mathbf{A}_K, \mathbf{B}_K, \mathbf{C}_K)$ are stabilizable and detectable, $(\mathbf{A}_c, \mathbf{B}_c, \mathbf{C}_c)$ is stabilizable and detectable. Thus from Theorem 5.6.1 we deduce that each eigenvalue λ of $\mathbf{A}_c$ with $\operatorname{Re}\lambda \geq 0$ is a pole of $\mathbf{F}(s)$. Since $\mathbf{F}(s)$ is assumed to be stable, $\mathbf{A}_c$ is stable.

In the sequel we denote by n_P and n_K be the number of poles of $\mathbf{P}_{22}(s)$ and $\mathbf{K}(s)$, respectively, lying in the closed RHP. Hereby, these poles are counted with respect to their multiplicities. The next theorem is the direct generalization of Theorem 3.2.1 to MIMO systems.

Theorem 6.5.3. *The general feedback system is well-posed and internally stable if and only if the following assertions hold:*

(i) $\mathbf{L}_o(s)$ *has exactly* $n_K + n_P$ *poles in the closed RHP.*

(ii) $\mathbf{S}_o(s) = (\mathbf{I} - \mathbf{P}_{22}(s)\mathbf{K}(s))^{-1}$ *exists and is stable.*

Proof. First we need a state space representation of $\mathbf{L}_o$ and $\mathbf{S}_o$. Equation (5.54) gives

$$\mathbf{P}_{22}\mathbf{K} = \left(\begin{array}{c|c} \mathbf{A}_o & \mathbf{B}_o \\ \hline \mathbf{C}_o & \mathbf{D}_o \end{array}\right) = \left(\begin{array}{cc|c} \mathbf{A} & \mathbf{B}_2\mathbf{C}_K & \mathbf{B}_2\mathbf{D}_K \\ \mathbf{0} & \mathbf{A}_K & \mathbf{B}_2 \\ \hline \mathbf{C}_2 & \mathbf{D}_{22}\mathbf{C}_K & \mathbf{D}_{22}\mathbf{D}_K \end{array}\right). \tag{6.66}$$

If $\mathbf{d}_1 = \mathbf{0}$, then $\hat{\mathbf{v}}_2 = \mathbf{S}_o\hat{\mathbf{d}}_2$ and (cf. Fig. 6.10)

$$\dot{\mathbf{x}} = \mathbf{A}\mathbf{x} + \mathbf{B}_2\mathbf{u}$$

$$\mathbf{v}_2 = \mathbf{C}_2\mathbf{x} + \mathbf{d}_2 + \mathbf{D}_{22}\mathbf{u} \ .$$

Thus the matrices describing the feedback system are in this case given by

$$\mathbf{B}_1 = \mathbf{0}, \quad \mathbf{C}_1 = \mathbf{C}_2, \quad \mathbf{D}_{11} = \mathbf{I}, \quad \mathbf{D}_{12} = \mathbf{D}_{22}, \quad \mathbf{D}_{21} = \mathbf{I} \, .$$

An application of the general formulas for the feedback loop yields

$$\mathbf{S}_o = \left(\begin{array}{c|c} \mathbf{A}_c & \mathbf{B}_c \\ \hline \mathbf{C}_c & \mathbf{D}_c \end{array} \right),$$

where $\mathbf{A}_c$ is as in (6.56) and

$$\mathbf{B}_c = \begin{pmatrix} \mathbf{B}_2\mathbf{D}_K \\ \mathbf{B}_K \end{pmatrix} \mathbf{R}^{-1}, \ \mathbf{C}_c = \mathbf{R}^{-1}\left(\mathbf{C}_2 \quad \mathbf{D}_{22}\mathbf{D}_K\right), \ \mathbf{D}_c = \mathbf{R}^{-1} .$$

The following assertions can easily be shown by applying the PHB test:

$$(\mathbf{A}_c, \mathbf{B}_c) \text{ is stabilizable} \Leftrightarrow (\mathbf{A}_o, \mathbf{B}_o) \text{ is stabilizable;} \tag{6.67}$$

$$(\mathbf{C}_c, \mathbf{A}_c) \text{ is detectable} \Leftrightarrow (\mathbf{C}_o, \mathbf{A}_o) \text{ is detectable.} \tag{6.68}$$

Moreover, we have:

$$\text{(i)} \Leftrightarrow (\mathbf{A}_o, \mathbf{B}_o) \text{ is stabilizable and } (\mathbf{C}_o, \mathbf{A}_o) \text{ is detectable} \, . \tag{6.69}$$

This follows directly from the Kalman decomposition (Theorem 5.4.4). If (i) holds, no eigenvalue λ of $\mathbf{A}_o$ with $\operatorname{Re}\lambda \geq 0$ can be eigenvalue of one of the matrices $\tilde{\mathbf{A}}_{22}, \tilde{\mathbf{A}}_{33}, \tilde{\mathbf{A}}_{44}$ because otherwise this would lead to a pole/zero cancelation when $\mathbf{L}_o$ is formed. Hence $(\mathbf{A}_o, \mathbf{B}_o)$ is stabilizable and $(\mathbf{C}_o, \mathbf{A}_o)$ is detectable. Vice versa, if the stabilizability and the detectability assumptions are satisfied, each unstable eigenvalue of $\mathbf{A}_o$ is eigenvalue of $\tilde{\mathbf{A}}_{11}$ and therefore also pole of $\mathbf{L}_o$ (cf. Theorem 5.6.1(b)). Consequently, (i) must hold.

We are now in a position to prove the theorem. The argument makes again use of Theorem 5.6.1.

"$\Leftarrow$": If (i) holds, then (6.67) – (6.69) imply that $(\mathbf{A}_c, \mathbf{B}_c)$ is stabilizable and $(\mathbf{C}_c, \mathbf{A}_c)$ is detectable. Thus each eigenvalue λ of $\mathbf{A}_c$ with $\operatorname{Re}\lambda \geq 0$ is a pole of $\mathbf{S}_o$. Since by (ii) $\mathbf{S}_o$ is assumed to be stable, there are no such poles, which proves the internal stability.

"$\Rightarrow$": If the general feedback system is internally stable, then $\mathbf{S}_o$ is stable. Moreover, in this case $(\mathbf{A}_c, \mathbf{B}_c)$ is stabilizable and $(\mathbf{C}_c, \mathbf{A}_c)$ is detectable. Thus by (6.67) and (6.68), $(\mathbf{A}_o, \mathbf{B}_o)$ is stabilizable and $(\mathbf{C}_o, \mathbf{A}_o)$ is detectable. Hence from (6.69) we see that (i) must hold.

It is possible Corollary 5.6.1(b) to $\mathbf{I} + \mathbf{L}_o$. By this way it is possible to characterize stability of $\mathbf{S}_o$ by the transmission zeros of $\mathbf{I} + \mathbf{L}_o$.

Corollary 6.5.1 $\mathbf{S}_o$ *is stable if and only if* $\mathbf{I}+\mathbf{L}_o$ *has all its transmission zeros in the open LHP.*

In order to prove a generalization of the Nyquist stability criterion to MIMO systems, the following lemma will be used, which is an immediate consequence of (5.58).

Lemma 6.5.4 (Hsu–Chen Theorem). *Let*

$$\mathbf{L}_o(s)=\left(\begin{array}{c|c}\mathbf{A}_0 & \mathbf{B}_0\\ \hline \mathbf{C}_0 & \mathbf{D}_0\end{array}\right) \tag{6.70}$$

be an arbitrary realization of $\mathbf{L}_o$ *and suppose that* $\mathbf{I}+\mathbf{D}_0$ *is invertible. Define*

$$c=\det(\mathbf{I}+\mathbf{D}_0)\,,\quad \mathbf{A}_c=\mathbf{A}_0-\mathbf{B}_0(\mathbf{I}+\mathbf{D}_0)^{-1}\mathbf{C}_0\,,$$

$$q(s)=\det(s\mathbf{I}-\mathbf{A}_0)\,,\quad p(s)=\det(s\mathbf{I}-\mathbf{A}_c)\,,\quad \Phi(s)=\det(\mathbf{I}+\mathbf{L}_o(s))\,.$$

Then the following formula holds:

$$\Phi(s)=c\frac{p(s)}{q(s)}\,.$$

Theorem 6.5.4 (Generalized Nyquist Stability Criterion). *Let the realization (6.70) be minimal and suppose that* $\mathbf{I}+\mathbf{D}_0$ *is invertible. Moreover, assume that* $\mathbf{I}+\mathbf{L}_o$ *has no transmission zero that is also a pole of this transfer function. Denote by* n_o *the number of poles of* $\mathbf{L}_o$. *Then* $\mathbf{S}_o$ *is stable if and only if the origin does not lie on the Nyquist plot* $\mathbf{C}_{p\,\Phi}$ *and the Nyquist plot encircles the origin* n_o *times in the mathematical positive sense.*

Proof. The assumption concerning the transmission zeros of $\mathbf{I}+\mathbf{L}_o$ guarantees that no pole/zero cancelation occurs when $\Phi(s)$ is formed (cf. (5.57) and Theorem 5.6.2). The result follows now from Corollary 3.4.1.

Notes and References

The idea of estimating the state from the measurement by an observer was introduced in Luenberger [65]. The solution of the pole placement for SISO systems appears in Popov [75] and for MIMO problems it was first solved in Wonham [106]. The robust eigenvalue and eigenstructure assignment presented in Sect. 6.4.1 can be found in Kautsky, Nichols, and Van Dooren [57]. The presentation of basic properties of general feedback systems (Sect. 6.5) is based on the presentation given in Zhou, Doyle, and Glover [112].

Chapter 7

Norms of Systems and Performance

In the preceding chapter, we developed systematically the structure of the general feedback configuration for multivariable systems. In comparison to the standard feedback loop for SISO systems, substantial generalizations were obtained, but as we saw, in particular in Sect. 6.5.2, many of them are based in a natural way on classical concepts.

Our next goal consists in finding the mathematical tools which are necessary to formulate performance specifications for feedback systems. The basic idea has already been given in Sect. 3.5.1, where we introduced the maximum of the magnitude plot of certain transfer functions to measure performance. These transfer functions made use of weighting functions and the maximum could be viewed as a norm. A transfer function, or more generally a transfer matrix, can be considered as an operator which maps one function space into another, and the norm of the operator measures in some sense its size.

We start our analysis by defining norms for vectors and matrices. The most important matrix norms are the spectral norm and the Frobenius norm. Both can be expressed by the singular values of the matrix. We will use them in a natural way to define norms for time- or frequency-dependent functions with matrices as values. Such functions are given by the impulse response or the transfer matrix of a linear system. This leads to the $\mathcal{H}_2$ or the $\mathcal{H}_\infty$ norm of a transfer matrix, respectively. In a control theory application, these norms are applied to the closed-loop system, and the controller has to be synthesized such that the norms are small. The plant is extended by suitable weights, which express the design goals. The definition and basic properties of the $\mathcal{H}_2$ and $\mathcal{H}_\infty$ norms are applied in the following chapters, in which optimal controllers and synthesis methods are studied.

Sections 7.4 and 7.5 are devoted to the question of how suitable requirements for the controller design can be imposed and which limitations arise. The chapter ends with a section about coprime factorization and inner functions, which will be used when $\mathcal{H}_\infty$ optimal controllers are investigated.

7.1 Norms of Vectors and Matrices

Definition. *A* ***norm*** $\|\cdot\|$ *on* $\mathbb{R}^n$ *is a real-valued function such that the following conditions are satisfied (for every* $\mathbf{x}, \mathbf{y} \in \mathbb{R}^n$, $\alpha \in \mathbb{R}$ *):*

(i) $\|\mathbf{x}\| \geq 0$,

(ii) $\|\mathbf{x}\| = 0 \Leftrightarrow \mathbf{x} = \mathbf{0}$,

(iii) $\|\alpha\mathbf{x}\| = |\alpha| \|\mathbf{x}\|$,

(iv) $\|\mathbf{x}+\mathbf{y}\| \leq \|\mathbf{x}\| + \|\mathbf{y}\|$ *(triangle inequality).*

A norm on $\mathbb{C}^n$ is completely analogously defined. An important example of a norm is the **Euclidean norm** (or **2-norm**). It is defined by

$$\|\mathbf{x}\|_2 = \sqrt{|x_1|^2 + \ldots + |x_n|^2}.$$

This definition is valid for real as well as for complex vectors. The following representation holds:

$$\|\mathbf{x}\|^2 = \langle \mathbf{x}, \mathbf{x} \rangle = \mathbf{x}^* \mathbf{x}. \tag{7.1}$$

Another important norm is the ∞ **-norm**:

$$\|\mathbf{x}\|_\infty = \max_{1 \leq k \leq n} |x_k|.$$

Since matrices may be viewed as vectors, norms are also defined for matrices. Let $\mathbf{A}$ be a $m \times n$ matrix. Then an example for a matrix norm is given by the so-called **induced norm**:

$$\|\mathbf{A}\| = \max_{\mathbf{x} \neq \mathbf{0}} \frac{\|\mathbf{A}\mathbf{x}\|}{\|\mathbf{x}\|}.$$

There are matrix norms, which are not induced by a vector norm. For an arbitrary $n \times n$ - m atrix $\mathbf{A}$ we define

$$\lambda_{\max}(\mathbf{A}) = \max\{\lambda \,|\, \lambda \text{ is an eigenvalue of } \mathbf{A}\}.$$

It can be shown that the following matrix norms are induced by the 2-norm and the ∞ -norm, respectively:

$$\|\mathbf{A}\|_2 = \sqrt{\lambda_{\max}(\mathbf{A}^*\mathbf{A})} \qquad \text{(spectral norm)},$$

$$\|\mathbf{A}\|_\infty = \max_{1 \leq i \leq m} \sum_{j=1}^{n} |a_{ij}| \qquad \text{(row sum norm)}.$$

For any induced matrix norm the following inequality holds:

$$\| \mathbf{AB} \| \leq \| \mathbf{A} \| \, \| \mathbf{B} \| .$$

Let $\mathbf{A}, \mathbf{B}$ be $m \times n$-matrices. Then, from

$$\langle \mathbf{A}, \mathbf{B} \rangle = \operatorname{trace}(\mathbf{A}^* \mathbf{B}) ,$$

a scalar product will be defined on the linear space of $m \times n$ matrices. The norm belonging to this scalar product is denoted as the **Frobenius norm**:

$$\| \mathbf{A} \|_F = \sqrt{\operatorname{trace}(\mathbf{A}^* \mathbf{A})} = \sqrt{\sum_{i=1}^{m} \sum_{j=1}^{n} | a_{ij} |^2} .$$

The Frobenius norm is not induced by a vector norm.

The spectral norm as well as the Frobenius norm can be calculated with the help of the **singular value decomposition** (SVD) of a matrix $\mathbf{A}$. This decomposition is described by the following theorem, which is a basic result of linear algebra. A proof can be found in many books; see, for example, Zhou, Doyle, and Glover [112], Theorem 2.11, where also the additional properties listed below are shown.

Theorem 7.1.1. *Let* $\mathbf{A}$ *be a (real or complex)* $m \times n$*-matrix. Then there is a unitary* $m \times m$*-matrix* $\mathbf{U}$ *and a unitary* $n \times n$*-matrix* $\mathbf{V}$ *such that*

$$\mathbf{A} = \mathbf{U} \boldsymbol{\Sigma} \mathbf{V}^* \quad \text{with} \quad \boldsymbol{\Sigma} = \begin{pmatrix} \boldsymbol{\Sigma}_1 & \mathbf{0} \\ \mathbf{0} & \mathbf{0} \end{pmatrix}$$

and a diagonal matrix

$$\boldsymbol{\Sigma}_1 = \operatorname{diag}(\sigma_1, \ldots, \sigma_p), \quad p = \min\{m, n\} .$$

Furthermore, the following inequality holds:

$$\sigma_1 \geq \sigma_2 \cdots \geq \sigma_p \geq 0 .$$

The numbers $\sigma_1, \sigma_2, \ldots, \sigma_p$ are called **singular values** of $\mathbf{A}$. If $\mathbf{U}, \mathbf{V}$ are written in the form

$$\mathbf{U} = (\mathbf{u}_1 \ \ \ldots \ \ \mathbf{u}_m) ,$$
$$\mathbf{V} = (\mathbf{v}_1 \ \ \ldots \ \ \mathbf{v}_n) ,$$

then

$$\mathbf{A} \mathbf{v}_i = \sigma_i \, \mathbf{u}_i ,$$
$$\mathbf{A}^* \mathbf{u}_i = \sigma_i \, \mathbf{v}_i .$$

This implies

$$\mathbf{A}^*\mathbf{A}\mathbf{v}_i = \sigma_i^2 \mathbf{v}_i ,$$
$$\mathbf{A}\mathbf{A}^*\mathbf{u}_i = \sigma_i^2 \mathbf{u}_i .$$

Hence the numbers σ_i^2 are the eigenvalues of $\mathbf{A}^*\mathbf{A}$ and $\mathbf{A}\mathbf{A}^*$.

Next we note some important properties of the SVD. Let

$$\sigma_1 \geq \cdots \geq \sigma_r > \sigma_{r+1} = \cdots = \sigma_p = 0 .$$

Moreover, we define

$$\bar{\sigma}(\mathbf{A}) = \sigma_1 \qquad \textit{(largest singular value of } \mathbf{A}\textit{)},$$
$$\underline{\sigma}(\mathbf{A}) = \sigma_p \qquad \textit{(smallest singular value of } \mathbf{A}\textit{)}.$$

Then the following assertions hold:

1. $r = \operatorname{rank}(\mathbf{A})$.
2. $\|\mathbf{A}\|_2 = \bar{\sigma}(\mathbf{A})$.
3. $\|\mathbf{A}\|_F^2 = \sigma_1^2 + \ldots + \sigma_n^2$.
4. $\underline{\sigma}(\mathbf{A}) = \min_{\|\mathbf{x}\|=1} \|\mathbf{A}\mathbf{x}\|$, if $m \geq n$.
5. $\mathbf{A}$ has the following dyadic expansion:

$$\mathbf{A} = \sum_{k=1}^{r} \sigma_k \mathbf{u}_k \mathbf{v}_k^* .$$

6. If $\mathbf{A}, \mathbf{B}$ are square matrices, then
   (i) $\underline{\sigma}(\mathbf{AB}) \geq \underline{\sigma}(\mathbf{A})\underline{\sigma}(\mathbf{B})$;
   (ii) $|\underline{\sigma}(\mathbf{A}+\mathbf{B}) - \underline{\sigma}(\mathbf{A})| \leq |\bar{\sigma}(\mathbf{B})|$;
   (iii) $\bar{\sigma}(\mathbf{A}^{-1}) = 1/\underline{\sigma}(\mathbf{A})$ if $\mathbf{A}$ is invertible.

If the norm has no index, in all that follows the 2-norm is meant. We can write

$$\|\mathbf{A}\| = \max_{\mathbf{x} \neq \mathbf{0}} \frac{\|\mathbf{A}\mathbf{x}\|}{\|\mathbf{x}\|} .$$

Consequently, $\mathbf{v}_1$ is the direction in which the matrix $\mathbf{A}$ has its largest amplification. If $\operatorname{rank}(\mathbf{A}) = \min\{m, n\}$, then all singular values are positive. In this case,

$$\kappa(\mathbf{A}) = \frac{\bar{\sigma}(\mathbf{A})}{\underline{\sigma}(\mathbf{A})}$$

is called the **condition number** of $\mathbf{A}$.

7.2 Norms of Systems

7.2.1 The Hardy-Space $\mathcal{H}_2$ and the Laplace Transform

We start with a linear system given in the time domain:

$$\dot{\mathbf{x}} = \mathbf{A}\,\mathbf{x} + \mathbf{B}\mathbf{u}$$
$$\mathbf{y} = \mathbf{C}\mathbf{x} + \mathbf{D}\mathbf{u}\,.$$

It can equivalently be described by a convolution operator F_g, which maps the space of input functions $\mathbf{u}$ to the space of output functions $\mathbf{y}$:

$$F_g(\mathbf{u})(t) = \int_0^t \mathbf{g}(t-\tau)\mathbf{u}(\tau)\,d\tau\,.$$

Here, the function $\mathbf{g}$ is the impulse response of the system (cf. Sect. 5.2). The induced operator norm is then a candidate for the norm of F_g. The question is now, how the function spaces can be suitably chosen. To answer this question, we need some basic facts from functional analysis. Readers who are not familiar with this material are referred to the appendix, in which a few very basic function spaces are defined. Function spaces which are more closely related to control theory will be introduced in the running text.

Another approach for defining an operator norm for a linear system starts with the frequency domain description of the system using the transfer matrix:

$$\hat{\mathbf{y}} = \mathbf{G}(s)\hat{\mathbf{u}}\,.$$

Here, we consider the transfer matrix as an operator, which maps the Laplace transforms of the inputs onto the Laplace transforms of the outputs and again the question arises how these function spaces have to be selected. As for the time domain approach, the operator norm could then be defined as an induced norm. Another possibility is to define the operator norm directly.

First, we make the concept of viewing the transfer matrix as operator more precise. This requires the definition of two function spaces. Let $\mathcal{L}_2^n(i\mathbb{R})$ be the space of all measurable functions $\mathbf{f}(i\omega)$ defined on the imaginary axis with values in $\mathbb{C}^n$ such that

$$\int_{-\infty}^{\infty} \|\mathbf{f}(i\omega)\|^2\,d\omega < \infty$$

(a short remark concerning measurable functions is made in the appendix). A scalar product on this space is given by

$$\langle \mathbf{f},\mathbf{h}\rangle = \frac{1}{2\pi}\int_{-\infty}^{\infty} \langle \mathbf{f}(i\omega),\mathbf{h}(i\omega)\rangle\,d\omega\,.$$

The corresponding norm is

$$\|\mathbf{f}\|_2^2 = \frac{1}{2\pi}\int_{-\infty}^{\infty} \|\mathbf{f}(i\omega)\|^2 \, d\omega .$$

Finally, let $\mathcal{L}_\infty^{l\times m}(i\mathbb{R})$ be the space of all matrix-valued functions $\mathbf{G}(i\omega)$, which are essentially bounded. It has the norm

$$\|\mathbf{G}\|_\infty = \operatorname*{ess\,sup}_{\omega\in\mathbb{R}} \bar{\sigma}(\mathbf{G}(i\omega)) .$$

The space $\mathcal{L}_\infty^n(i\mathbb{R})$ is similarly defined for functions. From now on, we drop the dimension in the notation of the function spaces if, from the context, it is clear what is meant.

Now we are in a position to define for a transfer matrix $\mathbf{G}\in\mathcal{L}_\infty(i\mathbb{R})$ a so-called **multiplication operator** by

$$M_\mathbf{G} : \mathcal{L}_2(i\mathbb{R}) \to \mathcal{L}_2(i\mathbb{R})$$
$$M_\mathbf{G}(\mathbf{f})(i\omega) = \mathbf{G}(i\omega)\mathbf{f}(i\omega) .$$

Clearly, this operator is well defined. We denote the induced operator norm simply by $\|\cdot\|$.

Lemma 7.2.1. *For every function $\mathbf{G}(i\omega)\in\mathcal{L}_\infty(i\mathbb{R})$, the norm of the multiplication operator induced by the 2-norm is equal to the ∞-norm of $\mathbf{G}(i\omega)$:*

$$\|M_\mathbf{G}\| = \|\mathbf{G}\|_\infty . \tag{7.2}$$

We omit the proof, since it is contained in the proof of a different version of this result, which will be proven later (Theorem 7.2.3).

The next question is how this operator can be translated into the time domain. This requires the Fourier transform, which in a natural manner is defined for absolutely integrable functions, but now is needed for square integrable functions. We define for a square integrable function $\mathbf{u}:\mathbb{R}\to\mathbb{C}^n$ the Fourier transform as follows:

$$\mathcal{F}(\mathbf{u})(i\omega) = \int_{-\infty}^{-\infty} e^{-i\omega t}\mathbf{u}(t)\,dt = \lim_{T\to\infty}\int_{-T}^{T} e^{-i\omega t}\mathbf{u}(t)\,dt .$$

If the limit does not exist, we set $\mathcal{F}(\mathbf{u})(i\omega)=0$. The next theorem known as the Plancherel theorem, says that the Fourier transform is a norm preserving mapping between two function spaces.

Theorem 7.2.1. *The Fourier transform is an invertible map between the spaces $\mathcal{L}_2(-\infty,\infty)$ and $\mathcal{L}_2(i\mathbb{R})$ and the following equation holds for all $\mathbf{u},\mathbf{v}$:*

$$\langle\mathcal{F}(\mathbf{u}),\mathcal{F}(\mathbf{v})\rangle = \langle\mathbf{u},\mathbf{v}\rangle .$$

The inverse of the Fourier transform, which is denoted by $\mathcal{F}^{-1}$, is also norm preserving. A state space description of the multiplication operator M_G is

$$G = \mathcal{F}^{-1} M_G \mathcal{F} .$$

This result is not precisely what we want, since the time dependent functions are not defined on $(-\infty, \infty)$, but on $[0, \infty)$. Hence we need a result similar to Theorem 7.2.1 for the Laplace transform. For functions $\mathbf{u} \in \mathcal{L}_2[0, \infty)$, the Laplace transform is defined by

$$\mathcal{L}(\mathbf{u})(s) = \int_0^\infty e^{-st}\mathbf{u}(t)\,dt = \lim_{T\to\infty}\int_0^T e^{-st}\mathbf{u}(t)\,dt .$$

Note that $\mathcal{L}(\mathbf{u})(i\omega) = \mathcal{F}(\mathbf{u})(i\omega)$ if $\mathbf{u}(t) = \mathbf{0}$ for $t < 0$. In order to describe the range of the Laplace transform, the definition of another function space is needed.

Remark. A function $\hat{\mathbf{w}}(s)$ is the Laplace transform of a real-valued function if and only if

$$\overline{\hat{\mathbf{w}}(s)} = \hat{\mathbf{w}}(\overline{s}) . \tag{7.3}$$

This follows directly from the formulas for the Laplace transform and the inverse Laplace transform.

Definition. *A function* $\mathbf{u} : \overline{\mathbb{C}}^+ \to \mathbb{C}^n$ *is in* $\mathcal{H}_2$ *if*

(a) $\mathbf{u}(s)$ *is analytic in the open right half-plane* $\mathbb{C}^+$ *;*

(b) for almost every real number ω *,*

$$\lim_{\sigma\to 0+} \mathbf{u}(\sigma + i\omega) = \mathbf{u}(i\omega) ;$$

(c) $\sup_{\sigma\geq 0} \frac{1}{2\pi}\int_{-\infty}^{\infty} |\mathbf{u}(\sigma + i\omega)|_2^2\, d\omega < \infty$.

It can be shown that for $\mathbf{u} \in \mathcal{H}_2$ the supremum in (c) is always achieved at $\sigma = 0$. Note that the functions of $\mathcal{H}_2$ are defined in the closed right half-plane. Moreover, it is possible to show that two functions of $\mathcal{H}_2$ which coincide on the imaginary axis, are identical on the whole half-plane $\overline{\mathbb{C}}^+$. Hence, there is a one-to-one correspondence between the functions of $\mathcal{H}_2$ and their restrictions to the imaginary axis. By identifying the function with its restriction, it is possible to consider $\mathcal{H}_2$ as a subspace of $\mathcal{L}_2(i\,\mathbb{R})$, which inherits the scalar product and the norm of this space. Thus, in particular, for $\mathbf{u}, \mathbf{v} \in \mathcal{H}_2$ we have

$$\langle \mathbf{u}, \mathbf{v} \rangle = \frac{1}{2\pi}\int_{-\infty}^{\infty} \mathbf{u}^*(i\omega)\mathbf{u}(i\omega)\,d\omega .$$

The next theorem states that the Laplace transform is a norm preserving one to one correspondence between $\mathcal{L}_2[0,\infty)$ and $\mathcal{H}_2$. Herein, (b) is the theorem of Paley-Wiener and (c) is the Theorem of Plancherel (cf. Rudin [77]).

Theorem 7.2.2 *The following assertions hold:*

(a) $\mathbf{u} \in \mathcal{L}_2[0,\infty)$ *implies* $\mathcal{L}(\mathbf{u}) \in \mathcal{H}_2$.

(b) For each $\mathbf{w} \in \mathcal{H}_2$ *there is a unique* $\mathbf{u} \in \mathcal{L}_2[0,\infty)$ *with* $\mathcal{L}(\mathbf{u}) = \mathbf{w}$.

(c) $\langle \mathcal{L}(\mathbf{u}), \mathcal{L}(\mathbf{v}) \rangle = \langle \mathbf{u}, \mathbf{v} \rangle$ *for all* $\mathbf{u}, \mathbf{v}$.

Assertion (c) states in particular that the Laplace transformation is norm preserving:

$$\| \mathcal{L}(\mathbf{u}) \|_2 = \| \mathbf{u} \|_2 .$$

7.2.2 The $\mathcal{H}_2$ Norm as an Operator Norm

In the next section, the results of the preceding section will be used to calculate the induced operator norm of the convolution operator F_g. Here, we first generalize the definition of $\mathcal{H}_2$ spaces to matrix-valued functions of a complex variable, i.e., to transfer matrices. To this end, we need the space $\mathcal{L}_2^{l\times m}(i\mathbb{R})$, which is the space of all matrix-valued functions $\mathbf{G}(i\omega)$ defined on the imaginary axis that are square integrable (where $\mathbf{G}(i\omega)$ is an $l \times m$-matrix). The inner product is

$$\langle \mathbf{F}, \mathbf{G} \rangle = \frac{1}{2\pi} \int_{-\infty}^{\infty} \operatorname{trace}(\mathbf{F}^*(i\omega)\mathbf{G}(i\omega))\, d\omega .$$

Then it is possible to define a $\mathcal{H}_2$ space of transfer matrices $\mathbf{G}(s)$ exactly as in the above definition with the only difference being that condition (c) must be replaced by

$$\sup_{\sigma \geq 0} \frac{1}{2\pi} \int_{-\infty}^{\infty} \operatorname{trace}(\mathbf{G}^*(\sigma + i\omega)\mathbf{G}(\sigma + i\omega))\, d\omega < \infty .$$

The $\mathcal{H}_2$ norm is not an induced operator norm. Despite this, it can also be used to measure performance of a feedback system, as we will see in the next chapter.

The norm on this space is defined by

$$\| \mathbf{G} \|_2^2 = \frac{1}{2\pi} \int_{-\infty}^{\infty} \operatorname{trace}(\mathbf{G}^*(i\omega)\mathbf{G}(i\omega))\, d\omega .$$

Moreover, the 2-norm can be expressed by the singular values as follows:

$$\|\mathbf{G}\|_2^2 = \frac{1}{2\pi}\int_{-\infty}^{\infty}\sum_{k=1}^{n}(\sigma_k(\mathbf{G}(i\omega)))^2\,d\omega\,.$$

Using Theorem 7.2.2(c) it can be seen that this norm can also be calculated in the time domain:

$$\|\mathbf{G}\|_2^2 = \int_0^{\infty}\text{trace}\,(\mathbf{g}^T(t)\mathbf{g}(t))\,dt\,.$$

The subspace of $\mathcal{H}_2$, which consists of all real rational transfer matrices in $\mathcal{H}_2$, will be denoted by $\mathcal{RH}_2$. It consists of all real rational stable and strictly proper transfer matrices.

The 2-norm has the following interesting interpretation. Let $\mathbf{G}(s)\in\mathcal{RH}_2$ and $\mathcal{L}(\mathbf{g})=\mathbf{G}$. Furthermore, suppose the system input $\mathbf{w}(t)$ is a vector of white noise having the following autocorrelation function:

$$E[\mathbf{w}(t)\mathbf{w}^T(t+\tau)] = \delta(\tau)\mathbf{I}\,.$$

Let $\mathbf{y}(t)$ be the system output for $\mathbf{w}$ as the input and define $\mathbf{y}_\infty = \lim_{t\to\infty}\mathbf{y}(t)$. Then

$$E[\mathbf{y}_\infty\mathbf{y}_\infty^T] = \int_0^{\infty}\mathbf{g}(t)\mathbf{g}^T(t)\,dt$$

(cf. Theorem 5.8.2). This yields

$$\begin{aligned} E[\mathbf{y}_\infty^T\mathbf{y}_\infty] &= \text{trace}\,(E[\mathbf{y}_\infty\mathbf{y}_\infty^T]) \\ &= \int_0^{\infty}\text{trace}\,(\mathbf{g}(t)\mathbf{g}^T(t))dt = \int_0^{\infty}\text{trace}\,(\mathbf{g}^T(t)\mathbf{g}(t))dt \end{aligned}$$

and therefore

$$E[\mathbf{y}_\infty^T\mathbf{y}_\infty] = \|\mathbf{G}\|_2^2\,.$$

The 2-norm has a second interpretation. Let (where $\mathbf{e}_1,\ldots,\mathbf{e}_m$ are columns of the identity matrix)

$$\mathbf{u}_k(t) = \delta(t)\mathbf{e}_k \quad \text{and} \quad \hat{\mathbf{y}}_k = \mathbf{G}(s)\hat{\mathbf{u}}_k\,.$$

Then

$$\begin{aligned} \|\mathbf{y}_k\|_2^2 &= \|\hat{\mathbf{y}}_k\|_2^2 = \|\mathbf{G}(i\omega)\mathbf{e}_k\hat{\delta}(i\omega)\|_2^2 \\ &= \|\mathbf{G}(i\omega)\mathbf{e}_k\|_2^2 = \int_{-\infty}^{\infty}\sum_{j=1}^{l}|G_{jk}(i\omega)|^2 d\omega \end{aligned}$$

and consequently

$$\| \mathbf{G} \|_2^2 = \sum_{k=1}^{m} \| \mathbf{y}_k \|_2^2 . \tag{7.4}$$

7.2.3 The Hardy-Space $\mathcal{H}_\infty$ and an Induced Operator Norm

As we have seen in the previous section, the suitable space for the signals $\hat{\mathbf{y}}(s)$ and $\hat{\mathbf{u}}(s)$ is $\mathcal{H}_2$. If we now define a multiplication operator by

$$M_{\mathbf{G}}(\mathbf{f})(s) = \mathbf{G}(s)\mathbf{f}(s),$$

we have to take care that for $\mathbf{f} \in \mathcal{H}_2$ the image $\mathbf{Gf}$ is also in $\mathcal{H}_2$. This requires that $\mathbf{G}(s)$ is analytic and bounded and leads us to a further function space to which $\mathbf{G}(s)$ must belong.

Definition. *A matrix-valued function* $\mathbf{G}: \overline{\mathbb{C}}^+ \to \mathbb{C}^{l \times m}$ *is in* $\mathcal{H}_\infty$ *if*

(a) $\mathbf{G}(s)$ *is analytic in the open right half-plane* $\mathbb{C}^+$;

(b) for almost every real number ω

$$\lim_{\sigma \to 0+} \mathbf{G}(\sigma + i\omega) = \mathbf{G}(i\omega);$$

(c) $\sup_{s \in \overline{\mathbb{C}}^+} \overline{\sigma}(\mathbf{G}(s)) < \infty$.

As in the case of the space $\mathcal{H}_2$, it can be shown that the functions of $\mathcal{H}_\infty$ are uniquely determined by their values on the imaginary axis. Hence $\mathcal{H}_\infty$ can be considered as a subspace of $\mathcal{L}_\infty(i\mathbb{R})$ which inherits its norm:

$$\| \mathbf{G} \|_\infty = \sup_{\omega \in \mathbb{R}} \overline{\sigma}(\mathbf{G}(i\omega)).$$

It should be mentioned that this supremum coincides with that of (b), as it can be shown with the boundary maximum principle of complex analysis.

The subspace of $\mathcal{H}_\infty$ consisting of all real rational transfer matrices in $\mathcal{H}_\infty$ will be denoted by $\mathcal{RH}_\infty$. It consists of all real rational stable and proper transfer matrices.

For $\mathbf{G} \in \mathcal{H}_\infty$, the multiplication operator is now well defined and we can draw the following diagram:

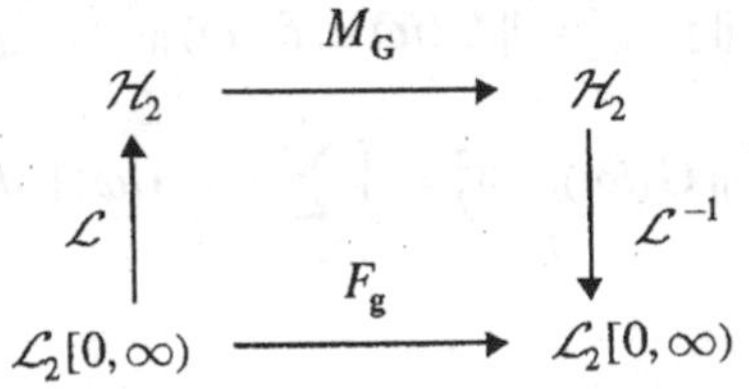

Hence in the time domain, the multiplication operator $M_{\mathbf{G}}$ is

$$F_{\mathbf{g}} = \mathcal{L}^{-1} M_{\mathbf{G}} \mathcal{L}\,.$$

It is also continuous as the following calculation shows:

$$\begin{aligned}\|\mathbf{Gf}\|_2^2 &= \frac{1}{2\pi}\int_{-\infty}^{\infty} \|\mathbf{G}(i\omega)\mathbf{f}(i\omega)\|_2^2\, d\omega \\ &\le \frac{1}{2\pi}\int_{-\infty}^{\infty} \|\mathbf{G}(i\omega)\|_2^2\, \|\mathbf{f}(i\omega)\|_2^2\, d\omega \\ &\le \|\mathbf{G}\|_\infty^2\, \|\mathbf{f}\|_2^2\,.\end{aligned}$$

The operator norm of $M_{\mathbf{G}}$ induced by the 2-norm is

$$\|M_{\mathbf{G}}\| = \sup\{\, \|M_{\mathbf{G}}(\mathbf{f})\|_2 \;\big|\; \mathbf{f} \in \mathcal{H}_2,\ \|\mathbf{f}\|_2 \le 1 \,\}\,.$$

The calculation shows that it is bounded from above by $\|\mathbf{G}\|_\infty$. The following theorems states that these norms are in fact equal.

Theorem 7.2.3. *For every function $\mathbf{G} \in \mathcal{H}_\infty$, the induced norm of the multiplication operator $M_{\mathbf{G}} : \mathcal{H}_2 \to \mathcal{H}_2$ is given by*

$$\|M_{\mathbf{G}}\| = \|\mathbf{G}\|_\infty\,.$$

If $\mathbf{G}$ is the Laplace transform of a real-valued function, then the supremum is achived by a real-valued function.

Proof. We prove the result only if (7.3) holds for $\mathbf{G}$, since this is the most interesting case with respect to applications. We already know that $\|M_{\mathbf{G}}\| \le \|\mathbf{G}\|_\infty$ and show that there is a function $\mathbf{f}(s)$, which is the Laplace transform of a real-valued function, and satisfies

$$\|\mathbf{f}\|_2 \le 1 \quad\text{and}\quad \|\mathbf{Gf}\|_2 = \|M_{\mathbf{G}}(\mathbf{f})\|_2 = \|\mathbf{G}\|_\infty\,.$$

Let now ω_0 be a frequency for which

$$\overline{\sigma}(\mathbf{G}(i\omega_0)) = \|\mathbf{G}\|_\infty$$

(with the possibility of $\omega_0 = \infty$). Applying the singular value decomposition, we can write

$$\mathbf{G}(i\omega_0) = \overline{\sigma}\mathbf{u}_1(i\omega_0)\mathbf{v}_1^*(i\omega_0) + \sum_{k=2}^{r} \sigma_k \mathbf{u}_k(i\omega_0)\mathbf{v}_k^*(i\omega_0)\,.$$

First, we suppose $\omega_0 < \infty$ and represent $\mathbf{v}_1(i\omega_0)$ in the form

$$\mathbf{v}_1(i\omega_0) = \begin{pmatrix} \gamma_1 e^{i\varphi_1} \\ \vdots \\ \gamma_m e^{i\varphi_m} \end{pmatrix}$$

with real numbers γ_k, which are chosen such that $\varphi_k \in (-\pi, 0]$ (m is the number of columns of $\mathbf{G}$). Let now numbers $\beta_k > 0$ be given with the property

$$\varphi_k = \arg \frac{\beta_k - i\omega_0}{\beta_k + i\omega_0},$$

which is always possible in the case of $\varphi_k < 0$. Let

$$\mathbf{w}(s) = \begin{pmatrix} \gamma_1 \dfrac{\beta_1 - s}{\beta_1 + s} \\ \vdots \\ \gamma_m \dfrac{\beta_m - s}{\beta_m + s} \end{pmatrix},$$

where for $\varphi_k = 0$ the fraction has to be replaced by 1. Choose now a scalar analytic function $h(s)$ such that the property (7.3) holds and

$$|h(i\omega)| = \begin{cases} c, & \text{if } |\omega - \omega_0| < \varepsilon \text{ or } |\omega + \omega_0| < \varepsilon, \\ 0, & \text{otherwise}. \end{cases}$$

Here, ε denotes a small positive number and c a constant, which still has to be defined. Let $\mathbf{f}(s) = h(s)\mathbf{w}(s)$. Then we have $\mathbf{f} \in \mathcal{H}_2$ and $\mathbf{f}(i\omega_0) = c\,\mathbf{v}_1(i\omega_0)$. Thus

$$\begin{aligned} \| \mathbf{Gf} \|_2^2 &= \frac{1}{2\pi} \int_{-\infty}^{\infty} \| \mathbf{G}(i\omega)\mathbf{f}(i\omega) \|_2^2 \, d\omega \\ &\approx \frac{\varepsilon}{\pi} \{ \| \mathbf{G}(-i\omega_0)\mathbf{f}(-i\omega_0) \|_2^2 + \| \mathbf{G}(i\omega_0)\mathbf{f}(i\omega_0) \|_2^2 \} \\ &= \frac{\varepsilon c^2}{\pi} \{ \| \mathbf{G}(-i\omega_0)\mathbf{v}_1(-i\omega_0) \|_2^2 + \| \mathbf{G}(i\omega_0)\mathbf{v}_1(i\omega_0) \|_2^2 \} \\ &= \frac{2\varepsilon c^2 \overline{\sigma}^2}{\pi}. \end{aligned}$$

The choice $c^2 = \pi / 2\varepsilon$ gives $\| \mathbf{Gf} \|_2^2 \approx \| \mathbf{G} \|_\infty^2$. The assertions follow for $\varepsilon \to 0$.

For $\omega_0 = \infty$, the assertion can be shown by slightly modifying the above argument.

By definition, the induced operator norm of the convolution operator is

$$\| F_g \| = \sup\{ \| F_g(\mathbf{u}) \|_2 \mid \mathbf{u} \in \mathcal{L}_2[0,\infty),\ \| \mathbf{u} \|_2 \le 1 \}.$$

From Theorem 7.2.3 and Theorem 7.2.2 we see now that this norm coincides with the $\mathcal{H}_\infty$ norm of the transfer matrix, which leads us to the next corollary.

Corollary 7.2.1. *The induced operator norm of the convolution operator* F_g *coincides with the* $\mathcal{H}_\infty$ *norm of the corresponding transfer matrix* $\mathbf{G}$ *:*

$$\| F_g \| = \| \mathbf{G} \|_\infty .$$

This formula shows how the ∞-norm can also be calculated in the time domain. It plays a key role in the theory of $\mathcal{H}_\infty$ control.

The $\mathcal{H}_\infty$ norm of a transfer matrix has the following interpretations. In applications, $\mathbf{G}(i\omega)$ is the frequency response of a transfer matrix $\mathbf{G}(s)$. For each frequency ω one has the singular value decomposition with frequency dependent singular values $\sigma_k(i\omega)$ and singular vectors $\mathbf{u}_k(i\omega)$, $\mathbf{v}_k(i\omega)$. Then the equation

$$\sigma_k(i\omega)\mathbf{u}_k(i\omega) = \mathbf{G}(i\omega)\mathbf{v}_k(i\omega)$$

holds. Since $\mathbf{u}_k(i\omega)$ and $\mathbf{v}_k(i\omega)$ have length 1, $\sigma_k(i\omega)$ is the system gain in direction $\mathbf{v}_k(i\omega)$. The gains $\sigma_k(i\omega)$ are the **principal gains** of the system. If they are plotted in dependency of the frequency, one obtains a generalization of the Bode plot to MIMO systems.

Similarly, the **condition number**

$$\kappa(\mathbf{G}(i\omega)) = \bar{\sigma}(\mathbf{G}(i\omega)) / \underline{\sigma}(\mathbf{G}(i\omega))$$

is frequency dependent. For $\kappa(\mathbf{G}(i\omega)) \approx 1$, the system is said to be **well-conditioned**. If $\kappa(\mathbf{G}(i\omega)) \gg 1$, the system is said to be **ill-conditioned**.

As we have seen above, $\| \mathbf{G} \|_\infty$ is the smallest number for which the inequality

$$\| \mathbf{Gf} \|_2 \le \| \mathbf{G} \|_\infty \| \mathbf{f} \|_2 \quad \text{for every } \mathbf{f} \in \mathcal{L}_2^m(i\mathbb{R})$$

holds. It gives an upper bound of the quadratic mean of the output signal with respect to the frequency.

7.3 Calculation of Operator Norms

Next, we investigate how the $\mathcal{H}_2$ norm and the $\mathcal{H}_\infty$ norm of a transfer matrix can be calculated.

Lemma 7.3.1. *Let the transfer matrix*

$$\mathbf{G}(s) = \left(\begin{array}{c|c} \mathbf{A} & \mathbf{B} \\ \hline \mathbf{C} & \mathbf{0} \end{array}\right)$$

with stable $\mathbf{A}$ *be given and denote by* $\mathbf{X}_o$ ($\mathbf{X}_c$) *the corresponding observability (controllability) gramian. Then*

$$\|\mathbf{G}\|_2^2 = \text{trace}\,(\mathbf{B}^T\mathbf{X}_o\mathbf{B}) = \text{trace}\,(\mathbf{C}\mathbf{X}_c\mathbf{C}^T)\,.$$

Proof. The following equation holds:

$$\|\mathbf{G}\|_2^2 = \int_0^\infty \text{trace}\,(\mathbf{g}^T(t)\mathbf{g}(t))\,dt = \int_0^\infty \text{trace}\,(\mathbf{B}^T e^{\mathbf{A}^T t}\mathbf{C}^T\mathbf{C}e^{\mathbf{A}t}\mathbf{B})\,dt$$

$$= \text{trace}\,(\mathbf{B}^T \int_0^\infty e^{\mathbf{A}^T t}\mathbf{C}^T\mathbf{C}e^{\mathbf{A}t}\,dt\,\mathbf{B}) = \text{trace}\,(\mathbf{B}^T\mathbf{X}_o\mathbf{B})\,.$$

The second equality follows completely analogously by dualization.

In the next theorem, a necessary and sufficient condition for

$$\operatorname*{ess\,sup}_{\omega\in\mathbb{R}} \|\mathbf{G}(i\omega\| < \gamma$$

will be given.

Theorem 7.3.1. *Let* $\gamma > 0$ *and the transfer matrix*

$$\mathbf{G}(s) = \left(\begin{array}{c|c} \mathbf{A} & \mathbf{B} \\ \hline \mathbf{C} & \mathbf{D} \end{array}\right)$$

be given. Put $\mathbf{R} = \gamma^2\mathbf{I} - \mathbf{D}^T\mathbf{D}$ *and suppose that* $\mathbf{A}$ *has no eigenvalues on the imaginary axis. Then the following assertions are equivalent:*

(i) $\|\mathbf{G}\|_\infty < \gamma$.

(ii) $\bar{\sigma}(\mathbf{D}) < \gamma$ *and the matrix*

$$\mathbf{H} = \begin{pmatrix} \mathbf{A} + \mathbf{B}\mathbf{R}^{-1}\mathbf{D}^T\mathbf{C} & \mathbf{B}\mathbf{R}^{-1}\mathbf{B}^T \\ -\mathbf{C}^T(\mathbf{I} + \mathbf{D}\mathbf{R}^{-1}\mathbf{D}^T)\mathbf{C} & -(\mathbf{A} + \mathbf{B}\mathbf{R}^{-1}\mathbf{D}^T\mathbf{C})^T \end{pmatrix}$$

has no eigenvalues on the imaginary axis.

Proof. Assertion (i) is equivalent to

$$\|\mathbf{G}(i\omega)\| < \gamma \text{ for every } \omega \quad \text{and} \quad \|\mathbf{D}\| = \bar{\sigma}(\mathbf{D}) < \gamma\,. \tag{7.5}$$

Define $\Phi(s) = \gamma^2 \mathbf{I} - \mathbf{G}^{\sim}(s)\mathbf{G}(s)$. Then (7.5) is equivalent to

$$\Phi(i\omega) > 0 \text{ for every } \omega \in \mathbb{R} \text{ and } \mathbf{R} > \mathbf{0}.$$

Because $\Phi(i\omega) \geq 0$ and $\mathbf{R} \geq \mathbf{0}$, this is equivalent to

$$\Phi(i\omega) \text{ is nonsingular for every } \omega \text{ and } \mathbf{R} \text{ is nonsingular.}$$

If $\mathbf{R}$ is nonsingular, using the formulas in Sect. 5.5, a short calculation yields

$$\Phi(s) = \left(\begin{array}{cc|c} \mathbf{A} & \mathbf{0} & -\mathbf{B} \\ -\mathbf{C}^T\mathbf{C} & -\mathbf{A}^T & \mathbf{C}^T\mathbf{D} \\ \hline \mathbf{D}^T\mathbf{C} & \mathbf{B}^T & \mathbf{R} \end{array}\right),$$

$$\Phi^{-1}(s) = \left(\begin{array}{cc|c} \multicolumn{2}{c|}{\mathbf{H}} & \begin{matrix} \mathbf{B}\mathbf{R}^{-1} \\ -\mathbf{C}^T\mathbf{D}\mathbf{R}^{-1} \end{matrix} \\ \hline \mathbf{R}^{-1}\mathbf{D}^T\mathbf{C} & \mathbf{R}^{-1}\mathbf{B}^T & \mathbf{R}^{-1} \end{array}\right). \tag{7.6}$$

Since $\mathbf{A}$ has no eigenvalues on the imaginary axis, $i\omega_0$ is a zero of $\det \Phi(s)$ if and only if $i\omega_0$ is an eigenvalue of $\mathbf{H}$, cf. (5.58). Hence the theorem is proven.

Remark. In the situation of the above theorem, each eigenvalue $i\omega_0$ of $\mathbf{H}$ is a pole of $\Phi^{-1}(s)$. This is true, since, as it can be seen from the PHB test, in general for an invertible system $\mathbf{G}$, an eigenvalue s_0 of $\mathbf{A} + \mathbf{B}\mathbf{D}^{-1}\mathbf{C}$, which is not an eigenvalue of $\mathbf{A}$, is a pole of $\mathbf{G}^{-1}$ (cf. Sect. 5.5).

This theorem is called the **bounded real lemma** and can be used to develop an algorithm for computing the $\mathcal{H}_\infty$ norm.

Example. We calculate the 2-norm and the ∞-norm for a second-order system. With $d > 0$ let

$$G(s) = \frac{\omega_0^2}{s^2 + 2d\omega_0 s + \omega_0^2}.$$

Then the system matrices are

$$\mathbf{A} = \begin{pmatrix} 0 & 1 \\ -\omega_0^2 & -2d\omega_0 \end{pmatrix}, \quad \mathbf{B} = \begin{pmatrix} 0 \\ \omega_0^2 \end{pmatrix}, \quad \mathbf{C} = \begin{pmatrix} 1 & 0 \end{pmatrix}, \quad \mathbf{D} = (0).$$

1. *Computation of the 2-norm.* With

$$\mathbf{X}_c = \begin{pmatrix} w_{11} & w_{12} \\ w_{12} & w_{22} \end{pmatrix},$$

the equation for the controllability gramian is

$$\begin{pmatrix} 2w_{12} & w_{22} - \omega_0^2 w_{11} - 2d\omega_0 w_{12} \\ w_{22} - \omega_0^2 w_{11} - 2d\omega_0 w_{12} & -2\omega_0^2 w_{12} - 4d\omega_0 w_{22} \end{pmatrix} = \begin{pmatrix} 0 & 0 \\ 0 & 0 \end{pmatrix}.$$

A short calculation gives $w_{12} = 0$ and

$$w_{11} = \frac{\omega_0}{4d}, \quad w_{22} = \frac{\omega_0^2}{4d}$$

and

$$\| \mathbf{G} \|_2^2 = \text{trace}\,(\mathbf{CWC}^T) = w_{11}.$$

Thus the 2-norm here is

$$\| \mathbf{G} \|_2 = \frac{1}{2}\sqrt{\frac{\omega_0}{d}}.$$

This can be interpreted as follows. If the second order system has white noise with spectral density 1 as the input, the stationary value y_∞ for the output is a normally distributed random variable with expected value 0 and standard deviation

$$\sigma_{y_\infty} = \| \mathbf{G} \|_2.$$

2. *Computation of the* ∞ *-norm.* Here we have

$$\mathbf{H} = \begin{pmatrix} 0 & 1 & 0 & 0 \\ -\omega_0^2 & -2d\omega_0 & 0 & \omega_0^4/\gamma^2 \\ -1 & 0 & 0 & \omega_0^2 \\ 0 & 0 & -1 & 2d\omega_0 \end{pmatrix}$$

and hence

$$\det(s\mathbf{I} - \mathbf{H}) = (s^2 - 2d\omega_0 s + \omega_0^2)(s^2 + 2d\omega_0 s + \omega_0^2) - \omega_0^4/\gamma^2.$$

Let s be an eigenvalue of $\mathbf{H}$ situated on the imaginary axis. Then $s = i\omega$ and ω is the solution of the equation (with $\gamma > 0$)

$$(1/\gamma^2 - 1)\omega_0^4 = f(\omega)$$

if one puts

$$f(\omega) = \omega^4 + 2(2d^2 - 1)\omega^2\omega_0^2.$$

We now look for the infumum γ_0 such that the above equation has no solution ω. This requires the minimization of $f(\omega)$ with respect to ω. We consider the following two cases.

Case 1: $2d^2 - 1 \geq 0$. Then $f(\omega)$ is a minimum for $\omega = 0$, which yields $\gamma_0 = 1$.

Case 2: $2d^2 - 1 < 0$. Then $f(\omega)$ is a minimum $\omega_m = \sqrt{1-2d^2}\,\omega_0$, which gives

$$f(\omega_m) = -(1-2d^2)^2\omega_0^4 .$$

This implies

$$\gamma_0 = \frac{1}{2d\sqrt{1-d^2}},$$

and consequently,

$$\|\mathbf{G}\|_\infty = \begin{cases} 1, & \text{if} \quad d \geq 1/\sqrt{2}, \\ \dfrac{1}{2d\sqrt{1-d^2}}, & \text{if} \quad d < 1/\sqrt{2}. \end{cases}$$

Of course, for this example it would have been much simpler to minimize the magnitude plot directly.

7.4 Specifications for Feedback Systems

In this section, we make some general observations concerning the specification of feedback systems. We discuss the most basic design goals and show that some of them are contradictory. Consequently, controller design consists always in finding a reasonable compromise for the contradictory specifications.

We analyze these problems for the feedback system shown in Fig. 7.1. It has four inputs, namely a reference signal $\mathbf{r}$, two disturbances $\mathbf{d}, \mathbf{d}_1$ and noise $\mathbf{n}$. In addition to the transfer functions defined in Sect. 6.5.2, the **input** and **output complementary sensitivity** functions are needed, which are defined as

$$\mathbf{T}_\mathrm{i} = \mathbf{I} - \mathbf{S}_\mathrm{i}, \quad \mathbf{T}_\mathrm{o} = \mathbf{I} - \mathbf{S}_\mathrm{o} .$$

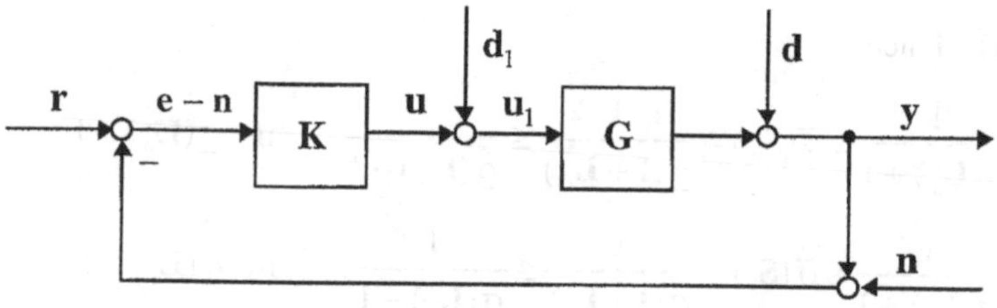

Fig. 7.1. Standard feedback configuration for MIMO systems

They can be written in the form

$$\mathbf{T}_i = \mathbf{L}_i(\mathbf{I}+\mathbf{L}_i)^{-1}, \quad \mathbf{T}_o = \mathbf{L}_o(\mathbf{I}+\mathbf{L}_o)^{-1}. \tag{7.7}$$

It is easy to see that the output $\mathbf{y}$ and the tracking error $\mathbf{e}=\mathbf{r}-\mathbf{y}$ can be calculated from the equations

$$\hat{\mathbf{y}} = \mathbf{T}_o\hat{\mathbf{r}} + \mathbf{S}_o\mathbf{G}\hat{\mathbf{d}}_1 + \mathbf{S}_o\hat{\mathbf{d}} - \mathbf{T}_o\hat{\mathbf{n}}, \tag{7.8}$$

$$\hat{\mathbf{e}} = \mathbf{S}_o\hat{\mathbf{r}} + \mathbf{T}_o\hat{\mathbf{n}} - \mathbf{S}_o\mathbf{G}\hat{\mathbf{d}}_1 - \mathbf{S}_o\hat{\mathbf{d}}. \tag{7.9}$$

The controller output $\mathbf{u}$ and the disturbed control $\mathbf{u}_1$ are given by

$$\hat{\mathbf{u}} = \mathbf{K}\mathbf{S}_o\hat{\mathbf{r}} - \mathbf{K}\mathbf{S}_o\hat{\mathbf{d}} - \mathbf{T}_i\hat{\mathbf{d}}_1 - \mathbf{K}\mathbf{S}_o\hat{\mathbf{n}}, \tag{7.10}$$

$$\hat{\mathbf{u}}_1 = \mathbf{K}\mathbf{S}_o\hat{\mathbf{r}} - \mathbf{K}\mathbf{S}_o\hat{\mathbf{d}} + \mathbf{S}_i\hat{\mathbf{d}}_1 - \mathbf{K}\mathbf{S}_o\hat{\mathbf{n}}. \tag{7.11}$$

The calculations in this section make use of the equations (cf. (6.63)):

$$\mathbf{S}_o\mathbf{G} = \mathbf{G}\mathbf{S}_i, \quad \mathbf{S}_i\mathbf{K} = \mathbf{K}\mathbf{S}_o.$$

In order to get a small control error $\mathbf{e}$, (7.9) leads to the following requirement.

R1: *In order to get a small control error with respect to the reference variable* $\mathbf{r}$ *and the disturbances* $\mathbf{d}, \mathbf{d}_1$, *the maximum singular values* $\bar{\sigma}(\mathbf{S}_o)$, $\bar{\sigma}(\mathbf{S}_o\mathbf{G})$ *have to be small.*

The requirement concerning the "size" of $\mathbf{S}_o$ can equivalently be expressed by the corresponding loop transfer function $\mathbf{L}_o$, as the next lemmma shows.

Lemma 7.4.1. *The following estimates hold:*

$$\bar{\sigma}(\mathbf{S}_o) \ll 1 \;\Leftrightarrow\; \underline{\sigma}(\mathbf{L}_o) \gg 1,$$

$$\bar{\sigma}(\mathbf{S}_i) \ll 1 \;\Leftrightarrow\; \underline{\sigma}(\mathbf{L}_i) \gg 1.$$

Proof. The inequalities

$$\underline{\sigma}(\mathbf{L}_o) - 1 \le \underline{\sigma}(\mathbf{I}+\mathbf{L}_o) \le \underline{\sigma}(\mathbf{L}_o) + 1,$$

$$\underline{\sigma}(\mathbf{L}_i) - 1 \le \underline{\sigma}(\mathbf{I}+\mathbf{L}_i) \le \underline{\sigma}(\mathbf{L}_i) + 1.$$

are obvious. Hence

$$\frac{1}{\underline{\sigma}(\mathbf{L}_o)+1} \le \bar{\sigma}(\mathbf{S}_o) = \frac{1}{\underline{\sigma}(\mathbf{I}+\mathbf{L}_o)} \le \frac{1}{\underline{\sigma}(\mathbf{L}_o)-1}, \quad \text{if } \underline{\sigma}(\mathbf{L}_o) > 1,$$

$$\frac{1}{\underline{\sigma}(\mathbf{L}_i)+1} \le \bar{\sigma}(\mathbf{S}_i) = \frac{1}{\underline{\sigma}(\mathbf{I}+\mathbf{L}_i)} \le \frac{1}{\underline{\sigma}(\mathbf{L}_i)-1}, \quad \text{if } \underline{\sigma}(\mathbf{L}_i) > 1.$$

The asserted inequalities are now directly seen.

Similarly, a second requirement can be derived from (7.11).

R2: *In order to get a good rejection of the disturbances* $\mathbf{d}, \mathbf{d}_1$ *with respect to the plant input* $\mathbf{u}_1$ *and in order to get a small influence of the reference signal and the noise on* $\mathbf{u}_1$, *the maximum singular values* $\bar{\sigma}(\mathbf{S}_i)$, $\bar{\sigma}(\mathbf{K}\mathbf{S}_o)$ *have to be small.*

The next requirement follows also from (7.9).

R3. *In order to get a good suppression of sensor noise with respect to the control error,* $\bar{\sigma}(\mathbf{T}_o)$ *has to be small.*

From the definition of $\mathbf{T}_o$ it is obvious that R3 is in contradiction to R1 and R2, but it has to be noted that these requirements have to be formulated with respect to the frequency. R1 and R2 are mainly relevant for low frequencies, whereas R3 is important only for high frequencies. As is seen from Lemma 7.4.1, $\bar{\sigma}(\mathbf{T}_o)$ is small if $\underline{\sigma}(\mathbf{L}_o)$ is small.

Finally, we have to analyze controller synthesis with respect to the magnitude of the control $\mathbf{u}$. Suppose there are frequencies such that the controller gain is large and the loop gain is small:

$$\bar{\sigma}(\mathbf{K}) \gg 1 \quad \text{and} \quad \bar{\sigma}(\mathbf{L}_i(i\omega)) \ll 1 .$$

Then $\mathbf{S}_i\mathbf{K} \approx \mathbf{K}$, $\mathbf{T}_i \approx \mathbf{0}$ and

$$\hat{\mathbf{u}} = \mathbf{S}_i\mathbf{K}(\hat{\mathbf{r}} - \hat{\mathbf{n}} - \hat{\mathbf{d}}) - \mathbf{T}_i\hat{\mathbf{d}}_1 \approx \mathbf{K}(\hat{\mathbf{r}} - \hat{\mathbf{n}} - \hat{\mathbf{d}}) .$$

Hence, in this case the reference signal, the disturbance $\mathbf{d}$ and the sensor noise will be also returned to the control in an amplified manner. Thus the following requirement may be formulated in order to avoid saturation of the actuators:

R4: *The controller should be designed such that there are no frequencies where the loop gain is small and the controller gain is very large.*

The requirements R1 – R4 are concerned with performance. Another important aspect comes into play when robustness is considered. Let the "true" plant be

$$\mathbf{G}_\Delta = (\mathbf{I} + \Delta)\mathbf{G} ,$$

where $\Delta = \Delta(s)$ is a stable transfer function. Suppose the nominal feedback system is stable. Using (7.7) and

$$\begin{aligned}\mathbf{I} + (\mathbf{I} + \Delta)\mathbf{L}_o &= \mathbf{I} + \mathbf{L}_o + \Delta\mathbf{L}_o \\ &= (\Delta\mathbf{L}_o(\mathbf{I} + \mathbf{L}_o)^{-1} + \mathbf{I})(\mathbf{I} + \mathbf{L}_o),\end{aligned}$$

we obtain

$$\det(\mathbf{I} + \mathbf{G}_\Delta\mathbf{K}) = \det(\mathbf{I} + \mathbf{L}_o)\det(\mathbf{I} + \Delta\mathbf{T}_o) .$$

This equation gives a relationship between the closed-loop poles of the nominal and the perturbed system. Hence the perturbed system is stable if $\bar{\sigma}(\Delta\mathbf{T}_o)$ is small. This is fulfilled if $\bar{\sigma}(\mathbf{T}_o)$ is small for those frequencies where $\bar{\sigma}(\Delta)$ is large. Typically, these are the high frequencies. Hence, a further requirement has to be imposed as follows.

R5: *In order to get a robustly stable feedback system,* $\bar{\sigma}(\mathbf{T}_o)$ *has to be small at high frequencies.*

Thus, in order to get a small control error, $\bar{\sigma}(\mathbf{L}_o)$ has to be made large, and in order to get a system which is insensitive with respect to sensor noise and model errors, $\bar{\sigma}(\mathbf{L}_o)$ has to be small. Again, these requirements have to be interpreted with respect to the frequency. For low frequencies, performance plays the bigger role and for high frequencies robustness is more important.

7.5 Performance Limitations

In this section, some fundamental performance limitations are discussed. We shall limit ourselves to the SISO case.

We suppose that S is stable and additionally, for the sake of simplicity, that L has no poles on the imaginary axis. Than S has no zeros on the imaginary axis. The **relative degree** of a transfer function is defined as the degree of the denominator minus the degree of the numerator. Denote the poles of L in the open RHP by $p_1, \ldots, p_l$. Then S_{ap} is given by (cf. Sect. 2.3.2)

$$S_{ap}(s) = \prod_{k=1}^{l} \frac{s - p_k}{s + \bar{p}_k}.$$

Lemma 7.5.1. *(a) For every* $s_0 = \sigma_0 + i\omega_0$ *with* $\sigma_0 > 0$ *we have*

$$\log|S_{mp}(s_0)| = \frac{1}{\pi}\int_{-\infty}^{\infty} \log|S(i\omega)| \frac{\sigma_0}{\sigma_0^2 + (\omega - \omega_0)^2} d\omega.$$

(b) If $z = \sigma_0 + i\omega_0$ *is a zero of the loop transfer function* L*, then*

$$\log|S_{ap}(z)^{-1}| = \frac{1}{\pi}\int_{-\infty}^{\infty} \log|S(i\omega)| \frac{\sigma_0}{\sigma_0^2 + (\omega - \omega_0)^2} d\omega.$$

Proof. (a) Since S_{mp} has no poles or zeros in the closed RHP, the function

$$F(s) = \ln S_{mp}(s)$$

is analytic in the closed RHP. For any $z \in \mathbb{C}$, the equation $w = \ln z$ implies $\operatorname{Re} w = \ln |z|$. Thus Lemma A.3.1 yields

$$\ln |S_{mp}(s_0)| = \frac{1}{\pi} \int_{-\infty}^{\infty} \ln |S_{mp}(i\omega)| \frac{\sigma_0}{\sigma_0^2 + (\omega - \omega_0)^2} d\omega .$$

The result follows by noting that $|S(i\omega)| = |S_{mp}(i\omega)|$ and $\log x = \log e \ln x$.
(b) The assumption $L(z) = 0$ implies $S(z) = 1$. This gives

$$S_{mp}(z) = S_{ap}(z)^{-1}$$

and the assertion follows directly from (a).

Assume that one requirement for the feedback loop is good tracking in an interval of low frequencies. This means that for a certain positive number M and a positive frequency ω_1 the following estimate has to hold:

$$|S(i\omega)| \le M \quad \text{for every} \quad -\omega_1 \le \omega \le \omega_1 . \tag{7.12}$$

Define now

$$c_1 = \frac{1}{\pi} \int_{-\omega_1}^{\omega_1} \frac{\sigma_0}{\sigma_0^2 + (\omega - \omega_0)^2} d\omega$$

$$c_2 = \frac{1}{\pi} \int_{-\infty}^{\omega_1} \frac{\sigma_0}{\sigma_0^2 + (\omega - \omega_0)^2} d\omega + \frac{1}{\pi} \int_{\omega_1}^{\infty} \frac{\sigma_0}{\sigma_0^2 + (\omega - \omega_0)^2} d\omega .$$

Then Lemma 7.5.1(b) leads immediately to the following estimate.

Corollary 7.5.1. *Let $z = \sigma_0 + i\omega_0$ be a zero of the loop transfer function L, and suppose that the estimate (7.12) holds. Then*

$$0 \le \log |S_{ap}(z)^{-1}| \le c_1 \log M + c_2 \log \| S \|_\infty .$$

Proof. It remains only to note that the first inequality holds, because $|S_{ap}(z)| \le 1$ by the maximum modulus theorem.

A good tracking capability is characterized by the inequality $M \ll 1$, whereas robust stability requires that $\| S \|_\infty$ is sufficiently small (cf. Sect. 4.1.1). Corollary 7.5.1 says that for a plant with a zero in the open RHP, it may become impossible to design a (SISO) controller with good tracking and robustness properties.

Example. Let a plant with a positive pole and a positive zero be given:

$$G(s) = \frac{s - z}{s - p} G_1(s) \quad \text{with } z > 0,\ p > 0,\ p \ne z.$$

Since an internally stabilizing controller does not cancel $s-p$, we conclude that p must be a zero of S. Hence

$$S_{ap} = \frac{s-p}{s+p} S_1,$$

where S_1 is all-pass. The maximum modulus theorem shows that $|S_1(z)| \le 1$, and thus

$$|S_{ap}(z)| \le \left| \frac{z-p}{z+p} \right|.$$

An application of Corollary 7.5.1 yields

$$\log \left| \frac{z+p}{z-p} \right| \le c_1 \log M + c_2 \log \| S \|_\infty .$$

If the zero z lies close to the pole p, the lower bound becomes very large and consequently the plant is difficult to control.

Theorem 7.5.1. *Suppose that the relative degree of* L *is at least 2. Then*

$$\int_0^\infty \log |S(i\omega)| d\omega = \pi (\log e) \sum_{k=1}^{l} \operatorname{Re} p_k . \tag{7.13}$$

Proof. An application of Lemma 7.5.1 with $\omega_0 = 0$ gives

$$\pi \sigma_0 \log |S_{mp}(s_0)| = \int_{-\infty}^{\infty} \log |S(i\omega)| \frac{\sigma_0^2}{\sigma_0^2 + \omega^2} d\omega .$$

This implies

$$\int_0^\infty \log |S(i\omega)| \frac{\sigma_0^2}{\sigma_0^2 + \omega^2} d\omega = \frac{\pi}{2} \sigma_0 \log |S_{mp}(\sigma_0)|$$

and therefore

$$\begin{aligned} \int_0^\infty \log |S(i\omega)| d\omega &= \lim_{\sigma \to \infty} \int_0^\infty \log |S(i\omega)| \frac{\sigma^2}{\sigma^2 + \omega^2} d\omega \\ &= \frac{\pi}{2} \lim_{\sigma \to \infty} (\sigma \log |S_{mp}(\sigma)|) \\ &= \frac{\pi}{2} \lim_{\sigma \to \infty} (\sigma \log |S(\sigma)| + \sigma \log |S_{ap}^{-1}(\sigma)|). \end{aligned}$$

We now show that

$$\lim_{\sigma\to\infty}(\sigma \log|S(\sigma)|)=0. \tag{7.14}$$

Since the relative degree of L is at least 2, we have

$$L(\sigma)\approx\frac{c}{\sigma^m}\quad\text{for large }\sigma$$

and some constant c and some integer $m\ge 2$. Hence for large σ, the following approximation holds:

$$\sigma\ln S(\sigma)=-\sigma\ln(1+L(\sigma))\approx-\sigma\ln\left(1+\frac{c}{\sigma^m}\right).$$

Since the series expansion of the $\ln$ function is

$$\ln(1+x)=x-\frac{x^2}{2}+\frac{x^3}{3}-\cdots,$$

we get the approximation

$$\sigma\ln S(\sigma)\approx-\sigma\left(\frac{c}{\sigma^m}-\ \cdots\right).$$

The right-hand side tends to zero as $\sigma\to\infty$ and (7.14) is shown. It remains to prove that

$$\lim_{\sigma\to\infty}\left(\frac{\sigma}{2}\ln|S_{ap}^{-1}(\sigma)|\right)=\sum_{k=1}^{l}\operatorname{Re}p_k.$$

Because

$$\ln|S_{ap}(\sigma)^{-1}|=\ln\prod_{k=1}^{l}\left|\frac{\sigma+\overline{p}_k}{\sigma-p_k}\right|=\sum_{k=1}^{l}\ln\left|\frac{\sigma+\overline{p}_k}{\sigma-p_k}\right|,$$

it suffices to show that

$$\lim_{\sigma\to\infty}\frac{\sigma}{2}\ln\left|\frac{\sigma+\overline{p}_k}{\sigma-p_k}\right|=\operatorname{Re}p_k. \tag{7.15}$$

Let $p_k=x+iy$. Then, using again the series expension of $\ln$, we get

$$\frac{\sigma}{2}\ln\left|\frac{\sigma+\overline{p}_k}{\sigma-p_k}\right|=\frac{\sigma}{2}\ln\left|\frac{1+\overline{p}_k\sigma^{-1}}{1-p_k\sigma^{-1}}\right|$$

$$=\frac{\sigma}{4}\ln\frac{(1+x\sigma^{-1})^2+(y\sigma^{-1})^2}{(1-x\sigma^{-1})^2+(y\sigma^{-1})^2}$$

$$= \frac{\sigma}{4}\{\ln[(1 + x\sigma^{-1})^2 + (y\sigma^{-1})^2] - \ln[(1 - x\sigma^{-1})^2 + (y\sigma^{-1})^2]\}$$

$$= \frac{\sigma}{4}\{4\frac{x}{\sigma} + \cdots\} .$$

This proves (7.15).

Corollary 7.5.2. *Suppose that L is stable and has a relative degree, which is at least 2. Then*

$$\int_0^\infty \log |S(i\omega)| d\omega = 0. \tag{7.16}$$

Remark. The results of this section remain valid if L has poles on the imaginary axis. In this case, the imaginary axis as the integration path has to be changed such that near the imaginary poles the integration path is a small half-circle with radius ε (cf. Sect. 3.4). The integrals over these half-circles tend to 0 as ε tends to 0.

Example. We consider a plant which is controlled by a purely proportional controller:

$$G(s) = \frac{1}{s(s+1)(s+2)}, \quad K(s) = k_R .$$

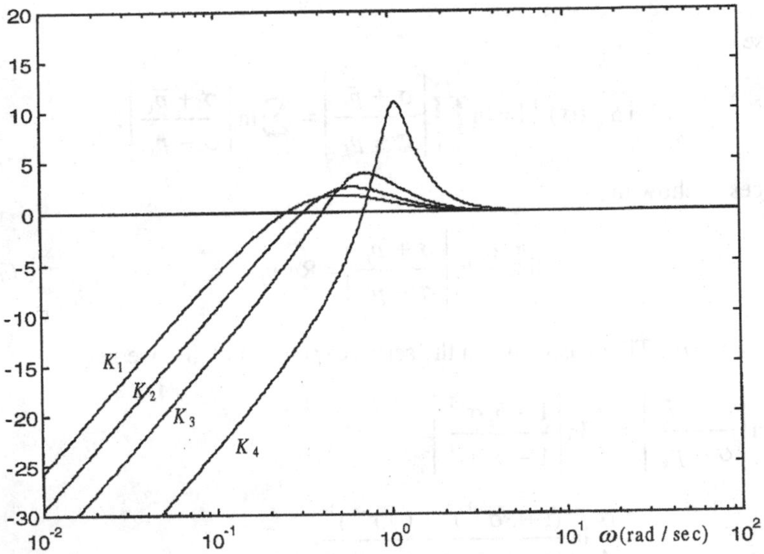

Fig. 7.2. $|S|_{dB}$ for various controller gains

Figure 7.2 shows $|S|_{dB}$ for $k_R = 0.39, 0.6, 1, 3$. (curves $K_1 - K_4$). The dampings of the feedback system are $d = 0.992, 0.75, 0.515, 0.155$. Equation (7.16) says that the area under the 0dB-line is equal to the area over the 0dB-line (for a *linear* frequency scale). The behavior which is expressed by (7.16) can be compared with that of a waterbed: pushing it down at one point, which reduces the water level locally, will result in an increased level somewhere else on the bed.

7.6 Coprime Factorization and Inner Functions

We conclude this chapter by presenting some basic results on coprime factorization and inner functions, which are required in Chap. 9 for the proof of the characterization theorem for $\mathcal{H}_\infty$ suboptimal controllers. Since the results are nowhere else needed in the book, we limit ourselves to a few basic facts concerning these topics. Nevertheless, it should be mentioned that coprime factorization is an interesting subject in itself, which plays an important role when controller parameterizations are considered.

Definition. *(a) Two matrices* $\mathbf{M} \in \mathcal{RH}_\infty$ *and* $\mathbf{N} \in \mathcal{RH}_\infty$ *are* ***right coprime over*** $\mathcal{RH}_\infty$ *if they have the same numbers of columns and if there exist matrices* $\mathbf{X}_r$ *and* $\mathbf{Y}_r$ *in* $\mathcal{RH}_\infty$ *such that*

$$(\mathbf{X}_r \quad \mathbf{Y}_r)\begin{pmatrix}\mathbf{M}\\ \mathbf{N}\end{pmatrix} = \mathbf{X}_r\mathbf{M} + \mathbf{Y}_r\mathbf{N} = \mathbf{I}. \tag{7.17}$$

(b) Similarly, two matrices $\tilde{\mathbf{M}}$ *and* $\tilde{\mathbf{N}}$ *are* ***left coprime over*** $\mathcal{RH}_\infty$ *if they have the same number of rows and if there exist matrices* $\mathbf{X}_l$ *and* $\mathbf{Y}_l$ *in* $\mathcal{RH}_\infty$ *such that*

$$(\tilde{\mathbf{M}} \quad \tilde{\mathbf{N}})\begin{pmatrix}\mathbf{X}_l\\ \mathbf{Y}_l\end{pmatrix} = \tilde{\mathbf{M}}\mathbf{X}_l + \tilde{\mathbf{N}}\mathbf{Y}_l = \mathbf{I}. \tag{7.18}$$

Let now $\mathbf{G}$ be a proper rational transfer matrix. Then a **right coprime factorization** of $\mathbf{G}$ is a factorization

$$\mathbf{G} = \mathbf{N}\mathbf{M}^{-1}, \qquad \mathbf{M}^{-1} \text{ is proper,} \tag{7.19}$$

where $\mathbf{M}$ and $\mathbf{N}$ are right coprime. Analogously, a **left coprime factorization** of $\mathbf{G}$ is a factorization

$$\mathbf{G} = \tilde{\mathbf{M}}^{-1}\tilde{\mathbf{N}}, \qquad \tilde{\mathbf{M}}^{-1} \text{ is proper,} \tag{7.20}$$

where and $\tilde{\mathbf{M}}$ and $\tilde{\mathbf{N}}$ are left coprime.

Remark. If $\mathbf{G}$ is stable, then the choice $\mathbf{N} = \mathbf{G}, \mathbf{M} = \mathbf{I}, \mathbf{X}_r = \mathbf{I}, \mathbf{Y}_r = \mathbf{0}$ leads to a right coprime factorization of $\mathbf{G}$. A left coprime factorization can be defined quite similarly. The equations (7.17) and (7.18) are denoted as **Bezout identities.**

Next we specialize these definitions to the SISO case. Then there is no need to distinguish between right and left coprimeness. To be specific, two transfer functions M and N are coprime over $\mathcal{RH}_\infty$ if there exist $X, Y \in \mathcal{RH}_\infty$ such that

$$XM + YN = 1. \tag{7.21}$$

Suppose s_0 is a common zero of M and N in the closed RHP. Since X and Y are stable, the linear factor $s - s_0$ cannot be cancelled when XM and YN are formed. Hence s_0 must be a zero of XM and YN, which is a direct contradiction of (7.21). This equation also implies that at least one of the transfer functions M and N must be proper, but not strictly proper. Thus (7.21) implies that M and N have no common zero in the closed RHP, or at $s = \infty$. It can be shown that this condition is also sufficient for (7.21).

As an example, let (with $a > 0, b > 0, a \neq b$)

$$G(s) = \frac{s - b}{s - a}.$$

Then a coprime factorization of G is given by

$$M(s) = \frac{s - a}{s + \alpha}, \quad N(s) = \frac{s - b}{s + \alpha} \quad \text{with } \alpha > 0.$$

The construction of X and Y is more complicated. The following lemma describes how a coprime factorization can be found for a general MIMO plant.

Lemma 7.6.1. *Let* $\mathbf{G}$ *be a proper rational transfer matrix and suppose that*

$$\mathbf{G} = \left(\begin{array}{c|c} \mathbf{A} & \mathbf{B} \\ \hline \mathbf{C} & \mathbf{D} \end{array}\right)$$

is a stabilizable and detectable realization. Let $\mathbf{F}, \mathbf{L}$ *be matrices such that*

$$\mathbf{A}_F = \mathbf{A} + \mathbf{BF}, \quad \mathbf{A}_L = \mathbf{A} + \mathbf{LC}$$

are stable. Define

$$\mathbf{C}_F = \mathbf{C} + \mathbf{DF}, \quad \mathbf{B}_L = \mathbf{B} + \mathbf{LD}$$

Then a right coprime factorization of $\mathbf{G}$ *is given by the transfer matrices*

$$\mathbf{M} = \left(\begin{array}{c|c} \mathbf{A}_F & \mathbf{B} \\ \hline \mathbf{F} & \mathbf{I} \end{array}\right), \mathbf{N} = \left(\begin{array}{c|c} \mathbf{A}_F & \mathbf{B} \\ \hline \mathbf{C}_F & \mathbf{D} \end{array}\right), \mathbf{X}_r = \left(\begin{array}{c|c} \mathbf{A}_L & -\mathbf{B}_L \\ \hline \mathbf{F} & \mathbf{I} \end{array}\right), \mathbf{Y}_r = \left(\begin{array}{c|c} \mathbf{A}_L & \mathbf{L} \\ \hline \mathbf{F} & \mathbf{0} \end{array}\right), \tag{7.22}$$

and a left coprime factorization of $\mathbf{G}$ *is given by*

$$\tilde{\mathbf{M}} = \left(\begin{array}{c|c} \mathbf{A}_L & \mathbf{L} \\ \hline \mathbf{C} & \mathbf{I} \end{array}\right), \ \tilde{\mathbf{N}} = \left(\begin{array}{c|c} \mathbf{A}_L & \mathbf{B}_L \\ \hline \mathbf{C} & \mathbf{D} \end{array}\right), \ \mathbf{X}_l = \left(\begin{array}{c|c} \mathbf{A}_F & -\mathbf{L} \\ \hline \mathbf{C}_F & \mathbf{I} \end{array}\right), \ \mathbf{Y}_l = \left(\begin{array}{c|c} \mathbf{A}_F & \mathbf{L} \\ \hline \mathbf{F} & \mathbf{0} \end{array}\right) \tag{7.23}$$

Proof. The state space realization of $\mathbf{G}$ can be written as

$$\dot{\mathbf{x}} = \mathbf{A}\mathbf{x} + \mathbf{B}\mathbf{u}$$
$$\mathbf{y} = \mathbf{C}\mathbf{x} + \mathbf{D}\mathbf{u}.$$

Applying state feedback using $\mathbf{F}$, a new variable can be introduced by defining

$$\mathbf{v} = \mathbf{u} - \mathbf{F}\mathbf{x}.$$

Then we get

$$\dot{\mathbf{x}} = (\mathbf{A} + \mathbf{B}\mathbf{F})\mathbf{x} + \mathbf{B}\mathbf{v}$$
$$\mathbf{u} = \mathbf{F}\mathbf{x} + \mathbf{v}$$
$$\mathbf{y} = (\mathbf{C} + \mathbf{D}\mathbf{F})\mathbf{x} + \mathbf{D}\mathbf{v}.$$

We denote the transfer matrix from $\mathbf{v}$ to $\mathbf{u}$ by $\mathbf{M}$ and the transfer matrix from $\mathbf{v}$ to $\mathbf{y}$ by $\mathbf{N}$:

$$\mathbf{u} = \mathbf{M}\mathbf{v}, \quad \mathbf{y} = \mathbf{N}\mathbf{v}. \tag{7.24}$$

It is directly seen that a state space realization of these transfer matrices is given by (7.22). Moreover, the state space realization of $\mathbf{M}$ directly shows that $\mathbf{M}^{-1}$ exists and is proper. Thus (7.22) implies (7.19).

The decomposition (7.20) can be shown in similar way by using a variable transform based on $\mathbf{L}$. The Bezout equations (7.17) and (7.18) are verified by direct calculation using the definition of $\mathbf{X}_r, \mathbf{Y}_r, \mathbf{X}_l, \mathbf{Y}_l$ given in (7.22) and (7.23).

Application: Controller Parameterization. First we consider again a SISO system with a coprime factorization described by M, N, X, Y as above. Suppose that Q is a transfer function in $\mathcal{RH}_\infty$ that makes the controller

$$K = \frac{Y + MQ}{X - NQ} \tag{7.25}$$

proper. Then it can easily be checked that

$$S = (X - NQ)M, \ KS = (Y + MQ)M,$$
$$GS = N(X - NQ), \ F_r = N(Y + QM).$$

Thus, the SISO feedback system is internally stable. It is an interesting and important fact that *all* proper and stabilizing controllers can be written in the form (7.25). Moreover, this result can be generalized to general MIMO plants (see, for example, Dullerud and Paganini [37], Theorem 5.13).

Definition. *A transfer matrix* $\mathbf{N}$ *in* $\mathcal{RH}_\infty$ *is called* ***inner*** *if* $\mathbf{N}^{\sim}\mathbf{N}=\mathbf{I}$ *and* ***co-inner*** *if* $\mathbf{N}\mathbf{N}^{\sim}=\mathbf{I}$.

If $\mathbf{N}$ is inner, then $\mathbf{N}(i\omega)^*\mathbf{N}(i\omega)=\mathbf{I}$ for all ω. This implies $\|\mathbf{N}(i\omega)\mathbf{v}\|=\|\mathbf{v}\|$ for all $\mathbf{v}\in\mathbb{C}^m$ and consequently

$$\|\mathbf{N}\mathbf{w}\|_2=\|\mathbf{w}\|_2 \quad \text{for all } \mathbf{w}\in\mathcal{L}_2.$$

Lemma 7.6.2. *Let* $\mathbf{N}\in\mathcal{RH}_\infty$ *with state space representation*

$$\mathbf{N}=\left(\begin{array}{c|c}\mathbf{A} & \mathbf{B}\\ \hline \mathbf{C} & \mathbf{D}\end{array}\right)$$

be given and let $\mathbf{X}\geq\mathbf{0}$ *be the solution of*

$$\mathbf{A}^T\mathbf{X}+\mathbf{X}\mathbf{A}+\mathbf{C}^T\mathbf{C}=\mathbf{0}. \tag{7.26}$$

Then $\mathbf{N}$ *is inner if*

$$\mathbf{D}^T\mathbf{C}+\mathbf{B}^T\mathbf{X}=\mathbf{0} \ \text{ and } \ \mathbf{D}^T\mathbf{D}=\mathbf{I}. \tag{7.27}$$

Proof. A state space representation of $\mathbf{N}^{\sim}\mathbf{N}$ is immediately calculated as

$$\mathbf{N}^{\sim}\mathbf{N}=\left(\begin{array}{cc|c}\mathbf{A} & \mathbf{0} & \mathbf{B}\\ -\mathbf{C}^T\mathbf{C} & -\mathbf{A}^T & -\mathbf{C}^T\mathbf{D}\\ \hline \mathbf{D}^T\mathbf{C} & \mathbf{B}^T & \mathbf{D}^T\mathbf{D}\end{array}\right).$$

Using the similarity transformation defined by

$$\mathbf{T}=\begin{pmatrix}\mathbf{I} & \mathbf{0}\\ -\mathbf{X} & \mathbf{I}\end{pmatrix},\quad \mathbf{T}^{-1}=\begin{pmatrix}\mathbf{I} & \mathbf{0}\\ \mathbf{X} & \mathbf{I}\end{pmatrix}$$

and the Lyapunov equation (7.26), it is easy to see that another state space representation of $\mathbf{N}^{\sim}\mathbf{N}$ is

$$\mathbf{N}^{\sim}\mathbf{N}=\left(\begin{array}{cc|c}\mathbf{A} & \mathbf{0} & \mathbf{B}\\ \mathbf{0} & -\mathbf{A}^T & -(\mathbf{X}\mathbf{B}+\mathbf{C}^T\mathbf{D})\\ \hline \mathbf{B}^T\mathbf{X}+\mathbf{D}^T\mathbf{C} & \mathbf{B}^T & \mathbf{D}^T\mathbf{D}\end{array}\right).$$

The assertion now follows easily.

Remark. If $\mathbf{A}$ is stable, $(\mathbf{A}, \mathbf{B})$ is controllable and $(\mathbf{C}, \mathbf{A})$ is observable, then the condition (7.27) is also necessary for $\mathbf{N}$ to be inner.

Notes and References

More detailed information about the function spaces introduced in Sects. 7.1 and 7.2 can be found in Rudin [77] and Duren [38] and the system theory interpretations of the norms and function spaces are analyzed in Desoer and Vidyasagar [29]. The discussion concerning the properties of feedback systems and their performance limitations can be found in greater detail in Zhou, Doyle, and Glover [112] and Doyle, Francis, and Tannenbaum [36]. The analysis of performance limitations has its roots in Bode [14]. Coprime factorizations were introduced in Vidyasagar [102].

Chapter 8

$\mathcal{H}_2$ Optimal Control

In Chap. 6 we studied the basic structure and fundamental properties of MIMO feedback systems. This chapter and the following two are concerned with controller synthesis. The idea is to consider the feedback system as a linear operator which maps the input $\mathbf{w}$ onto the output $\mathbf{z}$. The feedback loop is given as in Sect. 6.5.1, which means in particular that the plant may contain weights. These weights formulate the design goals and the controller has to be constructed in such a way that it minimizes the operator norm of the feedback system. Operator norms of linear systems governed by ordinary differential equations were introduced in the previous chapter. The most important norms for controller synthesis are the $\mathcal{H}_2$ norm and the $\mathcal{H}_\infty$ norm.

In this chapter, we are concerned with the synthesis of $\mathcal{H}_2$ optimal controllers. Our analysis starts by considering a certain quadratic optimal control problem. The resulting controller is denoted as linear quadratic regulator (LQR). It has the form $\mathbf{u} = \mathbf{F}\mathbf{x}$, where $\mathbf{x}$ is the state of the system. This approach is classical and leads in a natural way to the famous Riccati equation, which plays a key role in $\mathcal{H}_2$ and $\mathcal{H}_\infty$ controller synthesis. The most important properties of Riccati equations are studied in Sect. 8.1.2. In Sect. 8.2, the general $\mathcal{H}_2$ problem is solved in several steps by breaking down the original problem into several subproblems. The basic idea is to introduce a new variable by $\mathbf{v} = \mathbf{u} - \mathbf{F}\mathbf{x}$. The remaining problem in the variable $\mathbf{v}$ is a so-called output estimation problem, which is the dual of another kind of problem, namely a disturbance feedforward problem. This can be interpreted as a special case of a full information problem, which essentially is the LQR problem. The same concept will be used when the $\mathcal{H}_\infty$ problem is solved.

One application of the $\mathcal{H}_2$ theory is the Kalman filter, which is a special kind of observer (Sect. 8.3). We also give a typical application, namely the estimation of the distance and velocity of an aircraft, where noisy range measurements are made by a ground radar. If a linear quadratic controller and a Kalman filter are combined, then an LQG controller results. It may be viewed as solution of a $\mathcal{H}_2$ problem (Sect. 8.4.1). In Sect. 8.4.2 robustness properties similar to those in Sect.

6.3.2 for SISO problems are analyzed for LQR problems. Section 8.4.3 is concerned with the dependence of the positions of the closed-loop poles on the weightings for LQ controllers. These considerations may be helpful when reasonable weightings have to be chosen in a practical situation. The combination of a LQ controller and a Kalman filter may have a lack of robustness. This can be improved by a procedure called loop transfer recovery, which is also presented in Sect. 8.4.3. The chapter ends with a discussion of general optimal control problems and a brief overview of infinite dimensional systems.

8.1 LQ Controllers

8.1.1 Controller Design by Minimization of a Cost Functional

We start with a dynamical system of the form

$$\dot{\mathbf{x}} = \mathbf{A}\mathbf{x} + \mathbf{B}_2\mathbf{u}, \quad \mathbf{x}(0) = \mathbf{x}_0 \tag{8.1}$$

and want to design a controller with feedback of the full state vector:

$$\mathbf{u} = \mathbf{F}\mathbf{x}$$

(where $\mathbf{F}$ is a suitably dimensioned matrix). This is a pure stabilization problem similar to that considered at the beginning of Sect. 6.1.

We put the problem in a somewhat different framework where controller synthesis is done by minimizing a certain cost functional. More precisely, let

$$f(\mathbf{x},\mathbf{u}) = \int_0^\infty \begin{pmatrix} \mathbf{x}(t) \\ \mathbf{u}(t) \end{pmatrix}^T \begin{pmatrix} \mathbf{Q} & \mathbf{S} \\ \mathbf{S}^T & \mathbf{R} \end{pmatrix} \begin{pmatrix} \mathbf{x}(t) \\ \mathbf{u}(t) \end{pmatrix} dt\,, \tag{8.2}$$

with symmetric matrices $\mathbf{Q}, \mathbf{R}$ and a positive definite matrix $\mathbf{R}$. Moreover, the matrices $\mathbf{Q}, \mathbf{R}, \mathbf{S}$ are assumed to be given such that

$$\begin{pmatrix} \mathbf{Q} & \mathbf{S} \\ \mathbf{S}^T & \mathbf{R} \end{pmatrix} \geq 0\,. \tag{8.3}$$

The problem now is to minimize the functional (8.2) over all $\mathbf{u} \in \mathcal{L}_2[0,\infty)$ subject to (8.1). The positive definiteness of $\mathbf{R}$ implies that the control energy is finite. The matrices $\mathbf{Q}, \mathbf{S}, \mathbf{R}$ are the design parameters for the controller synthesis. For the common choice $\mathbf{S} = \mathbf{0}$ the functional (8.2) takes the particular form

$$f(\mathbf{x},\mathbf{u}) = \int_0^\infty (\mathbf{x}^T(t)\mathbf{Q}\mathbf{x}(t) + \mathbf{u}^T(t)\mathbf{R}\mathbf{u}(t))\, dt\,.$$

Since $\mathbf{R}$ is positive definite, $\mathbf{R}$ has a square root $\mathbf{R}^{1/2}$ and it is possible to introduce a new control $\mathbf{v}$ by $\mathbf{v}=\mathbf{R}^{1/2}\mathbf{u}$. Then $\mathbf{R}$ has to be replaced by the identity matrix and consequently, without loss of generality, we may assume $\mathbf{R}=\mathbf{I}$.

For our analysis it is helpful to rewrite the functional (8.2). To this end, the matrix in (8.3) can be decomposed as follows:

$$\begin{pmatrix} \mathbf{Q} & \mathbf{S} \\ \mathbf{S}^T & \mathbf{R} \end{pmatrix} = \begin{pmatrix} \mathbf{C}_1^T \\ \mathbf{D}_{12}^T \end{pmatrix} \begin{pmatrix} \mathbf{C}_1 & \mathbf{D}_{12} \end{pmatrix}.$$

Herein, $\mathbf{C}_1$ is a $l \times n$ and $\mathbf{D}_{12}$ is a $l \times m$ matrix. The decomposition can be derived by using the positive semidefiniteness of the matrix in (8.3) (cf. Appendix A.1.3). With the matrices $\mathbf{C}_1$ and $\mathbf{D}_{12}$, we obtain

$$f(\mathbf{x},\mathbf{u}) = \|\mathbf{z}\|_2^2 \quad \text{where} \quad \mathbf{z}=\mathbf{C}_1\mathbf{x}+\mathbf{D}_{12}\mathbf{u}.$$

It is useful to note that

$$\mathbf{S}=\mathbf{0} \;\Leftrightarrow\; \mathbf{C}_1^T\mathbf{D}_{12}=\mathbf{0} \quad \text{and} \quad \mathbf{R}=\mathbf{I} \;\Leftrightarrow\; \mathbf{D}_{12}^T\mathbf{D}_{12}=\mathbf{I}.$$

In many applications the weightings are chosen with $\mathbf{S}=\mathbf{0}$. In this case, with $\mathbf{C}_{10}^T\mathbf{C}_{10}=\mathbf{Q}$ and $\mathbf{R}=\mathbf{I}$, define

$$\mathbf{C}_1 = \begin{pmatrix} \mathbf{C}_{10} \\ \mathbf{0} \end{pmatrix}, \quad \mathbf{D}_{12} = \begin{pmatrix} \mathbf{0} \\ \mathbf{I} \end{pmatrix}.$$

The weighted output is now given by

$$\mathbf{z} = \begin{pmatrix} \mathbf{C}_0\mathbf{x} \\ \mathbf{u} \end{pmatrix}.$$

These considerations put us in a position to formulate the optimization problem as follows:

$$\begin{aligned} &\text{Minimize } \|\mathbf{C}_1\mathbf{x}+\mathbf{D}_{12}\mathbf{u}\|_2^2 \\ &\text{subject to } \mathbf{x},\mathbf{u} \text{ with } \mathbf{u}\in\mathcal{L}_2[0,\infty) \text{ and} \\ &\dot{\mathbf{x}}=\mathbf{A}\mathbf{x}+\mathbf{B}_2\mathbf{u}, \quad \mathbf{x}(0)=\mathbf{x}_0. \end{aligned}$$

This problem is denoted as the **linear quadratic regulator (LQR) problem.**

We start the analysis by considering a specialized version of the LQR problem where it is assumed that the controls are given by state feedback with constant gain: $\mathbf{u}=\mathbf{F}\mathbf{x}$. Then the minimization has to be taken over all matrices $\mathbf{F}$ of suitable dimension and the problem can be rewritten as follows. With

$$\mathbf{A}_F=\mathbf{A}+\mathbf{B}_2\mathbf{F}, \quad \mathbf{C}_F=\mathbf{C}_1+\mathbf{D}_{12}\mathbf{F}$$

we get

$$\mathbf{C}_1\mathbf{x} + \mathbf{D}_{12}\mathbf{u} = \mathbf{C}_F\mathbf{x}, \quad \dot{\mathbf{x}} = \mathbf{A}\mathbf{x} + \mathbf{B}_2\mathbf{u} = \mathbf{A}_F\mathbf{x}.$$

If $\mathbf{A}_F$ is stable, the equation

$$\| \mathbf{C}_F\mathbf{x} \|_2^2 = \mathbf{x}_0^T\mathbf{X}\mathbf{x}_0$$

follows, where $\mathbf{X}$ is the solution of the Lyapunov equation

$$\mathbf{A}_F^T\mathbf{X} + \mathbf{X}\mathbf{A}_F + \mathbf{C}_F^T\mathbf{C}_F = \mathbf{0} \tag{8.4}$$

(cf. Theorem 5.7.1). Minimizing $\mathbf{x}_0^T\mathbf{X}\mathbf{x}_0$ over all $\mathbf{F}$ (where $\mathbf{X}$ depends on $\mathbf{F}$) can now be done by calculating the gradient and putting it to zero. This yields

$$\mathbf{F} = -(\mathbf{B}_2^T\mathbf{X} + \mathbf{D}_{12}^T\mathbf{C}_1), \tag{8.5}$$

where $\mathbf{X}$ is the solution of

$$\begin{aligned}(\mathbf{A} - \mathbf{B}_2\mathbf{D}_{12}^T\mathbf{C}_1)^T\mathbf{X} + \mathbf{X}(\mathbf{A} - \mathbf{B}_2\mathbf{D}_{12}^T\mathbf{C}_1) - \mathbf{X}\mathbf{B}_2\mathbf{B}_2^T\mathbf{X}& \\ + \mathbf{C}_1^T(\mathbf{I} - \mathbf{D}_{12}\mathbf{D}_{12}^T)\mathbf{C}_1 = \mathbf{0}&.\end{aligned} \tag{8.6}$$

We do not carry out the calculations, which are somewhat technical but rather straightforward. Of course, additional assumptions are necessary to make the argument rigorous concerning the existence of an optimal solution. Equation (8.6) is called the **algebraic Riccati equation**. It plays a key role in $\mathcal{H}_2$ and $\mathcal{H}_\infty$ optimization.

As we want to show now, the LQR problem has the same solution as its specialized version, where only controls of the form $\mathbf{u} = \mathbf{F}\mathbf{x}$ are admitted. The main idea is to introduce a new variable $\mathbf{v}$ by

$$\mathbf{v} = \mathbf{u} - \mathbf{F}\mathbf{x}$$

and to calculate the cost functional as for the specialized problem by using the solution $\mathbf{X}$ of the Lyapunov equation (8.4). Let $\mathbf{u} \in \mathcal{L}_2[0,\infty)$ be given and let $\mathbf{x}$ be the corresponding solution of (8.1). Then $\mathbf{v} = \mathbf{u} - \mathbf{F}\mathbf{x}$ is also square-integrable, as we shall show somewhat later. Hence the LQR problem is equivalent to minimizing

$$\| \mathbf{C}_F\mathbf{x} + \mathbf{D}_{12}\mathbf{v} \|_2^2$$

subject to

$$\dot{\mathbf{x}} = \mathbf{A}_F\mathbf{x} + \mathbf{B}_2\mathbf{v}, \quad \mathbf{v} \in \mathcal{L}_2[0,\infty).$$

Assume

$$\mathbf{D}_{12}^T\mathbf{D}_{12} = \mathbf{I}. \tag{8.7}$$

Then the norm that has to be minimized can be computed as follows:

$$\| \mathbf{C}_F\mathbf{x} + \mathbf{D}_{12}\mathbf{v} \|_2^2 = \left\langle \mathbf{x}^T, \mathbf{C}_F^T\mathbf{C}_F\mathbf{x} \right\rangle + 2\left\langle \mathbf{x}^T, \mathbf{C}_F^T\mathbf{D}_{12}\mathbf{v} \right\rangle + \| \mathbf{v} \|_2^2$$

$$= -\left\langle \mathbf{x}^T, (\mathbf{A}_F^T\mathbf{X} + \mathbf{X}\mathbf{A}_F)\mathbf{x}\right\rangle + 2\left\langle \mathbf{x}^T, \mathbf{C}_F^T\mathbf{D}_{12}\mathbf{v}\right\rangle + \|\mathbf{v}\|_2^2$$

$$= -\left\langle \dot{\mathbf{x}}^T, \mathbf{X}\mathbf{x}\right\rangle + \left\langle \mathbf{v}^T, \mathbf{B}_2^T\mathbf{X}\mathbf{x}\right\rangle - \left\langle \mathbf{x}^T, \mathbf{X}\dot{\mathbf{x}}\right\rangle + \left\langle \mathbf{x}^T, \mathbf{X}\mathbf{B}_2\mathbf{v}\right\rangle$$
$$+ 2\left\langle \mathbf{x}^T, \mathbf{C}_F^T\mathbf{D}_{12}\mathbf{v}\right\rangle + \|\mathbf{v}\|_2^2$$

$$= -\int_0^\infty \frac{d}{dt}(\mathbf{x}^T\mathbf{X}\mathbf{x})\,dt + 2\left\langle \mathbf{x}^T, \mathbf{X}\mathbf{B}_2\mathbf{v} + \mathbf{C}_F^T\mathbf{D}_{12}\mathbf{v}\right\rangle + \|\mathbf{v}\|_2^2$$

$$= \mathbf{x}_0^T\mathbf{X}\mathbf{x}_0 + 2\left\langle \mathbf{x}^T, \mathbf{X}\mathbf{B}_2\mathbf{v} + \mathbf{C}_F^T\mathbf{D}_{12}\mathbf{v}\right\rangle + \|\mathbf{v}\|_2^2 .$$

Now choose now $\mathbf{F}$ according to (8.5). With this $\mathbf{F}$ the Lyapunov equation (8.4) becomes the Riccati equation (8.6) and the norm reduces to

$$\|\mathbf{C}_F\mathbf{x} + \mathbf{D}_{12}\mathbf{v}\|_2^2 = \mathbf{x}_0^T\mathbf{X}\mathbf{x}_0 + \|\mathbf{v}\|_2^2 . \tag{8.8}$$

Thus, the minimum is obtained for $\mathbf{v} = \mathbf{0}$, and again $\mathbf{u} = \mathbf{F}\mathbf{x}$ is the optimal control. In order to make a complete proof of this argument, it must be shown that the Riccati equation actually has a solution $\mathbf{X}$. This requires some additional assumptions, which are formulated in the next theorem. The theorem also makes use of a slight reformulation of the Riccati equation, which will be given now.

Because of (8.7), the columns of $\mathbf{D}_{12}$ are orthonormal. Thus $m \le l$ and it is possible to complete the columns to an orthonormal basis in $\mathbb{R}^l$. This can be formulated as follows. There is a matrix $\mathbf{D}_\perp$ such that $(\mathbf{D}_{12} \quad \mathbf{D}_\perp)$ is unitary:

$$\begin{pmatrix}\mathbf{D}_{12} & \mathbf{D}_\perp\end{pmatrix}\begin{pmatrix}\mathbf{D}_{12}^T \\ \mathbf{D}_\perp^T\end{pmatrix} = \mathbf{I} .$$

Consequently,

$$\mathbf{D}_\perp\mathbf{D}_\perp^T = \mathbf{I} - \mathbf{D}_{12}\mathbf{D}_{12}^T .$$

The final result is stated in the next theorem.

Theorem 8.1.1. *Let a dynamical system of the form*

$$\dot{\mathbf{x}} = \mathbf{A}\mathbf{x} + \mathbf{B}_2\mathbf{u},$$
$$\mathbf{z} = \mathbf{C}_1\mathbf{x} + \mathbf{D}_{12}\mathbf{u}$$

be given where the following assumptions are made for the system matrices.

(A1) $(\mathbf{A}, \mathbf{B}_2)$ *is stabilizable.*

(A2) The columns of $\mathbf{D}_{12}$ *form an orthonormal system and* $\mathbf{D}_{12}$ *is completed by a matrix* $\mathbf{D}_\perp$ *to a unitary matrix* $(\mathbf{D}_{12} \quad \mathbf{D}_\perp)$.

(A3) $(\mathbf{C}_1, \mathbf{A})$ *is detectable.*

(A4) The following matrix has for every ω full column rank:

$$\begin{pmatrix} \mathbf{A} - i\omega\mathbf{I} & \mathbf{B}_2 \\ \mathbf{C}_1 & \mathbf{D}_{12} \end{pmatrix}.$$

Then $\mathbf{u} = \mathbf{F}\mathbf{x}$ *is the unique solution of the LQR problem if* $\mathbf{F}$ *is chosen according to*

$$\mathbf{F} = -(\mathbf{B}_2^T\mathbf{X} + \mathbf{D}_{12}^T\mathbf{C}_1), \tag{8.9}$$

where $\mathbf{X}$ *is the solution of the Riccati equation*

$$(\mathbf{A} - \mathbf{B}_2\mathbf{D}_{12}^T\mathbf{C}_1)^T\mathbf{X} + \mathbf{X}(\mathbf{A} - \mathbf{B}_2\mathbf{D}_{12}^T\mathbf{C}_1) - \mathbf{X}\mathbf{B}_2\mathbf{B}_2^T\mathbf{X} + \mathbf{C}_1^T\mathbf{D}_\perp\mathbf{D}_\perp^T\mathbf{C}_1 = \mathbf{0}. \tag{8.10}$$

The optimal value is $\mathbf{x}_0^T\mathbf{X}\,\mathbf{x}_0$.

Proof. The main point that remains to prove is that under the given assumptions the Riccati equation (8.10) has a positive semidefinite solution $\mathbf{X}$. This requires a detailed analysis of Riccati equations, which will be carried out in the next section. One main result is Theorem 8.1.5. With this result, the existence of a positive semidefinite solution $\mathbf{X}$ of (8.10) can be guaranteed if the conditions (A1), (A3), (A4) hold. Moreover, by Theorem 8.1.2, the matrix $\mathbf{A}_F$ is stable if $\mathbf{F}$ is given by (8.5).

Next we have to show that the change of variables is feasible, i.e. that $\mathbf{u}, \mathbf{z}$ are square-integrable if and only if $\mathbf{v}, \mathbf{z}$ are square-integrable. Let $\mathbf{v}, \mathbf{z} \in \mathcal{L}_2[0,\infty)$, then $\mathbf{x} \in \mathcal{L}_2[0,\infty)$ since $\mathbf{A}_F$ is stable. Hence we also have $\mathbf{u} = \mathbf{v} + \mathbf{F}\mathbf{x} \in \mathcal{L}_2[0,\infty)$. Conversely, let $\mathbf{u}, \mathbf{z} \in \mathcal{L}_2[0,\infty)$. We must show that $\mathbf{x}$ is square-integrable. By assumption (A3) there is a matrix $\mathbf{L}$ such that $\mathbf{A} + \mathbf{L}\mathbf{C}_1$ is stable. We now define a state estimate $\tilde{\mathbf{x}}$ by

$$\dot{\tilde{\mathbf{x}}} = (\mathbf{A} + \mathbf{L}\mathbf{C}_1)\tilde{\mathbf{x}} + (\mathbf{L}\mathbf{D}_{12} + \mathbf{B}_2)\mathbf{u} - \mathbf{L}\mathbf{z}.$$

Because of $\mathbf{u}, \mathbf{z} \in \mathcal{L}_2[0,\infty)$ it follows that $\tilde{\mathbf{x}} \in \mathcal{L}_2[0,\infty)$. Let $\mathbf{e} = \mathbf{x} - \tilde{\mathbf{x}}$. Then

$$\dot{\mathbf{e}} = (\mathbf{A} + \mathbf{L}\mathbf{C}_1)\mathbf{e},$$

and consequently $\mathbf{e} \in \mathcal{L}_2[0,\infty)$ and therefore also $\mathbf{x} \in \mathcal{L}_2[0,\infty)$ and $\mathbf{v} \in \mathcal{L}_2[0,\infty)$.

With these matrices $\mathbf{X}, \mathbf{F}$, the representation (8.8) of the norm which has to be minimized holds and the assertions directly follow.

Remarks. **1.** Assumption (A2) is essentially a reformulation of the requirement $\mathbf{R} = \mathbf{I}$.
2. Assumptions (A1), (A2) and (A4) are needed to guarantee the existence of a positive semidefinite solution $\mathbf{X}$ of the Riccati equation (8.10) (cf. Theorem 8.1.5).
3. In the special case $\mathbf{S} = \mathbf{C}_1^T\mathbf{D}_{12} = \mathbf{0}$ the formula for $\mathbf{F}$ and the Riccati equation

reduce to

$$\mathbf{F} = -\mathbf{B}_2^T \mathbf{X},$$

$$\mathbf{A}^T \mathbf{X} + \mathbf{X}\mathbf{A} - \mathbf{X}\mathbf{B}_2 \mathbf{B}_2^T \mathbf{X} + \mathbf{C}_1^T \mathbf{C}_1 = \mathbf{0}.$$

It is easy to see that in this case (A4) can be replaced by the following assumption.

(A4') $(\mathbf{C}_1, \mathbf{A})$ has no unobservable eigenvalues on the imaginary axis.

This is in particular true if $(\mathbf{C}_1, \mathbf{A})$ is detectable. Hence assumption (A4) can be skipped in this case.

4. A slight modification of the above proof shows that the detectability of $(\mathbf{C}_1, \mathbf{A})$ is actually not necessary for the solvability of the LQR problem. The solution remains the same if this condition is omitted (cf. Zhou, Doyle, and Glover [112], Theorem 14.4). Note that if (A3) is omitted, (A4) cannot be skipped in the case $\mathbf{C}_1^T \mathbf{D}_{12} = \mathbf{0}$ but may be replaced by (A4').

5. Assumption (A4) can be equivalently formulated as follows.

The system $(\mathbf{A}, \mathbf{B}, \mathbf{C}, \mathbf{D})$ has no invariant zeros on the imaginary axis.

8.1.2 Algebraic Riccati Equation

The discussion in the preceding section leads us to a matrix equation of the following form:

$$\mathbf{A}^T \mathbf{X} + \mathbf{X}\mathbf{A} + \mathbf{X}\mathbf{R}\mathbf{X} + \mathbf{Q} = \mathbf{0}. \tag{8.11}$$

Here, $\mathbf{A}, \mathbf{R}, \mathbf{Q}$ are real $n \times n$-matrices and $\mathbf{R}, \mathbf{Q}$ are supposed to be symmetric. The equation (8.11) is called the **algebraic Riccati equation**. To each Riccati equation belongs the $2n \times 2n$-matrix

$$\mathbf{H} = \begin{pmatrix} \mathbf{A} & \mathbf{R} \\ -\mathbf{Q} & -\mathbf{A}^T \end{pmatrix}, \tag{8.12}$$

which is the so-called **Hamilton matrix**. It can be used to give a representation of the solution $\mathbf{X}$ of (8.11) as we shall see now.

First, we observe that $\mathbf{H}$ and $-\mathbf{H}^T$ are similar because of

$$\mathbf{J}^{-1} \mathbf{H} \mathbf{J} = -\mathbf{H}^T$$

with

$$\mathbf{J} = \begin{pmatrix} \mathbf{0} & -\mathbf{I} \\ \mathbf{I} & \mathbf{0} \end{pmatrix}$$

(note $\mathbf{J}^2 = \mathbf{I}$). Consequently, λ is an igenvalue of $\mathbf{H}$ if and only if $-\overline{\lambda}$ has this property.

We introduce now the following assumption:

(A1) Stability property: $\mathbf{H}$ *has no eigenvalues on the imaginary axis.*

Then $\mathbf{H}$ has n eigenvalues λ with $\operatorname{Re}\lambda < 0$ and n eigenvalues with $\operatorname{Re}\lambda > 0$. Let $\chi_-(\mathbf{H})$ be the subspace of $\mathbb{R}^{2n}$ spanned by the eigenvectors and principal vectors which belong to eigenvalues λ with $\operatorname{Re}\lambda < 0$. Then $\chi_-(\mathbf{H})$ is an n-dimensional $\mathbf{H}$ invariant subspace of $\mathbb{R}^{2n}$. Suppose the vectors $\mathbf{w}_1, \ldots, \mathbf{w}_n$ are a basis of $\chi_-(\mathbf{H})$. Then we can write

$$\begin{pmatrix} \mathbf{X}_1 \\ \mathbf{X}_2 \end{pmatrix} = \begin{pmatrix} \mathbf{w}_1 & \ldots & \mathbf{w}_n \end{pmatrix}.$$

Herein, $\mathbf{X}_1, \mathbf{X}_2$ are $n \times n$-matrices, which can be chosen as real matrices. Hence we have

$$\chi_-(\mathbf{H}) = \operatorname{Im}\begin{pmatrix} \mathbf{X}_1 \\ \mathbf{X}_2 \end{pmatrix}.$$

The nonsingularity of $\mathbf{X}_1$ is equivalent to

(A2) Complementarity Property: $\chi_-(\mathbf{H}) \cap \operatorname{Im}\begin{pmatrix} \mathbf{0} \\ \mathbf{I} \end{pmatrix} = \{\mathbf{0}\}$.

To see this, note that a vector $\mathbf{w}$ is in the left-hand set of (A2) if it has the representations

$$\mathbf{w} = \alpha_1 \mathbf{w}_1 + \ldots + \alpha_n \mathbf{w}_n, \qquad \alpha_1, \ldots, \alpha_n \in \mathbb{C},$$

$$\mathbf{w} = \begin{pmatrix} \mathbf{0} \\ \mathbf{v} \end{pmatrix}, \qquad \mathbf{v} \in \mathbb{C}^n.$$

These equations are equivalent to

$$\mathbf{X}_1 \begin{pmatrix} \alpha_1 \\ \vdots \\ \alpha_n \end{pmatrix} = \mathbf{0},$$

and the assertion now follows immediatly.

The fact that $\chi_-(\mathbf{H})$ is an n-dimensional $\mathbf{H}$-invariant subspace of $\mathbb{R}^{2n}$ can be formulated as follows. There is an $n \times n$-matrix $\mathbf{H}_-$ with

$$\mathbf{H}\begin{pmatrix}\mathbf{X}_1\\ \mathbf{X}_2\end{pmatrix}=\begin{pmatrix}\mathbf{X}_1\\ \mathbf{X}_2\end{pmatrix}\mathbf{H}_-. \tag{8.13}$$

Let (A2) be fulfilled and put

$$\mathbf{X}=\mathbf{X}_2\mathbf{X}_1^{-1}. \tag{8.14}$$

Multiplying (8.13) with $\mathbf{X}_1^{-1}$ from the right-hand side gives

$$\mathbf{H}\begin{pmatrix}\mathbf{I}\\ \mathbf{X}\end{pmatrix}=\begin{pmatrix}\mathbf{I}\\ \mathbf{X}\end{pmatrix}\mathbf{X}_1\mathbf{H}_-\mathbf{X}_1^{-1}. \tag{8.15}$$

If this is multiplied from the left-hand side with $(\mathbf{X} \quad -\mathbf{I})$, we get

$$(\mathbf{X} \quad -\mathbf{I})\mathbf{H}\begin{pmatrix}\mathbf{I}\\ \mathbf{X}\end{pmatrix}=\mathbf{0}.$$

Carrying out the multiplication leads directly to the Riccati equation (8.11).

Hence, by (8.14) a solution of the Riccati equation is defined. We must still show that the matrix $\mathbf{X}$ constructed by this way is unique. To this end we prove that the same matrix $\mathbf{X}$ will be obtained if a different basis of $\chi_-(\mathbf{H})$ is chosen. For a different basis

$$\begin{pmatrix}\mathbf{X}_1\mathbf{T}\\ \mathbf{X}_2\mathbf{T}\end{pmatrix} \text{ has to be replaced by } \begin{pmatrix}\mathbf{X}_1\\ \mathbf{X}_2\end{pmatrix},$$

where $\mathbf{T}$ is nonsingular. The assertion follows now from $(\mathbf{X}_2\mathbf{T})(\mathbf{X}_1\mathbf{T})^{-1}=\mathbf{X}$.

The considerations are the basis for the following definition.

Definition. *Let* $\mathbf{H}$ *be a Hamilton matrix such that the assumptions (A1), (A2) hold and let* $\mathbf{X}$ *be the uniquely constructed matrix according to (8.14). Then the Operator which maps* $\mathbf{H}$ *onto* $\mathbf{X}$ *will be denoted by* Ric, *i.e.* $\mathbf{X}=Ric(\mathbf{H})$. *This operator is defined on*

$$dom(Ric)=\{\,\mathbf{H}\in\mathbb{R}^{2n\times 2n}\mid \mathbf{H} \text{ is a Hamilton matrix and fulfills (A 1), (A 2)}\}.$$

Theorem 8.1.2. *Suppose* $\mathbf{H}\in dom(Ric)$ *and* $\mathbf{X}=Ric(\mathbf{H})$. *Then the following assertions hold.*

(a) $\mathbf{X}$ *fulfills the Riccati equation (8.11).*

(b) $\mathbf{X}$ *is real and symmetric.*

(c) $\mathbf{A}+\mathbf{R}\mathbf{X}$ *is stable.*

Proof. (a) This has already been shown above.

(b) The matrices $\mathbf{X}_1, \mathbf{X}_2$ can be supposed as real. We show that $\mathbf{X}_1^T\mathbf{X}_2$ is symmetric. Multiplication of (8.13) from the left-hand side with

$$\begin{pmatrix} \mathbf{X}_1 \\ \mathbf{X}_2 \end{pmatrix}^T \mathbf{J}$$

gives

$$\begin{pmatrix} \mathbf{X}_1 \\ \mathbf{X}_2 \end{pmatrix}^T \mathbf{J}\,\mathbf{H} \begin{pmatrix} \mathbf{X}_1 \\ \mathbf{X}_2 \end{pmatrix} = \begin{pmatrix} \mathbf{X}_1 \\ \mathbf{X}_2 \end{pmatrix}^T \mathbf{J} \begin{pmatrix} \mathbf{X}_1 \\ \mathbf{X}_2 \end{pmatrix} \mathbf{H}_- . \tag{8.16}$$

Since $\mathbf{J}\,\mathbf{H}$ is symmetric, this holds also for the left-hand side of (8.16) and therefore also for the right-hand side of this equation. Hence we get

$$\begin{aligned}(-\mathbf{X}_1^T\mathbf{X}_2 + \mathbf{X}_2^T\mathbf{X}_1)\mathbf{H}_- &= \mathbf{H}_-^T(-\mathbf{X}_1^T\mathbf{X}_2 + \mathbf{X}_2^T\mathbf{X}_1)^T \\ &= -\mathbf{H}_-^T(-\mathbf{X}_1^T\mathbf{X}_2 + \mathbf{X}_2^T\mathbf{X}_1).\end{aligned}$$

This is a Lyapunov equation for the matrix in the bracket. Because of (8.14), $\mathbf{H}_-$ is the matrix realization of $\mathbf{H}$ on $\chi_-(\mathbf{H})$. This shows the stability of $\mathbf{H}_-$. Consequently, this Lyapunov equation has $\mathbf{0}$ as the unique solution (cf. Theorem 5.7.1):

$$-\mathbf{X}_1^T\mathbf{X}_2 + \mathbf{X}_2^T\mathbf{X}_1 = \mathbf{0},$$

i.e. $\mathbf{X}_1^T\mathbf{X}_2$ is symmetric. The symmetry of $\mathbf{X}$ follows now from the representation

$$\mathbf{X} = (\mathbf{X}_1^{-1})^T(\mathbf{X}_1^T\mathbf{X}_2)\mathbf{X}_1^{-1}.$$

(c) Multiplication of (8.15) with $(\mathbf{I} \quad \mathbf{0})$ gives

$$\mathbf{A} + \mathbf{R}\,\mathbf{X} = \mathbf{X}_1\mathbf{H}_-\mathbf{X}_1^{-1}.$$

The stability of $\mathbf{A} + \mathbf{R}\,\mathbf{X}$ follows now from the stability of $\mathbf{H}_-$.

Theorem 8.1.3. *Suppose* $\mathbf{H}$ *has no imaginary eigenvalues and let* $\mathbf{R}$ *be either positive semidefinite or negative semidefinite. Then* $\mathbf{H} \in dom(Ric)$ *holds if and only if* $(\mathbf{A}, \mathbf{R})$ *is stabilizable.*

Proof. "$\Leftarrow$": We have to show that (A2) is fulfilled, i.e. that $\mathbf{X}_1$ is nonsingular. Herein $\mathbf{X}_1, \mathbf{X}_2, \mathbf{H}_-$ are defined as above. First we prove the invariance of $\operatorname{Ker}\mathbf{X}_1$ under $\mathbf{H}_-$. Let $\mathbf{x} \in \operatorname{Ker}\mathbf{X}_1$. Multiplication of (8.13) from the left-hand side with $(\mathbf{I} \quad \mathbf{0})$ gives

$$\mathbf{A}\,\mathbf{X}_1 + \mathbf{R}\,\mathbf{X}_2 = \mathbf{X}_1\mathbf{H}_- . \tag{8.17}$$

Multiplication of this equation from the left-hand side with $\mathbf{x}^T\mathbf{X}_2^T$ and from the right-hand side with $\mathbf{x}$ yields

$$\mathbf{x}^T \mathbf{X}_2^T \mathbf{R} \mathbf{X}_2 \mathbf{x} = \mathbf{x}^T \mathbf{X}_2^T \mathbf{X}_1 \mathbf{H}_- \mathbf{x}.$$

This implies

$$\mathbf{x}^T \mathbf{X}_2^T \mathbf{R} \mathbf{X}_2 \mathbf{x} = \mathbf{x}^T \mathbf{H}_-^T \mathbf{X}_1^T \mathbf{X}_2 \mathbf{x} = \mathbf{x}^T \mathbf{H}_-^T \mathbf{X}_2^T \mathbf{X}_1 \mathbf{x} = 0.$$

Here we have used the symmetry of $\mathbf{X}_1^T \mathbf{X}_2$, which was shown in the proof of Theorem 8.1.2. Since $\mathbf{R}$ is semidefinite, the equation $\mathbf{R} \mathbf{X}_2 \mathbf{x} = \mathbf{0}$ follows. Multiplication of (8.17) from the right-hand side with $\mathbf{x}$ gives $\mathbf{X}_1 \mathbf{H}_- \mathbf{x} = \mathbf{0}$, i.e. $\mathbf{H}_- \mathbf{x} \in \operatorname{Ker} \mathbf{X}_1$.

Assume now that $\operatorname{Ker} \mathbf{X}_1 \neq \mathbf{0}$. As we have shown, $\mathbf{H}_- |_{\operatorname{Ker} \mathbf{X}_1}$ is a map from $\operatorname{Ker} \mathbf{X}_1$ to $\operatorname{Ker} \mathbf{X}_1$. Hence it has an eigenvalue λ and a corresponding eigenvector $\mathbf{x}$ lying in $\operatorname{Ker} \mathbf{X}_1$:

$$\begin{aligned} &\mathbf{H}_- \mathbf{x} = \lambda \mathbf{x}, \quad \mathbf{x} \neq \mathbf{0}, \\ &\operatorname{Re} \lambda < 0, \quad \mathbf{x} \in \operatorname{Ker} \mathbf{X}_1 . \end{aligned} \tag{8.18}$$

Multiplication of (8.13) from the left-hand side with $(\mathbf{0} \quad \mathbf{I})$ gives

$$\mathbf{Q} \mathbf{X}_1 - \mathbf{A}^T \mathbf{X}_2 = \mathbf{X}_2 \mathbf{H}_- .$$

This equation and (8.18) show that

$$(\mathbf{A}^T + \lambda \mathbf{I}) \mathbf{X}_2 \mathbf{x} = \mathbf{0} .$$

As we have seen above, $\mathbf{x} \in \operatorname{Ker} \mathbf{X}_1$ implies $\mathbf{R} \mathbf{X}_2 \mathbf{x} = \mathbf{0}$. Consequently,

$$\mathbf{x}^T \mathbf{X}_2^T \left(\mathbf{A} + \bar{\lambda} \mathbf{I} \quad \mathbf{R} \right) = \mathbf{0}.$$

From the stabilizability of $(\mathbf{A}, \mathbf{R})$ we deduce that $\mathbf{X}_2 \mathbf{x} = \mathbf{0}$ (cf. Theorem 5.3.5). Together with $\mathbf{X}_1 \mathbf{x} = \mathbf{0}$ and the fact that $\begin{pmatrix} \mathbf{X}_1 \\ \mathbf{X}_2 \end{pmatrix}$ has full column rank, it follows that $\mathbf{x} = \mathbf{0}$, which is a contradiction. Hence $\mathbf{X}_1$ is nonsingular.

"$\Rightarrow$": If $\mathbf{H}$ is nonsingular, the application of Theorem 8.1.2 shows that $\mathbf{A} + \mathbf{R} \mathbf{X}$ is stable. Therefore, $(\mathbf{A}, \mathbf{R})$ is stabilizable.

Theorem 8.1.4. *Let* $(\mathbf{A}, \mathbf{B})$ *be stabilizable, and suppose that* $(\mathbf{C}, \mathbf{A})$ *has no unobservable eigenvalues on the imaginary axis. Let* $\mathbf{H}$ *be of the form*

$$\mathbf{H} = \begin{pmatrix} \mathbf{A} & -\mathbf{B}\mathbf{B}^T \\ -\mathbf{C}^T \mathbf{C} & -\mathbf{A}^T \end{pmatrix}.$$

Then $\mathbf{H} \in dom(Ric)$ *and* $\mathbf{X} = Ric(\mathbf{H})$ *is positive semidefinite. If* $(\mathbf{C}, \mathbf{A})$ *has no stable unobservable eigenvalues, then* $\mathbf{X}$ *is positive definite.*

Proof. Using Theorem 5.3.5 it is easily seen that the stabilizability of $(\mathbf{A},\mathbf{B})$ implies the stabilizability of $(\mathbf{A},-\mathbf{B}\mathbf{B}^T)$. Hence from Theorem 8.1.3 we get $\mathbf{H} \in dom(Ric)$, if $\mathbf{H}$ has no eigenvalues on the imaginary axis. Suppose that $\mathbf{H}$ has an eigenvalue on the imaginary axis. We show now that this leads to a contradiction of the assumption that $(\mathbf{C},\mathbf{A})$ has no unobservable eigenvalues on the imaginary axis.

Let now $i\omega$ be an eigenvalue of $\mathbf{H}$ with eigenvector $(\mathbf{x}^* \quad \mathbf{z}^*) \neq \mathbf{0}$. Then

$$\mathbf{A}\mathbf{x} - \mathbf{B}\mathbf{B}^T\mathbf{z} = i\omega\mathbf{x}$$

$$-\mathbf{C}\mathbf{C}^T\mathbf{x} - \mathbf{A}^T\mathbf{z} = i\omega\mathbf{z},$$

and

$$(\mathbf{A} - i\omega\mathbf{I})\mathbf{x} = \mathbf{B}\mathbf{B}^T\mathbf{z}$$

$$-(\mathbf{A} - i\omega\mathbf{I})^*\mathbf{z} = \mathbf{C}^T\mathbf{C}\mathbf{x}.$$

This implies

$$\mathbf{z}^T(\mathbf{A} - i\omega\mathbf{I})\mathbf{x} = \mathbf{z}^T\mathbf{B}\mathbf{B}^T\mathbf{z} = \|\mathbf{B}^T\mathbf{z}\|^2$$

$$-\mathbf{x}^T(\mathbf{A} - i\omega\mathbf{I})^*\mathbf{z} = \mathbf{x}^T\mathbf{C}\mathbf{C}^T\mathbf{x} = \|\mathbf{C}\mathbf{x}\|^2 .$$

Hence $\mathbf{x}^T(\mathbf{A} - i\omega\mathbf{I})\mathbf{x}$ is real and the equation

$$-\|\mathbf{C}\mathbf{x}\|^2 = \mathbf{x}^T(\mathbf{A} - i\omega\mathbf{I})^*\mathbf{z} = \overline{\mathbf{x}^T(\mathbf{A} - i\omega\mathbf{I})^*\mathbf{z}} = \mathbf{z}^T(\mathbf{A} - i\omega\mathbf{I})\mathbf{x} = \|\mathbf{B}^T\mathbf{z}\|^2$$

holds. Thus we have $\mathbf{C}\mathbf{x} = \mathbf{0}$ and $\mathbf{B}^T\mathbf{z} = \mathbf{0}$ and consequently,

$$(\mathbf{A} - i\omega\mathbf{I})\mathbf{x} = \mathbf{0}$$

$$-(\mathbf{A} - i\omega\mathbf{I})^*\mathbf{z} = \mathbf{0}.$$

Using the stabilizability of $(\mathbf{A},\mathbf{B})$ the equation $\mathbf{z} = \mathbf{0}$ follows. Since $(\mathbf{C},\mathbf{A})$ has no unobservable modes on the imaginary axis, we get also $\mathbf{x} = \mathbf{0}$, which is a contradiction.

Next we show that $\mathbf{X}$ is positive semidefinite. The Riccati equation

$$\mathbf{A}^T\mathbf{X} + \mathbf{X}\mathbf{A} - \mathbf{X}\mathbf{B}\mathbf{B}^T\mathbf{X} + \mathbf{C}^T\mathbf{C} = \mathbf{0}$$

can here equivalently be rewritten as

$$(\mathbf{A} - \mathbf{B}\mathbf{B}^T\mathbf{X})^T\mathbf{X} + \mathbf{X}(\mathbf{A} - \mathbf{B}\mathbf{B}^T\mathbf{X}) + \mathbf{X}\mathbf{B}\mathbf{B}^T\mathbf{X} + \mathbf{C}^T\mathbf{C} = \mathbf{0}. \qquad (8.19)$$

This can be viewed as a Lyapunov equation for $\mathbf{X}$ with $\mathbf{Q} = \mathbf{X}\mathbf{B}\mathbf{B}^T\mathbf{X} + \mathbf{C}^T\mathbf{C}$. Since $\mathbf{A} - \mathbf{B}\mathbf{B}^T\mathbf{X}$ is stable (cf. Theorem 8.1.2), the positive semidefiniteness of $\mathbf{X}$ follows from Theorem 5.7.1.

Suppose that $(\mathbf{C},\mathbf{A})$ has no stable unobservable eigenvalues and let $\mathbf{X}\mathbf{x} = \mathbf{0}$. Pre-multiplication of (8.19) by $\mathbf{x}^*$ and post-multiplication by $\mathbf{x}$ yields $\mathbf{C}\mathbf{x} = \mathbf{0}$.

Then post-multiply (8.19) again by $\mathbf{x}$ to get $\mathbf{XAx}=\mathbf{0}$, i.e. $\mathbf{A}(\ker\mathbf{X})\subset\ker\mathbf{X}$. If $\mathbf{X}$ is not positive definite, then $\ker\mathbf{X}\neq\{\mathbf{0}\}$. Thus, in this case there is $\mathbf{x}\neq\mathbf{0}$ in $\ker\mathbf{X}$ and a λ such that $\lambda\mathbf{x}=\mathbf{Ax}=(\mathbf{A}-\mathbf{BB}^T\mathbf{X})\mathbf{x}$. Because $\mathbf{A}-\mathbf{BB}^T\mathbf{X}$ is stable, $\mathrm{Re}\,\lambda<0$ follows. Since $(\mathbf{C},\mathbf{A})$ has no stable unobservable eigenvalues, we get a contradiction and $\mathbf{X}$ must be positive definite.

Remark. Let $(\mathbf{A},\mathbf{B})$ be stabilizable and $(\mathbf{C},\mathbf{A})$ be observable. Then, by the last theorem, $\mathbf{X}=Ric(\mathbf{H})$ is positive definite.

Theorem 8.1.5. *Suppose* $\mathbf{D}$ *satisfies* $\mathbf{D}^T\mathbf{D}=\mathbf{I}$. *Let* $\mathbf{H}$ *be of the form*

$$\mathbf{H}=\begin{pmatrix}\mathbf{A}-\mathbf{BD}^T\mathbf{C} & -\mathbf{BB}^T\\ -\mathbf{C}^T(\mathbf{I}-\mathbf{DD}^T)\mathbf{C} & -(\mathbf{A}-\mathbf{BD}^T\mathbf{C})^T\end{pmatrix}$$

and let $(\mathbf{A},\mathbf{B})$ *be stabilizable. Finally, assume that the matrix*

$$\begin{pmatrix}\mathbf{A}-i\omega\mathbf{I} & \mathbf{B}\\ \mathbf{C} & \mathbf{D}\end{pmatrix}$$

has full column rank for all ω. *Then* $\mathbf{H}\in dom(Ric)$ *and* $\mathbf{X}=Ric(\mathbf{H})$ *is positive semidefinite.*

The result is a consequence of Theorem 8.1.4. It remains to show that the condition concerning unobservable eigenvalues on the imaginary axis is satisfied. We omit the detailed proof and refer the reader to Zhou, Doyle, and Glover [112], Corollary 13.10.

This section ends with a reformulation of the bounded real lemma (Theorem 7.3.1).

Theorem 8.1.6. *Let* $\gamma>0$ *and a let transfer matrix*

$$\mathbf{G}(s)=\left(\begin{array}{c|c}\mathbf{A} & \mathbf{B}\\ \hline \mathbf{C} & \mathbf{D}\end{array}\right)\in\mathcal{RH}_\infty$$

be given. Put $\mathbf{R}=\gamma^2\mathbf{I}-\mathbf{D}^T\mathbf{D}$ *and define*

$$\mathbf{H}=\begin{pmatrix}\mathbf{A}+\mathbf{BR}^{-1}\mathbf{D}^T\mathbf{C} & \mathbf{BR}^{-1}\mathbf{B}^T\\ -\mathbf{C}^T(\mathbf{I}+\mathbf{DR}^{-1}\mathbf{D}^T)\mathbf{C} & -(\mathbf{A}+\mathbf{BR}^{-1}\mathbf{D}^T\mathbf{C})^T\end{pmatrix}.$$

Then the following assertions are equivalent:

(i) $\|\mathbf{G}\|_\infty<\gamma$.

(ii) $\bar{\sigma}(\mathbf{D})<\gamma$ *and* $\mathbf{H}$ *has no eigenvalues on the imaginary axis.*

(iii) $\bar{\sigma}(\mathbf{D}) < \gamma$ *and* $\mathbf{H} \in dom(Ric)$.

(iv) $\bar{\sigma}(\mathbf{D}) < \gamma$ *and* $\mathbf{H} \in dom(Ric)$ *and* $\mathbf{X} = Ric(\mathbf{H}) \geq \mathbf{0}$.

Proof. The equivalence of (i) and (ii) is Theorem 7.3.1 The fact that (iii) implies (ii) is in the definition of the Riccati operator. In order to show that (iii) follows from (ii) we apply Theorem 8.1.3. Hence we have to prove the stabilizability of $(\mathbf{A} + \mathbf{B}\mathbf{R}^{-1}\mathbf{D}^T\mathbf{C}, \mathbf{B}\mathbf{R}^{-1}\mathbf{B}^T)$. This will be done by applying the PHB test. Suppose

$$\mathbf{x}^*(\mathbf{A} + \mathbf{B}\mathbf{R}^{-1}\mathbf{D}^T\mathbf{C}) = \lambda\mathbf{x}^* \quad \text{with} \quad \operatorname{Re}\lambda \geq 0 \quad \text{and} \quad \mathbf{x}^*\mathbf{B}\mathbf{R}^{-1}\mathbf{B}^T = \mathbf{0}.$$

This implies that $\mathbf{x}^*\mathbf{B}\mathbf{R}^{-1/2} = \mathbf{0}$ and therefore $\mathbf{x}^*\mathbf{A} = \lambda\mathbf{x}^*$, i.e. $\mathbf{x} = \mathbf{0}$, since $\mathbf{A}$ is stable. It remains to show that $\mathbf{X} \geq \mathbf{0}$ if (iii) is fulfilled. The Riccati equation can be written in the form

$$\mathbf{A}^T\mathbf{X} + \mathbf{X}\mathbf{A} + (\mathbf{C}^T\mathbf{D} + \mathbf{X}\mathbf{B})\mathbf{R}^{-1}(\mathbf{B}^T\mathbf{X} + \mathbf{D}^T\mathbf{C}) + \mathbf{C}^T\mathbf{C} = \mathbf{0}.$$

This is a Lyapunov equation for $\mathbf{X}$ with

$$\mathbf{Q} = (\mathbf{C}^T\mathbf{D} + \mathbf{X}\mathbf{B})\mathbf{R}^{-1}(\mathbf{B}^T\mathbf{X} + \mathbf{D}^T\mathbf{C}) + \mathbf{C}^T\mathbf{C} \geq \mathbf{0}.$$

Since $\mathbf{A}$ is stable, we deduce $\mathbf{X} \geq \mathbf{0}$.

8.2 Characterization of $\mathcal{H}_2$ Optimal Controllers

8.2.1 Problem Formulation and Characterization Theorem

We start with a plant of the form

$$\begin{aligned}
\dot{\mathbf{x}} &= \mathbf{A}\,\mathbf{x} + \mathbf{B}_1\mathbf{w} + \mathbf{B}_2\mathbf{u} \\
\mathbf{z} &= \mathbf{C}_1\,\mathbf{x} + \mathbf{D}_{12}\mathbf{u} \\
\mathbf{y} &= \mathbf{C}_2\,\mathbf{x} + \mathbf{D}_{21}\mathbf{w}
\end{aligned}$$

i.e. written as transfer function

$$\mathbf{P}(s) = \left(\begin{array}{c|cc} \mathbf{A} & \mathbf{B}_1 & \mathbf{B}_2 \\ \hline \mathbf{C}_1 & \mathbf{0} & \mathbf{D}_{12} \\ \mathbf{C}_2 & \mathbf{D}_{21} & \mathbf{0} \end{array}\right).$$

With a controller

$$\hat{\mathbf{u}} = \mathbf{K}(s)\,\hat{\mathbf{y}}$$

the closed-loop transfer function is denoted as before by $\mathbf{F}_{zw}(s)$. We now formulate the following problem.

$\mathcal{H}_2$ Optimal Control Problem. *The $\mathcal{H}_2$ optimal control problem is to find a proper controller $\mathbf{K}(s)$ which internally stabilizes $\mathbf{P}(s)$ such that the $\mathcal{H}_2$ norm of the closed-loop transfer function $\mathbf{F}_{zw}(s)$ is minimized.*

We consider once again the LQR problem. Then, with $\mathbf{u} = \mathbf{F}_2\,\mathbf{y}$ and treating the initial value as an impulse, $\mathbf{w} = \mathbf{x}_0\delta(t)$, it is possible to write

$$\dot{\mathbf{x}} = (\mathbf{A} + \mathbf{B}_2\mathbf{F})\mathbf{x} + \mathbf{w}$$
$$\mathbf{z} = (\mathbf{C}_1 + \mathbf{D}_{12}\mathbf{F})\mathbf{x}.$$

Choose, in particular, $\mathbf{w} = \mathbf{e}_k$ and denote the corresponding output by $\mathbf{z}_k$. Then (7.4) implies

$$\| \mathbf{F}_{zw} \|_2^2 = \sum_{k=1}^{n} \| \mathbf{z}_k \|_2^2 .$$

The minimization of each summand $\| \mathbf{z}_k \|_2$ is achieved by the constant gain controller $\mathbf{F}$ described in the LQR theory. This controller is the same for all summands and therefore also the solution of the $\mathcal{H}_2$ problem. In this sense, it is possible to consider the LQR problem as a special $\mathcal{H}_2$ problem.

Next we formulate conditions under which the $\mathcal{H}_2$ optimal control problem is solvable:

(A1) $(\mathbf{A}, \mathbf{B}_2)$ *is stabilizable.*

(A2) $(\mathbf{C}_2, \mathbf{A})$ *is detectable.*

(A3) *The columns of $\mathbf{D}_{12}$ form an orthonormal system and $\mathbf{D}_{12}$ is completed by a matrix $\mathbf{D}_\perp$ to a unitary matrix $(\mathbf{D}_{12} \quad \mathbf{D}_\perp)$.*

(A4) *The rows of $\mathbf{D}_{21}$ form an orthonormal system and $\mathbf{D}_{21}$ is completed by a matrix $\overline{\mathbf{D}}_\perp$ to a unitary matrix*

$$\begin{pmatrix} \mathbf{D}_{21} \\ \overline{\mathbf{D}}_\perp \end{pmatrix}.$$

(A5) *The following matrix has full column rank for all ω:*

$$\begin{pmatrix} \mathbf{A} - i\omega\mathbf{I} & \mathbf{B}_2 \\ \mathbf{C}_1 & \mathbf{D}_{12} \end{pmatrix}.$$

(A6) *The following matrix has full row rank for all ω:*

$$\begin{pmatrix} \mathbf{A} - i\omega\mathbf{I} & \mathbf{B}_1 \\ \mathbf{C}_2 & \mathbf{D}_{21} \end{pmatrix}.$$

For the formulation of the solution, it is convenient to define the following Hamiltonian matrices:

$$\mathbf{H}_2 = \begin{pmatrix} \mathbf{A} - \mathbf{B}_2\mathbf{D}_{12}^T\mathbf{C}_1 & -\mathbf{B}_2\mathbf{B}_2^T \\ -\mathbf{C}_1^T\mathbf{D}_\perp\mathbf{D}_\perp^T\mathbf{C}_1 & -(\mathbf{A} - \mathbf{B}_2\mathbf{D}_{12}^T\mathbf{C}_1)^T \end{pmatrix} \tag{8.20}$$

$$\mathbf{J}_2 = \begin{pmatrix} \mathbf{A}^T - \mathbf{C}_2^T\mathbf{D}_{21}\mathbf{B}_1^T & -\mathbf{C}_2^T\mathbf{C}_2 \\ -\mathbf{B}_1\overline{\mathbf{D}}_\perp^T\overline{\mathbf{D}}_\perp\mathbf{B}_1^T & -(\mathbf{A}^T - \mathbf{C}_2^T\mathbf{D}_{21}\mathbf{B}_1^T)^T \end{pmatrix}. \tag{8.21}$$

From the assumptions and Theorem 8.1.5, it is seen that both matrices lie in the domain of the Riccati operator. Moreover, the matrices $\mathbf{X}_2 = Ric(\mathbf{H}_2)$ and $\mathbf{Y}_2 = Ric(\mathbf{J}_2)$ are positive semidefinite.

Theorem 8.2.1. *Let the assumptions (A1) – (A6) be fulfilled. With the Hamiltonian matrices $\mathbf{H}_2$ of (8.20) and $\mathbf{J}_2$ of (8.21) and*

$$\mathbf{X}_2 = Ric(\mathbf{H}_2), \quad \mathbf{Y}_2 = Ric(\mathbf{J}_2)$$

define

$$\mathbf{F}_2 = -(\mathbf{B}_2^T\mathbf{X}_2 + \mathbf{D}_{12}^T\mathbf{C}_1), \qquad \mathbf{L}_2 = -(\mathbf{Y}_2\mathbf{C}_2^T + \mathbf{B}_1\mathbf{D}_{21}^T)$$

$$\tilde{\mathbf{A}}_2 = \mathbf{A} + \mathbf{B}_2\mathbf{F}_2 + \mathbf{L}_2\mathbf{C}_2$$

$$\mathbf{A}_F = \mathbf{A} + \mathbf{B}_2\mathbf{F}_2, \quad \mathbf{C}_{1F} = \mathbf{C}_1 + \mathbf{D}_{12}\mathbf{F}_2$$

$$\mathbf{A}_L = \mathbf{A} + \mathbf{L}_2\mathbf{C}_2, \quad \mathbf{B}_{1L} = \mathbf{B}_1 + \mathbf{L}_2\mathbf{D}_{21}$$

$$\mathbf{G}_F(s) = \left(\begin{array}{c|c} \mathbf{A}_F & \mathbf{B}_1 \\ \hline \mathbf{C}_{1F} & \mathbf{0} \end{array}\right), \quad \mathbf{G}_L(s) = \left(\begin{array}{c|c} \mathbf{A}_L & \mathbf{B}_{1L} \\ \hline -\mathbf{F}_2 & \mathbf{0} \end{array}\right).$$

Then the $\mathcal{H}_2$ problem has the optimal controller

$$\mathbf{K}(s) = \left(\begin{array}{c|c} \tilde{\mathbf{A}}_2 & -\mathbf{L}_2 \\ \hline \mathbf{F}_2 & \mathbf{0} \end{array}\right)$$

as a unique solution. The optimal value is given by

$$\min \| \mathbf{F}_{zw} \|_2^2 = \| \mathbf{G}_F \|_2^2 + \| \mathbf{G}_L \|_2^2 .$$

Remarks. 1. The controller is observer based. In the time domain the equations for the observer and controller are given by

$$\dot{\tilde{\mathbf{x}}} = \tilde{\mathbf{A}}_2\tilde{\mathbf{x}} - \mathbf{L}_2\mathbf{y}$$

$$\mathbf{u} = \mathbf{F}_2\tilde{\mathbf{x}} .$$

Equivalently they can be written as (cf. (6.13))

$$\dot{\tilde{\mathbf{x}}} = \mathbf{A}\tilde{\mathbf{x}} + \mathbf{B}_2\mathbf{u} + \mathbf{L}_2(\mathbf{C}_2\tilde{\mathbf{x}} - \mathbf{y})$$
$$\mathbf{u} = \mathbf{F}_2\tilde{\mathbf{x}}\ .$$

Here $\tilde{\mathbf{x}}$ is the observer estimate of $\mathbf{x}$.

2. In the important special case $\mathbf{D}_{12}^T\mathbf{C}_1 = \mathbf{0}$ and $\mathbf{D}_{21}\mathbf{B}_1^T = \mathbf{0}$ the Riccati equations for the controller and the observer are

$$\mathbf{A}^T\mathbf{X}_2 + \mathbf{X}_2\mathbf{A} - \mathbf{X}_2\mathbf{B}_2\mathbf{B}_2^T\mathbf{X}_2 + \mathbf{C}_1^T\mathbf{C}_1 = \mathbf{0},$$
$$\mathbf{Y}_2\mathbf{A}^T + \mathbf{A}\mathbf{Y}_2 - \mathbf{Y}_2\mathbf{C}_2^T\mathbf{C}_2\mathbf{Y}_2 + \mathbf{B}_1\mathbf{B}_1^T = \mathbf{0}.$$

Note that in this case assumptions (A5) can be replaced by (A4') of Sect, 8.1.1 (and analogously for (A6)).

3. If the full state vector can be measured, then the observer is no longer necessary. In this case, the controller reduces to

$$\mathbf{K}(s) = \mathbf{F}_2\ .$$

8.2.2 State Feedback

One important idea of the proof of Theorem 8.2.1 is to introduce again a new variable by $\mathbf{v} = \mathbf{u} - \mathbf{F}_2\mathbf{x}$. With the help of this variable, it is possible to decompose the closed-loop transfer function $\mathbf{F}_{zw}$ in a certain way. To be more specific, define the system $\mathbf{G}_v$ by

$$\dot{\mathbf{x}} = \mathbf{A}\,\mathbf{x} + \mathbf{B}_1\mathbf{w} + \mathbf{B}_2\mathbf{u}$$
$$\mathbf{v} = -\mathbf{F}_2\mathbf{x} + \mathbf{u}$$
$$\mathbf{y} = \mathbf{C}_2\,\mathbf{x} + \mathbf{D}_{21}\mathbf{w}$$

and let $\mathbf{K}(s)$ be a stabilizing controller for the plant $\mathbf{G}_v$: $\hat{\mathbf{u}} = \mathbf{K}(s)\hat{\mathbf{y}}$. Closing the feedback loop leads to a transfer function $\mathbf{F}_{vw}$ (cf. Fig. 8.1). Then, with the above

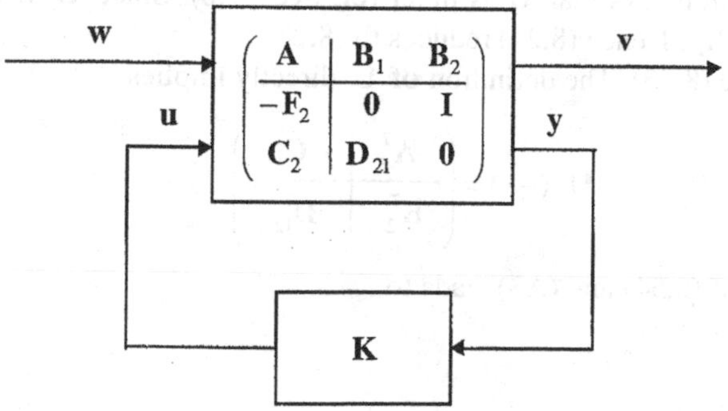

Fig. 8.1. Definition of $\mathbf{F}_{vw}$

definitions, the state equations of the original plant imply

$$\hat{\mathbf{z}} = \mathbf{F}_{zw}\hat{\mathbf{w}} = \mathbf{C}_{1F}(s\mathbf{I}-\mathbf{A}_F)^{-1}\mathbf{B}_1\hat{\mathbf{w}} + (\mathbf{C}_{1F}(s\mathbf{I}-\mathbf{A}_F)^{-1}\mathbf{B}_2 + \mathbf{D}_{12})\hat{\mathbf{v}}$$
$$= \mathbf{G}_F\hat{\mathbf{w}} + \mathbf{U}\mathbf{F}_{vw}\hat{\mathbf{w}},$$

where we have set

$$\mathbf{U}(s) = \left(\begin{array}{c|c} \mathbf{A}_F & \mathbf{B}_2 \\ \hline \mathbf{C}_{1F} & \mathbf{D}_{12} \end{array}\right).$$

The system $\mathbf{G}_v$ can be viewed as the plant of the transformed system which is stabilized by the controller $\mathbf{K}$. The following Lemma plays a key role in the proof of Theorem 8.2.1.

Lemma 8.2.1. *The following equations hold:*

$$\mathbf{F}_{zw} = \mathbf{G}_F + \mathbf{U}\mathbf{F}_{vw} \tag{8.22}$$

$$\|\mathbf{F}_{zw}\|_2^2 = \|\mathbf{G}_F\|_2^2 + \|\mathbf{F}_{vw}\|_2^2. \tag{8.23}$$

Proof. The first equation is clear from the previous calculations. Moreover, the following representation for the norm of $\mathbf{F}_{zw}$ is a direct consequence of (8.22):

$$\|\mathbf{F}_{zw}\|_2^2 = \|\mathbf{G}_F\|_2^2 + 2\langle\mathbf{G}_F, \mathbf{U}\mathbf{F}_{vw}\rangle + \|\mathbf{U}\mathbf{F}_{vw}\|_2^2. \tag{8.24}$$

The inner product can be written as (cf. Sect. 7.2.2)

$$\langle\mathbf{G}_F, \mathbf{U}\mathbf{F}_{vw}\rangle = \frac{1}{2\pi}\int_{-\infty}^{\infty} \text{trace}\,(\mathbf{F}_{vw}^T(-i\omega)\mathbf{U}^T(-i\omega)\mathbf{G}_F(i\omega)\,d\omega\,.$$

Next we will show the following relationships:

$$\mathbf{U}^T(-s)\mathbf{G}_F(s) \in \mathcal{H}_2^{\perp}, \quad \mathbf{U}^T(-s)\mathbf{U}(s) = \mathbf{I}\,. \tag{8.25}$$

The second equation says that $\mathbf{U}$ is inner (cf. Sect. 7.6). Since $\mathbf{K}$ is stabilizing, we have $\mathbf{F}_{vw} \in \mathcal{RH}_2$. Hence (8.24) reduces to (8.23).

We now prove (8.25). The definition of $\mathbf{U}$ directly implies

$$\mathbf{U}^T(-s) = \left(\begin{array}{c|c} -\mathbf{A}_F^T & -\mathbf{C}_{1F}^T \\ \hline \mathbf{B}_2^T & \mathbf{D}_{12}^T \end{array}\right).$$

The application of (5.54) and (A3) leads to

$$\mathbf{U}^T(-s)\mathbf{U}(s) = \left(\begin{array}{cc|c} -\mathbf{A}_F^T & -\mathbf{C}_{1F}^T\mathbf{C}_{1F} & -\mathbf{C}_{1F}^T\mathbf{D}_{12} \\ \mathbf{0} & \mathbf{A}_F & \mathbf{B}_2 \\ \hline \mathbf{B}_2^T & \mathbf{D}_{12}^T\mathbf{C}_F & \mathbf{I} \end{array}\right)$$

$$\mathbf{U}^T(-s)\mathbf{G}_F(s) = \left(\begin{array}{cc|c} -\mathbf{A}_F^T & -\mathbf{C}_{1F}^T\mathbf{C}_{1F} & \mathbf{0} \\ \mathbf{0} & \mathbf{A}_F & \mathbf{B}_1 \\ \hline \mathbf{B}_2^T & \mathbf{D}_{12}^T\mathbf{C}_{1F} & \mathbf{0} \end{array}\right).$$

We perform now a similarity transform with

$$\begin{pmatrix} \mathbf{I} & -\mathbf{X} \\ \mathbf{0} & \mathbf{I} \end{pmatrix}.$$

Here, as before, $\mathbf{X}$ denotes the solution of the Lyapunov equation (8.4). Using this equation and the definition of $\mathbf{F}_2$, it is easy to verify that

$$\mathbf{U}^T(-s)\mathbf{U}(s) = \left(\begin{array}{cc|c} -\mathbf{A}_F^T & \mathbf{0} & \mathbf{0} \\ \mathbf{0} & \mathbf{A}_F & \mathbf{B}_2 \\ \hline \mathbf{B}_2^T & \mathbf{0} & \mathbf{I} \end{array}\right) = \mathbf{I}$$

$$\mathbf{U}^T(-s)\mathbf{G}_F(s) = \left(\begin{array}{cc|c} -\mathbf{A}_F^T & \mathbf{0} & -\mathbf{X} \\ \mathbf{0} & \mathbf{A}_F & \mathbf{B}_2 \\ \hline \mathbf{B}_2^T & \mathbf{0} & \mathbf{0} \end{array}\right) = \left(\begin{array}{c|c} -\mathbf{A}_F^T & -\mathbf{X} \\ \hline \mathbf{B}_2^T & \mathbf{0} \end{array}\right).$$

As we have seen in the proof of Theorem 8.1.1, $\mathbf{A}_F$ is stable and hence

$$\mathbf{H}(s) = \mathbf{B}_2^T(s\mathbf{I} - \mathbf{A}_F^T)\mathbf{X} \in \mathcal{RH}_2.$$

The calculation shows that $\mathbf{U}^T(-s)\mathbf{G}_F(s) = \mathbf{H}(-s)$. By $\mathcal{L}_-$ we denote as usual the left-sided Laplace transform, i.e.

$$\mathcal{L}_-(\mathbf{f})(s) = \int_{-\infty}^{0} e^{-st}\mathbf{f}(t)\,dt$$

if $\mathbf{f} \in \mathcal{L}^2(-\infty, 0]$. Let now $\mathbf{h}$ be defined by $\mathcal{L}(\mathbf{h}) = \mathbf{H}$ and put $\mathbf{h}^-(t) = \mathbf{h}(-t)$ for $t \le 0$. Then $\mathbf{h}^- \in \mathcal{L}^2(-\infty, 0]$ and it is easy to see that $\mathcal{L}_-(\mathbf{h}^-)(s) = \mathbf{H}(-s)$. Let now an arbitrary $\mathbf{M} \in \mathcal{H}_2$ be given and define $\mathbf{m}$ by $\mathcal{L}(\mathbf{m}) = \mathbf{M}$. Then, in $\mathcal{L}^2(-\infty, \infty)$, the function $\mathbf{m}$ and $\mathbf{h}^-$ are orthogonal and the same holds for the Laplace transforms $\mathbf{M}(s)$ and $\mathbf{H}(-s)$ in $\mathcal{L}^2(i\mathbb{R})$. This shows the first equation of (8.25).

At first we consider the case where the controller is allowed to use the full state vector. Then the plant and the feedback law are

$$\begin{aligned}\dot{\mathbf{x}} &= \mathbf{A}\,\mathbf{x} + \mathbf{B}_1\mathbf{w} + \mathbf{B}_2\mathbf{u}\\ \mathbf{z} &= \mathbf{C}_1\,\mathbf{x} + \mathbf{D}_{12}\mathbf{u}\\ \mathbf{y} &= \mathbf{x}\\ \hat{\mathbf{u}} &= \mathbf{K}(s)\,\hat{\mathbf{x}}\,.\end{aligned}$$

This situation is similar to that in the LQR problem. The next theorem is a simple consequence of the preceding lemma.

Theorem 8.2.2. *Let the assumptions (A1), (A3), (A5) be fulfilled. Then the constant gain controller, which solves the LQR problem, is also optimal for the $\mathcal{H}_2$ problem with* ***state feedback:***

$$\mathbf{K}(s) = \mathbf{F}_2\,.$$

The optimal value is given by

$$\min \|\,\mathbf{F}_{zw}\,\|_2 = \|\,\mathbf{G}_F\,\|_2\,.$$

Proof. First, observe that $\mathbf{F}_{vw}$ is described by

$$\hat{\mathbf{v}} = (\mathbf{K}(s) - \mathbf{F}_2)\hat{\mathbf{x}}, \quad s\hat{\mathbf{x}} = (\mathbf{A} + \mathbf{B}_2\mathbf{K}(s))\hat{\mathbf{x}} + \mathbf{B}_1\hat{\mathbf{w}}.$$

Hence the minimum is attained for $\mathbf{K} = \mathbf{F}_2$, since for this choice we have $\mathbf{F}_{vw} = \mathbf{0}$. The formula for the optimal value is also a consequence of (8.23).

8.2.3 Proof of the Characterization Theorem

The idea is now as follows. The plant $\mathbf{G}_v$ of the transformed system has a special structure: it is a so-called output estimation problem. An output estimation problem is the dual of a disturbance feedforward problem, which can be viewed as a special full information problem. The precise definition for these problems will be given soon. The solution for a full information problem is essentially the same as for the LQR problem. So we have to solve all these subproblems, but in the inverse order, and the final result will be the general characterization theorem for $\mathcal{H}_2$ controllers.

Let a general feedback connection as in Fig. 6.8 be given. Then one defines

$$\mathcal{F}_l(\mathbf{P},\mathbf{K}) = \mathbf{F}_{zw}\,. \tag{8.26}$$

We will use this notation in the proof of the characterization theorem and also later in the book.

Full information problem. The first special case is described by

$$\begin{aligned} \dot{\mathbf{x}} &= \mathbf{A}\,\mathbf{x} + \mathbf{B}_1\mathbf{w} + \mathbf{B}_2\mathbf{u} \\ \mathbf{z} &= \mathbf{C}_1\,\mathbf{x} + \mathbf{D}_{12}\mathbf{u} \\ \mathbf{y} &= \begin{pmatrix} \mathbf{x} \\ \mathbf{w} \end{pmatrix}. \end{aligned}$$

In this case we have

$$\mathbf{C}_2 = \begin{pmatrix} \mathbf{I} \\ \mathbf{0} \end{pmatrix}, \quad \mathbf{D}_{21} = \begin{pmatrix} \mathbf{0} \\ \mathbf{I} \end{pmatrix},$$

and the controller can make use not only of the state vector $\mathbf{x}$ but also of the vector of disturbances $\mathbf{w}$. This problem is called the **full information (FI) problem.**

Theorem 8.2.3. *Let the assumptions (A1), (A3), (A5) be fulfilled. Then the optimal controller and the optimal value for the FI problem are given by*

$$\mathbf{K}(s) = \begin{pmatrix} \mathbf{F}_2 & \mathbf{0} \end{pmatrix},$$

$$\min \| \mathbf{F}_{zw} \|_2 = \| \mathbf{G}_F \|_2 .$$

Proof. The general controller is now

$$\hat{\mathbf{u}} = \begin{pmatrix} \mathbf{K}_1(s) & \mathbf{K}_2(s) \end{pmatrix} \begin{pmatrix} \hat{\mathbf{x}} \\ \hat{\mathbf{w}} \end{pmatrix},$$

and the transfer function $\mathbf{F}_{vw}$ is described by

$$\begin{aligned} s\hat{\mathbf{x}} &= (\mathbf{A} + \mathbf{B}_2\mathbf{K}_1)\hat{\mathbf{x}} + (\mathbf{B}_1 + \mathbf{B}_2\mathbf{K}_2)\hat{\mathbf{w}} \\ \hat{\mathbf{v}} &= (\mathbf{K}_1 - \mathbf{F}_2)\hat{\mathbf{x}} + \mathbf{K}_2\hat{\mathbf{w}}. \end{aligned}$$

As in the proof of Theorem 8.2.2, the function $\mathbf{F}_{vw}$ vanishes for the choice $\mathbf{K}_1 = \mathbf{F}_2$, $\mathbf{K}_2 = \mathbf{0}$ and the result follows from Lemma 8.2.1.

Remark. The controller for the FI problem does not make use of the disturbance and is the same as for the state feedback problem.

A second important special case is the **disturbance feedforward (DF) problem**. For a DF problem we have by definition $\mathbf{D}_{21} = \mathbf{I}$ and it has here the form

$$\begin{aligned} \dot{\mathbf{x}} &= \mathbf{A}\,\mathbf{x} + \mathbf{B}_1\mathbf{w} + \mathbf{B}_2\mathbf{u} \\ \mathbf{z} &= \mathbf{C}_1\,\mathbf{x} + \mathbf{D}_{12}\mathbf{u} \end{aligned}$$

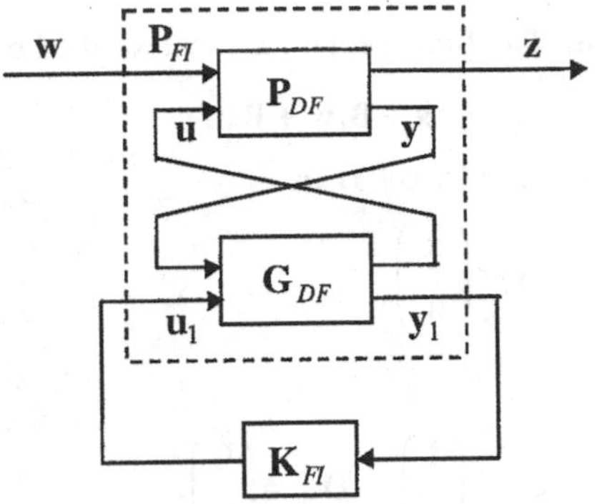

Fig. 8.2. Extension of a DF problem to a FI problem

$$\mathbf{y} = \mathbf{C}_2\mathbf{x} + \mathbf{w}.$$

This system will be denoted by $\mathbf{P}_{DF}$. It can be considered as a special case of the FI problem (and vice versa, as it can be seen). The idea is shown in Fig. 8.2. Here, $\mathbf{G}_{DF}$ is defined as follows:

$$\dot{\mathbf{x}}_1 = (\mathbf{A} - \mathbf{B}_1\mathbf{C}_2)\mathbf{x}_1 + \mathbf{B}_1\mathbf{y} + \mathbf{B}_2\mathbf{u}_1$$
$$\mathbf{u} = \mathbf{u}_1$$
$$\mathbf{y}_1 = \begin{pmatrix} \mathbf{x}_1 \\ -\mathbf{C}_2\mathbf{x}_1 + \mathbf{y} \end{pmatrix}.$$

We calculate now the transfer function $\mathbf{P}_{FI}$. Replacing the variable $\mathbf{x}_1$ by $\mathbf{e} = \mathbf{x} - \mathbf{x}_1$, it has the realization

$$\dot{\mathbf{x}} = \mathbf{A}\mathbf{x} + \mathbf{B}_1\mathbf{w} + \mathbf{B}_2\mathbf{u}_1$$
$$\dot{\mathbf{e}} = (\mathbf{A} - \mathbf{B}_1\mathbf{C}_2)\mathbf{e}$$
$$\mathbf{z} = \mathbf{C}_1\mathbf{x} + \mathbf{D}_{12}\mathbf{u}$$
$$\mathbf{y}_1 = \begin{pmatrix} \mathbf{x} - \mathbf{e} \\ \mathbf{C}_2\mathbf{e} + \mathbf{w} \end{pmatrix}.$$

We replace the second output equation by

$$\mathbf{y}_1 = \begin{pmatrix} \mathbf{x} \\ \mathbf{w} \end{pmatrix}.$$

This has no influence on the norm and the internal stability of the corresponding feedback system and we get a FI problem. Let $\mathbf{A} - \mathbf{B}_1\mathbf{C}_2$ be stable. Then, because of (A1), $\mathbf{P}_{FI}$ is stabilizable. Besides this, the following relationship holds (with the notation defined in (8.26):

$$\mathcal{F}_l(\mathbf{P}_{DF}, \mathcal{F}_l(\mathbf{G}_{DF}, \mathbf{K}_{FI})) = \mathcal{F}_l(\mathbf{P}_{FI}, \mathbf{K}_{FI}).$$

Hence the controller for $\mathbf{P}_{DF}$ is $\mathbf{K}_{DF} = \mathcal{F}_l(\mathbf{G}_{DF}, \mathbf{K}_{FI}))$. Applying now Theorem 8.2.3 to $\mathbf{P}_{FI}$ gives $\mathbf{K}_{FI}(s) = (\mathbf{F}_2 \quad \mathbf{0})$. Using the definition of $\mathbf{G}_{DF}$, the following representation of $\mathbf{K}_{DF}$ immediately follows:

$$\dot{\mathbf{x}}_1 = (\mathbf{A} - \mathbf{B}_1\mathbf{C}_2)\mathbf{x}_1 + \mathbf{B}_1\mathbf{y} + \mathbf{B}_2\mathbf{F}_2\mathbf{x}_1$$
$$\mathbf{u} = \mathbf{u}_1 = \mathbf{F}_2\mathbf{x}_1 .$$

Thus we have shown the following theorem.

Theorem 8.2.4. *Suppose that (A1), (A3), (A5) hold and let* $\mathbf{A} - \mathbf{B}_1\mathbf{C}_2$ *be stable. Then the* $\mathcal{H}_2$ *optimal controller for the DF problem is given by*

$$\mathbf{K}(s) = \left(\begin{array}{c|c} \mathbf{A} + \mathbf{B}_2\mathbf{F}_2 - \mathbf{B}_1\mathbf{C}_2 & \mathbf{B}_1 \\ \hline \mathbf{F}_2 & \mathbf{0} \end{array}\right).$$

The optimal value is

$$\min \| \mathbf{F}_{zw} \|_2 = \| \mathbf{G}_F \|_2 .$$

The $\mathcal{H}_2$ optimization problem for the plant

$$\dot{\mathbf{x}} = \mathbf{A}\,\mathbf{x} + \mathbf{B}_1\mathbf{w} + \mathbf{B}_2\mathbf{u}$$
$$\mathbf{z} = \mathbf{C}_1\mathbf{x} + \mathbf{u}$$
$$\mathbf{y} = \mathbf{C}_2\mathbf{x} + \mathbf{D}_{21}\mathbf{w}$$

is called the **output estimation (OE)** problem. Dualization gives a DF problem:

$$\dot{\mathbf{x}} = \mathbf{A}^T\mathbf{x} + \mathbf{C}_1^T\mathbf{w} + \mathbf{C}_2^T\mathbf{u}$$
$$\mathbf{z} = \mathbf{B}_1^T\mathbf{x} + \mathbf{D}_{21}^T\mathbf{u}$$
$$\mathbf{y} = \mathbf{B}_2^T\mathbf{x} + \mathbf{w} .$$

For a general plant $\mathbf{P}$ and a general controller $\mathbf{K}$ it can be checked that

$$\mathcal{F}_l(\mathbf{P}, \mathbf{K})) = \mathcal{F}_l(\mathbf{P}^T, \mathbf{K}^T)^T .$$

The norm and internal stability are not changed by transposition. An application of Theorem 8.2.4 leads immediately to the following result.

Theorem 8.2.5. *Suppose (A2), (A4), (A6) and let* $\mathbf{A} - \mathbf{B}_2\mathbf{C}_1$ *be stable. Then the* $\mathcal{H}_2$ *optimal controller for the OE problem is given by*

$$\mathbf{K}(s) = \left(\begin{array}{c|c} \mathbf{A} + \mathbf{L}_2\mathbf{C}_2 - \mathbf{B}_2\mathbf{C}_1 & \mathbf{L}_2 \\ \hline \mathbf{C}_1 & \mathbf{0} \end{array} \right).$$

The optimal value is

$$\min \| \mathbf{F}_{zw} \|_2 = \| \tilde{\mathbf{G}}_L \|_2 , \tag{8.27}$$

where

$$\tilde{\mathbf{G}}_L(s) = \left(\begin{array}{c|c} \mathbf{A}_L & \mathbf{B}_{1L} \\ \hline \mathbf{C}_1 & \mathbf{0} \end{array} \right).$$

Proof of Theorem 8.2.1. We use the decompositions (8.22) and (8.23). The transfer function $\mathbf{F}_{vw}$ describes a feedback system with plant $\mathbf{G}_v(s)$, which is of type OE. Thus, by (8.23) the problem reduces to solving the OE problem for

$$\mathbf{G}_v(s) = \left(\begin{array}{c|cc} \mathbf{A} & \mathbf{B}_1 & \mathbf{B}_2 \\ \hline -\mathbf{F}_2 & \mathbf{0} & \mathbf{I} \\ \mathbf{C}_2 & \mathbf{D}_{21} & \mathbf{0} \end{array} \right).$$

Here we note that a controller $\mathbf{K}$ for $\mathbf{F}_{zw}$ is admissible if and only if it is for $\mathbf{F}_{vw}$, since the plants for both feedback systems have the same $\mathbf{A}$ matrix. The solution of the OE problem for $\mathbf{G}_v(s)$ is, according to Theorem 8.2.5, given by

$$\mathbf{K}(s) = \left(\begin{array}{c|c} \mathbf{A} + \mathbf{L}_2\mathbf{C}_2 + \mathbf{B}_2\mathbf{F}_2 & -\mathbf{L}_2 \\ \hline \mathbf{F}_2 & \mathbf{0} \end{array} \right).$$

Assumptions (A2), (A4), (A6) are needed for the application of Theorem 8.2.5. The needed stability of $\mathbf{A} + \mathbf{F}_2\mathbf{C}_1$ is obvious from the construction of $\mathbf{F}_2$, which requires the assumptions (A1), (A3), (A4). The remaining assertion for the optimal value follows from (8.23) and (8.27).

8.3 Kalman Bucy Filter

8.3.1 Kalman Bucy Filter as a Special $\mathcal{H}_2$ State Estimator

We consider the following dynamical system:

$$\begin{aligned} \dot{\mathbf{x}} &= \mathbf{A}\,\mathbf{x} + \overline{\mathbf{B}}_1\overline{\mathbf{w}} \\ \mathbf{e} &= \mathbf{x} - \tilde{\mathbf{x}} \\ \mathbf{y} &= \mathbf{C}_2\mathbf{x} + \mathbf{D}_{21}\overline{\mathbf{w}}. \end{aligned}$$

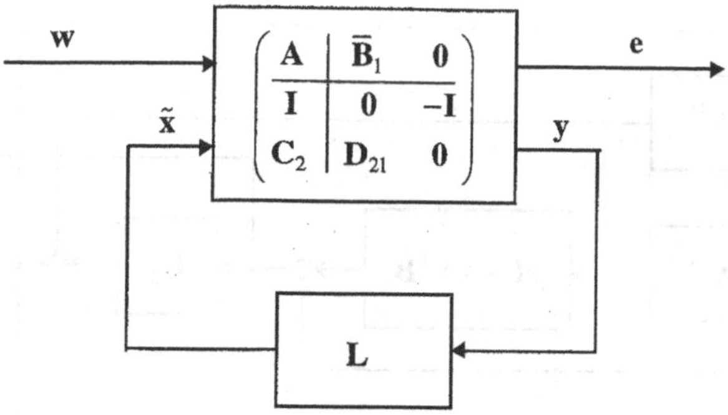

Fig. 8.3. State estimator

The task is to estimate the state vector $\mathbf{x}$ which is caused by the disturbance $\mathbf{w}$. The estimate is denoted by $\tilde{\mathbf{x}}$ and the estimator is assumed to be a dynamical system $\mathbf{L}(s)$ which has the measured output $\mathbf{y}$ of the system as its input:

$$\hat{\tilde{\mathbf{x}}} = \mathbf{L}(s)\hat{\mathbf{y}} .$$

Fig. 8.3 shows the corresponding feedback system.

The problem is to find the estimator which minimizes the 2-norm of $\mathbf{F}_{ew}$. This is an output estimation problem with $\mathbf{B}_2 = \mathbf{0}$ and consequently, Theorem 8.2.5 is not directly applicable if $\mathbf{A}$ is not stable. Nevertheless, when the OE problem is reformulated as a FI problem, it is possible to synthesize an internally stabilizing controller for the problem without the uncontrollable dynamics caused by $\mathbf{A}$. Since we are going to design a filter, it is feasible to delete this part of the system dynamics. Then the application of Theorem 8.2.5 gives

$$\mathbf{L}(s) = \left(\begin{array}{c|c} \mathbf{A}+\mathbf{L}_2\mathbf{C}_2 & -\mathbf{L}_2 \\ \hline \mathbf{I} & \mathbf{0} \end{array}\right).$$

Thus the state equation for the estimator is

$$\dot{\tilde{\mathbf{x}}} = (\mathbf{A}+\mathbf{L}_2\mathbf{C}_2)\tilde{\mathbf{x}} - \mathbf{L}_2\mathbf{y} . \tag{8.28}$$

It is easy to check that $\mathbf{X}_c = \mathbf{Y}_2$ if $\mathbf{X}_c$ denotes the controllability gramian of $\tilde{\mathbf{G}}_L$ (cf. Theorem 8.2.5). Thus, since $\mathbf{C}_1 = \mathbf{I}$, we obtain by Lemma 7.3.1 and Theorem 8.2.5

$$\min \|\mathbf{F}_{ew}\|_2 = \mathrm{trace}(\mathbf{Y}_2) . \tag{8.29}$$

Let $\tilde{\mathbf{g}}_L$ be the impulse response of $\tilde{\mathbf{G}}_L$. Then $\mathbf{X}_c = \mathbf{Y}_2$ implies (cf. Sect. 5.7)

$$\mathbf{Y}_2 = \int_0^\infty \tilde{\mathbf{g}}_L(t)\tilde{\mathbf{g}}_L^T(t)\,dt \tag{8.30}$$

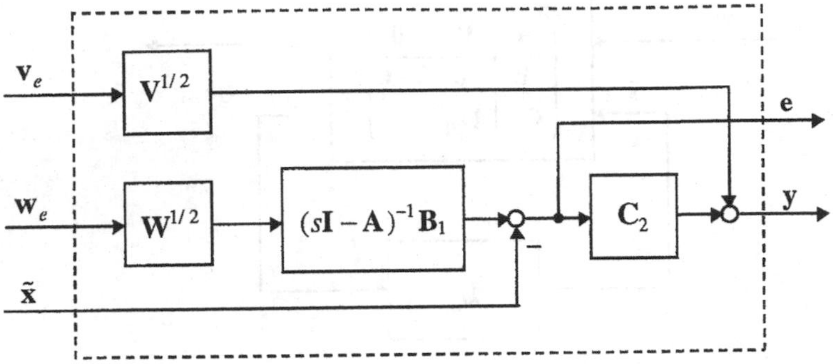

Fig. 8.4. Plant for the Kalman Bucy filter

We introduce now a somewhat different system by

$$\dot{\mathbf{x}} = \mathbf{A}\,\mathbf{x} + \mathbf{B}_1\mathbf{W}^{1/2}\mathbf{w}_e$$

$$\mathbf{y} = \mathbf{C}_2\mathbf{x} + \mathbf{V}^{1/2}\mathbf{v}_e\,.$$

Then the disturbance consists of two parts, namely a plant disturbance $\mathbf{w}$ and a measurement disturbance $\mathbf{v}$. The matrices $\mathbf{W}$ and $\mathbf{V}$ are the weights for the estimator design. We assume that they are symmetric and positive semidefinite and moreover that $\mathbf{V}$ is positive definite. Then we have

$$\overline{\mathbf{w}} = \begin{pmatrix} \mathbf{w} \\ \mathbf{v} \end{pmatrix}, \quad \overline{\mathbf{B}}_1 = (\mathbf{B}_1\mathbf{W}^{1/2} \quad \mathbf{0}), \quad \mathbf{D}_{21} = (\mathbf{0} \quad \mathbf{V}^{1/2})$$

and $\mathbf{D}_{21}\overline{\mathbf{B}}_1^T = \mathbf{0}$. The filter gain results as

$$\mathbf{L}_2 = -\mathbf{Y}_2\mathbf{C}_2^T\mathbf{V}^{-1}, \tag{8.31}$$

where $\mathbf{Y}_2$ is the solution of the following Riccati equation:

$$\mathbf{Y}_2\mathbf{A}^T + \mathbf{A}\mathbf{Y}_2 - \mathbf{Y}_2\mathbf{C}_2^T\mathbf{V}^{-1}\mathbf{C}_2\mathbf{Y}_2 + \mathbf{B}_1\mathbf{W}\mathbf{B}_1^T = \mathbf{0}\,. \tag{8.32}$$

The $\mathcal{H}_2$ estimator becomes of particular interest if the disturbances are assumed to be random processes. To be specific, we suppose that $\mathbf{w}_e$ and $\mathbf{v}_e$ are white noise processes, where the power spectral densities are identity matrices. Then $\mathbf{w}$ and $\mathbf{v}$ are white noise processes with power spectral densities $\mathbf{W}$ and $\mathbf{V}$:

$$E(\mathbf{w}(t)\mathbf{w}^T(\tau)) = \mathbf{W}\,\delta(t-\tau)$$

$$E(\mathbf{v}(t)\mathbf{v}^T(\tau)) = \mathbf{V}\,\delta(t-\tau)\,.$$

Using the interpretation of the $\mathcal{H}_2$ norm for systems driven by white noise which we have derived in Sect. 7.2.2 , Theorem 8.2.5 and (8.30) lead to the following result.

Theorem 8.3.1 (Kalman Bucy filter). *Let the system*

$$\dot{\mathbf{x}} = \mathbf{A}\,\mathbf{x} + \mathbf{B}_1\mathbf{w}$$

$$\mathbf{y} = \mathbf{C}_2\mathbf{x} + \mathbf{v}$$

be given. Here, $\mathbf{w}(t)$, $\mathbf{v}(t)$ *denote uncorrelated white noise processes whose power spectral densities are described by a positive semidefinite matrix* $\mathbf{W}$ *and a positive definite matrix* $\mathbf{V}$. *Suppose that* $(\mathbf{C}_2, \mathbf{A})$ *is detectable and that* $(\mathbf{A}, \mathbf{B}_1\mathbf{W}^{1/2})$ *has no uncontrollable eigenvalues on the imaginary axis. Let* $\mathbf{L}_2$ *be the matrix defined by (8.31) and (8.32). Then the linear system*

$$\dot{\tilde{\mathbf{x}}} = \mathbf{A}\tilde{\mathbf{x}} + \mathbf{L}_2(\mathbf{C}_2\,\tilde{\mathbf{x}} - \mathbf{y}) \tag{8.33}$$

defines a state estimator which minimizes the mean square value $E[\mathbf{e}_\infty^T \mathbf{e}_\infty]$ *of the stationary estimation error* $\mathbf{e}_\infty = \mathbf{x}_\infty - \tilde{\mathbf{x}}_\infty$. *The covariance matrix of the stationary estimation error is given by*

$$E[\mathbf{e}_\infty \mathbf{e}_\infty^T] = \mathbf{Y}_2\,.$$

Remark. It should be noted that the Kalman filter is designed with respect to an optimal stationary value of the error $\mathbf{e} = \mathbf{x} - \tilde{\mathbf{x}}$. With respect to other properties such as bandwidth, for example, the filter may have properties which are not so favorable. In such a situation it is often helpful to use the matrices $\mathbf{W}$ and $\mathbf{V}$ as design parameters (cf. Sect. 8.4.3).

8.3.2 Kalman Filtering with Non-White Noise

In many applications, the assumption of white system or measurement noise is too strong and one has to assume colored noise. This is obtained as the output of a linear system with white noise as the input.

Non-white process noise. Let the state space realization of the shaping filter be given by

$$\dot{\mathbf{x}}_w = \mathbf{A}_w\mathbf{x}_w + \mathbf{B}_w\mathbf{W}^{1/2}\mathbf{w}_e$$

$$\mathbf{w} = \mathbf{C}_w\mathbf{x}_w\,.$$

Herein, $\mathbf{w}_1$ denotes white noise and $\mathbf{A}_w$ is assumed to be stable. Adding the dynamics of the shaping filter to that of the system leads to

$$\begin{pmatrix} \dot{\mathbf{x}} \\ \dot{\mathbf{x}}_w \end{pmatrix} = \begin{pmatrix} \mathbf{A} & \mathbf{B}_1\mathbf{C}_w \\ \mathbf{0} & \mathbf{A}_w \end{pmatrix} \begin{pmatrix} \mathbf{x} \\ \mathbf{x}_w \end{pmatrix} + \begin{pmatrix} \mathbf{0} \\ \mathbf{B}_w \mathbf{W}^{1/2} \end{pmatrix} \mathbf{w}_e$$

$$\mathbf{y} = \begin{pmatrix} \mathbf{C}_2 & \mathbf{0} \end{pmatrix} \begin{pmatrix} \mathbf{x} \\ \mathbf{x}_w \end{pmatrix} + \mathbf{v}.$$

This is a system where the disturbance is white noise and the equations for the Kalman Bucy filter are directly applicable.

Non-white measurement noise. If the measurement noise is non-white, the situation is somewhat more complicated. We assume the shaping filter to be of the form:

$$\dot{\mathbf{v}} = \mathbf{A}_v \mathbf{v} + \mathbf{V}^{1/2} \mathbf{v}_1 .$$

As before, the filter dynamics has to be added to the plant dynamics. This yields

$$\begin{pmatrix} \dot{\mathbf{x}} \\ \dot{\mathbf{v}} \end{pmatrix} = \begin{pmatrix} \mathbf{A} & \mathbf{0} \\ \mathbf{0} & \mathbf{A}_v \end{pmatrix} \begin{pmatrix} \mathbf{x} \\ \mathbf{v} \end{pmatrix} + \begin{pmatrix} \mathbf{B}_1 \mathbf{W}^{1/2} & \mathbf{0} \\ \mathbf{0} & \mathbf{V}^{1/2} \end{pmatrix} \begin{pmatrix} \mathbf{w} \\ \mathbf{v}_1 \end{pmatrix}$$

$$\mathbf{y} = \begin{pmatrix} \mathbf{C}_2 & \mathbf{0} \end{pmatrix} \begin{pmatrix} \mathbf{x} \\ \mathbf{v} \end{pmatrix}.$$

For this plant we have $\mathbf{D}_{21} = \mathbf{0}$ and consequently, (A4) is violated. In order to find a problem formulation to which the $\mathcal{H}_2$ filtering theory is applicable, a new measurement equation will be introduced by

$$\mathbf{y}_1 = \dot{\mathbf{y}} - \mathbf{A}_v \mathbf{y} .$$

Then the measurement equation is given by

$$\mathbf{y} = \begin{pmatrix} \mathbf{C}_2\mathbf{A} - \mathbf{A}_v\mathbf{C}_2 & \mathbf{0} \end{pmatrix} \begin{pmatrix} \mathbf{x} \\ \mathbf{v} \end{pmatrix} + \begin{pmatrix} \mathbf{C}_2\mathbf{B}_1\mathbf{W}^{1/2} & \mathbf{V} \end{pmatrix} \begin{pmatrix} \mathbf{w} \\ \mathbf{v}_1 \end{pmatrix}.$$

It is easy to see that for this new plant the assumption (A2) is fulfilled if $(\mathbf{C}_2, \mathbf{A})$ is detectable and $\mathbf{A}_v$ is stable. If $\mathbf{V}$ is positive definite, (A4) can be satisfied by a suitable normalization procedure, which will be described in Sect. 9.5. Note that for this problem formulation, $\mathbf{D}_{21}\overline{\mathbf{B}}_1^T \neq \mathbf{0}$ holds although the physical plant and measurement noise are uncorrelated. Again an optimal estimator can be obtained by appliying Theorem 8.2.5, but we make no attempt to carry out the somewhat tedious and purely technical calculation which is necessary to describe it explicitely (cf. Burl [17], Sect. 7.4.2).

8.3.3 Range and Velocity Estimate of an Aircraft

We consider an example where a Kalman filter has to be designed to estimate the range r and the radial velocity v of an aircraft. The sensor is a radar system, which is able to measure the range but not the velocity. The acceleration w of the aircraft is considered as a stochastic process. In the first step, w is assumed to be white noise with power spectral density W. Of course, this stochastic model for the acceleration is not very realistic, but it puts us in a position to describe the Kalman filter as a transfer function by an explicit formula. By this way, it is possible to gain some insight by studying the influence of certain parameters. In the second step, a more realistic model for the aircraft will be used and some simulations are made.

In the first approach, the system is given by

$$\begin{pmatrix} \dot{x}_1 \\ \dot{x}_2 \end{pmatrix} = \begin{pmatrix} 0 & 1 \\ 0 & 0 \end{pmatrix} \begin{pmatrix} x_1 \\ x_2 \end{pmatrix} + \begin{pmatrix} 0 \\ 1 \end{pmatrix} w$$

$$r = (1 \quad 0) \begin{pmatrix} x_1 \\ x_2 \end{pmatrix} + v .$$

Denote the power spectral density of the measurement noise by V. The Riccati equation (8.32) can be solved by an easy computation and the result is

$$\mathbf{Y}_2 = V \begin{pmatrix} \sqrt{2}\,\omega_0 & \omega_0^2 \\ \omega_0^2 & \sqrt{2}\,\omega_0^3 \end{pmatrix} \qquad \text{with} \quad \omega_0 = W / V .$$

Hence the feedback matrix $\mathbf{L}_2$ and the system matrix of the Kalman filter are

$$\mathbf{L}_2 = -\begin{pmatrix} \sqrt{2}\,\omega_0 \\ \omega_0^2 \end{pmatrix}, \qquad \mathbf{A} + \mathbf{L}_2\mathbf{C}_2 = \begin{pmatrix} -\sqrt{2}\,\omega_0 & 1 \\ -\omega_0^2 & 0 \end{pmatrix}.$$

The characteristic polynomial is computed as

$$\det(s\mathbf{I} - (\mathbf{A} + \mathbf{L}_2\mathbf{C}_2)) = s^2 + \sqrt{2}\,\omega_0\, s + \omega_0^2 ,$$

and the poles are

$$s_{1/2} = \frac{1}{2}\sqrt{2}\,\omega_0\,(-1 \pm i) .$$

Their damping is $1/\sqrt{2}$, independent of ω_0. The state space description of the Kalman filter is

$$\begin{pmatrix} \dot{\tilde{r}} \\ \dot{\tilde{v}} \end{pmatrix} = \begin{pmatrix} -\sqrt{2}\,\omega_0 & 1 \\ -\omega_0^2 & 0 \end{pmatrix} \begin{pmatrix} \tilde{r} \\ \tilde{v} \end{pmatrix} + \begin{pmatrix} \sqrt{2}\,\omega_0 \\ \omega_0^2 \end{pmatrix} y .$$

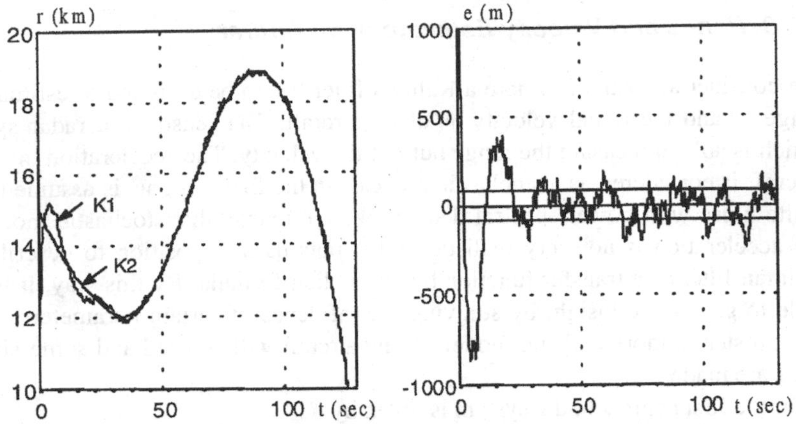

Fig. 8.5. Estimated range (K1) and true range (K2) (left-hand plot); estimation error (right-hand plot)

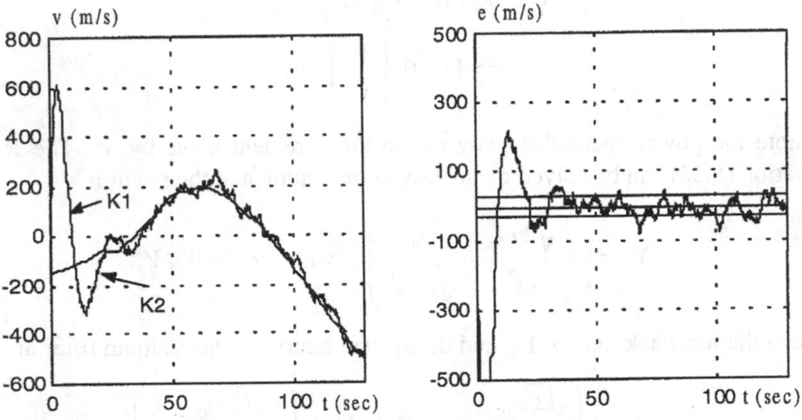

Fig. 8.6. Estimated velocity (K1) and true velocity (K2) (left-hand plot); estimation error (right-hand plot)

Hence the transfer functions from the measurement to the components of the estimated state vector are

$$\hat{r} = \frac{\sqrt{2}\omega_0\, s + \omega_0^2}{s^2 + \sqrt{2}\omega_0\, s + \omega_0^2}\,\hat{y}, \qquad \hat{v} = \frac{\omega_0^2\, s}{s^2 + \sqrt{2}\omega_0^2\, s + \omega_0^2}\,\hat{y}.$$

If the measurement error decreases, ω_0 tends to ∞ and in this limit case the equations $\hat{r} = y$, $\hat{v} = s\hat{y}$ hold, as is seen from the transfer functions. With respect to the velocity, the Kalman filter is then a pure differentiator. If the measurement error increases, the time constant of the filter becomes larger.

We now use a somewhat more realistic model for the aircraft and suppose that its acceleration is given by

$$T_F\,\dot{a} + a = w_1\,.$$

Here, w_1 denotes again white noise. The correlation time of the system noise is assumed to be $T_F = 20\ \text{s}$ and the standard deviation of the aircraft acceleration is chosen as $\sigma_F = 10\,\text{m}/\text{s}^2$. Then the power spectral density of the system noise w_1 must be (cf. Sect. 5.8.3)

$$W_1 = \sqrt{2T_F}\,\sigma_F\,.$$

Let the power spectral density of the radar noise be $V = 12000\ \text{m}^2/\text{Hz}$. Figures 8.5 and 8.6 show the simulation results for the initial estimates $\tilde{r}(0) = 12\,\text{km}$, $\tilde{v}(0) = 0$, $\tilde{a}_F(0) = 0$. It is seen that the Kalman filter reaches a stationary value after approximately $25\ \text{s}$. In both right-hand plots, the estimation error is depicted together with the standard deviation of the stationary estimation error.

8.4 LQG Controller and further Properties of LQ Controllers

8.4.1 LQG Controller as a Special $\mathcal{H}_2$ Controller

It is possible to combine a LQR Controller with a Kalman filter. This leads to the **linear quadratic Gaussian (LQG)** controller. For such a controller, the plant is assumed to be given by

$$\begin{aligned}\dot{\mathbf{x}} &= \mathbf{A}\,\mathbf{x} + \mathbf{B}_1\mathbf{w} + \mathbf{B}_2\mathbf{u}\\ \mathbf{y} &= \mathbf{C}_2\mathbf{x} + \mathbf{v}\,.\end{aligned}$$

As before, $\mathbf{w}, \mathbf{v}$ are white noise processes with power spectral densities $\mathbf{W}, \mathbf{V}$. The task is to find a controller $\hat{\mathbf{u}} = \mathbf{K}(s)\,\hat{\mathbf{y}}$ which minimizes the expected value

$$E[\mathbf{x}_\infty^T\,\mathbf{Q}\,\mathbf{x}_\infty + \mathbf{u}_\infty^T\,\mathbf{R}\,\mathbf{u}_\infty]\,.$$

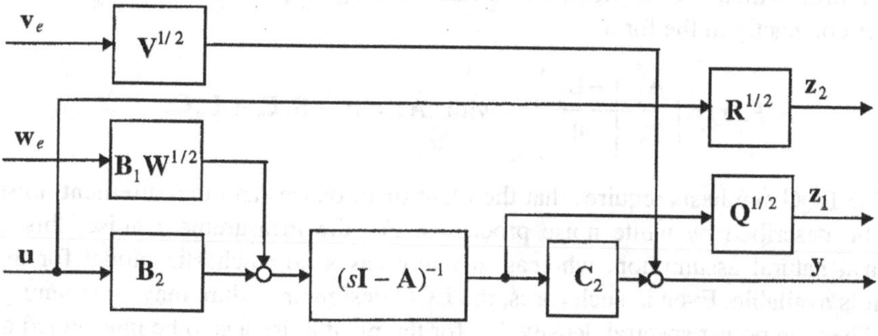

Fig. 8.7. Plant for LQG controller design

Here, $\mathbf{Q}, \mathbf{R}$ are matrices as in the LQR problem.

It is possible to interpret the LQG problem as a special $\mathcal{H}_2$ problem (cf. Fig. 8.7). The key is again the interpretation of the $\mathcal{H}_2$ norm given in Sect. 7.2. Then the characterization of the optimal controller follows directly from Theorem 8.2.1. The necessary assumptions are:

(A1) $(\mathbf{A}, \mathbf{B}_2)$ *is stabilizable and* $(\mathbf{C}_2, \mathbf{A})$ *is detectable.*

(A2) $(\mathbf{A}, \mathbf{B}_1\mathbf{W}^{1/2})$ *has no uncontrollable eigenvalues on the imaginary axis and* $(\mathbf{Q}^{1/2}, \mathbf{A})$ *has no unobservable eigenvalues on the imaginary axis.*

(A3) The matrices $\mathbf{R}$ *and* $\mathbf{V}$ *are positive definite and the matrices* $\mathbf{Q}$ *and* $\mathbf{W}$ *are positive semidefinite.*

Theorem 8.4.1. *Suppose that (A1)–(A3) hold and let* $\mathbf{X}_2, \mathbf{Y}_2$ *be the solutions of the following Riccati equations:*

$$\mathbf{A}^T\mathbf{X}_2 + \mathbf{X}_2\mathbf{A} - \mathbf{X}_2\mathbf{B}_2\mathbf{R}^{-1}\mathbf{B}_2^T\mathbf{X}_2 + \mathbf{Q} = \mathbf{0}$$
$$\mathbf{Y}_2\mathbf{A}^T + \mathbf{A}\mathbf{Y}_2 - \mathbf{Y}_2\mathbf{C}_2^T\mathbf{V}^{-1}\mathbf{C}_2\mathbf{Y}_2 + \mathbf{B}_1\mathbf{W}\mathbf{B}_1^T = \mathbf{0}\,.$$

Define

$$\mathbf{F}_2 = -\mathbf{R}^{-1}\mathbf{B}_2^T\mathbf{X}_2, \qquad \mathbf{L}_2 = -\mathbf{Y}_2\mathbf{C}_2^T\mathbf{V}^{-1},$$

Then the LQG *problem has the optimal controller*

$$\begin{aligned} \dot{\tilde{\mathbf{x}}} &= \mathbf{A}\tilde{\mathbf{x}} + \mathbf{B}_2\mathbf{u} + \mathbf{L}_2(\mathbf{C}_2\tilde{\mathbf{x}} - \mathbf{y}) \\ \mathbf{u} &= \mathbf{F}_2\tilde{\mathbf{x}}. \end{aligned} \tag{8.34}$$

as a unique solution.

As this theorem shows, the LQG controller consists of a Kalman filter and a constant gain controller that feeds back the estimate of the state vector. The Kalman filter has essentially the form (8.33). The only difference is the part caused by the control, which can be viewed as a bias. It is again possible to write the controller compactly in the form

$$\mathbf{K}(s) = \left(\begin{array}{c|c} \tilde{\mathbf{A}}_2 & -\mathbf{L}_2 \\ \hline \mathbf{F}_2 & \mathbf{0} \end{array}\right), \quad \text{with } \tilde{\mathbf{A}}_2 = \mathbf{A} + \mathbf{B}_2\mathbf{F}_2 + \mathbf{L}_2\mathbf{C}_2\,.$$

The LQG synthesis requires that the plant disturbances and measurement noise can be described by white noise processes. For the measurement noise, this is often a natural assumption, whereas in many cases no stochastic model for the plant is available. Even in such cases, the LQG design procedure may be meaningful. Then the power spectral density $\mathbf{W}$ for the plant noise has to be interpreted as a design parameter. Although the estimator possibly is not optimal from a stochastic viewpoint, the LQG controller may have good performance and robustness properties. We will return to this point in Sect. 8.4.3 (compare also the controller design for the inverse pendulum in Sect. 11.1).

8.4.2 An Estimate for the Input Return Difference Matrix

The robust stability analysis for SISO systems with state vector feedback bases mainly on the inequality (6.36). As we saw in Sect. 6.3.2, it is always possible to choose the closed-loop poles such that (6.36) is satisfied. In this section, we prove that (6.36) is also satisfied for an LQ controller with state fedback. Moreover, for MIMO systems it is possible to derive a generalized version of this inequaliy which will be used in Sect. 13.1.2 for the robust stability analysis of LQ controllers with state feedback.

The input loop transfer matrix $\mathbf{L}_i(s)$ for a controller with state feedback is given by (cf. Sect. 6.5.2)

$$\mathbf{L}_i(s) = -\mathbf{F}_2(s\mathbf{I}-\mathbf{A})^{-1}\mathbf{B}_2 .$$

With the nomenclature of Sect. 6.5.1 the following are equations valid:

$$\mathbf{P}_{12}(s) = \mathbf{C}_1(s\mathbf{I}-\mathbf{A})^{-1}\mathbf{B}_2 + \mathbf{D}_{12}$$

$$\mathbf{P}_{22}(s) = (s\mathbf{I}-\mathbf{A})^{-1}\mathbf{B}_2 .$$

We first need the following lemma.

Lemma 8.4.1. (a) *The following equation holds:*

$$(\mathbf{I}+\mathbf{L}_i^T(-s))(\mathbf{I}+\mathbf{L}_i(s)) = \mathbf{P}_{12}^T(-s)\mathbf{P}_{12}(s) . \tag{8.35}$$

(b) *Under the assumption* $\mathbf{D}_{12}^T\mathbf{C}_1 = \mathbf{0}$, *(8.35) can be written as*

$$(\mathbf{I}+\mathbf{L}_i^T(-s))(\mathbf{I}+\mathbf{L}_i(s)) = \mathbf{I}+\mathbf{B}_2^T\mathbf{P}_{22}^T(-s)\mathbf{C}_1^T\mathbf{C}_1\mathbf{P}_{22}(s)\mathbf{B}_2 . \tag{8.36}$$

Proof. (a) First, note that the Riccati equation (8.10) for $\mathbf{X}_2$ can be written as

$$\mathbf{A}^T\mathbf{X}_2 + \mathbf{X}_2\mathbf{A} - \mathbf{X}_2\mathbf{B}_2\mathbf{B}_2^T\mathbf{X}_2 - \mathbf{X}_2\mathbf{B}_2\mathbf{D}_{12}^T\mathbf{C}_1 - \mathbf{C}_1^T\mathbf{D}_{12}\mathbf{B}_2^T\mathbf{X}_2 - \mathbf{C}_1^T\mathbf{D}_{12}\mathbf{D}_{12}^T\mathbf{C}_1$$
$$+\mathbf{C}_1^T\mathbf{C}_1 = \mathbf{A}^T\mathbf{X}_2 + \mathbf{X}_2\mathbf{A} - \mathbf{F}_2^T\mathbf{F}_2 + \mathbf{C}_1^T\mathbf{C}_1 = \mathbf{0}.$$

With the transfer function

$$\mathbf{H}(s) = (s\mathbf{I}-\mathbf{A})^{-1}$$

we obtain

$$(\mathbf{I}+\mathbf{L}_i^T(-s))(\mathbf{I}+\mathbf{L}_i(s)) = (\mathbf{I}-\mathbf{B}_2^T\mathbf{H}^T(-s)\mathbf{F}_2^T)(\mathbf{I}-\mathbf{F}_2\mathbf{H}(s)\mathbf{B}_2)$$

$$= \mathbf{I} - \mathbf{F}_2 \mathbf{H}(s)\mathbf{B}_2 - \mathbf{B}_2^T \mathbf{H}^T(-s)\mathbf{F}_2^T + \mathbf{B}_2^T \mathbf{H}^T(-s)\mathbf{F}_2^T \mathbf{F}_2 \mathbf{H}(s)\mathbf{B}_2$$

$$= \mathbf{I} + (\mathbf{B}_2^T \mathbf{X}_2 + \mathbf{D}_{12}^T \mathbf{C}_1)\mathbf{H}(s)\mathbf{B}_2 + \mathbf{B}_2^T \mathbf{H}^T(-s)(\mathbf{X}_2 \mathbf{B}_2 + \mathbf{C}_1^T \mathbf{D}_{12})$$
$$+ \mathbf{B}_2^T \mathbf{H}^T(-s)(\mathbf{A}^T \mathbf{X}_2 + \mathbf{X}_2 \mathbf{A} + \mathbf{C}_1^T \mathbf{C}_1)\mathbf{H}(s)\mathbf{B}_2$$

$$= \mathbf{I} + \mathbf{B}_2^T \mathbf{X}_2 \mathbf{H}(s)\mathbf{B}_2 + \mathbf{D}_{12}^T \mathbf{C}_1 \mathbf{H}(s)\mathbf{B}_2 + \mathbf{B}_2^T \mathbf{H}^T(-s)\mathbf{X}_2 \mathbf{B}_2$$
$$+ \mathbf{B}_2^T \mathbf{H}^T(-s)\mathbf{C}_1^T \mathbf{D}_{12} + \mathbf{B}_2^T \mathbf{H}^T(-s)\mathbf{A}^T \mathbf{X}_2 \mathbf{H}(s)\mathbf{B}_2$$
$$+ \mathbf{B}_2^T \mathbf{H}^T(-s)\mathbf{X}_2 \mathbf{A}\mathbf{H}(s)\mathbf{B}_2 + \mathbf{B}_2^T \mathbf{H}^T(-s)\mathbf{C}_1^T \mathbf{C}_1 \mathbf{H}(s)\mathbf{B}_2$$

$$= \mathbf{P}_{12}^T(-s)\mathbf{P}_{12}(s) + \mathbf{I} + \mathbf{B}_2^T \mathbf{X}_2 \mathbf{H}(s)\mathbf{B}_2 + \mathbf{B}_2^T \mathbf{H}^T(-s)\mathbf{X}_2 \mathbf{B}_2$$
$$+ \mathbf{B}_2^T \mathbf{H}^T(-s)\mathbf{A}^T \mathbf{X}_2 \mathbf{H}(s)\mathbf{B}_2 + \mathbf{B}_2^T \mathbf{H}^T(-s)\mathbf{X}_2 \mathbf{A}\mathbf{H}(s)\mathbf{B}_2 - \mathbf{D}_{12}^T \mathbf{D}_{12}$$

$$= \mathbf{P}_{12}^T(-s)\mathbf{P}_{12}(s)$$
$$+ \mathbf{B}_2^T [\mathbf{X}_2 \mathbf{H}(s) + \mathbf{H}^T(-s)\mathbf{X}_2 + \mathbf{H}^T(-s)\mathbf{A}^T \mathbf{X}_2 \mathbf{H}(s) + \mathbf{H}^T(-s)\mathbf{X}_2 \mathbf{A}\mathbf{H}(s)]\mathbf{B}_2$$

$$= \mathbf{P}_{12}^T(-s)\mathbf{P}_{12}(s)$$
$$+ \mathbf{B}_2^T \mathbf{H}^T(-s)\,[\mathbf{H}^T(-s)^{-1}\mathbf{X}_2 + \mathbf{X}\mathbf{H}(s)^{-1} + \mathbf{A}^T \mathbf{X}_2 + \mathbf{X}_2 \mathbf{A}\,]\mathbf{H}(s)\mathbf{B}_2$$

$$= \mathbf{P}_{12}^T(-s)\mathbf{P}_{12}(s).$$

(b) Under the assumption $\mathbf{D}_{12}^T \mathbf{C}_1 = \mathbf{0}$, (8.35) implies

$$(\mathbf{I} + \mathbf{L}_i^T(-s))(\mathbf{I} + \mathbf{L}_i(s)) = \mathbf{I} + \mathbf{B}_2^T(-s\mathbf{I} - \mathbf{A}^T)^{-1}\mathbf{C}_1^T \mathbf{C}_1 (s\mathbf{I} - \mathbf{A})^{-1}\mathbf{B}_2,$$

and (8.36) is proven.

Remark. The transfer matrix $\mathbf{E} + \mathbf{L}_i$ occurring in Lemma 8.4.1 is called the **input return difference matrix.**

Theorem 8.4.2. *Suppose* $\mathbf{D}_{12}^T \mathbf{C}_1 = \mathbf{0}$. *Then the following inequality holds:*

$$(\mathbf{I} + \mathbf{L}_i(i\omega))^* (\mathbf{I} + \mathbf{L}_i(i\omega)) \geq \mathbf{I} \qquad (8.37)$$

For SISO systems this takes the following special form:

$$|1 + L(i\omega)| \geq 1.$$

Proof. Using Lemma 8.4.1, (8.37) follows from

$$(\mathbf{I} + \mathbf{L}_i(i\omega))^* (\mathbf{I} + \mathbf{L}_i(i\omega)) = (\mathbf{I} + \mathbf{L}_i^T(-i\omega))(\mathbf{I} + \mathbf{L}_i(i\omega))$$
$$= \mathbf{I} + \mathbf{B}_2^T \mathbf{P}_{22}^T(-i\omega)\mathbf{C}_1^T \mathbf{C}_1 \mathbf{P}_{22}(i\omega)\mathbf{B}_2$$
$$= \mathbf{I} + \mathbf{B}_2^T \mathbf{P}_{22}(i\omega)^* \mathbf{C}_1^T \mathbf{C}_1 \mathbf{P}_{22}(i\omega)\mathbf{B}_2$$
$$\geq \mathbf{I}.$$

8.4.3 Poles of LQ Feedback Systems and Loop Transfer Recovery

Poles of LQR feedback loops. One crucial point in LQG controller design is the selection of the weighting matrices $\mathbf{Q}, \mathbf{R}, \mathbf{W}, \mathbf{V}$. First we investigate this problem for $\mathbf{Q}, \mathbf{R}$, i.e. the matrices which describe the constant gain controller for the state feedback. It is normally not possible to translate performance requirements directly into the choice of $\mathbf{Q}, \mathbf{R}$, but in some sense this may be done indirectly by studying the dependence of $\mathbf{Q}, \mathbf{R}$ for the closed-loop poles.

The closed-loop poles of the LQ feedback system are the zeros of

$$p(s) = \det(s\mathbf{I} - \mathbf{A} - \mathbf{B}_2\mathbf{F}_2).$$

Our aim is to express this characteristic equation by means of the weighting matrices. Denote by $q(s)$ the characteristic polynomial of $\mathbf{A}$:

$$q(s) = \det(s\mathbf{I} - \mathbf{A}).$$

The transfer function $\mathbf{I} + \mathbf{L}_i$ has the state space representation

$$\mathbf{I} + \mathbf{L}_i(s) = \left(\begin{array}{c|c} \mathbf{A} & \mathbf{B}_2 \\ \hline -\mathbf{F}_2 & \mathbf{I} \end{array}\right).$$

Thus (5.58) implies

$$p(s) = q(s)\det(\mathbf{I} + \mathbf{L}_i(s))$$

and analogously

$$p(-s) = q(-s)\det(\mathbf{I} + \mathbf{L}_i^T(-s)).$$

Moreover, the equation

$$(\mathbf{I} + \mathbf{L}_i^T(-s))\mathbf{R}(\mathbf{I} + \mathbf{L}_i(s)) = \mathbf{R} + \mathbf{B}_2^T(-s\mathbf{I} - \mathbf{A}^T)^{-1}\mathbf{Q}(s\mathbf{I} - \mathbf{A})^{-1}\mathbf{B}_2.$$

holds. This is exactly (8.36) without the normalization $\mathbf{R} = \mathbf{I}$. Hence the following lemma is valid.

Lemma 8.4.2. *With the above notations the following equation holds:*

$$p(s)\det\mathbf{R}\,p(-s) = q(s)\,q(-s)\det(\mathbf{R} + \mathbf{B}_2^T(-s\mathbf{I} - \mathbf{A}^T)^{-1}\mathbf{Q}(s\mathbf{I} - \mathbf{A})^{-1}\mathbf{B}_2).$$

Note that with s the number $-s$ is also a zero of

$$r(s) = q(s)\,q(-s)\det(\mathbf{R} + \mathbf{B}_2^T(-s\mathbf{I} - \mathbf{A}^T)^{-1}\mathbf{Q}(s\mathbf{I} - \mathbf{A})^{-1}\mathbf{B}_2).$$

Thus the poles of the LQ feedback loop are the LHP zeros of $r(s)$. It is now possible to analyze the asymptotic behavior of the zeros of $r(s)$. Let $\mathbf{R} = \rho\,\mathbf{R}_0$ with a positive definite matrix $\mathbf{R}_0$ and a positive scalar parameter ρ. We consider the following cases.

Large weighting of the control. In this case the zeros are approximately the zeros of $r_\infty(s) = q(s)q(-s)$.

Small Weighting of the Control. For $\rho \to 0$ the polynomial $r(s)$ tends to

$$r_0(s) = q(s)q(-s)\det(\mathbf{B}_2^T(-s\mathbf{I} - \mathbf{A}^T)^{-1}\mathbf{Q}(s\mathbf{I} - \mathbf{A})^{-1}\mathbf{B}_2).$$

Let now the weighting matrix $\mathbf{Q}$ be decomposed as $\mathbf{Q} = \mathbf{C}_{10}^T\mathbf{C}_{10}$ (cf. Sect. 8.1.1), and define the transfer function

$$\mathbf{G}_1(s) = \left(\begin{array}{c|c} \mathbf{A} & \mathbf{B}_2 \\ \hline \mathbf{C}_{10} & \mathbf{0} \end{array}\right),$$

which is assumed to be square. Then we obtain

$$r_0(s) = q(s)q(-s)\det\mathbf{G}_1^T(-s)\det\mathbf{G}_1(s).$$

Finally, suppose that no transmission zero of $\mathbf{G}_1(s)$ is also pole of $\mathbf{G}_1(s)$. We can write $\mathbf{G}_1(s) = \mathbf{M}(s)/q(s)$, where the coefficients of $\mathbf{M}(s)$ are polynomials. Then the equation

$$\det\mathbf{G}_1(s) = \det(\frac{1}{q(s)}\mathbf{M}(s)) = \frac{1}{q(s)^n}\det\mathbf{M}(s)$$

implies that $r_0(s_0) = 0$ if and only if s_0 is a transmission zero of $\mathbf{G}_1(s)$ or if $-s_0$ is a transmission zero of $\mathbf{G}_1^T(-s)$. Based on these considerations, the following result on the limiting behavior of the closed-loop poles can be proven.

Theorem 8.4.3. *Let the weighting matrix $\mathbf{R}$ be given as $\mathbf{R} = \rho\mathbf{R}_0$ with a positive definite matrix $\mathbf{R}_0$ and a positive scalar parameter ρ. Then the following assertions hold.*

(a) For $\rho \to \infty$ the poles of the LQ feedback system tend to the stable plant poles and to the reflections of the instable plant poles about the imaginary axis.

(b) Let the weighting matrix $\mathbf{Q}$ be decomposed as $\mathbf{Q} = \mathbf{C}_{10}^T\mathbf{C}_{10}$ and suppose that the above defined system $\mathbf{G}_1(s)$ is square. Denote the number of transmisson zeros of $\mathbf{G}_1(s)$ by l. Then for $\rho \to 0$, l poles of the n closed-loop poles tend to the LHP plant zeros of $\mathbf{G}_1(s)$ or the reflections about the imaginary axis of the RHP zeros. The remaining $n - l$ closed-loop poles tend to infinity in the left half-plane.

(c) For a SISO system, the remaining $n - l$ closed-loop poles tend asymptotically to infinity near straight lines radiating from the origin. With the negative real axis these lines form angles of the size

$$\pm \nu \frac{\pi}{n-l}, \qquad \nu = 0, 1, \ldots, \frac{n-l-1}{2}, \quad \textit{if } n-l \textit{ is odd,}$$

$$\pm (\nu + \frac{1}{2}) \frac{\pi}{n-l}, \qquad \nu = 0, 1, \ldots, \frac{n-l}{2} - 1, \quad \textit{if } n-l \textit{ is even.}$$

Proof. Assertions (a) and (b) follow directly from our above calculations. Concerning (c), we refer the reader to Kwakernaak and Sivan [58], Sect. 3.8.

Remark. The poles form a so-called Butterworth configuration. For MIMO systems a similar result holds where several Butterworth configurations may result.

Formally, the Kalman filter is very similar to the LQR controller. This similarity may be used to make corresponding assertions concerning the asymptotic behavior of the filter poles, but we do not want to analyze this in detail (but compare Kwakernaak and Sivan [58], Sect. 4.4.4).

The assertions about the asymptotic behavior of the closed-loop poles can be used for the choice of the weighting matrices, since, as we have seen, there is a connection between these matrices and the damping of the poles and the bandwidth of the feedback system. This will be applied in the case study for the inverted pendulum.

Loop Transfer Recovery. For a low plant noise, the Kalman filter estimate is mainly based on the prediction of the plant state, while the measurement is almost ignored. Concerning the performance of the feedback system, this is not critical, but the good robustness properties of the LQR controller (cf. Theorem 8.4.2 and the discussion in Sect. 6.3.2) may get lost if a Kalman filter is used as estimator. Under certain assumptions, it is possible to recover the good robustness of the LQR controller by increasing the plant noise. In some cases, it may even be necessary to introduce additional synthetic noise. This method is called the **loop transfer recovery (LTR)**. The matrix, which models the power spectral density, then no longer describes its actual size and has to be interpreted as a design parameter. As a consequence, the estimate of the Kalman filter is less accurate and one sacrifices performance for the benefit of robustness.

If the plant noise is actuator noise (i.e. if it enters the plant with the control), then this noise can be increased to yield LTR. Otherwise, fictitious noise has to be added to the plant equation:

$$\dot{\mathbf{x}} = \mathbf{A}\,\mathbf{x} + \mathbf{B}_2\mathbf{u} + \mathbf{B}_1\mathbf{w} + \mathbf{B}_2\tilde{\mathbf{w}}\,.$$

Here, $\tilde{\mathbf{w}}$ denotes the additional noise. Denote its covariance matrix by $\tilde{\mathbf{W}}$ and suppose that the norm of $\tilde{\mathbf{W}}$ tends to ∞. Then the open-loop transfer function $\mathbf{L}_i(s)$ for the system with a Kalman filter tends to the corresponding transfer function for the LQR feedback loop if the following assumptions are fulfilled:

(a) The number of measurements is at least as large as the number of controls.

(b) The transfer function of the plant

$$\mathbf{C}_2(\mathbf{I}-\mathbf{A})^{-1}\mathbf{B}_2$$

is minimum phase, i.e. all its zeros are in the open LHP.

A proof and further details can be found in Burl [17], Sect. 8.3.1.

8.5 Related Problems

8.5.1 Optimal Control Problems

Linear quadratic optimal control problems with time-varying matrices. The starting point of LQR controller design was to define an optimization problem in which the functional (8.2) had to be minimized. We consider this problem again, but now for a finite time horizon. Then it causes no problem to allow the use of time-dependent matrices, and the functional may now be defined as (with $\mathbf{S}=\mathbf{0}$ for notational simplicity)

$$f(\mathbf{x},\mathbf{u})=\mathbf{x}^T(t_1)\mathbf{Q}_1\mathbf{x}(t_1)+\int_0^{t_1}(\mathbf{x}^T(t)\mathbf{Q}(t)\mathbf{x}(t)+\mathbf{u}^T(t)\mathbf{R}(t)\mathbf{u}(t))\,dt\,. \tag{8.38}$$

The system matrices are also allowed to be time-varying:

$$\dot{\mathbf{x}}=\mathbf{A}(t)\mathbf{x}+\mathbf{B}(t)\mathbf{u}, \quad \mathbf{x}(0)=\mathbf{x}_0\,. \tag{8.39}$$

The problem is now to minimize the functional (8.38) over all functions $\mathbf{x},\mathbf{u}$ belonging to a suitable function space under the constraint (8.39). The solution of this problem is similar to that of the LQR problem. The optimal control is given by

$$\mathbf{u}(t)=-\mathbf{R}(t)^{-1}\mathbf{B}^T(t)\mathbf{X}(t)\mathbf{x}(t)\,. \tag{8.40}$$

Here, $\mathbf{X}$ is the solution of the following Riccati differential equation:

$$-\dot{\mathbf{X}}=\mathbf{A}^T(t)\mathbf{X}+\mathbf{X}\mathbf{A}(t)-\mathbf{X}\mathbf{B}(t)\mathbf{R}^{-1}(t)\mathbf{B}^T(t)\mathbf{X}+\mathbf{Q}(t) \tag{8.41}$$

$$\mathbf{X}(t_1)=\mathbf{Q}_1\,.$$

If all matrices in (8.41) are constant, then the constant solutions of (8.41) are the solutions of the algebraic Riccati equation. The main difference is that the gains in the feedback law (8.40) are time-varying. The Kalman filter can also be generalized to the case of time-varying linear systems. Next we illustrate this result by an example.

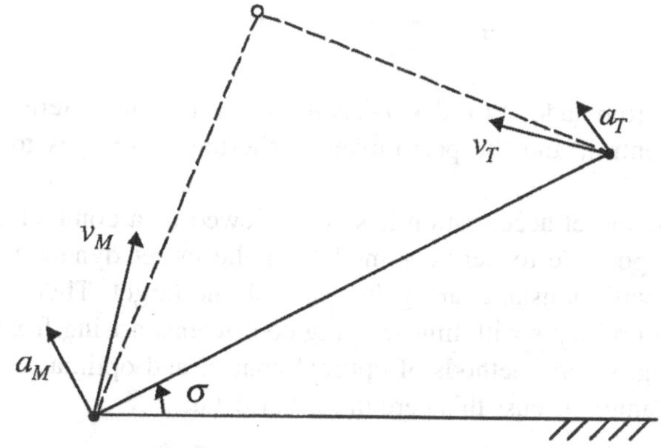

Fig. 8.8. Engagement geometry

Proportional navigation. Suppose a missile has to intercept a target as depicted in Fig. 8.8. Here, it is assumed that the missile hits the target after some time t_f if their velocities point along the dashed lines. Suppose now that the directions of the actual velocities are somewhat different from these nominal directions. Moreover, let a_M and a_T be the lateral accelerations with respect to the line of sight (LOS) of missile and target. We denote by $\lambda = \dot{\sigma}$ the rate of change of the direction σ of the LOS and, finally, let v_R be the relative velocity of the missile and target with respect to the LOS. Then it can be shown that

$$\dot{\lambda} = \frac{2}{t_f - t}\lambda - \frac{1}{v_R(t_f - t)} a_M + \frac{1}{v_R(t_f - t)} a_T, \tag{8.42}$$

and that the missile will hit the target if the LOS rate vanishes for some $t_1 \le t_f$, i.e. if $\lambda(t_1) = 0$.

We look now for a guidance law that prescribes the lateral acceleration a_M of the missile such that $\lambda(t_1) = 0$. To this end, an adequate optimal control problem will be formulated. For $t_1 < t_f$ define the functional

$$f(\lambda, a_M) = Q_1 \lambda(t_1)^2 + \int_{t_0}^{t_1} (Q\lambda^2 + a_M^2)\, d\tau \tag{8.43}$$

and minimize it over all square-integrable functions such that (8.42) holds (with $a_T = 0$). The solution of this problem can be calculated as $a_M(t) = v_R N(t) \lambda(t)$ with a certain function $N(t)$. Moreover, it is possible to pass to the limit: $t_1 \to t_f$. Then we get $N_\infty \ge 3$ as the limit of $N(t)$ and the gain is constant. This is the well-known proportional guidance law, which was discovered long before optimal control existed.

It is possible to develop much more refined guidance laws by optimal control, which do not use the constant gain feedback of the LOS rate. One approach is to take in a simple way the missile dynamics into account by using the equation

$$\dot{a}_M = -\frac{1}{T_M} a_M + \frac{1}{T_M} a_c \, .$$

This equation has to be added to (8.42). Then the commanded lateral acceleration a_c acts as the control, and the performance criterion (8.43) has to be suitably adapted.

Up to now, the target acceleration has been viewed as a completely unknown disturbance. It is possible to derive a model for the target dynamics and to use Kalman filtering with nonstationary gains to track the target. There exists a rich literature for such problems with time-varying coefficients arising from navigation and target tracking, where methods of optimal control and optimal estimation are applied. But we cannot discuss this here in further detail.

General optimal control problems. A much more general class of optimal control problems arises when nonquadratic cost functionals, nonlinear differential equations and additional constraints are admitted. The necessary conditions for optimality are formulated in the famous Pontryagin maximum principle.

A problem of this kind (which is not the most general) may be formulated as follows.

Minimize

$$\int_0^T f(t, \mathbf{x}(t), \mathbf{u}(t))\, dt$$

over all functions $\mathbf{x}, \mathbf{u}$ (belonging to a suitable function space) under the constraints

$$\dot{\mathbf{x}}(t) = g(t, \mathbf{x}(t), \mathbf{u}(t)), \quad \mathbf{x}(0) = \mathbf{x}_0$$
$$\mathbf{x}(T) = \mathbf{x}_T$$
$$\mathbf{u}(t) \in M \quad \text{for almost every } t \in [0, T]\, .$$

An important difference to the (time-varying) LQR problem is that an optimal solution can no longer be calculated from $\mathbf{x}$ by feedback. Despite this, the feedback law (8.40) can be obtained as a special case of the Pontryagin maximum principle. Next, we consider an example.

Re-entry of a space vehicle. An interesting optimal control problem arises when an aircraft returns from space. Since the vehicle is heated up considerably when it flies through the atmosphere, a flight trajectory has to be chosen such that the heating is minimized.

We first briefly discuss the dynamics of the space vehicle and model it simply as a point mass with mass m on which the gravitational force and an aerody-

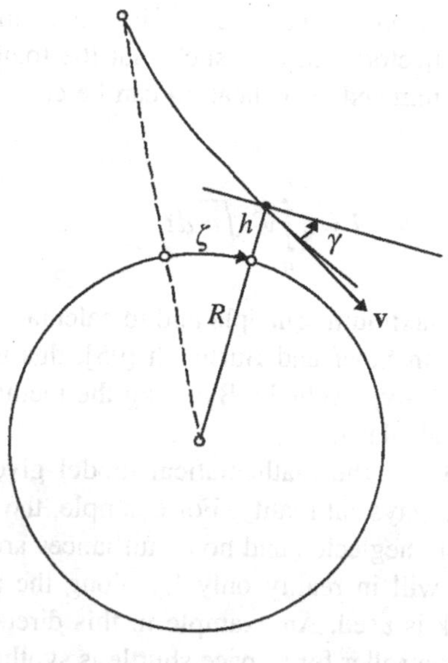

Fig. 8.9. States for the space vehicle reentry

namic force act. The component of the aerodynamic force in direction of **v** (velocity of the vehicle with absolute value v) is denoted by D (drag) and its component orthogonal to **v** is denoted by L (lift). The state variables are v, the flight-path angle γ, the normalized altitude $\xi = h/R$ (where R is the earth's radius) and the distance ζ along the earth's surface (cf. Fig. 8.9). Then it is not difficult to see that the following differential equations hold (note that the acceleration orthogonal to the flight path is given by $v\dot{\gamma}$):

$$\dot{v} = \frac{1}{m} D(\xi, v, \alpha) - \frac{g \sin\gamma}{(1+\xi)^2}$$

$$\dot{\gamma} = \frac{1}{mv} L(\xi, v, \alpha) + \frac{v \cos\gamma}{R(1+\xi)} - \frac{g \cos\gamma}{v(1+\xi)^2}$$

$$\dot{\xi} = \frac{v \sin\gamma}{R}$$

$$\dot{\zeta} = \frac{v}{1+\xi} \cos\gamma \, .$$

Here, α is the angle of attack, which acts as a control. The drag and lift can be expressed in terms of the atmospheric density ρ, the aerodynamical coefficients C_D, C_L and a reference area S by the expressions

$$D = \frac{S}{2} \rho(\xi) v^2 C_D(\alpha), \quad L = \frac{S}{2} \rho(\xi) v^2 C_L(\alpha) \, .$$

The task is to find a trajectory such that at the end of the maneuver the quantities $v(T), \gamma(T)$ and $\xi(T)$ attain prescribed values. The final time T is not fixed in advance. Moreover, the trajectory must be such that the total convective heating per unit area J will be minimized. This heating can be calculated by the formula (with a constant c)

$$J = c \int_0^T v^3 \sqrt{\rho}\, dt .$$

It is possible to apply the maximum principle and to calculate a solution by appropriate numerical methods. In Stoer and Burlirsch [95], this is discussed in detail for the re-entry of an Apollo-type vehicle. By using the methods applied there, an optimal trajectory can be calculated.

It must be emphasized that the mathematical model given above is a rather crude simplification of the physical reality. For example, the rotational motion of the spacecraft is completely neglected and no disturbances are taken into account. Therefore, the spacecraft will in reality only fly along the prescribed trajectory when control via feedback is used. An example in this direction can be found in Balas et. al. [7], where a controller for a space shuttle is synthesized.

8.5.2 Systems Described by Partial Differential Equations

We conclude this chapter by giving a very brief overview of problems whose dynamics is governed by a partial differential equation (PDE). It must be emphasized that many of our calculations are purely formal, but all the results presented here can be proven in a rigorous mathematical manner. There is a rich theory on this subject and the reader may consult for example the excellent introduction of Curtain and Zwart [28].

Temperature control of a thin metal bar. We start with the following example. The temperature distribution $z(t,x)$ of a thin metal bar of length one can be calculated from the following parabolic PDE:

$$\begin{aligned} &\frac{\partial z}{\partial t}(t,x) = \frac{\partial^2 z}{\partial x^2}(t,x) + w(t,x) \\ &\frac{\partial z}{\partial x}(t,0) = 0, \quad \frac{\partial z}{\partial x}(t,1) = 0, \quad z(0,x) = z_0(x). \end{aligned} \tag{8.44}$$

Here, x is the position and $z_0(x)$ is a given temperature distribution at time $t=0$. The function $w(t,x)$ describes the addition of heat along the bar. If suitable units are used for the physical quantities, it is possible to write the heat equation in this normalized form. Moreover, this system can be written in a form which is similar to (5.7) for linear ordinary differential equations (ODEs). To be specific, put $H = \mathcal{L}_2[0, 1]$ and define an operator by

$$D(A) = \{ h \in H \mid h, \frac{dh}{dx} \text{ are absolutely continuous}, \frac{d^2h}{dx^2} \in H, \frac{dh}{dx}(0) = \frac{dh}{dx}(1) = 0 \}$$

$$A : D(A) \to H, \quad Ah = \frac{d^2h}{dx^2}.$$

If B denotes the identity operator, (8.44) can be written in the form

$$\dot{z}(t) = A\,z(t) + Bu(t), \quad z(0) = z_0. \tag{8.45}$$

Our next aim is to derive a solution formula for this system which is analogous to (5.11). Thereby we start with a solution formula for the homogeneous system, i.e. for $w = 0$. The function $e^{\lambda t}v(x)$ is a solution of the homogeneous system if λ and v are a solution of the eigenvalue problem for the operator A:

$$\frac{d^2h}{dx^2}(x) = \lambda\,h(x), \quad h(0) = 0, \quad h(1) = 0.$$

The solutions of this eigenvalue problem are given by

$$\lambda_0 = 0, \quad v_0(x) = 1, \quad \lambda_k = -k^2\pi^2, \quad v_k(x) = \sqrt{2}\cos(k\pi x), \quad k \geq 1.$$

Since the functions $v_0, v_1, \ldots$ form an orthonormal basis of H, each initial value $z_0 \in H$ can be expressed by the series

$$z_0 = \sum_{k=0}^{\infty} \langle v_k, z_0 \rangle v_k.$$

Here, the bracket denotes the scalar product of H. Hence, the solution of the homogeneous system is given by

$$z(t,x) = \sum_{k=0}^{\infty} e^{\lambda_k t} \langle v_k, z_0 \rangle v_k(x).$$

We note that this equation is similar to the eigenvector expansion (5.19), but now the function values $z(t, \cdot)$ lie in the infinite dimensional space $H = \mathcal{L}_2[0, 1]$. For each $t \geq 0$ define an operator $S(t) : H \to H$ by

$$S(t)\,z_0 = \sum_{k=0}^{\infty} e^{\lambda_k t} \langle v_k, z_0 \rangle v_k. \tag{8.46}$$

Then we can write $z(t) = S(t)\,z_0$, and hence $S(t)$ generalizes the matrix exponential function. In the terminology of infinite dimensional systems theory, $S(t)$ is a strongly continuous semigroup. The desired solution formula is now

$$z(t) = S(t)\,z_0 + \int_0^t S(t-\tau)Bu(\tau)d\tau\,. \tag{8.47}$$

Of course, this argument is purely formal and not a proof, but for a very large class of linear PDEs it is possible to define semigroups such that (8.47) holds in a rigorous sense.

It is possible to evaluate (8.46) further, but again in all what follows, the calculations are only formal. Interchanging summation and integration, we obtain

$$\begin{aligned} S(t)\,z_0(x) &= \sum_{k=0}^{\infty} e^{\lambda_k t} \int_0^1 v_k(\xi) z_0(\xi)\, d\xi\, v_k(x) \\ &= \int_0^1 \sum_{k=0}^{\infty} e^{\lambda_k t}\, v_k(x) v_k(\xi) z_0(\xi)\, d\xi\,. \end{aligned}$$

Thus, defining the Green's function

$$\gamma(t,x,\xi) = 1 + \sum_{k=1}^{\infty} 2\,e^{-k^2\pi^2 t} \cos(k\pi x)\cos(k\pi\xi)\,,$$

the semigroup can also be represented as

$$S(t)\,z_0(x) = \int_0^1 \gamma(t,x,\xi) z_0(\xi)\, d\xi\,. \tag{8.48}$$

We now formulate a control problem for the heat equation. Suppose that control is exerted by using a heating element around the point x_0. At another point x_1, the temperature is measured. Then it is helpful to introduce two "shaping functions" by

$$b(x) = \begin{cases} 1/2\varepsilon & \text{for } x_0 - \varepsilon \le x \le x_0 + \varepsilon, \\ 0 & \text{elsewhere}\,, \end{cases}$$

$$c(x) = \begin{cases} 1/2\varepsilon & \text{for } x_1 - \nu \le x \le x_1 + \nu, \\ 0 & \text{elsewhere}\,, \end{cases}$$

with small positive constants $\varepsilon > 0, \nu > 0$. The input operator and the output operator are defined by

$$Bu = b(x)u$$

$$Cz = \int_0^1 c(x)z(x)dx\,.$$

The controlled system is then described by the state equation (8.45) and the output equation $y = Cx$.

The impulse response for this system is defined by (compare (5.12))

$$g(t) = CS(t)B.$$

For the system given here the impulse response can easily be calculated as

$$g(t) = 1 + \sum_{k=1}^{\infty} a_k \, e^{-k^2\pi^2 t},$$

where the coefficients are defined by

$$a_k = \frac{2\cos(k\pi x_0)\cos(k\pi x_1)\sin(k\pi\varepsilon)\sin(k\pi v)}{\varepsilon v k^2 \pi^2}.$$

Since the Laplace transform of the impulse response is the transfer function $G(s)$, the transfer function for the controlled heat equation is

$$G(s) = \frac{1}{s} + \sum_{k=1}^{\infty} \frac{a_k}{s + k^2\pi^2}.$$

This is a nonrational transfer function and the sum of an integrator and an infinite sum of first-order systems.

Supported beam. A simply supported beam is described by a PDE of second order in t:

$$\frac{\partial^2 z}{\partial t^2}(t,x) = -\frac{\partial^4 z}{\partial x^4}(t,x)$$

$$z(t,0) = 0, \quad z(t,1) = 0, \quad \frac{\partial^2 z}{\partial x^2}(t,0) = 0, \quad \frac{\partial^2 z}{\partial x^2}(t,1) = u(t) \tag{8.49}$$

$$z(0,x) = z_0(x), \quad \frac{\partial z}{\partial t}(0,x) = z_1(x).$$

Here, z denotes the displacement of the beam. The boundary conditions imply the following: at the ends the beam is fixed, and at the left-hand end, the bending moment is zero, whereas at the right-hand end the bending moment is prescribed by the control $u(t)$. The observation is the displacement at a position x_1:

$$y(t) = z(t, x_1).$$

The task is that the control suppresses vibrations at the beam position $x = x_1$ that are caused by a force acting on the beam at a different position (cf. Fig. 8.10).

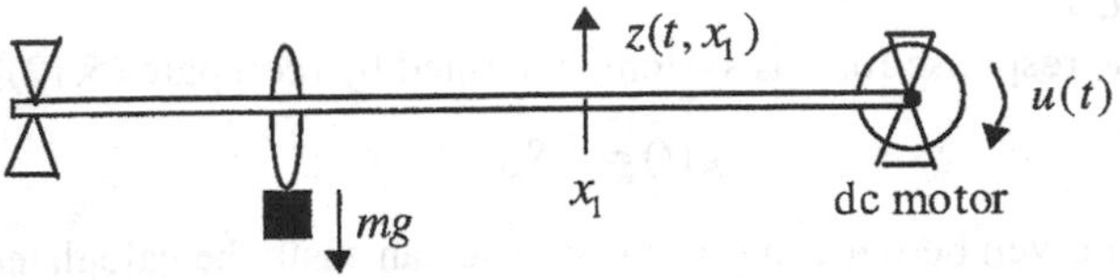

Fig. 8.10. Vibration control of a beam

It is possible to perform a similar analysis to that for the heat equation. The transfer function is

$$G(s) = \sum_{k=1}^{\infty} \frac{(-1)^k 2k\pi \sin(k\pi x_1)}{s^2 + (k\pi)^4}.$$

This is an infinite superposition of second-order transfer functions with zero damping. A similar problem can be formulated where the left-hand end is free. Then in the transfer function, a double integrator has to be added. For a rigid beam, this double integrator would describe the complete transfer function.

A further interesting point is that the transfer function can also be obtained by calculating the Laplace transform of the system directly. We use the relation

$$\mathcal{L}(\frac{\partial z}{\partial x}(\cdot, x))(s) = \frac{d}{dx}\mathcal{L}(z(\cdot, x))(s) = \frac{d}{dx}\hat{z}(s, x).$$

Then the Laplace transform leads to the following boundary value problem for $\hat{z}(s, x)$ with s as a parameter:

$$\frac{d^4 \hat{z}}{dx^4}(s, x) = -s^2 \hat{z}(s, x)$$

$$\hat{z}(s, 0) = 0, \quad \hat{z}(s, 1) = 0$$

$$\frac{d^2 \hat{z}}{dx^2}(s, 0) = 0, \quad \frac{d^2 \hat{z}}{dx^2}(s, 1) = \hat{u}(s).$$

This is a linear boundary value problem for the function $\hat{z}(s, \cdot)$, which can be solved explicitly and its solution at x_1 yields another formula for the transfer function. With the abbreviation $\lambda = \sqrt{s}/\sqrt{2}$, one obtains

$$G(s) = \frac{\cos\lambda \sinh\lambda \sin\lambda x_1 \cosh\lambda x_1 - \sin\lambda \cosh\lambda \cos\lambda x_1 \sinh\lambda x}{s(\sin^2\lambda + \sinh^2\lambda)}.$$

For such problems, all the questions concerning system theory and controller design can be formulated again, but it is far beyond the scope of this book to present the results.

Optimal control problems. Finally, we mention another type of problem, which resembles those formulated for systems driven by ODEs in Sect. 8.5.1. To give an idea of their nature, consider again the heat equation (8.44), but now in a finite time interval $[0, T]$ and with $w(t,x) = b(x)u(t)$. The task is to find a control which minimizes the norm

$$\int_0^T (z(T,x) - z_T(x))^2 dx$$

under the constraint

$$0 \leq u(t) \leq m .$$

In other words, at time T, a given temperature distribution $z_T(x)$ has to be reached as closely as possible. There is a vast literature on this and related problems. One of the first books dealing with them is Lions [64].

Notes and References

The theory of LQG control begins with the papers of Kalman [55] and of Kalman and Bucy [56]. A thorough analysis of this subject can be found in Anderson and Moore [2], and in Kwakernaak and Sivan [58]. The reformulation and generalization of the LQG problem as an $\mathcal{H}_2$ optimization problem is more recent, and the solution given here is based on the famous paper of Doyle, Glover, Khargonekar, and Francis [35]; see also Zhou, Doyle, and Glover [112]. The analysis of $\mathcal{H}_2$ optimal controllers uses results for the Riccati equation. An intensive treatment of this subject is Lancaster and Rodman [59]. Optimal control problems for nonlinear systems are treated in Lewis and Syrmos [60]. An introduction to infinite dimensional linear systems theory including linear quadratic control is Curtain and Zwart [28].

Chapter 9

$\mathcal{H}_\infty$ Optimal Control: Riccati-Approach

In the last chapter, we considered the problem of minimizing $\mathbf{F}_{zw}$ with respect to the $\mathcal{H}_2$ norm. The performance specifications were given in the time domain. As we have seen in Chap. 3 for SISO problems, for specifications in the frequency domain the $\mathcal{H}_\infty$ norm is an adequate tool. In this way we are naturally lead to the question of how controllers can be characterized in a way which minimizes the closed-loop transfer function $\mathbf{F}_{zw}$ with respect to the $\mathcal{H}_\infty$ norm. There are two important methods for solving this problem. One is based on two Riccati equations similar to those used in the $\mathcal{H}_2$ problem. It will be analyzed in this chapter, whereas the other method uses linear matrix inequalities and is presented in the next chapter.

As we will see, we are lead to the characterization of suboptimal controllers instead of optimal controllers. The basic idea for solving the characterization problem is, as for the $\mathcal{H}_2$ problem a change of variables of the kind $\mathbf{v} = \mathbf{u} - \mathbf{Fx}$, with a matrix $\mathbf{F}$ which is related to a Riccati equation. The resulting problem is again an output estimation problem, which can be reduced in several steps to a full information problem. The technical details are much more complicated as for the $\mathcal{H}_2$ problem, and the full proof of the characterization theorem essentially requires the whole of this chapter. Thereby the basic structure of the plant is similar to that in $\mathcal{H}_2$ control. In particular, the assumption $\mathbf{D}_{11} = \mathbf{0}$ is made again and the proof of the characterization theorem for problems fulfilling this requirement is given in Sects. 9.2– 9.4. This assumption is natural for $\mathcal{H}_2$ problems but restrictive for $\mathcal{H}_\infty$ problems, since the introduction of weights in order to get a certain closed-loop frequency response normally leads to generalized plants with $\mathbf{D}_{11} \neq \mathbf{0}$. It is possible to reduce problems with $\mathbf{D}_{11} \neq \mathbf{0}$ to ones with $\mathbf{D}_{11} = \mathbf{0}$ by a procedure called loop shifting, which is presented in Sect. 9.5. The introduction of frequency-dependent weights leads in many situation to a so-called mixed sensitivity problem. We discuss it together with problems based on other weighting schemes in Sect. 9.6 and finally present a result on pole-zero cancellations.

9.1 Formulation of the General $\mathcal{H}_\infty$ Problem

We start with a general plant of the form

$$\mathbf{P}(s) = \left(\begin{array}{c|cc} \mathbf{A} & \mathbf{B}_1 & \mathbf{B}_2 \\ \hline \mathbf{C}_1 & \mathbf{D}_{12} & \mathbf{D}_{12} \\ \mathbf{C}_2 & \mathbf{D}_{21} & \mathbf{D}_{22} \end{array}\right),$$

i.e.

$$\dot{\mathbf{x}} = \mathbf{A}\,\mathbf{x} + \mathbf{B}_1\mathbf{w} + \mathbf{B}_2\mathbf{u},$$

$$\mathbf{z} = \mathbf{C}_1\,\mathbf{x} + \mathbf{D}_{11}\mathbf{w} + \mathbf{D}_{12}\mathbf{u},$$

$$\mathbf{y} = \mathbf{C}_2\,\mathbf{x} + \mathbf{D}_{21}\mathbf{w} + \mathbf{D}_{22}\mathbf{u}.$$

A controller $\mathbf{K}(s)$ is denoted as **admissible**, if it is proper and if it stabilizes internally the system $\mathbf{F}_{zw}(s)$. We now formulate the following problem.

Optimal $\mathcal{H}_\infty$ problem. *Find all admissible controllers $\mathbf{K}(s)$ which minimize the $\mathcal{H}_\infty$ norm of the feedback system, i.e. all admissible controllers that minimize $\| \mathbf{F}_{zw} \|_\infty$.*

As we will see, for the minimization of the $\mathcal{H}_\infty$ norm it is more natural to ask for all suboptimal controllers. Finding an optimal controller is more difficult and, besides this, optimal controllers for the $\mathcal{H}_\infty$ problem are not unique. They can be viewed as the limit case for suboptimal controllers and are not explicitly constructed. Therefore we are led to the following problem.

Suboptimal $\mathcal{H}_\infty$ problem. *For a given $\gamma > 0$, find all admissible controllers $\mathbf{K}(s)$ with $\| \mathbf{F}_{zw} \|_\infty < \gamma$. Such a controller is denoted as **suboptimal**.*

We define

$$\gamma_{\text{opt}} = \inf\left\{ \| \mathbf{F}_{zw} \|_\infty \,\middle|\, \mathbf{K}(s) \text{ is admissible} \right\}.$$

Note that for $\gamma = \gamma_{\text{opt}}$, there are no suboptimal controllers. For numbers γ, which are greater than the optimal value γ_{opt}, there are always admissible controllers with $\| \mathbf{F}_{zw} \|_\infty < \gamma$. It is possible to characterize the suboptimal controllers belonging to $\gamma > \gamma_{\text{opt}}$ completely.

Example. $\mathcal{H}_\infty$ control of a first-order plant. Let the plant be given by

$$\dot{x} = -a\,x + u$$

$$y = x\,.$$

(a) *Performance specification in the time domain.* We inject standard white noise processes d and n into the plant by

$$\dot{x} = -a\,x + u + \sqrt{W}\,d$$
$$y = x + n$$

and define for controller design the output

$$\mathbf{z} = \begin{pmatrix} \sqrt{Q}\,x \\ u \end{pmatrix}.$$

With $\mathbf{w} = (d \quad n)^T$ this can be viewed as a typical LQG problem. The state space description of the generalized plant is given by

$$\mathbf{A} = (-a), \quad \mathbf{B}_1 = \begin{pmatrix} \sqrt{W} & 0 \end{pmatrix}, \quad \mathbf{B}_2 = (1), \quad \mathbf{w} = \begin{pmatrix} d \\ n \end{pmatrix},$$
$$\mathbf{C}_1 = \begin{pmatrix} \sqrt{Q} \\ 0 \end{pmatrix}, \quad \mathbf{D}_{11} = (0), \quad \mathbf{D}_{12} = \begin{pmatrix} 0 \\ 1 \end{pmatrix},$$
$$\mathbf{C}_2 = (1), \quad \mathbf{D}_{21} = (0 \quad 1), \quad \mathbf{D}_{22} = (0).$$

Of course, it would be possible to minimize $\mathbf{F}_{zw}$ also with respect to the $\mathcal{H}_\infty$ norm, but then the design parameters can not so naturally be related to the operator norm.

(b) *Specification in the frequency domain.* For a $\mathcal{H}_\infty$ controller design, it is more natural to specify the performance in the frequency domain. To this end, we introduce a (deterministic) disturbance at the plant output and obtain

$$\dot{x} = -a\,x + u$$
$$y = x + d\,.$$

Then

$$\hat{y} = S(s)\hat{d}\,,$$

where S is as usual the sensitivity function. With a weight $W_1(s)$ it is possible to specify the frequency response of S in the form

$$|\,S(i\omega)\,|\,|W_1(i\omega)\,| \le 1\,.$$

A reasonable choice for the weight is

$$W_1(s) = \frac{s + \omega_1}{s + \omega_\varepsilon}\,,$$

with $\omega_1 > \omega_\varepsilon > 0$. As we will see later, $\omega_\varepsilon = 0$ is not allowed. We define for the sake of controller design the output

$$\hat{z}_1 = W_1(s)S(s)\hat{d}$$
$$\hat{z}_2 = -w_2\hat{u} = -w_2 K(s)S(s)\hat{d}.$$

The number $w_2 > 0$ describes a second weight. The state space realization of the generalized plant is given by

$$\mathbf{A} = \begin{pmatrix} -a & 0 \\ 1 & -\omega_\varepsilon \end{pmatrix}, \quad \mathbf{B}_1 = \begin{pmatrix} 0 \\ 1 \end{pmatrix}, \quad \mathbf{B}_2 = \begin{pmatrix} 1 \\ 0 \end{pmatrix},$$

$$\mathbf{C}_1 = \begin{pmatrix} 1 & \omega_1 - \omega_\varepsilon \\ 0 & 0 \end{pmatrix}, \quad \mathbf{D}_{11} = \begin{pmatrix} 1 \\ 0 \end{pmatrix}, \quad \mathbf{D}_{12} = \begin{pmatrix} 0 \\ -w_2 \end{pmatrix},$$

$$\mathbf{C}_2 = (1 \quad 0), \quad \mathbf{D}_{21} = (1), \quad \mathbf{D}_{22} = (0).$$

In contrast to the specification in the time domain, we have $\mathbf{D}_{11} \neq \mathbf{0}$. This is typical for frequency-domain specifications and causes some difficulties. In the Riccati approach, the $\mathcal{H}_\infty$ problem is first solved for $\mathbf{D}_{11} = \mathbf{0}$. This (and some other assumptions) can later be relaxed by a procedure called loop shifting. Moreover, we have

$$\mathbf{D}_{12}^T\mathbf{C}_1 = (0), \quad \mathbf{B}_1\mathbf{D}_{21}^T = \begin{pmatrix} 0 \\ 1 \end{pmatrix},$$

and thus $\mathbf{B}_1\mathbf{D}_{21}^T \neq \mathbf{0}$. Even for this simple specification, mixed terms arise.

9.2 Characterization of $\mathcal{H}_\infty$ Suboptimal Controllers by Means of Riccati Equations

9.2.1 Characterization Theorem for Output Feedback

In this section, we describe suboptimal $\mathcal{H}_\infty$ controllers for problems with a special structure. The following assumptions are made.

(A1) $(\mathbf{A}, \mathbf{B}_1)$ *is stabilizable and* $(\mathbf{C}_1, \mathbf{A})$ *is detectable.*

(A2) $(\mathbf{A}, \mathbf{B}_2)$ *is stabilizable and* $(\mathbf{C}_2, \mathbf{A})$ *is detectable.*

(A3) $\mathbf{D}_{12}^T\mathbf{C}_1 = \mathbf{0}$ *and* $\mathbf{D}_{12}^T\mathbf{D}_{12} = \mathbf{I}$.

(A4) $\mathbf{B}_1\mathbf{D}_{21}^T = \mathbf{0}$ *and* $\mathbf{D}_{21}\mathbf{D}_{21}^T = \mathbf{I}$.

(A5) $\mathbf{D}_{11} = \mathbf{0}$ *and* $\mathbf{D}_{22} = \mathbf{0}$.

As we have seen from the previous example, these assumptions are too restrictive. In Sect. 9.5 it will be shown how they can be relaxed.

Our first question is under which conditions internal stability is equivalent to $\mathbf{F}_{zw} \in \mathcal{RH}_\infty$. It can be shown that the above assumptions imply those of Lemma 6.5.2. We omit the proof (but compare Zhou, Doyle, and Glover [112], Corollary 16.3).

Corollary 9.2.1. *The assumptions (A1), (A3), (A4) imply that the feedback loop is internally stable if and only if* $\mathbf{F}_{zw} \in \mathcal{RH}_\infty$.

For the next theorem, the following Hamilton matrices are used:

$$\mathbf{H}_\infty = \begin{pmatrix} \mathbf{A} & \gamma^{-2}\mathbf{B}_1\mathbf{B}_1^T - \mathbf{B}_2\mathbf{B}_2^T \\ -\mathbf{C}_1^T\mathbf{C}_1 & -\mathbf{A}^T \end{pmatrix}$$

$$\mathbf{J}_\infty = \begin{pmatrix} \mathbf{A}^T & \gamma^{-2}\mathbf{C}_1^T\mathbf{C}_1 - \mathbf{C}_2^T\mathbf{C}_2 \\ -\mathbf{B}_1\mathbf{B}_1^T & -\mathbf{A} \end{pmatrix}.$$

Theorem 9.2.1. *Suppose the assumptions (A1)–(A5) hold. Then there exists an admissible controller with* $\| \mathbf{F}_{zw} \|_\infty < \gamma$ *if and only if the following conditions are fulfilled:*

(i) $\mathbf{H}_\infty \in dom(Ric)$ *and* $\mathbf{X}_\infty = Ric(\mathbf{H}_\infty) \geq \mathbf{0}$;

(ii) $\mathbf{J}_\infty \in dom(Ric)$ *and* $\mathbf{Y}_\infty = Ric(\mathbf{J}_\infty) \geq \mathbf{0}$;

(iii) $\rho(\mathbf{X}_\infty \mathbf{Y}_\infty) < \gamma^2$.

If these conditions hold, such a controller is

$$\mathbf{K}_{sub}(s) = \left(\begin{array}{c|c} \hat{\mathbf{A}}_\infty & -\mathbf{Z}_\infty\mathbf{L}_\infty \\ \hline \mathbf{F}_\infty & \mathbf{0} \end{array}\right)$$

with

$$\hat{\mathbf{A}}_\infty = \mathbf{A} + \gamma^{-2}\mathbf{B}_1\mathbf{B}_1^T\mathbf{X}_\infty + \mathbf{B}_2\mathbf{F}_\infty + \mathbf{Z}_\infty\mathbf{L}_\infty\mathbf{C}_2$$

$$\mathbf{F}_\infty = -\mathbf{B}_2^T\mathbf{X}_\infty, \quad \mathbf{L}_\infty = -\mathbf{Y}_\infty\mathbf{C}_2^T, \quad \mathbf{Z}_\infty = (\mathbf{I} - \gamma^{-2}\mathbf{Y}_\infty\mathbf{X}_\infty)^{-1}.$$

It is possible to describe this controller with an observer. The controller can equivalently be written in the form

$$\dot{\tilde{\mathbf{x}}} = \mathbf{A}\tilde{\mathbf{x}} + \mathbf{B}_1\tilde{\mathbf{w}}_{\text{worst}} + \mathbf{B}_2\mathbf{u} + \mathbf{Z}_\infty\mathbf{L}_\infty(\mathbf{C}_2\tilde{\mathbf{x}} - \mathbf{y})$$

$$\mathbf{u} = \mathbf{F}_\infty \tilde{\mathbf{x}}, \quad \tilde{\mathbf{w}}_{\text{worst}} = \gamma^{-2} \mathbf{B}_1^T \mathbf{X}_\infty \tilde{\mathbf{x}}\,.$$

The first equation defines an observer. The term $\tilde{\mathbf{w}}_{\text{worst}} = \gamma^{-2}\mathbf{B}_1^T\mathbf{X}_\infty\tilde{\mathbf{x}}$ can be understood as an estimate of the disturbance $\mathbf{w}_{\text{worst}} = \gamma^{-2}\mathbf{B}_1^T\mathbf{X}_\infty\mathbf{x}$. In this way, one gets a controller-observer structure similar to that for the $\mathcal{H}_2$ problem (cf. 8.34). In contrast to this problem, the vector $\mathbf{B}_1$ enters in the $\mathcal{H}_\infty$ observer.

The $\mathcal{H}_\infty$ suboptimal controller has also the representation

$$\mathbf{K}_{sub}(s) = -\mathbf{Z}_\infty \mathbf{L}_\infty (s\mathbf{I} - \hat{\mathbf{A}}_\infty)^{-1} \mathbf{F}_\infty\,.$$

It has as many states as the (generalized) plant $\mathbf{P}(s)$ and is strictly proper. The Riccati equations for $\mathbf{X}_\infty$ and $\mathbf{Y}_\infty$ are

$$\mathbf{X}_\infty \mathbf{A} + \mathbf{A}^T \mathbf{X}_\infty - \mathbf{X}_\infty (\mathbf{B}_2 \mathbf{B}_2^T - \gamma^{-2} \mathbf{B}_1 \mathbf{B}_1^T) \mathbf{X}_\infty + \mathbf{C}_1^T \mathbf{C}_1 = \mathbf{0} \tag{9.1}$$

$$\mathbf{A}\mathbf{Y}_\infty + \mathbf{Y}_\infty \mathbf{A}^T - \mathbf{Y}_\infty (\mathbf{C}_2^T \mathbf{C}_2 - \gamma^{-2} \mathbf{C}_1^T \mathbf{C}_1) \mathbf{Y}_\infty + \mathbf{B}_1 \mathbf{B}_1^T = \mathbf{0}\,. \tag{9.2}$$

The proof of Theorem 9.2.1 is long and takes up the main part of this chapter. We follow essentially the presentation of Zhou, Doyle, and Glover [112].

9.2.2 Outline of the Proof

We start with the following consideration. By Corollary 7.2.1, it is seen that the inequality $\|\mathbf{F}_{zw}\|_\infty < \gamma$ can equivalently be written as

$$\|\mathbf{z}\|_2^2 - \gamma^2 \|\mathbf{w}\|_2^2 < 0 \quad \text{for all } \mathbf{w} \in \mathcal{L}_2[0,\infty), \quad \mathbf{w} \neq \mathbf{0}\,. \tag{9.3}$$

Here, $\mathbf{z}$ is given by

$$\dot{\mathbf{x}} = \mathbf{A}\mathbf{x} + \mathbf{B}_1\mathbf{w} + \mathbf{B}_2\mathbf{u}, \quad \mathbf{x}(0) = \mathbf{0}$$

$$\mathbf{z} = \mathbf{C}_1\mathbf{x} + \mathbf{D}_{12}\mathbf{u}\,,$$

where $\mathbf{u}$ is the controller output. We assume that the Riccati equation (9.1) has a solution $\mathbf{X}_\infty$. Using this equation and assumption (A3), the following calculation can be made:

$$\begin{aligned}
\frac{d}{dt}(\mathbf{x}^T\mathbf{X}_\infty\mathbf{x}) &= \dot{\mathbf{x}}^T\mathbf{X}_\infty\mathbf{x} + \mathbf{x}^T\mathbf{X}_\infty\dot{\mathbf{x}} \\
&= \mathbf{x}^T(\mathbf{A}^T\mathbf{X}_\infty + \mathbf{X}_\infty\mathbf{A})\mathbf{x} + 2\mathbf{w}^T\mathbf{B}_1^T\mathbf{X}_\infty\mathbf{x} + 2\mathbf{u}^T\mathbf{B}_2^T\mathbf{X}_\infty\mathbf{x} \\
&= \mathbf{x}^T(-\mathbf{C}_1^T\mathbf{C}_1 - \gamma^{-2}\mathbf{X}_\infty\mathbf{B}_1\mathbf{B}_1^T\mathbf{X}_\infty + \mathbf{X}_\infty\mathbf{B}_2\mathbf{B}_2^T\mathbf{X}_\infty)\mathbf{x} \\
&\quad + 2\mathbf{w}^T\mathbf{B}_1^T\mathbf{X}_\infty\mathbf{x} + 2\mathbf{u}^T\mathbf{B}_2^T\mathbf{X}_\infty\mathbf{x}
\end{aligned}$$

$$
\begin{aligned}
&= -\|\mathbf{C}_1\mathbf{x}\|^2 - \gamma^{-2}\|\mathbf{B}_1^T\mathbf{X}_\infty\mathbf{x}\|^2 + \|\mathbf{B}_2^T\mathbf{X}_\infty\mathbf{x}\|^2 \\
&\quad + 2\mathbf{w}^T\mathbf{B}_1^T\mathbf{X}_\infty\mathbf{x} + 2\mathbf{u}^T\mathbf{B}_2^T\mathbf{X}_\infty\mathbf{x} \\
&= -\|\mathbf{z}\|^2 + \gamma^2\|\mathbf{w}\|^2 - \gamma^2\|\mathbf{w} - \gamma^{-2}\mathbf{B}_1^T\mathbf{X}_\infty\mathbf{x}\|^2 + \|\mathbf{u} + \mathbf{B}_2^T\mathbf{X}_\infty\mathbf{x}\|^2 .
\end{aligned}
$$

Suppose that $\mathbf{x}(t)$ tends to $\mathbf{x}_\infty = \mathbf{0}$ as $t \to \infty$. Then the above equation can be integrated from 0 to ∞ and one obtains

$$
\|\mathbf{z}\|_2^2 - \gamma^2\|\mathbf{w}\|_2^2 = \|\mathbf{u} + \mathbf{B}_2^T\mathbf{X}_\infty\mathbf{x}\|_2^2 - \gamma^2\|\mathbf{w} - \gamma^{-2}\mathbf{B}_1^T\mathbf{X}_\infty\mathbf{x}\|_2^2 . \tag{9.4}
$$

If all states are available, the choice

$$
\mathbf{u} = -\mathbf{B}_2^T\mathbf{X}_\infty\mathbf{x} \tag{9.5}
$$

can be made. This leads to

$$
\|\mathbf{z}\|_2^2 - \gamma^2\|\mathbf{w}\|_2^2 = -\gamma^2\|\mathbf{w} - \gamma^{-2}\mathbf{B}_1^T\mathbf{X}_\infty\mathbf{x}\|_2^2 \quad \text{for every} \quad \mathbf{w} \in \mathcal{L}_2[0,\infty) .
$$

The difference on the right-hand side vanishes only for $\mathbf{x} = \mathbf{0}$ and $\mathbf{w} = \mathbf{0}$. Hence, with the controller (9.5), inequality (9.3) holds and therefore we have $\|\mathbf{F}_{zw}\|_\infty < \gamma$.

We need the following matrices:

$$
\mathbf{A}_{F_\infty} = \mathbf{A} + \mathbf{B}_2\mathbf{F}_\infty , \quad \mathbf{C}_{1F_\infty} = \mathbf{C}_1 + \mathbf{D}_{12}\mathbf{F}_\infty
$$

The next lemma is a first step in solving the FI problem for $\mathcal{H}_\infty$ optimal control.

Lemma 9.2.1. *Suppose* $\mathbf{H}_\infty \in dom(Ric)$ *and* $\mathbf{X}_\infty = Ric(\mathbf{H}_\infty) \geq \mathbf{0}$. *Then the inequality* $\|\mathbf{F}_{zw}\|_\infty < \gamma$ *is fulfilled if the controller is given by the constant matrix*

$$
\mathbf{K}(s) = -\mathbf{B}_2^T\mathbf{X}_\infty .
$$

Proof. The proof is essentially the preceding calculation. It only remains to show that $\mathbf{x} \in \mathcal{L}_2[0,\infty)$ and $\mathbf{x}_\infty = \mathbf{0}$. This requires the assumptions concerning $\mathbf{H}_\infty$ and $\mathbf{X}_\infty$. Define

$$
\mathbf{C}_e = \begin{pmatrix} \mathbf{C}_{1F_\infty} \\ \gamma^{-1}\mathbf{B}_1^T\mathbf{X}_\infty \end{pmatrix} .
$$

Then (9.1) can be rewritten as a Lyapunov equation for $\mathbf{X}_\infty$:

$$
\mathbf{A}_{F_\infty}^T\mathbf{X}_\infty + \mathbf{X}_\infty\mathbf{A}_{F_\infty} + \mathbf{C}_e^T\mathbf{C}_e = \mathbf{0} . \tag{9.6}
$$

Since $\mathbf{H}_\infty \in dom(Ric)$, Theorem 8.1.2 shows that the matrix

$$
\mathbf{A} + (\gamma^{-2}\mathbf{B}_1\mathbf{B}_1^T - \mathbf{B}_2\mathbf{B}_2^T)\mathbf{X}_\infty = \mathbf{A}_{F_\infty} + \gamma^{-2}\mathbf{B}_1\mathbf{B}_1^T\mathbf{X}_\infty
$$

is stable. Hence, by the PHB test the detectability of $(\mathbf{C}_e^T\mathbf{C}_e, \mathbf{A}_{F_\infty})$ follows. Because of $\mathbf{X}_\infty \geq \mathbf{0}$, Lemma 5.7.1 shows that the matrix $\mathbf{A}_{F_\infty}$ is stable. Thus, since $\mathbf{x}$ is the solution of

$$\dot{\mathbf{x}} = \mathbf{A}_{F_\infty}\mathbf{x} + \mathbf{B}_1\mathbf{w}$$

we obtain $\mathbf{x} \in \mathcal{L}_2[0,\infty)$ and also, as it is not very difficult to see, $\mathbf{x}_\infty = \mathbf{0}$.

The proof of the converse direction, namely the validity of assertion (i) in Theorem 9.2.1 under the assumption $\| \mathbf{F}_{zw} \|_\infty < \gamma$, is much more difficult to show. It is the content of Sect. 9.3.

Equation (9.4) is also the key to the solution of the $\mathcal{H}_\infty$ problem if not all states are available. Then this equation suggests to introduce the following variables:

$$\mathbf{v} = \mathbf{u} - \mathbf{F}_\infty\mathbf{x}, \qquad \mathbf{r} = \mathbf{w} - \gamma^{-2}\mathbf{B}_1^T\mathbf{X}_\infty\mathbf{x}.$$

With these definitions the equations

$$\begin{aligned}\dot{\mathbf{x}} &= \mathbf{A}_{F_\infty}\mathbf{x} + \mathbf{B}_1\mathbf{w} + \mathbf{B}_2\mathbf{v} \\ \mathbf{z} &= \mathbf{C}_{1F_\infty}\mathbf{x} + \mathbf{D}_{12}\mathbf{v} \\ \mathbf{r} &= -\gamma^{-2}\mathbf{B}_1^T\mathbf{X}_\infty\mathbf{x} + \mathbf{w}\end{aligned}$$

hold. Moreover, with

$$\mathbf{A}_{tmp} = \mathbf{A} + \gamma^{-2}\mathbf{B}_1\mathbf{B}_1^T\mathbf{X}_\infty$$

we get (using $\mathbf{B}_1\mathbf{D}_{21}^T = \mathbf{0}$)

$$\begin{aligned}\dot{\mathbf{x}} &= \mathbf{A}_{tmp}\mathbf{x} + \mathbf{B}_1\mathbf{r} + \mathbf{B}_2\mathbf{u} \\ \mathbf{v} &= -\mathbf{F}_\infty\mathbf{x} + \mathbf{u} \\ \mathbf{y} &= \mathbf{C}_2\mathbf{x} + \mathbf{D}_{21}\mathbf{r}.\end{aligned} \tag{9.7}$$

The transfer matrices belonging to these systems are denoted as follows:

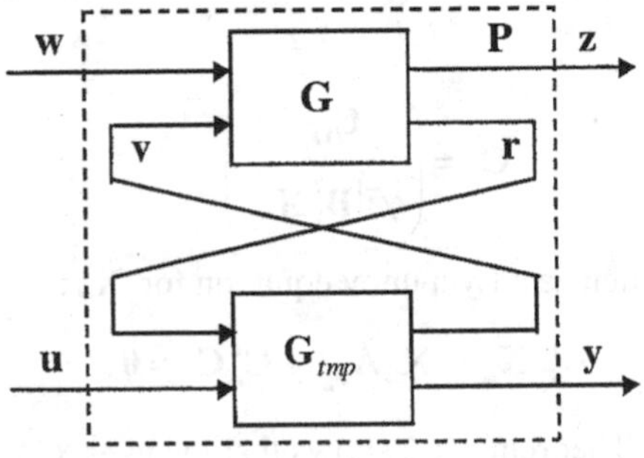

Fig. 9.1. Decomposition of $\mathbf{P}$ in $\mathbf{G}$ in $\mathbf{G}_{tmp}$

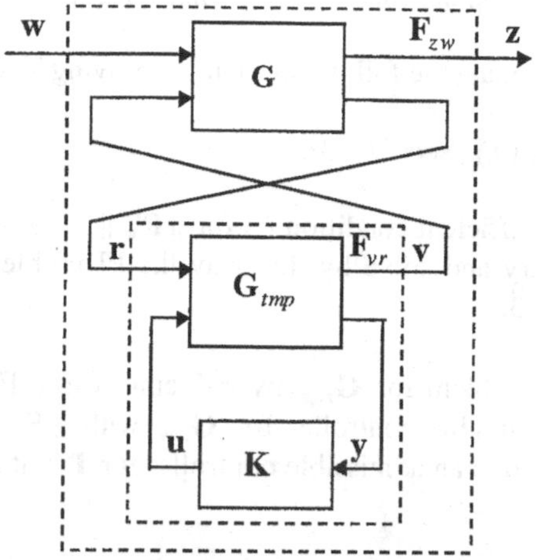

Fig. 9.2. Structure of $\mathbf{F}_{zw}$ and $\mathbf{F}_{vr}$

$$\mathbf{G} = \left(\begin{array}{c|cc} \mathbf{A}_{F_\infty} & \mathbf{B}_1 & \mathbf{B}_2 \\ \hline \mathbf{C}_{1F_\infty} & \mathbf{0} & \mathbf{D}_{12} \\ -\gamma^{-2}\mathbf{B}_1^T\mathbf{X}_\infty & \mathbf{I} & \mathbf{0} \end{array}\right), \qquad \mathbf{G}_{tmp} = \left(\begin{array}{c|cc} \mathbf{A}_{tmp} & \mathbf{B}_1 & \mathbf{B}_2 \\ \hline -\mathbf{F}_\infty & \mathbf{0} & \mathbf{I} \\ \mathbf{C}_2 & \mathbf{D}_{21} & \mathbf{0} \end{array}\right).$$

If $\mathbf{G}$ and $\mathbf{G}_{tmp}$ are composed as shown in Fig. 9.1, the original plant $\mathbf{P}$ results. The closed-loop system with plant $\mathbf{P}$ and controller $\mathbf{K}$ is $\mathbf{F}_{zw}$. The closed-loop system with plant $\mathbf{G}_{tmp}$ and controller $\mathbf{K}$ is denoted as $\mathbf{F}_{vr}$ (cf. Fig. 9.2).

From these considerations it is seen that a controller for $\mathbf{G}_{tmp}$ yields also a controller for $\mathbf{P}$. It is important to note that $\mathbf{G}_{tmp}$ is an OE problem (cf. Sect. 8.2.3). Lemma 9.2.1 is also the key for the solution of the $\mathcal{H}_\infty$ OE problem. The idea is now to proceed as in the $\mathcal{H}_2$ case by reducing the OE problem to a dual DF problem which can be interpreted as special FI problem. This strategy is only successful if there is a suitable relationship between the norms of $\mathbf{F}_{zw}$ and $\mathbf{F}_{vr}$. Because of (9.4) we have

$$\| \mathbf{z} \|_2^2 - \gamma^2 \| \mathbf{w} \|_2^2 = \| \mathbf{v} \|_2^2 - \gamma^2 \| \mathbf{r} \|_2^2 \tag{9.8}$$

and consequently

$$\| \mathbf{F}_{zw} \|_\infty \leq \gamma \quad \text{if and only if} \quad \| \mathbf{F}_{vr} \|_\infty \leq \gamma .$$

Since we want to characterize suboptimal controllers, a version of this equivalence where the inequalities are strict will be needed. More precisely, the following assertion can be shown:

(A) The controller $\mathbf{K}$ *is admissible for* $\mathbf{P}$ *and* $\| \mathbf{F}_{zw} \|_\infty < \gamma$ *holds if and only*

if the controller $\mathbf{K}$ *is admissible for* $\mathbf{G}_{tmp}$ *and* $\|\mathbf{F}_{vr}\|_\infty < \gamma$.

Summing up, we may formulate the following plan for proving Theorem 9.2.1.

Step 1: Prove assertion (A) (Sect. 9.2.3).

Step 2: Show that the sufficient condition (i) for $\|\mathbf{F}_{zw}\|_\infty < \gamma$ in Lemma 9.2.1 is also necessary and solve by this way the FI problem. This will be done in Sect. 9.3.

Step 3: Solve the OE problem for $\mathbf{G}_{tmp}$ by reducing it to a FI problem. This leads to an admissible controller for $\mathbf{G}_{tmp}$ with $\|\mathbf{F}_{vr}\|_\infty < \gamma$ and because of (A) also to an admissible controller for $\mathbf{P}$ with $\|\mathbf{F}_{zw}\|_\infty < \gamma$.

9.2.3 Contraction and Stability

We consider the feedback system shown in Fig. 9.3 and partition $\mathbf{G}$ adequately:

$$\mathbf{G} = \begin{pmatrix} \mathbf{G}_{11} & \mathbf{G}_{12} \\ \mathbf{G}_{21} & \mathbf{G}_{22} \end{pmatrix}.$$

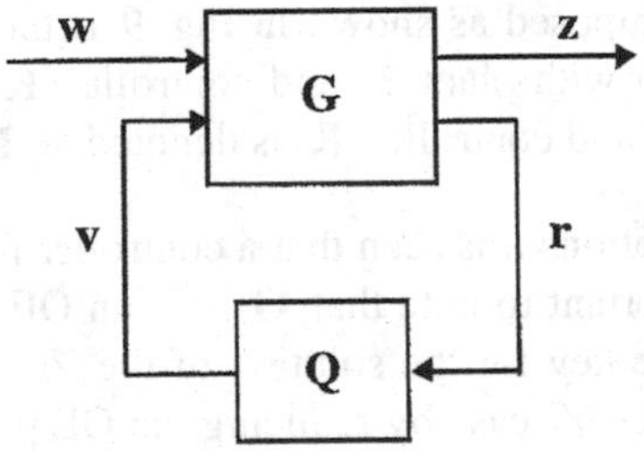

Fig. 9.3. Contraction and stability

Theorem 9.2.2. *Let* $\mathbf{G} \in \mathcal{RH}_\infty$, *suppose* $\mathbf{G}_{21}^{-1}$ *exists and assume* $\mathbf{G}_{21}^{-1} \in \mathcal{RH}_\infty$, $\mathbf{G}^{\sim}\mathbf{G} = \mathbf{I}$. *Then, with a proper rational transfer matrix* $\mathbf{Q}$, *the following assertions are equivalent:*

(i) The feedback system in Fig. 9.3 is well-posed, internally stable and $\|\mathbf{F}_{zw}\|_\infty < 1$.

(ii) $\mathbf{Q} \in \mathcal{RH}_\infty$ *and* $\|\mathbf{Q}\|_\infty < 1$.

Proof. *(ii)* $\Rightarrow$ *(i)*: Since $\mathbf{G}$ is inner, we have $\|\mathbf{G}\|_\infty = 1$, and this implies $\|\mathbf{G}_{22}\|_\infty \le 1$. Because of $\mathbf{G}, \mathbf{Q} \in \mathcal{RH}_\infty$ and $\|\mathbf{Q}\|_\infty < 1$, well-posedness and internal stability follow from the small gain theorem (Theorem 13.1.1). Thus, in particular $(\mathbf{I}-\mathbf{G}_{22}\mathbf{Q})^{-1} \in \mathcal{RH}_\infty$ holds.

In order to show $\|\mathbf{F}_{zw}\|_\infty < 1$, we first note that

$$\mathbf{F}_{wr} = \mathbf{G}_{21}^{-1}(\mathbf{I}-\mathbf{G}_{22}\mathbf{Q}) \in \mathcal{RH}_\infty .$$

Let $\gamma = \|\mathbf{Q}\|_\infty$, $\delta = \|\mathbf{F}_{wr}\|_\infty$. We have $\|\mathbf{z}\|^2 + \|\mathbf{r}\|^2 = \|\mathbf{w}\|^2 + \|\mathbf{v}\|^2$, since $\mathbf{G}$ is inner (with $\mathbf{r},\mathbf{v},\mathbf{w},\mathbf{z} \in \mathcal{H}_2$). Because of $\gamma < 1$ and $\|\mathbf{w}\| \le \delta\|\mathbf{r}\|$, the inequality

$$\|\mathbf{z}\|^2 \le \|\mathbf{w}\|^2 - (1-\gamma^2)\|\mathbf{r}\|^2 \le [1-(1-\gamma^2)\delta^{-2}]\|\mathbf{w}\|^2$$

follows; hence $\|\mathbf{F}_{zw}\|_\infty < 1$ (cf. Theorem 7.2.3).

$(i) \Rightarrow (ii)$: We first suppose $\|\mathbf{Q}\|_\infty \ge 1$ and generate a contradiction. This assumption implies the existence of $\omega_0 \in \mathbb{R} \cup \{\infty\}$ and of a constant vector $\mathbf{r}$ such that $\|\mathbf{Q}(i\omega_0)\mathbf{r}\| \ge \|\mathbf{r}\|$. Defining

$$\mathbf{v} = \mathbf{Q}\mathbf{r}, \quad \mathbf{w} = \mathbf{G}_{21}^{-1}(\mathbf{I}-\mathbf{G}_{22}\mathbf{Q})\mathbf{r}$$

we obtain $\mathbf{v} = \mathbf{F}_{vw}\mathbf{w}$. Since $\mathbf{G}$ is inner, the inequality

$$\begin{aligned}\|\mathbf{z}(i\omega_0)\|^2 + \|\mathbf{r}(i\omega_0)\|^2 &= \|\mathbf{w}(i\omega_0)\|^2 + \|\mathbf{v}(i\omega_0)\|^2 \\ &\ge \|\mathbf{w}(i\omega_0)\|^2 + \|\mathbf{r}(i\omega_0)\|^2\end{aligned}$$

follows, and therefore $\|\mathbf{z}(i\omega_0)\| \ge \|\mathbf{w}(i\omega_0)\|$. This contradicts $\|\mathbf{F}_{zw}\|_\infty < 1$ and $\|\mathbf{Q}\|_\infty < 1$ is shown.

We now prove $\mathbf{Q} \in \mathcal{RH}_\infty$. Let $\mathbf{M},\mathbf{N},\mathbf{X},\mathbf{Y} \in \mathcal{RH}_\infty$ be a right coprime factorization of $\mathbf{Q}$. Then $\mathbf{Q} = \mathbf{N}\mathbf{M}^{-1}$ and $\mathbf{X}\mathbf{M}+\mathbf{Y}\mathbf{N} = \mathbf{I}$. We show $\mathbf{M}^{-1} \in \mathcal{RH}_\infty$. The internal stability implies

$$\mathbf{Q}(\mathbf{I}-\mathbf{G}_{22}\mathbf{Q})^{-1} = \mathbf{N}(\mathbf{M}-\mathbf{G}_{22}\mathbf{N})^{-1} \in \mathcal{RH}_\infty$$

and

$$(\mathbf{I}-\mathbf{G}_{22}\mathbf{Q})^{-1} = \mathbf{M}(\mathbf{M}-\mathbf{G}_{22}\mathbf{N})^{-1} \in \mathcal{RH}_\infty .$$

Hence we get

$$(\mathbf{M}-\mathbf{G}_{22}\mathbf{N})^{-1} = \mathbf{Y}\mathbf{Q}(\mathbf{I}-\mathbf{G}_{22}\mathbf{Q})^{-1} + \mathbf{X}(\mathbf{I}-\mathbf{G}_{22}\mathbf{Q})^{-1} \in \mathcal{RH}_\infty .$$

Let C be a Nyquist contour and define

$$f_\alpha(s) = \det(\mathbf{M}(s) - \alpha\mathbf{G}_{22}(s)\mathbf{N}(s)) = \frac{p_\alpha(s)}{q_\alpha(s)} \quad \text{with} \quad 0 \le \alpha \le 1 .$$

Since $\mathbf{M}-\mathbf{G}_{22}\mathbf{N}$ and $(\mathbf{M}-\mathbf{G}_{22}\mathbf{N})^{-1}$ are stable, p_1 and q_1 have no zeros in $\overline{\mathbb{C}}_+$. Therefore, we get $n(C_{f1},0) = 0$ (with the notation of Sect. 3.4). The index does not change with $0 \le \alpha \le 1$. Otherwise, there would exist an α such that the origin lies on $C_{f\alpha}$. Assume this is the case. Then there exists a frequency ω with

$f_\alpha(i\omega) = 0$ and consequently

$$\det(\mathbf{I} - \alpha \mathbf{G}_{22}(i\omega)\mathbf{Q}(i\omega)) = \det(\mathbf{M}(i\omega) - \alpha \mathbf{G}_{22}(i\omega)\mathbf{N}(i\omega)) \det \mathbf{M}^{-1}(i\omega) = 0 .$$

This cannot be valid, since the matrix $(\mathbf{I} - \alpha \mathbf{G}_{22}(i\omega)\mathbf{Q}(i\omega))$ is nonsingular (because of $\| \mathbf{G}_{22} \|_\infty \le 1$, $\| \mathbf{Q} \|_\infty < 1$ and $0 \le \alpha \le 1$). Thus, we get $n(C_{f0}, 0) = 0$. Denote by k_{p0} (resp. k_{q0}) the number of zeros of p_0 (resp. q_0) in the closed RHP. Then we have $n(C_{f0}, 0) = k_{p0} - k_{q0}$. Since $\mathbf{M}$ is stable, it follows that $k_{q0} = 0$ and therefore also $k_{p0} = 0$. Hence $\mathbf{M}^{-1}$ is stable and hence $\mathbf{Q}$ is also stable (cf. Corollary 5.6.1). This proves the theorem.

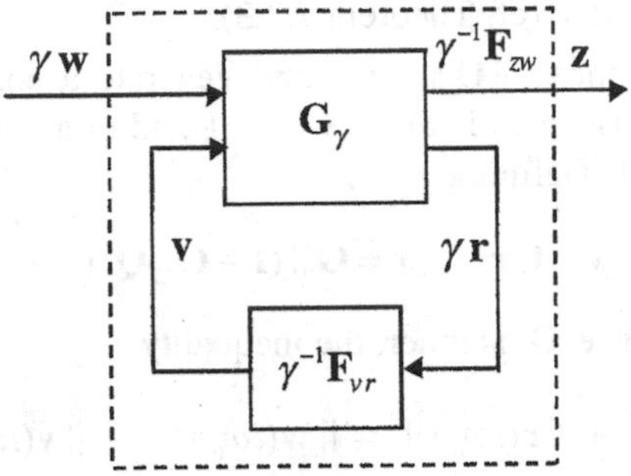

Fig. 9.4. Reformulation of the block diagram of Fig. 9.2

The transfer function $\mathbf{G}$ introduced in Sect 9.2.1 is not inner, but the following equation (which is a consequence of (9.8)

$$\| \mathbf{z} \|_2^2 + \| \gamma \mathbf{r} \|_2^2 = \| \mathbf{v} \|_2^2 + \| \gamma \mathbf{w} \|_2^2$$

suggests that

$$\mathbf{G}_\gamma = \left(\begin{array}{c|cc} \mathbf{A}_{F_\infty} & \gamma^{-1}\mathbf{B}_1 & \mathbf{B}_2 \\ \hline \mathbf{C}_{1F_\infty} & \mathbf{0} & \mathbf{D}_{12} \\ -\gamma^{-1}\mathbf{B}_1^T \mathbf{X}_\infty & \mathbf{I} & \mathbf{0} \end{array} \right),$$

might be inner because of

$$\begin{pmatrix} \hat{\mathbf{z}} \\ \gamma \hat{\mathbf{r}} \end{pmatrix} = \mathbf{G}_\gamma \begin{pmatrix} \gamma \hat{\mathbf{w}} \\ \hat{\mathbf{v}} \end{pmatrix}.$$

We replace now the block diagram of Fig. 9.2 by the block diagram of Fig. 9.4. Denote by $\mathbf{F}_{z(\gamma w)}$ the transfer function which results when in the feedback system $\mathbf{F}_{zw}$ the plant $\mathbf{G}$ is replaced by $\mathbf{G}_\gamma$. Then we have $\mathbf{F}_{z(\gamma w)} = \gamma^{-1}\mathbf{F}_{zw}$. Moreover, let $\mathbf{F}_{v(\gamma r)} = \gamma^{-1}\mathbf{F}_{vr}$. Then $\| \mathbf{F}_{zw} \|_\infty < \gamma$ is equivalent to $\| \mathbf{F}_{z(\gamma w)} \|_\infty < 1$ and the inequality $\| \mathbf{F}_{vr} \|_\infty < \gamma$ is equivalent to $\| \mathbf{F}_{v(\gamma r)} \|_\infty < 1$. In this way we arrive at the situation of Theorem 9.2.2.

Lemma 9.2.2. *Suppose* $\mathbf{H}_\infty \in dom(Ric)$ *and* $\mathbf{X}_\infty = Ric(\mathbf{H}_\infty) \geq \mathbf{0}$. *Then* $\mathbf{G}_\gamma$ *is in* $\mathcal{RH}_\infty$ *and inner, and* $\mathbf{G}_{\gamma 21}^{-1} \in \mathcal{RH}_\infty$.

Proof. We have already shown in the proof of Lemma 9.2.1 that $\mathbf{A}_{F_\infty}$ is stable. Since $\mathbf{A}_{F_\infty}$ is stable, $\mathbf{G}_\gamma$ is in $\mathcal{RH}_\infty$. From (9.6) it is seen that the observability gramian of $\mathbf{G}_\gamma$ is $\mathbf{X}_\infty$. An application of Lemma 7.5.2 shows that $\mathbf{G}_\gamma$ is inner (condition (7.27) is easily checked in this situation). Since the system matrix of $\mathbf{G}_{\gamma 21}^{-1}$ is $\mathbf{A}_{F_\infty} + \gamma^{-2}\mathbf{B}_1\mathbf{B}_1^T\mathbf{X}_\infty$, it follows that $\mathbf{G}_{\gamma 21}^{-1} \in \mathcal{RH}_\infty$ (compare again the proof of Lemma 9.2.1).

Lemma 9.2.3. *Suppose* $\mathbf{H}_\infty \in dom(Ric)$ *and* $\mathbf{X}_\infty = Ric(\mathbf{H}_\infty) \geq \mathbf{0}$. *Then* $\mathbf{K}$ *is admissible for* $\mathbf{P}$ *and* $\| \mathbf{F}_{zw} \|_\infty < \gamma$ *if and only if* $\mathbf{K}$ *is admissible for* $\mathbf{G}_{tmp}$ *and* $\| \mathbf{F}_{vr} \|_\infty < \gamma$.

Proof. Let the realization of $\mathbf{K}$ be stabilizable and detectable. We show that the conditions of Lemma 6.5.2 are fulfilled for the feedback system $\mathbf{F}_{vr}$. This is the case since the matrix

$$\begin{pmatrix} \mathbf{A}_{tmp} - \lambda\mathbf{I} & \mathbf{B}_1 \\ \mathbf{C}_2 & \mathbf{D}_{21} \end{pmatrix} = \begin{pmatrix} \mathbf{A} - \lambda\mathbf{I} & \mathbf{B}_1 \\ \mathbf{C}_2 & \mathbf{D}_{21} \end{pmatrix} \begin{pmatrix} \mathbf{I} & \mathbf{0} \\ \gamma^{-2}\mathbf{B}_1^T\mathbf{X}_\infty & \mathbf{I} \end{pmatrix}$$

has full rank for all $\operatorname{Re}\lambda \geq 0$ and also since

$$\det \begin{pmatrix} \mathbf{A}_{tmp} - \lambda\mathbf{I} & \mathbf{B}_2 \\ -\mathbf{F}_\infty & \mathbf{I} \end{pmatrix} = \det(\mathbf{A}_{tmp} + \mathbf{B}_2\mathbf{F}_\infty - \lambda\mathbf{I}) \neq 0$$

for all $\operatorname{Re}\lambda \geq 0$ since $\mathbf{A}_{tmp} + \mathbf{B}_2\mathbf{F}_\infty$ is stable, as we have seen in the proof of Lemma 9.2.1. Thus $\mathbf{F}_{vr} \in \mathcal{RH}_\infty$ implies internal stability for the system $\mathbf{F}_{vr}$. The result follows from Lemma 9.2.2 and Theorem 9.2.2 with $\mathbf{G} = \mathbf{G}_\gamma$ and $\mathbf{Q} = \gamma^{-1}\mathbf{F}_{vr}$.

9.3 $\mathcal{H}_\infty$ Control with Full Information

9.3.1 Mixed Hankel-Toeplitz Operators

Concerning the solution of the FI problem, we still have to show that the existence of a feasible controller with $\| \mathbf{F}_{zw} \|_\infty < \gamma$ implies condition (i) of Theorem 9.2.1. This will be done in the proof of Theorem 9.3.1 by reformulating the inequality $\| \mathbf{F}_{zw} \|_\infty < \gamma$ equivalently. Thereby a certain operator Λ arises in a natural way and has to be investigated in some detail. Since the necessary analysis is rather complicated, it will be separated from the proof of Theorem 9.2.1 and presented in this section.

We start with a stable system

$$\dot{\mathbf{x}} = \mathbf{A}\,\mathbf{x} + \mathbf{B}\mathbf{u}$$

and define the so-called **controllability operator** $\Psi_c : \mathcal{L}_2(-\infty, 0] \to \mathbb{R}^n$ by

$$\Psi_c(u) = \int_{-\infty}^{0} e^{-\mathbf{A}\tau}\mathbf{B}\mathbf{u}(\tau)\,d\tau\,.$$

Let now the following transfer matrix with $\mathbf{B} = (\mathbf{B}_1 \quad \mathbf{B}_2)$ be given:

$$\mathbf{G}(s) = (\mathbf{G}_1(s) \quad \mathbf{G}_2(s)) = \left(\begin{array}{c|cc} \mathbf{A} & \mathbf{B}_1 & \mathbf{B}_2 \\ \hline \mathbf{C} & \mathbf{0} & \mathbf{0} \end{array}\right), \tag{9.9}$$

where the matrix $\mathbf{A}$ is also is assumed to be stable. Define the space

$$\mathcal{W} = \mathcal{L}_2(-\infty, 0] \times \mathcal{L}_2(-\infty, \infty)\,,$$

and for $\mathbf{w} = (\mathbf{w}_1, \mathbf{w}_2) \in \mathcal{W}$ put

$$\begin{aligned} \dot{\mathbf{x}}(t) &= \mathbf{A}\,\mathbf{x}(t) + \mathbf{B}_2\mathbf{w}_2(t) \qquad t \geq 0 \\ \mathbf{x}(0) &= \mathbf{x}_0 = \Psi_c(\mathbf{w}) \\ \mathbf{z} &= \mathbf{C}\mathbf{x}\,. \end{aligned} \tag{9.10}$$

We define now an operator by

$$\Lambda : \mathcal{W} \to \mathcal{L}_2[0, \infty)\,, \qquad \Lambda\mathbf{w} = \mathbf{z}\,.$$

For our analysis, it is necessary to give another description of Λ. Define a convolution operator $F : \mathcal{L}_2(-\infty, \infty) \to \mathcal{L}_2(-\infty, \infty)$ by

$$F(u)(t) = \int_{-\infty}^{t} \mathbf{C}\,e^{\mathbf{A}(t-\tau)}\mathbf{B}\mathbf{w}(\tau)\,d\tau$$

and similarly define F_2 with $\mathbf{B}_2$ instead of $\mathbf{B}$. Moreover, for $\mathbf{z} \in \mathcal{L}_2(-\infty, \infty)$ let

$$P_+(z)(t) = \begin{cases} z(t), & \text{if } t \geq 0, \\ 0, & \text{otherwise} \end{cases}, \qquad P_-(z)(t) = \begin{cases} z(t), & \text{if } t < 0, \\ 0, & \text{otherwise}. \end{cases}$$

Then it is immediately verified that ("$\circ$" denotes the composition of two mappings)

$$\Lambda\mathbf{w} = P_+ \circ F \circ P_-\mathbf{w} + P_+ \circ F_2 \circ P_+\mathbf{w}_2\,. \tag{9.11}$$

The first summand is denoted as a Hankel operator and the second as a Toeplitz operator. These operators are an interesting topic in themselves, but here we investigate only those properties which are needed for the proof of Theorem 9.3.1.

We need to describe Λ together with its adjoint in the frequency domain. Let $\mathcal{H}_2^\perp$ be the orthogonal complement of $\mathcal{H}_2$ in $\mathcal{L}_2(i\mathbb{R})$. Then the operators P_+, P_- translated in the frequency domain are

$$\pi^+ : \mathcal{L}_2(i\mathbb{R}) \to \mathcal{H}_2, \quad \pi^+ = \mathcal{L} \circ P_+ \circ \mathcal{F}^{-1}$$
$$\pi^- : \mathcal{L}_2(i\mathbb{R}) \to \mathcal{H}_2^\perp, \quad \pi^- = \mathcal{L}_- \circ P_- \circ \mathcal{F}^{-1}.$$

Here, $\mathcal{L}_-$ denotes the left-sided Laplace transform:

$$\mathcal{L}_-(f)(s) = \int_{-\infty}^{t} e^{-st} f(t)\,dt\,.$$

Note that P_+, P_-, π_+, π_- are projections. In particular, we have

$$z = P_+ z + P_- z, \quad \hat{z} = \pi_+ \hat{z} + \pi_- \hat{z}\,.$$

It is now easy to see that in the frequency domain the operator Λ is given by

$$\Gamma : \mathcal{H}_2^\perp \times \mathcal{L}_2(i\mathbb{R}) \to \mathcal{H}_2$$
$$\Gamma \hat{\mathbf{w}} = \pi_+ \mathbf{G} \pi_- \hat{\mathbf{w}} + \pi_+ \mathbf{G}_2 \pi_+ \hat{\mathbf{w}}_2\,.$$

To calculate the adjoint $\Gamma^* : \mathcal{H}_2 \to \mathcal{H}_2^\perp \times \mathcal{L}_2(i\mathbb{R})$ of Γ, let $\hat{\mathbf{w}} \in \mathcal{H}_2^\perp \times \mathcal{L}_2(i\mathbb{R})$ and $\hat{\mathbf{v}} \in \mathcal{H}_2$ be given. We note that in general the adjoint of a multiplication operator $\mathbf{G}$ is given by $\mathbf{G}^\sim$. Then (the bracket denotes the inner product of $\mathcal{H}_2$)

$$\begin{aligned}
\langle \hat{\mathbf{v}}, \Gamma \hat{\mathbf{w}} \rangle &= \langle \hat{\mathbf{v}}, \pi_+ (\mathbf{G}_1 \hat{\mathbf{w}}_1 \quad \mathbf{G}_2 \pi_- \hat{\mathbf{w}}_2) \rangle + \langle \hat{\mathbf{v}}, \pi_+ \mathbf{G}_2 \pi_+ \hat{\mathbf{w}}_2 \rangle \\
&= \langle \hat{\mathbf{v}}, \pi_+ \mathbf{G}_1 \hat{\mathbf{w}}_1 \rangle + \langle \hat{\mathbf{v}}, \pi_+ \mathbf{G}_2 \hat{\mathbf{w}}_2 \rangle \\
&= \langle \hat{\mathbf{v}}, \mathbf{G}_1 \hat{\mathbf{w}}_1 \rangle + \langle \hat{\mathbf{v}}, \mathbf{G}_2 \hat{\mathbf{w}}_2 \rangle \\
&= \left\langle \mathbf{G}_1^\sim \hat{\mathbf{v}}, \hat{\mathbf{w}}_1 \right\rangle + \left\langle \mathbf{G}_2^\sim \hat{\mathbf{v}}, \hat{\mathbf{w}}_2 \right\rangle \\
&= \left\langle \pi_- \mathbf{G}_1^\sim \hat{\mathbf{v}}, \hat{\mathbf{w}}_1 \right\rangle + \left\langle \mathbf{G}_2^\sim \hat{\mathbf{v}}, \hat{\mathbf{w}}_2 \right\rangle .
\end{aligned}$$

This implies

$$\Gamma^* = \begin{pmatrix} \pi_- \mathbf{G}_1^\sim \\ \mathbf{G}_2^\sim \end{pmatrix}.$$

The remaining task is to give equivalent conditions for $\|\Gamma\| < 1$. This will be done in several steps and we first have to take a closer look at the operator Ψ_c introduced at the beginning of this section.

Lemma 9.3.1. *Let the system* $(\mathbf{A}, \mathbf{B})$ *be controllable. Then the following assertions hold:*

(a) The matrix $\mathbf{X}_c = \Psi_c \Psi_c^*$ *is the controllability gramian of* $(\mathbf{A}, \mathbf{B})$ *and nonsingular.*

(b) *For* $\mathbf{x}_0 \in \mathbb{R}^n$ *define the set*

$$\{\mathbf{u} \in \mathcal{L}_2(-\infty,0] \,|\, \Psi_c(\mathbf{u}) = \mathbf{x}_0\}.$$

Then the function $\mathbf{u}_m = \Psi_c^* \mathbf{X}_c^{-1} \mathbf{x}_0$ *belongs to this set and is the element which minimizes the norm.*

(c) *The norm of the minimizing element is*

$$\| \mathbf{u}_m \|_2^2 = \mathbf{x}_0^T \mathbf{X}_c^{-1} \mathbf{x}_0 .$$

Proof. It is easy to see that the adjoint operator Ψ_c^* is given by

$$\Psi_c^*(\xi) = \begin{cases} \mathbf{B}^T e^{-\mathbf{A}^T t} \xi, & \text{if } t \le 0, \\ \mathbf{0}, & \text{otherwise}. \end{cases}$$

This implies

$$\mathbf{X}_c = \Psi_c \Psi_c^* = \int_{-\infty}^{0} e^{-\mathbf{A}\tau} \mathbf{B}\mathbf{B}^T e^{-\mathbf{A}^T \tau} \, d\tau = \int_{0}^{\infty} e^{\mathbf{A}t} \mathbf{B}\mathbf{B}^T e^{\mathbf{A}^T t} \, dt .$$

This is the controllability gramian. Hence $\mathbf{X}_c > \mathbf{0}$ since $(\mathbf{A},\mathbf{B})$ is controllable and (a) is shown.

For the proof of (b), we note first that $\mathbf{u}_m$ lies in the defined set because of

$$\Psi_c \mathbf{u}_m = \Psi_c \Psi_c^* \mathbf{X}_c^{-1} \mathbf{x}_0 = \mathbf{x}_0 .$$

Define now the operator $P = \Psi_c^* \mathbf{X}_c^{-1} \Psi_c$. Obviously $P^2 = P$ and therefore (with an arbitrary $\mathbf{u} \in \mathcal{L}_2(-\infty,0]$)

$$\begin{aligned} \| \mathbf{u} \|_2^2 &= \| P\mathbf{u} + (\mathbf{u} - P\mathbf{u}) \|_2^2 = \| P\mathbf{u} \|_2^2 + \| \mathbf{u} - P\mathbf{u} \|_2^2 + 2\langle P\mathbf{u}, \mathbf{u} - P\mathbf{u} \rangle \\ &= \| P\mathbf{u} \|_2^2 + \| \mathbf{u} - P\mathbf{u} \|_2^2 \ge \| P\mathbf{u} \|_2^2 . \end{aligned}$$

Let $\mathbf{u} \in \mathcal{L}_2(-\infty,0]$ with $\Psi_c(\mathbf{u}) = \mathbf{x}_0$ be given. Then

$$P\mathbf{u} = \Psi_c^* \mathbf{X}_c^{-1} \Psi_c \mathbf{u} = \Psi_c^* \mathbf{X}_c^{-1} \mathbf{x}_0 = \mathbf{u}_m$$

and the preceding inequality shows $\| \mathbf{u} \|_2 \ge \| \mathbf{u}_0 \|_2$.

Assertion (c) follows from

$$\begin{aligned} \| \mathbf{u}_m \|_2^2 &= \langle \mathbf{X}_c^{-1} \mathbf{x}_0, \Psi_c \Psi_c^* \mathbf{X}_c^{-1} \mathbf{x}_0 \rangle \\ &= \langle \mathbf{X}_c^{-1} \mathbf{x}_0, \Psi_c \Psi_c^* \mathbf{X}_c^{-1} \mathbf{x}_0 \rangle \\ &= \mathbf{x}_0^T \mathbf{X}_c^{-1} \mathbf{x}_0 . \end{aligned}$$

Remark. If $(\mathbf{A},\mathbf{B})$ is not controllable, then it can be shown by using the decomposition of Theorem 5.3.4 that the infimum in (b) is achieved by $\mathbf{x}_0^T\mathbf{y}_0$ if $\mathbf{y}_0$ is a solution of $\mathbf{X}_c\mathbf{y}_0=\mathbf{x}_0$ (supposed that the set is not empty).

The operator $\Psi_o:\mathbb{R}^n\to\mathcal{L}_2[0,\infty)$

$$\Psi_o(\mathbf{x}_0)=\begin{cases}\mathbf{C}\,e^{\mathbf{A}t}\mathbf{x}_0, & \text{if } t\geq 0,\\ \mathbf{0}, & \text{otherwise},\end{cases}$$

is denoted as **observability operator**. The observability gramian is

$$\mathbf{X}_o=\int_0^\infty e^{\mathbf{A}^T\tau}\mathbf{C}^T\mathbf{C}e^{\mathbf{A}\tau}d\tau=\Psi_o^*\Psi_o.$$

We consider now the first summand in (9.11), which is a Hankel operator:

$$\Lambda_H=P_+\circ F\circ P_-.$$

From the above considerations it follows immediately that Λ_H can be decomposed as :

$$\Lambda_H=\Psi_o\Psi_c.$$

Let now $\mathbf{z}(\mathbf{w})=\mathbf{z}$ be the output of the system (with a stable matrix $\mathbf{A}$)

$$\dot{\mathbf{x}}=\mathbf{A}\mathbf{x}+\mathbf{B}\mathbf{w},\quad \mathbf{x}(0)=\mathbf{x}_0$$
$$\mathbf{z}=\mathbf{C}\mathbf{x},$$

let $\mathbf{G}(s)$ be its transfer matrix and let $\mathbf{H}$ be the following Hamiltonian matrix:

$$\mathbf{G}(s)=\left(\begin{array}{c|c}\mathbf{A}&\mathbf{B}\\\hline\mathbf{C}&\mathbf{0}\end{array}\right),\quad \mathbf{H}=\begin{pmatrix}\mathbf{A}&\mathbf{B}\mathbf{B}^T\\-\mathbf{C}^T\mathbf{C}&-\mathbf{A}^T\end{pmatrix}.$$

Since $\mathbf{A}$ is stable, we can apply Theorem 8.1.6 and the following lemma holds.

Lemma 9.3.3. *With the notations defined above let* $\mathbf{X}=Ric(\mathbf{H})$ *and suppose* $\|\mathbf{G}\|_\infty<1$. *Then*

$$\sup_{\mathbf{w}\in\mathcal{L}_2[0,\infty)}\left(\|\mathbf{z}(\mathbf{w})\|_2^2-\|\mathbf{w}\|_2^2\right)=\mathbf{x}_0^T\mathbf{X}\mathbf{x}_0.$$

Proof. Using the Riccati equation, we get

$$\frac{d}{dt}(\mathbf{x}^T(t)\mathbf{X}\mathbf{x}(t))=\dot{\mathbf{x}}^T(t)\mathbf{X}\mathbf{x}(t)+\mathbf{x}^T(t)\mathbf{X}\dot{\mathbf{x}}(t)$$

$$=\mathbf{x}^T(t)(\mathbf{A}^T\mathbf{X}+\mathbf{X}\mathbf{A})\mathbf{x}(t)+2\mathbf{w}^T(t)\mathbf{B}^T\mathbf{X}\mathbf{x}(t)$$

$$= -\mathbf{x}^T(t)(\mathbf{X}\mathbf{B}\mathbf{B}^T\mathbf{X} + \mathbf{C}^T\mathbf{C})\mathbf{x}(t) + 2\mathbf{w}^T(t)\mathbf{B}^T\mathbf{X}\mathbf{x}(t)$$
$$= -\| \mathbf{z}(\mathbf{w})(t) \|^2 + \| \mathbf{w}(t) \|^2 - \| \mathbf{w}(t) - \mathbf{B}^T\mathbf{X}\mathbf{x}(t) \|^2 .$$

Because $\mathbf{A}$ is stable, $\mathbf{x}$ is in $\mathcal{L}_2[0,\infty)$ and hence this equation can be integrated from 0 to ∞:

$$\| \mathbf{z}(\mathbf{w}) \|_2^2 - \| \mathbf{w} \|_2^2 = \mathbf{x}_0^T\mathbf{X}\mathbf{x}_0 - \| \mathbf{w} - \mathbf{B}^T\mathbf{X}\mathbf{x} \|_2^2 \le \mathbf{x}_0^T\mathbf{X}\mathbf{x}_0 . \tag{9.12}$$

Define $\mathbf{x}$ as the solution of

$$\dot{\mathbf{x}} = (\mathbf{A} + \mathbf{B}\mathbf{B}^T\mathbf{X})\mathbf{x}, \quad \mathbf{x}(0) = \mathbf{x}_0 .$$

Then $\mathbf{x}$ can be interpreted as the solution of the homogeneous system for the disturbance $\mathbf{w} = \mathbf{B}^T\mathbf{X}\mathbf{x}$. By Theorem 8.1.2, $\mathbf{A} + \mathbf{B}\mathbf{B}^T\mathbf{X}$ is stable. Therefore, $\mathbf{x} \in \mathcal{L}_2[0,\infty)$ and thus $\mathbf{w}$ is a feasible perturbation which leads to equality in (9.12) and the lemma is shown.

In the proof of the next lemma, we use the following notation:

$$\mathcal{L}_{2+} = \mathcal{L}_2[0,\infty), \quad \mathcal{L}_{2-} = \mathcal{L}_2(-\infty,0] .$$

Lemma 9.3.4. *Let* $\mathbf{G}(s)$ *be as in* (9.9). *Then* $\| \Lambda \| < 1$ *holds if and only if the following conditions are fulfilled:*

(i) $\mathbf{H}_W \in dom(Ric)$ and $\mathbf{W} = Ric(\mathbf{H}_W) \ge \mathbf{0}$ with

$$\mathbf{H}_W = \begin{bmatrix} \mathbf{A} & \mathbf{B}_2\mathbf{B}_2^T \\ -\mathbf{C}^T\mathbf{C} & -\mathbf{A}^T \end{bmatrix} .$$

(ii) $\rho(\mathbf{W}\mathbf{X}_c) < 1$.

Proof. We suppose first that $(\mathbf{A},\mathbf{B})$ is controllable. From the definition of $\mathbf{G}_2$ the inequality $\| \mathbf{G}_2 \|_\infty < \| \Lambda \|$ is immediate. Hence $\| \Lambda \| < 1$ implies $\| \mathbf{G}_2 \|_\infty < 1$ and by Theorem 8.1.6 assertion (i). It remains to show that if (i) holds, then $\| \Lambda \| < 1$ is equivalent to (ii). We use the representation (9.10) of $\mathbf{z} = \Lambda\mathbf{w}$. Applying Lemma 9.3.1 and Lemma 9.3.3 we obtain

$$\| \Gamma \| < 1 \quad \Leftrightarrow \quad 0 > \sup\left\{ \| \mathbf{z} \|_{\mathcal{L}_{2+}} - \| \mathbf{w} \|_{\mathcal{W}} \;\middle|\; \mathbf{w} \in \mathcal{W} \right\}$$
$$= \sup\left\{ \| \mathbf{z} \|_{\mathcal{L}_{2+}} - \| P_+\mathbf{w} \|_{\mathcal{L}_{2+}} - \| P_-\mathbf{w} \|_{\mathcal{L}_{2-}} \;\middle|\; \mathbf{w} \in \mathcal{W} \right\}$$
$$= \sup\Big\{ \| \mathbf{z} \|_{\mathcal{L}_{2+}} - \| \mathbf{w}_{2+} \|_{\mathcal{L}_{2+}} - \| \mathbf{w}_- \|_{\mathcal{L}_{2-}} \;\Big|\; \mathbf{w}_{2+} \in \mathcal{L}_{2+}, \mathbf{w}_- \in \mathcal{L}_{2-},$$
$$\mathbf{x}_0 \in \mathbb{R}^n, \ \mathbf{x}_0 = \Psi_c(\mathbf{w}_-) \Big\}$$

$$= \sup_{\mathbf{x}_0 \in \mathbb{R}^n} \Big\{ \sup\{ \| \mathbf{z} \|_{\mathcal{L}_{2+}} - \| \mathbf{w}_{2+} \|_{\mathcal{L}_{2+}} \mid \mathbf{w}_{2+} \in \mathcal{L}_{2+} \}$$

$$- \inf \{ \| \mathbf{w}_- \|_{\mathcal{L}_{2-}} \mid \mathbf{w}_- \in \mathcal{L}_{2-}, \mathbf{x}_0 = \Psi_c(\mathbf{w}) \} \Big\}$$

$$= \sup_{\mathbf{x}_0 \in \mathbb{R}^n} \{ \mathbf{x}_0^T \mathbf{W} \mathbf{x}_0 - \mathbf{x}_0^T \mathbf{X}_c^{-1} \mathbf{x}_0 \}$$

$$= \sup_{\mathbf{x}_1 \in \mathbb{R}^n} \{ \mathbf{x}_1^T \mathbf{X}_c^{1/2} \mathbf{W} \mathbf{X}_c^{1/2} \mathbf{x}_1 - \mathbf{x}_1^T \mathbf{x}_1 \}.$$

We have

$$\rho(\mathbf{W}\mathbf{X}_c) = \rho(\mathbf{X}_c^{1/2}\mathbf{W}\mathbf{X}_c^{1/2}).$$

Hence, condition (ii) is equivalent to

$$\sup_{\mathbf{x}_1 \in \mathbb{R}^n} \{ \mathbf{x}_1^T \mathbf{X}_c^{1/2} \mathbf{W} \mathbf{X}_c^{1/2} \mathbf{x}_1 - \mathbf{x}_1^T \mathbf{x}_1 \} < 0,$$

which shows the equivalence of $\| \Lambda \| < 1$ and condition (ii). If $(\mathbf{A}, \mathbf{B})$ is not controllable, then in the above calculation only those $\mathbf{x}_0$ occur which are an element of $\operatorname{Im} \Psi_c$ and it is possible to generalize the proof to the uncontrollable case.

Remark. Lemma 9.3.4 gives also a condition for $\| \Gamma^* \| < 1$, since we have

$$\| \Lambda \| = \| \Gamma \| = \| \Gamma^* \|. \tag{9.13}$$

9.3.2 Proof of the Characterization Theorem for Full Information

As in Sect. 8.2.3, we consider the following FI problem:

$$\dot{\mathbf{x}} = \mathbf{A}\mathbf{x} + \mathbf{B}_1\mathbf{w} + \mathbf{B}_2\mathbf{u}$$
$$\mathbf{z} = \mathbf{C}_1\mathbf{x} + \mathbf{D}_{12}\mathbf{u}$$
$$\mathbf{y} = \begin{pmatrix} \mathbf{x} \\ \mathbf{w} \end{pmatrix}.$$

Then

$$\mathbf{C}_2 = \begin{pmatrix} \mathbf{I} \\ \mathbf{0} \end{pmatrix}, \quad \mathbf{D}_{21} = \begin{pmatrix} \mathbf{0} \\ \mathbf{I} \end{pmatrix}.$$

The remaining assumptions are

(A1') $(\mathbf{C}_1, \mathbf{A})$ *is detectable.*

(A2') $(\mathbf{A}, \mathbf{B}_2)$ *is stabilizable.*

(A3) $\mathbf{D}_{12}^T\mathbf{C}_1 = \mathbf{0}$ *and* $\mathbf{D}_{12}^T\mathbf{D}_{12} = \mathbf{I}$.

For the moment, we assume that $(\mathbf{C}_1, \mathbf{A})$ is even observable. Put

$$\mathbf{H}_2 = \begin{pmatrix} \mathbf{A} & -\mathbf{B}_2\mathbf{B}_2^T \\ -\mathbf{C}_1^T\mathbf{C}_1 & -\mathbf{A}^T \end{pmatrix}, \quad \mathbf{H}_\infty = \begin{pmatrix} \mathbf{A} & \mathbf{B}_1\mathbf{B}_1^T - \mathbf{B}_2\mathbf{B}_2^T \\ -\mathbf{C}_1^T\mathbf{C}_1 & -\mathbf{A}^T \end{pmatrix}.$$

Then our assumptions imply that $\mathbf{H}_2 \in dom(Ric)$ and $\mathbf{X}_2 = Ric(\mathbf{H}_2) > \mathbf{0}$ (cf. Theorem 8.1.4). Let $\mathbf{D}_\perp$ be such that the matrix $(\mathbf{D}_{12} \;\; \mathbf{D}_\perp)$ is orthogonal. Define the matrices

$$\mathbf{F}_2 = -\mathbf{B}_2^T\mathbf{X}_2, \quad \mathbf{A}_F = \mathbf{A} + \mathbf{B}_2\mathbf{F}_2, \quad \mathbf{C}_{1F} = \mathbf{C}_1 + \mathbf{D}_{12}\mathbf{F}_2,$$

$$\mathbf{G}_c(s) = \left(\begin{array}{c|c} \mathbf{A}_F & \mathbf{I} \\ \hline \mathbf{C}_{1F} & \mathbf{0} \end{array}\right), \quad \mathbf{U}(s) = \left(\begin{array}{c|c} \mathbf{A}_F & \mathbf{B}_2 \\ \hline \mathbf{C}_{1F} & \mathbf{D}_{12} \end{array}\right), \quad \mathbf{U}_\perp(s) = \left(\begin{array}{c|c} \mathbf{A}_F & -\mathbf{X}_2^{-1}\mathbf{C}_1^T\mathbf{D}_\perp \\ \hline \mathbf{C}_{1F} & \mathbf{D}_\perp \end{array}\right).$$

Then, as in Sect. 8.2.2, the variable transform

$$\mathbf{v} = \mathbf{u} - \mathbf{F}_2\mathbf{x}$$

yields the representation

$$\hat{\mathbf{z}} = \mathbf{G}_c\mathbf{B}_1\hat{\mathbf{w}} + \mathbf{U}\hat{\mathbf{v}}. \tag{9.14}$$

The reason for introducing the transfer function $\mathbf{U}_\perp(s)$ will become clear in the proof of Theorem 9.3.1.

Lemma 9.3.5. $(\mathbf{U} \;\; \mathbf{U}_\perp)$ *is square and inner and*

$$\mathbf{G}_c^\sim(\mathbf{U} \;\; \mathbf{U}_\perp) = \left(\begin{array}{c|cc} \mathbf{A}_F & \mathbf{B}_2 & \mathbf{X}_2^{-1}\mathbf{C}_1^T\mathbf{D}_\perp \\ \hline \mathbf{X}_2 & \mathbf{0} & \mathbf{0} \end{array}\right) \in \mathcal{RH}_2.$$

Proof. This is shown by a direct, but somewhat tedious computation using the formulas for systems connections of Sect. 5.5 and by applying Lemma 7.6.2. First, the series connections have to be transformed to block diagonal form by means of the similarity transforms

$$\begin{pmatrix} \mathbf{I} & \mathbf{0} \\ \mathbf{X}_2 & \mathbf{I} \end{pmatrix}, \quad \begin{pmatrix} \mathbf{I} & \mathbf{0} \\ -\mathbf{X}_2 & \mathbf{I} \end{pmatrix}.$$

Here, one has to use the property that $\mathbf{X}_2$ is a solution of a Riccati equation. Then the uncontrollable states have to be eliminated.

In the proof of the next theorem, we need the following set:

$$\mathcal{BH}_2 = \{\mathbf{w} \in \mathcal{H}_2 \mid \| \mathbf{w} \|_2 \le 1\}.$$

Theorem 9.3.1. *Let the assumptions (A1'), (A2') and (A3) be fulfilled. Then there exists an admissible controller for the FI problem with* $\| \mathbf{F}_{zw} \|_\infty < \gamma$ *if and only if the following condition holds:*

(i) $\mathbf{H}_\infty \in dom(Ric)$ *and* $\mathbf{X}_\infty = Ric(\mathbf{H}_\infty) \ge \mathbf{0}$.

If (i) is true, then such a controller is given by

$$\mathbf{K}(s) = \mathbf{F}_\infty \quad \textit{with} \quad \mathbf{F}_\infty = -\mathbf{B}_2^T \mathbf{X}_\infty .$$

Proof. „$\Rightarrow$": Without restriction of generality, suppose $\gamma = 1$. Moreover, without loss of generality it is also possible to assume that $(\mathbf{C}_1, \mathbf{A})$ is observable instead of detectable. To see this, one has to use a change of coordinates using the normal form of Theorem 5.4.2 (cf. Zhou, Doyle, and Glover [112], p. 426).

According to Theorem 7.2.3, the condition $\| \mathbf{F}_{zw} \|_\infty < 1$ is equivalent to

$$\sup_{\hat{\mathbf{w}} \in \mathcal{BH}_2} \| \hat{\mathbf{z}} \|_2 = \sup_{\hat{\mathbf{w}} \in \mathcal{BH}_2} \| \mathbf{F}_{zw} \hat{\mathbf{w}} \|_2 < 1 .$$

This implies

$$\sup_{\hat{\mathbf{w}} \in \mathcal{BH}_2} \inf_{\hat{\mathbf{u}} \in \mathcal{H}_2} \| \hat{\mathbf{z}} \|_2 < 1 ,$$

since each control $\mathbf{u}$ which is generated by an internally stabilizing controller is in $\mathcal{L}_2[0,\infty)$. With the variable transform $\mathbf{v} = \mathbf{u} - \mathbf{F}_2 \mathbf{x}$ using (9.14), we can write

$$\sup_{\hat{\mathbf{w}} \in \mathcal{BH}_2} \inf_{\hat{\mathbf{v}} \in \mathcal{H}_2} \| \hat{\mathbf{z}} \|_2 < 1 . \tag{9.15}$$

The idea is now to calculate the infimum in (9.15) for fixed $\hat{\mathbf{w}}$. This would be trivial if we had in (9.14) $\mathbf{I}$ instead of $\mathbf{U}$. Multiplication of (9.14) with $\mathbf{U}^\sim$ isolates $\hat{\mathbf{v}}$, but $\mathbf{U}^\sim$ is not norm preserving. So we apply $(\mathbf{U} \quad \mathbf{U}_\perp)$ to (9.14), since $(\mathbf{U} \quad \mathbf{U}_\perp)$ is norm preserving: $\| \hat{\mathbf{z}} \|_2 = \| (\mathbf{U} \quad \mathbf{U}_\perp)^\sim \hat{\mathbf{z}} \|_2$. This is true because by Lemma 9.3.5, $(\mathbf{U} \quad \mathbf{U}_\perp)$ is square and inner. Then the following equation results:

$$(\mathbf{U} \quad \mathbf{U}_\perp)^\sim \hat{\mathbf{z}} = \begin{pmatrix} \mathbf{U}^\sim \mathbf{G}_c \mathbf{B}_1 \hat{\mathbf{w}} + \hat{\mathbf{v}} \\ \mathbf{U}_\perp^\sim \mathbf{G}_c \mathbf{B}_1 \hat{\mathbf{w}} \end{pmatrix} = \begin{pmatrix} \pi_-(\mathbf{U}^\sim \mathbf{G}_c \mathbf{B}_1 \hat{\mathbf{w}}) + \pi_+(\mathbf{U}^\sim \mathbf{G}_c \mathbf{B}_1 \hat{\mathbf{w}} + \hat{\mathbf{v}}) \\ \mathbf{U}_\perp^\sim \mathbf{G}_c \mathbf{B}_1 \hat{\mathbf{w}} \end{pmatrix}.$$

Hence

$$\sup_{\hat{\mathbf{w}} \in \mathcal{BH}_2} \inf_{\hat{\mathbf{v}} \in \mathcal{H}_2} \| \hat{\mathbf{z}} \|_2 = \sup_{\hat{\mathbf{w}} \in \mathcal{BH}_2} \inf_{\hat{\mathbf{v}} \in \mathcal{H}_2} \left\| \begin{pmatrix} \pi_-(\mathbf{U}^\sim \mathbf{G}_c \mathbf{B}_1 \hat{\mathbf{w}}) + \pi_+(\mathbf{U}^\sim \mathbf{G}_c \mathbf{B}_1 \hat{\mathbf{w}} + \hat{\mathbf{v}}) \\ \mathbf{U}_\perp^\sim \mathbf{G}_c \mathbf{B}_1 \hat{\mathbf{w}} \end{pmatrix} \right\|_2 .$$

For fixed $\hat{\mathbf{w}}$, the infimum on the right-hand side is attained by the function

$\hat{\mathbf{v}} = -\pi_+(\mathbf{U}^\sim \mathbf{G}_c \mathbf{B}_1 \hat{\mathbf{w}})$. With

$$\mathbf{G}_1 = \mathbf{B}_1^T \mathbf{G}_c^\sim \mathbf{U}, \quad \mathbf{G}_2 = \mathbf{B}_1^T \mathbf{G}_c^\sim \mathbf{U}_\perp,$$

we obtain

$$\begin{aligned} \sup_{\hat{\mathbf{w}} \in \mathcal{BH}_2} \inf_{\hat{\mathbf{v}} \in \mathcal{H}_2} \|\hat{\mathbf{z}}\|_2 &= \sup_{\hat{\mathbf{w}} \in \mathcal{BH}_2} \left\| \begin{pmatrix} \pi_-(\mathbf{U}^\sim \mathbf{G}_c \mathbf{B}_1 \hat{\mathbf{w}}) \\ \mathbf{U}_\perp^\sim \mathbf{G}_c \mathbf{B}_1 \hat{\mathbf{w}} \end{pmatrix} \right\|_2 \\ &= \sup_{\hat{\mathbf{w}} \in \mathcal{BH}_2} \left\| \begin{pmatrix} \pi_- \mathbf{G}_1^\sim \\ \mathbf{G}_2^\sim \end{pmatrix} \hat{\mathbf{w}} \right\|_2 \\ &= \sup_{\hat{\mathbf{w}} \in \mathcal{BH}_2} \|\Gamma^* \hat{\mathbf{w}}\|_2 = \|\Gamma^*\|. \end{aligned}$$

Here, Γ is the map defined in Sect. 9.3.1. Thus we get $\|\Lambda\| < 1$. The Hamilton matrix of Lemma 9.3.4 is given by

$$\mathbf{H}_W = \begin{pmatrix} \mathbf{A}_F & \mathbf{X}_2^{-1} \mathbf{C}_1^T \mathbf{C}_1 \mathbf{X}_2^{-1} \\ -\mathbf{X}_2 \mathbf{B}_1 \mathbf{B}_1^T \mathbf{X}_2 & -\mathbf{A}_F^T \end{pmatrix}$$

(note that $\mathbf{C}_1^T \mathbf{D}_\perp \mathbf{D}_\perp^T \mathbf{C}_1 = \mathbf{C}_1^T \mathbf{C}_1$). Lemma 9.3.4 gives now $\mathbf{H}_W \in dom(Ric)$ and $\mathbf{W} = Ric(\mathbf{H}_W) \geq \mathbf{0}$ and moreover $\rho(\mathbf{W}\mathbf{X}_c) < 1$, if $\mathbf{X}_c$ is the controllability Gramian of $(\mathbf{A}_F, \mathbf{B})$ (with $\mathbf{B} = (\mathbf{B}_2 \quad \mathbf{X}_2^{-1} \mathbf{C}_1^T \mathbf{D}_\perp)$). Using the Riccati equation for $\mathbf{X}_2$, it is easily verified that

$$\mathbf{A}_F \mathbf{X}_2^{-1} + \mathbf{X}_2^{-1} \mathbf{A}_F^T + \mathbf{B}\mathbf{B}^T = \mathbf{0}.$$

Hence $\mathbf{X}_c = \mathbf{X}_2^{-1}$ and therefore $\rho(\mathbf{W}\mathbf{X}_2^{-1}) < 1$. This in turn implies $\mathbf{X}_2 > \mathbf{W}$. Define

$$\mathbf{T} = \begin{pmatrix} -\mathbf{I} & \mathbf{X}_2^{-1} \\ -\mathbf{X}_2 & \mathbf{0} \end{pmatrix}.$$

With the Riccati equation for $\mathbf{X}_2$ it can be shown that the Hamiltonian matrices $\mathbf{H}_W$ and $\mathbf{H}_\infty$ are similar: $\mathbf{H}_\infty = \mathbf{T}\mathbf{H}_W\mathbf{T}^{-1}$. Moreover,

$$\chi_-(\mathbf{H}_\infty) = \mathbf{T}\chi_-(\mathbf{H}_W) = \mathbf{T}\,\mathrm{Im}\begin{pmatrix} \mathbf{I} \\ \mathbf{W} \end{pmatrix} = \mathrm{Im}\begin{pmatrix} \mathbf{I} - \mathbf{X}_2^{-1}\mathbf{W} \\ \mathbf{X}_2 \end{pmatrix}.$$

This implies $\mathbf{H}_\infty \in dom(Ric)$. Consequently, $\mathbf{X}_\infty = Ric(\mathbf{H}_\infty)$ is given by

$$\mathbf{X}_\infty = \mathbf{X}_2(\mathbf{I} - \mathbf{X}_2^{-1}\mathbf{W})^{-1} = \mathbf{X}_2(\mathbf{X}_2 - \mathbf{W})^{-1}\mathbf{X}_2 > \mathbf{0}.$$

„$\Leftarrow$": This is Lemma 9.2.1.

Remark. As in the $\mathcal{H}_2$ case, the FI controller does not make use of $\mathbf{w}$.

9.4 Proof of the Characterization Theorem for Output Feedback

Next, we consider, as in the $\mathcal{H}_2$ case, a disturbance feedforward problem:

$$\begin{aligned}\dot{\mathbf{x}} &= \mathbf{A}\,\mathbf{x} + \mathbf{B}_1\mathbf{w} + \mathbf{B}_2\mathbf{u}\\ \mathbf{z} &= \mathbf{C}_1\,\mathbf{x} + \mathbf{D}_{12}\mathbf{u}\\ \mathbf{y} &= \mathbf{C}_2\mathbf{x} + \mathbf{w}.\end{aligned}$$

As in Sect. 8.2.3, $\mathbf{A} - \mathbf{B}_1\mathbf{C}_2$ is assumed to be stable to get internal stability.

Theorem 9.4.1. *Let the assumptions (A1'), (A2') and (A3) of Sect. 9.3.2 be fulfilled and suppose* $\mathbf{A} - \mathbf{B}_1\mathbf{C}_2$ *to be stable. Then there exists an admissible controller for the DF problem with* $\|\mathbf{F}_{zw}\|_\infty < \gamma$ *if and only if the following condition holds:*

(i) $\quad \mathbf{H}_\infty \in dom(Ric)$ *and* $\mathbf{X}_\infty = Ric(\mathbf{H}_\infty) \geq \mathbf{0}$.

If (i) is true, then such a controller is given by

$$\mathbf{K}(s) = \left(\begin{array}{c|c} \mathbf{A} + \mathbf{B}_2\mathbf{F}_\infty - \mathbf{B}_1\mathbf{C}_2 & \mathbf{B}_1 \\ \hline \mathbf{F}_\infty & \mathbf{0} \end{array}\right).$$

Proof. Let $\|\mathbf{F}_{zw}\|_\infty < \gamma$. Because of (cf. Sect. 8.2.3)

$$\mathcal{F}_l(\mathbf{P}_{FI}, \mathbf{K}_{FI}) = \mathcal{F}_l(\mathbf{P}_{DF}, \mathbf{K}_{DF}) = \mathbf{F}_{zw},$$

$\|\mathcal{F}_l(\mathbf{P}_{FI}, \mathbf{K}_{FI})\| < \gamma$ follows and Theorem 9.3.1 gives (i). Let now (i) hold. Then $\mathbf{K}_{FI}(s) = \mathbf{F}_\infty$ is an admissible controller for $\mathbf{P}_{FI}$ with $\|\mathcal{F}_l(\mathbf{P}_{FI}, \mathbf{K}_{FI})\| < \gamma$, and $\mathbf{K}_{DF} = \mathcal{F}_l(\mathbf{G}_{DF}, \mathbf{K}_{FI})$ is an admissible controller for the DF problem with $\|\mathcal{F}_l(\mathbf{P}_{DF}, \mathbf{K}_{DF})\| < \gamma$. The representation of $\mathbf{K}_{DF}$ follows directly from the definition of $\mathbf{G}_{DF}$.

Next, we investigate the $\mathcal{H}_\infty$ output estimation problem, i.e. the $\mathcal{H}_\infty$ problem for the plant

$$\begin{aligned}\dot{\mathbf{x}} &= \mathbf{A}\,\mathbf{x} + \mathbf{B}_1\mathbf{w} + \mathbf{B}_2\mathbf{u}\\ \mathbf{z} &= \mathbf{C}_1\mathbf{x} + \mathbf{u}\\ \mathbf{y} &= \mathbf{C}_2\mathbf{x} + \mathbf{D}_{21}\mathbf{w}.\end{aligned}$$

The assumptions for OE are:

(A1'') $\quad (\mathbf{A}, \mathbf{B}_1)$ *is stabilizable and* $\mathbf{A} - \mathbf{B}_2\mathbf{C}_1$ *is stable.*

(A2') $\quad (\mathbf{C}_2, \mathbf{A})$ *is detectable.*

(A4) $\mathbf{B}_1\mathbf{D}_{21}^T = \mathbf{0}$ *and* $\mathbf{D}_{21}\mathbf{D}_{21}^T = \mathbf{I}$.

Taking into account that the OE problem is dual to a DF problem which is given by (cf. Sect. 8.2.3)

$$\dot{\mathbf{x}} = \mathbf{A}^T\mathbf{x} + \mathbf{C}_1^T\mathbf{w} + \mathbf{C}_2^T\mathbf{u}$$
$$\mathbf{z} = \mathbf{B}_1^T\mathbf{x} + \mathbf{D}_{21}^T\mathbf{u}$$
$$\mathbf{y} = \mathbf{B}_2^T\mathbf{x} + \mathbf{u},$$

the application of Theorem 9.4.1 directly gives the solution of the OE problem.

Theorem 9.4.2. *Let the assumptions (A1''), (A2') and (A4) be fulfilled. Then there exists an admissible controller for the OE problem with* $\|\mathbf{F}_{zw}\|_\infty < \gamma$ *if and only if the following condition holds:*

(i) $\mathbf{J}_\infty \in dom(Ric)$ *and* $\mathbf{Y}_\infty = Ric(\mathbf{J}_\infty) \geq \mathbf{0}$.

If (i) is true, then such a controller is given by

$$\mathbf{K}(s) = \left(\begin{array}{c|c} \mathbf{A} + \mathbf{L}_\infty\mathbf{C}_2 - \mathbf{B}_2\mathbf{C}_1 & \mathbf{L}_\infty \\ \hline \mathbf{C}_1 & \mathbf{0} \end{array}\right).$$

The OE problem has an interesting application in the special case $\mathbf{B}_2 = \mathbf{0}$, $\mathbf{D}_{12} = \mathbf{0}$. We write

$$\dot{\mathbf{x}} = \mathbf{A}\mathbf{x} + \mathbf{B}_1\mathbf{w}$$
$$\mathbf{e} = \mathbf{C}_1\mathbf{x} - \tilde{\mathbf{z}}$$
$$\mathbf{y} = \mathbf{C}_2\mathbf{x} + \mathbf{D}_{21}\mathbf{w}.$$

The corresponding $\mathcal{H}_\infty$ problem is denoted as the $\mathcal{H}_\infty$ filter problem (compare the $\mathcal{H}_2$ estimator in Sect. 8.3.1). It may be formulated as follows (with $\mathbf{z} = \mathbf{C}_1\mathbf{x}$).

For given $\gamma > 0$ *find a filter* $\mathbf{F}(s) \in \mathcal{RH}_\infty$ *such that with* $\hat{\mathbf{z}} = \mathbf{F}(s)\hat{\mathbf{y}}$

$$\sup_{\mathbf{w}\in\mathcal{L}_2[0,\infty)} \frac{\|\mathbf{z} - \tilde{\mathbf{z}}\|_2}{\|\mathbf{w}\|_2} < \gamma.$$

The formula of Theorem 9.4.2 gives the following representation of $\mathbf{F}(s)$:

$$\dot{\tilde{\mathbf{x}}} = \mathbf{A}\tilde{\mathbf{x}} + \mathbf{L}_\infty(\mathbf{C}_2\tilde{\mathbf{x}} - \mathbf{y})$$
$$\tilde{\mathbf{z}} = \mathbf{C}_1\tilde{\mathbf{x}}.$$

In is possible to add a control to the plant dynamics. Then with the plant

$$\begin{aligned}\dot{\mathbf{x}} &= \mathbf{A}\,\tilde{\mathbf{x}} + \mathbf{B}_1\mathbf{w} + \mathbf{B}_2\mathbf{u}\\ \mathbf{z} &= \mathbf{C}_1\tilde{\mathbf{x}}\\ \mathbf{y} &= \mathbf{C}_2\tilde{\mathbf{x}} + \mathbf{D}_{21}\mathbf{w}\,,\end{aligned}$$

there is associated the $\mathcal{H}_\infty$ observer

$$\begin{aligned}\dot{\tilde{\mathbf{x}}} &= \mathbf{A}\,\tilde{\mathbf{x}} + \mathbf{B}_2\mathbf{u} + \mathbf{L}_\infty(\mathbf{C}_2\tilde{\mathbf{x}} - \mathbf{y})\\ \tilde{\mathbf{z}} &= \mathbf{C}_1\tilde{\mathbf{x}}\,.\end{aligned}$$

For the proof of Theorem 9.2.1, we need the following lemma.

Lemma 9.4.1. *Suppose there exists an admissible controller such that* $\|\mathbf{F}_{zw}\|_\infty < \gamma$ *holds. Then (i) and (ii) of Theorem 9.2.1 are valid.*

Proof. Let $\mathbf{K}$ be an admissible controller for which $\|\mathbf{F}_{zw}\|_\infty < \gamma$. We can write $\mathbf{u} = \mathbf{K}(\mathbf{C}_2\mathbf{x} + \mathbf{D}_{21}\mathbf{w})$. Thus the controller $\mathbf{K}_{FI} = \mathbf{K}(\mathbf{C}_2 \quad \mathbf{D}_{21})$ solves an FI problem and (i) follows from Theorem (9.3.1).

Define now the plant

$$\begin{aligned}\dot{\mathbf{x}} &= \mathbf{A}\mathbf{x} + \mathbf{B}_1\mathbf{w} + \mathbf{u}_1\\ \mathbf{z} &= \mathbf{C}_1\mathbf{x} + \mathbf{u}_2\\ \mathbf{y} &= \mathbf{C}_2\mathbf{x} + \mathbf{D}_{21}\mathbf{w}\,.\end{aligned}$$

This is a so-called **full control (FC)** problem. It is dual to a FI problem and vice versa. The controller $\mathbf{K}$ gives also a controller for the FC problem:

$$\begin{pmatrix}\mathbf{u}_1\\ \mathbf{u}_2\end{pmatrix} = \begin{pmatrix}\mathbf{B}_2\\ \mathbf{D}_{12}\end{pmatrix}\mathbf{K}\,\mathbf{y}\,.$$

By passing to the dual problem, it is possible to apply again Theorem 9.3.1 and this leads to (ii).

We are now in a position to prove Theorem 9.2.1. The main idea is to apply Theorem 9.4.2 to $\mathbf{G}_{tmp}$, which is an OE problem. Here we use the fact that under the assumption (i) the controller $\mathbf{K}$ is feasible for $\mathbf{P}$ and $\|\mathbf{F}_{zw}\|_\infty < \gamma$ if and only if $\mathbf{K}$ is feasible for $\mathbf{G}_{tmp}$ and $\|\mathbf{F}_{vr}\|_\infty < \gamma$ (Lemma 9.2.3).

Proof of Theorem 9.2.1. Since $\mathbf{G}_{tmp}$ is an OE problem, we have to apply Theorem 9.4.2. Hence the assumptions of this theorem have to be verified. These are:

(a1) $(\mathbf{A}_{tmp}, \mathbf{B}_1)$ is stabilizable and $\mathbf{A}_{tmp} + \mathbf{B}_2\mathbf{F}_\infty$ is stable.

(a2) $(\mathbf{C}_2, \mathbf{A}_{tmp})$ is detectable.

(A4) $\mathbf{B}_1\mathbf{D}_{21}^T = \mathbf{0}$ and $\mathbf{D}_{21}\mathbf{D}_{21}^T = \mathbf{I}$.

Assumption (A4) and the stabilizability of $(\mathbf{A}_{tmp}, \mathbf{B}_1)$ directly follow from the corresponding assumptions of Theorem 9.2.1. The stability of

$$\mathbf{A}_{tmp} + \mathbf{B}_2\mathbf{F}_\infty = \mathbf{A} + (\gamma^{-2}\mathbf{B}_1\mathbf{B}_1^T - \mathbf{B}_2\mathbf{B}_2^T)\mathbf{X}_\infty$$

is a consequence of Theorem 8.1.2 if (i) holds. Note that the OE Hamilton matrix for $\mathbf{G}_{tmp}$ is given by

$$\mathbf{J}_{tmp} = \begin{pmatrix} \mathbf{A}_{tmp}^T & \gamma^{-2}\mathbf{F}_\infty^T\mathbf{F}_\infty - \mathbf{C}_2^T\mathbf{C}_2 \\ -\mathbf{B}_1\mathbf{B}_1^T & -\mathbf{A}_{tmp} \end{pmatrix}.$$

The following assertions are true:

(*) If $\mathbf{J}_{tmp} \in dom(Ric)$ and $\mathbf{Y}_{tmp} = Ric(\mathbf{J}_{tmp}) \geq \mathbf{0}$, then $(\mathbf{C}_2, \mathbf{A}_{tmp})$ is detectable.

(**) If there is a feasible controller for $\mathbf{G}_{tmp}$, then $(\mathbf{C}_2, \mathbf{A}_{tmp})$ is detectable.

To prove (*), we argue similar as in the proof of Lemma 9.2.1. With

$$\tilde{\mathbf{A}}_{tmp} = \mathbf{A}_{tmp} - \mathbf{Y}_{tmp}\mathbf{C}_2^T\mathbf{C}_2, \quad \mathbf{B}_e = \begin{pmatrix} \mathbf{B}_1 - \mathbf{Y}_{tmp}\mathbf{C}_2^T\mathbf{D}_{21} & \gamma^{-1}\mathbf{Y}_{tmp}\mathbf{F}_\infty^T \end{pmatrix}$$

the Riccati equation for $\mathbf{Y}_{tmp}$ can be written as a Lyapunov equation:

$$\mathbf{Y}_{tmp}\tilde{\mathbf{A}}_{tmp}^T + \tilde{\mathbf{A}}_{tmp}\mathbf{Y}_{tmp} + \mathbf{B}_e\mathbf{B}_e^T = \mathbf{0}.$$

Moreover, since the matrix

$$\mathbf{A}_{tmp}^T + (\gamma^{-2}\mathbf{F}_\infty^T\mathbf{F}_\infty - \mathbf{C}_2^T\mathbf{C}_2)\mathbf{Y}_{tmp} = \tilde{\mathbf{A}}_{tmp}^T + \gamma^{-2}\mathbf{F}_\infty^T\mathbf{F}_\infty\mathbf{Y}_{tmp}$$

is stable, the system $(\tilde{\mathbf{A}}_{tmp}, \mathbf{B}_e)$ is stabilizable. This implies that $\tilde{\mathbf{A}}_{tmp}$ is stable and consequently, $(\mathbf{C}_2, \mathbf{A}_{tmp})$ is detectable.

Assertion (**) is a direct consequence of Theorem 6.5.1.

We now show the asserted equivalence of Theorem 9.2.1.

„⇐“: Let (i) – (iii) hold. From the above discussion we conclude that assumption (a1) is fulfilled. We now prove (a2). Let

$$\mathbf{T} = \begin{pmatrix} \mathbf{I} & -\gamma^{-2}\mathbf{X}_\infty \\ \mathbf{0} & \mathbf{I} \end{pmatrix}.$$

Then using the Riccati equation for $\mathbf{X}_\infty$, we obtain $\mathbf{J}_\infty = \mathbf{T}^{-1}\mathbf{J}_{tmp}\mathbf{T}$. Thus $\mathbf{J}_{tmp}$ and $\mathbf{J}_\infty$ are similar. Consequently,

$$\chi_-(\mathbf{J}_{tmp}) = \mathbf{T}\chi_-(\mathbf{J}_\infty) = \mathbf{T}\,\mathrm{Im}\begin{pmatrix} \mathbf{I} \\ \mathbf{Y}_\infty \end{pmatrix} = \mathrm{Im}\begin{pmatrix} \mathbf{I}-\gamma^{-2}\mathbf{X}_\infty\mathbf{Y}_\infty \\ \mathbf{Y}_\infty \end{pmatrix}. \tag{9.16}$$

Hence from (iii) we get $\mathbf{I}-\gamma^{-2}\mathbf{X}_\infty\mathbf{Y}_\infty > \mathbf{0}$ and therefore $\mathbf{J}_{tmp} \in dom(Ric)$. For $\mathbf{Y}_{tmp} = Ric(\mathbf{J}_{tmp})$, the equation

$$\mathbf{Y}_{tmp} = \mathbf{Y}_\infty(\mathbf{I}-\gamma^{-2}\mathbf{X}_\infty\mathbf{Y}_\infty)^{-1} = (\mathbf{I}-\gamma^{-2}\mathbf{Y}_\infty\mathbf{X}_\infty)^{-1}\mathbf{Y}_\infty = \mathbf{Z}_\infty\mathbf{Y}_\infty \geq \mathbf{0}$$

follows. Hence (*) shows that $(\mathbf{C}_2, \mathbf{A}_{tmp})$ is detectable. Consequently, all assumptions for the OE problem $\mathbf{G}_{tmp}$ are fulfilled and by Theorem 9.4.2 there is a controller $\mathbf{K}$ which internally stabilizes $\mathbf{G}_{tmp}$ such that $\|\mathbf{F}_{vr}\|_\infty < \gamma$. It stabilizes also $\mathbf{P}$ with $\|\mathbf{F}_{zw}\|_\infty < \gamma$. One such controller is

$$\mathbf{K}(s) = \left(\begin{array}{c|c} \mathbf{A}+\gamma^{-2}\mathbf{B}_1\mathbf{B}_1^T\mathbf{X}_\infty - \mathbf{Y}_{tmp}\mathbf{C}_2^T\mathbf{C}_2 + \mathbf{B}_2\mathbf{F}_\infty & \mathbf{Y}_{tmp}\mathbf{C}_2^T \\ \hline \mathbf{F}_\infty & \mathbf{0} \end{array}\right)$$

which coincides with the controller of Theorem 9.2.1.

„⇒“: Let now $\mathbf{K}$ be a feasible controller with $\|\mathbf{F}_{zw}\|_\infty < \gamma$. Then by Lemma 9.4.1 we get $\mathbf{H}_\infty \in dom(Ric)$, $\mathbf{X}_\infty = Ric(\mathbf{H}_\infty) \geq \mathbf{0}$ and $\mathbf{J}_\infty \in dom(Ric)$, $\mathbf{Y}_\infty = Ric(\mathbf{J}_\infty) \geq \mathbf{0}$. Moreover, the detectability of $(\mathbf{C}_2, \mathbf{A}_{tmp})$ follows from (**). Hence the OE assumptions for $\mathbf{G}_{tmp}$ are fulfilled and from Theorem 9.4.2 we get $\mathbf{J}_{tmp} \in dom(Ric)$ and $\mathbf{Y}_{tmp} = Ric(\mathbf{J}_{tmp}) \geq \mathbf{0}$. Thus (9.16) shows now the invertibility of $\mathbf{I}-\gamma^{-2}\mathbf{X}_\infty\mathbf{Y}_\infty$ and it follows that

$$(\mathbf{I}-\gamma^{-2}\mathbf{Y}_\infty\mathbf{X}_\infty)^{-1}\mathbf{Y}_\infty = \mathbf{Y}_{tmp} \geq \mathbf{0}.$$

We have to show that this implies (iii).

Case 1: $\mathbf{Y}_\infty$ is invertible. Then the last equation implies $(\mathbf{I}-\gamma^{-2}\mathbf{Y}_\infty\mathbf{X}_\infty)^{-1} > \mathbf{0}$, which gives $\mathbf{I}-\gamma^{-2}\mathbf{Y}_\infty\mathbf{X}_\infty > \mathbf{0}$ and also $\mathbf{I}-\gamma^{-2}\mathbf{Y}_\infty^{1/2}\mathbf{X}_\infty\mathbf{Y}_\infty^{1/2} > \mathbf{0}$. Hence

$$\rho(\mathbf{X}_\infty\mathbf{Y}_\infty) = \rho(\mathbf{Y}_\infty^{1/2}\mathbf{X}_\infty\mathbf{Y}_\infty^{1/2}) < \gamma^2.$$

Case 2: $\mathbf{Y}_\infty$ is singular. Then there is an orthogonal matrix $\mathbf{U}$ such that

$$\mathbf{Y}_\infty = \mathbf{U}^T\begin{pmatrix} \mathbf{Y}_{11} & \mathbf{0} \\ \mathbf{0} & \mathbf{0} \end{pmatrix}\mathbf{U},$$

with $\mathbf{Y}_{11} > \mathbf{0}$. Let $\mathbf{U}\mathbf{X}_\infty\mathbf{U}^T$ be correspondingly partitioned:

$$\mathbf{U}\mathbf{X}_\infty\mathbf{U}^T = \begin{pmatrix} \mathbf{X}_{11} & \mathbf{X}_{12} \\ \mathbf{X}_{21} & \mathbf{X}_{22} \end{pmatrix}.$$

Then

$$\mathbf{Y}_{tmp} = \mathbf{U}^T \begin{pmatrix} \mathbf{I} - \gamma^{-2}\mathbf{Y}_{11}\mathbf{X}_{11} & -\mathbf{Y}_{11}\mathbf{X}_{12} \\ \mathbf{0} & \mathbf{I} \end{pmatrix}^{-1} \begin{pmatrix} \mathbf{Y}_{11} & \mathbf{0} \\ \mathbf{0} & \mathbf{0} \end{pmatrix} \mathbf{U}$$

$$= \mathbf{U}^T \begin{pmatrix} (\mathbf{I} - \gamma^{-2}\mathbf{Y}_{11}\mathbf{X}_{11})^{-1}\mathbf{Y}_{11} & \mathbf{0} \\ \mathbf{0} & \mathbf{0} \end{pmatrix} \mathbf{U} \geq \mathbf{0}.$$

This implies $(\mathbf{I} - \gamma^{-2}\mathbf{Y}_{11}\mathbf{X}_{11})^{-1}\mathbf{Y}_{11} \geq \mathbf{0}$ and, as in case 1, $\rho(\mathbf{X}_{11}\mathbf{Y}_{11}) < \gamma^2$ follows (note $\mathbf{Y}_{11} > \mathbf{0}$). Because of $\rho(\mathbf{X}_\infty \mathbf{Y}_\infty) = \rho(\mathbf{X}_{11}\mathbf{Y}_{11})$, the inequality $\rho(\mathbf{X}_\infty \mathbf{Y}_\infty) < \gamma^2$ again holds.

Remark. In the preceding theorems, we have always described only the so-called central controller. It is also possible to characterize all suboptimal controllers (cf. Zhou, Doyle, and Glover [112]).

9.5 General $\mathcal{H}_\infty$ Problem: Scaling and Loop Shifting

In this section, we want to show how the assumptions of Theorem 9.2.1 can be relaxed. First, we note that the stabilizability of $(\mathbf{A}, \mathbf{B}_1)$ and the detectability of $(\mathbf{C}_1, \mathbf{A})$ are not required, but if these conditions are skipped, then two assumptions concerning the invariant zeros of two subsystems have to be added (compare the remarks at the end of Sect. 8.1.1). It is also possible to skip the orthogonality relations in (A3) and (A4). In this way we are lead to the following weakened assumptions (they may be compared with that for the $\mathcal{H}_2$ problem in Sect. 8.2.1):

(A1') $(\mathbf{A}, \mathbf{B}_2)$ *is stabilizable and* $(\mathbf{C}_2, \mathbf{A})$ *is detectable.*

(A2') *The following matrix has full column rank for every* ω *:*

$$\begin{pmatrix} \mathbf{A} - i\omega\mathbf{I} & \mathbf{B}_2 \\ \mathbf{C}_1 & \mathbf{D}_{12} \end{pmatrix}.$$

(A3') *The following matrix has full row rank for every* ω *:*

$$\begin{pmatrix} \mathbf{A} - i\omega\mathbf{I} & \mathbf{B}_1 \\ \mathbf{C}_2 & \mathbf{D}_{21} \end{pmatrix}.$$

(A4)* $\mathbf{D}_{12} = \begin{pmatrix} \mathbf{0} \\ \mathbf{I} \end{pmatrix}$ *and* $\mathbf{D}_{21} = (\mathbf{0} \quad \mathbf{I})$.

(A5) $\mathbf{D}_{11} = \mathbf{0}$ *and* $\mathbf{D}_{22} = \mathbf{0}$.

Assumption (A4*) can be fulfilled by a procedure called **scaling** as we see in what follows and is used for the loop shifting procedure. The scaling procedure requires that the following rank condition holds.

(A4') $\mathbf{D}_{12}$ *has full column rank and* $\mathbf{D}_{21}$ *has full row rank.*

Remark. Assumptions (A2') and (A3') can alternatively be formulated as follows. The systems

$$\left(\begin{array}{c|c} \mathbf{A} & \mathbf{B}_2 \\ \hline \mathbf{C}_1 & \mathbf{D}_{12} \end{array}\right), \quad \left(\begin{array}{c|c} \mathbf{A} & \mathbf{B}_1 \\ \hline \mathbf{C}_2 & \mathbf{D}_{21} \end{array}\right)$$

have no invariant zeros on the imaginary axis.

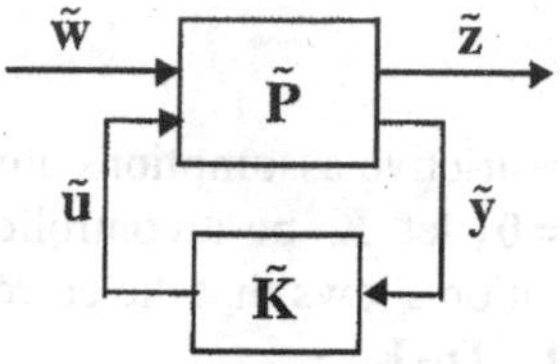

Fig. 9.5. Unscaled feedback loop

Scaling. We consider the feedback loop depicted in Fig. 9.5. Denote its realization by

$$\tilde{\mathbf{P}}(s) = \left(\begin{array}{c|cc} \tilde{\mathbf{A}} & \tilde{\mathbf{B}}_1 & \tilde{\mathbf{B}}_2 \\ \hline \tilde{\mathbf{C}}_1 & \tilde{\mathbf{D}}_{11} & \tilde{\mathbf{D}}_{12} \\ \tilde{\mathbf{C}}_2 & \tilde{\mathbf{D}}_{21} & \tilde{\mathbf{D}}_{22} \end{array}\right). \tag{9.17}$$

Here we assume that $\tilde{\mathbf{D}}_{12}$ has full column rank and that $\tilde{\mathbf{D}}_{21}$ has full row rank. It is easy to see that the singular value decomposition may be used to get decompositions of the form

$$\tilde{\mathbf{D}}_{12} = \mathbf{U}_1 \begin{pmatrix} \mathbf{0} \\ \mathbf{I} \end{pmatrix} \mathbf{R}_1, \qquad \tilde{\mathbf{D}}_{21} = \mathbf{R}_2 \begin{pmatrix} \mathbf{0} & \mathbf{I} \end{pmatrix} \mathbf{U}_2,$$

with (real) unitary matrices $\mathbf{U}_1, \mathbf{U}_2$ and (real) invertible matrices $\mathbf{R}_1, \mathbf{R}_2$. We now introduce new variables by

$$\tilde{\mathbf{z}} = \mathbf{U}_1 \mathbf{z}, \quad \tilde{\mathbf{w}} = \mathbf{U}_2^T \mathbf{w}, \quad \tilde{\mathbf{y}} = \mathbf{R}_2 \mathbf{y}, \quad \tilde{\mathbf{u}} = \mathbf{R}_1 \mathbf{u}$$

and define a new plant and a new controller by

$$\mathbf{P}(s) = \left(\begin{array}{c|cc} \mathbf{A} & \mathbf{B}_1 & \mathbf{B}_2 \\ \hline \mathbf{C}_1 & \mathbf{D}_{11} & \mathbf{D}_{12} \\ \mathbf{C}_2 & \mathbf{D}_{21} & \mathbf{D}_{22} \end{array}\right) = \begin{pmatrix} \mathbf{U}_1^T & \mathbf{0} \\ \mathbf{0} & \mathbf{R}_2^{-1} \end{pmatrix} \tilde{\mathbf{P}}(s) \begin{pmatrix} \mathbf{U}_2^T & \mathbf{0} \\ \mathbf{0} & \mathbf{R}_1^{-1} \end{pmatrix}$$

$$\mathbf{K}(s) = \mathbf{R}_1 \tilde{\mathbf{K}}(s) \mathbf{R}_2 .$$

Using the formulas of Sect. 5.5, a state space representation of $\mathbf{P}(s)$ with $\mathbf{D}_{12}$ and $\mathbf{D}_{21}$ as in (A4*) results.

Moreover, since $\mathbf{U}_1$, $\mathbf{U}_2$ are unitary, the transformation is norm preserving:

$$\| \mathcal{F}_l(\tilde{\mathbf{P}}, \tilde{\mathbf{K}}) \|_\infty = \| \mathbf{U}_1 \mathcal{F}_l(\mathbf{P}, \mathbf{K}) \mathbf{U}_2 \|_\infty = \| \mathcal{F}_l(\mathbf{P}, \mathbf{K}) \|_\infty .$$

Thus we are allowed to replace the plant $\tilde{\mathbf{P}}$ by $\mathbf{P}$ before the $\mathcal{H}_\infty$ optimization is performed. After the optimization process, the controller for $\mathbf{P}$ has to be transformed back.

The remaining, and possibly restrictive assumptions are $\mathbf{D}_{11} = \mathbf{0}$ and $\mathbf{D}_{22} = \mathbf{0}$. To remove the assumption $\mathbf{D}_{22} = \mathbf{0}$, let $\mathbf{K}$ be a controller for the plant $\mathbf{P}$ with $\mathbf{D}_{22}$ set to $\mathbf{0}$. Then a short calculation shows that the corresponding controller for the original plant is given by $\mathbf{K}(\mathbf{I} + \mathbf{D}_{22}\mathbf{K})^{-1}$.

To remove the assumption $\mathbf{D}_{11} = \mathbf{0}$ is more complicated and needs a procedure called loop shifting, which will be described next.

Loop Shifting. We start with a plant of the form

$$\mathbf{P}(s) = \left(\begin{array}{c|cc} \mathbf{A} & \mathbf{B}_1 & \mathbf{B}_2 \\ \hline \mathbf{C}_1 & \mathbf{D}_{11} & \begin{pmatrix} \mathbf{0} \\ \mathbf{I} \end{pmatrix} \\ \mathbf{C}_2 & (\mathbf{0} \ \ \mathbf{I}) & \mathbf{0} \end{array}\right) \tag{9.18}$$

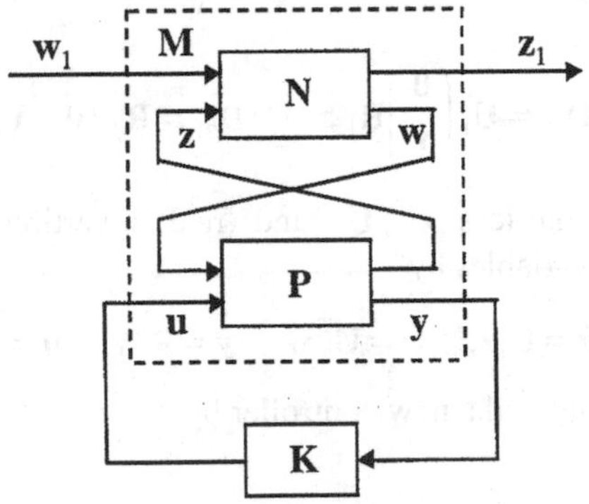

Fig. 9.6. Loop Shifting

and represent $\mathbf{D}_{11}$ as a 2×2-block matrix:

$$\mathbf{D}_{11}=\begin{pmatrix}\mathbf{D}_{1111} & \mathbf{D}_{1112}\\ \mathbf{D}_{1121} & \mathbf{D}_{1122}\end{pmatrix}.$$

The plant will be extended as shown in Fig. 9.6. Here, $\mathbf{N}$ denotes a constant transfer function:

$$\mathbf{N}=\begin{pmatrix}\mathbf{N}_{11} & \mathbf{N}_{12}\\ \mathbf{N}_{21} & \mathbf{N}_{22}\end{pmatrix}.$$

The objective is to construct $\mathbf{N}$ in such a way that that the $\mathbf{D}_{11}$-block $\mathbf{D}_{M11}$ of the transfer function $\mathbf{M}$, which is made up of $\mathbf{P}$ and $\mathbf{N}$, vanishes. The feedback connection of $\mathbf{M}$ and $\mathbf{K}$ will normally not have the same norm as that of $\mathbf{P}$ and $\mathbf{K}$, but we try to ensure that the change of the plant preserves suboptimality with the same γ. For the sake of simplicity, let γ be normalized to 1. Then we want to have the following:

$$\begin{array}{lcl}\mathbf{K}\text{ stabilizes }\mathbf{P}\text{ internally} & \Leftrightarrow & \mathcal{F}_l(\mathbf{P},\mathbf{K})\text{ stabilizes }\mathbf{N}\text{ internally}\\ \text{and } \|\mathcal{F}_l(\mathbf{P},\mathbf{K})\|_\infty<1 & & \text{and } \|\mathcal{F}_l(\mathbf{N},\mathcal{F}_l(\mathbf{P},\mathbf{K}))\|_\infty<1.\end{array}$$

This is true if Theorem 9.2.2 can be applied. The assumptions of Theorem 9.2.2 are fulfilled if $\mathbf{N}$ is orthonormal and if $\mathbf{N}_{21}$ is nonsingular.

Suppose $\|\mathbf{D}_{11}\|<\gamma$ and put

$$\mathbf{N}=\begin{pmatrix}-\mathbf{D}_{11} & \tilde{\mathbf{R}}^{1/2}\\ \mathbf{R}^{1/2} & \mathbf{D}_{11}^T\end{pmatrix},\quad \mathbf{R}=\mathbf{I}-\mathbf{D}_{11}^T\mathbf{D}_{11},\quad \tilde{\mathbf{R}}=\mathbf{I}-\mathbf{D}_{11}\mathbf{D}_{11}^T.$$

It is easy to verify that $\mathbf{N}$ is orthogonal with $\mathbf{N}_{12}$ and $\mathbf{N}_{21}$ nonsingular. Then we have in particular that

$$\tilde{\mathbf{R}}^{1/2}\mathbf{D}_{11}=\mathbf{D}_{11}\mathbf{R}^{1/2}.$$

From the formulas for the state space representation for a feedback system (cf. Sect. 6.5.1) we see immediately that the $\mathbf{D}_{11}$-matrix of $\mathbf{M}$ actually vanishes:

$$\begin{aligned}\mathbf{D}_{M11}&=-\mathbf{D}_{11}+\tilde{\mathbf{R}}^{1/2}\mathbf{D}_{11}\mathbf{R}^{-1}\mathbf{R}^{1/2}=-\mathbf{D}_{11}+\tilde{\mathbf{R}}^{1/2}\mathbf{D}_{11}\mathbf{R}^{-1/2}\\ &=-\mathbf{D}_{11}+\mathbf{D}_{11}\mathbf{R}^{1/2}\mathbf{R}^{-1/2}=0.\end{aligned}$$

Before we proceed further, we have to discuss the condition $\|\mathbf{D}_{11}\|<1$. Since it cannot simply be assumed that this condition holds, we try to replace the plant $\mathbf{P}$ by a plant $\tilde{\mathbf{P}}$ such that the $\mathbf{D}_{11}$-matrix of $\tilde{\mathbf{P}}$ has a norm less than 1. This will be achieved by introducing a new controller by

$$\tilde{\mathbf{K}}(s)=\mathbf{K}(s)-\mathbf{D}_\infty\,, \tag{9.19}$$

where the matrix $\mathbf{D}_\infty$ still has to be specified. We also define a new plant by

$$\tilde{\mathbf{P}}(s)=\left(\begin{array}{c|cc}\mathbf{A}+\mathbf{B}_2\mathbf{D}_\infty\mathbf{C}_2 & \mathbf{B}_1+\mathbf{B}_2\mathbf{D}_\infty\mathbf{D}_{21} & \mathbf{B}_2\\ \hline \mathbf{C}_1+\mathbf{D}_{12}\mathbf{D}_\infty\mathbf{C}_2 & \tilde{\mathbf{D}}_{11} & \begin{pmatrix}\mathbf{0}\\ \mathbf{I}\end{pmatrix}\\ \mathbf{C}_2 & (\mathbf{0}\ \ \mathbf{I}) & \mathbf{0}\end{array}\right),\quad \tilde{\mathbf{D}}_{11}=\begin{pmatrix}\mathbf{D}_{1111} & \mathbf{D}_{1112}\\ \mathbf{D}_{1121} & \mathbf{D}_{1122}+\mathbf{D}_\infty\end{pmatrix}.$$

It is easy to see with the formulas of Sect. 6.5.1 that

$$\mathcal{F}_l(\mathbf{P},\mathbf{K})=\mathcal{F}_l(\tilde{\mathbf{P}},\tilde{\mathbf{K}}).$$

Moreover, the $\mathbf{D}$-block of $\mathcal{F}_l(\tilde{\mathbf{P}},\tilde{\mathbf{K}})$ is given by

$$\mathcal{F}_l(\tilde{\mathbf{P}},\tilde{\mathbf{K}})(\infty)=\begin{pmatrix}\mathbf{D}_{1111} & \mathbf{D}_{1112}\\ \mathbf{D}_{1121} & \mathbf{D}_{1122}+\mathbf{D}_K\end{pmatrix}.$$

The next task is to construct $\mathbf{D}_\infty$ such that $\|\tilde{\mathbf{D}}_{11}\|<1$. Let

$$\gamma_0=\min_{D_\infty}\left\|\begin{pmatrix}\mathbf{D}_{1111} & \mathbf{D}_{1112}\\ \mathbf{D}_{1121} & \mathbf{D}_{1122}+\mathbf{D}_\infty\end{pmatrix}\right\|.$$

The minimum can be calculated by Parrott's Theorem (see Appendix, Theorem A.1.4) and has the value

$$\gamma_0=\max\left\{\bar{\sigma}((\mathbf{D}_{1111}\ \ \mathbf{D}_{1112})),\bar{\sigma}\left((\mathbf{D}_{1111}^T\ \ \mathbf{D}_{1121}^T)\right)\right\}.$$

We have $\gamma_0<1$ since

$$\gamma_0\le\left\|\begin{pmatrix}\mathbf{D}_{1111} & \mathbf{D}_{1112}\\ \mathbf{D}_{1121} & \mathbf{D}_{1122}+\mathbf{D}_K\end{pmatrix}\right\|=\|\mathcal{F}_l(\tilde{\mathbf{P}},\tilde{\mathbf{K}})(\infty)\|\le\|\mathcal{F}_l(\mathbf{P},\mathbf{K})\|_\infty<1.$$

From the Appendix, Corollary A.1.2 it follows that $\|\tilde{\mathbf{D}}_{11}\|<1$ is achieved for

$$\mathbf{D}_\infty=-\mathbf{D}_{1122}-\mathbf{D}_{1121}(\mathbf{I}-\mathbf{D}_{1111}^T\mathbf{D}_{1111})^{-1}\mathbf{D}_{1111}^T\mathbf{D}_{1112}. \tag{9.20}$$

Now the above procedure of loop shifting can be applied to the new plant $\tilde{\mathbf{P}}$.

It is not difficult to calculate an explicit formula for the new plant $\mathbf{M}$ (cf. Zhou, Doyle, and Glover [112], Sect. 17.2). From our considerations it follows that $\mathbf{K}$ stabilizes $\mathbf{P}$ internally and $\|\mathcal{F}_l(\mathbf{P},\mathbf{K})\|_\infty<1$ if and only if $\tilde{\mathbf{K}}$ stabilizes $\mathbf{M}$ internally and $\|\mathcal{F}_l(\mathbf{M},\tilde{\mathbf{K}})\|_\infty<1$.

The preceding consideration shows that $\|\mathcal{F}_l(\mathbf{P},\mathbf{K})\|_\infty<\gamma$ is only possible if (with γ not normalized)

$$\max\left\{\bar{\sigma}((\mathbf{D}_{1111}\ \ \mathbf{D}_{1112})),\bar{\sigma}\left((\mathbf{D}_{1111}^T\ \ \mathbf{D}_{1121}^T)\right)\right\}<\gamma. \tag{9.21}$$

Let now a general plant

$$\mathbf{P}(s)=\left(\begin{array}{c|cc}\mathbf{A} & \mathbf{B}_1 & \mathbf{B}_2 \\ \hline \mathbf{C}_1 & \mathbf{D}_{11} & \mathbf{D}_{12} \\ \mathbf{C}_2 & \mathbf{D}_{21} & \mathbf{D}_{22}\end{array}\right)$$

be given and let the assumptions (A1') - (A4') and (A5) hold. The procedure of scaling and loop shifting can be summarized as follows.

Step 1: Set $\mathbf{D}_{22}=\mathbf{0}$ (and calculate after the synthesis the controller for $\mathbf{D}_{22}\neq\mathbf{0}$ as described above).

Step 2: Scale the plant such that it has the form (9.18).

Step 3: Calculate a new plant $\tilde{\mathbf{P}}$ by means of $\mathbf{D}_\infty$ (cf. (9.20)) as described in the loop shifting procedure and apply the loop shifting procedure to $\tilde{\mathbf{P}}$.

Step 4: The resulting plant $\mathbf{M}$ of step 3 normally has $\mathbf{D}_{22}\neq\mathbf{0}$ (and is not normalized). If $\mathbf{D}_{22}$ is set to $\mathbf{0}$, it is possible to calculate the explicit formulas for all suboptimal controllers $\tilde{\mathbf{K}}$ for $\mathbf{M}$, i.e. all suboptimal controllers $\mathbf{K}$ for $\mathbf{P}$ (cf. Zhou, Doyle, and Glover [112], Sect. 17.1).

Remark 1. It is clear that (A1') is necessary for the existence of an optimal controller. If the rank condition (A4') is relaxed, then a singular optimal control problem arises.

Remark 2. The simple example at the beginning of this chapter shows that it is actually necessary to consider the general case with $\mathbf{D}_{11}\neq\mathbf{0}$.

9.6 Mixed Sensitivity Design

9.6.1 Weighting Schemes

In designing a $\mathcal{H}_\infty$ controller, it is necessary to extend the original plant $\mathbf{G}$, which is the mathematical model of the physical plant by weights. These weights contain the design goals, and the extended plant is denoted by $\mathbf{P}$. The functions that can be shaped are all the transfer functions belonging to a feedback loop, as introduced in Sect. 7.4. For example, if $\mathbf{S}_o$ and $\mathbf{K}\mathbf{S}_o$ have to be shaped, $\mathbf{P}$ is

$$\mathbf{P}=\begin{pmatrix}\mathbf{W}_1 & \mathbf{W}_1\mathbf{G} \\ \mathbf{0} & -\mathbf{W}_2 \\ \mathbf{I} & \mathbf{G}\end{pmatrix}.$$

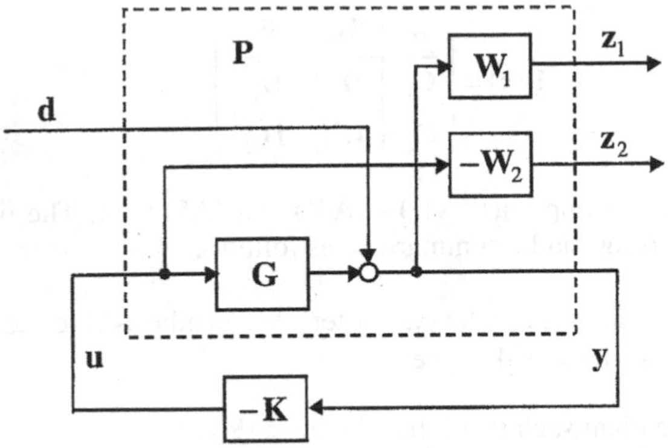

Fig. 9.7. S/KS mixed-sensitivity optimization

Using (7.8) and (7.10), it is seen that closed-loop transfer function results as

$$\mathcal{F}_l(\mathbf{P},\mathbf{K}) = \begin{pmatrix} \mathbf{W}_1\mathbf{S}_o \\ \mathbf{W}_2\mathbf{K}\mathbf{S}_o \end{pmatrix}.$$

Another possibility is to shape the sensitivity function $\mathbf{S}_o$ and the complementary sensitivity function $\mathbf{T}_o$ (cf. Fig. 9.8). Then $\mathbf{W}_1$ has the control error $\hat{\mathbf{e}} = \mathbf{S}_o\hat{\mathbf{r}}$ as the input and, similarly the input to $\mathbf{W}_3$ is the measurement $\hat{\mathbf{y}} = \mathbf{T}_o\hat{\mathbf{r}}$. The extended plant is

$$\mathbf{P} = \begin{pmatrix} \mathbf{W}_1 & -\mathbf{W}_1\mathbf{G} \\ \mathbf{0} & \mathbf{W}_3\mathbf{G} \\ \mathbf{I} & -\mathbf{G} \end{pmatrix}$$

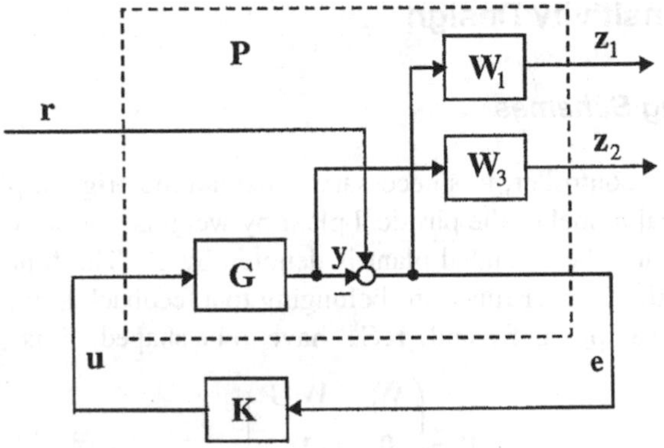

Fig. 9.8. S/T mixed-sensitivity control

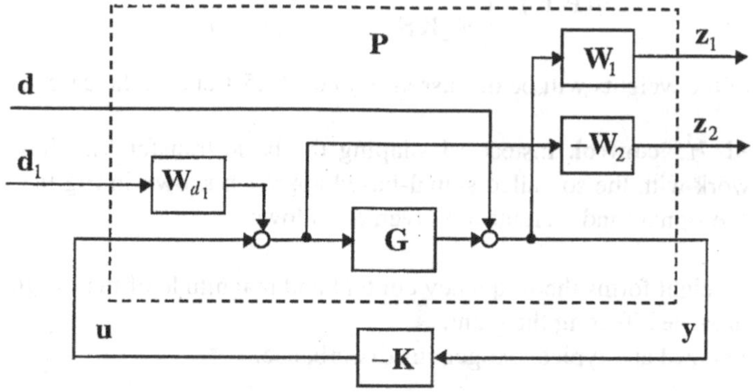

Fig. 9.9. S/KS mixed-sensitivity control with additional disturbance at the plant input

and the closed-loop transfer function is

$$\mathcal{F}_l(\mathbf{P},\mathbf{K}) = \begin{pmatrix} \mathbf{W}_1\mathbf{S}_o \\ \mathbf{W}_3\mathbf{T}_o \end{pmatrix}.$$

Of course, it is possible to combine both approaches. Then the closed-loop transfer function of the extended plant is

$$\mathcal{F}_l(\mathbf{P},\mathbf{K}) = \begin{pmatrix} \mathbf{W}_1\mathbf{S}_o \\ \mathbf{W}_2\mathbf{K}\mathbf{S}_o \\ \mathbf{W}_3\mathbf{T}_o \end{pmatrix}.$$

Often it is also necessary to weight the transfer function $\mathbf{S}_o\mathbf{G}$. This requires an additional disturbance at the plant input (cf. Fig. 9.9). The closed-loop transfer function of the extended plant is now

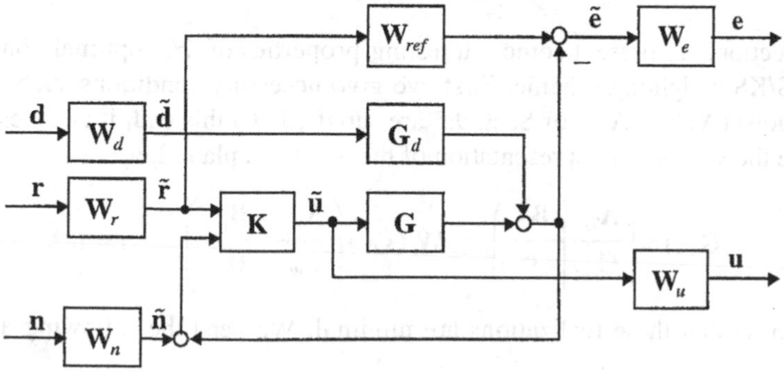

Fig. 9.10. Signal-based $\mathcal{H}_\infty$ control

$$\mathcal{F}_l(\mathbf{P},\mathbf{K}) = \begin{pmatrix} \mathbf{W}_1\mathbf{S}_o & \mathbf{W}_1\mathbf{S}_o\mathbf{G}\mathbf{W}_{d_1} \\ \mathbf{W}_2\mathbf{K}\,\mathbf{S}_o & -\mathbf{W}_2\mathbf{S}_i\mathbf{W}_{d_1} \end{pmatrix}.$$

The choice of the weights will be discussed in Sect. 10.5.1 and in the case studies.

Signal-Based $\mathcal{H}_\infty$ control. Instead of shaping the basic transfer functions, it is possible to work with the so-called signal-based approach shown in Fig 9.10. The meaning of the signals and weights are given as follows.

$\mathbf{W}_d$: This weight forms the frequency content and magnitude of the exogenous disturbance affecting the plant.
$\mathbf{d}, \tilde{\mathbf{d}}$: Normalized and typical exogenous disturbance.

$\mathbf{W}_r$: This weight shapes the magnitude and frequency of a reference command.
$\mathbf{r}, \tilde{\mathbf{r}}$: Normalized and typical reference command.

$\mathbf{W}_n$: This weight represents the frequency-domain models of sensor noise.
$\mathbf{n}, \tilde{\mathbf{n}}$: Normalized and typical sensor noise.

$\mathbf{W}_{ref}$: This weight represents the desired model for the closed-loop system with tracking.

$\mathbf{W}_e$: This weight shapes the tracking error
$\mathbf{e}, \tilde{\mathbf{e}}$: Weighted and actual tracking error.

$\mathbf{W}_u$: This weight forms the frequency content and magnitude of the control signal use.
$\mathbf{u}, \tilde{\mathbf{u}}$: Weighted and actual control signal.

9.6.2 Pole-Zero Cancellations

In this section, we present some interesting properties of $\mathcal{H}_\infty$ optimal controllers for the S/KS weighting scheme. First, we give necessary conditions such that the assumptions (A1') – (A4') of Sect. 9.5 are satisfied. To this end, it is necessary to calculate the state space representation of the extended plant. Let

$$\mathbf{G}(s) = \left(\begin{array}{c|c} \mathbf{A}_G & \mathbf{B}_G \\ \hline \mathbf{C}_G & \mathbf{0} \end{array}\right), \quad \mathbf{W}_i(s) = \left(\begin{array}{c|c} \mathbf{A}_{W_i} & \mathbf{B}_{W_i} \\ \hline \mathbf{C}_{W_i} & \mathbf{D}_{W_i} \end{array}\right) \qquad i = 1, 2,$$

and suppose that these realizations are minimal. We need the following assumptions:

(a1') $(\mathbf{A}_G, \mathbf{B}_G)$ is stabilizable, $(\mathbf{C}_G, \mathbf{A})$ is detectable and $\mathbf{W}_1, \mathbf{W}_2$ are stable.

(a2') The original plant $\mathbf{G}(s)$ has no poles on the imaginary axis.

(a3') The weights $\mathbf{W}_1, \mathbf{W}_2$ have no zeros on the imaginary axis.

(a4') The matrices $\mathbf{D}_{W_1}, \mathbf{D}_{W_2}$ are square and nonsingular.

The extended plant is given by

$$\mathbf{P}(s) = \left(\begin{array}{ccc|cc} \mathbf{A}_{W_1} & \mathbf{B}_{W_1}\mathbf{C}_G & \mathbf{0} & \begin{pmatrix} \mathbf{B}_{W_1} \\ \mathbf{0} \\ \mathbf{0} \end{pmatrix} & \begin{pmatrix} \mathbf{0} \\ \mathbf{B}_G \\ -\mathbf{B}_{W_2} \end{pmatrix} \\ \mathbf{0} & \mathbf{A}_G & \mathbf{0} & & \\ \mathbf{0} & \mathbf{0} & \mathbf{A}_{W_2} & & \\ \hline \begin{pmatrix} \mathbf{C}_{W_1} & \mathbf{D}_{W_1}\mathbf{C}_G & \mathbf{0} \\ \mathbf{0} & \mathbf{0} & \mathbf{C}_{W_2} \end{pmatrix} & & & \begin{pmatrix} \mathbf{D}_{W_1} \\ \mathbf{0} \end{pmatrix} & \begin{pmatrix} \mathbf{0} \\ \mathbf{D}_{W_2} \end{pmatrix} \\ (\mathbf{0} & \mathbf{C}_G & \mathbf{0}) & \mathbf{I} & \mathbf{0} \end{array}\right). \tag{9.22}$$

First, we note that the $\mathbf{D}_{11}$-matrix of the generalized plant is nonvanishing if $\mathbf{D}_{W_1}$ has this property. As we will see soon, this is normally the case. The rank condition requires that $\mathbf{D}_{W_2}$ has full row rank. It is now an easy exercise to see that conditions (A1') – (A4') hold if (a1') – (a4') hold.

Example. In order to get an idea of how the weightings may be chosen and which properties a S/KS mixed-sensitivity controller might have, we consider a simple but illustrative example. Let $a > 0$.
(a) Stable plant. Let

$$G(s) = \frac{1}{s+a}.$$

We want to design a controller such that the sensitivity function is

$$S = \frac{s}{s+\omega_1},$$

with a prescribed corner frequency ω_1. This leads to the choice $W_1 = 1/S$. Since the weights must be stable, this is not allowed and we define

$$W_1 = \frac{s+\omega_1}{s+\omega_\varepsilon}$$

with a small positive number ω_ε. Let W_2 be constant: $W_2 = d_2$. Since we do not want W_2 to become active, put $d_2 = 0$. Because of (a4'), this choice is excluded,

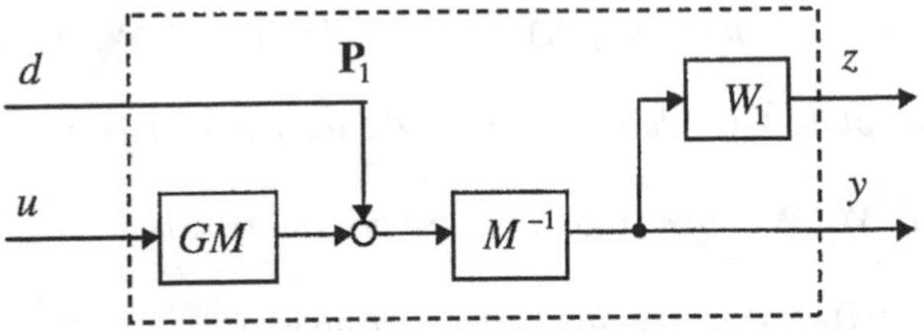

Fig. 9.11. Modified extended plant

but we ignore this for the moment. Then $\mathcal{F}_l(\mathbf{P}, K) = W_1 S$ and $\gamma = 1$ is achieved for the controller

$$K = (\omega_1 - \omega_\varepsilon)\frac{s+a}{s+\omega_\varepsilon}. \tag{9.23}$$

It cancels the denominator of the plant and has the same pole as W_1.

(b) Unstable plant. Let

$$G(s) = \frac{1}{s-a}.$$

The controller (9.23) with $-a$ instead of a would again lead to $S = 1/W_1$, but it no longer internally stabilizes the plant. Thus, our method has to be modified. Let a coprime factorization $G = NM^{-1}$ of G be given such that M is inner, which means $M(-s)M(s) = 1$. For this simple example such a factorization is easily constructed as

$$M(s) = \frac{s-a}{s+a}, \quad N = \frac{1}{s+a}.$$

We now replace the extended plant by the plant depicted in Fig. 9.11 and denote it by $\mathbf{P}_1$. For a given controller K, the relationship between the feedback loop with $\mathbf{P}$ and the feedback loop with $\mathbf{P}_1$ is given by $\mathcal{F}_l(\mathbf{P}, K) = M\mathcal{F}_l(\mathbf{P}_1, K)$. Thus, if K internally stabilizes $\mathbf{P}_1$, it also stabilizes $\mathbf{P}$, and since M is inner, the norms of both feedback loops are the same. For the modified feedback loop, we have $\mathcal{F}_l(\mathbf{P}_1, K) = W_1 S M^{-1}$. Thus, $\gamma = 1$ is achieved for

$$S = M W_1^{-1} = \frac{(s+\omega_\varepsilon)(s-a)}{(s+\omega_1)(s+a)}.$$

The optimal controller is immediately calculated as

$$K = \frac{(2a+\omega_1)s + a(\omega_1+\omega_\varepsilon)}{s+\omega_\varepsilon}.$$

The sensitivity function now has $-a$ as a pole and the controller has again the same pole as W_1.

It is a remarkable fact that the basic ideas of the previous example can be translated to the general situation. First, one has to note that the plant (9.22) exactly describes a disturbance feedforward problem. Moreover, if $\mathbf{G}$ is stable, the matrix $\mathbf{A}-\mathbf{B}_1\mathbf{C}_2$ is stable and it would be possible to apply Theorem 9.4.1 if $\mathbf{D}_{11}=\mathbf{0}$. If $\mathbf{G}$ is not stable, one has to transform the plant via coprime factorization with $\mathbf{M}$ inner. For the transformed plant, the matrix $\mathbf{A}-\mathbf{B}_1\mathbf{C}_2$ is stable and it is possible to calculate a state space representation of the controller $\mathbf{K}$ and of $\mathbf{L}_o=\mathbf{GK}$. This matrix is block diagonal with $\mathbf{A}_G$ as one block. It can now be shown that all stable eigenvalues of $\mathbf{A}_G$ are uncontrollable eigenvalues of $\mathbf{L}_o$. The problem is that we cannot assume $\mathbf{D}_{11}$ to vanish and thus we have to consider the general DF problem, a fact that makes the proof technically complicated. It can be found in Sefton and Glover [85]. The precise result is the following theorem.

Theorem 9.6.1. *Suppose (a1') – (a4') hold and assume additionally that* $\mathbf{W}_1^{-1},\mathbf{W}_2^{-1}$ *are stable. Then the following assertions hold for the* $\mathcal{H}_\infty$ *suboptimal controller* $\mathbf{K}$ *of the S/KS mixed-sensitivity problem.*

(i) *Each stable pole of the plant* $\mathbf{G}$ *is a transmission zero of* $\mathbf{K}$.

(ii) *Let* s_0 *be an unstable pole of* $\mathbf{G}$. *Then* $-s_0^*$ *is a pole of* $\mathbf{S}_o$.

(iii) *Each pole of* $\mathbf{W}_1$ *is a pole of* $\mathbf{K}$ *and each pole of* $\mathbf{W}_2$ *is a transmission zero of* $\mathbf{K}$.

Remarks. 1. The cancellation properties in the S/KS mixed-sensitivity design may be rather unfavorable if the plant has poles with a small positive or negative damping, because these poles (eventually reflected at the imaginary axis) appear internally in the feedback loop. A possible remedy against this is to use the S/KS mixed-sensitivity control with an additional disturbance at the plant input. Then (at least normally) no more cancellations occur and, as we will see in the examples and case studies, the poles of the feedback systems are well damped. Another possibility is to use the LMI approach with a prescribed region for the closed-loop poles.

2. It should be noted that for the S/T mixed-sensitivity problem the rank condition (A4') of Sect. 9.5 is not fulfilled (since the $\mathbf{u}$ does not enter the output equations for $\mathbf{z}_1,\mathbf{z}_2$). (A2') requires in this case that $\mathbf{G}$ has no zero on the imaginary axis. If the S/KS and the S/T design are combined (for example, with a small constant weight $\mathbf{W}_2$), the rank condition is fulfilled and $\mathbf{G}$ is allowed to have zeros on the imaginary axis.

Notes and References

The $\mathcal{H}_\infty$ control problem was originally formulated by Zames [111]. Different approaches to solve this problem were made and the first $\mathcal{H}_\infty$ optimal controllers tended to have a high state dimension. The solution presented here is based on Doyle, Glover, Khargonekar, and Francis [35], Glover, Limebeer, Doyle, Kasanally, and Safonov [48] and Safonov, Limebeer and Chiang [79]. Compare also the book by Zhou, Doyle and Glover [112]. The results on pole-zero cancellations are derived in Sefton and Glover [85].

Chapter 10

$\mathcal{H}_\infty$ Optimal Control: LMI - Approach and Applications

In this chapter, we consider again $\mathcal{H}_\infty$ optimal control, but we pursue a different approach. The starting point is a revised version of the bounded real lemma in which for a transfer matrix $\mathbf{G}$ the property $\| \mathbf{G} \|_\infty < 1$ is characterized by a linear matrix inequality (LMI). Then the idea for solving the $\mathcal{H}_\infty$ optimal control problem is quite simple: Apply this characterization to the closed-loop transfer function $\mathbf{F}_{zw}$ in order to get a description of suboptimal controllers. In doing so, the problem is that the characterization is not convex, but this property is needed to get necessary and sufficient optimality conditions. Some tricky algebra is required to get a convex characterization by three LMIs (Sect. 10.1).

In Sect. 10.2, we derive some properties of $\mathcal{H}_\infty$ suboptimal controllers. Thereby we first compare the optimality conditions of the Riccati and the LMI approach. It will be seen that two of the three LMIs correspond in some sense to the Riccati equations for $\mathbf{X}_\infty$ and $\mathbf{Y}_\infty$ (cf. Theorem 9.2.1) and that the third one corresponds to the condition $\rho(\mathbf{X}_\infty \mathbf{Y}_\infty) < \gamma^2$. Then the limiting behavior of the suboptimal controllers for $\gamma \to \gamma_{\text{opt}}$ is analyzed. The next section is devoted to LMI controller synthesis for problems where the closed-loop poles have to lie in a prescribed region. With the LMI method, it is possible to give a theoretical justification of gain-scheduling, which is a common praxis in classical control (Sect. 10.4). The chapter concludes with some numerical design experiments for a SISO plant of order 2. By this way, the effect of the weights will be studied for a simple example and the $\mathcal{H}_\infty$ synthesis can be compared with classical PID controller synthesis.

10.1 Characterization of $\mathcal{H}_\infty$ Suboptimal Controllers by Linear Matrix Inequalities

10.1.1 Bounded Real Lemma

The following lemma states the equivalence of two matrix inequalities and will be often used afterwards.

Lemma 10.1.1 (Schur Complement). *Let* $\mathbf{Q}, \mathbf{M},$ *and* $\mathbf{R}$ *be (real or complex) matrices and suppose that* $\mathbf{M}$ *and* $\mathbf{Q}$ *are self-adjoint. Then the following are equivalent:*

(i) $\quad \mathbf{Q} > \mathbf{0}$ *and* $\mathbf{M} - \mathbf{R}\mathbf{Q}^{-1}\mathbf{R}^* > \mathbf{0}$.

(ii) $\quad \begin{pmatrix} \mathbf{M} & \mathbf{R} \\ \mathbf{R}^* & \mathbf{Q} \end{pmatrix} > \mathbf{0}$.

Proof. Assertion (i) is equivalent to

$$\begin{pmatrix} \mathbf{M} - \mathbf{R}\mathbf{Q}^{-1}\mathbf{R}^* & \mathbf{0} \\ \mathbf{0} & \mathbf{Q} \end{pmatrix} > \mathbf{0}.$$

Since the matrix

$$\begin{pmatrix} \mathbf{I} & \mathbf{R}\mathbf{Q}^{-1} \\ \mathbf{0} & \mathbf{I} \end{pmatrix}$$

is nonsingular, the last inequality is equivalent to

$$\begin{pmatrix} \mathbf{M} & \mathbf{R} \\ \mathbf{R}^* & \mathbf{Q} \end{pmatrix} = \begin{pmatrix} \mathbf{I} & \mathbf{R}\mathbf{Q}^{-1} \\ \mathbf{0} & \mathbf{I} \end{pmatrix} \begin{pmatrix} \mathbf{M} - \mathbf{R}\mathbf{Q}^{-1}\mathbf{R}^* & \mathbf{0} \\ \mathbf{0} & \mathbf{Q} \end{pmatrix} \begin{pmatrix} \mathbf{I} & \mathbf{0} \\ \mathbf{Q}^{-1}\mathbf{R}^* & \mathbf{I} \end{pmatrix} > \mathbf{0}.$$

Remark. In a similar way it can be shown that (i) is also equivalent to

$$\begin{pmatrix} \mathbf{Q} & \mathbf{R}^* \\ \mathbf{R} & \mathbf{M} \end{pmatrix} > \mathbf{0}.$$

The following theorem holds for all transfer matrices of the form

$$\mathbf{G}(s) = \left(\begin{array}{c|c} \mathbf{A} & \mathbf{B} \\ \hline \mathbf{C} & \mathbf{D} \end{array}\right) \tag{10.1}$$

and is a revision of Theorem 7.3.1.

Theorem 10.1.1 (Bounded Real Lemma). *For an arbitrary transfer matrix of the form (10.1), the following assertions are equivalent:*

(i) $\mathbf{A}$ *is stable and*

$$\|\mathbf{G}\|_\infty < 1.$$

(ii) There is a symmetric matrix $\mathbf{X} > \mathbf{0}$ *such that*

$$\begin{pmatrix} \mathbf{A}^T\mathbf{X} + \mathbf{X}\mathbf{A} + \mathbf{C}^T\mathbf{C} & \mathbf{X}\mathbf{B} + \mathbf{C}^T\mathbf{D} \\ \mathbf{D}^T\mathbf{C} + \mathbf{B}^T\mathbf{X} & \mathbf{D}^T\mathbf{D} - \mathbf{I} \end{pmatrix} < \mathbf{0}. \tag{10.2}$$

Proof. "$\Rightarrow$": Put $\mathbf{R} = \mathbf{I} - \mathbf{D}^T\mathbf{D}$ and define

$$\mathbf{H} = \begin{pmatrix} \mathbf{A} + \mathbf{B}\mathbf{R}^{-1}\mathbf{D}^T\mathbf{C} & \mathbf{B}\mathbf{R}^{-1}\mathbf{B}^T \\ -\mathbf{C}^T(\mathbf{I} + \mathbf{D}\mathbf{R}^{-1}\mathbf{D}^T)\mathbf{C} & -(\mathbf{A} + \mathbf{B}\mathbf{R}^{-1}\mathbf{D}^T\mathbf{C})^T \end{pmatrix}.$$

Then Theorem 8.1.6 implies $\mathbf{H} \in dom(Ric)$ and $\mathbf{X}_0 = Ric(\mathbf{H}) \geq \mathbf{0}$. With

$$\mathbf{A}_H = \mathbf{A} + \mathbf{B}\mathbf{R}^{-1}\mathbf{D}^T\mathbf{C},$$

the Riccati equation for $\mathbf{X}_0$ can be written as

$$\mathbf{A}_H^T\mathbf{X}_0 + \mathbf{X}_0\mathbf{A}_H + \mathbf{X}_0\mathbf{B}\mathbf{R}^{-1}\mathbf{B}^T\mathbf{X}_0 + \mathbf{C}^T(\mathbf{I} + \mathbf{D}\mathbf{R}^{-1}\mathbf{D}^T)\mathbf{C} = \mathbf{0}. \tag{10.3}$$

By Theorem 8.1.2, the matrix $\mathbf{A}_H + \mathbf{B}\mathbf{R}^{-1}\mathbf{B}^T\mathbf{X}_0$ is stable. Thus by Lemma 5.7.2 the solution $\mathbf{X}_1$ of the Lyapunov equation

$$(\mathbf{A}_H + \mathbf{B}\mathbf{R}^{-1}\mathbf{B}^T\mathbf{X}_0)^T\mathbf{X}_1 + \mathbf{X}_1(\mathbf{A}_H + \mathbf{B}\mathbf{R}^{-1}\mathbf{B}^T\mathbf{X}_0) = -\mathbf{I} \tag{10.4}$$

is positive definite: $\mathbf{X}_1 > \mathbf{0}$. Consequently, the matrix $\mathbf{X} = \mathbf{X}_0 + \varepsilon\mathbf{X}_1$ is also positive definite for every $\varepsilon > 0$. Using (10.3) and (10.4), we obtain

$$\mathbf{A}_H^T\mathbf{X} + \mathbf{X}\mathbf{A}_H + \mathbf{X}\mathbf{B}\mathbf{R}^{-1}\mathbf{B}^T\mathbf{X} + \mathbf{C}^T(\mathbf{I} + \mathbf{D}\mathbf{R}^{-1}\mathbf{D}^T)\mathbf{C} = -\varepsilon\mathbf{I} + \varepsilon^2\mathbf{X}_1\mathbf{B}\mathbf{R}^{-1}\mathbf{B}^T\mathbf{X}_1.$$

Thus for sufficiently small $\varepsilon > 0$ the following Riccati inequality results:

$$\mathbf{A}^T\mathbf{X} + \mathbf{X}\mathbf{A} + \mathbf{C}^T\mathbf{C} - (\mathbf{X}\mathbf{B} + \mathbf{C}^T\mathbf{D})(\mathbf{D}^T\mathbf{D} - \mathbf{I})^{-1}(\mathbf{B}^T\mathbf{X} + \mathbf{D}^T\mathbf{C}) < \mathbf{0}. \tag{10.5}$$

The application of Lemma 10.1.1 shows that (10.5) is equivalent to the asserted inequality (10.2).

"$\Leftarrow$": Vice versa, let (10.2) be fulfilled with $\mathbf{X} > \mathbf{0}$. Then, in particular,

$$\mathbf{A}^T\mathbf{X} + \mathbf{X}\mathbf{A} + \mathbf{C}^T\mathbf{C} < \mathbf{0}.$$

holds. Hence, $\mathbf{A}$ is stable by Corollary 5.7.1. Moreover, (10.2) shows the existence of $0 < \varepsilon < 1$ with

$$\begin{pmatrix} \mathbf{A}^T\mathbf{X}+\mathbf{X}\mathbf{A}+\mathbf{C}^T\mathbf{C} & \mathbf{X}\mathbf{B}+\mathbf{C}^T\mathbf{D} \\ \mathbf{D}^T\mathbf{C}+\mathbf{B}^T\mathbf{X} & \mathbf{D}^T\mathbf{D}-(1-\varepsilon)\mathbf{I} \end{pmatrix} < \mathbf{0}. \tag{10.6}$$

Let $\hat{\mathbf{y}} = \mathbf{G}(s)\hat{\mathbf{u}}$ with $\mathbf{u} \in \mathcal{L}_2$. The state space description of the last equation is

$$\dot{\mathbf{x}} = \mathbf{A}\,\mathbf{x} + \mathbf{B}\mathbf{u}, \quad \mathbf{x}(0) = \mathbf{0}$$

$$\mathbf{y} = \mathbf{C}\mathbf{x} + \mathbf{D}\mathbf{u}\,.$$

By Corollary 7.2.1, $\|\mathbf{G}\|_\infty \leq 1-\varepsilon < 1$ is true if we have for all such pairs $\mathbf{y},\mathbf{u}$

$$\|\mathbf{y}\|_2 \leq (1-\varepsilon)\,\|\mathbf{u}\|_2\,.$$

Multiplication of (10.6) from the left-hand side by $(\mathbf{x}^T \quad \mathbf{u}^T)$ and from the right-hand side by $(\mathbf{x}^T \quad \mathbf{u}^T)^T$ leads to

$$\|\mathbf{y}(t)\|^2 + 2\mathbf{x}^T(t)\mathbf{X}(\mathbf{A}\mathbf{x}(t)+\mathbf{B}\mathbf{u}(t)) \leq (1-\varepsilon)\,\|\mathbf{u}(t)\|^2\,. \tag{10.7}$$

Define the function $p(t) = \mathbf{x}^T(t)\mathbf{X}\mathbf{x}(t)$. Then $\dot{p}(t) = 2\mathbf{x}^T(t)\mathbf{X}\dot{\mathbf{x}}(t)$, and (10.7) implies

$$\|\mathbf{y}(t)\|^2 + \dot{p}(t) \leq (1-\varepsilon)\|\mathbf{u}(t)\|^2\,.$$

Integration gives

$$p(T) + \int_0^T \|\mathbf{y}(t)\|^2\,dt \leq (1-\varepsilon)\int_0^T \|\mathbf{u}(t)\|^2\,dt\,.$$

Because of $p(T) \geq 0$, passing to the limit $T \to \infty$ yields

$$\|\mathbf{y}\|_2^2 \leq (1-\varepsilon)\,\|\mathbf{u}\|_2^2\,.$$

10.1.2 Nonconvex Characterization of $\mathcal{H}_\infty$ Suboptimal Controllers

The basic idea is now to apply Theorem 10.1.1 to the closed-loop system. Then (10.2) turns out to be a nonconvex matrix inequality and the remaining task is to replace it by a set of convex matrix inequalities. Fortunately, this is possible.

Let a plant $\mathbf{P}$ and a controller $\mathbf{K}$ with the following state space description be given:

$$\mathbf{P}(s) = \left(\begin{array}{c|cc} \mathbf{A} & \mathbf{B}_1 & \mathbf{B}_2 \\ \hline \mathbf{C}_1 & \mathbf{D}_{11} & \mathbf{D}_{12} \\ \mathbf{C}_2 & \mathbf{D}_{21} & \mathbf{0} \end{array}\right), \qquad \mathbf{K}(s) = \left(\begin{array}{c|c} \mathbf{A}_K & \mathbf{B}_K \\ \hline \mathbf{C}_K & \mathbf{D}_K \end{array}\right).$$

Here, $\mathbf{A}$ is an $n\times n$-matrix and $\mathbf{A}_K$ is an $n_K \times n_K$-matrix. Note that we do not require any normalizations concerning the matrices $\mathbf{D}_{11}, \mathbf{D}_{12}, \mathbf{D}_{21}$. The feedback system has the state space representation (cf. Sect. 6.5.1)

$$\mathbf{F}_{zw} = \left(\begin{array}{c|c} \mathbf{A}_c & \mathbf{B}_c \\ \hline \mathbf{C}_c & \mathbf{D}_c \end{array}\right)$$

$$= \left(\begin{array}{cc|c} \mathbf{A} + \mathbf{B}_2\mathbf{D}_K\mathbf{C}_2 & \mathbf{B}_2\mathbf{C}_K & \mathbf{B}_1 + \mathbf{B}_2\mathbf{D}_K\mathbf{D}_{21} \\ \mathbf{B}_K\mathbf{C}_2 & \mathbf{A}_K & \mathbf{B}_K\mathbf{D}_{21} \\ \hline \mathbf{C}_1 + \mathbf{D}_{12}\mathbf{D}_K\mathbf{C}_2 & \mathbf{D}_{12}\mathbf{C}_K & \mathbf{D}_{11} + \mathbf{D}_{12}\mathbf{D}_K\mathbf{D}_{21} \end{array}\right).$$

Let the controller be described by the matrix

$$\mathbf{J} = \begin{pmatrix} \mathbf{A}_K & \mathbf{B}_K \\ \mathbf{C}_K & \mathbf{D}_K \end{pmatrix}.$$

With this matrix, the matrices of the feedback system can be parameterized in terns of $\mathbf{J}$ if we define

$$\mathbf{A}_e = \begin{pmatrix} \mathbf{A} & \mathbf{0} \\ \mathbf{0} & \mathbf{0} \end{pmatrix}, \quad \mathbf{B}_{1e} = \begin{pmatrix} \mathbf{B}_1 \\ \mathbf{0} \end{pmatrix}, \quad \mathbf{B}_{2e} = \begin{pmatrix} \mathbf{0} & \mathbf{B}_2 \\ \mathbf{I} & \mathbf{0} \end{pmatrix}$$

$$\mathbf{C}_{1e} = \begin{pmatrix} \mathbf{C}_1 & \mathbf{0} \end{pmatrix}, \quad \mathbf{C}_{2e} = \begin{pmatrix} \mathbf{0} & \mathbf{I} \\ \mathbf{C}_2 & \mathbf{0} \end{pmatrix}$$

$$\mathbf{D}_{12e} = \begin{pmatrix} \mathbf{0} & \mathbf{D}_{12} \end{pmatrix}, \quad \mathbf{D}_{21e} = \begin{pmatrix} \mathbf{0} \\ \mathbf{D}_{21} \end{pmatrix}.$$

The desired parameterization is now

$$\mathbf{A}_c = \mathbf{A}_e + \mathbf{B}_{2e}\mathbf{J}\,\mathbf{C}_{2e}, \quad \mathbf{B}_c = \mathbf{B}_{1e} + \mathbf{B}_{2e}\mathbf{J}\,\mathbf{D}_{21e}$$
$$\mathbf{C}_c = \mathbf{C}_{1e} + \mathbf{D}_{12e}\mathbf{J}\,\mathbf{C}_{2e}, \quad \mathbf{D}_c = \mathbf{D}_{11} + \mathbf{D}_{12e}\mathbf{J}\,\mathbf{D}_{21e}.$$

Notice that the closed-loop state matrices depend affinely on the controller matrix $\mathbf{J}$. The transfer matrix for the closed-loop system is

$$\mathbf{F}_{zw}(s) = \left(\begin{array}{c|c} \mathbf{A}_c & \mathbf{B}_c \\ \hline \mathbf{C}_c & \mathbf{D}_c \end{array}\right). \tag{10.8}$$

The next step is to apply the bounded real lemma to this feedback system. For the sake of simplicity, we suppose γ to be normalized to one.

Corollary 10.1.1. *The following assertions are equivalent:*

(i) $\mathbf{A}_c$ *is stable and* $\|\mathbf{F}_{zw}\|_\infty < 1$.

(ii) There is a symmetric matrix $\mathbf{Z} > \mathbf{0}$ *such that*

$$\begin{pmatrix} \mathbf{A}_c^T\mathbf{Z} + \mathbf{Z}\mathbf{A}_c & \mathbf{Z}\mathbf{B}_c & \mathbf{C}_c^T \\ \mathbf{B}_c^T\mathbf{Z} & -\mathbf{I} & \mathbf{D}_c^T \\ \mathbf{C}_c & \mathbf{D}_c & -\mathbf{I} \end{pmatrix} < \mathbf{0}. \qquad (10.9)$$

Proof. This follows directly from Theorem 10.1.1 and an application of Lemma 10.1.1 to the inequality (10.2).

In principle, the inequality (10.9) offers already the possibility of calculating a suboptimal controller such that (i) holds. The problem is that (10.9) is affine in $\mathbf{J}$ and $\mathbf{Z}$ individually, but not jointly affine in both variables. Therefore, we try to reformulate this condition to get an affine characterization for the existence of $\mathbf{Z}$ which does not make use of $\mathbf{J}$.

Our first step is to insert the formulas for the closed-loop state matrices into (10.9). Then the following matrices arise:

$$\mathbf{P}_Z = \begin{pmatrix} \mathbf{B}_{2e}^T\mathbf{Z} & \mathbf{0} & \mathbf{D}_{12e}^T \end{pmatrix}, \quad \mathbf{Q} = \begin{pmatrix} \mathbf{C}_{2e} & \mathbf{D}_{21e} & \mathbf{0} \end{pmatrix},$$

$$\mathbf{H}_Z = \begin{pmatrix} \mathbf{A}_e^T\mathbf{Z} + \mathbf{Z}\mathbf{A}_e & \mathbf{Z}\mathbf{B}_{1e} & \mathbf{C}_{1e}^T \\ \mathbf{B}_{1e}^T\mathbf{Z} & -\mathbf{I} & \mathbf{D}_{11}^T \\ \mathbf{C}_{1e} & \mathbf{D}_{11} & -\mathbf{I} \end{pmatrix}.$$

It is easily checked that (10.9) is equivalent to

$$\mathbf{H}_Z + \mathbf{P}_Z^T\mathbf{J}\mathbf{Q} + \mathbf{Q}^T\mathbf{J}^T\mathbf{P}_Z < \mathbf{0}. \qquad (10.10)$$

Next, we give the necessary and sufficient conditions for the existence of a controller matrix $\mathbf{J}$ such that (10.10) holds. Therefore, the following lemma will be needed.

Lemma 10.1.2. *Let* $\mathbf{P}, \mathbf{Q}$ *be matrices with linearly independent columns. Then for every matrix* $\mathbf{Y}$ *there is a matrix* $\mathbf{J}$ *such that*

$$\mathbf{P}^T\mathbf{J}\mathbf{Q} = \mathbf{Y}.$$

This can be easily proved by using the singular value decompositions of $\mathbf{P}$ and $\mathbf{Q}$. If $\mathbf{H}$ is symmetric, then there is a matrix $\mathbf{Y}$ such that $\mathbf{H} + \mathbf{Y} + \mathbf{Y}^T < \mathbf{0}$. Thus, if $\mathbf{P}$ and $\mathbf{Q}$ have linearly independent columns, then by Lemma 10.1.2 there is a matrix $\mathbf{J}$ such that $\mathbf{H} + \mathbf{P}^T\mathbf{J}^T\mathbf{Q} + \mathbf{Q}^T\mathbf{J}\,\mathbf{P} < \mathbf{0}$. In the next lemma, the necessary and sufficient conditions for the existence of a matrix $\mathbf{J}$ fulfilling this inequality are formulated for the case where $\mathbf{P}$ and $\mathbf{Q}$ have nontrivial kernels.

Lemma 10.1.3. *Let* $\mathbf{P}, \mathbf{Q}$ *and* $\mathbf{H}$ *be matrices and suppose* $\mathbf{H}$ *to be symmetric. Furthermore, assume that* $\mathbf{N}_P, \mathbf{N}_Q$ *are full-rank matrices satisfying*

$$\operatorname{Im} \mathbf{N}_P = \operatorname{Ker} \mathbf{P}, \quad \operatorname{Im} \mathbf{N}_Q = \operatorname{Ker} \mathbf{Q} .$$

Then there exists a matrix $\mathbf{J}$ *such that*

$$\mathbf{H} + \mathbf{P}^T \mathbf{J}^T \mathbf{Q} + \mathbf{Q}^T \mathbf{J} \mathbf{P} < 0 \qquad (10.11)$$

if and only if the following inequalities hold:

$$\mathbf{N}_P^T \mathbf{H} \mathbf{N}_P < \mathbf{0} \quad \textit{and} \quad \mathbf{N}_Q^T \mathbf{H} \mathbf{N}_Q < \mathbf{0} . \qquad (10.12)$$

Remark. If $\operatorname{Ker} \mathbf{P} = \mathbf{0}$ and $\operatorname{Ker} \mathbf{Q} = \mathbf{0}$, then the inequalities (10.12) cannot of course be fulfilled. In this case, they have to be skipped and the existence of $\mathbf{J}$ with (10.11) follows as discussed above. If $\operatorname{Ker} \mathbf{P} = \mathbf{0}$ or $\operatorname{Ker} \mathbf{Q} = \mathbf{0}$, then the corresponding inequality of (10.12) has to be omitted and the proof is a simplified version of the following proof, in which it is assumed that none of the kernels consists only of the zero vector.

Proof. Suppose the following:

$$\{\mathbf{v}_1, \ldots, \mathbf{v}_k\} \text{ is a basis of } \operatorname{Ker} \mathbf{P} \cap \operatorname{Ker} \mathbf{Q} ,$$
$$\{\mathbf{v}_1, \ldots, \mathbf{v}_k, \mathbf{v}_{k+1}, \ldots, \mathbf{v}_l\} \text{ is a basis of } \operatorname{Ker} \mathbf{P} ,$$
$$\{\mathbf{v}_1, \ldots, \mathbf{v}_k, \mathbf{v}_{l+1}, \ldots, \mathbf{v}_m\} \text{ is a basis of } \operatorname{Ker} \mathbf{Q}$$

Moreover, define the matrices

$$\mathbf{V}_1 = \begin{pmatrix} \mathbf{v}_1 & \cdots & \mathbf{v}_k \end{pmatrix}, \quad \mathbf{V}_2 = \begin{pmatrix} \mathbf{v}_{k+1} & \cdots & \mathbf{v}_l \end{pmatrix}, \quad \mathbf{V}_3 = \begin{pmatrix} \mathbf{v}_{l+1} & \cdots & \mathbf{v}_m \end{pmatrix}$$

and let $\mathbf{V}_4$ be a matrix such that

$$\mathbf{V} = \begin{pmatrix} \mathbf{V}_1 & \mathbf{V}_2 & \mathbf{V}_2 & \mathbf{V}_4 \end{pmatrix}$$

is quadratic and nonsingular. Then (10.11) is equivalent to

$$\mathbf{V}^T \mathbf{H} \mathbf{V} + \mathbf{V}^T \mathbf{P}^T \mathbf{J}^T \mathbf{Q} \mathbf{V} + \mathbf{V}^T \mathbf{Q}^T \mathbf{J} \mathbf{P} \mathbf{V} < 0 . \qquad (10.13)$$

According to the construction of $\mathbf{V}$ we have

$$\mathbf{P}\mathbf{V} = \begin{pmatrix} \mathbf{0} & \mathbf{0} & \mathbf{P}_1 & \mathbf{P}_2 \end{pmatrix}, \quad \mathbf{Q}\mathbf{V} = \begin{pmatrix} \mathbf{0} & \mathbf{Q}_1 & \mathbf{0} & \mathbf{Q}_2 \end{pmatrix}.$$

with suitable matrices $\mathbf{P}_1, \mathbf{P}_2, \mathbf{Q}_1, \mathbf{Q}_2$. Write $\mathbf{V}^T \mathbf{H} \mathbf{V}$ in the form

$$\mathbf{V}^T \mathbf{H} \mathbf{V} = \begin{pmatrix} \mathbf{H}_{11} & \mathbf{H}_{12} & \mathbf{H}_{13} & \mathbf{H}_{14} \\ \mathbf{H}_{12}^T & \mathbf{H}_{22} & \mathbf{H}_{23} & \mathbf{H}_{24} \\ \mathbf{H}_{13}^T & \mathbf{H}_{23}^T & \mathbf{H}_{33} & \mathbf{H}_{34} \\ \mathbf{H}_{14}^T & \mathbf{H}_{24}^T & \mathbf{H}_{34}^T & \mathbf{H}_{44} \end{pmatrix}.$$

Furthermore, define a matrix $\mathbf{Y}$ with a matrix $\mathbf{J}$ that still has to be specified by

$$\mathbf{Y}=\begin{pmatrix}\mathbf{Y}_{11} & \mathbf{Y}_{12}\\ \mathbf{Y}_{21} & \mathbf{Y}_{22}\end{pmatrix}=\begin{pmatrix}\mathbf{P}_1^T\\ \mathbf{P}_2^T\end{pmatrix}\mathbf{J}^T\begin{pmatrix}\mathbf{Q}_1 & \mathbf{Q}_2\end{pmatrix}.$$

By the above construction we have

$$\operatorname{Ker}\begin{pmatrix}\mathbf{P}_1 & \mathbf{P}_2\end{pmatrix}=\mathbf{0},\quad \operatorname{Ker}\begin{pmatrix}\mathbf{Q}_1 & \mathbf{Q}_2\end{pmatrix}=\mathbf{0}.$$

Then it follows from Lemma 10.1.2 that each matrix $\mathbf{Y}$ can be written in the described way with an appropriate matrix $\mathbf{J}$.

Using these relationships, a short calculation shows that (10.13) can be written as follows:

$$\begin{pmatrix}\mathbf{H}_{11} & \mathbf{H}_{12} & \mathbf{H}_{13} & \mathbf{H}_{14}\\ \mathbf{H}_{12}^T & \mathbf{H}_{22} & \mathbf{H}_{23}+\mathbf{Y}_{11}^T & \mathbf{H}_{24}+\mathbf{Y}_{21}^T\\ \mathbf{H}_{13}^T & \mathbf{H}_{23}^T+\mathbf{Y}_{11} & \mathbf{H}_{33} & \mathbf{H}_{34}+\mathbf{Y}_{12}\\ \mathbf{H}_{14}^T & \mathbf{H}_{24}^T+\mathbf{Y}_{21} & \mathbf{H}_{34}^T+\mathbf{Y}_{12}^T & \mathbf{H}_{44}+\mathbf{Y}_{22}+\mathbf{Y}_{22}^T\end{pmatrix}<\mathbf{0}.$$

Thus it remains to show that (10.12) holds if and only if there is a matrix $\mathbf{Y}$ such that the latter inequality is valid. Lemma 10.1.1 shows that this inequality is equivalent to the following two inequalities:

$$\bar{\mathbf{H}}=\begin{pmatrix}\mathbf{H}_{11} & \mathbf{H}_{12} & \mathbf{H}_{13}\\ \mathbf{H}_{12}^T & \mathbf{H}_{22} & \mathbf{H}_{23}+\mathbf{Y}_{11}^T\\ \mathbf{H}_{13}^T & \mathbf{H}_{23}^T+\mathbf{Y}_{11} & \mathbf{H}_{33}\end{pmatrix}<\mathbf{0}, \tag{10.14}$$

$$\mathbf{H}_{44}+\mathbf{Y}_{22}+\mathbf{Y}_{22}^T-\begin{pmatrix}\mathbf{H}_{14}\\ \mathbf{H}_{24}+\mathbf{Y}_{21}^T\\ \mathbf{H}_{34}+\mathbf{Y}_{12}\end{pmatrix}^T\bar{\mathbf{H}}^{-1}\begin{pmatrix}\mathbf{H}_{14}\\ \mathbf{H}_{24}+\mathbf{Y}_{21}^T\\ \mathbf{H}_{34}+\mathbf{Y}_{12}\end{pmatrix}<\mathbf{0}.$$

The latter inequality can always be fulfilled with adequately chosen $\mathbf{Y}_{22}, \mathbf{Y}_{12}, \mathbf{Y}_{21}$. Hence, (10.13) holds if there is a matrix $\mathbf{Y}_{11}$ such that (10.14) is true. We now apply Lemma 10.1.1 to $\bar{\mathbf{H}}$ and see in this way that $\bar{\mathbf{H}}<\mathbf{0}$ holds if and only if $\mathbf{H}_{11}<\mathbf{0}$ and

$$\tilde{\mathbf{H}}=\begin{pmatrix}\mathbf{H}_{22}-\mathbf{H}_{12}^T\mathbf{H}_{11}^{-1}\mathbf{H}_{12} & \mathbf{Y}_{11}^T+\mathbf{W}^T\\ \mathbf{Y}_{11}+\mathbf{W} & \mathbf{H}_{33}-\mathbf{H}_{13}^T\mathbf{H}_{11}^{-1}\mathbf{H}_{13}\end{pmatrix}<\mathbf{0}$$

with $\mathbf{W}=\mathbf{H}_{23}^T-\mathbf{H}_{13}^T\mathbf{H}_{11}^{-1}\mathbf{H}_{13}$.

If $\tilde{\mathbf{H}}<\mathbf{0}$, then also all the diagonal entries of $\tilde{\mathbf{H}}$ are negative definite. Vice versa, if these diagonal entries are negative definite, then $\mathbf{Y}_{11}$ can be chosen such that $\hat{\mathbf{H}}$ is negative definite. Hence $\mathbf{Y}_{11}$ can be chosen such that $\hat{\mathbf{H}}$ is negative definite if and only if the diagonal entries are negative definite. Another application of Lemma 10.1.1 shows that this is the case if and only if

$$\begin{pmatrix} \mathbf{H}_{11} & \mathbf{H}_{12} \\ \mathbf{H}_{12}^T & \mathbf{H}_{22} \end{pmatrix} < \mathbf{0} \text{ and } \begin{pmatrix} \mathbf{H}_{11} & \mathbf{H}_{13} \\ \mathbf{H}_{13}^T & \mathbf{H}_{33} \end{pmatrix} < \mathbf{0}. \tag{10.15}$$

Put now

$$\mathbf{N}_P = \begin{pmatrix} \mathbf{V}_1 & \mathbf{V}_2 \end{pmatrix}, \quad \mathbf{N}_Q = \begin{pmatrix} \mathbf{V}_1 & \mathbf{V}_3 \end{pmatrix}.$$

Then the left-hand sides of (10.12) and (10.15) coincide and the lemma is proved.

The next lemma is an immediate consequence of Corollary 10.1.1 and Lemma 10.1.3.

Lemma 10.1.4. *With the preceding definitions, the following assertions are equivalent.*

(i) $\mathbf{A}_c$ *is stable and* $\| \mathbf{F}_{zw} \|_\infty < 1$.

(ii) There is a symmetric matrix $\mathbf{Z} > \mathbf{0}$ *such that*

$$\mathbf{N}_{P_Z}^T \mathbf{H}_Z \mathbf{N}_{P_Z} < \mathbf{0} \quad \text{and} \quad \mathbf{N}_Q^T \mathbf{H}_Z \mathbf{N}_Q < \mathbf{0}.$$

Here, $\mathbf{N}_{P_Z}$ *and* $\mathbf{N}_Q$ *are defined as in Lemma 10.1.3.*

In the following lemma, the matrix $\mathbf{P}_Z$ is replaced by a matrix $\mathbf{P}$, which does not depend on $\mathbf{Z}$. With a matrix $\mathbf{Z} > \mathbf{0}$ we define

$$\mathbf{P} = \begin{pmatrix} \mathbf{B}_{2e}^T & \mathbf{0} & \mathbf{D}_{12e}^T \end{pmatrix},$$

$$\mathbf{F}_Z = \begin{pmatrix} \mathbf{A}_e \mathbf{Z}^{-1} + \mathbf{Z}^{-1} \mathbf{A}_e^T & \mathbf{B}_{1e} & \mathbf{Z}^{-1} \mathbf{C}_{1e}^T \\ \mathbf{B}_{1e}^T & -\mathbf{I} & \mathbf{D}_{11}^T \\ \mathbf{C}_{1e} \mathbf{Z}^{-1} & \mathbf{D}_{11} & -\mathbf{I} \end{pmatrix}.$$

Lemma 10.1.5. *With the above definitions, the following are equivalent:*

(i) $\mathbf{A}_c$ *is stable and* $\| \mathbf{F}_{zw} \|_\infty < 1$.

(ii) There is a symmetric matrix $\mathbf{Z} > \mathbf{0}$ *such that*

$$\mathbf{N}_P^T \mathbf{F}_Z \mathbf{N}_P < \mathbf{0} \quad \text{and} \quad \mathbf{N}_Q^T \mathbf{H}_Z \mathbf{N}_Q < \mathbf{0}$$

(with $\mathbf{N}_P, \mathbf{N}_Q$ *as in Lemma 10.1.3).*

Proof. The assertion is a consequence of the preceding lemma if the equivalence

$$\mathbf{N}_{P_Z}^T \mathbf{H}_Z \mathbf{N}_{P_Z} < \mathbf{0} \quad \Leftrightarrow \quad \mathbf{N}_P^T \mathbf{F}_Z \mathbf{N}_P < \mathbf{0}$$

is demonstrated. Put

$$\mathbf{S} = \begin{pmatrix} \mathbf{Z} & \mathbf{0} & \mathbf{0} \\ \mathbf{0} & \mathbf{I} & \mathbf{0} \\ \mathbf{0} & \mathbf{0} & \mathbf{I} \end{pmatrix}.$$

Then $\mathbf{P}_Z = \mathbf{P}\mathbf{S}$ and

$$\text{Ker}\, \mathbf{P}_Z = \mathbf{S}^{-1}\, \text{Ker}\, \mathbf{P}\,.$$

Taking into account the definitions of $\mathbf{N}_{P_Z}$ and $\mathbf{N}_P$, we may define $\mathbf{N}_{P_Z} = \mathbf{S}^{-1}\mathbf{N}_P$. Hence (note that $\mathbf{S}^{-1}$ is symmetric)

$$\mathbf{N}_{P_Z}^T \mathbf{H}_Z \mathbf{N}_{P_Z} < \mathbf{0} \quad \Leftrightarrow \quad \mathbf{N}_P^T \mathbf{S}^{-1} \mathbf{H}_Z \mathbf{S}^{-1} \mathbf{N}_P < \mathbf{0}$$

and the assertion follows because of $\mathbf{S}^{-1}\mathbf{H}_Z\mathbf{S}^{-1} = \mathbf{F}_Z$.

10.1.3 Convex Characterization of $\mathcal{H}_\infty$ Suboptimal Controllers

The matrix $\mathbf{Z}$ in Lemma 10.1.5(ii) is symmetric, positive definite and of dimension $n + n_K$. We write $\mathbf{Z}$ and $\mathbf{Z}^{-1}$ in the form

$$\mathbf{Z} = \begin{pmatrix} \mathbf{X} & \mathbf{X}_2 \\ \mathbf{X}_2^T & \mathbf{X}_3 \end{pmatrix}, \quad \mathbf{Z}^{-1} = \begin{pmatrix} \mathbf{Y} & \mathbf{Y}_2 \\ \mathbf{Y}_2^T & \mathbf{Y}_3 \end{pmatrix},$$

with $n \times n$ - matrices $\mathbf{X}$ and $\mathbf{Y}$. With these matrices it is possible to exploit the structure $\mathbf{F}_Z$ and $\mathbf{H}_Z$ and to give an explicit formulation of the matrix products in the inequalities of Lemma 10.1.5(ii). In particular, it will be seen that they depend only on $\mathbf{X}$ and $\mathbf{Y}$.

Lemma 10.1.6. *Let* $\mathbf{N}_o$ *and* $\mathbf{N}_c$ *be full-rank matrices such that*

$$\text{Im}\, \mathbf{N}_o = \text{Ker}\begin{pmatrix} \mathbf{C}_2 & \mathbf{D}_{21} \end{pmatrix},$$
$$\text{Im}\, \mathbf{N}_c = \text{Ker}\begin{pmatrix} \mathbf{B}_2^T & \mathbf{D}_{12}^T \end{pmatrix}.$$

Then with the above definitions, the inequalities

$$\mathbf{N}_Q^T \mathbf{H}_Z \mathbf{N}_Q < \mathbf{0} \quad \text{and} \quad \mathbf{N}_P^T \mathbf{F}_Z \mathbf{N}_P < \mathbf{0}$$

are equivalent to

$$\begin{pmatrix} \mathbf{N}_o & \mathbf{0} \\ \mathbf{0} & \mathbf{I} \end{pmatrix}^T \begin{pmatrix} \mathbf{A}^T\mathbf{X}+\mathbf{X}\mathbf{A} & \mathbf{X}\mathbf{B}_1 & \mathbf{C}_1^T \\ \mathbf{B}_1^T\mathbf{X} & -\mathbf{I} & \mathbf{D}_{11}^T \\ \mathbf{C}_1 & \mathbf{D}_{11} & -\mathbf{I} \end{pmatrix} \begin{pmatrix} \mathbf{N}_o & \mathbf{0} \\ \mathbf{0} & \mathbf{I} \end{pmatrix} < \mathbf{0},$$

$$\begin{pmatrix} \mathbf{N}_c & \mathbf{0} \\ \mathbf{0} & \mathbf{I} \end{pmatrix}^T \begin{pmatrix} \mathbf{A}\mathbf{Y}+\mathbf{Y}\mathbf{A}^T & \mathbf{Y}\mathbf{C}_1^T & \mathbf{B}_1 \\ \mathbf{C}_1\mathbf{Y} & -\mathbf{I} & \mathbf{D}_{11} \\ \mathbf{B}_1^T & \mathbf{D}_{11}^T & -\mathbf{I} \end{pmatrix} \begin{pmatrix} \mathbf{N}_c & \mathbf{0} \\ \mathbf{0} & \mathbf{I} \end{pmatrix} < \mathbf{0}.$$

Proof. We show that the second inequality of the first assertion is equivalent to the second inequality of the second assertion. The proof of the remaining asserted equivalence is analogous. The definitions directly yield

$$\mathbf{F}_Z = \begin{pmatrix} \mathbf{A}\mathbf{Y}+\mathbf{Y}\mathbf{A}^T & \mathbf{A}\mathbf{Y}_2 & \mathbf{B}_1 & \mathbf{Y}\mathbf{C}_1^T \\ \mathbf{Y}_2^T\mathbf{A}^T & \mathbf{0} & \mathbf{0} & \mathbf{Y}_2^T\mathbf{C}_1^T \\ \mathbf{B}_1^T & \mathbf{0} & -\mathbf{I} & \mathbf{D}_{11}^T \\ \mathbf{C}_1\mathbf{Y} & \mathbf{C}_1\mathbf{Y}_2 & \mathbf{D}_{11} & -\mathbf{I} \end{pmatrix}$$

$$\mathbf{P} = \begin{pmatrix} \mathbf{0} & \mathbf{I} & \mathbf{0} & \mathbf{0} \\ \mathbf{B}_2^T & \mathbf{0} & \mathbf{0} & \mathbf{D}_{12}^T \end{pmatrix}.$$

Write $\mathbf{N}_c$ in the form

$$\mathbf{N}_c = \begin{pmatrix} \mathbf{V}_1 \\ \mathbf{V}_2 \end{pmatrix}.$$

The kernel of $\mathbf{P}$ is spanned by the columns of

$$\mathbf{N}_P = \begin{pmatrix} \mathbf{V}_1 & \mathbf{0} \\ \mathbf{0} & \mathbf{0} \\ \mathbf{0} & \mathbf{I} \\ \mathbf{V}_2 & \mathbf{0} \end{pmatrix},$$

i.e. $\operatorname{Im}\mathbf{N}_P = \operatorname{Ker}\mathbf{P}$. Since the second block-row of $\mathbf{N}_P$ vanishes, the second block-row and the second block-column of $\mathbf{F}_Z$ do not enter into the matrix product $\mathbf{N}_P^T\mathbf{F}_Z\mathbf{N}_P$, the equation

$$\mathbf{N}_P^T\mathbf{F}_Z\mathbf{N}_P = \begin{pmatrix} \mathbf{V}_1 & \mathbf{0} \\ \mathbf{0} & \mathbf{I} \\ \mathbf{V}_2 & \mathbf{0} \end{pmatrix}^T \begin{pmatrix} \mathbf{A}\mathbf{Y}+\mathbf{Y}\mathbf{A}^T & \mathbf{B}_1 & \mathbf{Y}\mathbf{C}_1^T \\ \mathbf{B}_1^T & -\mathbf{I} & \mathbf{D}_{11}^T \\ \mathbf{C}_1\mathbf{Y} & \mathbf{D}_{11} & -\mathbf{I} \end{pmatrix} \begin{pmatrix} \mathbf{V}_1 & \mathbf{0} \\ \mathbf{0} & \mathbf{I} \\ \mathbf{V}_2 & \mathbf{0} \end{pmatrix}.$$

follows. Taking into account that

$$\begin{pmatrix} \mathbf{V}_1 & \mathbf{0} \\ \mathbf{0} & \mathbf{I} \\ \mathbf{V}_2 & \mathbf{0} \end{pmatrix} = \begin{pmatrix} \mathbf{I} & \mathbf{0} & \mathbf{0} \\ \mathbf{0} & \mathbf{0} & \mathbf{I} \\ \mathbf{0} & \mathbf{I} & \mathbf{0} \end{pmatrix} \begin{pmatrix} \mathbf{N}_c & \mathbf{0} \\ \mathbf{0} & \mathbf{I} \end{pmatrix},$$

the assertion is shown.

Lemma 10.1.7. *Let* $\mathbf{X}$ *and* $\mathbf{Y}$ *be symmetric positive definite* $n \times n$ *- matrices. Then there are* $n \times n_K$ *- matrices* $\mathbf{X}_2, \mathbf{Y}_2$ *and symmetric* $n_K \times n_K$ *- matrices* $\mathbf{X}_3, \mathbf{Y}_3$ *with*

$$\begin{pmatrix} \mathbf{X} & \mathbf{X}_2 \\ \mathbf{X}_2^T & \mathbf{X}_3 \end{pmatrix} > \mathbf{0} \quad \text{and} \quad \begin{pmatrix} \mathbf{X} & \mathbf{X}_2 \\ \mathbf{X}_2^T & \mathbf{X}_3 \end{pmatrix}^{-1} = \begin{pmatrix} \mathbf{Y} & \mathbf{Y}_2 \\ \mathbf{Y}_2^T & \mathbf{Y}_3 \end{pmatrix}$$

if and only if

$$\begin{pmatrix} \mathbf{X} & \mathbf{I} \\ \mathbf{I} & \mathbf{Y} \end{pmatrix} \geq \mathbf{0} \quad \text{and} \quad \operatorname{rank} \begin{pmatrix} \mathbf{X} & \mathbf{I} \\ \mathbf{I} & \mathbf{Y} \end{pmatrix} \leq n + n_K . \tag{10.16}$$

Proof. "$\Rightarrow$": The equation

$$\begin{pmatrix} \mathbf{X} & \mathbf{X}_2 \\ \mathbf{X}_2^T & \mathbf{X}_3 \end{pmatrix} \begin{pmatrix} \mathbf{Y} & \mathbf{Y}_2 \\ \mathbf{Y}_2^T & \mathbf{Y}_3 \end{pmatrix} = \begin{pmatrix} \mathbf{I} & \mathbf{0} \\ \mathbf{0} & \mathbf{I} \end{pmatrix}$$

implies

$$\mathbf{X}\mathbf{Y} + \mathbf{X}_2\mathbf{Y}_2^T = \mathbf{I}, \quad \mathbf{Y}\mathbf{X} + \mathbf{Y}_2\mathbf{X}_2^T = \mathbf{I} \quad \text{and} \quad \mathbf{X}_2^T\mathbf{Y} + \mathbf{X}_3\mathbf{Y}_2^T = \mathbf{0}.$$

This yields

$$\mathbf{0} \leq \begin{pmatrix} \mathbf{I} & \mathbf{0} \\ \mathbf{Y} & \mathbf{Y}_2 \end{pmatrix} \begin{pmatrix} \mathbf{X} & \mathbf{X}_2 \\ \mathbf{X}_2^T & \mathbf{X}_3 \end{pmatrix} \begin{pmatrix} \mathbf{I} & \mathbf{Y} \\ \mathbf{0} & \mathbf{Y}_2 \end{pmatrix} = \begin{pmatrix} \mathbf{X} & \mathbf{I} \\ \mathbf{I} & \mathbf{Y} \end{pmatrix}.$$

Furthermore, we have

$$\begin{pmatrix} \mathbf{X} & \mathbf{I} \\ \mathbf{I} & \mathbf{Y} \end{pmatrix} = \begin{pmatrix} \mathbf{I} & \mathbf{Y}^{-1} \\ \mathbf{0} & \mathbf{I} \end{pmatrix} \begin{pmatrix} \mathbf{X} - \mathbf{Y}^{-1} & \mathbf{0} \\ \mathbf{0} & \mathbf{Y}^T \end{pmatrix} \begin{pmatrix} \mathbf{I} & \mathbf{0} \\ \mathbf{Y}^{-1} & \mathbf{I} \end{pmatrix}. \tag{10.17}$$

This implies

$$\operatorname{rank} \begin{pmatrix} \mathbf{X} & \mathbf{I} \\ \mathbf{I} & \mathbf{Y} \end{pmatrix} = n + \operatorname{rank}(\mathbf{X} - \mathbf{Y}^{-1}) = n + \operatorname{rank}(\mathbf{X}\mathbf{Y} - \mathbf{I}) \leq n + n_K .$$

The second inequality of (10.16) follows now from $\mathbf{I} - \mathbf{X}\mathbf{Y} = \mathbf{X}_2\mathbf{Y}_2^T$ and the fact that $\mathbf{X}_2$ and $\mathbf{Y}_2$ are $n \times n_K$ -matrices.

"$\Leftarrow$": We may conclude from (10.17) that

$$\mathbf{X}-\mathbf{Y}^{-1}\geq \mathbf{0} \quad \text{and} \quad \operatorname{rank}(\mathbf{X}-\mathbf{Y}^{-1})\leq n_K .$$

Hence, there is a $n \times n_K$-matrix $\mathbf{X}_2$ with

$$\mathbf{X}-\mathbf{Y}^{-1}=\mathbf{X}_2\mathbf{X}_2^T \geq \mathbf{0} .$$

This leads to $\mathbf{X}-\mathbf{X}_2\mathbf{X}_2^T > \mathbf{0}$. An application of Lemma 10.1.1 yields

$$\begin{pmatrix} \mathbf{X} & \mathbf{X}_2 \\ \mathbf{X}_2^T & \mathbf{I} \end{pmatrix} > \mathbf{0} .$$

The inverse of the left-hand matrix is

$$\begin{pmatrix} \mathbf{X} & \mathbf{X}_2 \\ \mathbf{X}_2^T & \mathbf{I} \end{pmatrix}^{-1} = \begin{pmatrix} \mathbf{Y} & -\mathbf{Y}\mathbf{X}_2 \\ -\mathbf{X}_2^T\mathbf{Y} & \mathbf{X}_2^T\mathbf{Y}\mathbf{X}_2+\mathbf{I} \end{pmatrix}.$$

Hence, there are matrices $\mathbf{X}_2, \mathbf{X}_3, \mathbf{Y}_2, \mathbf{Y}_3$ with the desired properties (with $\mathbf{X}_3=\mathbf{I}$).

We now are in a position to formulate the desired existence condition by a set of LMIs. Note that the rank condition of the preceding lemma can be skipped if controllers of order $n_K \geq n$ are admitted.

Theorem 10.1.2 (Existence). *There is a stabilizing controller with*

$$\| \mathbf{F}_{zw} \|_\infty < 1 \tag{10.18}$$

if and only if there are symmetric matrices $\mathbf{X}>\mathbf{0}$ and $\mathbf{Y}>\mathbf{0}$ such that the following matrix inequalities are fulfilled:

$$\begin{pmatrix} \mathbf{N}_o & \mathbf{0} \\ \mathbf{0} & \mathbf{I} \end{pmatrix}^T \begin{pmatrix} \mathbf{A}^T\mathbf{X}+\mathbf{X}\mathbf{A} & \mathbf{X}\mathbf{B}_1 & \mathbf{C}_1^T \\ \mathbf{B}_1^T\mathbf{X} & -\mathbf{I} & \mathbf{D}_{11}^T \\ \mathbf{C}_1 & \mathbf{D}_{11} & -\mathbf{I} \end{pmatrix} \begin{pmatrix} \mathbf{N}_o & \mathbf{0} \\ \mathbf{0} & \mathbf{I} \end{pmatrix} < \mathbf{0}, \tag{10.19}$$

$$\begin{pmatrix} \mathbf{N}_c & \mathbf{0} \\ \mathbf{0} & \mathbf{I} \end{pmatrix}^T \begin{pmatrix} \mathbf{A}\mathbf{Y}+\mathbf{Y}\mathbf{A}^T & \mathbf{Y}\mathbf{C}_1^T & \mathbf{B}_1 \\ \mathbf{C}_1\mathbf{Y} & -\mathbf{I} & \mathbf{D}_{11} \\ \mathbf{B}_1^T & \mathbf{D}_{11}^T & -\mathbf{I} \end{pmatrix} \begin{pmatrix} \mathbf{N}_c & \mathbf{0} \\ \mathbf{0} & \mathbf{I} \end{pmatrix} < \mathbf{0}, \tag{10.20}$$

$$\begin{pmatrix} \mathbf{X} & \mathbf{I} \\ \mathbf{I} & \mathbf{Y} \end{pmatrix} \geq \mathbf{0}. \tag{10.21}$$

Here $\mathbf{N}_o$ and $\mathbf{N}_c$ are matrices with full rank such that

$$\operatorname{Im}\mathbf{N}_o = \operatorname{Ker}\begin{pmatrix} \mathbf{C}_2 & \mathbf{D}_{21} \end{pmatrix}, \quad \operatorname{Im}\mathbf{N}_c = \operatorname{Ker}\begin{pmatrix} \mathbf{B}_2^T & \mathbf{D}_{12}^T \end{pmatrix}.$$

Proof. " $\Rightarrow$ ": If there is a stabilizing controller with (10.18), then condition (ii) of Lemma 10.1.5 holds. Thus, by Lemma 10.1.6, (10.19) and (10.20) are valid and Lemma 10.1.7 proves (10.21).
" $\Leftarrow$ ": We choose $n_K = n$. Then the rank condition of (10.16) is automatically fulfilled and thus there is a matrix

$$\mathbf{Z} = \begin{pmatrix} \mathbf{X} & \mathbf{X}_2 \\ \mathbf{X}_2^T & \mathbf{X}_3 \end{pmatrix}$$

with the properties described in Lemma 10.1.7. Then, using (10.19) and (10.20), the proof is completed by using Lemma 10.1.5 and Lemma 10.1.6.

The next theorem is an immediate consequence of the results obtained up to now.

Theorem 10.1.3 (Synthesis). *Let* $\mathbf{X}, \mathbf{Y}$ *be matrices, which fulfill the conditions of the preceding theorem. Moreover, let* $\mathbf{X}_2$ *be a* $n \times n$ *- matrix which satisfies*

$$\mathbf{X} - \mathbf{Y}^{-1} = \mathbf{X}_2 \mathbf{X}_2^T \geq \mathbf{0}$$

and put

$$\mathbf{Z} = \begin{pmatrix} \mathbf{X} & \mathbf{X}_2 \\ \mathbf{X}_2^T & \mathbf{I} \end{pmatrix}.$$

Finally, let $\mathbf{J}$ *be a solution of the linear matrix inequality*

$$\begin{pmatrix} \mathbf{A}_c^T \mathbf{Z} + \mathbf{Z} \mathbf{A}_c & \mathbf{Z} \mathbf{B}_c & \mathbf{C}_c^T \\ \mathbf{B}_c^T \mathbf{Z} & -\mathbf{I} & \mathbf{D}_c^T \\ \mathbf{C}_c & \mathbf{D}_c & -\mathbf{I} \end{pmatrix} < \mathbf{0}. \tag{10.22}$$

Then $\mathbf{J}$ *is a stabilizing controller such that condition (i) of Corollary 10.1.1 holds.*

Remarks. 1. Theorems 10.1.2, 10.1.3 only need the assumption $\mathbf{D}_{22} = \mathbf{0}$, which is no restriction (cf. Sect. 9.5).
2. If γ is not normed to 1, then in Theorem 10.1.2 the identity matrices in the 3×3-block matrices have to be replaced by $\gamma \mathbf{I}$ and in Theorem 10.1.3 the identity matrix in the second block row has to be replaced by $\gamma^2 \mathbf{I}$.

10.2 Properties of $\mathcal{H}_\infty$ Suboptimal Controllers

10.2.1 Connection between the Riccati- and the LMI-Approaches

In this section, we want to compare the results of the Riccati - and the LMI - approaches. Let the assumptions (A3), (A4) and (A5) of Theorem 9.2.1 be fulfilled. Then, with $\mathbf{D}_\perp$ as in Theorem 8.1.1, we have

$$\begin{pmatrix}\mathbf{B}_2^T & \mathbf{D}_{12}^T\end{pmatrix}\begin{pmatrix}\mathbf{I} & \mathbf{0}\\ -\mathbf{D}_{12}\mathbf{B}_2^T & \mathbf{D}_\perp\end{pmatrix}=\begin{pmatrix}\mathbf{0} & \mathbf{0}\end{pmatrix}.$$

Hence, we can choose (note that $\mathbf{D}_\perp$ has full column rank)

$$\mathbf{N}_c=\begin{pmatrix}\mathbf{I} & \mathbf{0}\\ -\mathbf{D}_{12}\mathbf{B}_2^T & \mathbf{D}_\perp\end{pmatrix}.$$

Carrying out the multiplication in (10.20) shows that (10.20) is equivalent to

$$\begin{pmatrix}\mathbf{AY}+\mathbf{YA}^T-\mathbf{B}_2\mathbf{B}_2^T & \mathbf{YC}_1^T\mathbf{D}_\perp & \mathbf{B}_1\\ \mathbf{D}_\perp^T\mathbf{C}_1\mathbf{Y} & -\mathbf{I} & \mathbf{0}\\ \mathbf{B}_1^T & \mathbf{0} & -\mathbf{I}\end{pmatrix}<\mathbf{0}.$$

Using $\mathbf{D}_\perp\mathbf{D}_\perp^T=\mathbf{I}-\mathbf{D}_{12}\mathbf{D}_{12}^T$, an application of Lemma 10.1.1 leads to the equivalent condition

$$\mathbf{AY}+\mathbf{YA}^T-\mathbf{B}_2\mathbf{B}_2^T+\mathbf{YC}_1^T\mathbf{C}_1\mathbf{Y}+\mathbf{B}_1\mathbf{B}_1^T<\mathbf{0}.$$

Defining $\tilde{\mathbf{X}}=\mathbf{Y}^{-1}$, this inequality can be rewritten as

$$\mathbf{A}^T\tilde{\mathbf{X}}+\tilde{\mathbf{X}}\mathbf{A}+\tilde{\mathbf{X}}(\mathbf{B}_1\mathbf{B}_1^T-\mathbf{B}_2\mathbf{B}_2^T)\tilde{\mathbf{X}}+\mathbf{C}_1^T\mathbf{C}_1<\mathbf{0}. \tag{10.23}$$

Similarly, (10.19) is equivalent to

$$\mathbf{A}\tilde{\mathbf{Y}}+\tilde{\mathbf{Y}}\mathbf{A}^T+\tilde{\mathbf{Y}}(\mathbf{C}_1^T\mathbf{C}_1-\mathbf{C}_2^T\mathbf{C}_2)\tilde{\mathbf{Y}}+\mathbf{B}_1\mathbf{B}_1^T<\mathbf{0} \tag{10.24}$$

if one puts $\tilde{\mathbf{Y}}=\mathbf{X}^{-1}$. Using the equation

$$\begin{pmatrix}\mathbf{X} & \mathbf{I}\\ \mathbf{I} & \mathbf{Y}\end{pmatrix}=\begin{pmatrix}\mathbf{I} & \mathbf{0}\\ \mathbf{X}^{-1} & \mathbf{I}\end{pmatrix}\begin{pmatrix}\mathbf{X} & \mathbf{0}\\ \mathbf{0} & \mathbf{Y}-\mathbf{X}^{-1}\end{pmatrix}\begin{pmatrix}\mathbf{I} & \mathbf{X}^{-1}\\ \mathbf{0} & \mathbf{I}\end{pmatrix},$$

it is easy to see that (10.21) is equivalent to

$$\rho(\tilde{\mathbf{X}}\tilde{\mathbf{Y}})\le 1. \tag{10.25}$$

Let now additionally assumptions (A1) and (A2) of Theorem 9.2.1 be true. Then it is possible to show that if the LMIs (10.19) – (10.21) are solvable, then the conditions (i) – (iii) of Theorem 9.2.2 are fulfilled and vice versa (cf. Sanchez-Pena and Snaier [80] for further details).

10.2.2 Limiting Behavior

The number γ_{opt} is defined as the infimum over all γ such that an admissible controller with $\|\mathbf{F}_{zw}\|_\infty < \gamma$ exist. It is therefore natural to ask what happens when γ tends to γ_{opt}. Due to Theorem 9.2.1, if (A1)–(A5) hold, we may equivalently characterize γ_{opt} as the infimum over all γ such that (i)–(iii) of this theorem hold. An optimal controller is an admissible controller such that $\|\mathbf{F}_{zw}\|_\infty = \gamma_{\text{opt}}$. For an optimal controller, at least one of the conditions (i)–(iii) are not fulfilled.

We begin our discussion with the case that (iii) fails first when γ tends to γ_{opt}. Thus for $\gamma = \gamma_{\text{opt}}$, (i) and (ii) still hold, whereas $\rho(\mathbf{X}_\infty \mathbf{Y}_\infty) = \gamma^2$. Since $\mathbf{X}_\infty \mathbf{Y}_\infty$ has only nonnegative eigenvalues, the matrix

$$\mathbf{I} - \gamma_{\text{opt}}^2 \mathbf{X}_\infty \mathbf{Y}_\infty \tag{10.26}$$

is singular. Thus the representation for the suboptimal controller of Theorem 9.2.1 does not hold for the optimal controller. To get a description for this case, we multiply the state space representation for the suboptimal controller by the matrix (10.26). Using the Riccati equation for $\mathbf{X}_\infty$, a short calculation shows that the suboptimal controller can be written in the form

$$\begin{aligned}(\mathbf{I} - \gamma^2 \mathbf{X}_\infty \mathbf{Y}_\infty)\dot{\tilde{\mathbf{x}}} &= \tilde{\mathbf{A}}_\infty \tilde{\mathbf{x}} - \mathbf{L}_\infty \mathbf{y} \\ \mathbf{u} &= \mathbf{F}_\infty \tilde{\mathbf{x}}\end{aligned} \tag{10.27}$$

where

$$\tilde{\mathbf{A}}_\infty = \mathbf{A} + \mathbf{B}_2\mathbf{F}_\infty + \mathbf{L}_\infty \mathbf{C}_2 + \gamma^{-2}\mathbf{Y}_\infty \mathbf{A}^T \mathbf{X}_\infty + \gamma^{-2}\mathbf{B}_1\mathbf{B}_1^T\mathbf{X}_\infty + \gamma^{-2}\mathbf{Y}_\infty \mathbf{C}_1^T \mathbf{C}_1 .$$

The state space description (10.27) is also valid in the limiting case $\gamma = \gamma_{\text{opt}}$. Since the matrix (10.26) is singular, at least one state of the controller can be eliminated, i.e. the optimal controller is at most of order $n-1$. This limiting behavior can easily observed for some simple numerical examples in Sect. 10.5.

If condition (i) or (ii) fails first, the limit case is more complicated and we only give the results (but compare Zhou, Doyle, and Glover [112] for further details). By duality, it suffices to discuss the case where (i) fails before (iii). It can be shown that the conditions $\mathbf{X}_\infty \geq \mathbf{0}$ or $\mathbf{Y}_\infty \geq \mathbf{0}$ cannot fail prior to (iii). This means that if (i) fails, then $\mathbf{H}_\infty$ is not an element of $dom(Ric)$ because this matrix has not the stability property or not the complementary property (cf. Sect. 8.1.2).

If $\mathbf{H}_\infty$ does not have the stability property, then $\mathbf{H}_\infty \notin dom(Ric)$, but it is possible to extend the Riccati operator to obtain $\mathbf{X}_\infty$ for $\gamma = \gamma_{\text{opt}}$. Then again the controller $\mathbf{u} = -\mathbf{B}_2^T\mathbf{X}_\infty\mathbf{x}$ stabilizes the plant and yields $\|\mathbf{F}_{zw}\|_\infty = \gamma_{\text{opt}}$. Besides this, it can be shown that if the stability property is violated first, then the infimum over all stabilizing controllers is the same as the infimum over all controllers.

If the complementary property does not hold for $\mathbf{H}_\infty$ at $\gamma = \gamma_{\text{opt}}$, then $\mathbf{H}_\infty \in dom(Ric)$ is again possible for $\gamma < \gamma_{\text{opt}}$, so that in this case the controller $\mathbf{u} = -\mathbf{B}_2^T\mathbf{X}_\infty\mathbf{x}$ is defined. It gives $\|\mathbf{F}_{zw}\|_\infty < \gamma$, but is not stabilizing. If the complementary property fails first, but (iii) still holds, it is also possible derive formulas for the optimal controller. If condition (iii) also fails (or is not present, as for the FI problem), it may happen that no optimal controller exists; see the example in Zhou, Doyle, and Glover [112], Sect 16.9.

As a conclusion, it may be said that for a reasonable specification of the feedback loop, it can be expected that condition (iii) fails first. In this case, an optimal controller exists and has the state space representation (10.27) as described above.

10.3 $\mathcal{H}_\infty$ Synthesis with Pole Placement Constraints

10.3.1 LMI Regions

In this section, we discuss how $\mathcal{H}_\infty$ synthesis is possible under the additional condition that the closed-loop poles have to lie in a prescribed region. In order to be compatible with the LMI approach, these regions must have a certain structure, which will be described now.

Definition. *A subset* $\mathcal{D}$ *of the open LHP is called an* ***LMI region*** *if there exists a symmetric* $m \times m$ *-matrix* $\mathbf{L}$ *and an arbitrary* $m \times m$ *-matrix* $\mathbf{M}$ *such that*

$$\mathcal{D} = \{\, z \in \mathbb{C} \mid \mathbf{L} + z\mathbf{M} + \bar{z}\mathbf{M}^T < \mathbf{0} \}.$$

The matrix-valued function

$$f_{\mathcal{D}}(z) = \mathbf{L} + z\mathbf{M} + \bar{z}\mathbf{M}^T$$

is denoted as the ***characteristic function*** *of* $\mathcal{D}$.

We define the coefficients of $\mathbf{L}$ and $\mathbf{M}$ by

$$\mathbf{L} = (\lambda_{kl}), \quad \mathbf{M} = (\mu_{kl}).$$

Then the characteristic function can also be written as

$$f_{\mathcal{D}}(z) = (\lambda_{kl} + \mu_{kl}\, z + \mu_{lk}\,\bar{z}).$$

Of course, in the definition of LMI regions it is not necessary to assume that $\mathcal{D}$ is a subset of the open LHP, but most of the following theorems require this, and,

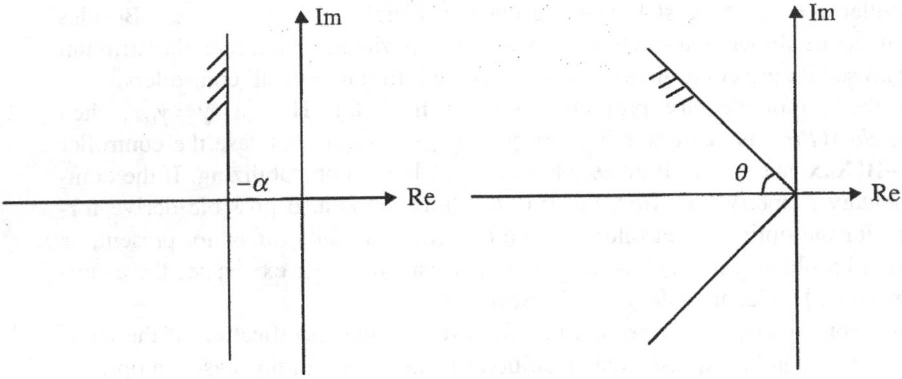

Fig. 10.1. LMI regions: half-plane and conic sector

since $\mathcal{D}$ is the set where the closed-loop poles are intended to lie, it is reasonable to consider only sets having this property. Typical regions for closed-loop poles are (with $\alpha > 0$, $0 \le \theta < \pi/2$)

$$\mathcal{D}_1 = \{ z \mid \operatorname{Re} z < -\alpha \} \qquad \text{(half-plane)},$$

$$\mathcal{D}_2 = \{ z \mid \operatorname{Re} z < 0, \frac{|\operatorname{Im} z|}{|\operatorname{Re} z|} < \tan\theta \} \qquad \text{(conic sector)},$$

compare Fig. 10.1. The characteristic functions are

$$f_{\mathcal{D}_1}(z) = 2\alpha + z + \bar{z},$$

$$f_{\mathcal{D}_2}(z) = \begin{pmatrix} \sin\theta & \cos\theta \\ -\cos\theta & \sin\theta \end{pmatrix} z + \begin{pmatrix} \sin\theta & -\cos\theta \\ \cos\theta & \sin\theta \end{pmatrix} \bar{z}.$$

This can easily be seen by calculating the values of z for which the eigenvalues of the characteristic function are negative. Moreover, it is not difficult to prove that LMI regions are convex and symmetric with respect to the real axis. An intersection of two LMI regions is again an LMI region. More precisely, let $f_{\mathcal{D}_1}$, $f_{\mathcal{D}_2}$ be the characteristic functions of $\mathcal{D}_1, \mathcal{D}_2$, respectively. Then the characteristic function of $\mathcal{D}_1 \cap \mathcal{D}_2$ is

$$f_{\mathcal{D}_1 \cap \mathcal{D}_2}(z) = \begin{pmatrix} f_{\mathcal{D}_1}(z) & \mathbf{0} \\ \mathbf{0} & f_{\mathcal{D}_2}(z) \end{pmatrix}.$$

Definition. *A matrix* **A** *is called* ***$\mathcal{D}$-stable*** *if all its eigenvalues lie in* $\mathcal{D}$.

In this sense, the $\mathcal{D}$-stability of linear systems is also defined. When $\mathcal{D}$ is the entire open LHP, then the $\mathcal{D}$-stability coincides with the usual stability. As we already know from the results in Sect. 5.7, $\mathbf{A}$ is stable if and only if there exists a symmetric matrix $\mathbf{X} > \mathbf{0}$ such that

$$\mathbf{A}\mathbf{X} + \mathbf{X}\mathbf{A}^T < \mathbf{0}.$$

We define now the $m \times m$ -block matrix

$$\mathbf{M}_{\mathcal{D}}(\mathbf{A}, \mathbf{X}) = \left(\lambda_{kl}\mathbf{X} + \mu_{kl}\,\mathbf{A}\mathbf{X} + \mu_{lk}\mathbf{X}\mathbf{A}^T\right).$$

If $\mathcal{D}$ is the open LHP, then this matrix consists only of a single block and is given by

$$\mathbf{M}_{\mathcal{D}}(\mathbf{A}, \mathbf{X}) = \mathbf{A}\mathbf{X} + \mathbf{X}\mathbf{A}^T.$$

For the conic sector, this matrix is

$$\mathbf{M}_{\mathcal{D}}(\mathbf{A}, \mathbf{X}) = \begin{pmatrix} \sin\theta(\mathbf{A}\mathbf{X} + \mathbf{X}\mathbf{A}^T) & \cos\theta(\mathbf{A}\mathbf{X} - \mathbf{X}\mathbf{A}^T) \\ \cos\theta(-\mathbf{A}\mathbf{X} + \mathbf{X}\mathbf{A}^T) & \sin\theta(\mathbf{A}\mathbf{X} + \mathbf{X}\mathbf{A}^T) \end{pmatrix}.$$

We can now state the Lyapunov criterion for stability as follows: $\mathbf{A}$ is stable if and only if and only if there exists a symmetric matrix $\mathbf{X} > \mathbf{0}$ such that $\mathbf{M}_{\mathcal{D}}(\mathbf{A}, \mathbf{X}) < \mathbf{0}$. This result can be generalized to the general $\mathcal{D}$-stability for LMI regions, as can be seen from the next theorem. A proof of this result is given in Chilali and Gahinet [26] (Theorem 2.2). The main point is that $\mathcal{D}$-stability can be described by an LMI.

Theorem 10.3.1. *The matrix* $\mathbf{A}$ *is* $\mathcal{D}$*-stable if and only if there exists a symmetric matrix* $\mathbf{X} > \mathbf{0}$ *such that*

$$\mathbf{M}_{\mathcal{D}}(\mathbf{A}, \mathbf{X}) < \mathbf{0}. \tag{10.28}$$

10.3.2 Design of $\mathcal{D}$-stable $\mathcal{H}_\infty$ Optimal Controllers

The basic idea is to apply Theorem 10.3.1 to the system matrix of the feedback loop and to combine this with $\mathcal{H}_\infty$ synthesis (in terms of LMIs). We start with the simple case of state feedback, i.e.

$$\mathbf{u} = \mathbf{F}\mathbf{x}.$$

Then the feedback system is given by

$$\mathbf{F}_{zw} = \left(\begin{array}{c|c} \mathbf{A} + \mathbf{B}_2\mathbf{F} & \mathbf{B}_1 \\ \hline \mathbf{C}_1 + \mathbf{D}_{12}\mathbf{F} & \mathbf{D}_{11} \end{array}\right).$$

We apply now Theorem 10.3.1 to $\mathbf{A} + \mathbf{B}_2\mathbf{F}$. Then equation (10.28) is equivalent to the existence of a symmetric matrix $\mathbf{Z} > \mathbf{0}$ such that

$$\left(\lambda_{kl}\mathbf{Z} + \mu_{kl}(\mathbf{A}+\mathbf{B}_2\mathbf{F})\mathbf{Z} + \mu_{lk}\mathbf{Z}(\mathbf{A}+\mathbf{B}_2\mathbf{F})^T\right) < \mathbf{0}.$$

Thus because of the multiplication of $\mathbf{Z}$ and $\mathbf{F}$, Theorem 10.3.1 leads to a non-convex condition for these variables. We now define a new variable by

$$\mathbf{W} = \mathbf{F}\mathbf{Z}.$$

In the variables $\mathbf{Z}$ and $\mathbf{W}$, the condition is convex. This leads us to the following theorem.

Theorem 10.3.2. *It is possible to find a feedback law of the form* $\mathbf{u} = \mathbf{F}\mathbf{x}$ *such that the closed-loop poles lie in* $\mathcal{D}$ *if and only if there is a symmetric matrix* $\mathbf{Z} > \mathbf{0}$ *and a matrix* $\mathbf{W}$ *such that*

$$\left(\lambda_{kl}\mathbf{Z} + \mu_{kl}(\mathbf{A}\mathbf{Z}+\mathbf{B}_2\mathbf{W}) + \mu_{lk}(\mathbf{Z}\mathbf{A}^T + \mathbf{W}^T\mathbf{B}_2^T)\right) < \mathbf{0}. \qquad (10.29)$$

If $\mathbf{Z}$ *and* $\mathbf{W}$ *have these properties, then such a constant gain matrix is given by*

$$\mathbf{F} = \mathbf{W}\mathbf{Z}^{-1}.$$

It is possible to combine this kind of pole positioning with $\mathcal{H}_\infty$ optimization. To this end, it is convenient to replace the condition (10.9) of Corollary 10.1.1 by the equivalent inequality

$$\begin{pmatrix} \mathbf{A}_c\mathbf{Z}+\mathbf{Z}\mathbf{A}_c^T & \mathbf{B}_c & \mathbf{Z}\mathbf{C}_c^T \\ \mathbf{B}_c^T & -\mathbf{I} & \mathbf{D}_c^T \\ \mathbf{C}_c\mathbf{Z} & \mathbf{D}_c & -\gamma^2\mathbf{I} \end{pmatrix} < \mathbf{0}. \qquad (10.30)$$

Here, γ is not normed to 1. Then, with the same variable transformation as above, it is an easy task to give the necessary and sufficient conditions for the existence of a constant gain controller $\mathbf{u} = \mathbf{F}\mathbf{x}$ such that the closed-loop system is $\mathcal{D}$-stable and in addition has $\mathcal{H}_\infty$ performance smaller than γ (Chilali and Gahinet [26], Theorem 3.1). Since constant-gain controllers are not so natural for $\mathcal{H}_\infty$ optimal control because of the (normally) dynamic weightings, we do not state the theorem here explicitly.

We now turn to the case of $\mathcal{H}_\infty$ output feedback. Again, one can try to combine the LMI condition for $\mathcal{D}$-stability and (10.30), but this is now much harder, since the simple variable transform from above no longer works. In this case, the controller parameterization is much more involved. We state only the main result and refer the reader to Chilali and Gahinet [26], Theorem 4.3. It is only necessary to suppose $\mathbf{D}_{22} = \mathbf{0}$, which is no restriction. Let the controller be given by

$$\mathbf{K} = \left(\begin{array}{c|c} \mathbf{A}_K & \mathbf{B}_K \\ \hline \mathbf{C}_K & \mathbf{D}_K \end{array}\right). \qquad (10.31)$$

It is assumed that the controller order is n. We now define a controller parameterization in terms of these matrices and the unknown Lyapunov matrix $\mathbf{X}$. Suppose that $\mathbf{X}$ and its inverse are partitioned in the form

$$\mathbf{Z} = \begin{pmatrix} \mathbf{X} & \mathbf{X}_2 \\ \mathbf{X}_2^T & \mathbf{I} \end{pmatrix}, \quad \mathbf{Z}^{-1} = \begin{pmatrix} \mathbf{Y} & -\mathbf{Y}\mathbf{X}_2 \\ -\mathbf{X}_2^T\mathbf{Y} & \mathbf{X}_2^T\mathbf{Y}\mathbf{X}_2 + \mathbf{I} \end{pmatrix}, \qquad \mathbf{X}, \mathbf{Y} \in \mathbb{R}^{n \times n}.$$

With $\mathbf{Y}_2 = -\mathbf{Y}\mathbf{X}_2$ define new controller variables as

$$\mathcal{B}_K = \mathbf{Y}_2\mathbf{B}_K + \mathbf{Y}\mathbf{B}_2\mathbf{D}_K \tag{10.32}$$

$$\mathcal{C}_K = \mathbf{C}_K\mathbf{X}_2^T + \mathbf{D}_K\mathbf{C}_2\mathbf{X} \tag{10.33}$$

$$\mathcal{A}_K = \mathbf{Y}_2\mathbf{A}_K\mathbf{X}_2^T + \mathbf{Y}_2\mathbf{B}_K\mathbf{C}_2\mathbf{X} + \mathbf{Y}\mathbf{B}_2\mathbf{C}_K\mathbf{X}_2^T + \mathbf{Y}(\mathbf{A} + \mathbf{B}_2\mathbf{D}_K\mathbf{C}_2)\mathbf{X}. \tag{10.34}$$

In the next theorem, the following matrices are used as abbreviations:

$$\Phi = \begin{pmatrix} \mathbf{A}\mathbf{X} + \mathbf{B}_2\mathcal{C}_K & \mathbf{A} + \mathbf{B}_2\mathbf{D}_K\mathbf{C}_2 \\ \mathcal{A}_K & \mathbf{Y}\mathbf{A} + \mathcal{B}_K\mathbf{C}_2 \end{pmatrix}$$

$$\Psi_{11} = \begin{pmatrix} \mathbf{A}\mathbf{X} + \mathbf{X}\mathbf{A}^T + \mathbf{B}_2\mathcal{C}_K + \mathcal{C}_K^T\mathbf{B}_2^T & \mathbf{B}_1 + \mathbf{B}_2\mathbf{D}_K\mathbf{D}_{21} \\ (\mathbf{B}_1 + \mathbf{B}_2\mathbf{D}_K\mathbf{D}_{21})^T & -\gamma\mathbf{I} \end{pmatrix}$$

$$\Psi_{21} = \begin{pmatrix} \mathcal{A}_K + (\mathbf{A} + \mathbf{B}_2\mathbf{D}_K\mathbf{C}_2)^T & \mathbf{Y}\mathbf{B}_1 + \mathcal{B}_K\mathbf{D}_{21} \\ \mathbf{C}_1\mathbf{X} + \mathbf{D}_{12}\mathcal{C}_K & \mathbf{D}_{11} + \mathbf{D}_{12}\mathbf{D}_K\mathbf{D}_{21} \end{pmatrix}$$

$$\Psi_{22} = \begin{pmatrix} \mathbf{A}^T\mathbf{Y} + \mathbf{Y}\mathbf{A} + \mathcal{B}_K\mathbf{C}_2 + \mathbf{C}_2^T\mathcal{B}_K^T & (\mathbf{C}_1 + \mathbf{D}_{12}\mathbf{D}_K\mathbf{C}_2)^T \\ \mathbf{C}_1 + \mathbf{D}_{12}\mathbf{D}_K\mathbf{C}_2 & -\gamma\mathbf{I} \end{pmatrix}.$$

As before, let $\mathbf{A}_c$ be the $\mathbf{A}$-matrix of the feedback system.

Theorem 10.3.3. *It is possible to find a controller* $\mathbf{K}$ *of the form (10.31) such that*

$$\mathbf{A}_c \textit{ is } \mathcal{D}\textit{-stable and } \|\mathbf{F}_{zw}\|_\infty < \gamma$$

if and only if there are matrices $\mathcal{A}_K$, $\mathcal{B}_K$, $\mathcal{C}_K$, $\mathbf{D}_K$ *and symmetric* $n \times n$*-matrices* $\mathbf{X}$ *and* $\mathbf{Y}$ *such that the following set of LMIs holds:*

$$\begin{pmatrix} \mathbf{X} & \mathbf{I} \\ \mathbf{I} & \mathbf{Y} \end{pmatrix} > \mathbf{0}$$

$$\left(\lambda_{kl}\begin{pmatrix} \mathbf{X} & \mathbf{I} \\ \mathbf{I} & \mathbf{Y} \end{pmatrix} + \mu_{kl}\Phi + \mu_{lk}\Phi^T \right) < \mathbf{0}$$

$$\begin{pmatrix} \boldsymbol{\Psi}_{11} & \boldsymbol{\Psi}_{21}^T \\ \boldsymbol{\Psi}_{21} & \boldsymbol{\Psi}_{22} \end{pmatrix} < \mathbf{0}.$$

Synthesis. If the matrices of the preceding theorem are found, a controller can be calculated as follows. Note that $\mathbf{X} - \mathbf{Y}^{-1} > \mathbf{0}$.

Step 1. Calculate via singular value decomposition a factorization

$$\mathbf{X}_2 \mathbf{X}_2^T = \mathbf{X} - \mathbf{Y}^{-1}$$

with a square (and invertible) matrix $\mathbf{X}_2$ and define $\mathbf{Y}_2 = -\mathbf{Y}\,\mathbf{X}_2$.

Step 2. Solve the systems (10.32), (10.33) (10.34) to get $\mathbf{B}_K, \mathbf{C}_K, \mathbf{A}_K$.

Remarks. **1.** With this approach, it is also possible to synthesize a $\mathcal{D}$-stable controller with a mixed $\mathcal{H}_\infty / \mathcal{H}_2$ performance specification.

2. This kind of $\mathcal{H}_\infty$ synthesis is implemented in the Matlab LMI Control Toolbox [47]. In Sect. 10.5.3, we consider a simple numerical example.

10.4 Gain Scheduling

In most practical problems, the plant is nonlinear and will be linearized before a synthesis method can be applied. It often happens that during the operation the system does not stay in the neighborhood of the equilibrium point. Then it may turn out to become necessary to adapt the controller to a new equilibrium point. As an example, consider an aircraft where the plant strongly depends on the velocity. Then the plant is a so-called **linear parameter-varying system (LPV system)**. Such a system is of the form

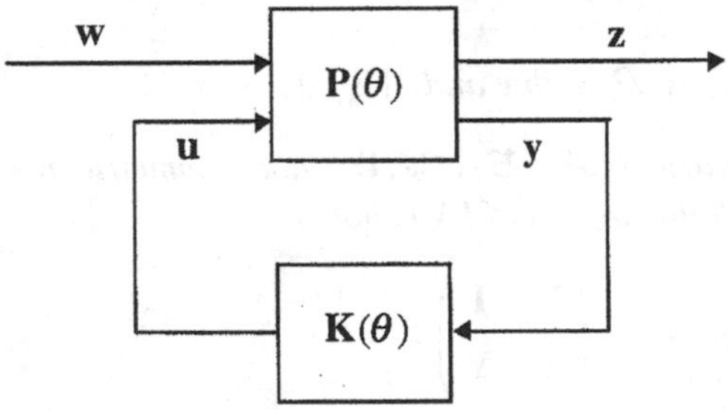

Fig. 10.2. LPV system

$$\begin{aligned}\dot{\mathbf{x}} &= \mathbf{A}(\theta(t))\mathbf{x} + \mathbf{B}_1(\theta(t))\mathbf{w} + \mathbf{B}_2(\theta(t))\mathbf{u}\\ \mathbf{z} &= \mathbf{C}_1(\theta(t))\mathbf{x} + \mathbf{D}_{11}(\theta(t))\mathbf{w} + \mathbf{D}_{12}(\theta(t))\mathbf{u}\\ \mathbf{y} &= \mathbf{C}_2(\theta(t))\mathbf{x} + \mathbf{D}_{21}(\theta(t))\mathbf{w} + \mathbf{D}_{22}(\theta(t))\mathbf{u}\,.\end{aligned} \tag{10.35}$$

The system matrices are now time-dependent. An important point is that the parameters are not known in advance, but result when the time t is reached by measuring the parameter vector θ. The idea is to adapt the controller in dependence on θ to the actual plant. This is known as **gain scheduling** and is well known in the engineering praxis. Here, the controller is designed at discrete points θ_k, and for a general parameter vector θ a certain interpolation procedure is applied. The approach appears to be plausible, but has no theoretical basis. Using the LMI design, it is possible give a theoretical foundation of gain scheduling which guarantees stability and performance independently of θ.

In order to get results which are of practical use in the design process, some assumptions concerning the dependence of the parameter are necessary.

(GS1) *The parameter region* Π *is the convex hull of the points* $\theta_1,\ldots,\theta_r \in \mathbb{R}^l$:

$$\Pi = co\{\theta_1,\ldots,\theta_r\} = \left\{ \sum_{k=1}^{r} v_k\theta_k \;\middle|\; v_k \geq 0,\ \sum_{k=1}^{r} v_k = 1 \right\}.$$

(GS2) *The matrices of the system (10.35) depend affinely on the parameter* θ .

The idea is now to construct controllers for the vertices of the polytop and to use for an arbitrary point of the convex polytop a convex combination of these controllers. The controllers are written in the form

$$\mathbf{J}(\theta) = \begin{pmatrix} \mathbf{A}_K(\theta) & \mathbf{B}_K(\theta) \\ \mathbf{C}_K(\theta) & \mathbf{D}_K(\theta) \end{pmatrix}.$$

They are assumed to have the order n of the plant (this is always possible). A realization of the feedback system $\mathbf{F}_{zw}(\theta)$ results from the formulas of Sect. 10.1.2. Later we will need the property that the matrices $\mathbf{F}_{zw}(\theta)$ also depend affinely on θ . This leads to the following assumption:

(GS3) *The matrices* $\mathbf{B}_2, \mathbf{C}_2, \mathbf{D}_{12}, \mathbf{D}_{21}$ *do not depend on* θ .

At first sight, this assumption seems to be restrictive. As we will see at the end of this section it can be always fulfilled without a severe restriction of generality. Moreover, as before, suppose $\mathbf{D}_{22} = \mathbf{0}$. Then we get

$$\begin{pmatrix} \mathbf{A}_c(\theta) & \mathbf{B}_c(\theta) \\ \mathbf{C}_c(\theta) & \mathbf{D}_{11}(\theta) \end{pmatrix} = \begin{pmatrix} \mathbf{A}_e(\theta) & \mathbf{B}_e(\theta) \\ \mathbf{C}_e(\theta) & \mathbf{D}_{11e}(\theta) \end{pmatrix} + \begin{pmatrix} \mathbf{B}_{2e} \\ \mathbf{D}_{12e} \end{pmatrix} \mathbf{J}(\theta) \begin{pmatrix} \mathbf{C}_{2e} & \mathbf{D}_{21e} \end{pmatrix}.$$

Define for $k = 1,\ldots,r$

$$\begin{pmatrix} \mathbf{A}_{ck} & \mathbf{B}_{ck} \\ \mathbf{C}_{ck} & \mathbf{D}_{ck} \end{pmatrix} = \begin{pmatrix} \mathbf{A}_c(\theta_k) & \mathbf{B}_c(\theta_k) \\ \mathbf{C}_c(\theta_k) & \mathbf{D}_c(\theta_k) \end{pmatrix},$$

$$\mathbf{J}_k = \begin{pmatrix} \mathbf{A}_K(\theta_k) & \mathbf{B}_K(\theta_k) \\ \mathbf{C}_K(\theta_k) & \mathbf{D}_K(\theta_k) \end{pmatrix}.$$

Let $\theta \in \Pi$ be arbitrarily given as

$$\theta = \sum_{k=1}^{r} v_k \theta_k \quad \text{with} \quad v_k \geq 0, \ \sum_{k=1}^{r} v_k = 1. \tag{10.36}$$

Then for the parameter θ define the controller

$$\mathbf{J}(\theta) = \sum_{k=1}^{r} v_k \mathbf{J}_k \,. \tag{10.37}$$

Because of (GS2) and (GS3), the dependence

$$\theta \to \begin{pmatrix} \mathbf{A}_c(\theta) & \mathbf{B}_c(\theta) \\ \mathbf{C}_c(\theta) & \mathbf{D}_c(\theta) \end{pmatrix}$$

is affine. We introduce now the map

$$\mathcal{B}(\mathbf{A},\mathbf{B},\mathbf{C},\mathbf{D},\mathbf{Z}) = \begin{pmatrix} \mathbf{A}^T\mathbf{Z}+\mathbf{Z}\mathbf{A} & \mathbf{Z}\mathbf{B} & \mathbf{C}^T \\ \mathbf{B}^T\mathbf{Z} & -\mathbf{I} & \mathbf{D}^T \\ \mathbf{C} & \mathbf{D} & -\mathbf{I} \end{pmatrix}.$$

It is affin in the first four parameters but not in all variables. Let the condition

$$\mathcal{B}(\mathbf{A}_{ck},\mathbf{B}_{ck},\mathbf{C}_{ck},\mathbf{D}_{ck},\mathbf{Z}) < \mathbf{0} \quad \text{for } k = 1,\ldots,r\,. \tag{10.38}$$

be fulfilled with a matrix $\mathbf{Z}$ which is independent of k. Thus for θ as in (10.36) we obtain

$$\mathcal{B}(\mathbf{A}_c(\theta),\mathbf{B}_c(\theta),\mathbf{C}_c(\theta),\mathbf{D}_c(\theta),\mathbf{Z}) = \sum_{k=1}^{r} v_k \mathcal{B}(\mathbf{A}_{ck},\mathbf{B}_{ck},\mathbf{C}_{ck},\mathbf{D}_{ck},\mathbf{Z})$$

and consequently

$$\mathcal{B}(\mathbf{A}_c(\theta),\mathbf{B}_c(\theta),\mathbf{C}_c(\theta),\mathbf{D}_c(\theta),\mathbf{Z}) < \mathbf{0} \quad \text{for every } \theta \in \Pi\,.$$

Corollary 10.1.1 shows that $\| \mathbf{F}_{zw}(\theta) \|_\infty < 1$ for every $\theta \in \Pi$. The matrix $\mathbf{Z}$ is now constructed as in Sect. 10.1.3 with the only difference that the inequalities have to be fulfilled commonly for all parameters θ_k with matrices $\mathbf{X}$ and $\mathbf{Y}$. We summarize our considerations in the following theorem.

Theorem 10.4.1 (Gain Scheduling, Existence). *Let the assumptions (GS1) – (GS3) be fulfilled. Then there is a stabilizing controller with*

$$\| \mathbf{F}_{zw}(\theta) \|_\infty < 1 \qquad \text{for every } \theta \in \Pi \tag{10.39}$$

if there are symmetric matrices $\mathbf{X} > \mathbf{0}$ and $\mathbf{Y} > \mathbf{0}$ such that the following matrix inequalities hold for every $k = 1, \ldots, r$:

$$\begin{pmatrix} \mathbf{N}_o & \mathbf{0} \\ \mathbf{0} & \mathbf{I} \end{pmatrix}^T \begin{pmatrix} \mathbf{A}_k^T \mathbf{X} + \mathbf{X}\mathbf{A}_k & \mathbf{X}\mathbf{B}_{1k} & \mathbf{C}_{1k}^T \\ \mathbf{B}_{1k}^T \mathbf{X} & -\mathbf{I} & \mathbf{D}_{11k}^T \\ \mathbf{C}_{1k} & \mathbf{D}_{11k} & -\mathbf{I} \end{pmatrix} \begin{pmatrix} \mathbf{N}_o & \mathbf{0} \\ \mathbf{0} & \mathbf{I} \end{pmatrix} < \mathbf{0}$$

$$\begin{pmatrix} \mathbf{N}_c & \mathbf{0} \\ \mathbf{0} & \mathbf{I} \end{pmatrix}^T \begin{pmatrix} \mathbf{A}_k \mathbf{Y} + \mathbf{Y}\mathbf{A}_k^T & \mathbf{Y}\mathbf{C}_{1k}^T & \mathbf{B}_{1k} \\ \mathbf{C}_{1k} \mathbf{Y} & -\mathbf{I} & \mathbf{D}_{11k} \\ \mathbf{B}_{1k}^T & \mathbf{D}_{11k}^T & -\mathbf{I} \end{pmatrix} \begin{pmatrix} \mathbf{N}_c & \mathbf{0} \\ \mathbf{0} & \mathbf{I} \end{pmatrix} < \mathbf{0},$$

$$\begin{pmatrix} \mathbf{X} & \mathbf{I} \\ \mathbf{I} & \mathbf{Y} \end{pmatrix} \geq \mathbf{0}.$$

Here $\mathbf{N}_o$ and $\mathbf{N}_c$ are matrices of full rank with the property

$$\mathbf{N}_o = \operatorname{Ker}\begin{pmatrix} \mathbf{C}_2 & \mathbf{D}_{21} \end{pmatrix}$$
$$\mathbf{N}_c = \operatorname{Ker}\begin{pmatrix} \mathbf{B}_2^T & \mathbf{D}_{12}^T \end{pmatrix}.$$

Gain Scheduling, Synthesis. *If matrices $\mathbf{X}, \mathbf{Y}$ with the above properties are found, then a matrix $\mathbf{Z}$ as in Theorem 10.1.3 has to be constructed. The next step is to solve matrix inequality (10.38) for $k = 1, \ldots, r$ in order to get $\mathbf{J}_k$. The controller belonging to a parameter $\theta \in \Pi$ is given by (10.36) and (10.37) and has the property that $\mathbf{A}_c(\theta)$ is stable and that the closed-loop system fulfills (10.39).*

If the requirement (GS3) does not hold, it is possible to extend the original problem such that for the extended system (GS3) holds. To this end, define a new input $\tilde{\mathbf{u}}$ and a new measured output by adding two new systems as follows:

$$\dot{\mathbf{x}}_u = \mathbf{A}_u \mathbf{x}_u + \mathbf{B}_2 \tilde{\mathbf{u}}$$
$$\mathbf{u} = \mathbf{C}_u \mathbf{x}_u$$
$$\dot{\mathbf{x}}_y = \mathbf{A}_y \mathbf{x}_y + \mathbf{B}_y \mathbf{y}$$
$$\tilde{\mathbf{y}} = \mathbf{C}_y \mathbf{x}_y .$$

Replacing $\mathbf{u}$ by $\tilde{\mathbf{u}}$ and $\mathbf{y}$ by $\tilde{\mathbf{y}}$ in the plant (10.35) yields

$$\begin{pmatrix} \dot{\mathbf{x}} \\ \dot{\mathbf{x}}_u \\ \dot{\mathbf{x}}_y \end{pmatrix} = \begin{pmatrix} \mathbf{A}(\theta(t)) & \mathbf{B}_2(\theta(t))\mathbf{C}_u & \mathbf{0} \\ \mathbf{0} & \mathbf{A}_u & \mathbf{0} \\ \mathbf{B}_2\mathbf{C}_2(\theta(t)) & \mathbf{0} & \mathbf{A}_y \end{pmatrix} \begin{pmatrix} \mathbf{x} \\ \mathbf{x}_u \\ \mathbf{x}_y \end{pmatrix}$$

$$+ \begin{pmatrix} \mathbf{B}_1(\theta(t)) \\ \mathbf{0} \\ \mathbf{B}_y\mathbf{D}_{21}(\theta(t)) \end{pmatrix} \mathbf{w} + \begin{pmatrix} \mathbf{0} \\ \mathbf{B}_u \\ \mathbf{0} \end{pmatrix} \tilde{\mathbf{u}}$$

$$\begin{pmatrix} \mathbf{z} \\ \tilde{\mathbf{y}} \end{pmatrix} = \begin{pmatrix} \mathbf{C}_1(\theta(t)) & \mathbf{D}_{12}(\theta(t))\mathbf{C}_u & \mathbf{0} \\ \mathbf{0} & \mathbf{0} & \mathbf{C}_y \end{pmatrix} \begin{pmatrix} \mathbf{x} \\ \mathbf{x}_u \\ \mathbf{x}_y \end{pmatrix} + \begin{pmatrix} \mathbf{D}_{11}(\theta(t)) & \mathbf{0} \\ \mathbf{0} & \mathbf{0} \end{pmatrix} \begin{pmatrix} \mathbf{w} \\ \tilde{\mathbf{u}} \end{pmatrix}.$$

For this extended plant, (GS3) holds. The additional systems are chosen to be stable with a sufficient damping and such that they have a bandwidth, which is considerably larger then that of the original plant. Then the new and the old plant have a similar dynamic behavior. In many cases, an artificial extension of the plant is not necessary, because it suffices to model the actuator and the sensor dynamics. Their behavior typically does not depend on the equilibrium point.

Example. We consider a very simple example. Let the plant be given by

$$\begin{aligned} \dot{x}_1 &= a_1 x_1 + x_2 \\ \dot{x}_2 &= a_2 x_2 + u \\ y &= x_1 . \end{aligned}$$

The parameters a_1 and a_2 are supposed to vary within certain known bounds: $a_{1l} \le a_1 \le a_{1u}$, $a_{2l} \le a_2 \le a_{2u}$. Moreover, we assume that a_1 and a_2 are functions of a quantity that can be measured, for example, the measurement y itself: $a_1 = a_1(y)$, $a_2 = a_2(y)$. Then we define as parameter vector and as vertices

$$\theta = \begin{pmatrix} a_1 \\ a_2 \end{pmatrix}, \quad \theta_1 = \begin{pmatrix} a_{1l} \\ a_{2l} \end{pmatrix}, \quad \theta_2 = \begin{pmatrix} a_{1u} \\ a_{2l} \end{pmatrix}, \quad \theta_3 = \begin{pmatrix} a_{1l} \\ a_{2u} \end{pmatrix}, \quad \theta_4 = \begin{pmatrix} a_{1u} \\ a_{2u} \end{pmatrix}.$$

With the projections $p_1(\mathbf{x}) = x_1$ and $p_2(\mathbf{x}) = x_2$, the parameter dependency of $\mathbf{A}$ can be written as

$$\mathbf{A}(\theta) = \begin{pmatrix} p_1(\theta) & 1 \\ p_2(\theta) & 0 \end{pmatrix}.$$

An elaborate example for a gain scheduled controller can be found in Apkarian, Gahinet and Becker [3].

10.5 $\mathcal{H}_\infty$ Optimal Control for a Second-Order Plant

10.5.1 Choice of the Weights

In this section, we design a $\mathcal{H}_\infty$ controller for several second-order plants. In this way, we get familiar with this kind of controller synthesis and with the basic properties of $\mathcal{H}_\infty$ controllers. We choose simple weights so that it will be possible to compare these controllers with conventional PIDT$_1$ controllers.

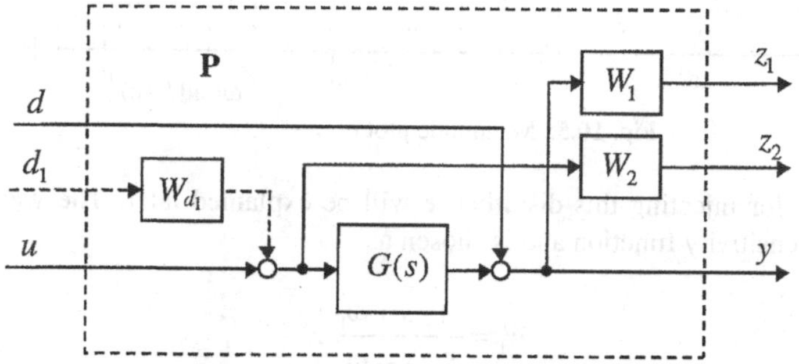

Fig. 10.3. Mixed-sensitivity specification for the second order plant

Let the second-order plant of the form

$$G(s) = \frac{1}{s^2 + as + b}$$

be given and specify the controller by a mixed-sensitivity approach as depicted in Fig. 10.3. In comparison to the mixed-sensitivity specifications in Sect. 9.6.1, there is a possible additional disturbance d_1 at the plant input with weight W_{d1}.

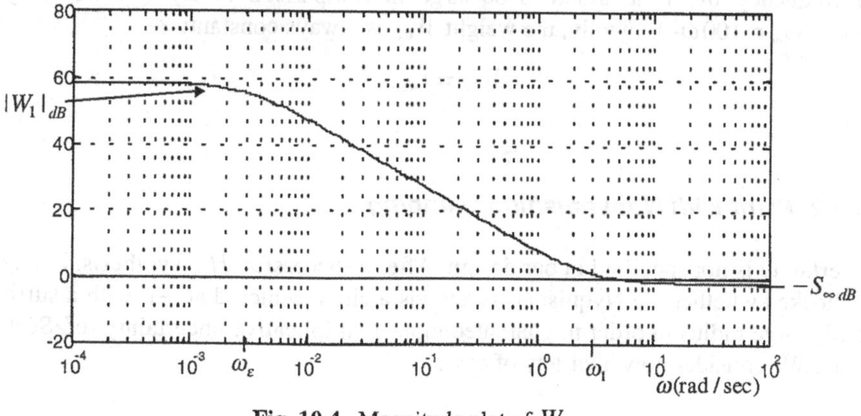

Fig. 10.4. Magnitude plot of W_1

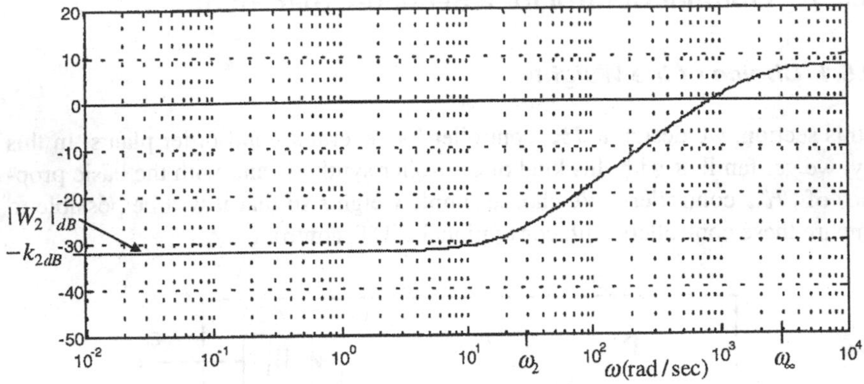

Fig. 10.5. Magnitude plot of W_2

The reason for injecting this disturbance will be explained later. The weight W_1 forms the sensitivity function and is chosen as

$$W_1 = \frac{1}{S_\infty} \frac{s+\omega_1}{s+\omega_\varepsilon}.$$

Then ω_1 specifies the bandwidth of the feedback system. Ideally, ω_ε would be equal to zero, but in order to get a stable weight, let $0 < \omega_\varepsilon \ll \omega_1$. Here we take in most cases $\omega_\varepsilon = 0.001\omega_1$. The second weight forms KS and in the simplest case it may be constant. If a limited actuator bandwidth has to be taken account, it is reasonable to define

$$W_2 = \frac{\omega_\infty}{k_2\omega_2} \frac{s+\omega_2}{s+\omega_\infty}.$$

Here, k_2 and ω_2 reflect the limited gain and bandwidth of the actuator. The corner frequency ω_∞ is assumed to be large in comparison to ω_2 (we normally choose $\omega_\infty = 100\,\omega_2$). Finally, the weight W_{d_1} is always constant:

$$W_{d_1} = k_{d_1}.$$

10.5.2 Plant with Nonvanishing Damping

Uncertainty is not specified in our design. After a successful $\mathcal{H}_\infty$ synthesis, it may be checked whether the Nyquist curve avoids a circle centered at -1 with a sufficiently large radius in order to guarantee input-multiplicative uncertainty (cf. Sect. 4.1.1). We consider now a variety of cases.

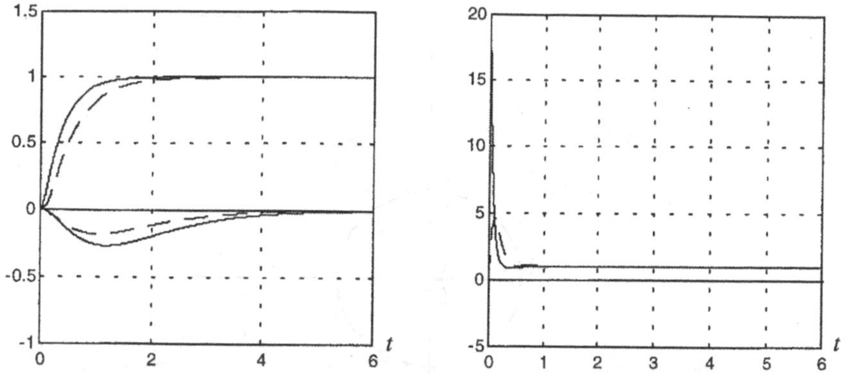

Fig. 10.6. Step response for feedback system 1; solid line: constant W_2, dashed line: dynamic W_2

Plant 1. $G(s)$ *is a series connection of two first-order plants.* Let

$$G(s) = \frac{1}{(1+T_1 s)(1+T_2 s)} \quad \text{with} \quad T_1 = 1,\ T_2 = 0.4$$

and

$$\omega_1 = 3, \quad S_\infty = 1.2, \quad W_2 = 1/k_2, \quad k_2 = 40\,.$$

The resulting controller is

$$K_\infty(s) = \frac{61992\,(s+2.5)\,(s+1)}{(s+3014)\,(s+20.16)\,(s+0.003)} \approx \frac{20.57\,(s+2.5)\,(s+1)}{s\,(s+20.16)}$$

with optimal value $\gamma_{\text{opt}} = 0.979$. As predicted by the theory (cf. Sect. 9.6.2), it cancels the linear factors belonging to the two plant poles and is approximately a PIDT$_1$ controller. Fig. 10.6 (left-hand plot) shows the step responses for the reference signal r and a disturbance at the plant input (cf. Fig. 3.3). In the right-hand plot, the control u is plotted for a unit step of r. The same holds for all other figures in the sequel where step responses are plotted. We observe that the magnitude of the control is very high for small times. This is quite typical for PIDT$_1$ controllers.

The peak can be avoided if the limited bandwidth of the actuator is taken into account by a properly dimensioned dynamical weight W_2. For this reason, we design a second controller by specifying

$$\omega_1 = 2, \quad S_\infty = 1.2, \quad k_2 = 30, \quad \omega_2 = 6\,.$$

In order to get an optimal value near 1, it was necessary to relax the requirement for the bandwidth slightly. A suboptimal controller is (with $\gamma_{\text{opt}} = 1.046$)

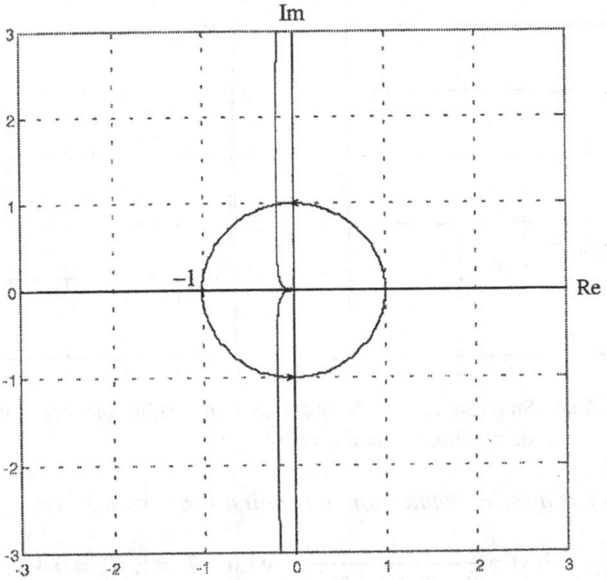

Fig.10.7. Nyquist plot of $L(s)$ for the controller with dynamic weight W_2

$$K_\infty(s) = \frac{178\,(s+600)\,(s+2.5)\,(s+1)}{(s+942)\,(s+0.002)\,(s^2+20.21s+178.7)} \approx \frac{0.189\,(s+600)\,(s+2.5)\,(s+1)}{s\,(s^2+20.21s+178.7)}$$

and the simulation result is also depicted in Fig. 10.6 (dashed lines). As can be seen from the figures, the actuator dynamics is much less required by this controller, and the step responses are somewhat slower. Due to the dynamical weight, the controller order has increased by 1. From Fig. 10.7, it is seen that the feedback system has very good robustness properties in terms of phase reserve and gain margin.

Plant 2. $G(s)$ *is a second-order plant with good (positive) damping.* Let

$$G(s) = \frac{\omega_0^2}{s^2 + 2d\omega_0 s + \omega_0^2} \quad \text{with } \omega_0 = 1,\ d = 0.7$$

and

$$\omega_1 = 3, \quad S_\infty = 1.2, \quad W_2 = 1/k_2, \quad k_2 = 80\,.$$

A suboptimal controller is now

$$K_\infty(s) = \frac{15287\,(s^2+1.4s+1)}{(s+341.3)\,(s+18)\,(s+0.003)} \approx \frac{15287\,(s^2+1.4s+1)}{s\,(s+18)}\,.$$

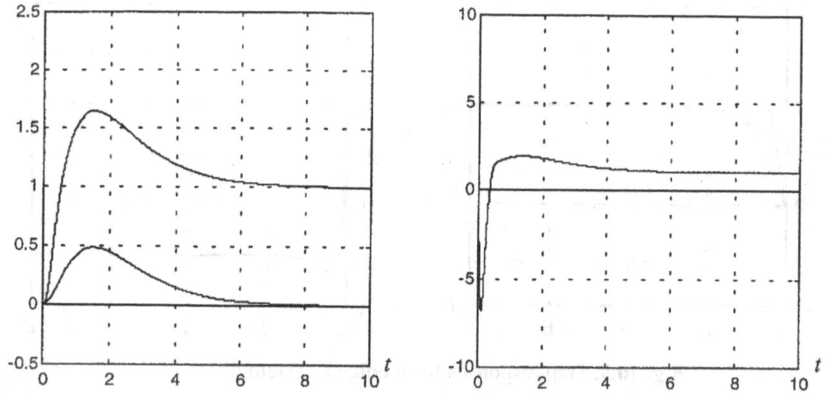

Fig. 10.8. Step responses for feedback system 3

This is also a PIDT_1 controller, and again the plant is inverted, but the quadratic numerator can no longer be decomposed into two real linear factors. The step responses resemble those of the previous example for the constant weight W_2 and are not shown here.

Plant 3. *$G(s)$ is a series connection of a stable and an unstable first-order plant:*

$$G(s) = \frac{1}{(1 - T_1 s)(1 + T_2 s)} \quad \text{with} \quad T_1 = 1,\ T_2 = 0.4\,.$$

The parameters for the weights are selected as

$$\omega_1 = 1, \quad S_\infty = 1.4, \quad k_2 = 40, \quad \omega_2 = 2\,.$$

A suboptimal controller is (with dynamic weight W_2)

$$K_\infty(s) = \frac{-189\,(s + 600)\,(s + 2.5)\,(s + 0.2795)}{(s + 632.8)\,(s + 0.001)\,(s^2 + 20.11s + 180.4)}\,.$$

The linear factor belonging to the stable pole is cancelled by the controller. The reflected unstable pole is a pole of the sensitivity function S. The step responses are shown in Fig. 10.8. The step response for the reference signal has a large overshoot, which could be reduced by a design with a prefilter, but we do not make any attempt in this direction.

Plant 4. *$G(s)$ is a series connection of a pure integrator and a first-order plant.* Then we have $b = 0$ and the plant can be written in the form

$$G(s) = \frac{1}{s\,(s + a)}\,.$$

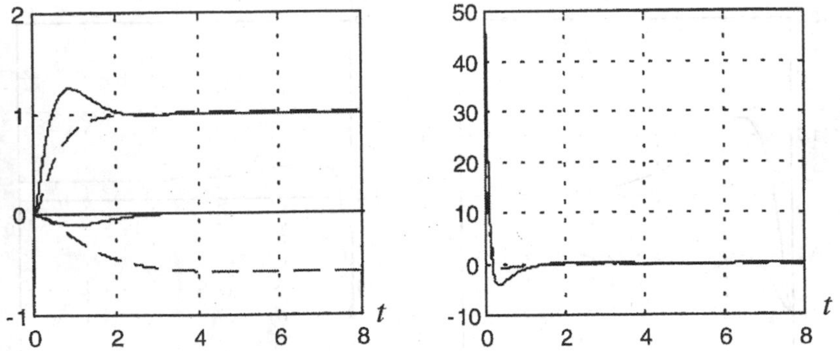

Fig. 10.9. Step responses for feedback system 4

Choose $a = 1$ and let the numerical values of the weights be given by

$$\omega_1 = 2, \quad S_\infty = 1.2, \quad k_2 = 40\,.$$

Since the plant has a pole on the imaginary axis, direct application of the software leads to an error (cf. Sect. 9.6.2). A simple way to solve this problem is to perturb the plant slightly (with a small $\varepsilon > 0$):

$$G(s) = \frac{1}{(s+\varepsilon)(s+a)}\,.$$

The simulation result is depicted in Fig. 10.9 (dashed lines). As can be seen, the disturbance at the plant input does not tend to zero. If we repeat the synthesis with

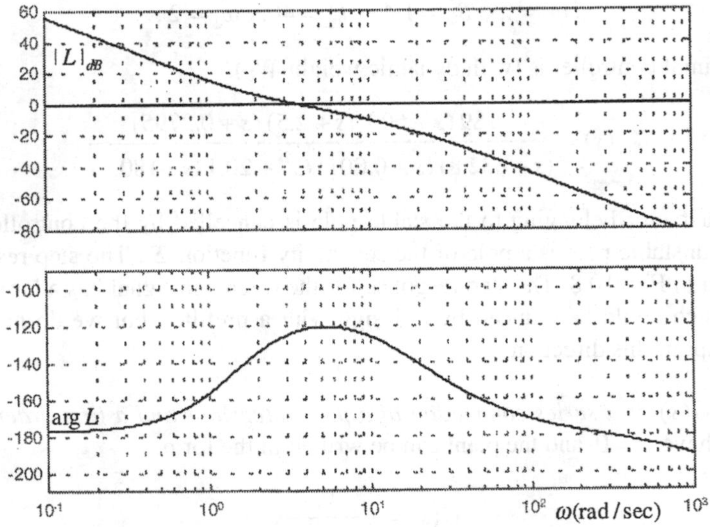

Fig. 10.10. Bode plot of $L(s)$ for feedback system 4

an additional perturbation d_1 as shown in Fig. 10.3 (and $k_{d1} = 4$, $k_2 = 100$), the disturbance will be rejected (Fig. 10.9, solid lines). A suboptimal controller is (with $\gamma_{opt} = 1.06$)

$$K_\infty(s) = \frac{61254\,(s^2 + 2.391s + 1.82)}{(s + 935.9)\,(s + 18.93)\,(s + 0.002)} \approx \frac{65.45\,(s^2 + 2.391s + 1.82)}{s\,(s + 18.93)}.$$

The bode plot of $L(s)$ for this system (Fig. 10.10) has some similarity with that which is obtained by applying the symmetrical optimum (Fig. 4.7).

Plant 5. $G(s)$ *is a double integrator:*

$$G(s) = \frac{1}{s^2}.$$

Again, for controller design the plant has to be perturbed:

$$G(s) = \frac{1}{(s+\varepsilon)^2}.$$

The results are quite similar to that of the previous example and therefore not shown here.

10.5.3 Plant with a Pole Pair on the Imaginary Axis

We continue our study for the case where the plant has a pole pair on the imaginary axis, a case which is excluded by the $\mathcal{H}_\infty$ theory (cf. Sect. 9.6.2). As for plants with an integrator, we slightly perturb such a pole pair such that the damping becomes a small positive number. Then, with the S/KS scheme, these poles remain poles of the feedback loop. If an additional perturbation at the plant input

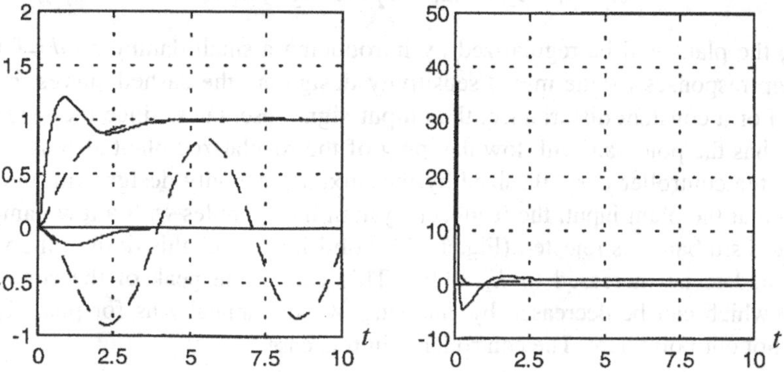

Fig. 10.11. Step responses for feedback system 6 (Riccati Design)

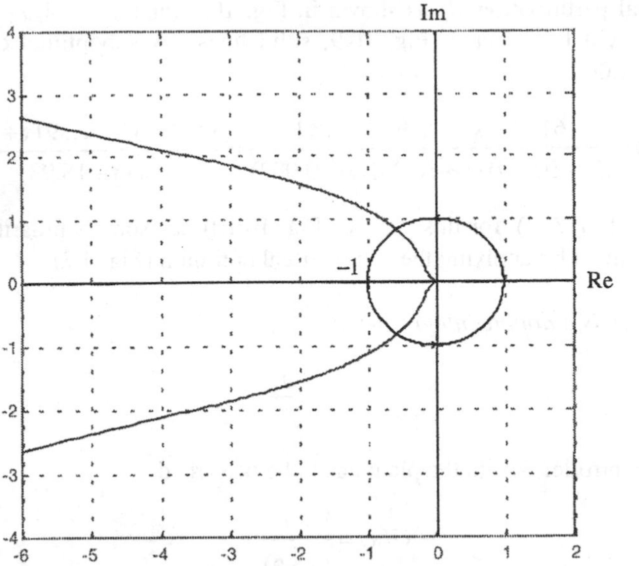

Fig.10.12. Nyquist plot of $L(s)$ for the controller (10.40)

is added, no cancellations occur and the poles of the feedback loop are well damped.

Plant 6. $G(s)$ *is a second-order plant with damping zero:*

$$G(s) = \frac{\omega_0^2}{s^2 + \omega_0^2} \quad \text{with } \omega_0 = 1.$$

(a) Mixed-sensitivity design. We again use the Riccati approach for designing the $\mathcal{H}_\infty$ controller, but in this case, we additionally apply LMIs for controller synthesis. Let

$$\omega_1 = 1, \quad S_\infty = 1.2, \quad W_2 = 1/k_2, \quad k_2 = 8.$$

Again, the plant will be regularized by introducing a small damping ($d = 0.01$). The step responses for the mixed sensitivity design are the dashed curves in Fig. 10.11. For a constant disturbance, the output signal oscillates, since the feedback system has the pole pair with low damping of the regularized plant also as a pole pair. If the controller is synthesized by the mixed-sensitivity design with the disturbance at the plant input, the feedback system has no poles with a low damping and the disturbance is rejected (Fig. 10.11, solid lines). For this design, the value of k_2 had to be increased to $k_2 = 110$. This causes the peak of the control at $t = 0$, which can be decreased by choosing W_2 dynamically as for plant 1, but this is not our point here. The controller is in this case

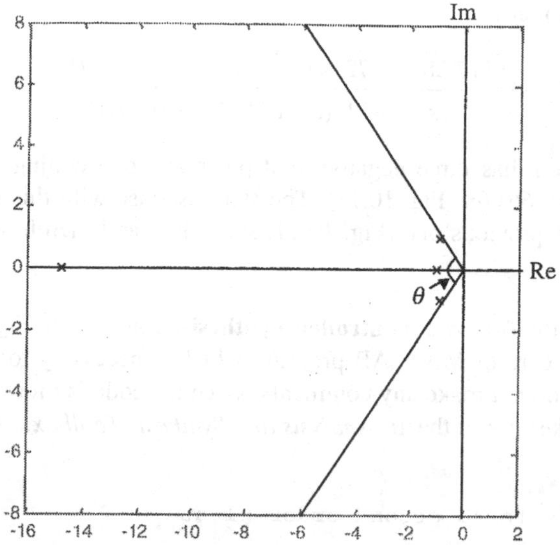

Fig. 10.13. Closed-loop poles and pole placement region for the LMI controller (system 6)

$$K_\infty(s) \approx \frac{61.58(s^2 + 1.525s + 1.109)}{s\,(s + 19.98)}. \tag{10.40}$$

The Nyquist plot of $L(s)$ with this controller is depicted in Fig. 10.12. The picture shows that the feedback system is robust with respect to input-multiplicative perturbations. Thus, with respect to our criteria, the design is successful.

(b) Mixed-sensitivity design with LMI synthesis. We now design for this plant a further controller via the LMI approach. Since the LMI theory does not exclude poles on the imaginary axis, it is interesting to see how the software handles this case. We use the same weights as before (with $k_2 = 8$). The LMI software actually works for this case and gives the controller

$$K_\infty(s) \approx \frac{6552\,(s^2 + 1)}{(s + 1457)\,(s + 5.385)\,(s + 0.000987)}.$$

It cancels the quadratic denominator of the plant. Due to Theorem 3.2.1 this necessarily leads to an unstable feedback loop. The LMI synthesis gives a closed-loop system having one eigenvalue with $d = 1.55 \times 10^{-6}$, whereas the true value is $d = 0$.

(c) LMI synthesis with pole placement. We now use ability of the LMI design to prescribe a region were the closed-loop poles have to lie. Let this region be a conic sector centered at the origin and with inner angle θ. In this way, the damping of

the closed-loop poles is at least $\cos(\theta/2)$. We require $d = 0.6$, i.e. $\theta = 106°$. The resulting controller is

$$K_\infty(s) = \frac{17.38\,(s+7215)(s^2+0.853\,s+0.614)}{(s+2832)\,(s+17.58)\,(s+0.001)}.$$

The feedback system has three negative real poles and one conjugate pole pair with damping $d = 0.661$ (cf. Fig. 10.13). The step response with this controller is quite similar to the previous one (Fig. 10.11, solid line) and therefore not shown here.

MATLAB program for $\mathcal{H}_\infty$ controller synthesis. Finally, in Fig. 10.14 we show the essential core of MATLAB program which is necessary to produce the above results. We do not make any comments, since the code is more or less self-explanatory. It makes use of the μ *-Analysis and Synthesis Toolbox.*

```
%Hinf synthesis for a second-order plant

%Data:
om0=1;om1=1;Sinf=1.2;umax =110;wd=4;
%Plant:
G=nd2sys([om0^2],[1 0 om0^2] );
Greg=nd2sys([om0^2],[1 2*0.05*om0 om0^2]);%regularized plant
%Weights:
omeps=0.01*om1;
W1=nd2sys([1,om1],[Sinf,Sinf*omeps]);
W2=nd2sys([1/umax],[1]);
Wd=nd2sys([wd],[1]);

%Extended plant
systemnames='Greg W1 W2 Wd';
inputvar='[d;d1;u]';
outputvar='[W1;W2;-Greg-d]';
input_to_Greg='[u+Wd]';
input_to_W1='[Greg+d]';
input_to_W2='[u+Wd]';
input_to_Wd='[d1]';
sysoutname='P';
cleanupsysic='yes';
sysic;

%Hinf synthesis
nmeas=1;ncon=1;gmin=0.1;gmax=200;tol=0.01;
[Kinf,ginf,gopt]=hinfsyn(P,nmeas,ncon,gmin,gmax,tol);

%Feedback system
systemnames='G Kinf';
inputvar='[r;d1]';
outputvar='[r-G;Kinf]';
```

```
input_to_G='[Kinf+d1]';
input_to_Kinf='[r-G]';
sysoutname='GRKsys';
cleanupsysic='yes';
sysic;

%Determination of the basic transfer functions
[a,b,c,d]=unpck(GRKsys);
GRKtf=tf(ss(a,b,c,d));
S=minreal(GRKtf(1,1));
Fr=-minreal(GRKtf(2,2));
KS=minreal(GRKtf(2,1));
SG=minreal(GRKtf(1,2));

%Plot of a step response
[y,t]=step(Fr,0:0.001:10);
plot(t,y);
```

Fig.10.14. MATLAB program for $\mathcal{H}_\infty$ controller synthesis

For controller synthesis with the *LMI Control Toolbox* the two lines concerning controller synthesis have to be replaced by

```
r=[0 1 1];
obj=[0 0 1 0];
region=lmireg;
[gopt,h2opt,Kinf,R,S]=hinfmix(P,r,obj,region);
```

Here, `lmireg` is an interactive function to fix the pole placement region.

Conclusions. A suboptimal mixed-sensitivity $\mathcal{H}_\infty$ controller for a second-order plant is of order 3 if the weights W_2, W_{d_1} are constant and if the weight W_1 is of order 1. If W_1 is chosen such that a reference signal has to be tracked, the suboptimal controller has a pole near 0 (cf. Theorem 9.6.1(iii)). For $\gamma \to \gamma_{\mathrm{opt}}$, the suboptimal controller can be very well-approximated by a controller of order 2. The resulting controller then has exactly the same structure as the classical PIDT_1 controller. Therefore it possible to compare directly $\mathcal{H}_\infty$ controller design with classical design for this simple, but nontrivial case. As can be seen from the discussed examples, with the $\mathcal{H}_\infty$ synthesis it is much simpler to design a controller that precisely fulfills the performance criteria than by direct application of the Nyquist criterion. Nevertheless, the full power of the $\mathcal{H}_\infty$ method will only be seen from the advanced case studies in Chaps. 11 and 14.

Notes and References

The solution of the $\mathcal{H}_\infty$ optimal control problem via LMIs was found after the solution method based on Riccati equations. A basic reference for the LMI approach is Gahinet and Apkarin [46]; compare also Packard [73] and the book of Dullerud and Paganini [37]. The LMI approach avoids the rank conditions of the Riccati approach and offers the possibility of incorporating additional requirements for the feedback system such as pole placement constraints. Concerning the last point, compare Chilali and Gahinet [26]. The LMI method is also flexible enough to handle gain scheduling; see in particular Apkarin and Gahinet [4].

Chapter 11

Case Studies for $\mathcal{H}_2$ and $\mathcal{H}_\infty$ Optimal Control

In this chapter, we apply the methods of the previous chapters to a series of classical control problems. The first one is the inverted pendulum, which is considered in almost every textbook on control. As for all the other examples, we start with a rather careful analysis of the plant dynamics. Then an LQ and an LQG controller are designed. The latter controller turns out to be not robust and the robustness properties are then improved by applying the LTR procedure.

The next example is a nonisothermal continuously stirred tank, which is an unstable MIMO system. Our approach is to feed back the state vector and to design the controller by pole positioning. One of the four states cannot be measured and this state will be estimated by a reduced observer.

The third case study is devoted to the lateral motion of an aircraft. The analysis of the plant shows that one eigenvalue is badly damped and the first task is to improve its damping. We start with a classical approach, which uses a so-called washout filter. Next an LQ controller for the lateral motion is designed. Performance of the closed-loop system is mainly expressed by its pole positions. It is assumed that all states can be measured by an inertial navigation system. In this way, it is possible to shift the badly damped eigenvalue to a region where the damping is better. This controller is compared with a $\mathcal{H}_\infty$ controller, which is designed by a mixed sensitivity approach. As expected from the general theory, this controller has no satisfactory disturbance rejection properties. A controller with satisfactory performance is then obtained by applying the mixed sensitivity method to the prestabilized plant.

11.1 Control of an Inverted Pendulum

11.1.1 Analysis of the Plant Dynamics

Our first example consists in stabilizing an inverted pendulum, which is a classical problem of modern control theory. The task is to hold a pendulum in a perpendicular position, where control is affected by a motor-driven cart (cf. Fig. 11.1). If the cart position is also prescribed, the control problem consists of stabilization and tracking.

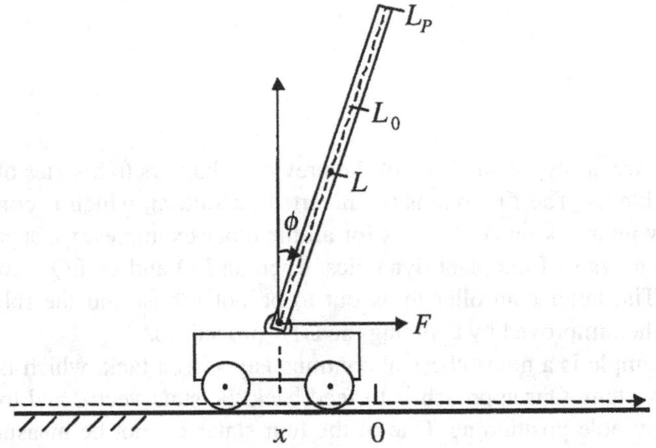

Fig. 11.1. Inverted pendulum

Derivation of the equations of motion. The physical parameters are the mass m_W of the cart, the mass m_P of the pendulum and the inertia J of the pendulum with respect to the center of gravity and its distance L to the cart. The length of the pendulum is denoted by L_P. The dynamical quantities are the cart position x, the angle ϕ of the pendulum and the force F generated by the motor.

Let F_x and F_z be the x - and z - components, respectively, of the reaction force caused by the motion of the pendulum acting on the cart and let x_P be the x - coordinate and z_P be the z - coordinate of the center of gravity of the pendulum. Then the translational and rotational dynamics of the pendulum is described by the equations

$$\begin{aligned} m_P \ddot{x}_P &= F_x \\ m_P \ddot{z}_P &= F_z - m_P g \\ J\ddot{\phi} &= LF_z \sin\phi - LF_x \cos\phi . \end{aligned}$$

The translational dynamics of the cart is given by

$$m_W \ddot{x} = F - F_x .$$

Elimination of F_x and F_z gives

$$m_W \ddot{x} + m_P \ddot{x}_P = F$$
$$J\ddot{\phi} = Lm_P \ddot{z}_P \sin\phi - Lm_P \ddot{x}_P \cos\phi + Lm_P g \sin\phi .$$

Moreover, the following geometric relationships hold:

$$x_P = x + L\sin\phi , \quad z_P = L\cos\phi .$$

Substitution of x_P and z_P and performing the differentiations yields

$$\ddot{x}\cos\phi + L_0\ddot{\phi} = g\sin\phi \tag{11.1}$$

$$m\ddot{x} + m_P L\ddot{\phi}\cos\phi = m_P L\dot{\phi}^2 \sin\phi + F , \tag{11.2}$$

if one puts

$$m = m_W + m_P , \quad L_0 = \frac{J + L^2 m_P}{Lm_P} .$$

The differential equations (11.1) and (11.2) describe the system dynamics. They can be linearized according to the procedure described in Sect. 5.1. Since an equilibrium point is given by $x_0 = 0$, $\phi_0 = 0$, we denote the deviations by the same symbol as they are used for the original quantities. Linearization of (11.1) and (11.2) yields

$$\ddot{x} + L_0\ddot{\phi} = g\,\phi \tag{11.3}$$

$$m\ddot{x} + m_P L\ddot{\phi} = F . \tag{11.4}$$

The force F is exerted by a dc motor. If its inductivity is neglected, the motor dynamics is described by the equations

$$J_M \dot{\omega} = c_A i_A - \rho F$$

$$R_A i_A = -c_A \omega + u_A . \tag{11.5}$$

Here, J_M denotes the inertia of the motor and the wheels of the cart (where the gears has to be taken into account), and ρ is the diameter of the wheels. Then we have $\dot{x} = \rho\omega$ and (11.4) has to be replaced by

$$\left(\frac{J_M}{\rho^2} + m\right)\ddot{x} + m_P L\ddot{\phi} = \frac{c_A}{\rho} i_A . \tag{11.6}$$

Concerning the choice of the control, there exist two possibilities.

Voltage u_A as control. If i_A is eliminated from (11.5), then (11.6) has to be replaced by

$$(\frac{J_M}{\rho^2}+m)\,\ddot{x}+m_P L\ddot{\phi}=-\frac{c_A^2}{\rho^2 R_A}\,\dot{x}+\frac{c_A}{\rho R_A}u_A\,. \tag{11.7}$$

Current i_A as control. In this case, a cascaded structure of the feedback loop has to be designed in which the current in the inner loop is controlled. The basic idea is as follows. Feedback of the current is done by a purely integrating controller (or by a PI controller, if the dynamics induced by the motor inductivity is taken into account):

$$\hat{u}_A=\frac{k_i}{s}(\hat{i}_{Ac}-\hat{i}_A)\,.$$

Hence, the inner loop for the current is given by

$$\hat{i}_A=\frac{1}{1+T_i s}\hat{i}_{Ac}-\frac{k_\omega s}{1+T_i s}\hat{\omega}\quad\text{with}\quad T_i=\frac{R_A}{k_i},\quad k_\omega=c_A/k_i\,.$$

The angular velocity $\omega=\rho^{-1}\dot{x}$ now plays the role of a disturbance. Since it only varies slowly in comparison to the fast inner loop, it will not seriously effect the dynamics of the inner loop. Hence, the current can be viewed as a control.

In what follows, the current is chosen as the control. Then the inverted pendulum is described by (11.3) and (11.6). We define

$$\tilde{m}=\frac{J_M}{\rho^2}+m,\quad \lambda=\tilde{m}-\frac{L}{L_0}m_P,\quad \gamma=\frac{c_A}{\rho\lambda},$$

$$\omega_1=\sqrt{\frac{m_P g}{L_0\lambda}},\quad \omega_2=\sqrt{\frac{\tilde{m}\,g}{L_0\lambda}}\,.$$

After eliminating $\ddot{x}$, $\ddot{\phi}$ from (11.3) and (11.6), the system dynamics becomes

$$\ddot{x}=-L\omega_1^2\phi+\gamma i_A$$

$$\ddot{\phi}=\omega_2^2\phi-\frac{\gamma}{L_0}i_A\,.$$

The state space description is $\dot{\mathbf{x}}=\mathbf{A}\mathbf{x}+\mathbf{b}_2\mathbf{u}$ with

$$\mathbf{A}=\begin{pmatrix}0&1&0&0\\0&0&-L\omega_1^2&0\\0&0&0&1\\0&0&\omega_2^2&0\end{pmatrix},\quad \mathbf{b}_2=\begin{pmatrix}0\\ \gamma\\ 0\\ -\gamma/L_0\end{pmatrix},\quad \mathbf{x}=\begin{pmatrix}x\\ \dot{x}\\ \phi\\ \dot{\phi}\end{pmatrix}.$$

The transfer functions are easily calculated as

$$\hat{x} = \gamma \frac{s^2 - \omega_2^2 + (L/L_0)\omega_1^2}{s^2(s^2 - \omega_2^2)} \hat{i}_A \tag{11.8}$$

$$\hat{\phi} = -\frac{\gamma}{L_0} \frac{1}{s^2 - \omega_2^2} \hat{i}_A . \tag{11.9}$$

We define now for $0 \le d \le L_P$ the position $x_d = x + d\,\phi$. It has the transfer function

$$\hat{x}_d = \gamma \frac{(1 - d/L_0)s^2 - \omega_2^2 + (L/L_0)\omega_1^2}{s^2(s^2 - \omega_2^2)} \hat{i}_A .$$

Hence with respect to x and x_d, the plant is a nonminimum-phase transfer function if $d < L_0$ holds. This is caused by the horizontal force component, which acts back to the cart. For $d = L_0$, this transfer function has no zero and is given by

$$\hat{x}_d = -\gamma \frac{\omega_2^2 - (L/L_0)\omega_1^2}{s^2(s^2 - \omega_2^2)} \hat{i}_A .$$

For $d > L_0$ it has two conjugate complex zeros on the imaginary axis. If the mass of the pendulum can be neglected with respect to that of the cart, the nonminimum phase behavior with respect to x disappears because of $\omega_1 = 0$.

Controllability, observability. Using the controllability matrix, it is easy to see that the system is controllable.

If only ϕ is measured, the system is not observable. It becomes observable when ϕ and x are measured. We still consider the interesting case where the position x_d is measured. Then (cf. Sect. 8.1.1)

$$\mathbf{C}_{10} = \begin{pmatrix} 1 & 0 & d & 0 \end{pmatrix} . \tag{11.10}$$

and the observability matrix is given by

$$\mathbf{O}_{C_{10},A} = \begin{pmatrix} 1 & 0 & d & 0 \\ 0 & 1 & 0 & d \\ 0 & 0 & \beta & 0 \\ 0 & 0 & 0 & \beta \end{pmatrix}, \qquad \beta = d\omega_2^2 - L\omega_1^2 .$$

Thus, if only x_d is measured, the system $(\mathbf{C}_{10}, \mathbf{A})$ is observable if and only if

$$d \neq L_0 \frac{m_P}{\tilde{m}} .$$

It should be noted that for a small mass of the pendulum, β is small. Thus, there is almost a pole/zero cancellation in the corresponding transfer function, and control of the inverted pendulum by measuring only x_d will not work in this case in a physical realization. If d is sufficiently large, this restriction no longer holds.

Values of the parameters. We suppose that the numerical values of the parameters are given as follows:

$$L_P = 1\,m, \quad m_P = 0.2\,kg, \quad m_W = 0.5\,kg, \quad J_M = 0,$$
$$\rho = 0.02\,m, \quad c_A = 0.116\,Vs, \quad R_A = 0.366\,\Omega, \quad T_m = 0.05\,s.$$

Remark. The linear differential equations for the crane positioning system can be obtained when the system (11.1), (11.2) is linearized at $\phi_0 = \pi$ (cf. Sect. 5.2.1).

11.1.2. Design of an LQ Controller

LQ Design. We want to design different controllers for the inverted pendulum and start with an LQ controller. Let $\mathbf{Q} = \mathbf{C}_{10}^T \mathbf{C}_{10}$ with $\mathbf{C}_{10}$ as in (11.10) and $\mathbf{R} = (\rho^{-1})$, where ρ is an arbitrary positive number. Then the assumptions of Theorem 8.1.1 are fulfilled and the controller design is particularly simple, since the controller coefficients depend only on the single parameter ρ. For the output which has to be weighted, we consider the following three different cases (here, $G(s)$ denotes the transfer function from i_A to x_d):

1. *Weighting of the cart position ($d = 0$):*

$$G(s) = 4.06\frac{(s-3.84)(s+3.84)}{s^2(s-4.33)(s+4.33)}.$$

2. *Weighting of the position $x_d = x + L_0\phi$:*

$$G(s) = -\frac{24.63}{s^2(s-4.33)(s+4.33)}.$$

3. *Weighting of the position $x_d = x + 1.2\,L_0\phi$:*

$$G(s) = -2.81\frac{s^2 + 8.58^2}{s^2(s-4.33)(s+4.33)}.$$

Figure 11.2 shows the root locus for $\rho > 0$ as a parameter. If the pendulum position $x_d = x + L_0\phi$ is weighted, the plant has no zeros, and for $\rho \to \infty$ the closed-loop poles tend along straight lines to infinity. They have the asymptotic behavior discussed in Sect. 8.4.3. The situation changes if x_d with $d \neq L_0$ is weighted, since in this case the plant zeros limit the bandwidth of the feedback system. The situation may become critical for $d > L_0$ because then the plant has a pair of zeros on the imaginary axis. This case is considered in Fig. 11.2(b) for $x_d = x + 1.2 L_0 \phi$.

Hence it is reasonable to optimize our controller for $x_d = x + L_0\phi$. With $\rho = 2500$ the feedback system has the poles

$$s_{1/2} = -4.03 \pm 8.89i, \quad s_{3/4} = -9.75 \pm 3.97i$$

with damping 0.41 and 0.94. Figure 11.3 shows the responses for x, ϕ, x_d and i_A for the initial deflection $\phi_0 = 5°$. Here it is assumed that all states are ideally measured. The initial cart position is reached again after approximately 0.7 s, and after 1 s the pendulum is in a perpendicular position. The required current is within acceptable bounds. Increasing the bandwidth of the closed-loop system leads to increased consumption of current and to an increased deflection of the pendulum. In particular the latter effect is undesired.

We assume now that the velocities are obtained from noisy measurements of the cart position x and the angle ϕ by using a differentiator of the form

$$G_D(s) = \frac{T_D s}{1 + T_D s}$$

with an appropriately chosen time constant T_D. The noise processes are supposed to be white with power spectral densities of $10^{-4} \, m^2/s$ for x and of $1 \, deg^2/s$ for ϕ. The corresponding simulation results are depicted in Fig. 11.4. As can be seen, the fluctuation of the cart position is in the region of several centimeters whereas the fluctuation of the pendulum position is very small.

The performance degradation due to sensor noise can be reduced if a smaller closed-loop bandwidth is chosen (by decreasing ρ). Then the pendulum deflec-

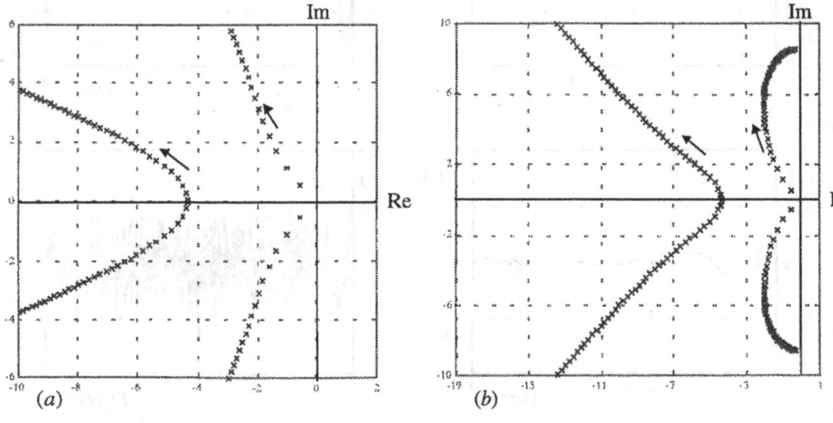

Fig. 11.2. Root locus for the LQ controller ($\rightarrow$: ρ increasing):
(a) Weighting of the position $x_d = x + L_0\phi$,
(b) Weighting of the position $x_d = x + 1.2 L_0\phi$

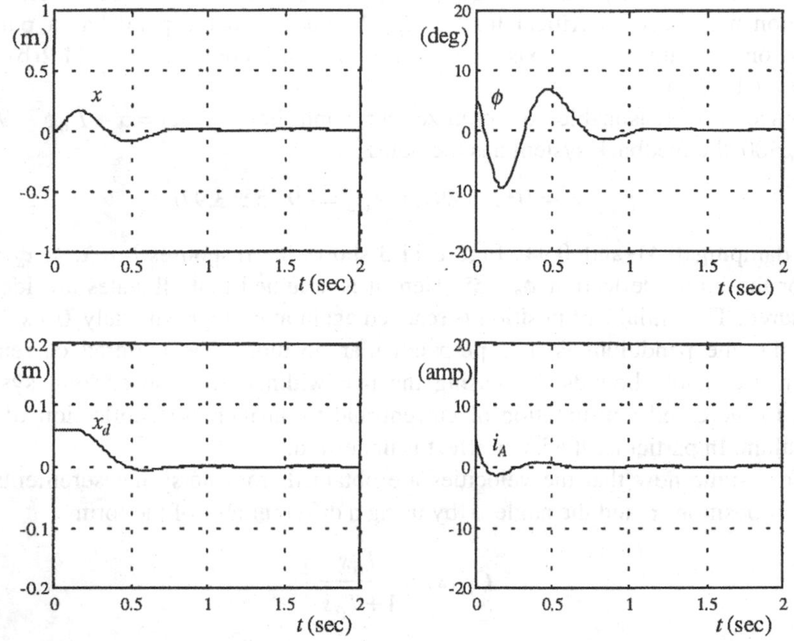

Fig. 11.3. Responses of x, ϕ, x_d, i_A with the LQ controller

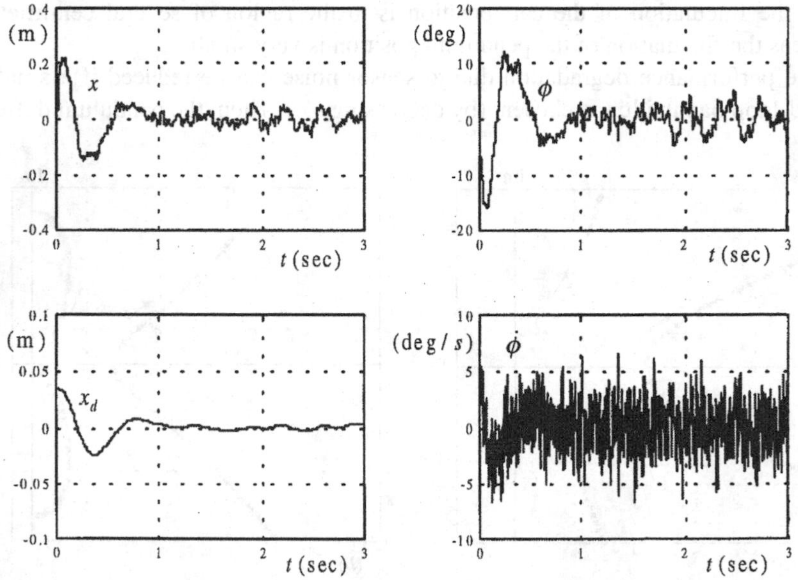

Fig. 11.4. Responses of $x, \phi, x_d, \dot{\phi}$ with the LQ controller in the presence of noise

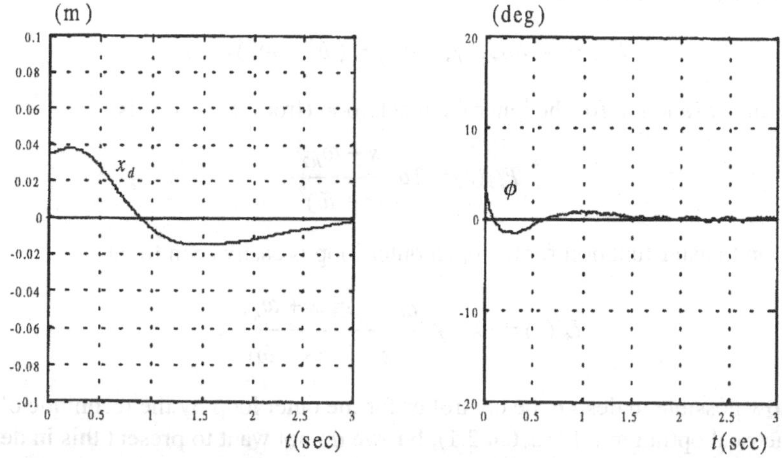

Fig. 11.5. Responses of x and ϕ with the LQ controller in the presence of noise and for a smaller closed-loop bandwidth

tions become considerably smaller, but it takes a much longer time before the pendulum position reaches its final value (cf. Fig 11.5).

Classical design. It is also possible to develop a controller by using classical methods. We only sketch one possible approach in which the basic idea is to use a cascaded structure. For simplicity, suppose $m_P = 0$ (then we also have $\omega_1 = 0$). If the current is eliminated from (11.8) and (11.9), then

$$\hat{x} = L_0 \frac{\omega_2^2 - s^2}{s^2} \hat{\phi} = G_1(s)\hat{\phi} .$$

If the transfer function from i_A to ϕ is denoted by $G_2(s)$ (cf. (11.9)), then the structure depicted in Fig. 11.6 results. The inner controller as well as the outer one is a PD controller:

$$K_1(s) = k_{R1}(s + \omega_{R1}), \quad K_2(s) = k_{R2}(s + \omega_{R2}) .$$

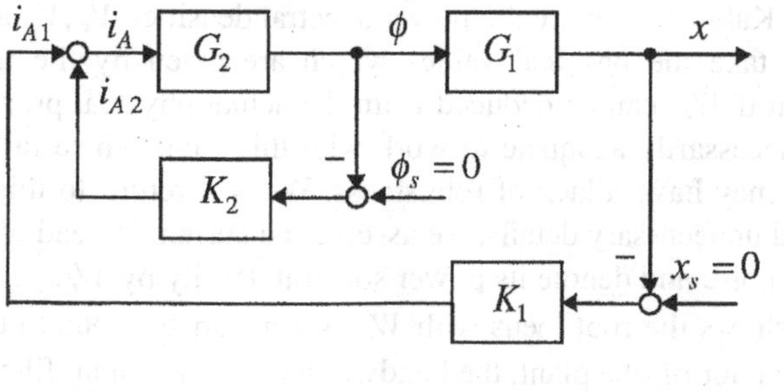

Fig. 11.6. Cascaded controller structure for the inverted pendulum

If we choose for a given $\tilde{\omega}$

$$k_{R2} = -2\tilde{\omega}L_0/\gamma, \quad \omega_{R2} = (\tilde{\omega}^2 + \omega_2^2)/2\tilde{\omega},$$

then transfer function for the inner feedback loop (from ϕ_s to ϕ) is

$$F_{r2}(s) = 2\tilde{\omega}\frac{s+\omega_{R2}}{(s+\tilde{\omega})^2}.$$

The loop transfer function for the open outer loop is easily seen to be

$$L_1(s) = -k_{R1}\gamma\frac{\omega_2^2 - s^2}{s^2}\frac{s+\omega_{R1}}{(s+\tilde{\omega})^2}.$$

It is now possible to design the controller for the outer loop by the technique of the symmetrical optimum (cf. Sect. 4.2.1), but we do not want to present this in detail. Both controllers can be written as one:

$$i_A = -k_{R1}\omega_{R1}x - k_{R1}\dot{x} - k_{R2}\omega_{R2}\phi - k_{R2}\dot{\phi}$$

which has the same structure as the LQ controller with full state feedback. We mention that not all meaningful combinations of gain coefficients can be generated in this way. In this sense, the LQ approach is more powerful. In addition, it can more easily be handled.

11.1.3. Design of an LQG Controller

Let again x and ϕ be the measured quantities, but instead of differentiating x and ϕ to get $\dot{x}$ and $\dot{\phi}$, we design a Kalman filter to estimate the whole state vector. As in the last section, x and ϕ are assumed to be corrupted by white noise. The Kalman filter and the LQG design procedure require that the plant also has a noisy input. In this case it may be assumed that the roughness of the surface causes a force acting as an additive input which behaves as white noise. It is also possible to treat this input as an artificial quantity for which the power spectral density W_F of the process serves as a design parameter. Consequently, the parameters for the Kalman filter are the power spectral densities V_x, V_ϕ and W_F. For V_x and V_ϕ we take the physical values which are given by the measurement equipment. Even if W_F can be deduced from the actual physical properties of the plant, it is not necessarily adequate to work with this value, since in this way the LQG controller may have a lack of robustness. We will return to this point later. In order to avoid unnecessary details, we assume the current instead of the force to be corrupted by noise and denote its power spectral density by W_i.

Figure 11.7 shows the root locus with W_i as a parameter. Due to the nonminimum-phase behavior of the plant, the bandwidth of the Kalman filter is limited.

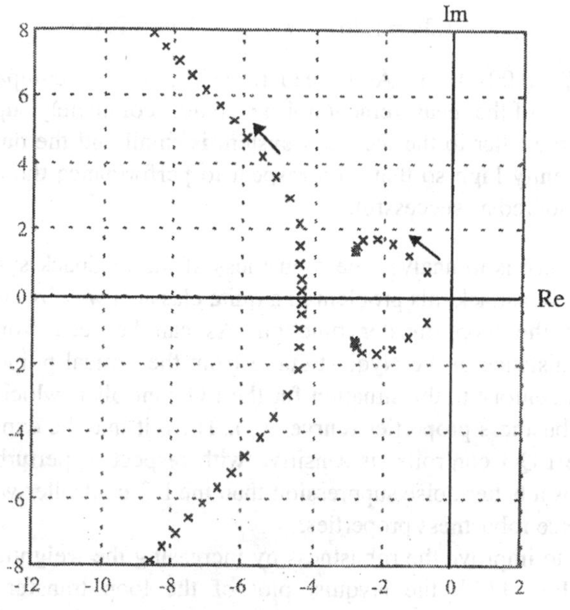

Fig. 11.7. Root locus of the Kalman filter as a function of W_i ($\rightarrow$: W_i increasing)

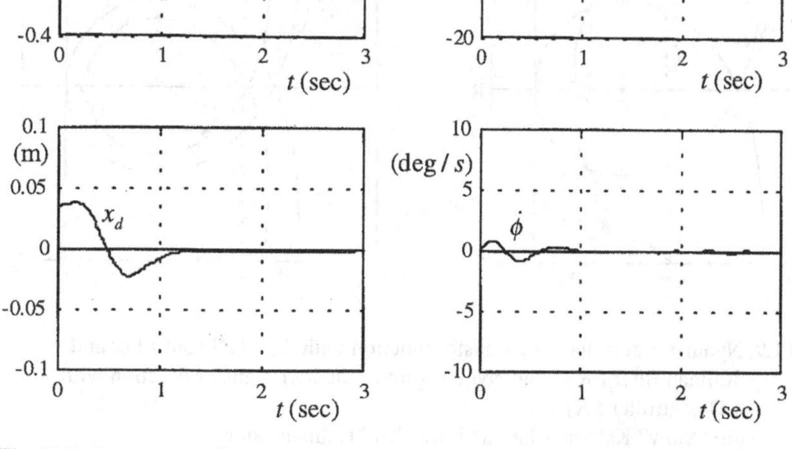

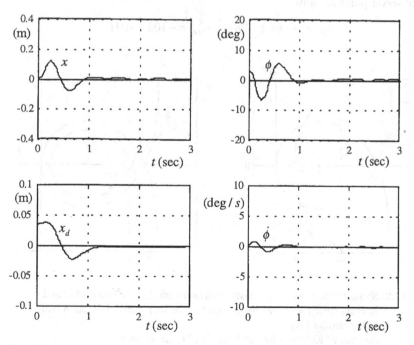

Fig.11.8. Responses of the states in the presence of noise with a "slow" Kalman filter

We consider an estimator with poles

$$s_{1/2} = -1.34 \pm 1.26i, \quad s_{3/4} = -4.34 \pm 0.58i\,.$$

It results for $W_i = 0.001\,A^2/s$. As is seen from Fig. 11.8 in comparison to Fig. 11.4, the influence of the measurement noise is almost completely suppressed. The delay added by the filter to the feedback system is small and the dampings of its poles are sufficiently high so that with respect to performance the controller design can be considered as successful.

Our next concern is to analyze the robustness of the feedback system with the LQG controller. We attack this problem in a quite classical way by considering the Nyquist plot of the loop transfer function. As can be seen from Fig. 11.9a (curve K_2), the distance of the Nyquist curve from the critical point -1 is rather small. This is in contrast to the situation for the LQ controller, which in this sense has excellent robustness properties (curve K_1). Thus, it may be expected that the system with the LQG controller is sensitive with respect to perturbations at the plant input. It has a better noise suppression than the LQ controller with the differentiators but worse robustness properties.

It is possible to improve the robustness by increasing the weighting of the system noise. In Fig. 11.9b the Nyquist plot of the loop transfer function for $W_i = 10^4\,A^2/s$ is depicted. The distance of the Nyquist plot from -1 has considerably increased and now the feedback system has good gain and phase margins. The observer poles are now

$$s_{1/2} = -2.84 \pm 1.28i, \quad s_{3/4} = -103 \pm 103i\,.$$

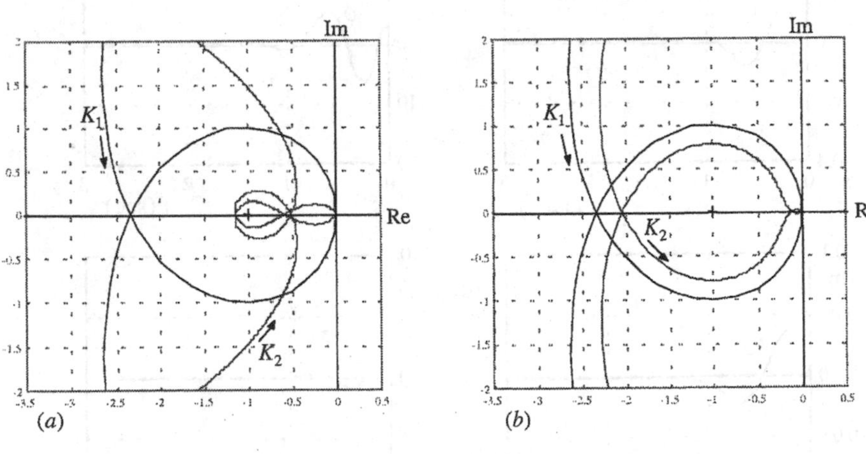

Fig. 11.9. Nyquist plot of the loop transfer function with the LQG controller and Kalman filter (K_2) and Nyquist plot of the loop transfer function with LQ controller (K_1):
(a) "Slow" Kalman filter, and (b) "fast" Kalman filter

The effect of the noise on the states is only very moderate and therefore not shown here. The feedback loop with the LQG controller now has good robustness and performance properties.

As we have seen above, it is in principle possible to design a controller with x_d as the only measurement (which is of particular interest if x_d is the cart position). A Kalman filter as in the case where x and ϕ are fed back can be constructed such that the feedback system again has good performance properties. It turns out that the closed-loop is then extremely sensitive with respect to perturbations at the plant input. As in the previous case, the robustness can be increased by loop transfer recovery, but at the price that extremely large deflections of the pendulum leading deeply into the nonlinear region of pendulum dynamics occur. This could probably be improved by incorporating ϕ into the weighting matrix $\mathbf{C}_{10}$, but we shall not investigate this further here. It should be noted that through the dependence on the sensor position x_d, some major performance limitations may arise. Compare in particular the transfer function for $d = 0$ and the first example in Sect. 7.5.

11.2 Control of a Continuously Stirred Tank

11.2.1 Analysis of the Plant Dynamics

In this section, we want to control a chemical reaction which takes place in a continuously stirred tank (CSTR); see Luyben [66], where several plants from chemical industries are described in detail. We assume that a component A reacts irreversibly to form a product component B and the reaction is supposed to be exothermic. To remove the heat of reaction, a cooling jacket surrounds the reactor. Cooling water is added to the jacket at a volumetric flow rate q_K. This flow rate as well as the flow rate q of the reacting component in the tank can be controlled by a pump. These two quantities are the controls of the system.

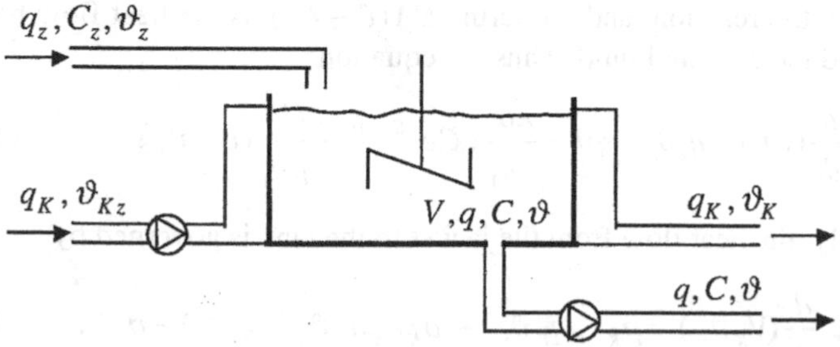

Fig. 11.10. Nonisothermal CSTR

The temperature of the incoming cooling liquid is assumed to be constant and is denoted by ϑ_{Kz}. In the tank, the liquid is supposed to be perfectly mixed. Then the temperature ϑ of the liquid depends only on the time. The same is assumed for the temperature ϑ_K of the liquid in the jacket.

The temperature ϑ of the substance in the tank and its volume V are the variables which have to be controlled. We denote the temperature of the substance which gets into the tank from outside by ϑ_z. The substance contains the component A with concentration C_z. Let q_z be the flow rate at the inlet and denote finally the concentration of component A in the tank by C. The situation is visualized in Fig. 11.10.

Our first task is to derive the differential equations, which govern the CSTR. The equation of continuity for the whole reactor is

$$\frac{dV}{dt} = q_z - q\,. \tag{11.11}$$

The equation of continuity for component A is given by

$$\frac{d}{dt}(VC) = C_z q_z - Cq - Vk(\vartheta)C\,.$$

Here, $k(\vartheta) = \alpha\, e^{-E/R\vartheta}$ is the Arrhenius term (with gas constant R, activation energy E and a proportionality factor α). The quantity $Vk(\vartheta)C$ describes the portion of the flow rate of component A which reacts to form B. Thus we have

$$\frac{d}{dt}(VC) = C_z q_z - Cq - \alpha VCe^{-E/R\vartheta}\,. \tag{11.12}$$

The heat flow in the reactor is described by the equation

$$\rho c_0 \frac{d}{dt}(V\vartheta) = \rho c_0 q_z \vartheta_z - \rho c_0 q\vartheta - \lambda V k(\vartheta) C - UA(\vartheta - \vartheta_K)\,.$$

In this equation, ρ is the density and c_0 the heat capacity of the liquid in the tank. U denotes the overall heat transfer coefficient with respect to the tank and jacket and A is the heat transfer area. The term $\lambda V k(\vartheta) C$ describes the heat flow caused by the reaction, and the term $UA(\vartheta - \vartheta_K)$ is the heat flow between the process and the cooling liquid. Thus the equation

$$\frac{d}{dt}(V\vartheta) = q_z\vartheta_z - q\vartheta - \frac{\lambda\alpha}{\rho c_0} VCe^{-E/R\vartheta} - \frac{UA}{\rho c_0}(\vartheta - \vartheta_K) \tag{11.13}$$

holds. Similarly, the heat flow from the jacket to the tank is governed by

$$\rho_K c_K \frac{d}{dt}(V_K \vartheta_K) = \rho_K c_K q_K \vartheta_{Kz} - \rho_K c_K q_K \vartheta_K + UA(\vartheta - \vartheta_K)\,. \tag{11.14}$$

Here, V_K is the (constant) volume of the cooling liquid and ρ_K and c_K are its density and heat capacity, respectively. The equations (11.11)–(11.14) describe the dynamical behavior of the reactor. The dynamical variables can be classified as follows: $V, C, \vartheta, \vartheta_K$ are the states, q, q_K are the controls and q_z, C_z, ϑ_z are the disturbances. All other quantities are constants.

We define

$$k_1 = \frac{\lambda\alpha}{\rho c_0},\quad k_2 = \frac{UA}{\rho c_0},\quad k_3 = \frac{UA}{\rho_K c_K V_K},\quad k_4 = \frac{1}{V_K},\quad k_5 = \frac{\vartheta_{Kz}}{V_K},\quad \mu = \frac{E}{R}$$

and

$$x_1 = V,\quad x_2 = C,\quad x_3 = \vartheta,\quad x_4 = \vartheta_K,$$
$$u_1 = q,\quad u_2 = q_K,$$
$$d_1 = q_z,\quad d_2 = C_z,\quad d_3 = \vartheta_z.$$

Then the system can be reformulated as

$$\dot{x}_1 = d_1 - u_1$$
$$\dot{x}_2 = \frac{d_1}{x_1}(d_2 - x_2) - \alpha x_2 e^{-\mu/x_3}$$

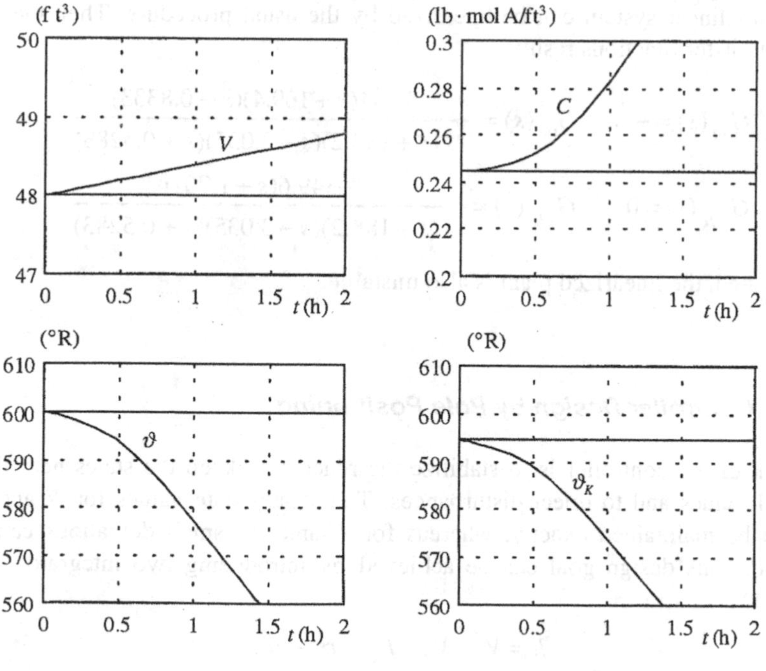

Fig. 11.11. States in dependence of time for the uncontrolled CSTR

$$\dot{x}_3 = \frac{d_1}{x_1}(d_3 - x_3) - \frac{k_2}{x_1}(x_3 - x_4) - k_1 x_2 e^{-\mu / x_3}$$

$$\dot{x}_4 = (k_5 - k_4 x_4) u_2 + k_3 (x_3 - x_4) \ .$$

The numerical values of the constants are assumed to be given by (cf. Luyben [66], Sect. 5.3):

$$\alpha = 7.08 \times 10^{10}\, h^{-1}, \quad E = 30000\, Btu / lb \cdot mol, \quad R = 1.99\, Btu / lb \cdot mol\, °R,$$

$$\lambda = -30000\, Btu / lb \cdot mol, \quad \rho = 50\, lb_m / ft^3, \quad \rho_K = 62.3\, lb_m / ft^3,$$

$$V_K = 3.85\, ft^3, \quad c_0 = 0.75\, Btu / lb_m\, °R, \quad c_K = 1.0\, Btu / lb_m\, °R,$$

$$A = 250\, ft^2, \quad U = 150\, Btu / h\, ft^2\, °R, \quad \vartheta_{Kz} = 530\, °R\, .$$

The equilibrium point is chosen as

$$V_0 = 48\, ft^3, \quad C_0 = 0.245 lb \cdot mol A / ft^3, \quad \vartheta_0 = 600°R, \quad \vartheta_{K0} = 594.6°R,$$

$$q_0 = 40\, ft^3 / h, \quad q_{K0} = 49.9\, ft^3 / h, \quad C_{z0} = 0.5 lb \cdot mol A / ft^3, \quad \vartheta_{z0} = 530°R.$$

Figure 11.11 shows the states for this equilibrium point when the disturbances q_z, C_z, ϑ_z are slightly perturbed. From these plots, it may be expected that the nonlinear system is unstable. The straight lines show the ideal solutions for the equilibrium point.

The nonlinear system can be linearized by the usual procedure. Then the following transfer functions result:

$$G_{Vq}(s) = -\frac{1}{s}, \quad G_{\vartheta q}(s) = -\frac{3.541(s+169.4)(s+0.8333)}{s(s+188.2)(s-3.035)(s+0.5283)},$$

$$G_{Vq_K}(s) = 0, \quad G_{\vartheta q_K}(s) = -\frac{-349.6(s+1.701)}{(s+188.2)(s-3.035)(s+0.5283)}.$$

As expected, the linearized plant is also unstable.

11.2.2 Controller Design by Pole Positioning

The task of the controller is to stabilize the reactor, to keep the states near their nominal values and to reject disturbances. The steady state values for V and ϑ have to be maintained exactly, whereas for C and ϑ_K small deviations can be tolerated. This design goal can be achieved by introducing two integrators (cf. Sect. 6.1):

$$\dot{h}_1 = V_c - V, \quad \dot{h}_2 = \vartheta_c - \vartheta\, .$$

Here, the choice $V_c = V_0$, $\vartheta_c = \vartheta_0$ is made. The extended linearized plant has the poles

$$s_1^0 = s_2^0 = s_3^0 = 0, \quad s_4^0 = -188.2, \quad s_5^0 = 3.035, \quad s_6^0 = -0.528$$

and the system matrices are of the form

$$\mathbf{A}_e = \begin{pmatrix} 0 & 0 & 0 & 0 & 0 & 0 \\ a_{21} & a_{22} & a_{23} & 0 & 0 & 0 \\ a_{31} & a_{32} & a_{33} & a_{34} & 0 & 0 \\ 0 & 0 & a_{43} & a_{44} & 0 & 0 \\ -1 & 0 & 0 & 0 & 0 & 0 \\ 0 & 0 & -1 & 0 & 0 & 0 \end{pmatrix}, \qquad \mathbf{B}_{2e} = \begin{pmatrix} b_{11}^2 & 0 \\ 0 & 0 \\ 0 & 0 \\ 0 & b_{42}^2 \\ 0 & 0 \\ 0 & 0 \end{pmatrix}.$$

Since the plant is controllable, the method of pole positioning is applicable. It is quite natural to assume that the states V, ϑ and ϑ_K are measurable, whereas the concentration C has to be estimated.

Although the problem has multiple inputs and outputs, we adopt some of the ideas developed for SISO systems in Sect. 6.3. We start with the unstable pole s_5^0 and replace it for the closed-loop system by $s_5 = -6$. The two stable poles are made somewhat "quicker" and prescribed as $s_4 = -200$, $s_6 = -1.5$. Since the coefficients for the gain matrix will be calculated by the numerical method described in Sect. 6.4.2, the stable counterparts of the integrator poles are not allowed to coincide. We choose $s_1 = -7.5$, $s_2 = -6.5$, $s_3 = -4.5$. Compared with s_5, this seems to be a reasonable choice.

Our next step is to construct an observer. Since three of the four states can be measured (namely x_1, x_3, x_4), it is natural to design a reduced observer. We proceed as described in Sect. 6.2.2. The description of the systems (6.23) and (6.24) is here very simple since it suffices to interchange some equations and to rename the corresponding variables. This yields

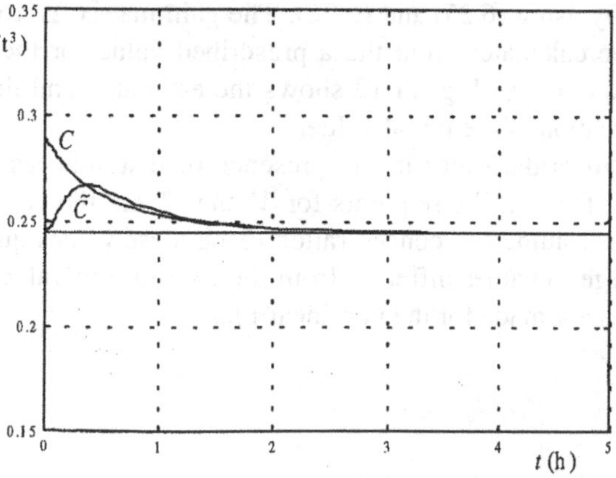

Fig. 11.12. True and estimated concentration

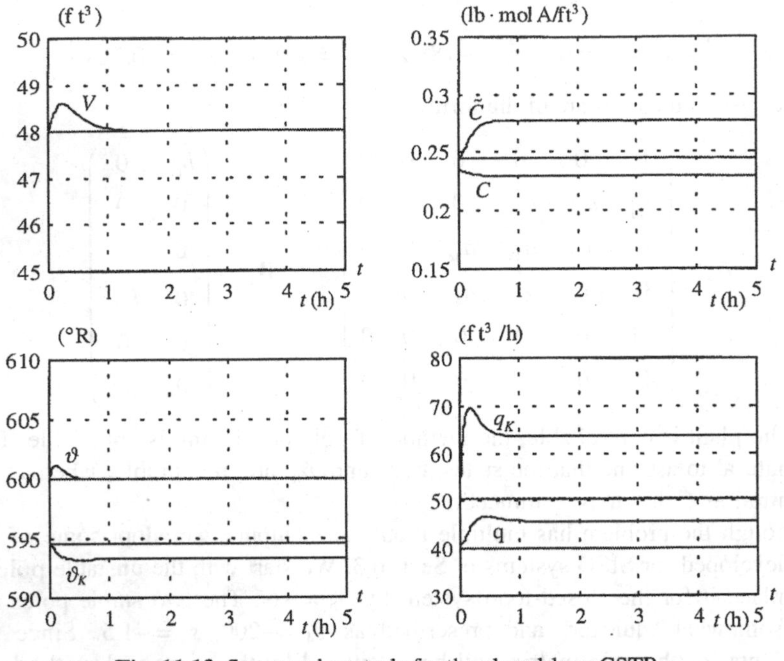

Fig. 11.13. States and controls for the closed-loop CSTR

$$\mathbf{A}_{11} = \begin{pmatrix} 0 & 0 & 0 \\ a_{31} & a_{33} & a_{34} \\ 0 & a_{43} & a_{44} \end{pmatrix}, \qquad \mathbf{A}_{12} = \begin{pmatrix} 0 \\ a_{32} \\ 0 \end{pmatrix}, \qquad \mathbf{B}_1 = \begin{pmatrix} b_{11}^2 & 0 \\ 0 & 0 \\ 0 & b_{42}^2 \end{pmatrix}$$

$$\mathbf{A}_{21} = \begin{pmatrix} a_{21} & a_{23} & 0 \end{pmatrix}, \qquad \mathbf{A}_{22} = \begin{pmatrix} a_{22} \end{pmatrix}, \qquad \mathbf{B}_2 = \begin{pmatrix} 0 & 0 \end{pmatrix}.$$

Equation (6.27) for the reduced observer, which here is of order 1, can easily be written explicitly by using (6.25) and (6.26). The gain matrix $\mathbf{L}$ for the measurement vector can be calculated from the a prescribed value for the observer pole s_{OB}. We choose $s_{OB} = -10$. Fig. 11.12 shows the estimated and the true concentration for a perturbation of the initial value.

Figure 11.13 shows the states in the presence of disturbances and for a perturbed initial value for C. The setpoints for V and ϑ are the corresponding stationary values. The volume and concentration reach these values quickly, whereas C and ϑ_K converge to values different from the equilibrium values. We mention that the simulations are made for the nonlinear plant.

11.3 Control of Aircraft

11.3.1 Nonlinear and Linearized Dynamics

Equations of Motion. An aircraft is, at least to a good approximation, a rigid body with gravitational, aerodynamic, and propulsive forces and moments acting on it. The differential equations describing the dynamics can be divided into two parts: one governs the translational motion and the other the rotational motion. To be more specific, denote by u, v, w the components of the velocity vector and by p, q, r the components of the angular velocity vector. In a suitable body-fixed coordinate system the equations of motion then can be formulated as follows:

$$\begin{aligned} m\dot{u} + m(qw - rv) &= F_x \\ m\dot{v} + m(ru - pw) &= F_y \\ m\dot{w} + m(pv - qu) &= F_z \end{aligned} \tag{11.15}$$

$$\begin{aligned} I_x\dot{p} - I_{xz}\dot{r} + (I_z - I_y)qr - I_{xz}pq &= M_x \\ I_y\dot{q} + (I_x - I_z)pr + I_{xz}(p^2 - r^2) &= M_y \\ I_z\dot{r} - I_{xz}\dot{p} + (I_y - I_x)pq + I_{xz}qr &= M_z . \end{aligned} \tag{11.16}$$

Here, m is the mass of the aircraft and I_x, I_y, I_{xz} are the nonvanishing components of the inertia tensor:

$$\mathbf{I} = \begin{pmatrix} I_x & 0 & -I_{xz} \\ 0 & I_y & 0 \\ -I_{xz} & 0 & I_z \end{pmatrix}.$$

The exterior force and the exterior moment are the vectors

$$\mathbf{F} = (F_x \quad F_y \quad F_z)^T, \quad \mathbf{M} = (M_x \quad M_y \quad M_z)^T .$$

The force vector is the sum $\mathbf{F} = \mathbf{F}_A + \mathbf{F}_G + \mathbf{F}_T$, where $\mathbf{F}_G$ is the gravitational force and where $\mathbf{F}_A$ and $\mathbf{F}_T$ is the aerodynamic and propulsion force, respectively. We assume that the propulsion induces no moment. Then the moment is of a purely aerodynamic nature. The aerodynamic forces and moments are as usual denoted as

$$\mathbf{F}_A = (X \quad Y \quad Z)^T, \quad \mathbf{M} = (L \quad M \quad N)^T ,$$

where L is the roll, M the pitch and N the yaw moment. The aerodynamic forces and moments depend on the state vector and its derivative and the deflections of the actuators. More precisely, the following dependences are typical:

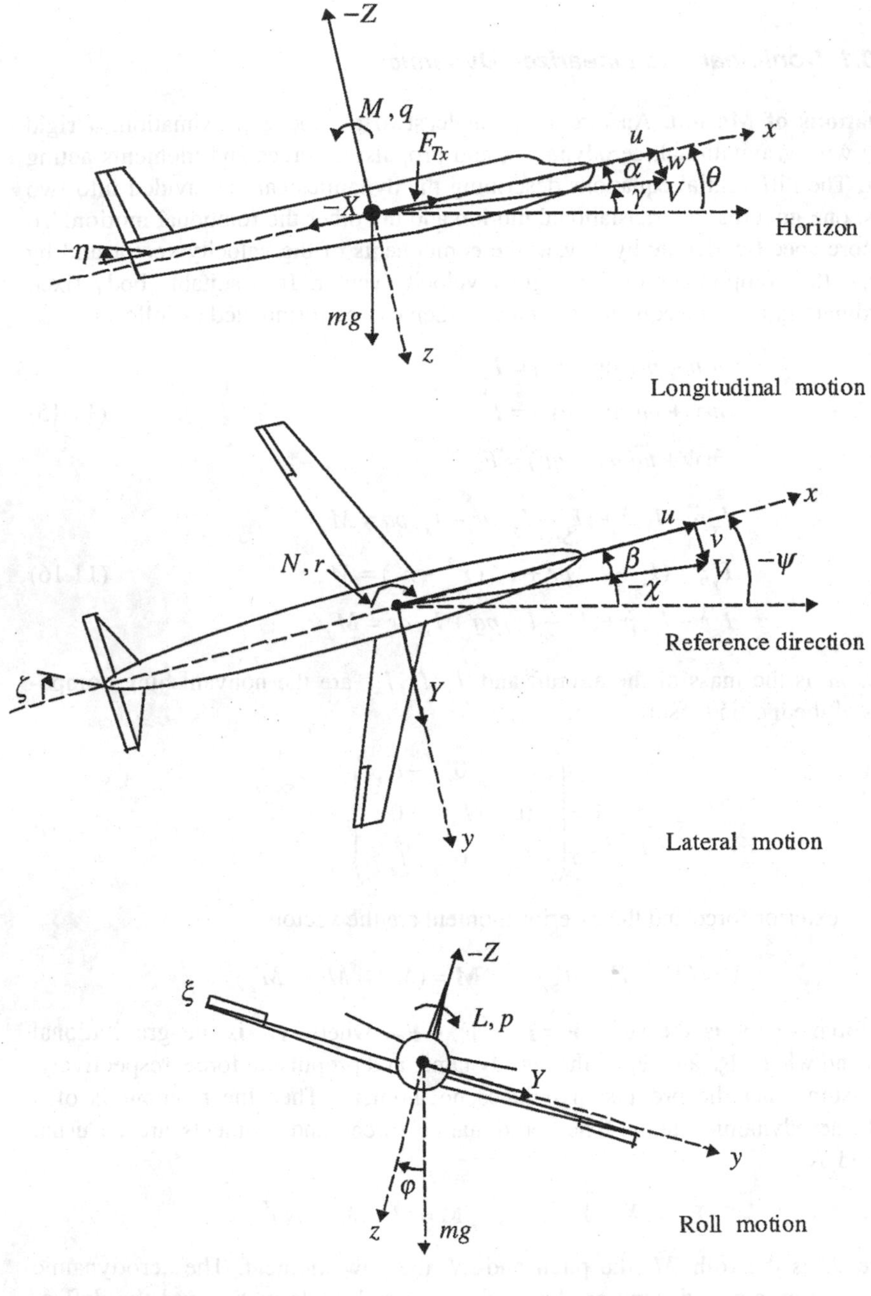

Fig. 11.14. Coordinates, states and controls for the aircraft dynamics

$$X = X(u, w, \eta), \qquad Y = Y(v, p, r, \zeta, \xi), \qquad Z = Z(u, w, \dot{w}, q, \eta),$$
$$L = L(v, p, r, \zeta, \xi), \quad M = M(u, w, \dot{w}, q, \eta), \quad N = N(v, p, r, \zeta, \xi).$$

Here, ξ (ζ, η) denotes the aileron (rudder, elevator) deflection. Note that $\dot{w}$ enters in Z and M. The moment M depends also on the center of gravity, which is a function of the time since the aircraft loses fuel. The forces and moments vary also with the air pressure and therefore with the altitude. Other exterior forces and moments are caused by the wind, which acts as a disturbance. The dependence on forces and moments induced by the wind is complicated and will only be described for the linearized model. The thrust vector has the representation

$$\mathbf{F}_T = (F_{Tx}(u, w, \delta) \quad 0 \quad -F_{Tz}(u, w, \delta))^T,$$

where the throttle deflection δ is a further control of the aircraft.

The attitude of the aircraft is described by the Euler angles θ, ϕ and ψ. The gravitational force $\mathbf{F}_G$ is a function of these angles:

$$\mathbf{F}_G = m\,g \begin{pmatrix} -\sin\theta \\ \cos\theta\sin\phi \\ \cos\theta\cos\phi \end{pmatrix}.$$

The differential equations for θ, ϕ and ψ are

$$\begin{aligned} \dot{\phi} &= p + q\sin\phi\tan\theta + r\cos\phi\tan\theta \\ \dot{\theta} &= q\cos\phi - r\sin\phi \\ \dot{\psi} &= q\frac{\sin\phi}{\cos\theta} + r\frac{\cos\phi}{\cos\theta}. \end{aligned} \tag{11.17}$$

Trim points. An equilibrium point for the system (11.15)–(11.17) is said to be a trim point. We adopt the notation of Sect. 5.1. From (11.17) it follows directly that $p_0 = q_0 = r_0 = 0$ for a trim point and then (11.16) implies that $L_0 = M_0 = N_0 = 0$. With the choice $\phi_0 = 0$, the equations (11.15) can now be written as

$$\begin{aligned} &X_0 + F_{Tx0} - mg\sin\theta_0 = 0 \\ &Y_0 = 0 \\ &Z_0 - F_{Tz0} + mg\cos\theta_0 = 0. \end{aligned}$$

They lead to conditions for u_0, v_0, w_0 and $\delta_0, \xi_0, \zeta_0, \eta_0$. It is reasonable to assume that the stationary yaw velocity also vanishes: $v_0 = 0$.

Linearization. We now linearize the equations (11.15) and (11.16) at the described trim point. Thereby it is convenient to linearize the aerodynamic forces and moments in a second step separately for the longitudinal and lateral motion. Moreover, the velocities w and v will be replaced by the angle of attack α and the sideslip angle β. For small velocity increments Δw and Δv, the following equations approximately hold:

$$\Delta\alpha = \frac{\Delta w}{u_0}, \qquad \Delta\beta = \frac{\Delta v}{u_0}.$$

The linearized version of (11.15) and (11.16) is now easily obtained as (here, the symbol "Δ" is omitted if 0 is the equilibrium point):

$$\begin{aligned}
& m\,\Delta\dot{u} + m u_0 \alpha_0 q = \Delta X - mg\cos\theta_0\,\Delta\theta + \Delta F_{Tx} \\
& m u_0 \dot{\beta} + m u_0 r - m u_0 \alpha_0 p = \Delta Y + mg\cos\theta_0\,\phi \\
& m u_0 \Delta\dot{\alpha} - m u_0 q = \Delta Z - mg\sin\theta_0\,\Delta\theta - \Delta F_{Tz} \\
& I_x \dot{p} - I_{xz}\dot{r} = \Delta L \\
& I_y \dot{q} = \Delta M \\
& I_z \dot{r} - I_{xz}\dot{p} = \Delta N .
\end{aligned} \tag{11.18}$$

Linearized longitudinal dynamics. For the longitudinal dynamics the forces and moments $\Delta X, \Delta Z$ and ΔM have to be further evaluated. Define

$$X_\alpha = \frac{1}{m}\frac{\partial X}{\partial \alpha}, \qquad M_\alpha = \frac{1}{I_y}\frac{\partial M}{\partial \alpha}$$

and corresponding quantities for $\dot{\alpha}, q, \eta, \delta$ and the remaining forces and moments. Then we obtain

$$\begin{aligned}
\Delta\dot{u} &= -u_0\alpha_0 q - \frac{1}{m}\Delta X - g\cos\theta_0\,\Delta\theta + \frac{1}{m}\Delta F_{Tx} \\
&= (X_u + F_{Txu})\Delta u + (X_\alpha + F_{Tx\alpha})\Delta\alpha - u_0\alpha_0 q - g\cos\theta_0\,\Delta\theta + X_\eta\Delta\eta + F_{Tx\delta}\Delta\delta \\
u_0\Delta\dot{\alpha} &= u_0 q + \frac{1}{m}\Delta Z - g\sin\theta_0\,\Delta\theta - \frac{1}{m}\Delta F_{Tz} \\
&= Z_{\dot{\alpha}}\Delta\dot{\alpha} + (Z_u - F_{Tzu})\Delta u + (Z_\alpha - F_{Tz\alpha})\Delta\alpha + (u_0 + Z_q)q \\
&\qquad - g\sin\theta_0\,\Delta\theta + Z_\eta\Delta\eta - F_{Tz\delta}\Delta\delta \\
\dot{q} &= M_u\Delta u + M_\alpha\Delta\alpha + M_{\dot{\alpha}}\Delta\dot{\alpha} + M_q q + M_\eta\Delta\eta .
\end{aligned}$$

The angle $\Delta\theta$ results from the equation $\Delta\dot{\theta} = q$. We now define vectors for the states and the controls by

$$\mathbf{x} = (\Delta u \quad \Delta\alpha \quad q \quad \Delta\theta)^T, \quad \mathbf{u} = (\Delta\eta \quad \Delta\delta)^T .$$

Thus, the linearized system dynamics is given by

$$\mathbf{E}_0\dot{\mathbf{x}} = \mathbf{A}_0\mathbf{x} + \mathbf{B}_0\mathbf{u}$$

with matrices

$$\mathbf{A}_0 = \begin{pmatrix} X_u + F_{Txu} & X_\alpha + F_{Tx\alpha} & -u_0\alpha_0 & -g\cos\theta_0 \\ Z_u - F_{Tzu} & Z_\alpha - F_{Tz\alpha} & u_0 + Z_q & -g\sin\theta_0 \\ M_u & M_\alpha & M_q & 0 \\ 0 & 0 & 1 & 0 \end{pmatrix},$$

$$\mathbf{E}_0 = \begin{pmatrix} 1 & 0 & 0 & 0 \\ 0 & u_0 - Z_{\dot\alpha} & 0 & 0 \\ 0 & -M_{\dot\alpha} & 1 & 0 \\ 0 & 0 & 0 & 1 \end{pmatrix}, \quad \mathbf{B}_0 = \begin{pmatrix} X_\eta & F_{Tx\delta} \\ Z_\eta & -F_{Tz\delta} \\ M_\eta & 0 \\ 0 & 0 \end{pmatrix}.$$

The matrix $\mathbf{E}_0$ is invertible. Putting $\mathbf{A}_L = \mathbf{E}_0^{-1}\mathbf{A}_0$, $\mathbf{B}_L = \mathbf{E}_0^{-1}\mathbf{B}_0$, the linearized longitudinal dynamics is given by the following system of order 4:

$$\dot{\mathbf{x}} = \mathbf{A}_L\,\mathbf{x} + \mathbf{B}_L\,\mathbf{u}.$$

Linearized lateral dynamics. For the lateral dynamics, the forces and moments $\Delta Y, \Delta L$ and ΔN have to be further evaluated. In a way similar to that for the longitudinal dynamics, we set

$$Y_\beta = \frac{1}{m}\frac{\partial Y}{\partial \beta}, \quad L_\beta = \frac{1}{I_x}\frac{\partial L}{\partial \beta}, \quad N_\beta = \frac{1}{I_y}\frac{\partial N}{\partial \beta}$$

and analogously for the other quantities. Then we get

$$\begin{aligned} u_0\dot\beta &= \frac{1}{m}\Delta Y - u_0 r + u_0\alpha_0 p + g\cos\theta_0\,\phi \\ &= Y_\beta\beta + (u_0\alpha_0 + Y_p)p + (Y_r - u_0)r + g\cos\theta_0\,\phi + Y_\zeta\zeta + Y_\xi\xi \end{aligned}$$

$$\dot p - \frac{I_{xz}}{I_x}\dot r = L_\beta\beta + L_p p + L_r r + L_\zeta\zeta + L_\xi\xi$$

$$-\frac{I_{xz}}{I_z}\dot p + \dot r = N_\beta\beta + N_p p + N_r r + N_\zeta\zeta + N_\xi\xi.$$

The roll angle ϕ results from the equation

$$\dot\phi = p + \tan\theta_0\, r.$$

The state vector and the control vector are now

$$\mathbf{x} = (\beta \quad p \quad r \quad \phi)^T, \quad \mathbf{u} = (\zeta \quad \xi)^T,$$

and the system matrices are calculated as

$$\mathbf{E}_1 = \begin{pmatrix} u_0 & 0 & 0 & 0 \\ 0 & 1 & -I_{xz}/I_x & 0 \\ 0 & -I_{xz}/I_z & 1 & 0 \\ 0 & 0 & 0 & 1 \end{pmatrix},$$

$$\mathbf{A}_1 = \begin{pmatrix} Y_\beta & \alpha_0 u_0 + Y_p & Y_r - u_0 & g\cos\theta_0 \\ L_\beta & L_p & L_r & 0 \\ N_\beta & N_p & N_r & 0 \\ 0 & 1 & \tan\theta_0 & 0 \end{pmatrix}, \quad \mathbf{B}_1 = \begin{pmatrix} Y_\zeta & Y_\xi \\ L_\zeta & L_\xi \\ N_\zeta & N_\xi \\ 0 & 0 \end{pmatrix}.$$

Since the matrix $\mathbf{E}_1$ is invertible, we can put $\mathbf{A}_S = \mathbf{E}_1^{-1}\mathbf{A}_1$, $\mathbf{B}_S = \mathbf{E}_1^{-1}\mathbf{B}_1$. Hence, the lateral dynamics is also governed by a system of order 4:

$$\dot{\mathbf{x}} = \mathbf{A}_S\, \mathbf{x} + \mathbf{B}_S\, \mathbf{u}\,.$$

Wind disturbance. The wind causes a force as well as a moment acting on the aircraft. A moment arises when the wind force depends on the position where it attacks the aircraft. If wind is present, the aerodynamic forces depend on $\mathbf{v} - \mathbf{v}_W$ instead of $\mathbf{v}$. Here, $\mathbf{v}_W$ denotes the wind velocity in the body-fixed coordinate system. Thus, in the linearized equations $\mathbf{v}$ has to be replaced by $\Delta\mathbf{v} - \mathbf{v}_W$ and $\Delta u, \Delta\alpha, \Delta\beta$ by $\Delta u - u_W, \Delta\alpha - \alpha_W, \Delta\beta - \beta_W$, respectively. A similar substitution has to be made for the vector of the angular velocities. Normally, only the component p_W is of significance. Then in the linearized equations, only the component p has to be replaced by $p - p_W$.

Thus for the *longitudinal motion* we put

$$\mathbf{B}_{0W} = \begin{pmatrix} -X_u & -X_\alpha \\ -Z_u & -Z_\alpha \\ -M_u & -M_\alpha \\ 0 & 0 \end{pmatrix}, \quad \mathbf{B}_{LW} = \mathbf{E}_0^{-1}\mathbf{B}_{0W}, \quad \mathbf{d} = \begin{pmatrix} u_W \\ \alpha_W \end{pmatrix}.$$

This yields the perturbed system

$$\dot{\mathbf{x}} = \mathbf{A}_L\, \mathbf{x} + \mathbf{B}_L\, \mathbf{u} + \mathbf{B}_{LW}\, \mathbf{d}\,.$$

For the *lateral motion* we define analogously

$$\mathbf{B}_{1W} = \begin{pmatrix} -Y_\beta & -\alpha_0 u_0 - Y_p \\ -L_\beta & -L_p \\ -N_\beta & -N_p \\ 0 & -1 \end{pmatrix}, \quad \mathbf{B}_{SW} = \mathbf{E}_1^{-1}\mathbf{B}_{1W}, \quad \mathbf{d} = \begin{pmatrix} \beta_W \\ p_W \end{pmatrix}$$

and get the system

$$\dot{\mathbf{x}} = \mathbf{A}_S\,\mathbf{x} + \mathbf{B}_S\,\mathbf{u} + \mathbf{B}_{SW}\,\mathbf{d}$$

There are mathematical models which describe the stochastic properties of the wind (cf. Sect. 5.8.3).

11.3.2 Analysis of the Linearized Plant Dynamics

Longitudinal dynamics. In this section we analyze the linearized longitudinal motion for a typical example, namely a B747 aircraft at an altitude of 40 kft at $Ma = 0.8$ (cf. Bryson [16], Sect. 10.2). Then the $\mathbf{A}$-matrix is

$$\mathbf{A}_L = \begin{pmatrix} -0.003 & 0.039 & 0 & -0.322 \\ -0.065 & -0.319 & 7.74 & 0 \\ 0.02 & -0.101 & -0.429 & 0 \\ 0 & 0 & 1 & 0 \end{pmatrix}.$$

This matrix has two pairs of conjugate complex poles. If we describe them by frequency and damping, they are given by

$$\omega_1 = 0.0674\,s^{-1},\quad d_1 = 0.0068,\quad \omega_2 = 0.958\,s^{-1},\quad d_2 = 0.39\,.$$

The numerical values show that there is an eigenvalue with a high frequency and rather good damping (the short-period eigenvalue) and an eigenvalue with a low frequency and very small damping (the long-period or phugoid eigenvalue). In Fig. 11.15 the time dependency of the angle of attack is depicted for the case, where the initial value lies in the corresponding eigenspace.

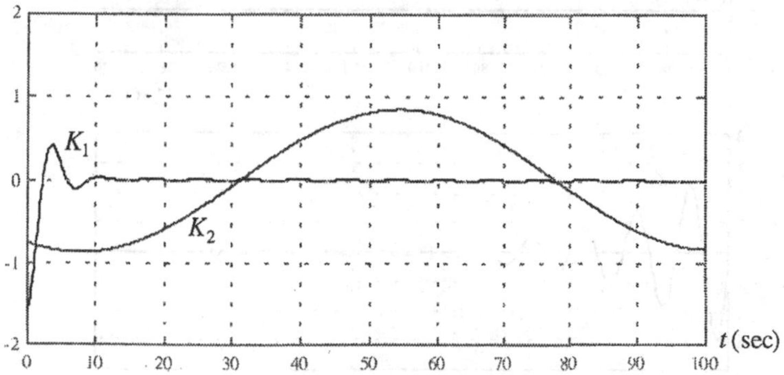

Fig. 11.15. Pure short period (K_1) and pure phugoid motion (K_2) of a B747 aircraft (angle of attack)

Lateral dynamics. The lateral dynamics is analyzed in a similar way. We take as an example a DC-8 aircraft at an altitude of 33000 ft at a velocity of 825 ft/s (cf. Schmidt [83], Sect. 7.3). Then the system matrices are

$$\mathbf{A}_S = \begin{pmatrix} -0.0869 & 0 & -1.0 & 0.039 \\ -4.424 & -1.184 & 0.335 & 0 \\ 2.148 & -0.021 & -0.228 & 0 \\ 0 & 1 & 0 & 0 \end{pmatrix}, \quad \mathbf{B}_S = \begin{pmatrix} 0.0223 & 0 \\ 0.547 & 2.12 \\ -1.169 & 0.065 \\ 0 & 0 \end{pmatrix}.$$

This matrix has two real eigenvalues and one pair of complex-conjugate eigenvalues, which all lie in the open left half-plane. If these eigenvalues are described by their frequencies, damping and time constants, the numerical values are

$$T_1 = 0.795s, \quad T_2 = 250.3\,s, \quad \omega_0 = 1.5s^{-1}, \quad d = 0.079, \quad T_S = 4.19s\,.$$

Here, T_S is the period associated with the complex eigenvalue. The eigenvalue associated with the small time constant is the so-called roll mode and the other real eigenvalue is the spiral mode. The complex eigenvalue is denoted as the dutch roll mode. This situation with two real eigenvalues (one with a small time constant and the other with a large time constant) is typical for an aircraft. The damping of the complex eigenvalue may also be negative. In this case, the lateral dynamics is unstable. Figure 11.16 shows the time dependence of the sideslip angle, where the initial values lie in the corresponding eigenspace.

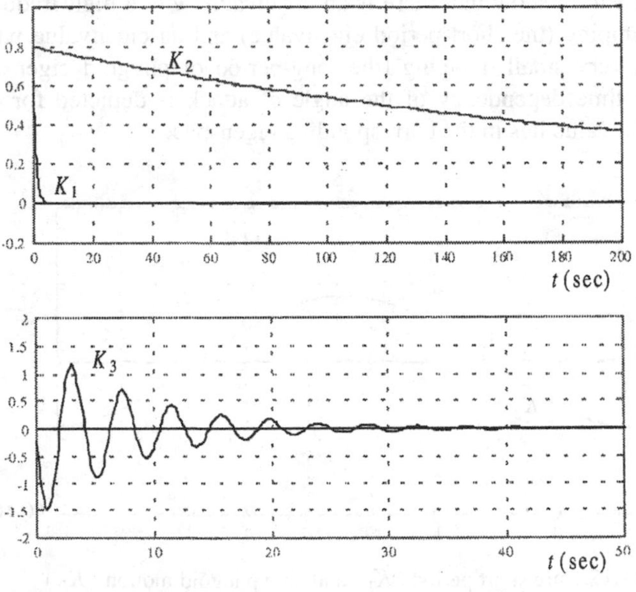

Fig. 11.16. Pure roll motion (K_1), pure spiral motion (K_2) and pure dutch roll motion for a DC-8 aircraft (sideslip angle)

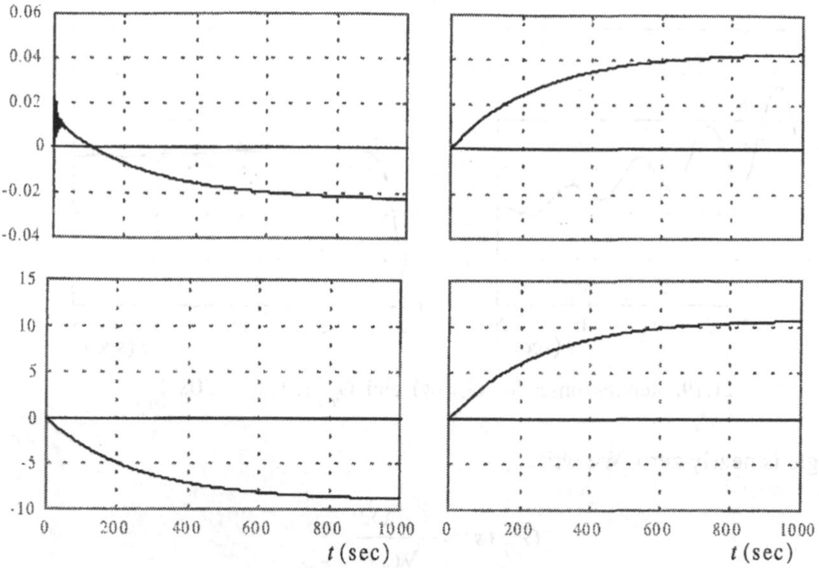

Fig. 11.17. Step responses for $G_{\beta\zeta}(s)$, $G_{\beta\xi}(s)$ (upper figures) and $G_{\phi\zeta}(s)$, $G_{\phi\xi}(s)$ (lower figures), $t_e = 1000s$

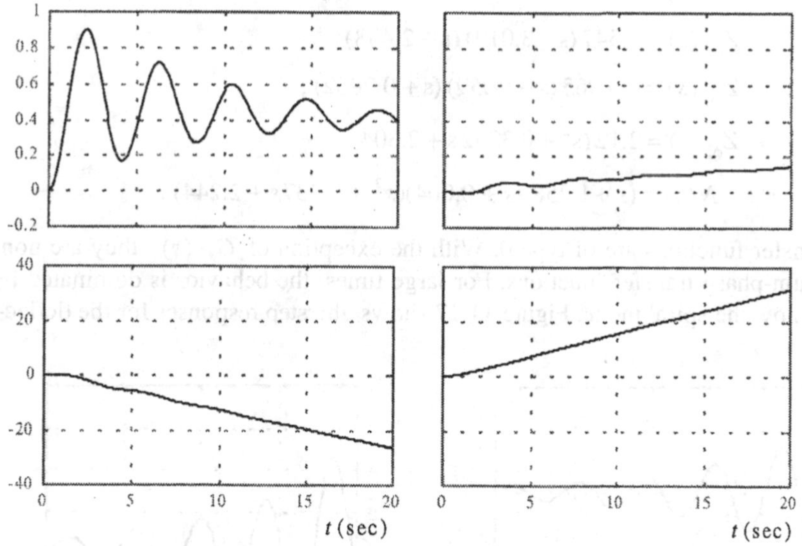

Fig. 11.18. Step responses for $G_{\beta\zeta}(s)$, $G_{\beta\xi}(s)$ (upper figures) and $G_{\phi\zeta}(s)$, $G_{\phi\xi}(s)$ (lower figures), $t_e = 20s$

Next, we study the transfer functions for the lateral motion. The outputs are the sideslip angle β and the roll angle ϕ. The latter serves in a certain sense as "control " for the turning of the aircraft. The turning is desired to be such that the side-

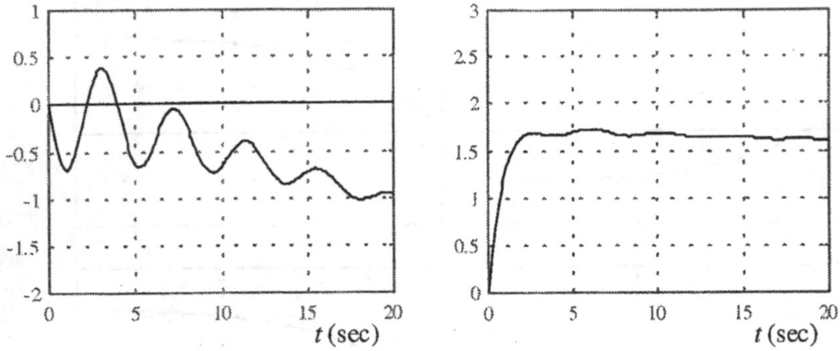

Fig. 11.19. Step responses for $G_{r\zeta}(s)$ and $G_{p\xi}(s)$, $t_e = 20s$

slip angle is nearly zero. We write

$$G_{\beta\zeta}(s) = \frac{Z_{\beta\zeta}(s)}{N(s)}$$

and similar expressions for the other transfer functions. We obtain

$$\begin{aligned}
Z_{\beta\zeta}(s) &= 0.0223\,(s-0.00727)\,(s+1.22)\,(s+52.62)\,,\\
Z_{\phi\zeta}(s) &= 0.547\,(s-3.019)\,(s+2.438)\,,\\
Z_{\beta\xi}(s) &= -0.065\,(s-1.059)\,(s+0.2862)\,,\\
Z_{\phi\xi}(s) &= 2.12\,(s^2+0.3252\,s+2.304)\,,\\
N(s) &= (s+1.258)(s+0.004)(s^2+0.237s+2.244)\,.
\end{aligned}$$

All transfer functions are of type 0. With the exception of $G_{\phi\xi}(s)$, they are non-minimum-phase transfer functions. For large times, the behavior is dominated by T_2, i.e. by the spiral mode. Figure 11.17 shows the step responses for the deflec-

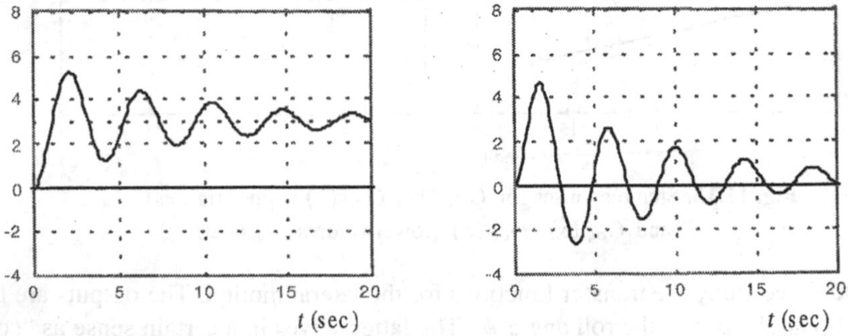

Fig. 11.20. Step responses of $G_{\beta\beta_W}(s)$ and $G_{\phi\beta_W}(s)$

tions $\zeta = \xi = 0.025°$. The nonminimum-phase property is only of significance for $G_{\beta\zeta}(s)$. For this transfer function, the effect of the dutch roll mode is clearly visible.

From the viewpoint of the pilot, the behavior of the step responses for small times is most significant. As can be seen from Fig. 11.18, over a short time interval the transfer functions $G_{\beta\xi}(s), G_{\phi\zeta}(s), G_{\phi\xi}(s)$ are nearly integrating. We also observe that the small damping of the dutch roll mode has almost no influence on $G_{\phi\xi}(s)$. The reason is that the quadratic factor in the denominator of $G_{\phi\xi}(s)$ is with a slight modification also a factor in the numerator. Figure 11.19 shows the step responses for the angular velocities, i.e. the step responses of $G_{r\zeta}(s)$ and $G_{p\xi}(s)$. Again, the effect of the badly damped dutch roll mode can be observed.

Finally, we study the effect of gust. Suppose that it causes the steady sideslip angle $\beta_W = 3°$, but induces no roll moment. Then the responses depicted in Fig. 11.20 result. $G_{\beta\beta_W}(s)$ is of type zero with gain 1, whereas $G_{\phi\beta_W}(s)$ is differentiating. Both properties can be observed in the figures.

11.3.3 Improvement of the Lateral Dynamics by State Feedback

As we have seen in the last section, the aircraft dynamics is mainly caused by the durch roll mode, in some aspects rather unfavorably. It can be improved by a suitable state feedback. The classical way is to increase the damping of the dutch roll mode by feeding back the yaw rate r to the rudder. In this way, the dynamic behavior of the aircraft as a plant is improved. In a second step, an outer feedback

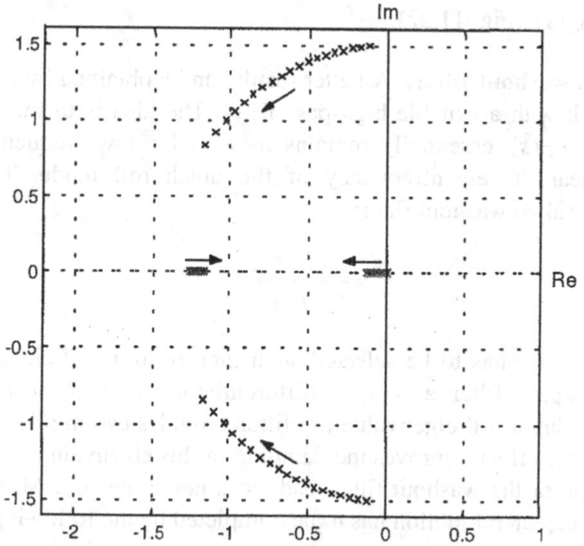

Fig. 11.21. Root locus for a purely proportional feedback of the pitch rate to the rudder

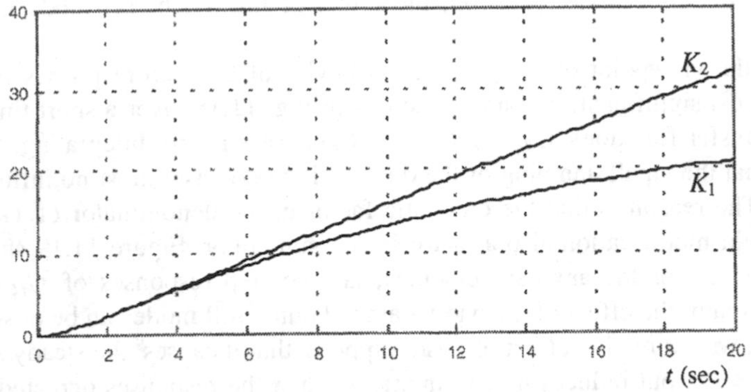

Fig. 11.22. Step response of $G_{\phi\xi}(s)$: K_1, with feedback of r; K_2, without feedback of r

loop is constructed in which reference commands are tracked. This loop also uses the rate r.

Proportional feedback. The simplest feedback is purely proportional: $\zeta = k\,r$. In Fig. 11.21 the system poles for this kind of feedback are depicted (with the gains $k = 0, 0.1, \ldots, 1.8$). The figure shows that the dutch roll eigenvalues move into the left half-plane such that the damping becomes high if the feedback gain is sufficiently large. The choice $k = 1$ leads to the damping $d = 0.467$ of the dutch roll mode. The time constant of the spiral mode reduces to $T_2 = 15.9\,s$. Thus for small times the systems acts no longer as an integrator. This is also observed from the step response $G_{\phi\xi}(s)$ (Fig. 11.22).

Application of a washout filter. A better result can be obtained by replacing the constant feedback with a suitable high-pass filter. The idea is to introduce feedback such that $G_{r\zeta}(s)$ essentially remains as it is for low frequencies and is changed only near the eigenfrequency of the dutch roll mode. This can be achieved by a so called washout filter:

$$\hat{\zeta} = \frac{k_W T_W s}{1 + T_W s}\,\hat{r}.$$

The time constant T_W has to be selected such that up to a sufficiently high frequency the high-pass filter acts as a differentiator, whereas near the eigenfrequency of the dutch roll eigenvalue the filter provides essentially proportional feedback. The latter effect improves the damping of this eigenvalue.

The application of the washout filter leads to a new state x_W. More precisely, the dynamics of the lateral motion has to be completed by the following equations:

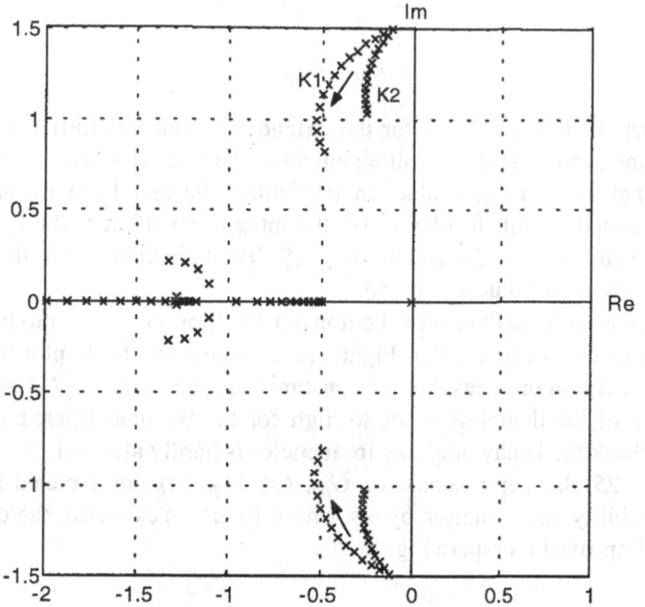

Fig. 11.23. Root locus for the plant with stability augmentation (washout filter), lateral dynamics for $T_W = 2\,s$ (K_1) and $T_W = 0.8\,s$ (K_2)

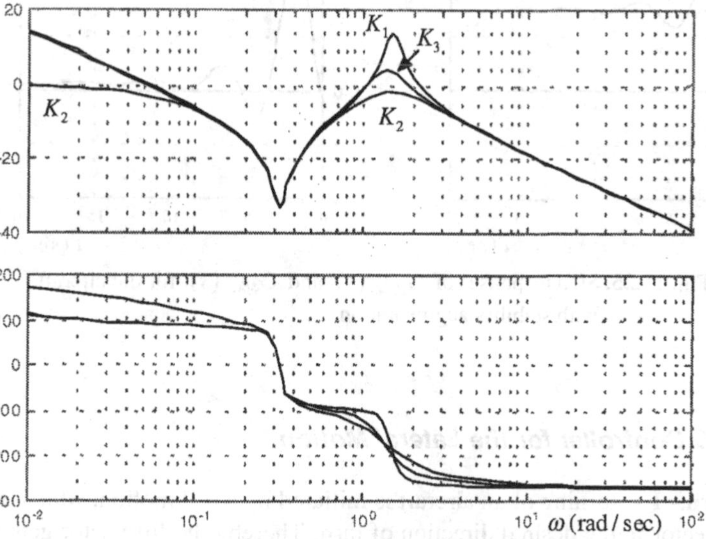

Fig. 11.24. Bode plot of $G_{r\zeta}(s)$: K_1, original plant;
K_2, proportional feedback of r;
K_3, feedback with washout filter

$$\dot{x}_W = -\frac{1}{T_W} x_W + \frac{k_W}{T_W} r$$

$$\zeta = k_W r - x_W .$$

Fig. 11.23 shows the root locus for the extended system and different values of T_W. The damping of the dutch roll eigenvalue increases if $1/T_W$ approaches the eigenfrequency of this eigenvalue. In particular, the spiral eigenvalue remains nearly unaffected by this feedback, i.e. the integrating effect of the system is not changed. We choose $T_W = 2s$ and $k_W = -0.5$. Then the damping of the dutch roll mode increases to the value $d = 0.256$.

Finally we want to analyze how the transfer function $G_{r\zeta}(s)$ is modified by the introduction of the washout filter. Figure 11.24 shows the Bode plot for the considered cases. Again one sees that for constant gain feedback $G_{r\zeta}(s)$ is of type 0. The increase of the damping is not so high for the washout filter, but with this kind of feedback the behavior at low frequencies is hardly affected.

In Fig. 11.25, the step response of $G_{\beta\beta_W}(s), G_{\phi\beta_W}(s)$ are depicted for the aircraft with stability augmentation by a washout filter. As expected, the disturbance rejection is improved (compare Fig. 11.20).

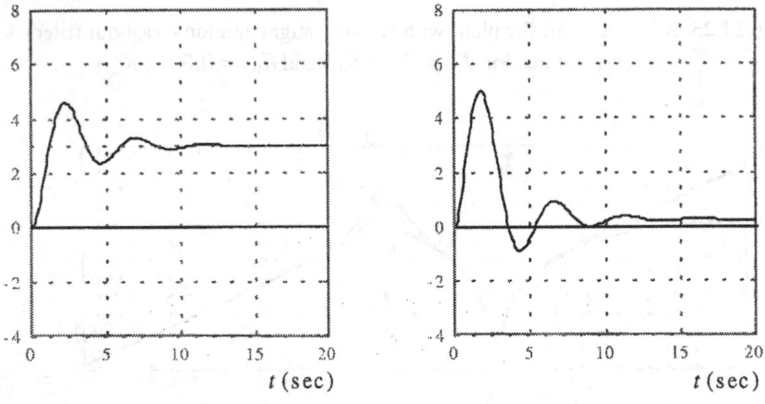

Fig. 11.25. Step responses of $G_{\beta\beta_W}(s)$ and $G_{\phi\beta_W}(s)$ for the aircraft with stability augmentation

11.3.4 LQ Controller for the Lateral Motion

Steady turn. The turning of an aircraft is initiated by a roll motion, which tilts the wing-lift vector in the desired direction of turn. Thereby the lift vector gets a non-vanishing component in the horizontal plane. This force component forces the aircraft to fly in a curve (cf. Fig. 11.26). We assume the angular velocity vector to be vertical with respect to the horizon. Then we have

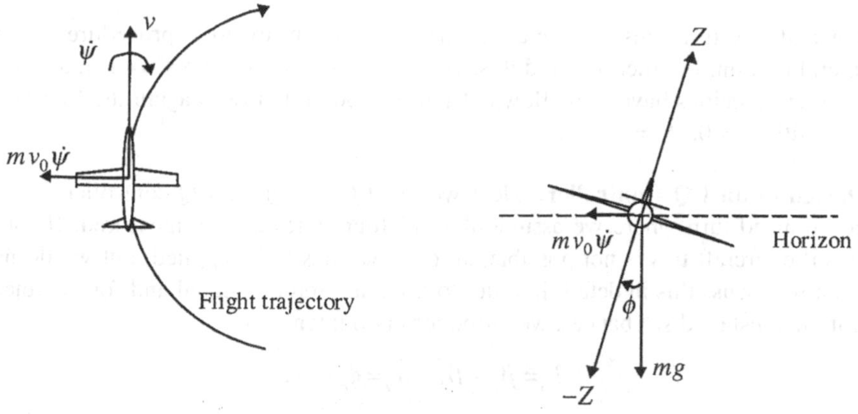

Fig. 11.26. Dynamic equilibrium for a steady turn at a constant altitude

$$\Omega = (\dot{\psi} \quad 0 \quad 0)^T .$$

For passenger comfort it is required that the lateral acceleration a_y in body-fixed coordinates is zero. Such a turn is denoted as **coordinated**. In a first-order approximation, we have $a_y = c_\beta \beta + c_\zeta \zeta$ with certain constants c_β, c_ζ. So we try to ensure that the stationary sideslip angle is zero. In what follows, we assume for simplicity $\theta_0 = 0$, but the considerations can be generalized to the case $\theta_0 \neq 0$.

For $\beta = 0$, the aircraft x-direction is tangential to the flight trajectory and we have $\chi = \psi$ and the situation depicted in Fig. 11.26 (with a constant absolute value v_0 of the velocity). Moreover, the stationary bank angle ϕ_0 has to be chosen according to

$$\tan \phi_0 = \frac{v_0 \dot{\psi}}{g} . \tag{11.19}$$

From this equation, the roll angle needed for the turn can be calculated. The stationary body-fixed angular velocities follow easily from solving (11.17):

$$p_0 = 0, \quad q_0 = \dot{\psi} \sin \phi_0, \quad r_0 = \dot{\psi} \cos \phi_0 .$$

Thus for a steady turn, the equations (11.16) imply

$$\begin{aligned} (I_z - I_y) q_0 r_0 &= L(u_0 \beta_0, 0, r_0, \zeta_0, \xi_0) \\ - I_{xz} q_0 r_0 &= N(u_0 \beta_0, 0, r_0, \zeta_0, \xi_0) . \end{aligned}$$

Here, q_0, r_0 are given and the sideslip angle is prescribed as $\beta_0 = 0$. Then ζ_0, ξ_0 can be calculated in a unique manner from these equations. Hence, by coordinating the rudder and aileron deflection in this way, a stationary turn with prescribed roll angle and sideslip angle zero is possible. The equilibrium point then is

$$(\beta_0 \quad p_0 \quad r_0 \quad \phi_0)^T = (0 \quad 0 \quad \dot{\psi}\cos\phi_0 \quad \phi_0)^T \quad \text{with } \phi_0 \text{ as in (11.19).}$$

For a steady turn, this point must be used in the linearization procedure for the lateral motion. In order to avoid this, we assume that only curves with small pitch angular velocities have to be flown. Then it is possible to use again the linearization with $r_0 = 0$, $\phi_0 = 0$.

Design of an LQ Controller. Next we want to design an LQ controller for the coordinated turn where we assume that all four states can be measured. If for a specific aircraft this is not possible, an observer has to be applied, but we do not want to discuss this in detail. In order to track a reference signal and for the rejection of constant disturbances, we introduces two integrators:

$$\dot{h}_1 = \beta_c - \beta, \quad \dot{h}_2 = \phi_c - \phi .$$

The extended system can be represented as described in Sect. 6.1. For the LQ controller the weighting matrices $\mathbf{Q}, \mathbf{S}, \mathbf{R}$ are the design parameters of the system. We choose them to be as simple as possible. In particular, let $\mathbf{S} = \mathbf{0}$ and choose $\mathbf{Q}$ and $\mathbf{R}$ diagonal:

$$\mathbf{Q} = \mathrm{diag}(\gamma, 0, 0, \gamma, \delta, \delta), \quad \mathbf{R} = \rho^{-1}\,\mathrm{diag}(1, 1).$$

Then the sideslip and roll angle are weighted by the same factor γ and the integrators and control errors are weighted by the same factor δ. The penalty for the controls decreases with ρ.

We start with a very small value for ρ. Then the control is almost not effective and the closed-loop poles essentially coincide with that of the plant, where of course the two poles caused by the integrators have to be added. Besides this, let

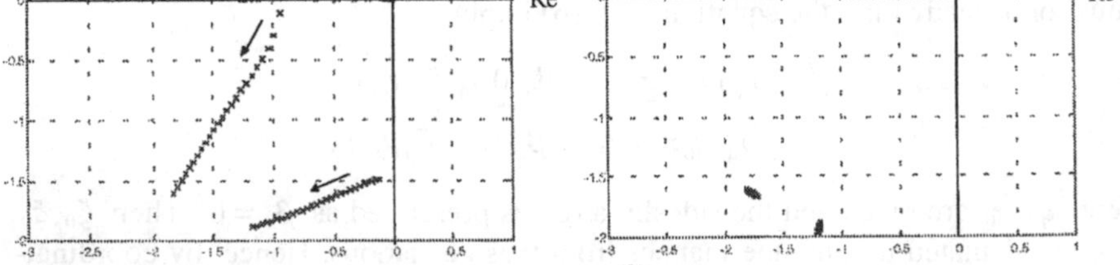

(a) (b)

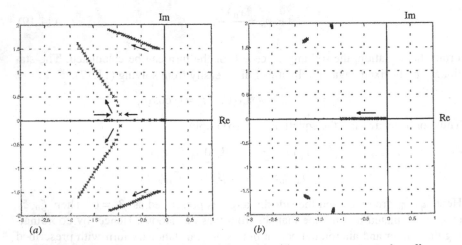

Fig. 11.27. Root locus of the feedback loop: (a) with ρ as parameter and small weighting of the integrators; (b): with δ as parameter and $\rho = 0.09$

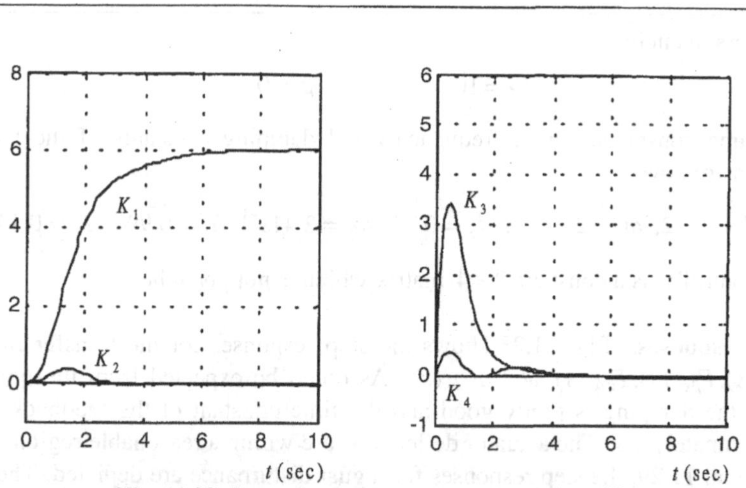

Fig. 11.28. Step response for a command of $\phi_c = 6°$
$(K_1 : \phi, K_2 : \beta, K_3 : \xi, \ K_4 : \zeta)$

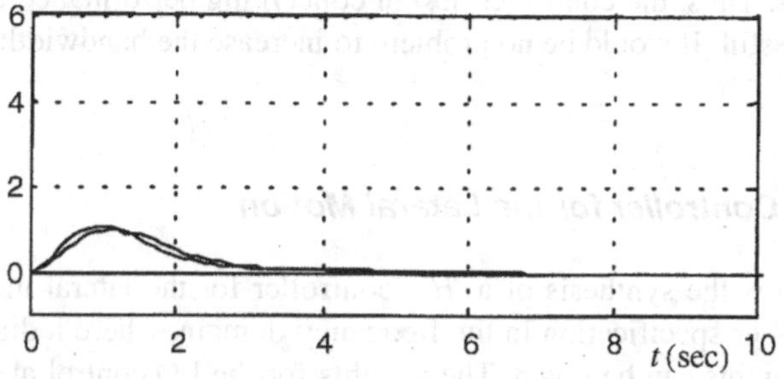

Fig. 11.29. Step responses (ϕ and β) in the presence of a gust

$\gamma = 10$ and $\delta = 0.5$. Then the root locus as a function of ρ depicted in Fig. 11.27(a) results. For small ρ, the only slightly modified roll and dutch roll eigenvalues can be recognized. With increasing ρ the roll eigenvalue moves along the real axis in the right-hand direction and one of the integrator eigenvalues moves in the left-hand direction. If ρ is further increased, they form a conjugate-complex pole pair, which moves along a straight line to ∞ for $\rho \to \infty$. The conjugate-complex dutch roll eigenvalue also moves along a straight line to ∞ (with increasing damping). The other integrator eigenvalue and the spiral eigenvalue are only very moderately changed by ρ.

These considerations show that for sufficiently large ρ a good damping d_1 of the dutch roll eigenvalue is achievable. The choice $\rho = 0.09$ gives $d_1 = 0.53$. The feedback loop now has good damping, but only a small bandwidth. It can be increased by increasing the weighting factor δ of the integrators. Fig 11.27(b) shows the corresponding root locus for $0.5 \le \delta \le 10$. It is seen that with an increasing value of δ the two eigenvalues near 0 are shifted along the real axis into the left half-plane, whereas the other four eigenvalues essentially remain unchanged. These considerations lead to the final choice of the weighting parameters, namely

$$\gamma = 10, \quad \delta = 10, \quad \rho = 0.09 .$$

The time constants, corner frequencies and damping constants of the feedback system are now

$$d_1 = 0.52,\ \omega_1 = 2.31 s^{-1},\ d_2 = 0.72,\ \omega_1 = 2.41 s^{-1},\ T_1 = 0.99 s,\ T_2 = 0.14 s .$$

The controller is a constant 2×4 matrix which is not given here.

Step responses. Fig. 11.28 shows the step responses for the transfer functions $F_{\phi\phi_c}(s), F_{\beta\phi_c}(s), F_{\xi\phi_c}(s)$ and $F_{\zeta\phi_c}(s)$. As could be expected from the pole positions, the damping is pretty good and the time constant of the feedback loop is approximately $2\,s$. The actuator deflections are within a reasonable region.

In Fig. 11.29, the step responses for a gust disturbance are depicted. The figure shows that disturbance rejection has not only improved with respect to the uncontrolled aircraft, but also with respect to the aircraft, which is prestabilized by a washout filter. Thus, the controller design concerning performance can be considered as successful. It would be no problem to increase the bandwidth moderately.

11.3.5 $\mathcal{H}_\infty$ Controller for the Lateral Motion

Our next step is the synthesis of a $\mathcal{H}_\infty$ controller for the lateral motion. This allows a controller specification in the frequency domain, where a direct interpretation of the weights can be given. The weights for the LQ control are in our example only indirectly specified from the position of the closed-loop poles. More precisely, we want to carry out a mixed-sensitivity approach as described in Sect. 9.6. In contrast to the LQ controller, we assume that only the sideslip angle and roll angle are measured. These are also the variables to be controlled. Consequently, the transfer function for the command and the output complementary sensitivity function coincide, a fact that makes the choice of the weights somewhat easier. Of course, the $\mathcal{H}_\infty$ design can also be done when other states are measured or in particular, when all states are measured. Since the LQ approach requires an observer if only these two angles are measured, the results of this section and the foregoing one are not directly comparable.

We apply the S/T weighting scheme (cf. Fig. 9.8) and choose as weights

$$w_1(s) = \gamma \frac{s + \omega_1}{s + \omega_2}, \quad w_3(s) = \frac{\omega_4^2}{\omega_3^2} \frac{s^2}{s^2 + 2\omega_4 s + \omega_4^2},$$

$$\mathbf{W}_j(s) = \operatorname{diag}(w_j(s), w_j(s)), \quad j = 1, 3,$$

with the numerical values

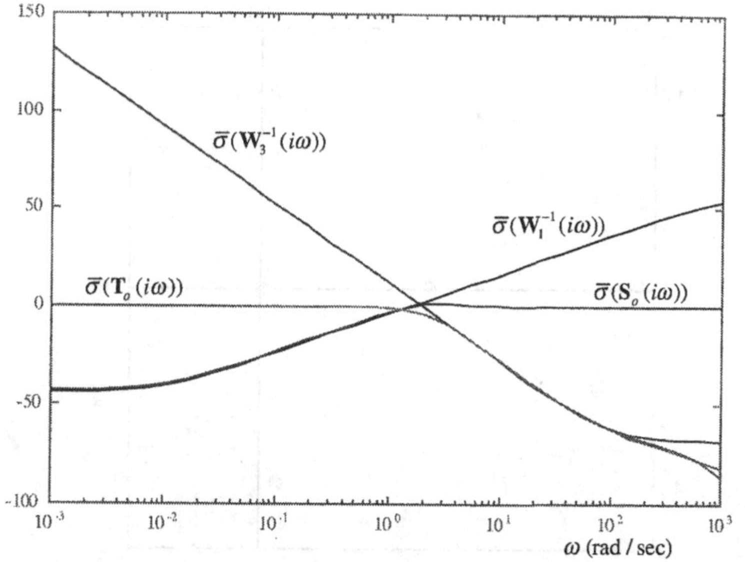

Fig. 11.30. Frequency response of the inverse weights and transfer functions for $\mathcal{H}_\infty$ synthesis

$$\gamma = 0.0015, \quad \omega_1 = 1000\,s^{-1}, \quad \omega_2 = 0.01s^{-1}, \quad \omega_3 = 2s^{-1}, \quad \omega_4 = 100s^{-1}.$$

For the inverse weights, the following approximations hold:

$$|w_1^{-1}(i\omega)| \approx \left|\frac{i\omega}{\gamma\omega_1}\right|, \quad \text{if } \omega_2 \ll \omega \ll \omega_1, \quad |w_3^{-1}(i\omega)| \approx \left|\frac{\omega_3^2}{(i\omega)^2}\right|, \quad \text{if } \omega \ll \omega_4$$

The frequency response of the inverse weights is depicted in Fig. 11.30.

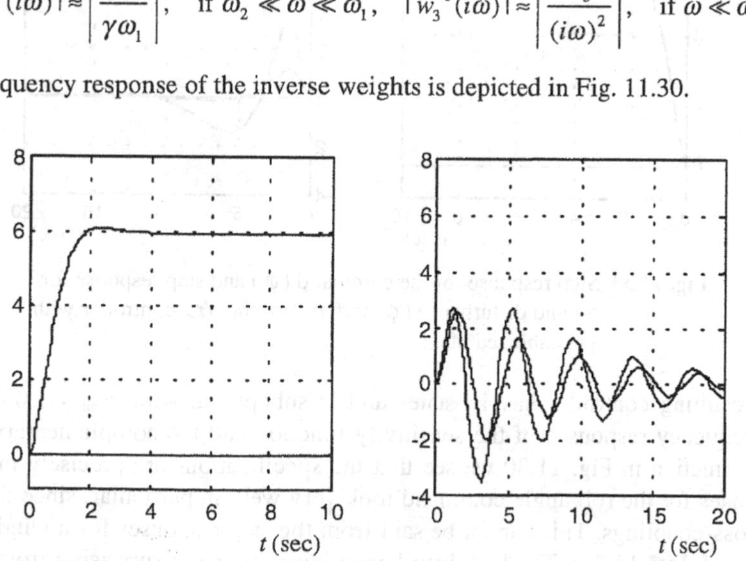

Fig. 11.31. Step response for the command (ϕ) and step response for a wind disturbance (ϕ and β) for the $\mathcal{H}_\infty$ controller

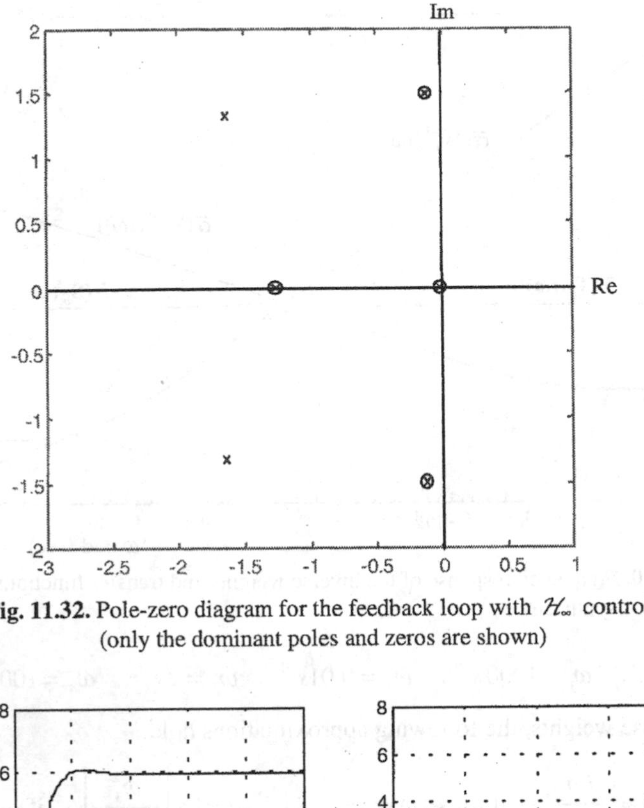

Fig. 11.32. Pole-zero diagram for the feedback loop with $\mathcal{H}_\infty$ controller (only the dominant poles and zeros are shown)

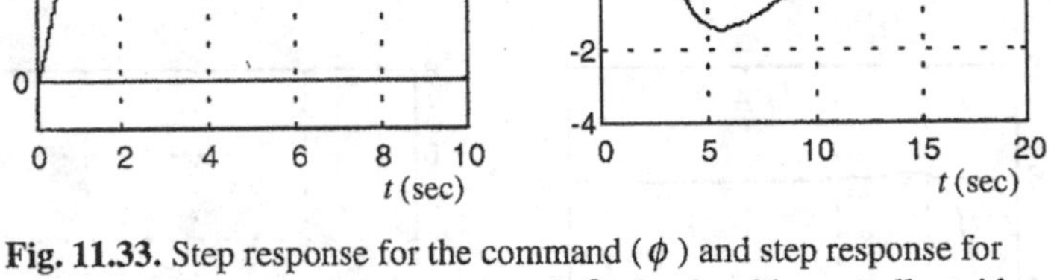

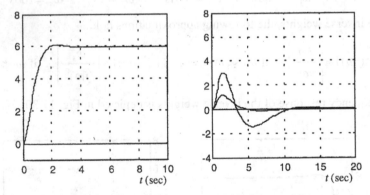

Fig. 11.33. Step response for the command (ϕ) and step response for a wind disturbance (ϕ and β) for the $\mathcal{H}_\infty$ controller with prestabilized plant

The resulting controller has 10 states and is suboptimal with $\gamma_{\min} = 1.09$. From the frequency responses of the sensitivity function and the complementary sensitivity function in Fig. 11.30 we see that the specifications are precisely met. The responses for the roll angle command look very well, in particular, since there are no cross-couplings. This cannot be said from the step responses for a wind disturbance (cf. Fig. 11.31). The bad disturbance rejection is not surprising from theory (cf. Sect. 9.6.2) since we know that the mixed sensitivity $\mathcal{H}_\infty$ controller cancels all stable plant poles. Hence these poles are still present in the system matrix of the closed-loop system and they appear as poles for the transfer function for the disturbance. In particular, the dutch roll eigenvalue is one of these poles. Fig. 11.32 shows the pole-zero plot for the closed-loop transfer function from (β_c, ϕ_c) to (β, ϕ) (here, only the dominant poles are depicted).

There are several possible ways to design a controller with good disturbance rejection properties. One of them is to make the feedback loop more robust by introducing uncertainties. We will return to this point in Sect. 14.1. A simple, but very natural approach is to apply the $\mathcal{H}_\infty$ design to the prestabilized plant. As before, prestabilization may be done by a high-pass filter. Then the $\mathcal{H}_\infty$ synthesis is specified with exactly the same weights which we have used for our first synthesis. For this design, the frequency response is as in Fig. 11.30 and again the stable plant poles are cancelled by the controller, but this is no longer critical. The step response for the command shown in Fig. 11.33 can be compared with that of Figs. 11.28 and 11.29 for the LQ controller and, concerning the wind disturbance, also with that of Fig. 11.25 for the washout filter. The step response for the wind disturbance is now well damped and with respect to their damping is similar to that for the prestabilized plant without tracking. One observes that the disturbance rejection for this $\mathcal{H}_\infty$ controller is not quite as good as for the LQ controller.

Chapter 12

Representation of Uncertainty

As we have seen, $\mathcal{H}_\infty$ optimization is an adequate tool for controller design if the specification of the feedback loop is formulated in the frequency domain. On the other hand, the $\mathcal{H}_\infty$ optimal controller is not necessarily robust. A robust controller design can be achieved by the so-called D-K iteration, which will be introduced in the next chapter. The application of this method requires a precise mathematical framework of system robustness, which will be given now. One source of uncertainty is neglected high-frequency dynamics. This will be described in Sect. 12.2, after the basic definitions concerning robust stability and robust performance are introduced (Sect. 12.1). Then uncertainty is modeled by one unknown transfer function together with some weights and is called unstructured uncertainty. Another type of uncertainty arises if one or more parameters of the plant are not known precisely. This leads to structured uncertainty, whose mathematical description is given in Sect. 12.3. Of course it may occur that parameter uncertainty arises in combination with uncertain dynamics. A mathematical framework covering all kinds of uncertainty will be described in Sect. 12.4. It is the basis for the analysis and design methods for robust feedback loops, which we will develop in Chap. 13. The chapter ends with discussing uncertainty for a simple SISO second-order plant, which prepares the way for the complex case studies in Chap. 15.

12.1 Model Uncertainty

Each control theory design method needs a mathematical model of the plant. It may result from a mathematical model building process using physical laws or by an identification procedure or a mixture of both methods. There are some fields in technology where considerable effort is made to get an accurate model, whereas in others the control engineer has to carry out his design based on a rather crude model. For example, in the aircraft or aerospace industries normally very good models of the plant are available. In the chemical industry, the models are often highly simplified or if they are not, they may be too complicated to handle for a controller design procedure. One possible reason for this is that the plant is governed by a partial differential equation. Even if the model is very accurate, there is one inherent difficulty due to the linearization procedure: the design methods are made for the linear model and therefore in some sense only applicable if the states and the controls remain in the vicinity of the equilibrium point. If this is not the case, the linear model is inaccurate even if the nonlinear model is very good. An extreme example of this kind is the missile studied in Chap. 14. Thus, in any case, controller design is based on a mathematical model which is only a simplified image of the physical reality.

There are two basic types of uncertainty, namely **parameter uncertainty** and **unmodeled dynamics**. A typical example for the first kind are the aerodynamical coefficients which describe the aerodynamical forces and moments which act on an aircraft or a missile. Depending on the equilibrium point, their measurement is only possible up to a more or less large amount of uncertainty. Not modeled dynamics may arise if certain cross-couplings in a MIMO system are neglected or if only simplified actuator or sensor models are available. A possible source of the neglected dynamics of a mechanical plant may be its structure, which is assumed as inelastic in the model but is flexible in reality. Examples for plants having such a behavior are a telescope, which has to be pointed very exactly, or a thin missile, which tends to vibrate if it has to fly in a sharp curve.

Thus, normally the **high-frequency dynamics** of the plant is inaccurate. This means that at high frequencies the gain and phase lie within a band that gets broader if the frequencies increase. The situation is depicted schematically in Fig 12.1 for a multiplicative uncertainty, which will be introduced in the next section.

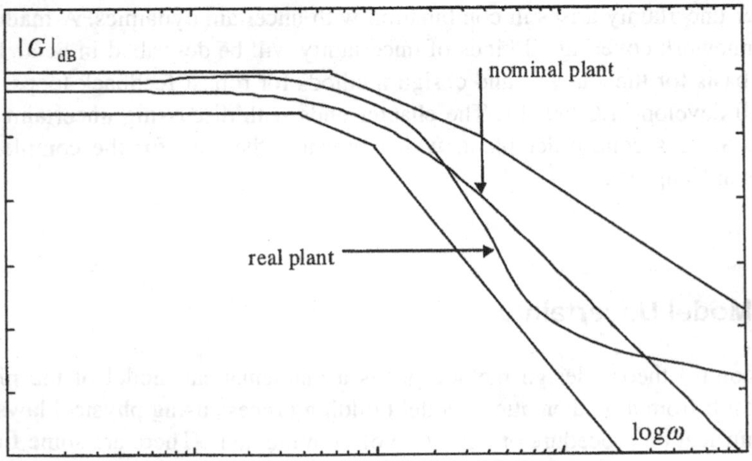

Fig. 12.1. Typical model uncertainty in the frequency domain

From these considerations, we conclude that a model of the plant consists of the following two parts:

(i) *a model of the nominal plant,*

(ii) *a model of the uncertainties.*

Then all possible plants covered by this modeling are elements of a certain set Π. The nominal plant as well as the "real" plant are elements of Π. In order to give our design goals a precise meaning, the following definition is of fundamental importance.

Definition 12.1. *Let a set* Π *of plants be given such that the nominal plant* $\mathbf{G}$ *is an element of* Π. *Moreover, let a set of performance objectives be given and suppose that* $\mathbf{K}$ *is a controller which meets these objectives. Then the feedback system is said to have*

nominal stability (NS), *if* $\mathbf{K}$ *internally stabilizes the nominal plant* $\mathbf{G}$;

robust stability (RS), *if* $\mathbf{K}$ *internally stabilizes every plant belonging to* Π;

nominal performance *(NP), if the performance objectives are satisfied for the nominal plant* $\mathbf{G}$;

robust performance *(RP), if the performance objectives are satisfied for every plant belonging to* Π.

Our aim is to design controllers, which are robustly stable and have robust performance.

12.2 Unstructured Uncertainties

We start our analysis by considering uncertainty caused by not modeled dynamics and assume that the real plant is obtained by combining the nominal plant $\mathbf{G}(s)$ with a transfer function of the kind $\mathbf{W}_1(s)\Delta(s)\mathbf{W}_2(s)$. Such a kind of uncertainty is denoted as **unstructured**. The transfer functions $\mathbf{W}_1, \mathbf{W}_2$ are given and specify the uncertainty, whereas Δ is unknown. It is only assumed that Δ is stable and has a bounded magnitude: $\|\Delta\|_\infty \le 1$. Later, the weights $\mathbf{W}_1, \mathbf{W}_2$ will be modeled as parts of the (extended) plant. Then the feedback system consists of the extended plant, the uncertainty Δ and the controller $\mathbf{K}$.

Fig. 12.2 shows several possible ways of connecting $\mathbf{W}_1\Delta\mathbf{W}_2$ with the plant. The perturbed plant is denoted as $\mathbf{G}_P$. An **additive uncertainty** is given by

$$\mathbf{G}_P = \mathbf{G} + \mathbf{W}_1\Delta\mathbf{W}_2 .$$

It can be used to model additive uncertainties. **Input-multiplicative uncertainties**

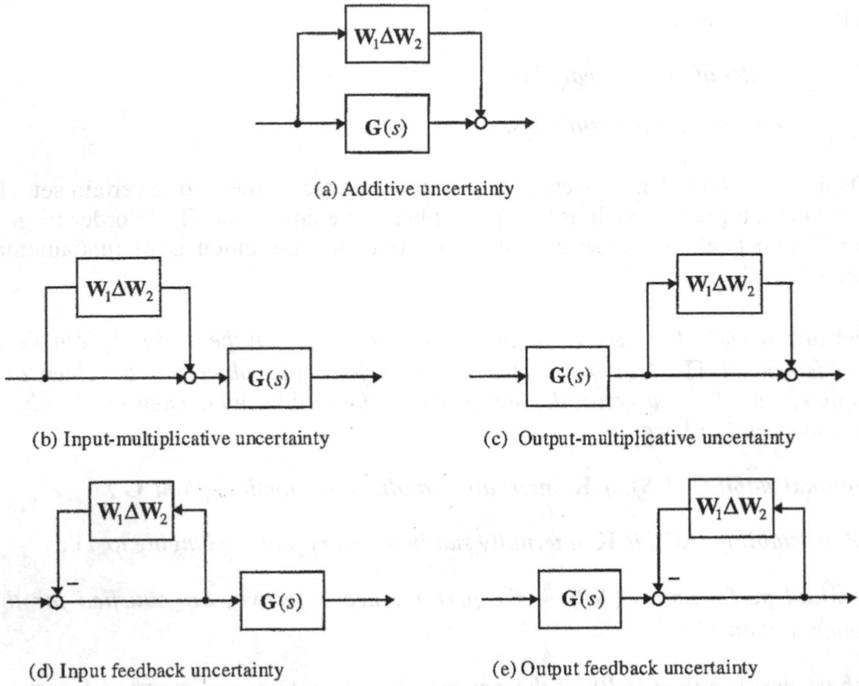

Fig. 12.2. Types of unstructured uncertainty

are described by

$$\mathbf{G}_P = \mathbf{G}(\mathbf{I} + \mathbf{W}_1 \Delta \mathbf{W}_2)\,.$$

They can be used to model actuator errors or neglected high-frequency dynamics. Similarly, **output-multiplicative perturbed systems** are modeled in the form

$$\mathbf{G}_P = (\mathbf{I} + \mathbf{W}_1 \Delta \mathbf{W}_2)\mathbf{G}\,.$$

They are used to model sensor errors and neglected high-frequency dynamics. Finally, there are **input feedback uncertainties**,

$$\mathbf{G}_P = \mathbf{G}(\mathbf{I} + \mathbf{W}_1 \Delta \mathbf{W}_2)^{-1}\,,$$

and similarly **output feedback uncertainties**, which model low-frequency parameter errors.

Examples. 1. A typical weighting function for modeling scalar input-multiplicative perturbed systems is given by

$$W(s) = \frac{k\omega_2}{\omega_1}\frac{s+\omega_1}{s+\omega_2} \quad \text{with } \omega_1 < \omega_2\,.$$

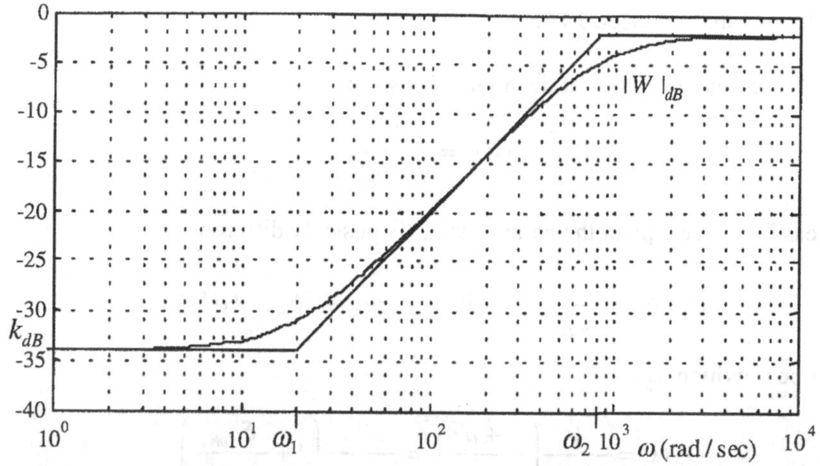

Fig. 12.3. Magnitude plot of $W(s)$

For frequencies $\omega \le \omega_1$, the percentage of the gain error is $\le k$, and for frequencies $\omega \ge \omega_2$ it is limited from above by $k\omega_2/\omega_1$ (cf. Fig. 12.3). Examples for input-multiplicative uncertainties are given in Sect. 12.5.

2. Suppose that a moment acts on a thin elastic body and let this moment be the control. Then the rotational dynamics is given by

$$\hat{\varphi} = G_0(s)\hat{u} \quad \text{with} \quad G_0(s) = \frac{k}{s^2}.$$

The task is to control the angle φ which describes the angular position of the body. If the elastic behavior is taken into account, the transfer function G_0 has to be replaced by

$$G(s) = \frac{k}{s^2} + G_{el}(s)$$

where

$$G_{el}(s) = \sum_{\nu=1}^{\infty} \frac{k_\nu}{1 + 2d_\nu \omega_\nu s + \omega_\nu s^2}$$

with small damping coefficients $d_\nu > 0$ (compare the example of the supported beam in Sect. 8.5.2). We now model G_{el} as an additive model uncertainty. To this end, let a stable weight W_1 with stable inverse be given such that

$$\| G_{el} \|_\infty \le \| W_1 \|_\infty .$$

We define now a set of plants by $G_P = G_0 + W_1 \Delta$ with $\| \Delta \|_\infty \le 1$. Then G is an element of this set (put $\Delta = W_1^{-1} G_{el}$). A similar approach will be used in the missile case study in Sect. 14.3.

3. In the second example, let the nominal plant be

$$G(s) = \frac{1}{s + a_0},$$

whereas for the real plant the parameter a_0 is possibly different:

$$G_P(s) = \frac{1}{s + a} \qquad \text{with} \qquad a = a_0 + \delta\delta_{\mathrm{m}}, \quad |\delta| \leq 1.$$

It can be rewritten as

$$G_P(s) = \frac{1}{s + a_0}\left(\frac{s + a}{s + a_0}\right)^{-1} = \frac{1}{s + a_0}\left(1 + \frac{\delta\delta_m}{s + a_0}\right)^{-1}.$$

By defining

$$W_1(s) = \delta_{\mathrm{m}} G(s), \quad W_2(s) = 1,$$

the uncertainty of the plant parameter a can be modeled as feedback uncertainty. Note that in this approach the set of uncertain plants consists of all plants which can be written in the form $G_P = G(1 + \Delta W_1)^{-1}$, where Δ is a stable transfer function such that $\|\Delta\|_\infty \leq 1$. In particular, Δ may be a constant $\Delta = \delta$ with $|\delta| \leq 1$. These constant transfer functions model the uncertain paramater.

We still mention that the input-multiplicative uncertainty of Fig. 12.2(b) can equivalently be represented as shown in Fig. 12.4. A similar reformulation is possible for the other unstructured uncertainties of Fig 12.2.

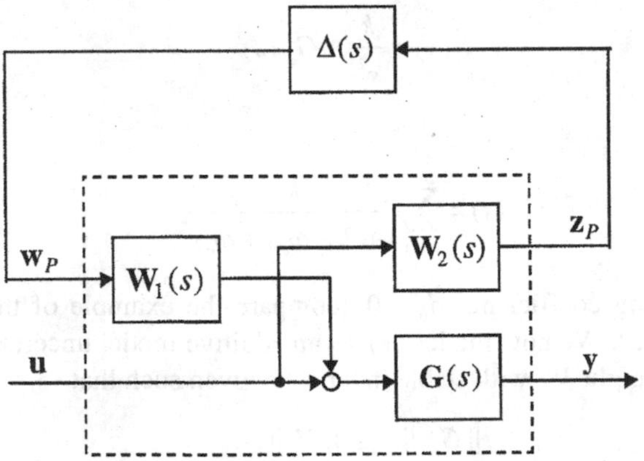

Fig. 12.4. Multiplicative uncertainties with a separated normed part Δ

12.3 Structured Model Uncertainties

12.3.1 Introductory Examples

Plant of order 1. We consider again a first-order plant as in the preceding section, but now we represent the plant and the uncertainty in the time domain:

$$\dot{x} = -(a_0 + \delta\delta_m)x + u \qquad \text{with } |\delta| < 1 .$$

In this way, uncertainty in the parameter a is modeled. Figure 12.5(a) shows the plant where the uncertainty occurs in a proportional block with gain $a + \delta\delta_m$. In the next step (Fig. 12.5(b)), one block consisting of an unknown gain δ is isolated. The magnitude of δ is limited by $|\delta| < 1$. In the second and final step, this gain is separated from the plant, which is now extended by a parameter which models the magnitude of the uncertainty and one additional input and output (Fig. 12.5(c)). The state space description of the extended plant is given by

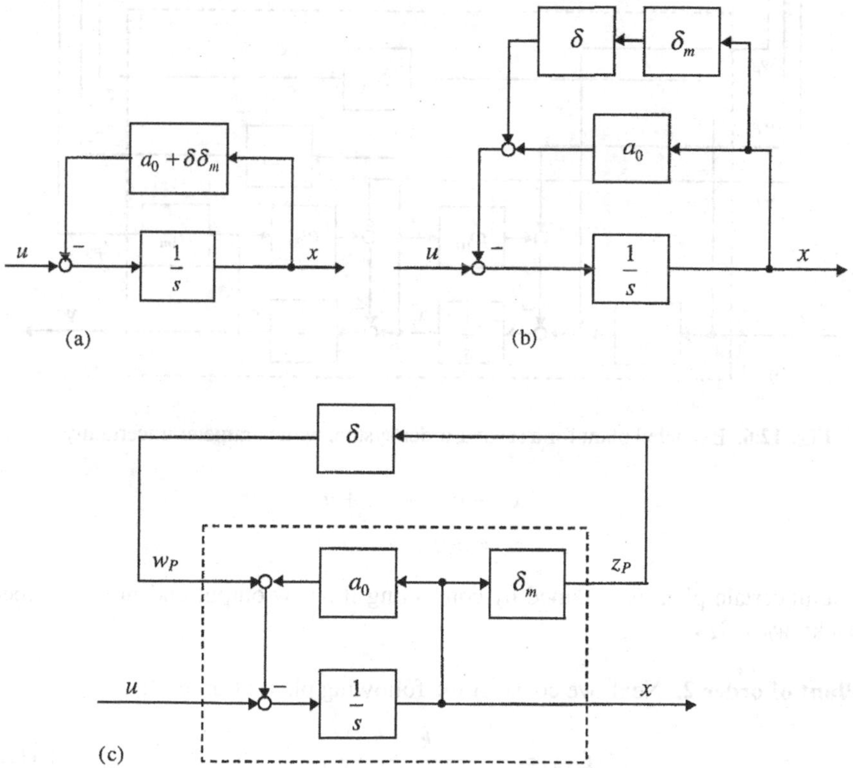

Fig. 12.5. Extended plant for a first-order system with parameter uncertainty

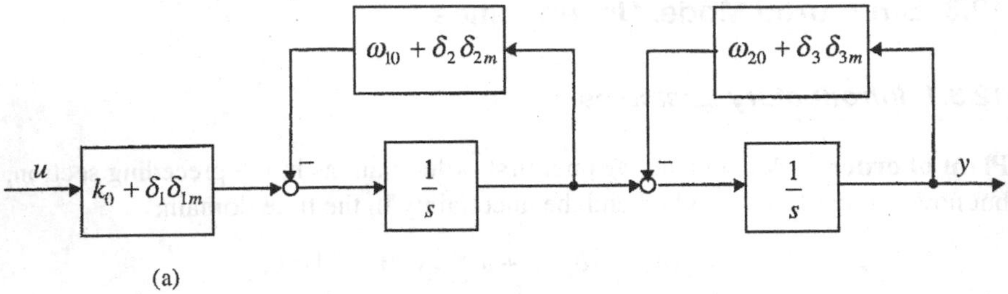

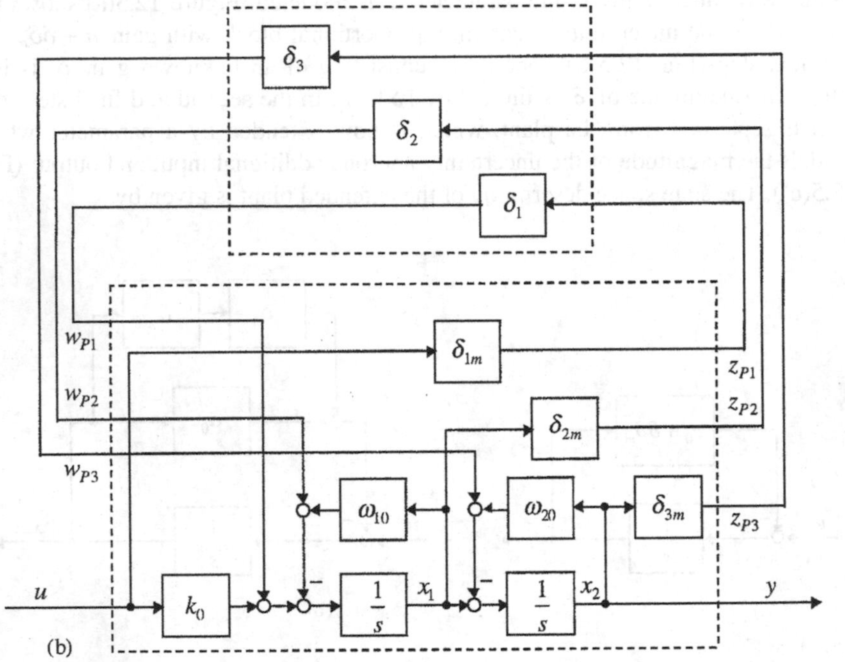

Fig. 12.6. Extended plant for a second-order system with parameter uncertainty

$$\dot{x} = -a_0 x - w_P + u$$
$$z_P = \delta_m x .$$

The uncertain plant is obtained by connecting the new output and input via feedback: $w_P = \delta z_P$.

Plant of order 2. Next, we consider the following plant of order 2:

$$G(s) = \frac{k}{(s+\omega_1)(s+\omega_2)} . \tag{12.1}$$

Let the gain as well as the corner frequencies be uncertain:

$$k = k_0 + \delta_1 \delta_{1m}, \quad \omega_1 = \omega_{10} + \delta_2 \delta_{2m}, \quad \omega_2 = \omega_{20} + \delta_3 \delta_{3m}.$$

As in the case of the first-order system we assume that $|\delta_i| \leq 1$ for $i = 1, 2, 3$. Figure 12.6(a) shows the block diagram for the plant with uncertainties. In Fig. 12.6(b), the unknown, but normed uncertainties $\delta_1, \delta_2, \delta_3$ are isolated, where for each constant one proceeds exactly as in the previous example, compare the block diagram of Fig. 12.5(c). It is evident that the two block diagrams of Fig. 12.6 are equivalent.

The state space description of the extended plant is

$$\begin{pmatrix} \dot{x}_1 \\ \dot{x}_2 \end{pmatrix} = \begin{pmatrix} -\omega_{10} & 0 \\ 1 & -\omega_{20} \end{pmatrix} \begin{pmatrix} x_1 \\ x_2 \end{pmatrix} + \begin{pmatrix} k_0 \\ 0 \end{pmatrix} u + \begin{pmatrix} 1 & -1 & 0 \\ 0 & 0 & -1 \end{pmatrix} \begin{pmatrix} w_{P1} \\ w_{P2} \\ w_{P3} \end{pmatrix}$$

$$y = \begin{pmatrix} 0 & 1 \end{pmatrix} \begin{pmatrix} x_1 \\ x_2 \end{pmatrix}$$

$$\begin{pmatrix} z_{P1} \\ z_{P2} \\ z_{P3} \end{pmatrix} = \begin{pmatrix} 0 & 0 \\ \delta_{2m} & 0 \\ 0 & \delta_{3m} \end{pmatrix} \begin{pmatrix} x_1 \\ x_2 \end{pmatrix} + \begin{pmatrix} \delta_{1m} \\ 0 \\ 0 \end{pmatrix} u .$$

The block with the uncertainties can be written as

$$\begin{pmatrix} w_{P1} \\ w_{P2} \\ w_{P3} \end{pmatrix} = \begin{pmatrix} \delta_1 & 0 & 0 \\ 0 & \delta_2 & 0 \\ 0 & 0 & \delta_3 \end{pmatrix} \begin{pmatrix} z_{P1} \\ z_{P2} \\ z_{P3} \end{pmatrix} .$$

The feedback system consisting of the extended plant and the constant-gain block with the uncertainties gives again the uncertain second-order plant (12.1). These uncertain parameters form a structured model uncertainty.

12.3.2 Structured State Space Uncertainties

Next we want to generalize the ideas of the previous section to a general linear system in which the uncertainty occurs in the parameters of the linear differential equation. Let an uncertain system of the form

$$\begin{aligned} \dot{\mathbf{x}} &= \mathbf{A}\mathbf{x} + \mathbf{B}\mathbf{u} \\ \mathbf{y} &= \mathbf{C}\mathbf{x} + \mathbf{D}\mathbf{u} \end{aligned} \tag{12.2}$$

be given and assume that the uncertainty is additive with respect to the system matrices:

$$\mathbf{A} = \mathbf{A}_0 + \mathbf{A}_\delta, \quad \mathbf{B} = \mathbf{B}_0 + \mathbf{B}_\delta, \quad \mathbf{C} = \mathbf{C}_0 + \mathbf{C}_\delta, \quad \mathbf{D} = \mathbf{D}_0 + \mathbf{D}_\delta .$$

The nominal plant is described by the matrices $\mathbf{A}_0, \mathbf{B}_0, \mathbf{C}_0, \mathbf{D}_0$. The other matrices $\mathbf{A}_\delta, \mathbf{B}_\delta, \mathbf{C}_\delta, \mathbf{D}_\delta$ are supposed to depend linearly on uncertain parameters $\delta_1, \delta_2, \ldots, \delta_k$ as follows:

$$\begin{pmatrix} \mathbf{A}_\delta & \mathbf{B}_\delta \\ \mathbf{C}_\delta & \mathbf{D}_\delta \end{pmatrix} = \sum_{i=1}^{k} \delta_i \begin{pmatrix} \mathbf{A}_i & \mathbf{B}_i \\ \mathbf{C}_i & \mathbf{D}_i \end{pmatrix}.$$

The task is to separate the parameters $\delta_1, \delta_2, \ldots, \delta_k$ from the plant. To this end, we extend the plant as in the examples by additional input and output variables:

$$\begin{aligned} \dot{\mathbf{x}} &= \mathbf{A}_0 \mathbf{x} + \mathbf{B}_0 \mathbf{u} + \mathbf{B}_2 \mathbf{w}_P \\ \mathbf{y} &= \mathbf{C}_0 \mathbf{x} + \mathbf{D}_0 \mathbf{u} + \mathbf{D}_{12} \mathbf{w}_P \\ \mathbf{z}_P &= \mathbf{C}_2 \mathbf{x} + \mathbf{D}_{21} \mathbf{u} . \end{aligned} \tag{12.3}$$

Here, $\mathbf{z}_P$ and $\mathbf{w}_P$ are assumed to be adequately partitioned:

$$\mathbf{z}_P = \begin{pmatrix} \mathbf{z}_{P1} \\ \vdots \\ \mathbf{z}_{Pk} \end{pmatrix}, \quad \mathbf{w}_P = \begin{pmatrix} \mathbf{w}_{P1} \\ \vdots \\ \mathbf{w}_{Pk} \end{pmatrix}.$$

The matrices $\mathbf{B}_2, \mathbf{C}_2, \mathbf{D}_{12}, \mathbf{D}_{21}, \mathbf{D}_{22}$ have to be constructed such that the uncertain plant is obtained from the extended plant by using the feedback law

$$\mathbf{w}_{Pi} = \delta_i \mathbf{z}_{Pi} \quad \text{for } i = 1, \ldots, k \tag{12.4}$$

with suitably dimensioned vectors $\mathbf{w}_{Pi}$ and $\mathbf{z}_{Pi}$. To construct the desired representation explicitly, let

$$q_i = \operatorname{rank} \begin{pmatrix} \mathbf{A}_i & \mathbf{B}_i \\ \mathbf{C}_i & \mathbf{D}_i \end{pmatrix},$$

and introduce decompositions

$$\begin{pmatrix} \mathbf{A}_i & \mathbf{B}_i \\ \mathbf{C}_i & \mathbf{D}_i \end{pmatrix} = \begin{pmatrix} \mathbf{L}_i \\ \mathbf{W}_i \end{pmatrix} \begin{pmatrix} \mathbf{R}_i^T & \mathbf{Z}_i^T \end{pmatrix} \tag{12.5}$$

such that $\mathbf{L}_i \in \mathbb{R}^{n \times qi}, \mathbf{W}_i \in \mathbb{R}^{l \times qi}, \mathbf{R}_i \in \mathbb{R}^{n \times qi}, \mathbf{Z}_i \in \mathbb{R}^{m \times qi}$. Let now $\mathbf{w}_{Pi}, \mathbf{z}_{Pi} \in \mathbb{R}^{qi}$ and define

$$\mathbf{E}_\delta = \begin{pmatrix} \delta_1 \mathbf{I}_{r_1} & & \\ & \ddots & \\ & & \delta_k \mathbf{I}_{r_k} \end{pmatrix}.$$

Then (12.4) can be written as

$$\mathbf{w}_P = \mathbf{E}_\delta \mathbf{z}_P .$$

If this vector $\mathbf{w}_P$ is inserted into (12.3), the original plant (12.2) has to result. Then the equations

$$\mathbf{A}_\delta = \mathbf{B}_2\mathbf{E}_\delta\mathbf{C}_2, \quad \mathbf{B}_\delta = \mathbf{B}_2\mathbf{E}_\delta\mathbf{D}_{21},$$
$$\mathbf{C}_\delta = \mathbf{D}_{12}\mathbf{E}_\delta\mathbf{C}_2, \quad \mathbf{D}_\delta = \mathbf{D}_{12}\mathbf{E}_\delta\mathbf{D}_{21},$$

must hold. These four equations can be written as one equation:

$$\begin{pmatrix} \mathbf{A}_\delta & \mathbf{B}_\delta \\ \mathbf{C}_\delta & \mathbf{D}_\delta \end{pmatrix} = \begin{pmatrix} \mathbf{B}_2 \\ \mathbf{D}_{12} \end{pmatrix} \mathbf{E}_\delta \begin{pmatrix} \mathbf{C}_2 & \mathbf{D}_{21} \end{pmatrix},$$

and it is easily seen that the matrices

$$\mathbf{B}_2 = \begin{pmatrix} \mathbf{L}_1 & \cdots & \mathbf{L}_k \end{pmatrix}, \quad \mathbf{D}_{12} = \begin{pmatrix} \mathbf{W}_1 & \cdots & \mathbf{W}_k \end{pmatrix},$$
$$\mathbf{C}_2 = \begin{pmatrix} \mathbf{R}_1^T \\ \vdots \\ \mathbf{R}_k^T \end{pmatrix}, \quad \mathbf{D}_{21} = \begin{pmatrix} \mathbf{Z}_1^T \\ \vdots \\ \mathbf{Z}_k^T \end{pmatrix}.$$

have the desired property.

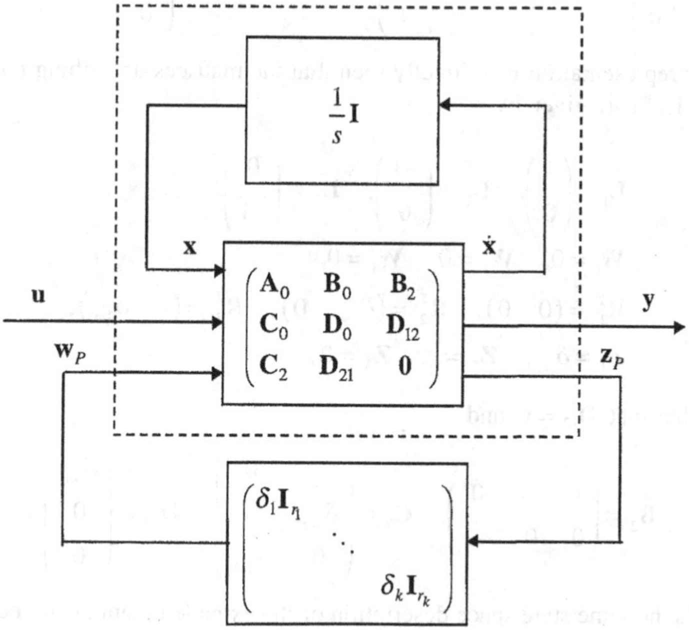

Fig. 12.7. Structured State Space Uncertainties

In this way, one obtains as result the block diagram depicted in Fig. 12.7. In particular, we have shown how the uncertain parameters $\delta_1, \delta_2, \ldots, \delta_k$ can be separated such that the uncertain plant can be viewed as a feedback system.

For Example 2 of the previous section, the uncertain plant matrices are

$$\mathbf{A}_0 = \begin{pmatrix} -\omega_{10} & 0 \\ 1 & -\omega_{20} \end{pmatrix}, \quad \mathbf{A}_\delta = \begin{pmatrix} -\delta_2\delta_{2m} & 0 \\ 0 & -\delta_3\delta_{3m} \end{pmatrix}$$

$$\mathbf{B}_0 = \begin{pmatrix} k_0 \\ 0 \end{pmatrix}, \quad \mathbf{B}_\delta = \begin{pmatrix} \delta_1\delta_m \\ 0 \end{pmatrix}$$

$$\mathbf{C}_0 = \begin{pmatrix} 0 & 1 \end{pmatrix}, \quad \mathbf{C}_\delta = \begin{pmatrix} 0 & 0 \end{pmatrix}, \quad \mathbf{D}_0 = 0, \quad \mathbf{D}_\delta = 0 .$$

This implies

$$\begin{pmatrix} \mathbf{A}_\delta & \mathbf{B}_\delta \\ \mathbf{C}_\delta & \mathbf{D}_\delta \end{pmatrix} = \delta_1 \begin{pmatrix} 0 & 0 & \delta_{1m} \\ 0 & 0 & 0 \\ 0 & 0 & 0 \end{pmatrix} + \delta_2 \begin{pmatrix} -\delta_{2m} & 0 & 0 \\ 0 & 0 & 0 \\ 0 & 0 & 0 \end{pmatrix} + \delta_3 \begin{pmatrix} 0 & 0 & 0 \\ 0 & -\delta_{3m} & 0 \\ 0 & 0 & 0 \end{pmatrix}$$

$$= \begin{pmatrix} 1 \\ 0 \\ 0 \end{pmatrix} \delta_1 \begin{pmatrix} 0 & 0 & \delta_{1m} \end{pmatrix} + \begin{pmatrix} -1 \\ 0 \\ 0 \end{pmatrix} \delta_2 \begin{pmatrix} \delta_{2m} & 0 & 0 \end{pmatrix} + \begin{pmatrix} 0 \\ -1 \\ 0 \end{pmatrix} \delta_3 \begin{pmatrix} 0 & \delta_{3m} & 0 \end{pmatrix} .$$

From this representation it is directly seen that the matrices describing the decomposition (12.5) are given by

$$\mathbf{L}_1 = \begin{pmatrix} 1 \\ 0 \end{pmatrix}, \quad \mathbf{L}_2 = \begin{pmatrix} -1 \\ 0 \end{pmatrix}, \quad \mathbf{L}_3 = \begin{pmatrix} 0 \\ -1 \end{pmatrix},$$

$$\mathbf{W}_1 = 0, \quad \mathbf{W}_2 = 0, \quad \mathbf{W}_3 = 0,$$

$$\mathbf{R}_1^T = \begin{pmatrix} 0 & 0 \end{pmatrix}, \quad \mathbf{R}_2^T = \begin{pmatrix} \delta_{2m} & 0 \end{pmatrix}, \quad \mathbf{R}_3^T = \begin{pmatrix} 0 & \delta_{3m} \end{pmatrix},$$

$$\mathbf{Z}_1 = \delta_{1m}, \quad \mathbf{Z}_2 = 0, \quad \mathbf{Z}_3 = 0 .$$

This implies that $\mathbf{D}_{12} = \mathbf{0}$ and

$$\mathbf{B}_2 = \begin{pmatrix} 1 & -1 & 0 \\ 0 & 0 & -1 \end{pmatrix}, \quad \mathbf{C}_2 = \begin{pmatrix} 0 & 0 \\ \delta_{2m} & 0 \\ 0 & \delta_{3m} \end{pmatrix}, \quad \mathbf{D}_{21} = \begin{pmatrix} \delta_{1m} \\ 0 \\ 0 \end{pmatrix},$$

and this is the same state space description of the extended plant as in Sect. 12.3.1.

12.3.3 Parameter Uncertainty for Transfer Functions

The main step in reformulating a system with uncertainties is to separate the uncertainties from the system ("pulling out the Δ s"). This was demonstrated in the previous sections for some examples and in detail for a general system in the state space representation. In this section, we want to describe this procedure for transfer functions and limit ourselves to one example. For other transfer functions, one proceeds in a quite similar way.

The transfer function from i_A to x for the crane-positioning system can be written in the form (cf. Sect. 11.1.1)

$$G(s) = k\frac{s^2 + a}{s^2(s^2 + b)}.$$

Let all three parameters be uncertain: $k = k_0 + \delta_1$, $a = a_0 + \delta_2$, $b = b_0 + \delta_3$. In order to pull out the Δ s, one has to draw a block diagram where again the uncertain parameters $\delta_1, \delta_2, \delta_3$ are isolated in the form of additional blocks. With each

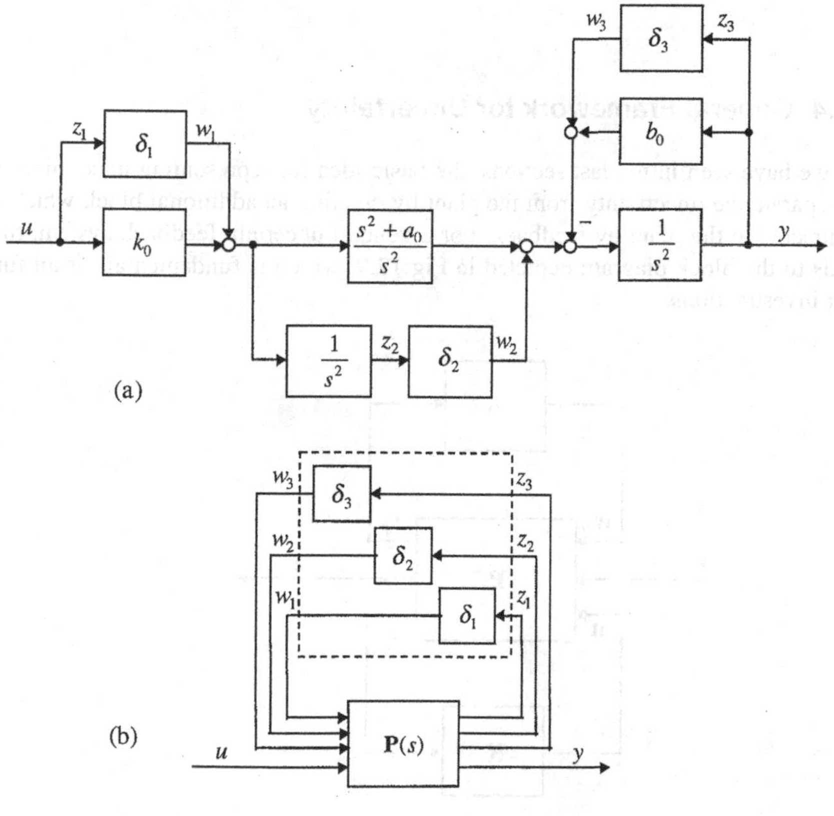

Fig. 12.8. Pulling out the Δ s for the crane positioning system

of these blocks, additional input variables w_1, w_2, w_3 and output variables z_1, z_2, z_3 are associated. In this way, we obtain the block diagram depicted in Fig. 12.8(a).

It is easy to see that the following equation holds:

$$\begin{pmatrix} \hat{z}_1 \\ \hat{z}_2 \\ \hat{z}_3 \\ \hat{y} \end{pmatrix} = \begin{pmatrix} 0 & 0 & 0 & 1 \\ \dfrac{1}{s^2} & 0 & 0 & \dfrac{k_0}{s^2} \\ \dfrac{s^2+a_0}{s^2(s^2+b_0)} & \dfrac{1}{s^2+b_0} & \dfrac{-1}{s^2+b_0} & k_0\dfrac{s^2+a_0}{s^2(s^2+b_0)} \\ \dfrac{s^2+a_0}{s^2(s^2+b_0)} & \dfrac{1}{s^2+b_0} & \dfrac{-1}{s^2+b_0} & k_0\dfrac{s^2+a_0}{s^2(s^2+b_0)} \end{pmatrix} \begin{pmatrix} \hat{w}_1 \\ \hat{w}_2 \\ \hat{w}_3 \\ \hat{u} \end{pmatrix}. \tag{12.6}$$

If this transfer matrix is denoted by $\mathbf{P}(s)$, the block diagram in Fig. 12.8(b) results. Closing the feedback loop in this diagram leads to the original uncertain plant in Fig. 12.8(a).

12.4 General Framework for Uncertainty

As we have seen in the last sections, the basic idea for representing uncertainty is to separate the uncertainty from the plant by creating an additional block which is connected to the plant by feedback. For a general uncertain feedback system, this leads to the block diagram depicted in Fig 12.9, which is fundamental for all further investigations.

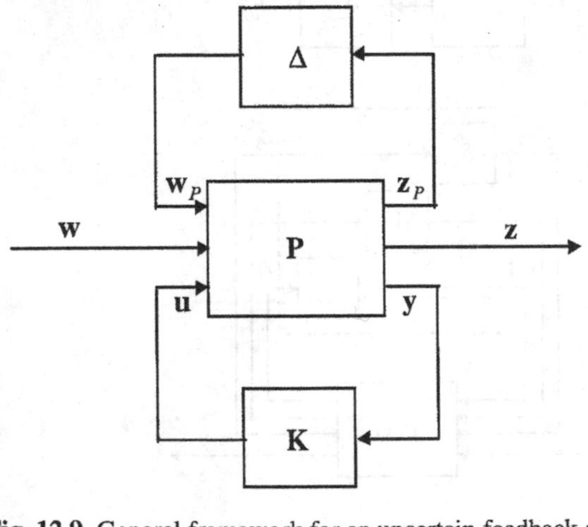

Fig. 12.9. General framework for an uncertain feedback system

The extended plant is described by nine blocks:

$$\begin{pmatrix} \mathbf{z}_P \\ \mathbf{z} \\ \mathbf{y} \end{pmatrix} = \mathbf{P}(s) \begin{pmatrix} \mathbf{w}_P \\ \mathbf{w} \\ \mathbf{u} \end{pmatrix} = \begin{pmatrix} \mathbf{P}_{11}(s) & \mathbf{P}_{12}(s) & \mathbf{P}_{13}(s) \\ \mathbf{P}_{21}(s) & \mathbf{P}_{22}(s) & \mathbf{P}_{23}(s) \\ \mathbf{P}_{31}(s) & \mathbf{P}_{32}(s) & \mathbf{P}_{33}(s) \end{pmatrix} \begin{pmatrix} \mathbf{w}_P \\ \mathbf{w} \\ \mathbf{u} \end{pmatrix}.$$

The uncertain plant is

$$\mathbf{G} = \mathcal{F}_u(\mathbf{P}, \Delta) .$$

Here, $\mathcal{F}_u(\mathbf{P}, \Delta)$ denotes the feedback connection of $\mathbf{P}$ and Δ (similar to $\mathcal{F}_l(\mathbf{P}, \mathbf{K})$, cf. Sect. 8.2.3), and the uncertain feedback system is

$$\mathbf{M}(s) = \mathcal{F}_l(\mathbf{P}(s), \mathbf{K}(s)) = \begin{pmatrix} \mathbf{M}_{11}(s) & \mathbf{M}_{12}(s) \\ \mathbf{M}_{21}(s) & \mathbf{M}_{22}(s) \end{pmatrix}.$$

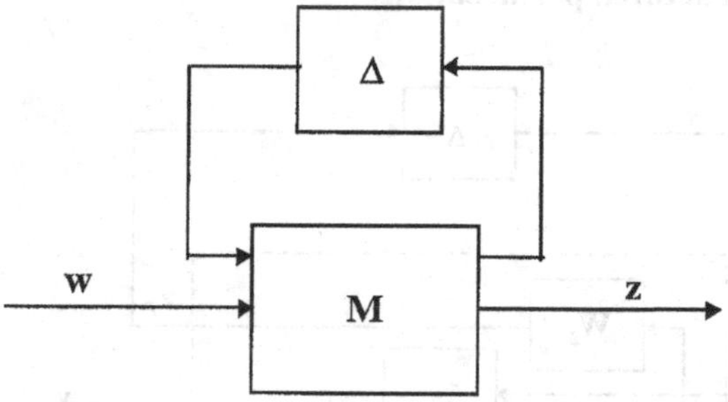

Fig. 12.10. Uncertain feedback system

The transfer function Δ is assumed to be unknown but stable. Thus, the uncertainty is separated from the feedback loop (cf. Fig 12.10). Suppose now that all such perturbations are considered with $\|\Delta\|_\infty \le \gamma$, where γ is some constant. Our task is to determine the maximum value of γ such that the feedback system is internally stable.

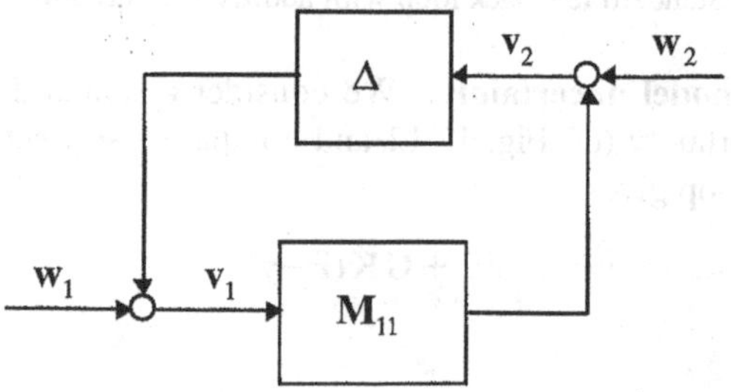

Fig. 12.11. Internal stability diagram in the presence of perturbations

Concerning internal stability, the situation is similar to that in Sect. 3.5.2: The feedback system in Fig. 12.10 is internally stable if and only if the feedback system depicted in Fig. 12.11 is internally stable (cf. Lemma 6.5.3). Thus, for internal stability it suffices to consider the block $\mathbf{M}_{11}(s)$. Compare also the corresponding discussion in Sect. 6.3.2.

The uncertainty Δ can be composed of several components. Normally we assume that it has the following block diagonal form:

$$\Delta = \operatorname{diag}\left(\delta_1 \mathbf{I}_{r_1}, \ldots, \delta_S \mathbf{I}_{r_S}, \Delta_1, \ldots, \Delta_F\right)$$

with $\Delta_i(s) \in \mathcal{RH}_\infty$. The quantities δ_i may be real numbers or transfer functions, $\delta_i(s) \in \mathcal{RH}_\infty$. This will be discussed in the next chapter. By suitable weights which have to be added to the plant, we may assume that

$$|\delta_i| \le 1 \;\; (\text{or } \|\delta_i\|_\infty \le 1) \;\text{ and }\; \|\Delta_i\|_\infty \le 1 .$$

Thus we consider **structured perturbations.**

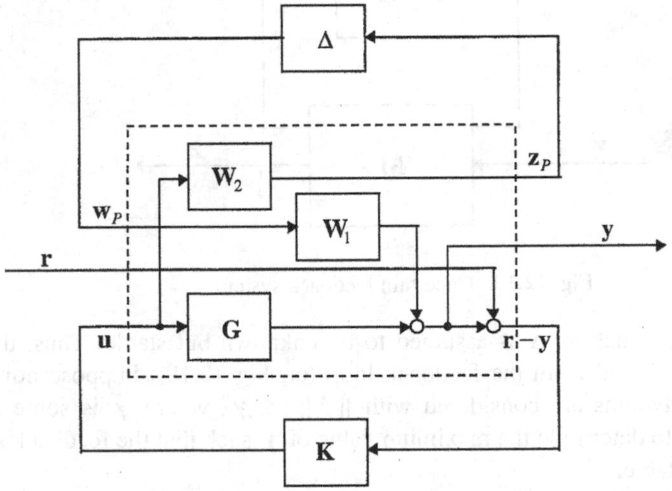

Fig. 12.12. Standard feedback loop with additive perturbation

Example: Additive model uncertainty. We consider a standard feedback loop with an additive uncertainty (cf. Fig. 12.12 and compare also Sect. 7.4). Closing the lower part of the loop gives

$$\hat{\mathbf{y}} = \mathbf{W}_1 \hat{\mathbf{w}}_P + \mathbf{G}\mathbf{K}(\hat{\mathbf{r}} - \hat{\mathbf{y}}) .$$

This leads to

$$\hat{\mathbf{y}} = (\mathbf{I} + \mathbf{G}\mathbf{K})^{-1}(\mathbf{W}_1 \hat{\mathbf{w}}_P + \mathbf{G}\mathbf{K}\hat{\mathbf{r}}) .$$

Hence we get

$$\hat{\mathbf{z}}_P = \mathbf{W}_2\mathbf{K}(\hat{\mathbf{r}} - \hat{\mathbf{y}}) = \mathbf{W}_2\mathbf{K}(\hat{\mathbf{r}} - (\mathbf{I} + \mathbf{G}\mathbf{K})^{-1}(\mathbf{W}_1\hat{\mathbf{w}}_P + \mathbf{G}\mathbf{K}\,\hat{\mathbf{r}}))$$
$$= -\mathbf{W}_2\mathbf{K}(\mathbf{I} + \mathbf{G}\mathbf{K})^{-1}\mathbf{W}_1\hat{\mathbf{w}}_P + \mathbf{W}_2\mathbf{K}(\mathbf{I} - (\mathbf{I} + \mathbf{G}\mathbf{K})^{-1}\mathbf{G}\mathbf{K})\hat{\mathbf{r}}$$

and

$$\mathbf{M}_{11} = -\mathbf{W}_2\mathbf{K}(\mathbf{I} + \mathbf{G}\mathbf{K})^{-1}\mathbf{W}_1 .$$

12.5 Example

We consider a simple example which will be continued in the next chapter. Let the plant nominally be a second-order plant of the form

$$G(s) = \frac{k_0\,\omega_0^2}{s^2 + 2\,d_0\omega_0 s + \omega_0^2} .$$

The damping d of the real plant is assumed to be uncertain:

$$d = d_0 + \delta\,\delta_m \quad \text{with } |\,\delta\,| \le 1 .$$

Additionally, let there be unmodeled high-frequency dynamics, which we take into account by an input-multiplicative uncertainty with weight W_Δ. The uncertain plant can be represented as shown in Fig. 12.13.

The uncertainty now has the form

$$\Delta(s) = \text{diag}\,(\delta_1 \quad \delta_2\,(s)) ,$$

with a real number with $|\,\delta_1\,| \le 1$ and a stable transfer function such that $\|\,\delta_2\,\|_\infty \le 1$. The perturbed plant can be written as

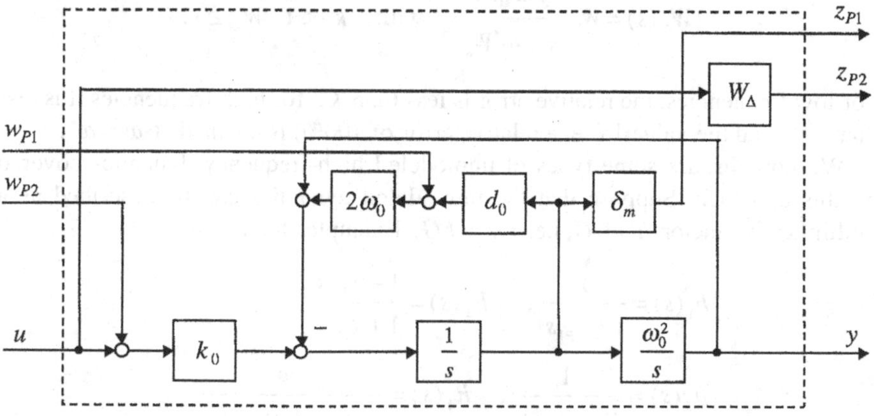

Fig. 12.13. Second-order plant with an uncertain parameter and uncertain dynamics

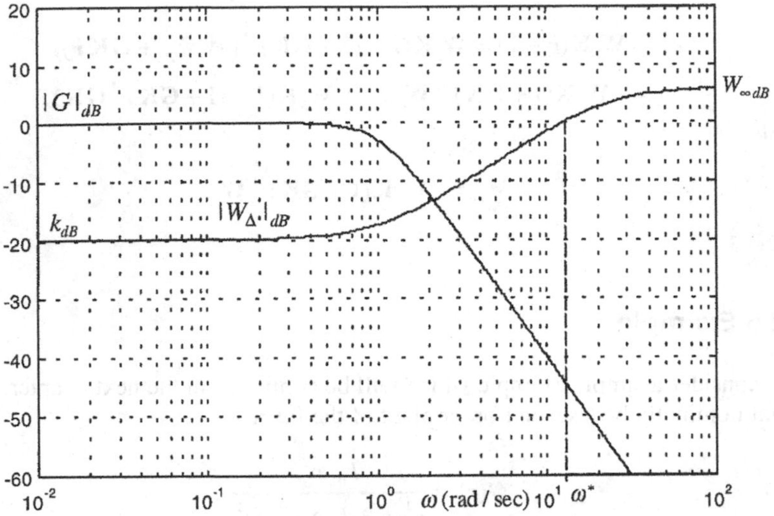

Fig. 12.14. Magnitude plot of G and W_Δ

$$G_P(s) = (1 + W_\Delta(s)\delta_2(s)) \frac{k\omega_0^2}{s^2 + 2d\omega_0 s + \omega_0^2}.$$

Specifying in this equation $d = d_0$, the requirement $\| \delta_2 \|_\infty \le 1$ is equivalent to

$$\left| \frac{G_P(i\omega) - G(i\omega)}{G(i\omega)} \right| \le |W_\Delta(i\omega)| \quad \text{for every } \omega. \tag{12.7}$$

Thus $|W_\Delta(i\omega)|$ describes the relative error of the plant model for each frequency ω. A typical choice for the uncertainty weight W_Δ is

$$W_\Delta(s) = W_\infty \frac{s + \omega^* k}{s + \omega^* W_\infty} \quad \text{with} \quad k \ll 1,\ W_\infty \ge 1.$$

For low frequencies, the relative error is less than k, for high frequencies it is less than W_∞ and the value 1 (i.e. a relative error of 100%) is attained at $\omega = \omega^*$.

We now discuss some types of unmodeled high-frequency dynamics covered by this approach. Suppose that the unmodeled dynamics can be described as a multiplicative factor h of G, i.e. $G_P = hG$. Examples for h are

$$h_1(s) = \frac{1}{1 + T_{P1}s}, \quad h_2(s) = \frac{1 - T_{P2}s}{1 + T_{P2}s},$$

$$h_3(s) = \frac{1}{(1 + T_{P3}s)^l}, \quad h_4(s) = \frac{\omega_P^2}{s^2 + 2d_P\omega_P s + \omega_P^2}.$$

Then (12.7) is equivalent to

$$|h(i\omega)-1| \leq |W_\Delta(i\omega)| \quad \text{for every } \omega . \tag{12.8}$$

The numerical values for the uncertain plant are chosen as

$$k_0 = 1,\ \omega_0 = 1s^{-1},\ d_0 = 0.7,\ \ d = 0.35,\ \ \delta_m = 0.5 .$$

The uncertainty weight W_Δ is described by

$$k = 0.1,\ W_\infty = 2,\ \omega^* = 11s^{-1}.$$

Moreover, let $T_{P3} = 0.018\,s$, $\omega_P = 80s^{-1}$. We now ask how large T_{P1}, T_{P2}, l and how small d_P are allowed to be such that (12.8) holds. The desired extreme values are

$$T_{P1} = 0.1s,\ T_{P2} = 0.045s\ ,\ l = 5,\ d_P = 0.32 .$$

Finally, we show that the plant of Fig. 12.13 can be implemented in Matlab as follows:

```
%Extended Plant in Fig. 12.13

%Nominal plant
k0=1;om0=1;d0=0.7;

%Weight Wdel:
k=0.1; omst=11; winf=2;
Wdel=nd2sys([winf winf*omst*k],[1 winf*omst]);

%Parametric uncertainty:
d00=0.35; delm=0.5;

OM0=2*om0;D=d00;Delm=delm;K0=k0;
Int1=nd2sys([1],[1 0]);
Int2=nd2sys([om0^2],[1 0]);

%Extended Plant P
systemnames='Int1 Int2 K0 OM0 D Delm Wdel';
inputvar='[wp1;wp2;u]';
outputvar='[Delm;Wdel;Int2]';
input_to_K0='[u+wp2]';
input_to_Int1='[K0-Int2-OM0]';
input_to_Int2='[Int1]';
input_to_OM0='[wp1+D]';
input_to_D='[Int1]';
input_to_Delm='[Int1]';
input_to_Wdel='[u]';
sysoutname='P';
cleanupsysic='yes';
sysic;
```

Chapter 13

Synthesis of Robust Controllers

In the last chapter, we introduced the mathematical framework for the description of model uncertainty. As we have seen, the basic idea is to separate the uncertainty Δ from the plant such that the uncertain plant can be interpreted as a feedback system consisting of $\mathbf{P}$ and Δ. Herein, $\mathbf{P}$ denotes the nominal plant, which results from the original physical plant $\mathbf{G}$ by adding weights for performance and uncertainty. The basic block diagram which visualizes this situation is Fig. 12.9. The nominal feedback system is denoted by $\mathbf{M}$ and consists of $\mathbf{P}$ and the controller $\mathbf{K}$. The uncertain feedback system is obtained if the input $\mathbf{w}_P$ is connected by feedback with the output $\mathbf{z}_P$ using the uncertainty Δ (cf. Fig 12.10). As we have seen, concerning internal stability it suffices to use the block $\mathbf{M}_{11}$ so that Fig. 12.11 becomes the fundamental block diagram with respect to robust stability. The question is now how large in the sense of the $\mathcal{H}_\infty$ norm the uncertainty may be so that the feedback system is stable. This question can completely be answered and is essentially the small gain theorem (cf. Fig. 12.11, where the uncertainty Δ can be formally viewed as a controller for the "plant" $\mathbf{M}_{11}$). This result together with some applications is the content of Sect. 13.1 and Sect. 13.2. A sufficient condition for robust performance for such a type of uncertainty is also given in Sect. 13.2.

Up to now, the uncertainty is an arbitrary element of $\mathcal{RH}_\infty$ and the size of the perturbation Δ as well as the size of the system $\mathbf{M}_{11}$ is expressed by their maximum singular value $\bar{\sigma}$. In many applications, this set of feasible perturbations is too large and has to be restricted to a smaller set. Then the perturbations belong to a structured set. This leads to the structured singular value μ which measures the size of the system $\mathbf{M}_{11}$ for this more general type of uncertainty and generalizes $\bar{\sigma}$. In general, it is not possible to give computational formulas for μ, but it is possible to derive upper and lower bounds which lead to rather precise estimates of μ. The basic theory concerning μ is derived in Sect. 13.3.

The next subsection is concerned with robust stability and robust performance for perturbations belonging to a structured set. In this way, the small gain theorem will be generalized and a necessary and sufficient condition for robust stability and also for robust performance are given. The latter result requires the structured singular value μ even if the set of feasible perturbations is unstructured.

Based on these results, it is possible to derive an algorithm, the so-called D-K iteration, which synthesizes a controller with robust performance and robust stability. Synthesis of several $\mathcal{H}_\infty$-optimal controllers are intermediate steps in the procedure. Up to now, convergence of this iteration could not be shown, but for the known examples, a few iterations always led to a satisfactory reduction of the objective function, which expresses the performance and robustness design goals by the structured singular value for the closed-loop system with respect to a certain structured set of perturbations. This method is described in Sect. 13.5. Since such μ-optimal controllers are typically of high order, it is desirable to replace them by controllers of lower orders. In Sect. 13.6, we describe how this can be done. The chapter ends with the construction of robust controllers for a plant of order 2.

13.1 Small Gain Theorem

We start with the general feedback system of Sect. 6 and consider in particular the feedback loop depicted in Fig. 6.10 and the loop transfer function $\mathbf{L}_o = -\mathbf{P}_{22}\mathbf{K}$.

Theorem 13.1.1 (Small Gain Theorem). *Let* $\mathbf{P}_{22} \in \mathcal{RH}_\infty$, $\mathbf{K} \in \mathcal{RH}_\infty$ *and additionally* $\|\mathbf{L}_o\|_\infty < 1$. *Then the feedback system is well-posed and internally stable.*

Proof. Theorem 6.5.3 shows that the feedback system is well-posed and internally stable if $\mathbf{S}_o(s)$ exists and is stable. Because of $\|\mathbf{L}_o\|_\infty < 1$, the inequality $\|\mathbf{D}_{22}\mathbf{D}_K\| < 1$ holds and hence $\mathbf{I} - \mathbf{D}_{22}\mathbf{D}_K$ is invertible. The stability of $\mathbf{S}_o(s)$ is shown if $\mathbf{I} + \mathbf{L}_o(s)$ has no transmission zero in $\overline{\mathbb{C}}_+$. This property is fulfilled if

$$\inf_{s \in \overline{\mathbb{C}}_+} \underline{\sigma}(\mathbf{I} + \mathbf{L}_o(s)) \neq 0 \tag{13.1}$$

because the following equation holds:

$$\underline{\sigma}(\mathbf{I} + \mathbf{L}_o(s)) = \min_{\mathbf{x} \neq \mathbf{0}} \frac{\|\mathbf{x} + \mathbf{L}_o(s)\mathbf{x}\|}{\|\mathbf{x}\|}. \tag{13.2}$$

We now prove (13.1). Because of (13.2) and since $\|\mathbf{L}_o(s)\| < 1$ for every s with $\operatorname{Re} s \geq 0$, we have

$$\underline{\sigma}(\mathbf{I} + \mathbf{L}_o(s)) \geq \min_{\mathbf{x} \neq \mathbf{0}} \frac{|\|\mathbf{x}\| - \|\mathbf{L}_o(s)\mathbf{x}\||}{\|\mathbf{x}\|}$$

$$= \min_{\mathbf{x} \neq \mathbf{0}} \left| 1 - \frac{\| \mathbf{L}_o(s)\mathbf{x} \|}{\| \mathbf{x} \|} \right|$$

$$\geq \min_{\mathbf{x} \neq \mathbf{0}} \left(1 - \frac{\| \mathbf{L}_o(s)\mathbf{x} \|}{\| \mathbf{x} \|} \right)$$

$$\geq 1 - \bar{\sigma}(\mathbf{L}_o(s))$$

$$\geq 1 - \sup_{s \in \bar{\mathbb{C}}_+} \bar{\sigma}(\mathbf{L}_o(s)) > 0 .$$

This implies (13.1).

The theorem gives a sufficient condition for stabilizing a stable plant with a stable controller. The condition makes use only of the magnitude plot of $\mathbf{L}_o$; the phase plot is not taken into account. Therefore, the small gain theorem gives only a sufficient condition for internal stability, but this condition is not necessary. As we will see from the following example, the assumption is in fact too restrictive.

Example. (a) Let with $T > 0$

$$P_{22}(s) = \frac{1}{1+Ts}, \qquad K(s) = k .$$

Then the small gain theorem requires $|k| < 1$ for stability. In contrast to this, the feedback system is stable for all $k < 1$.
(b) Define now with $0 < \alpha \leq 1$

$$P_{22}(s) = \frac{1 - \alpha Ts}{1+Ts}, \qquad K(s) = k .$$

Then we get

$$S(s) = \frac{1+Ts}{1 - k + (\alpha k + 1)Ts} .$$

Suppose for simplicity $k < 0$, which is the interesting case. Because of $\| k P_{22}(s) \|_\infty = |k|$, the small gain theorem again requires $|k| < 1$. $S(s)$ is stable if and only if $k > -1/\alpha$. For $\alpha = 1$, the sufficient condition of the theorem is also necessary. If α decreases, the condition becomes more and more conservative.

From this example it is seen that the sufficient condition of the small gain theorem may be very conservative, but cannot be relaxed if only the size of $\| \mathbf{L} \|_\infty$ is taken into account. Therefore, concerning controller synthesis the small gain theorem plays no important role.

In contrast to this, for estimating the destabilizing effect of model uncertainties the small gain theorem is the key. Note that no phase information for the uncer-

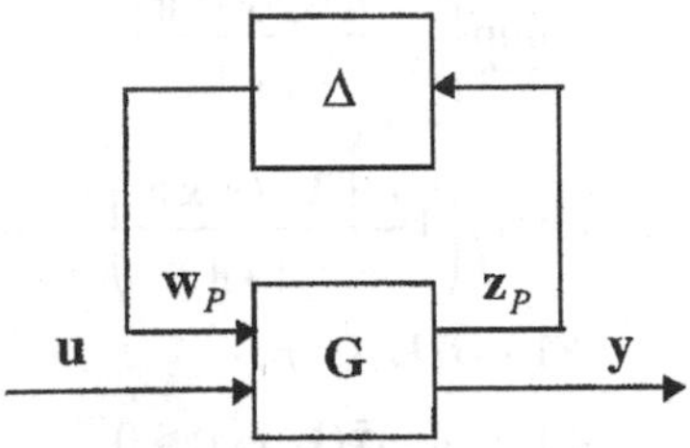

Fig. 13.1. Unstructured uncertainty for a linear system

uncertainties Δ is available. We now consider an uncertain system $\mathbf{G}$ and assume that the uncertainty is exactly modeled as described in the last chapter (cf. Fig. 13.1). It is now natural to ask the following question. Let $\mathbf{G}$ be stable and let Δ be any stable perturbation such that $\|\Delta\|_\infty \le 1/\gamma$. How large is $\mathbf{G}$ allowed to be in the $\mathcal{H}_\infty$ norm such that the system in Fig. 13.1 is internally stable?

To answer this question, we first look at the transfer function from $\mathbf{u}$ to $\mathbf{y}$ for the perturbed system. It is easily calculated as

$$\mathcal{F}_u(\mathbf{G},\Delta) = \mathbf{G}_{22} + \mathbf{G}_{21}\Delta(\mathbf{I}-\mathbf{G}_{11}\Delta)^{-1}\mathbf{G}_{12}\,.$$

Thus, since $\mathbf{G}$ and Δ are stable, $\mathcal{F}_u(\mathbf{G},\Delta)$ is stable if $(\mathbf{I}-\mathbf{G}_{11}\Delta)^{-1}$ is stable. If we assume that $\|\mathbf{G}_{11}\|_\infty < \gamma$ and $\|\Delta\|_\infty \le 1/\gamma$, then the submultiplicity of the $\mathcal{H}_\infty$ norm yields

$$\|\mathbf{G}\Delta\|_\infty \le \|\mathbf{G}\|_\infty \|\Delta\|_\infty < 1$$

and Theorem 13.1.1 implies that $(\mathbf{I}-\mathbf{G}_{11}\Delta)^{-1}$ is stable. Hence, the condition $\|\mathbf{G}_{11}\|_\infty < \gamma$ is sufficient for the stability of $\mathcal{F}_u(\mathbf{G},\Delta)$ if Δ is an arbitrary stable perturbation satisfying $\|\Delta\|_\infty \le 1/\gamma$. The next theorem tells us that this condition is also necessary.

Theorem 13.1.2. *Let $\mathbf{G}_{11} \in \mathcal{RH}_\infty$ and $\gamma > 0$. Then the following assertions are equivalent:*

(i) $\|\mathbf{G}_{11}\|_\infty < \gamma$.

(ii) *The transfer function $(\mathbf{I}-\mathbf{G}_{11}\Delta)^{-1}$ exists and is stable for all perturbations $\Delta \in \mathcal{RH}_\infty$ with*

$$\|\Delta\|_\infty \le 1/\gamma\,.$$

Proof. Without restriction of generality it may be assumed that $\gamma = 1$. We already know that (i) implies (ii) and show the opposite direction by contradiction.

To this end, suppose $\|\mathbf{G}_{11}\|_\infty \ge 1$. Then there is a frequency $\omega_0 \in \mathbb{R}_+ \cup \{\infty\}$ such that $\bar{\sigma}(\mathbf{G}_{11}(i\omega_0)) \ge 1$. Let

$$\mathbf{G}_{11}(i\omega_0) = \mathbf{U}(i\omega_0)\Sigma(i\omega_0)\mathbf{V}^*(i\omega_0)$$

be the singular value decomposition with

$$\mathbf{U}(i\omega_0) = \begin{pmatrix}\mathbf{u}_1 & \mathbf{u}_2 & \dots & \mathbf{u}_l\end{pmatrix}$$
$$\mathbf{V}(i\omega_0) = \begin{pmatrix}\mathbf{v}_1 & \mathbf{v}_2 & \dots & \mathbf{v}_m\end{pmatrix},$$

where $\Sigma(i\omega_0)$ has the entries $\sigma_1,\dots,\sigma_p$ such that $\sigma_1 = \overline{\sigma}(\mathbf{G}_{11}(i\omega_0))$. We now construct $\Delta \in \mathcal{RH}_\infty$ with

$$\Delta(i\omega_0) = \frac{1}{\sigma_1}\mathbf{v}_1\mathbf{u}_1^* \quad \text{and} \quad \|\Delta\|_\infty \le 1 .$$

This leads to

$$\begin{aligned}(\mathbf{I} - \mathbf{G}_{11}(i\omega_0)\Delta(i\omega_0))\mathbf{u}_1 &= \mathbf{u}_1 - \mathbf{G}_{11}(i\omega_0)\sigma_1^{-1}\mathbf{v}_1 \\ &= \mathbf{u}_1 - \mathbf{U}(i\omega_0)\Sigma(i\omega_0)\mathbf{V}^*(i\omega_0)\sigma_1^{-1}\mathbf{v}_1 \\ &= \mathbf{u}_1 - \mathbf{U}(i\omega_0)\Sigma(i\omega_0)\sigma_1^{-1}\mathbf{e}_1 = \mathbf{0} .\end{aligned}$$

Hence $\mathbf{I} - \mathbf{G}_{11}(s)\Delta(s)$ has a transmission zero on the imaginary axis and the system is unstable (if $\omega_0 \in \mathbb{R}_+$) or not well-posed (if $\omega_0 = \infty$), cf. Theorem 6.5.3. We consider now two cases.

Case 1: $\omega_0 = 0$ or $\omega_0 = \infty$. Then $\mathbf{U}$ and $\mathbf{V}$ are real and $\Delta(s)$ can be chosen as follows:

$$\Delta(s) = \frac{1}{\sigma_1}\mathbf{v}_1\mathbf{u}_1^* .$$

Case 2: $0 < \omega_0 < \infty$. In this case we write the vectors $\mathbf{u}_1$ and $\mathbf{v}_1$ in the form

$$\mathbf{u}_1^* = \begin{pmatrix}u_1e^{i\theta_1} & u_2e^{i\theta_2} & \dots & u_le^{i\theta_l}\end{pmatrix}, \quad \mathbf{v}_1 = \begin{pmatrix}v_1e^{i\psi_1} \\ v_2e^{i\psi_2} \\ \vdots \\ v_me^{i\psi_m}\end{pmatrix}.$$

Here, the numbers u_k, v_j are real and θ_k, ψ_l lie in the interval $[-\pi, 0)$. Choose now $\beta_k \ge 0$ and $\alpha_j \ge 0$ such that

$$\frac{\beta_k - i\omega_0}{\beta_k + i\omega_0} = e^{i\theta_k}, \quad \frac{\alpha_k - i\omega_0}{\alpha_k + i\omega_0} = e^{i\psi_j} .$$

Define the perturbation

$$\Delta(s) = \frac{1}{\sigma_1} \begin{pmatrix} v_1 \dfrac{\alpha_1 - s}{\alpha_1 + s} \\ \vdots \\ v_m \dfrac{\alpha_m - s}{\alpha_m + s} \end{pmatrix} \begin{pmatrix} u_1 \dfrac{\beta_1 - s}{\beta_1 + s} & \cdots & u_l \dfrac{\beta_l - s}{\beta_l + s} \end{pmatrix}.$$

Then Δ has the desired properties:

$$\Delta \in \mathcal{RH}_\infty, \quad \|\Delta\|_\infty = \frac{1}{\sigma_1} \le 1, \quad \Delta(i\omega_0) = \frac{1}{\sigma_1} \mathbf{v}_1 \mathbf{u}_1^* .$$

Remark. An analogous result holds where "<" in (i) and "≤" in (ii) are interchanged.

Next we ask for internal stability of the perturbed feedback system depicted in Fig. 13.1. Since $\mathbf{G}$ and Δ and are stable, we obtain from Theorem 6.5.3 that the system in Fig. 13.1 is internally stable if and only if $(\mathbf{I} - \mathbf{G}_{11}\Delta)^{-1}$ exists and is stable. This fact together with Theorem 13.1.2 yields the following result.

Corollary 13.1.1. *Let* $\mathbf{G} \in \mathcal{RH}_\infty$ *and* $\gamma > 0$. *Then the following assertions are equivalent:*

(i) $\quad \|\mathbf{G}_{11}\|_\infty < \gamma$.

(ii) *The perturbed system in Fig. 13.1 is well-posed and internallly stable for all all perturbations* $\Delta \in \mathcal{RH}_\infty$ *with*

$$\|\Delta\|_\infty \le 1/\gamma .$$

13.2 Robust Stability Under Stable Unstructured Uncertainties

Our next step is to apply Corollary 13.1.1 to the general uncertain feedback system shown in Fig. 12.9.

Corollary 13.2.1. *Let the nominal feedback system* $\mathbf{M} = \mathcal{F}_l(\mathbf{P}, \mathbf{K})$ *be internally stable and let* $\gamma > 0$. *Then the following assertions are equivalent:*

(i) $\quad \|\mathbf{M}_{11}\|_\infty \le \gamma$.

(ii) *The perturbed feedback system in Fig. 12.9 is well-posed and internally stable for all perturbations* $\Delta \in \mathcal{RH}_\infty$ *with*

$$\|\Delta\|_\infty < 1/\gamma .$$

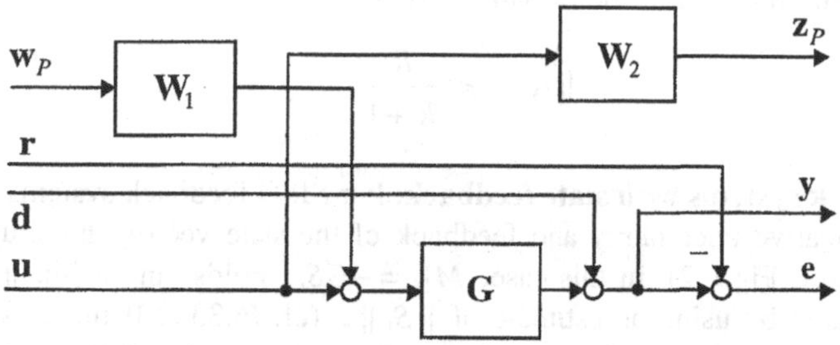

Fig. 13.2. Extended plant for an input-multiplicative perturbation

In Corollary 13.2.1, robust stability is characterized for a general feedback connection. Next we want to specialize this result to the special unstructured uncertainties introduced in Sect. 12.2. The most popular one is the input-multiplicative perturbation (cf. Fig. 12.4):

$$\mathcal{F}_u(\mathbf{G},\Delta) = \mathbf{G}(\mathbf{I} + \mathbf{W}_1\Delta\mathbf{W}_2)\,.$$

We combine this with a standard feedback configuration as shown in Fig. 13.2. Using (7.10), we obtain

$$\mathbf{M}_{11} = -\mathbf{W}_2\mathbf{T}_i\mathbf{W}_1\,.$$

For this specialized situation, Corollary 13.2.1 can be formulated as follows.

Corollary 13.2.2. *Let Δ be an input-multiplicative perturbation which is weighted by stable weights $\mathbf{W}_1$ and $\mathbf{W}_2$. Assume that the nominal feedback system $\mathcal{F}_l(\mathbf{G},\mathbf{K})$ is internally stable and let $\gamma > 0$. Then the following assertions are equivalent:*

(i) $\| \mathbf{W}_2\mathbf{T}_i\mathbf{W}_1 \|_\infty \le \gamma$.

(ii) The perturbed feedback system is well-posed and internally stable for all perturbations $\Delta \in \mathcal{RH}_\infty$ with

$$\|\Delta\|_\infty < 1/\gamma.$$

Next we consider several interesting special cases in which it is possible to estimate the $\mathcal{H}_\infty$ norm of $\mathbf{M}_{11}$.

SISO feedback systems with output feedback. Assume that Fig. 13.2 describes a classical SISO feedback system. Then $M_{11} = -W_2F_r$ holds (with $W_1 = 1$) and Theorem 3.5.1 proves to be a special case of Corollary 13.2.2. For $W_2 = 1$ it follows that $M_{11} = -F_r = S - 1$. If the curve $L(i\omega)$ avoids a circle with radius R and center -1 (i.e. if (4.1) holds), then $\| S \|_\infty \le R^{-1}$ and $\| M_{11} \|_\infty$ can also be estimated as described at the beginning of Sect. 4.1.1. The feedback system then is internally stable for all stable perturbations Δ with

$$\|\Delta\|_\infty < \frac{R}{R+1}.$$

SISO feedback systems with state feedback. For SISO feedback systems with an input-multiplicative uncertainty and feedback of the state vector, the situation is quite similar (cf. Fig 6.2). In this case, $M_{11} = -L_i S_i$ holds, and again $\| M_{11} \|_\infty$ can be estimated by using an estimate of $\| S_i \|_\infty$ (cf. (6.33)). If the closed-loop poles are adequately chosen, the above inequality can even be fulfilled with $R = 1$ (cf. Lemma 6.3.1). This means that the feedback system is stable for all perturbations Δ with $\|\Delta\|_\infty < 1/2$. If actuator and sensor dynamics is taken into account, the radius R becomes smaller as discussed in Sect. 6.3.2.

LQ feedback loops. We now consider the feedback system with an input-multiplicative uncertainty and feedback of the state vector: $\mathbf{u} = \mathbf{F}_2\mathbf{x}$. No external inputs and outputs occur so that the only task is to stabilize the plant. The weights are chosen as identity matrices. We have

$$\mathbf{M}_{11} = -(\mathbf{I}+\mathbf{L}_i)^{-1}\mathbf{L}_i = -\mathbf{S}_i\mathbf{L}_i$$

and consequently

$$\|\mathbf{M}_{11}\|_\infty = \|-(\mathbf{I}+\mathbf{L}_i)^{-1}\mathbf{L}_i\|_\infty = \|(\mathbf{I}+\mathbf{L}_i)^{-1} - \mathbf{I}\|_\infty .$$

Hence the estimate

$$\|\mathbf{M}_{11}\|_\infty \le \|\mathbf{S}_i\|_\infty + 1$$

follows. We now suppose that this system is stabilized by an LQ controller. Then

$$\mathbf{L}_i(s) = -\mathbf{F}_2(s\mathbf{I}-\mathbf{A})^{-1}\mathbf{B}_2 ,$$

and by Theorem 8.4.2, the following inequality holds:

$$(\mathbf{I}+\mathbf{L}_i(i\omega))^*(\mathbf{I}+\mathbf{L}_i(i\omega)) \ge \mathbf{I} .$$

Thus $1 \le \underline{\sigma}(\mathbf{I}+\mathbf{L}_i(i\omega))$ follows and therefore $\overline{\sigma}((\mathbf{I}+\mathbf{L}_i(i\omega))^{-1}) \le 1$ for every ω. Consequently, we have $\|(\mathbf{I}+\mathbf{L}_i)^{-1}\|_\infty \le 1$ and $\|\mathbf{M}_{11}\|_\infty \le 2$. An application of Corollary 13.2.1 gives now the following result:

A feedback system with an LQ controller and an input-multiplicative uncertainty is robustly stable for all stable perturbations Δ which fulfill

$$\|\Delta\|_\infty < 1/2 .$$

Similar results can be formulated for the other unstructured perturbations discussed in Fig. 12.2. The corresponding conditions for internal stability are summa-

	Uncertain plant	Criterion for robust stability
Additive uncertainty	$\mathbf{G}_P = \mathbf{G} + \mathbf{W}_1 \Delta \mathbf{W}_2$	$\lVert \mathbf{W}_2 \mathbf{K} \mathbf{S}_o \mathbf{W}_1 \rVert_\infty \le 1$
Input-multiplicative uncertainty	$\mathbf{G}_P = \mathbf{G}(\mathbf{I} + \mathbf{W}_1 \Delta \mathbf{W}_2)$	$\lVert \mathbf{W}_2 \mathbf{T}_i \mathbf{W}_1 \rVert_\infty \le 1$
Output-multiplicative uncertainty	$\mathbf{G}_P = (\mathbf{I} + \mathbf{W}_1 \Delta \mathbf{W}_2)\mathbf{G}$	$\lVert \mathbf{W}_2 \mathbf{T}_o \mathbf{W}_1 \rVert_\infty \le 1$
Input feedback uncertainty	$\mathbf{G}_P = \mathbf{G}(\mathbf{I} + \mathbf{W}_1 \Delta \mathbf{W}_2)^{-1}$	$\lVert \mathbf{W}_2 \mathbf{S}_i \mathbf{W}_1 \rVert_\infty \le 1$
Output feedback uncertainty	$\mathbf{G}_P = (\mathbf{I} + \mathbf{W}_1 \Delta \mathbf{W}_2)^{-1}\mathbf{G}$	$\lVert \mathbf{W}_2 \mathbf{S}_o \mathbf{W}_1 \rVert_\infty \le 1$

Table 13.1. Robust Stability Tests

rized in Table 13.1.

Finally, we want to analyze robustness under stable unstructured uncertainties where we limit ourselves to output-multiplicative uncertainties. Then the uncertain plant is $\mathbf{G}_P = (\mathbf{I} + \mathbf{W}_1 \Delta \mathbf{W}_2)\mathbf{G}$. Fig. 13.3 describes the situation. The requirement of certain performance is now a requirement for the $\mathcal{H}_\infty$ norm of $\mathbf{F}_{ed_1}$. By an adequate normalization it may be formulated in the form

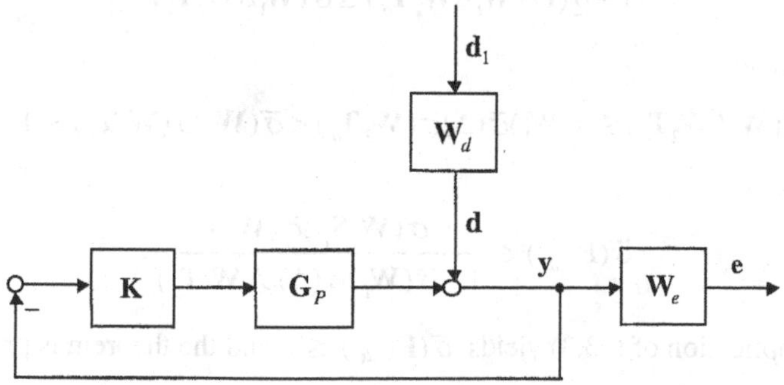

Fig. 13.3. Block diagram for robust performance

$$\| \mathbf{F}_{ed_1} \|_\infty \le 1 \qquad \text{(for the nominal plant).}$$

The feedback loop has robust performance if this also holds for all plants which are perturbed by an admissible perturbation. Additionally, it has to be required that the feedback system is robustly stable. Using the adequate robust stability test, it is seen that robust stability and robust performance are given if and only if the following conditions hold:

$$\| \mathbf{W}_2 \mathbf{T}_o \mathbf{W}_1 \|_\infty \le 1,$$

$$\| \mathbf{F}_{\Delta e d_1} \|_\infty \le 1 \quad \text{for all } \Delta \in \mathcal{RH}_\infty \text{ with } \| \Delta \|_\infty < 1 .$$

Here, $\mathbf{F}_{\Delta e d_1}$ denotes the transfer function from d_1 to e for the perturbed plant and is easily calculated as

$$\mathbf{F}_{\Delta e d_1} = \mathbf{W}_e \mathbf{S}_o (\mathbf{I} + \mathbf{W}_1 \Delta \mathbf{W}_2 \mathbf{T}_o)^{-1} \mathbf{W}_d .$$

Theorem 13.2.1. *Let* $\mathbf{G}_P = (\mathbf{I} + \mathbf{W}_1 \Delta \mathbf{W}_2)\mathbf{G}$ *with* $\Delta \in \mathcal{RH}_\infty$, $\| \Delta \|_\infty < 1$ *and let* $\mathbf{K}$ *be an internally stabilizing controller for* $\mathbf{G}$. *Then robust performance is guaranteed if the following condition holds:*

for each frequency ω,

$$\bar{\sigma}(\mathbf{W}_d(i\omega))\,\bar{\sigma}(\mathbf{W}_e(i\omega)\mathbf{S}_o(i\omega)) + \bar{\sigma}(\mathbf{W}_1(i\omega))\,\bar{\sigma}(\mathbf{W}_2(i\omega)\mathbf{T}_o(i\omega)) \le 1 . \qquad (13.3)$$

Proof. First we note that (13.3) implies $\| \mathbf{W}_2 \mathbf{T}_o \mathbf{W}_1 \|_\infty \le 1$. For each frequency ω we have

$$\bar{\sigma}(\mathbf{F}_{\Delta e d_1}) \le \bar{\sigma}(\mathbf{W}_e \mathbf{S}_o)\bar{\sigma}((\mathbf{I} + \mathbf{W}_1 \Delta \mathbf{W}_2 \mathbf{T}_o)^{-1})\bar{\sigma}(\mathbf{W}_d)$$
$$= \frac{\bar{\sigma}(\mathbf{W}_e \mathbf{S}_o)\bar{\sigma}(\mathbf{W}_d)}{\underline{\sigma}(\mathbf{I} + \mathbf{W}_1 \Delta \mathbf{W}_2 \mathbf{T}_o)} .$$

From the general results on singular values presented in Sect. 7.1 it follows that

$$1 - \underline{\sigma}(\mathbf{I} + \mathbf{W}_1 \Delta \mathbf{W}_2 \mathbf{T}_o) \le \bar{\sigma}(\mathbf{W}_1 \Delta \mathbf{W}_2 \mathbf{T}_o)$$

and (13.3) yields

$$\bar{\sigma}(\mathbf{W}_1 \Delta \mathbf{W}_2 \mathbf{T}_o) \le \bar{\sigma}(\mathbf{W}_1)\bar{\sigma}(\Delta)\bar{\sigma}(\mathbf{W}_2 \mathbf{T}_o) < \bar{\sigma}(\mathbf{W}_1)\bar{\sigma}(\mathbf{W}_2 \mathbf{T}_o) \le 1 .$$

This implies

$$\bar{\sigma}(\mathbf{F}_{\Delta e d_1}) \le \frac{\bar{\sigma}(\mathbf{W}_e \mathbf{S}_o)\bar{\sigma}(\mathbf{W}_d)}{1 - \bar{\sigma}(\mathbf{W}_1)\bar{\sigma}(\Delta)\bar{\sigma}(\mathbf{W}_2 \mathbf{T}_o)} .$$

A further application of (13.3) yields $\bar{\sigma}(\mathbf{F}_{\Delta e d_1}) \le 1$ and the theorem is proven.

It is possible to show that for SISO systems the condition (13.7) is also necessary for robust performance. Necessary and sufficient conditions for robust performance of MIMO systems are investigated in Sect. 13.4.

13.3 Structured Singular Value μ

13.3.1 Basic Idea and Definition

We consider again the feedback system depicted in Fig. 12.11 and write for simplicity $\mathbf{M}$ instead of $\mathbf{M}_{11}$, where $\mathbf{M}$ is an arbitrary stable $m \times n$ transfer matrix. As in the situation of the small gain theorem, we ask how large in the sense of the $\mathcal{H}_\infty$ norm a stable perturbation Δ is allowed to be so that the perturbed system remains stable. Since for s with $\operatorname{Re} s \geq 0$ the poles of the feedback system in Fig. 12.11 are the zeros of $\det(\mathbf{I} - \mathbf{M}(s)\Delta(s))$, stability is given if and only if this function has no zero in $\overline{\mathbb{C}}_+$. Up to now, all perturbations $\Delta \in \mathcal{RH}_\infty$ were admitted. For many situations, this set is too large, so that it is reasonable to consider only perturbations which belong to a certain smaller set which will be denoted as $\mathcal{M}(\underline{\Delta})$. This leads to so-called **structured perturbations**. Again, we ask for a criterion for stability. The set of feasible perturbations is defined as follows:

$$\mathcal{M}(\underline{\Delta}) = \left\{ \Delta \in \mathcal{RH}_\infty \,\middle|\, \Delta(s) \in \underline{\Delta} \quad \text{for every } s \in \overline{\mathbb{C}}_+ \right\}.$$

Here, $\underline{\Delta}$ denotes an arbitrary closed subset of $\mathbb{C}^{n \times m}$ which later has to be specialized in a certain way.

We start with the following consideration. Let $\gamma > 0$ be given and suppose

$$\min\left\{ \overline{\sigma}(\Delta) \,\middle|\, \Delta \in \underline{\Delta},\, \det(\mathbf{I} - \mathbf{M}(s)\Delta) = 0 \right\} \geq \frac{1}{\gamma} \quad \text{for every } s \in \overline{\mathbb{C}}_+ .$$

If the set whose minimum has to be formed is empty, the minimum is ∞ by definition. Let now $\Delta(s)$ be a perturbation with $\Delta \in \mathcal{M}(\underline{\Delta})$ and $\|\Delta\|_\infty < 1/\gamma$. Then in particular $\overline{\sigma}(\Delta(s)) < 1/\gamma$ and consequently

$$\det(\mathbf{I} - \mathbf{M}(s)\Delta(s)) \neq 0 \quad \text{for every } s \in \overline{\mathbb{C}}_+ .$$

This can equivalently be formulated as follows: if

$$\frac{1}{\min\left\{ \overline{\sigma}(\Delta) \,\middle|\, \Delta \in \underline{\Delta},\, \det(\mathbf{I} - \mathbf{M}(s)\Delta) = 0 \right\}} \leq \gamma \quad \text{for every } s \in \overline{\mathbb{C}}_+ ,$$

the feedback system is stable for all perturbations $\Delta \in \mathcal{M}(\underline{\Delta})$ with $\|\Delta\|_\infty < 1/\gamma$.

These considerations justify the following definition.

Definition. *For an arbitrary real or complex* $m \times n$ *- matrix* $\mathbf{M}$ *the* ***structured singular value*** *is defined by*

$$\mu_{\underline{\Delta}}(\mathbf{M}) = \frac{1}{\min\{\overline{\sigma}(\Delta) \mid \Delta \in \underline{\Delta},\ \det(\mathbf{I} - \mathbf{M}\Delta) = 0\}},$$

if there is a perturbation $\Delta \in \underline{\Delta}$ *such that* $\det(\mathbf{I} - \mathbf{M}\Delta) = 0$. *Otherwise let* $\mu_{\underline{\Delta}}(\mathbf{M}) = 0$.

As we will see in the next section, the structured singular value is a generalization of the ordinary singular value. Our considerations show that the following lemma holds.

Lemma 13.3.1. *Let* $\gamma > 0$. *Then the feedback loop shown in Fig. 12.10 is internally stable for all* $\Delta \in \mathcal{M}(\underline{\Delta})$ *with* $\|\Delta\|_\infty < 1/\gamma$ *if*

$$\mu_{\underline{\Delta}}(\mathbf{M}_{11}(s)) \le \gamma \quad \textit{for every } s \in \overline{\mathbb{C}}_+. \tag{13.4}$$

Note that this lemma is nothing else but a skillful reformulation of the definition of robust stability in the presence of structured uncertainties and therefore it has up to now no real mathematical content. It becomes of importance only if the following questions can be positively answered:

1. Is the sufficient condition (13.4) for robust stability also necessary?
2. Are there practicable methods to calculate the structured singular value?

For an arbitrary set $\underline{\Delta}$, there are no satisfactory answers to these questions. However, if this set has an adequate structure (namely the structure we found in Sect. 12.3 when structured model uncertainties were modeled), the condition for robust stability is also necessary and it is possible to find calculable upper and lower bounds for the structured singular value which in most cases yield a sufficiently accurate estimate of it.

13.3.2 Basic Properties of the Structured Singular Value μ

As already mentioned, the structured singular value μ can only be evaluated if the set $\underline{\Delta}$ has a particular structure. We start with the simplest case in which $\underline{\Delta}$ is the whole space of complex $n \times m$ - matrices. Then μ coincides with the maximum singular value $\overline{\sigma}$. Consequently, the structured singular value actually generalizes the maximum singular value of a matrix.

Lemma 13.3.2. *Let* $\underline{\Delta} = \mathbb{C}^{n \times m}$. *Then* $\mu_{\underline{\Delta}}(\mathbf{M}) = \overline{\sigma}(\mathbf{M})$.

Proof. First we prove $\mu_{\underline{\Delta}}(\mathbf{M}) \le \bar{\sigma}(\mathbf{M})$. If $\mu_{\underline{\Delta}}(\mathbf{M}) = 0$, this inequality is certainly fulfilled. Suppose now $\mu_{\underline{\Delta}}(\mathbf{M}) \ne 0$ and let Δ_0 be a matrix such that the minimum is attained. Then there is a vector $\mathbf{x}_0 \ne \mathbf{0}$ with $\mathbf{M}\Delta\mathbf{x}_0 = \mathbf{x}_0$. This implies

$$\bar{\sigma}(\mathbf{M})\bar{\sigma}(\Delta_0) \ge \bar{\sigma}(\mathbf{M}\Delta_0) = \sup_{\mathbf{x}\ne 0}\frac{\|\mathbf{M}\Delta_0\mathbf{x}\|}{\|\mathbf{x}\|} \ge 1$$

and consequently

$$\mu_{\underline{\Delta}}(\mathbf{M}) = \frac{1}{\bar{\sigma}(\Delta_0)} \le \bar{\sigma}(\mathbf{M}).$$

Define now with the corresponding singular vectors of $\mathbf{M}$ (with $\sigma_1 = \bar{\sigma}(\mathbf{M})$)

$$\Delta_1 = \frac{1}{\sigma_1}\mathbf{v}_1\mathbf{u}_1^*.$$

From the construction of Δ_1 the equation $\bar{\sigma}(\Delta_1) = 1/\bar{\sigma}(\mathbf{M})$ is immediately obtained. Because of $\mathbf{M}\mathbf{v}_1 = \sigma_1\mathbf{u}_1$ we get

$$(\mathbf{I} - \mathbf{M}\Delta_1)\mathbf{u}_1 = \mathbf{u}_1 - \frac{1}{\sigma_1}\mathbf{M}\mathbf{v}_1\mathbf{u}_1^*\mathbf{u}_1 = \mathbf{0}.$$

Hence $\det(\mathbf{I} - \mathbf{M}\Delta_1) = \mathbf{0}$ and therefore

$$\bar{\sigma}(\mathbf{M}) = \frac{1}{\bar{\sigma}(\Delta_1)} \le \mu_{\underline{\Delta}}(\mathbf{M}).$$

This proves the asserted equality. Additionally, we have constructed a matrix Δ (namely Δ_1) for which the minimum is attained.

Let now $\underline{\Delta}$ be the following set of $n \times m$-matrices:

$$\underline{\Delta} = \left\{\operatorname{diag}\left(\delta_1\mathbf{I}_{r_1},\ldots,\delta_S\mathbf{I}_{r_S},\Delta_1,\ldots,\Delta_F\right)\middle|\,\delta_i \in \mathbb{C},\ \Delta_j \in \mathbb{C}^{n_j \times m_j}\right\}. \quad (13.5)$$

The dimensions must be given such that

$$\sum_{i=1}^{S} r_i + \sum_{j=1}^{F} n_j = n, \quad \sum_{i=1}^{S} r_i + \sum_{j=1}^{F} m_j = m.$$

In the definition of $\underline{\Delta}$ two different types of blocks occur, namely **repeated scalar blocks** and **full blocks**. For this set $\underline{\Delta}$, uncertainty is always expressed by transfer functions. If the control system has uncertain parameters, repeated real blocks have to be allowed, i.e. blocks of the form $\delta\mathbf{I}$ with a real parameter δ.

The structured singular value is no norm. Nevertheless, the following equation holds for every $\alpha \in \mathbb{C}$:

$$\mu_{\underline{\Delta}}(\alpha \mathbf{M}) = |\alpha| \mu_{\underline{\Delta}}(\mathbf{M}).$$

Hence, μ measures in a certain sense the size of a matrix. Moreover, it is possible to show that for a transfer matrix $\mathbf{M}(s) \in \mathcal{RH}_\infty$ the following equation is valid (cf. Zhou, Doyle and Glover [112], Lemma 11.1):

$$\sup_{s \in \overline{\mathbb{C}}_+} \mu_{\underline{\Delta}}(\mathbf{M}(s)) = \sup_{\omega \in \mathbb{R}} \mu_{\underline{\Delta}}(\mathbf{M}(i\omega)). \tag{13.6}$$

For a single repeated scalar block it is also possible to calculate the structured singular value:

Lemma 13.3.3. *Let* $m = n$ *and* $\underline{\Delta} = \{ \delta \mathbf{I}_n \mid \delta \in \mathbb{C} \}$. *Then*

$$\mu_{\underline{\Delta}}(\mathbf{M}) = \rho(\mathbf{M}).$$

Proof. We have (with $\mathbf{I}_n = \mathbf{I}$)

$$\begin{aligned}
\mu_{\underline{\Delta}}(\mathbf{M})^{-1} &= \min \{ \overline{\sigma}(\delta \mathbf{I}) \mid \delta \in \mathbb{C},\ \det(\mathbf{I} - \delta \mathbf{M}) = 0 \} \\
&= \min \{ |\delta| \mid \delta \in \mathbb{C},\ \det(\mathbf{I} - \delta \mathbf{M}) = 0 \} \\
&= \min \{ |\delta| \mid \delta \in \mathbb{C},\ 1/\delta \text{ is an eigenvalue of } \mathbf{M} \} \\
&= \rho(\mathbf{M})^{-1}.
\end{aligned}$$

For a structured set $\underline{\Delta}$ of the form (13.5) with $m = n$, the largest singular value and the spectral radius can be used to give an upper and a lower bound of the structured singular value of $\mathbf{M}$.

Lemma 13.3.4. *Let* $m = n$ *and* $\underline{\Delta}$ *as in (13.5). Then the following estimate holds:*

$$\rho(\mathbf{M}) \le \mu_{\underline{\Delta}}(\mathbf{M}) \le \overline{\sigma}(\mathbf{M}). \tag{13.7}$$

Proof. We have $\underline{\Delta}_1 = \{ \delta \mathbf{I}_n \mid \delta \in \mathbb{C} \} \subset \underline{\Delta}$. Hence Lemma 13.3.3 implies

$$\begin{aligned}
\mu_{\underline{\Delta}}(\mathbf{M})^{-1} &= \min \{ \overline{\sigma}(\Delta) \mid \Delta \in \underline{\Delta},\ \det(\mathbf{I} - \mathbf{M}\Delta) = 0 \} \\
&\le \min \{ \overline{\sigma}(\Delta) \mid \Delta \in \underline{\Delta}_1,\ \det(\mathbf{I} - \mathbf{M}\Delta) = 0 \} \\
&= \rho(\mathbf{M})^{-1}
\end{aligned}$$

(again with $\mathbf{I}_n = \mathbf{I}$). Lemma 13.3.2 yields

$$\begin{aligned}
\overline{\sigma}(\mathbf{M})^{-1} &= \min \{ \overline{\sigma}(\Delta) \mid \Delta \in \mathbb{C}^{n \times n},\ \det(\mathbf{I} - \mathbf{M}\Delta) = \mathbf{0} \} \\
&\le \min \{ \overline{\sigma}(\Delta) \mid \Delta \in \underline{\Delta},\ \det(\mathbf{I} - \mathbf{M}\Delta) = \mathbf{0} \} \\
&= \mu_{\underline{\Delta}}(\mathbf{M})^{-1}.
\end{aligned}$$

Remark. The preceding lemma can be generalized to the nonsquare case. For $m \neq n$, we define a new structured set $\underline{\Delta}_e$ by replacing all nonsquare blocks in $\underline{\Delta}$ by appropriately sized square blocks. The matrix $\mathbf{M}$ is substituted by a new matrix $\mathbf{M}_e$, which is square and generated from $\mathbf{M}$ by adding zero columns or rows such that

$$\det(\mathbf{I} + \mathbf{M}_e \Delta_e) = \det(\mathbf{I} + \mathbf{M}\Delta)$$

for any $\Delta_e \in \underline{\Delta}_e$. Then it can be shown that

$$\mu_{\underline{\Delta}_e}(\mathbf{M}_e) = \mu_{\underline{\Delta}}(\mathbf{M}).$$

Consequently, in (13.7), $\rho(\mathbf{M})$ has to be replaced by $\rho(\mathbf{M}_e)$:

$$\rho(\mathbf{M}_e) \le \mu_{\underline{\Delta}}(\mathbf{M}).$$

For example, let $\mathbf{M} \in \mathbb{C}^{2\times 3}$ and let $\Delta \in \mathbb{C}^{3\times 2}$ have the block structure

$$\Delta = \begin{pmatrix} \Delta_1 & \mathbf{0} \\ \mathbf{0} & \Delta_2 \end{pmatrix} = \begin{pmatrix} \delta_{11} & 0 \\ \delta_{12} & 0 \\ 0 & \delta_{32} \end{pmatrix}.$$

Then

$$\mathbf{M}_e = \begin{pmatrix} m_{11} & m_{12} & m_{13} \\ 0 & 0 & 0 \\ m_{31} & m_{32} & m_{33} \end{pmatrix}, \quad \Delta_e = \begin{pmatrix} \delta_{11} & \delta_{12} & 0 \\ \delta_{21} & \delta_{22} & 0 \\ 0 & 0 & \delta_{33} \end{pmatrix}.$$

The inequalities (13.7) may be arbitrarily conservative. Nevertheless, in the next section we will use them to derive good upper and lower bounds for $\mu_{\underline{\Delta}}(\mathbf{M})$.

13.3.3 Estimates for μ

The idea in using (13.7) to get sharper bounds is to replace $\mathbf{M}$ by a matrix which has the same structured singular value, but a different largest singular value and a different spectral radius. We start with the case $m = n$ and introduce the following sets:

$$\mathcal{U} = \left\{ \mathbf{U} \in \underline{\Delta} \,\middle|\, \mathbf{U}^*\mathbf{U} = \mathbf{I} \right\},$$

$$\mathcal{D} = \Big\{ \operatorname{diag}\left(\mathbf{D}_1, \ldots, \mathbf{D}_S, d_1\mathbf{I}_{m_1}, \ldots, d_F\mathbf{I}_{m_F}\right) \,\Big|\, \mathbf{D}_i \in \mathbb{C}^{r_i\times r_i},\ \mathbf{D}_i = \mathbf{D}_i^* > \mathbf{0},\ d_j > 0 \Big\}.$$

For every $\Delta \in \underline{\Delta}$, $\mathbf{U} \in \mathcal{U}$, $\mathbf{D} \in \mathcal{D}$, the following equations hold:

$$\mathbf{U}\mathcal{U} = \mathcal{U} = \mathcal{U}\mathbf{U}, \quad \bar{\sigma}(\mathbf{U}\Delta) = \bar{\sigma}(\Delta\mathbf{U}) = \bar{\sigma}(\Delta),$$
$$\mathbf{D}\Delta = \Delta\mathbf{D}. \tag{13.8}$$

Lemma 13.3.5. *For every* $\mathbf{U} \in \mathcal{U}$, $\mathbf{D} \in \mathcal{D}$ *and the set* $\underline{\Delta}$ *defined in (13.5) the following assertions hold:*

(a) $\mu_{\underline{\Delta}}(\mathbf{MU}) = \mu_{\underline{\Delta}}(\mathbf{UM}) = \mu_{\underline{\Delta}}(\mathbf{M})$,

(b) $\mu_{\underline{\Delta}}(\mathbf{DMD}^{-1}) = \mu_{\underline{\Delta}}(\mathbf{M})$.

Proof. (a) We have

$$\begin{aligned}
\mu_{\underline{\Delta}}(\mathbf{MU})^{-1} &= \min\left\{\bar{\sigma}(\Delta) \,\middle|\, \Delta \in \underline{\Delta},\ \det(\mathbf{I} - \mathbf{MU}\Delta) = 0\right\} \\
&= \min\left\{\bar{\sigma}(\mathbf{U}^*\Delta) \,\middle|\, \Delta \in \underline{\Delta},\ \det(\mathbf{I} - \mathbf{MUU}^*\Delta) = 0\right\} \\
&= \min\left\{\bar{\sigma}(\Delta) \,\middle|\, \Delta \in \underline{\Delta},\ \det(\mathbf{I} - \mathbf{M}\Delta) = 0\right\} \\
&= \mu_{\underline{\Delta}}(\mathbf{M})^{-1}.
\end{aligned}$$

The second equation of (a) can be proven completely analogously.
(b) Using (13.8), assertion (b) follows from

$$\begin{aligned}
\mu_{\underline{\Delta}}(\mathbf{DMD}^{-1})^{-1} &= \min\left\{\bar{\sigma}(\Delta) \,\middle|\, \Delta \in \underline{\Delta},\ \det(\mathbf{I} - \mathbf{DMD}^{-1}\Delta) = 0\right\} \\
&= \min\left\{\bar{\sigma}(\Delta) \,\middle|\, \Delta \in \underline{\Delta},\ \det(\mathbf{I} - \mathbf{DMD}^{-1}(\mathbf{D}\Delta\mathbf{D}^{-1})) = 0\right\} \\
&= \min\left\{\bar{\sigma}(\Delta) \,\middle|\, \Delta \in \underline{\Delta},\ \det(\mathbf{I} - \mathbf{DM}\Delta\mathbf{D}^{-1}) = 0\right\} \\
&= \min\left\{\bar{\sigma}(\Delta) \,\middle|\, \Delta \in \underline{\Delta},\ \det(\mathbf{I} - \mathbf{M}\Delta) = 0\right\} \\
&= \mu_{\underline{\Delta}}(\mathbf{M})^{-1}.
\end{aligned}$$

Corollary 13.3.1. *For the set* $\underline{\Delta}$ *defined in (13.5) the following inequality holds:*

$$\sup_{\mathbf{U} \in \mathcal{U}} \rho(\mathbf{UM}) \le \mu_{\underline{\Delta}}(\mathbf{M}) \le \inf_{\mathbf{D} \in \mathcal{D}} \bar{\sigma}(\mathbf{DMD}^{-1}). \tag{13.9}$$

Proof. This follows directly from the preceding lemma and the Lemma 13.3.4:

$$\rho(\mathbf{UM}) \le \mu_{\underline{\Delta}}(\mathbf{UM}) = \mu_{\underline{\Delta}}(\mathbf{M}) = \mu_{\underline{\Delta}}(\mathbf{DMD}^{-1}) \le \bar{\sigma}(\mathbf{DMD}^{-1}).$$

It can even be shown that the lower bound is sharp (Doyle [32]):

$$\max_{\mathbf{U} \in \mathcal{U}} \rho(\mathbf{UM}) = \mu_{\underline{\Delta}}(\mathbf{M}).$$

For the upper bound we have

$$\mu_{\underline{\Delta}}(\mathbf{M}) = \inf_{\mathbf{D}\in\mathcal{D}} \overline{\sigma}(\mathbf{DMD}^{-1}), \quad \text{if } 2S+F \le 3$$

(see Packard and Doyle [72]). For block structures with $2S+F>3$ it is possible to construct examples where the infimum on the right-hand side is strictly larger than the structured singular value, but the deviations are so small that it is in any case reasonable to work with the infimum.

Remark. The proof of Corollary 13.3.6 does not make use of the property that the scaling matrices of $\mathcal{D}$ are self-adjoint and positive definite. For this reason it may be suspected that the infimum in (13.9) becomes smaller if this property is skipped in the definition of $\mathcal{D}$. It can be shown that this is not the case (cf. Zhou, Doyle, and Glover [112], Remark 11.2).

We consider now the case where $\mathbf{M}$ is nonsquare and limit ourselves for the sake of simplicity to structured sets $\underline{\Delta}$ without repeated scalar blocks. Then $\underline{\Delta}$ can be written as

$$\underline{\Delta} = \left\{ \operatorname{diag}(\Delta_1,\ldots,\Delta_F) \,\middle|\, \Delta_j \in \mathbb{C}^{n_j \times m_j} \right\}. \tag{13.10}$$

We first try to get an upper bound similar to that in (13.9) and define scaling matrices by

$$\mathbf{D}_L = \begin{pmatrix} d_1\mathbf{I}_{n_1} & \cdots & \mathbf{0} \\ \vdots & \ddots & \vdots \\ \mathbf{0} & \cdots & d_F\mathbf{I}_{n_F} \end{pmatrix}, \qquad \mathbf{D}_R = \begin{pmatrix} d_1\mathbf{I}_{m_1} & \cdots & \mathbf{0} \\ \vdots & \ddots & \vdots \\ \mathbf{0} & \cdots & d_F\mathbf{I}_{m_F} \end{pmatrix}$$

with parameters $d_1>0,\ldots,d_F>0$ which are denoted as **D-scales.** Note that with these matrices the commutivity property (13.8) is replaced by

$$\mathbf{D}_L\Delta = \Delta\mathbf{D}_R.$$

Then with completely analogous arguments as in the proof of Corollary 13.3.6, the following lemma can be shown.

Lemma 13.3.7. *The following inequality holds for the set* $\underline{\Delta}$ *defined in (13.3.7):*

$$\mu_{\underline{\Delta}}(\mathbf{M}) \le \inf\left\{ \overline{\sigma}(\mathbf{D}_R\mathbf{M}\mathbf{D}_L^{-1}) \,\middle|\, d_i>0, \ i=1,\ldots,F \right\}.$$

Remarks. 1. The proof of the preceding Lemma does not make use of the sign of the numbers d_k. Therefore it may be questioned whether the minimum becomes smaller when arbitrary complex numbers are admitted, but it can be shown that this is not the case (compare the remark after Corollary 13.3.1).

2. The matrix

$$\mathbf{D}_R \mathbf{M} \mathbf{D}_L^{-1}$$

contains the d_k only in form of the quotients d_k / d_j. Consequently, without restriction of generality, one of these numbers may be fixed, so that only $F-1$ free parameters remain when the minimum is calculated. A typical choice is $d_F = 1$.

3. The lower bound of $\mu_\Delta(\mathbf{M})$ given in Corollary 13.3.1 can also be adapted to this modified choice of $\underline{\Delta}$. With the notation introduced at the end of the previous section, we have

$$\sup\left\{\rho(\mathbf{U}\mathbf{M}_e) \middle| \mathbf{U} \in \underline{\Delta}_e,\ \mathbf{U}^*\mathbf{U} = \mathbf{I}\right\} \leq \mu_{\underline{\Delta}}(\mathbf{M}).$$

From a practical point of view, it is important to admit also *scalar real* or *repeated scalar real* blocks, compare the examples given in Chap. 12. Then the *mixed real/complex structured singular value* has to be computed. As an extreme case, an uncertainty structure with only real blocks can occur. We cannot give the theory for this more general structured set, but mention that there is commercial software which is able to compute bounds for μ in this situation (cf. Balas et al. [7]). There are also some short remarks concerning the mixed case given.

13.4 Structured Robust Stability and Performance

13.4.1 Robust Stability

We start with the block diagram depicted in Fig. 12.9. Here, as before, $\mathbf{P}$ denotes the extended plant, i.e. the original "physical" plant $\mathbf{G}$ with the weights for performance and robustness. The uncertainties Δ are isolated from the plant as discussed in detail in Chapter 12. We assume that they belong to the structured set $\mathcal{M}(\underline{\Delta})$ with $\underline{\Delta}$ as in (13.5). The feedback system without model uncertainties is again denoted by $\mathbf{M}$, i.e. $\mathbf{M}(s) = \mathcal{F}_l(\mathbf{P},\mathbf{K})(s)$ (cf. Fig. 13.4)). We ask under

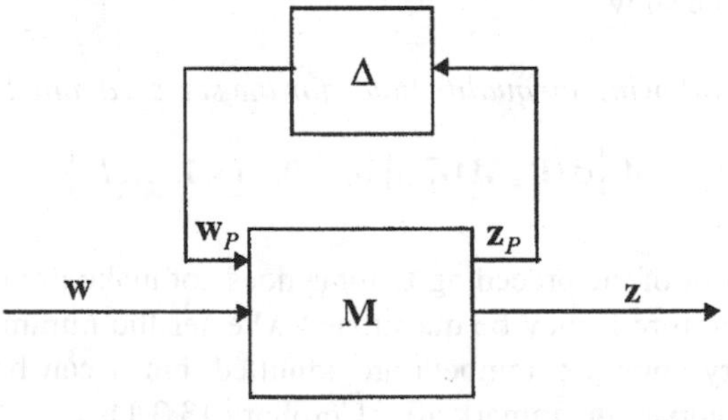

Fig. 13.4. Feedback system with perturbations

which conditions this feedback loop with structured model uncertainties is robustly stable. The answer is given in the next theorem, which generalizes Corollary 13.2.1.

For the proof of this result it will be helpful to note that the minimal norm matrix Δ in the definition of μ can be chosen dyads. For the sake of simplicity, assume that all blocks are full blocks (i.e. $S=0$) and square. Let $\Delta \in \underline{\Delta}$ be the smallest perturbation such that $\mathbf{I}-\mathbf{M}\Delta$ is singular. Then there exists an $\mathbf{x} \in \mathbb{C}^m$ such that $\mathbf{M}\Delta\mathbf{x}=\mathbf{x}$. Now partition $\mathbf{x}$ compatibly with Δ:

$$\mathbf{x}=\begin{pmatrix} \mathbf{x}_1 \\ \vdots \\ \mathbf{x}_F \end{pmatrix}, \quad \mathbf{x}_j \in \mathbb{C}^{m_j}, j=1,\ldots,F$$

and define for $j=1,\ldots,F$

$$\tilde{\Delta}_j = \begin{cases} \dfrac{1}{\|\mathbf{x}_j\|^2}\Delta_j \mathbf{x}_j \mathbf{x}_j^*, & \text{if } \mathbf{x}_j \neq \mathbf{0}, \\ \mathbf{0}, & \text{if } \mathbf{x}_j = \mathbf{0}. \end{cases}$$

Define

$$\tilde{\Delta} = \operatorname{diag}(\tilde{\Delta}_1,\ldots,\tilde{\Delta}_F).$$

Then $\tilde{\Delta}\mathbf{x}=\Delta\mathbf{x}$ and $\tilde{\Delta}$ has rank 1. Moreover, we have

$$\bar{\sigma}(\tilde{\Delta}) \le \bar{\sigma}(\Delta) \quad \text{and} \quad (\mathbf{I}-\mathbf{M}\tilde{\Delta})\mathbf{x}=(\mathbf{I}-\mathbf{M}\Delta)\mathbf{x}=\mathbf{0}.$$

Since Δ is a minimum norm perturbation for which $\mathbf{I}-\mathbf{M}\Delta$ is singular, the maximum singular values of Δ and $\tilde{\Delta}$ must coincide.

Theorem 13.4.1. *Let the nominal feedback system* $\mathbf{M}=\mathcal{F}_l(\mathbf{P},\mathbf{K})$ *be internally stable and let* $\gamma>0$. *Then the following assertions are equivalent:*

(i) $\mu_{\underline{\Delta}}(\mathbf{M}_{11}(s)) \le \gamma$ *for every* $s \in \overline{\mathbb{C}}_+$. *(13.11)*

(ii) The perturbed feedback system in Fig. 13.4 is well-posed and internally stable for all perturbations $\Delta \in \mathcal{M}(\underline{\Delta})$ *with*

$\|\Delta\|_\infty < 1/\gamma$.

Proof. "$\Leftarrow$": This is exactly Lemma 13.3.1.
"$\Rightarrow$": Suppose (13.11) is not true. Consequently, applying (13.6), we can conclude that there is a frequency $\omega_0 \ge 0$ such that $\mu_\Delta(\mathbf{M}_{11}(i\omega_0)) > \gamma$. Using the preceding consideration we get the existence of a matrix $\Delta_0 \in \underline{\Delta}$ such that each full block has rank 1 and

$$\bar{\sigma}(\Delta_0) < 1/\gamma, \quad \det(\mathbf{I} - \mathbf{M}_{11}(i\omega_0)\Delta_0) = 0 .$$

As in the proof of Theorem 13.1.2 it is now possible to construct a transfer function $\Delta \in \mathcal{M}(\underline{\Delta})$ such that $\Delta(i\omega_0) = \Delta_0$ and $\|\Delta\|_\infty = \bar{\sigma}(\Delta_0) < 1/\gamma$. Hence

$$\det(\mathbf{I} - \mathbf{M}_{11}(i\omega_0)\Delta(i\omega_0)) = 0$$

which means that Δ destabilizes the system and leads us to a contradiction.

Remarks. 1. As we see from Theorem 13.4.1, condition (13.4) is not only sufficient but also necessary for robust stability. We also note that Corollary 13.2.1 is a special case of Theorem 13.4.1 (put $\underline{\Delta} = \mathbb{C}^{n\times m}$).

2. Because of (13.6), condition (13.11) is equivalent to

$$\sup_{\omega \in \mathbb{R}} \mu_{\underline{\Delta}}(\mathbf{M}_{11}(i\omega)) \le \gamma .$$

Hence, if γ denotes the peak of the μ -plot of $\mathbf{M}_{11}$, then $1/\gamma$ is the largest magnitude of the perturbations for which the feedback loop is robustly stable.

13.4.2 Robust Performance

Next we want to derive a criterion for robust performance. Performance is measured by the $\mathcal{H}_\infty$ norm of the feedback system. As defined in Sect. 12.1, the system is said to have robust performance if this norm stays under a certain limit for all admissible uncertainties. The transfer function of the feedback system including the uncertainty is given by $\mathcal{F}_u(\mathbf{M}, \Delta)$. Hence the system has robust performance if $\|\mathcal{F}_u(\mathbf{M}, \Delta)\|_\infty \le \gamma$ for all stable transfer functions Δ, whose norm is bounded by a certain constant.

This problem can be elegantly handled by treating performance also as a kind of uncertainty. The basic idea is visualized in Fig. 13.5. The uncertain system $\mathcal{F}_u(\mathbf{M}, \Delta)$ (which still has $\mathbf{w}$ as input and $\mathbf{z}$ as output) is made into a feedback system with an additional uncertainty Δ_1 by putting $\mathbf{w} = \Delta_1 \mathbf{z}$ (Fig. 13.5(a)). Here, the uncertainty Δ_1 is assumed to be unstructured. In the next step, the uncertainties Δ, Δ_1 are unified to a single block (Fig. 13.5(b)). As we will show in the following theorem, robust stability of the system depicted in Fig. 13.5(b) is equivalent to robust stability and robust performance of the original uncertain feedback system. Thus define (if $\mathbf{z} \in \mathbb{R}^{p_2}$ and $\mathbf{w} \in \mathbb{R}^{q_2}$)

$$\underline{\Delta}_P = \left\{ \begin{pmatrix} \Delta & \mathbf{0} \\ \mathbf{0} & \Delta_1 \end{pmatrix} \middle| \, \Delta \in \underline{\Delta}, \ \Delta_1 \in \mathbb{C}^{q_2 \times p_2} \right\}.$$

Moreover, let $\underline{\Delta}_1 = \mathbb{C}^{q_2 \times p_2}$. Then have

$$\mathbf{w}_P = \Delta \mathbf{z}_P, \quad \mathbf{w} = \Delta_1 \mathbf{z} .$$

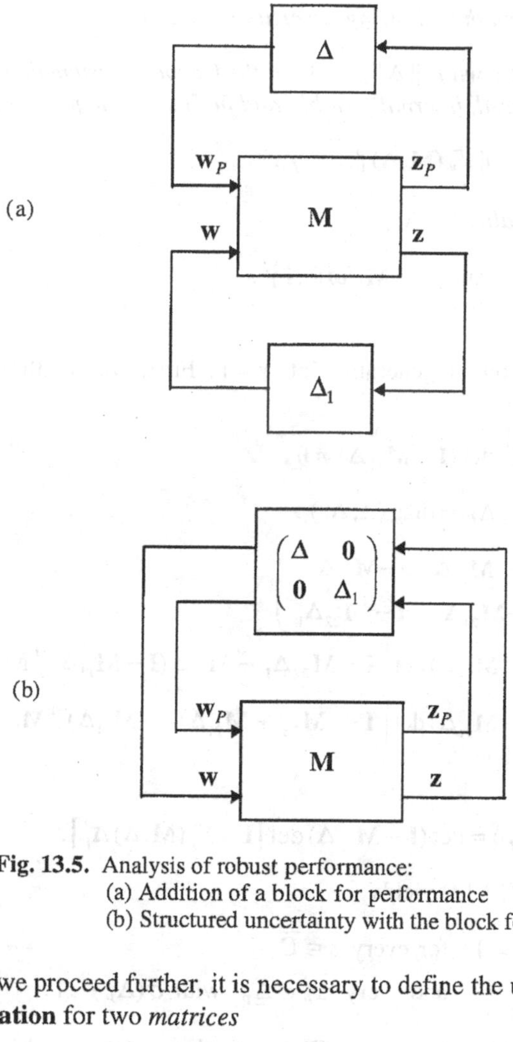

Fig. 13.5. Analysis of robust performance:
(a) Addition of a block for performance
(b) Structured uncertainty with the block for performance

Before we proceed further, it is necessary to define the **upper linear fractional transformation** for two *matrices*

$$\mathbf{M} = \begin{pmatrix} \mathbf{M}_{11} & \mathbf{M}_{12} \\ \mathbf{M}_{21} & \mathbf{M}_{22} \end{pmatrix} \in \mathbb{C}^{(p_1+p_2)\times(q_1+q_2)}, \quad \Delta \in \mathbb{C}^{q_1 \times p_1}.$$

It is defined by

$$\mathcal{F}_u(\mathbf{M}, \Delta) = \mathbf{M}_{22} + \mathbf{M}_{21}\Delta(\mathbf{I} - \mathbf{M}_{11}\Delta)^{-1}\mathbf{M}_{12},$$

provided that the inverse $(\mathbf{I} - \mathbf{M}_{11}\Delta)^{-1}$ exists. It is easy to see that for two transfer matrices the following equation holds:

$$\mathcal{F}_u(\mathbf{M}, \Delta)(s) = \mathcal{F}_u(\mathbf{M}(s), \Delta(s)).$$

Theorem 13.4.2. *Let the nominal feedback system* $\mathbf{M} = \mathcal{F}_l(\mathbf{P}, \mathbf{K})$ *be internally stable and let* $\gamma > 0$. *Then the following assertions are equivalent.*

(i) For every $\Delta \in \mathcal{M}(\underline{\Delta})$ *with* $\|\Delta\|_\infty < 1/\gamma$ *the feedback system depicted in Fig. 13.4 is well-posed, internally stable and fulfills the inequality*

$$\|\mathcal{F}_u(\mathbf{M}, \Delta)\|_\infty \leq \gamma. \tag{13.12}$$

(ii) The following inequality holds:

$$\sup_{\omega \in \mathbb{R}} \mu_{\underline{\Delta}_P}(\mathbf{M}(i\omega)) \leq \gamma. \tag{13.13}$$

Proof. Without restriction of generality let $\gamma = 1$. First, we fix the following. Under the assumption

$$\det(\mathbf{I} - \mathbf{M}_{11}\Delta) \neq \mathbf{0}, \tag{13.14}$$

we have for matrices $\mathbf{M}$, $\Delta_P = \text{diag}(\Delta, \Delta_1)$:

$$\begin{aligned}
\det(\mathbf{I} - \mathbf{M}\Delta_P) &= \det\begin{pmatrix} \mathbf{I} - \mathbf{M}_{11}\Delta & -\mathbf{M}_{12}\Delta_1 \\ -\mathbf{M}_{21}\Delta & \mathbf{I} - \mathbf{M}_{22}\Delta_1 \end{pmatrix} \\
&= \det(\mathbf{I} - \mathbf{M}_{11}\Delta)\det\left[\mathbf{I} - \mathbf{M}_{22}\Delta_1 - \mathbf{M}_{21}\Delta(\mathbf{I} - \mathbf{M}_{11}\Delta)^{-1}\mathbf{M}_{12}\Delta_1\right] \\
&= \det(\mathbf{I} - \mathbf{M}_{11}\Delta)\det\left[\mathbf{I} - (\mathbf{M}_{22} + \mathbf{M}_{21}\Delta(\mathbf{I} - \mathbf{M}_{11}\Delta)^{-1}\mathbf{M}_{12})\Delta_1\right].
\end{aligned}$$

Hence (13.14) implies

$$\det(\mathbf{I} - \mathbf{M}\Delta_P) = \det(\mathbf{I} - \mathbf{M}_{11}\Delta)\det\left[\mathbf{I} - \mathcal{F}_u(\mathbf{M}, \Delta)\Delta_1\right]. \tag{13.15}$$

"(ii) $\Rightarrow$ (i)": Condition (13.13) yields

$$\det(\mathbf{I} - \mathbf{M}(s)\Delta_P) \neq 0 \quad \text{for every } s \in \overline{\mathbb{C}}_+ \text{ and every } \Delta_P \in \underline{\Delta}_P \text{ with } \bar{\sigma}(\Delta_P) < 1. \tag{13.16}$$

Let now $\Delta \in \underline{\Delta}$ with $\bar{\sigma}(\Delta) < 1$ be given. Then one obtains (choose $\Delta_1 = \mathbf{0}$)

$$\det(\mathbf{I} - \mathbf{M}_{11}(s)\Delta) \neq 0 \quad \text{for every } s \in \overline{\mathbb{C}}_+. \tag{13.17}$$

This implies

$$\mu_{\underline{\Delta}}(\mathbf{M}_{11}(s)) \leq 1 \quad \text{for every } s \in \overline{\mathbb{C}}_+, \tag{13.18}$$

i.e. according to Theorem 13.4.1, the feedback loop is well-posed and robustly internally stable. Besides this, because of (13.17) we are allowed to apply (13.15) and obtain

$$\det\left[\mathbf{I}-(\mathcal{F}_u(\mathbf{M}(s),\Delta)\Delta_1\right]\neq 0 \quad \text{for every } s\in\overline{\mathbb{C}}_+$$
$$\text{and every } \Delta_P\in\underline{\Delta}_P \text{ with } \bar{\sigma}(\Delta_P)<1.$$

Suppose now $\Delta(s)\in\mathcal{M}(\underline{\Delta})$ with $\|\Delta\|_\infty<1$ and let $\Delta_1\in\underline{\Delta}_1$ with $\bar{\sigma}(\Delta_1)<1$. Then because of

$$\bar{\sigma}\begin{pmatrix}\Delta(s) & \mathbf{0}\\ \mathbf{0} & \Delta_1\end{pmatrix}=\max\{\bar{\sigma}(\Delta(s)),\bar{\sigma}(\Delta_1)\}<1$$

the following relation holds:

$$\det\left[\mathbf{I}-(\mathcal{F}_u(\mathbf{M}(s),\Delta(s))\Delta_1\right]\neq 0 \quad \text{for every } s\in\overline{\mathbb{C}}^+$$
$$\text{and every } \Delta_1\in\underline{\Delta}_1 \text{ with } \bar{\sigma}(\Delta_1)<1. \tag{13.19}$$

The definition of μ leads now to the desired inequality:

$$\sup_{s\in\overline{\mathbb{C}}^+}\bar{\sigma}(\mathcal{F}_u(\mathbf{M},\Delta)(s))=\sup_{s\in\overline{\mathbb{C}}^+}\mu_{\underline{\Delta}_1}(\mathcal{F}_u(\mathbf{M},\Delta)(s))\leq 1.$$

Here, we have applied Lemma 13.3.2.
"(i) $\Rightarrow$ (ii)": Suppose that (13.13) is not true. Then there is a frequency $\omega_0\geq 0$ such that $\mu_{\Delta_P}(\mathbf{M}(i\omega_0))>1$. Hence there exists a matrix Δ_{P0} of the form

$$\Delta_{P0}=\begin{pmatrix}\Delta_0 & \mathbf{0}\\ \mathbf{0} & \Delta_{10}\end{pmatrix} \quad \text{with } \Delta_0\in\underline{\Delta},\ \Delta_{10}\in\underline{\Delta}_1$$

such that

$$\bar{\sigma}(\Delta_{P0})<1 \quad \text{and} \quad \det(\mathbf{I}-\mathbf{M}(i\omega_0)\Delta_{P0})=0.$$

As in the proof of Theorem 13.4.1 there exists a transfer function $\Delta_P\in\mathcal{M}(\underline{\Delta}_P)$ such that $\Delta_P(i\omega_0)=\Delta_{P0}$ and $\|\Delta_P\|_\infty=\bar{\sigma}(\Delta_{P0})<1$. This implies $\|\Delta\|_\infty<1$, $\|\Delta_1\|_\infty<1$ and, since the system is internally stable, it follows from (13.15) and Theorem 13.4.1 that

$$\begin{aligned}0&=\det(\mathbf{I}-\mathbf{M}(i\omega_0)\Delta_P(i\omega_0))\\&=\det(\mathbf{I}-\mathbf{M}_{11}(i\omega_0)\Delta(i\omega_0))\det\left[\mathbf{I}-\mathcal{F}_u(\mathbf{M}(i\omega_0),\Delta(i\omega_0))\Delta_1(i\omega_0)\right].\end{aligned}$$

Because of $\det(\mathbf{I}-\mathbf{M}_{11}(i\omega_0)\Delta(i\omega_0))\neq 0$ we obtain

$$\det\left[\mathbf{I}-\mathcal{F}_u(\mathbf{M}(i\omega_0),\Delta(i\omega_0))\Delta_1(i\omega_0)\right]=0.$$

This cannot be since (13.12) implies

$$\begin{aligned}\mu_{\underline{\Delta}_1}(\mathcal{F}_u(\mathbf{M}(i\omega_0),\Delta(i\omega_0)))&=\bar{\sigma}(\mathcal{F}_u(\mathbf{M}(i\omega_0),\Delta(i\omega_0)))\\&=\bar{\sigma}(\mathcal{F}_u(\mathbf{M},\Delta)(i\omega_0))\leq 1.\end{aligned}$$

Remark. Hence the peak of the μ -plot of $\mathbf{M}$ indicates the magnitude of the perturbations for which the feedback loop is robustly stable and has additionally robust performance. This can be compared with the corresponding remark concerning robust stability and the peak of the μ -plot of $\mathbf{M}_{11}$ at the end of Sect. 13.4.1.

13.5. D-K Iteration

Basic idea. We consider the feedback loop depicted in Fig 13.6 with $\mathbf{P}$ as described in Sect. 13.4.2 and an input $\mathbf{w}_P$ and an output $\mathbf{z}_P$ for incorporating uncertainty. As before, let $\mathbf{M} = \mathcal{F}_l(\mathbf{P},\mathbf{K})$ and suppose that the controller $\mathbf{K}$ has already been designed. Then the largest value of $\|\Delta\|_\infty$ such that the feedback system is robustly stable and has robust performance is $1/\gamma$, where γ is the supremum in (13.13). Now we want to find a controller $\mathbf{K}$ such that this supremum is as small as possible. This leads us to the following optimization problem:

Determine a controller $\mathbf{K}$ *such that the supremum*

$$\sup_{\omega \in \mathbb{R}} \mu_{\Delta_P} (\mathcal{F}_l(\mathbf{P},\mathbf{K})(i\omega)) \le \gamma \tag{13.20}$$

is as small as possible.

According to Theorem 13.4.2, such a controller is optimal with respect to robust stability as well as to robust performance. Of course, this optimality is related to the specified weightings for performance and the uncertainty structure which is implicitly given in the extended plant $\mathbf{P}$. It is only possible to solve this optimization approximately and we want to show now in some detail how this can be done.

We first describe the algorithm for the case where $\mathcal{F}_l(\mathbf{P},\mathbf{K})$ is a square transfer

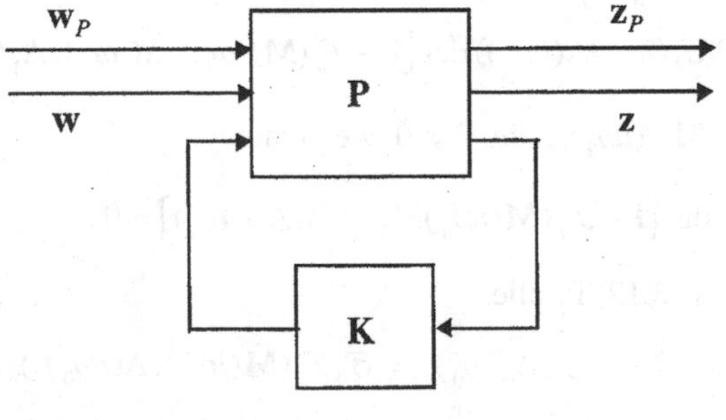

Fig. 13.6. Feedback system with input and output for model uncertainties

matrix. The first step is to synthesize a $\mathcal{H}_\infty$ controller $\mathbf{K}$. With this controller, the structured singular value $\mu_{\Delta_P}(\mathcal{F}_l(\mathbf{P},\mathbf{K})(i\omega))$ has to be calculated for each frequency ω. Here, the upper bound of Corollary 13.3.1 has to be used. This Corollary gives for each ω a scaling matrix $\mathbf{D}_\omega$ which is a solution of the optimization problem

$$\inf_{\mathbf{D}_\omega \in \mathcal{D}} \bar{\sigma}(\mathbf{D}_\omega \mathcal{F}_l(\mathbf{P},\mathbf{K})(i\omega)\mathbf{D}_\omega^{-1}) .$$

In a practical calculation, it is of course not possible to perform this minimization for each frequency, so that finitely many discrete frequencies $\omega_1,\ldots,\omega_J$ have to be selected where this is done. Then a rational transfer matrix $\mathbf{D}(s)$ has to be determined such that $\mathbf{D},\mathbf{D}^{-1} \in \mathcal{H}_\infty$ and

$$\mathbf{D}(i\omega_j) = \mathbf{D}_{\omega_j} .$$

This implies

$$\begin{aligned}\sup_j \inf_{\mathbf{D}_\omega \in \mathcal{D}} \bar{\sigma}(\mathbf{D}_{\omega_j} \mathcal{F}_l(\mathbf{P},\mathbf{K})(i\omega_j)\mathbf{D}_{\omega_j}^{-1}) &= \sup_j \bar{\sigma}(\mathbf{D}(i\omega_j)\mathcal{F}_l(\mathbf{P},\mathbf{K})(i\omega_j)\mathbf{D}(i\omega_j)^{-1}) \\ &\approx \|\mathbf{D}\mathcal{F}_l(\mathbf{P},\mathbf{K})\mathbf{D}^{-1}\|_\infty = \|\mathcal{F}_l(\hat{\mathbf{P}},\mathbf{K})\|_\infty .\end{aligned}$$

Here, $\hat{\mathbf{P}}$ denotes the scaled plant, which is obtained from $\mathbf{P}$ by the scaling transfer functions $\mathbf{D}$, $\mathbf{D}^{-1}$ (cf. Fig 13.7). Then a new $\mathcal{H}_\infty$ optimal controller has to be calculated for $\hat{\mathbf{P}}$. In the block diagram of Fig. 13.7, the inputs $\mathbf{w}_P, \mathbf{w}$ and the outputs $\mathbf{z}_P, \mathbf{z}$ are written as one quantity.

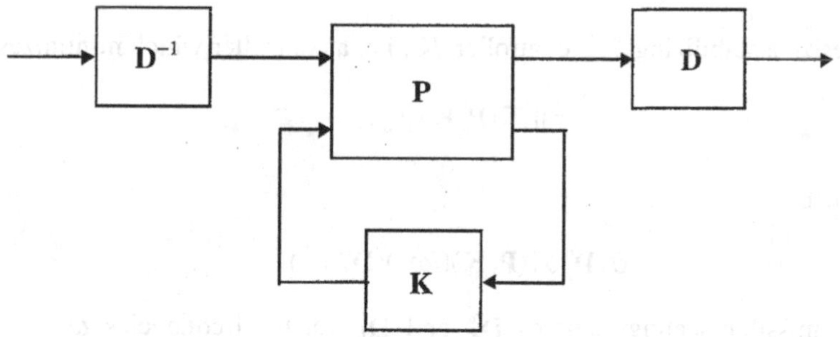

Fig. 13.7. Scaled plant

Essentially, the algorithm consists of two steps, which can roughly be described as follows.

Step 1: Given a scaling matrix $\mathbf{D}$, calculate the $\mathcal{H}_\infty$ controller $\mathbf{K}$ for the plant $\hat{\mathbf{P}}$ which has been scaled by $\mathbf{D}$ and $\mathbf{D}^{-1}$.

Step 2: Minimize $\bar{\sigma}(\mathbf{D}_\omega \mathcal{F}_l(\mathbf{P},\mathbf{K})(i\omega)\mathbf{D}_\omega^{-1})$ for each frequency ω (here $\mathbf{P}$ is the original unscaled plant). This leads to the scaling matrix $\mathbf{D}$.

This algorithm is denoted as the **D-K iteration.** Up to now, it was not possible to prove its convergence. However, in practical applications the methods needs only a few steps to calculate a controller which is nearly optimal.

Detailed description of the method for a special case. We want to describe the algorithm in greater detail for the case where the set $\underline{\Delta}$ has no repeated scalar blocks and where the full blocks are not required to be quadratic, i.e. $\underline{\Delta}$ has the structure (13.10). In this case, left and right scaling matrices of the form

$$\mathbf{D}_L^\omega = \text{diag}\left(d_1^\omega \mathbf{I}_{n_1}, \ldots, d_{F-1}^\omega \mathbf{I}_{n_{F-1}}, \mathbf{I}_{n_F}\right),$$

$$\mathbf{D}_R^\omega = \text{diag}\left(d_1^\omega \mathbf{I}_{m_1}, \ldots, d_{F-1}^\omega \mathbf{I}_{m_{F-1}}, \mathbf{I}_{m_F}\right)$$

are required (cf. Sect. 13.3.3). Here $\mathcal{F}_l(\mathbf{P}, \mathbf{K})$ is supposed to be an $m \times n$ transfer matrix.

Then the D-K iteration consists of the following steps.

1. For given scaling matrices $\mathbf{D}_L(s)$, $\mathbf{D}_R(s)$ calculate a new plant by

$$\hat{\mathbf{P}}(s) = \begin{pmatrix} \mathbf{D}_R(s) & \mathbf{0} \\ \mathbf{0} & \mathbf{I} \end{pmatrix} \mathbf{P}(s) \begin{pmatrix} \mathbf{D}_L(s)^{-1} & \mathbf{0} \\ \mathbf{0} & \mathbf{I} \end{pmatrix}$$

and determine a state space representation. The algorithm can be initialized by choosing $\mathbf{D}_L = \mathbf{I}$, $\mathbf{D}_R = \mathbf{I}$.

2. Synthesize a stabilizing $\mathcal{H}_\infty$ controller $\mathbf{K}$, i.e. a controller which minimizes

$$\| \mathcal{F}_l(\hat{\mathbf{P}}, \mathbf{K}) \|_\infty .$$

3. Minimize

$$\bar{\sigma}(\mathbf{D}_R^\omega \mathcal{F}_l(\mathbf{P}, \mathbf{K})(i\omega_j)(\mathbf{D}_L^\omega)^{-1})$$

over all admissible scaling matrices $\mathbf{D}_R^\omega$ and $\mathbf{D}_L^\omega$ for the frequencies $\omega_1, \ldots, \omega_J$. As a result, one obtains positive numbers

$$d_k^{\omega_1}, \ldots, d_k^{\omega_J} \quad \text{for } k = 1, \ldots, F-1 .$$

4. Determine $k = 1, \ldots, F-1$ transfer functions $d_k \in \mathcal{RH}_\infty$ with $d_k^{-1} \in \mathcal{RH}_\infty$ such that the points $d_k^{\omega_j}$ are approximately interpolated in the following sense:

$$| d_k(i\omega_j) | = d_k^{\omega_j} \quad \text{for } j = 1, \ldots, J .$$

Then put

$$\mathbf{D}_L(s) = \operatorname{diag}\left(d_1(s)\mathbf{I}_{n_1}, \ldots, d_{F-1}(s)\mathbf{I}_{n_{F-1}}, \mathbf{I}_{n_F}\right),$$

$$\mathbf{D}_R(s) = \operatorname{diag}\left(d_1(s)\mathbf{I}_{m_1}, \ldots, d_{F-1}(s)\mathbf{I}_{m_{F-1}}, \mathbf{I}_{m_F}\right).$$

5. Compare the numbers $d_k^{\omega_1}, \ldots, d_k^{\omega_J}$ with those of the previous step. If they are approximately the same, stop the algorithm. Otherwise, continue with step 1.

Remarks. 1. The synthesis of $\mathcal{H}_\infty$ controllers is an essential part of the D-K iteration.

2. The minimization in step 3 is a convex optimization problem, which can be solved with adequate numerical algorithms.

3. We want to describe the calculation of the transfer functions of step 4 in greater detail. Let values $d_j > 0$ for $j = 1, \ldots, J$ be given. For an arbitrary minimum-phase stable transfer function $G(s)$ the phase can be calculated from the gain by the following formula (cf. Doyle, Francis, and Tannenbaum [36], Sect. 7.2):

$$\arg G(i\omega_0) = \frac{2\omega_0}{\pi} \int_0^\infty \frac{\ln|G(i\omega)| - \ln|G(i\omega_0)|}{\omega^2 - \omega_0^2} d\omega .$$

This formula is evaluated numerically at the given frequencies. In this way, one gets discrete phases φ_j for the frequencies ω_j. As a result, a complex D-scale $\tilde{d}_j = d_j e^{i\varphi_j}$ is obtained. The remaining task is to find a proper and stable minimum-phase transfer function $G(s)$ which is not strictly proper such that

$$G(i\omega_j) \approx \tilde{d}_j .$$

4. If repeated scalar blocks occur, then the scaling matrices contain full blocks and the construction of the scaling transfer functions $\mathbf{D}_L, \mathbf{D}_R$ becomes more complicated. We refer the reader to Balas et al. [7], where additional information is given.

5. It is obvious that the D-K iteration is a highly complicated procedure, which can even in the simplest cases only be solved numerically. Every single step of the algorithm is by no means trivial. Fortunately, there is software which fully automates its application. We will demonstrate this later for a simple example.

13.6 Reduction of the Controller Order

13.6.1 Balanced Realizations

The order of a controller resulting from μ synthesis is typically high. Therefore, a natural question is whether it is possible to replace the original controller by a controller of lower order which only moderately degrades the robustness and performance properties of the feedback system. There are actually methods to achieve this and one of them we describe now. It requires a special kind of system realization, which is our next topic.

In Sect. 9.3.1, we introduced the observability operator $\Psi_o : \mathbb{R}^n \to \mathcal{L}_2[0,\infty)$. It has the property $\mathbf{X}_o = \Psi_o^*\Psi_o$, where $\mathbf{X}_o$ is the observability gramian. Thus the norm of $\mathbf{y} = \Psi_o\mathbf{x}_0$ is given by

$$\| \mathbf{y} \|^2 = \mathbf{x}_0^T \mathbf{X}_o \mathbf{x}_0 . \tag{13.21}$$

Let now

$$\eta_1 \geq \eta_2 \geq \ldots \geq \eta_n \geq 0$$

be the eigenvalues of $\mathbf{X}_o^{1/2}$ and $\mathbf{v}_1,\ldots,\mathbf{v}_n$ the corresponding eigenvectors. An application of (13.21) to the eigenvectors $\mathbf{v}_k$ yields

$$\| \mathbf{y} \| = \eta_k \quad \text{for} \quad \mathbf{y} = \Psi_o \mathbf{v}_k .$$

This tells us that for $k > l$ the state $\mathbf{v}_k$ is in some sense "more observable" than the state $\mathbf{v}_l$.

In a similar way, it is possible to introduce states which are "more controllable" than others. To this end, the controllability operator $\Psi_c : \mathcal{L}_2(-\infty,0] \to \mathbb{R}^n$ (cf. Sect. 9.3.1) will be needed. We already know that $\mathbf{X}_c = \Psi_c\Psi_c^*$, where $\mathbf{X}_c$ is the controllability gramian, cf. Lemma 9.3.1, in which the basic properties of Ψ_c are formulated. In the sequel, the following set is required:

$$R = \{ \Psi_c \mathbf{u} \mid \mathbf{u} \in \mathcal{L}_2(-\infty,0] \text{ and } \| \mathbf{u} \| \leq 1 \} . \tag{13.22}$$

As the next lemma says, it can be characterized by the controllability gramian.

Lemma 13.6.1. *The set R defined in (13.22) can equivalently be written as*

$$R = \{ \mathbf{X}_c^{1/2} \mathbf{x}_c \mid \mathbf{x}_c \in \mathbb{R}^n \text{ and } \| \mathbf{x}_c \| \leq 1 \} . \tag{13.23}$$

Proof. Let $\mathbf{u} \in \mathcal{L}_2(-\infty,0]$ with $\| \mathbf{u} \| \leq 1$ and put

$$\mathbf{x}_c = \mathbf{X}_c^{-1/2} \Psi_c \mathbf{u} .$$

The norm of $\mathbf{x}_c$ can be estimated by (note that $P = \Psi_c^* \mathbf{X}_c^{-1} \Psi_c$ is a projection operator; see Sect. 9.3.1)

$$\| \mathbf{x}_c \|^2 = \left\langle \mathbf{X}_c^{-1/2} \Psi_c \mathbf{u}, \mathbf{X}_c^{-1/2} \Psi_c \mathbf{u} \right\rangle = \left\langle \mathbf{u}, \Psi_c^* \mathbf{X}_c^{-1} \Psi_c \mathbf{u} \right\rangle$$
$$= \left\langle \mathbf{u}, P\mathbf{u} \right\rangle = \| P\mathbf{u} \|^2 \leq \|\mathbf{u}\|^2 \leq 1 .$$

Hence $\Psi_c \mathbf{u} = \mathbf{X}_c^{1/2} \mathbf{x}_c$ is in the set defined in the right-hand side of (13.23).

Let now $\mathbf{x}_c \in \mathbb{R}^n$ be given with $\| \mathbf{x}_c \| \leq 1$. Put $\mathbf{u}_m = \Psi_c^* \mathbf{X}_c^{-1} \mathbf{x}_0$, where $\mathbf{x}_0 = \mathbf{X}_c^{1/2} \mathbf{x}_c$. Then $\Psi_c \mathbf{u}_m = \mathbf{x}_0 = \mathbf{X}_0^{1/2} \mathbf{x}_c$ and Lemma 9.3.1 implies

$$\| \mathbf{u}_m \|^2 = \mathbf{x}_0^T \mathbf{X}_c^{-1} \mathbf{x}_0 = \mathbf{x}_c^T \mathbf{x}_c \leq 1 .$$

Hence $\mathbf{X}_c^{1/2} \mathbf{x}_c$ is an element of R defined in (13.22) and the lemma is proven.

The lemma says that all the states reachable by a control $\mathbf{u}$ with $\| \mathbf{u} \| \leq 1$ is given by all vectors $\mathbf{X}_c^{1/2} \mathbf{x}_c$, where $\| \mathbf{x}_c \| \leq 1$. Let

$$\mu_1 \geq \mu_2 \geq \ldots \geq \mu_n \geq 0$$

be the eigenvalues of $\mathbf{X}_c^{1/2}$ and $\mathbf{w}_1, \ldots, \mathbf{w}_n$ the corresponding eigenvectors. Then

$$\| \mathbf{x}_0 \| = \mu_k \quad \text{for} \quad \mathbf{x}_0 = \mathbf{X}_c^{1/2} \mathbf{w}_k .$$

Thus it can be said that for $k > l$ the state $\mathbf{w}_k$ is “more observable” than the state $\mathbf{w}_l$.

The idea is now to simplify a system by deleting the states which are only slightly observable or controllable. But, of course, this only makes sense when for $\mathbf{X}_o^{1/2}$ and $\mathbf{X}_c^{1/2}$ a common system of eigenvectors can be found.

Let a stable system of the form

$$\begin{aligned} \dot{\mathbf{x}} &= \mathbf{A}\mathbf{x} + \mathbf{B}\mathbf{u} \\ \mathbf{y} &= \mathbf{C}\mathbf{x} \end{aligned} \tag{13.24}$$

be given. We perform a state space transformation by a nonsingular matrix $\mathbf{T}$. Then

$$\tilde{\mathbf{A}} = \mathbf{T}\mathbf{A}\mathbf{T}^{-1}, \quad \tilde{\mathbf{B}} = \mathbf{T}\mathbf{B}, \quad \tilde{\mathbf{C}} = \mathbf{C}\mathbf{T}^{-1}$$

and the controllability gramian associated with this new realization is

$$\tilde{\mathbf{X}}_c = \int_0^\infty e^{\tilde{\mathbf{A}}t} \tilde{\mathbf{B}} \tilde{\mathbf{B}}^T e^{\tilde{\mathbf{A}}^T t} \, dt$$

$$= \int_0^\infty \mathbf{T} e^{\mathbf{A}t} \mathbf{T}^{-1} \mathbf{T} \mathbf{B} \mathbf{B}^T \mathbf{T}^T (\mathbf{T}^T)^{-1} e^{\mathbf{A}^T t} \mathbf{T}^T \, dt$$

$$= \mathbf{T}\mathbf{X}_c\mathbf{T}^T .$$

In a similar way, it can be shown that the transformed observability gramian is

$$\tilde{\mathbf{X}}_o = (\mathbf{T}^T)^{-1}\mathbf{X}_o\mathbf{T}^{-1} .$$

It is now possible to apply Theorem A1.3 to the gramians and the result is the next Theorem.

Theorem 13.6.1. *Let* $\mathbf{A}$ *be stable and suppose that* $(\mathbf{A},\mathbf{B},\mathbf{C})$ *is a controllable and observable realization of a system. Then there is a state transformation* $\mathbf{T}$ *such that the gramians of the transformed system* $(\tilde{\mathbf{A}},\tilde{\mathbf{B}},\tilde{\mathbf{C}})$ *fulfill*

$$\tilde{\mathbf{X}}_c = \tilde{\mathbf{X}}_o = \mathbf{D}$$

with a positive definite diagonal matrix $\mathbf{D}$.

A state space realization such that the controllability and observability gramians are equal and diagonal is denoted as **balanced realization.** Let $\mathbf{G}(s)$ be a transfer function in $\mathcal{RH}_\infty$. Then its minimal realization will necessarily have a stable matrix $\mathbf{A}$ and, from the previous theorem, we deduce that $\mathbf{G}(s)$ has a balanced realization.

Hence for a balanced realization, the gramians can be written as

$$\mathbf{X}_c = \mathbf{X}_o = \operatorname{diag}(\sigma_1,\ldots,\sigma_n) \text{ with } \sigma_1 \geq \sigma_2 \geq \ldots \geq \sigma_n > 0 . \qquad (13.25)$$

The numbers $\sigma_1,\ldots,\sigma_n$ are denoted as **Hankel singular values.** They are the nonzero eigenvalues of the operator $\Gamma_g^*\Gamma_g$, where $\Gamma_g = P_+ \circ F \circ P_-$ is the Hankel operator associated with the system (cf. (9.11)). The proof of this result can be found in Zhou, Doyle, and Glover [112], Sect. 8.1.

13.6.2 Balanced Truncation

We assume that (13.25) is a balanced realization of a transfer function. Then the gramians can be written in the form (13.25). We try to approximate the system by another system which has a lower order than n. To this end we select a number r such that

$$\sigma_r > \sigma_{r+1}$$

and compatibly partition the realization in the form

$$\mathbf{A} = \begin{pmatrix} \mathbf{A}_{11} & \mathbf{A}_{12} \\ \mathbf{A}_{21} & \mathbf{A}_{22} \end{pmatrix}, \quad \mathbf{B} = \begin{pmatrix} \mathbf{B}_1 \\ \mathbf{B}_2 \end{pmatrix}, \quad \mathbf{C} = \begin{pmatrix} \mathbf{C}_1 & \mathbf{C}_2 \end{pmatrix}, \qquad (13.26)$$

where $\mathbf{A}_{11}$ is an $r \times r$-matrix. Then the lower-order approximation of the original system is defined as

$$\mathbf{G}_r(s) = \left(\begin{array}{c|c} \mathbf{A}_{11} & \mathbf{B}_1 \\ \hline \mathbf{C}_1 & \mathbf{0} \end{array}\right).$$

In the next theorem, some basic properties of this reduced order model are formulated.

Theorem 13.6.2. *The approximation (13.26) of the system (13.24) has the following properties.*

(a) The matrix $\mathbf{A}_{11}$ *is stable.*

(b) The realization $(\mathbf{A}_{11}, \mathbf{B}_1, \mathbf{C}_1)$ *is balanced with Hankel singular values* $\sigma_1, \ldots, \sigma_r$.

Proof. Define

$$\mathbf{D}_1 = \operatorname{diag}(\sigma_1, \ldots, \sigma_r), \quad \mathbf{D}_2 = \operatorname{diag}(\sigma_{r+1}, \ldots, \sigma_n).$$

Then, using the above partition, the Lyapunov equations for the gramians can be written as

$$\begin{pmatrix} \mathbf{A}_{11}^T & \mathbf{A}_{21}^T \\ \mathbf{A}_{12}^T & \mathbf{A}_{22}^T \end{pmatrix}\begin{pmatrix} \mathbf{D}_1 & \mathbf{0} \\ \mathbf{0} & \mathbf{D}_2 \end{pmatrix}+\begin{pmatrix} \mathbf{D}_1 & \mathbf{0} \\ \mathbf{0} & \mathbf{D}_2 \end{pmatrix}\begin{pmatrix} \mathbf{A}_{11} & \mathbf{A}_{12} \\ \mathbf{A}_{21} & \mathbf{A}_{22} \end{pmatrix}+\begin{pmatrix} \mathbf{C}_1^T \\ \mathbf{C}_2^T \end{pmatrix}\begin{pmatrix} \mathbf{C}_1 & \mathbf{C}_2 \end{pmatrix}=\mathbf{0}$$

$$\begin{pmatrix} \mathbf{A}_{11} & \mathbf{A}_{12} \\ \mathbf{A}_{21} & \mathbf{A}_{22} \end{pmatrix}\begin{pmatrix} \mathbf{D}_1 & \mathbf{0} \\ \mathbf{0} & \mathbf{D}_2 \end{pmatrix}+\begin{pmatrix} \mathbf{D}_1 & \mathbf{0} \\ \mathbf{0} & \mathbf{D}_2 \end{pmatrix}\begin{pmatrix} \mathbf{A}_{11}^T & \mathbf{A}_{21}^T \\ \mathbf{A}_{12}^T & \mathbf{A}_{22}^T \end{pmatrix}+\begin{pmatrix} \mathbf{B}_1 \\ \mathbf{B}_2 \end{pmatrix}\begin{pmatrix} \mathbf{B}_1^T & \mathbf{B}_2^T \end{pmatrix}=\mathbf{0}.$$

Hence we obtain

$$\mathbf{A}_{11}^T\mathbf{D}_1 + \mathbf{D}_1\mathbf{A}_{11} + \mathbf{C}_1^T\mathbf{C}_1 = \mathbf{0} \tag{13.27}$$

$$\mathbf{A}_{11}\mathbf{D}_1 + \mathbf{D}_1\mathbf{A}_{11}^T + \mathbf{B}_1\mathbf{B}_1^T = \mathbf{0}. \tag{13.28}$$

Consequently, if $\mathbf{A}_{11}$ is stable, then $\mathbf{D}_1$ is a balanced gramian for the truncated system and (b) is shown. Thus it remains to prove (a).

Let λ be an eigenvalue of $\mathbf{A}_{11}$ and let $\mathbf{V}$ be a matrix whose columns are linearly independent and generate the eigenvector space. Then

$$\operatorname{Ker}(\mathbf{A}_{11} - \lambda\mathbf{I}) = \operatorname{Im}(\mathbf{V}).$$

Multiplying (13.27) on the left by $\mathbf{V}^*$ and on the right by $\mathbf{V}$ yields

$$(\bar{\lambda} + \lambda)\mathbf{V}^*\mathbf{D}_1\mathbf{V} + \mathbf{V}^*\mathbf{C}_1^T\mathbf{C}_1\mathbf{V} = \mathbf{0}.$$

Because of $\mathbf{V}^*\mathbf{D}_1\mathbf{V} > \mathbf{0}$ and $\mathbf{V}^*\mathbf{C}_1^*\mathbf{C}_1\mathbf{V} \geq \mathbf{0}$, we get $\operatorname{Re}\lambda \leq 0$. Thus it remains to exclude the possibility of a purely imaginary eigenvalue.

Let now $\lambda = i\omega$ be an eigenvalue of $\mathbf{A}_{11}$. Then the last equation implies

$$\mathbf{C}_1\mathbf{V} = \mathbf{0}.$$

Multiplication of (13.27) on the right by $\mathbf{V}$ shows that

$$\mathbf{A}_{11}^T(\mathbf{D}_1\mathbf{V}) = -i\omega\,\mathbf{D}_1\mathbf{V}. \tag{13.29}$$

Next, we apply (13.28): Using (13.29), multiplication on the left by $\mathbf{V}^*\mathbf{D}_1$ and multiplication on the right by $\mathbf{D}_1\mathbf{V}$ yields

$$\mathbf{B}_1^T\mathbf{D}_1\mathbf{V} = \mathbf{0}.$$

Then multiplication of (13.28) on the right by $\mathbf{D}_1\mathbf{V}$ and applying again (13.29) yields

$$\mathbf{A}_{11}(\mathbf{D}_1^2\mathbf{V}) = i\omega\,\mathbf{D}_1^2\mathbf{V}.$$

Thus we have

$$\operatorname{Im}(\mathbf{D}_1^2\mathbf{V}) \subset \operatorname{Ker}(\mathbf{A} - i\omega\,\mathbf{I}) = \operatorname{Im}(\mathbf{V}).$$

Hence $\operatorname{Im}(\mathbf{V})$ is an invariant subspace of $\mathbf{D}_1^2$. Thus there must be an eigenvalue $\mu > 0$ of $\mathbf{D}_1^2$ and an eigenvector $\mathbf{v} \in \operatorname{Im}(\mathbf{V})$:

$$\mathbf{D}_1^2\mathbf{v} = \mu\,\mathbf{v}.$$

We also have $\mathbf{A}_{11}\mathbf{v} = i\omega\mathbf{v}$ and from the considerations above

$$\mathbf{C}_1\mathbf{v} = \mathbf{0} \quad \text{and} \quad \mathbf{B}_1^T\mathbf{D}_1\mathbf{v} = \mathbf{0}. \tag{13.30}$$

Next we show that

$$\begin{pmatrix} \mathbf{A}_{11} & \mathbf{A}_{12} \\ \mathbf{A}_{21} & \mathbf{A}_{22} \end{pmatrix}\begin{pmatrix} \mathbf{v} \\ \mathbf{0} \end{pmatrix} = i\omega\begin{pmatrix} \mathbf{v} \\ \mathbf{0} \end{pmatrix}.$$

This contradicts the stability of $\mathbf{A}$, and consequently $\mathbf{A}_{11}$ cannot have a purely complex eigenvalue.

Thus we have to show that $\mathbf{A}_{21}\mathbf{v} = \mathbf{0}$. We consider again the Lyapunov equations for the gramians and multiply them, respectively, by

$$\begin{pmatrix} \mathbf{v} \\ \mathbf{0} \end{pmatrix} \quad \text{and} \quad \begin{pmatrix} \mathbf{D}_1\mathbf{v} \\ \mathbf{0} \end{pmatrix}.$$

Calculating for each equation the second row and using (13.30) leads to

$$\mathbf{A}_{12}^T\mathbf{D}_1\mathbf{v} + \mathbf{D}_2\mathbf{A}_{21}\mathbf{v} = 0$$

$$\mu\,\mathbf{A}_{21}\mathbf{v} + \mathbf{D}_2\mathbf{A}_{12}^T\mathbf{D}_1\mathbf{v} = 0.$$

From these equations we conclude

$$\mathbf{D}_2^2\mathbf{A}_{21}\mathbf{v} = \mu\,\mathbf{A}_{21}\mathbf{v}.$$

Hence μ is an eigenvalue of $\mathbf{D}_1^2$ and $\mathbf{D}_2^2$, which is impossible, since these matrices have no eigenvalue in common by construction.

The above technique of constructing $\mathbf{G}_r$ from $\mathbf{G}$ is called **balanced truncation**. The next task is to estimate the error $\mathbf{G}-\mathbf{G}_r$ in the $\mathcal{H}_\infty$ norm. To formulate the corresponding result, we have to extract the different singular values from the tail

$$\sigma_{r+1} \ge \sigma_{r+2} \ge \ldots \ge \sigma_n .$$

They are denoted by $\sigma_1^t, \sigma_2^t, \ldots, \sigma_k^t$ and also assumed to be ordered:

$$\sigma_1^t > \sigma_2^t > \ldots > \sigma_k^t .$$

Theorem 13.6.3. *Let the system $\mathbf{G}_r$ be of order r and obtained from the stable system $\mathbf{G}$ by balanced truncation. Then, with the notation introduced above, the following estimate holds:*

$$\| \mathbf{G}-\mathbf{G}_r \|_\infty \le 2(\sigma_1^t + \cdots + \sigma_k^t).$$

For a proof of this result, we refer the reader to Dullerud and Paganini [37], Sect. 4.6.4. Compare also Zhou, Doyle, and Glover [112], Chap. 8, where the problem of balanced model reduction is analyzed in detail.

Application. Balanced truncation may be applied to any stable system, for the plant as well as for the controller, but we are mainly interested in using it for the controller. A $\mathcal{H}_\infty$ or μ optimal controller must not necessarily be stable, so that balanced truncation can possibly not directly be applied. If the controller $\mathbf{K}$ is not stable, a reasonable approach is to decompose it into a sum $\mathbf{K} = \mathbf{K}_+ + \mathbf{K}_-$, where $\mathbf{K}_+$ has all its poles in the open left half-plane and $\mathbf{K}_-$ has all its poles in the closed right half-plane. Then a balanced realization for $\mathbf{K}_+$ is calculated. It yields the Hankel singular values as in (13.25). If there is a singular value σ_r such that $\sigma_r \gg \sigma_{r+1}$, one approximates $\mathbf{K}_+$ by $\mathbf{K}_{+a}$, where $\mathbf{K}_{+a}$ is the truncated system as described in Theorem 13.6.3 and uses $\mathbf{K}_a = \mathbf{K}_{+a} + \mathbf{K}_-$ as a new controller. It must be emphasized that with $\mathbf{K}_a$ the feedback system may have reduced robust stability and performance properties; it may even be unstable. For this reason it has to be checked whether with this new controller the stability and performance properties are not degraded too much.

13.7 Robust Control of a Second-Order Plant

In Sect. 10.5, we developed $\mathcal{H}_\infty$ controllers for several second-order plants. Although most of them where robust in classical terms, they are not necessarily robust in the sense of this chapter since no uncertainty structure of the plant was specified. We design now a robust controller for the second-order plant with the uncertainty structure defined in Sect. 12.5 and compare it with a PIDT$_1$ controller. Performance is specified by an S/KS weighting scheme as in Sect. 10.5. Thus the extended plant has the structure depicted in Fig. 13.8. We adopt the numerical values defined in Sect. 12.5 for the plant and the uncertainty weights. The numerical values for the performance weights still have to be specified.

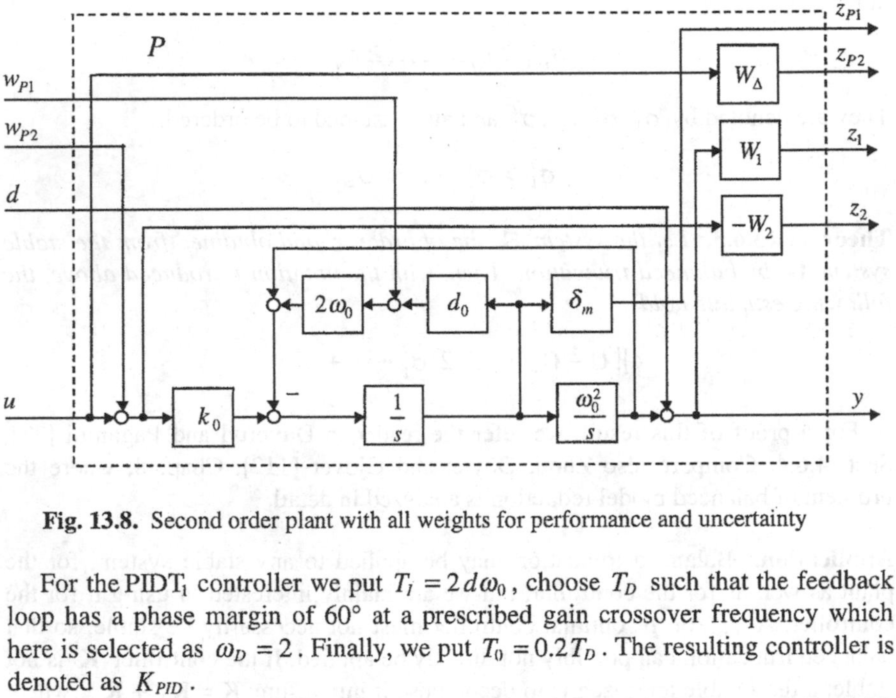

Fig. 13.8. Second order plant with all weights for performance and uncertainty

For the PIDT$_1$ controller we put $T_I = 2d\omega_0$, choose T_D such that the feedback loop has a phase margin of 60° at a prescribed gain crossover frequency which here is selected as $\omega_D = 2$. Finally, we put $T_0 = 0.2T_D$. The resulting controller is denoted as K_{PID}.

At first, we want to analyze the stability properties of the PIDT$_1$ controller with respect to the uncertainty structure defined in Sect. 12.5. To this end, the uncertain plant of Fig.12.13 has to be closed with the PIDT$_1$ controller and the structured singular value μ has to be calculated for this feedback connection (cf. Fig. 13.9). Fig. 13.10 shows the corresponding μ-plot. The maximum value is approximately $\mu_{\max} = 1.03$, thus robust stability is almost achieved for K_{PID}. Due to Theorem 13.4.1, the feedback loop is stable for perturbations such that $\|\Delta\|_\infty < 0.971$. Using μ tools, it is possible to calculate a destabilizing perturbation. For this perturbation we get (with $\Delta = \mathrm{diag}(\delta_1, \delta_2)$)

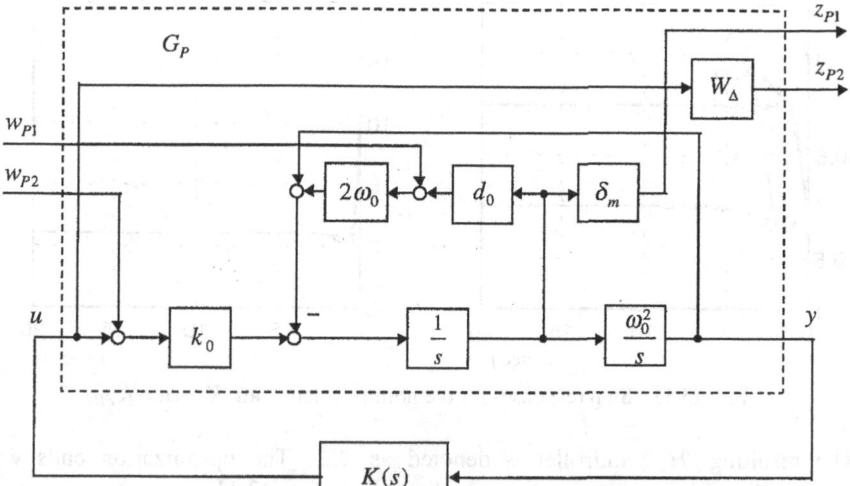

Fig. 13.9. Analysis of robust stability for a given controller

$$\delta_1 = -0.958, \quad 1 + W_\Delta(s)\delta_2(s) = \frac{2.92\,(s^2 - 36.5\,s + 974)}{(s + 142.8)\,(s + 22)}. \tag{13.31}$$

The next step is to find performance specifications such that the resulting $\mathcal{H}_\infty$ controller is comparable to the PIDT_1 controller. We use a dynamical weight W_2 and put (with the notations of Sect. 10.5.1)

$$\omega_1 = 2, \; S_\infty = 1.5, \; \omega_\varepsilon = 0.01\omega_1, \; k_2 = 50, \; \omega_2 = 6.$$

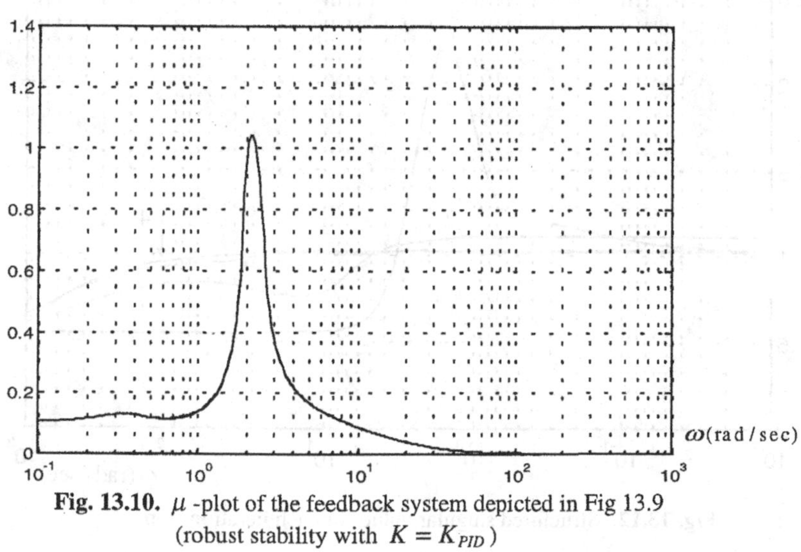

Fig. 13.10. μ -plot of the feedback system depicted in Fig 13.9 (robust stability with $K = K_{PID}$)

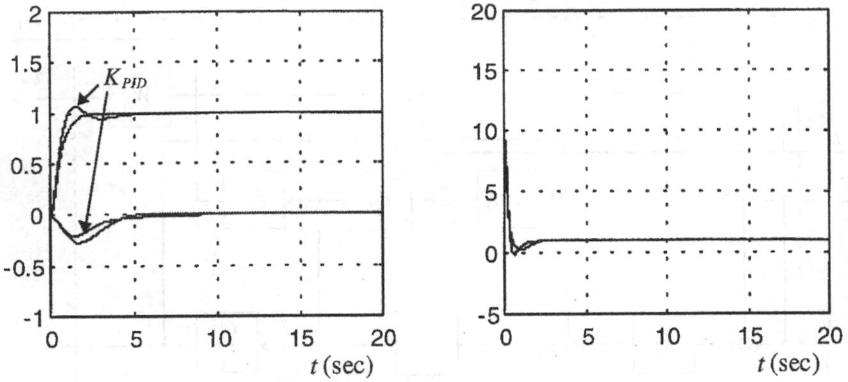

Fig. 13.11. Step responses for the nominal plant with K_∞ and K_{PID}

The resulting $\mathcal{H}_\infty$ controller is denoted as K_∞. The optimization ends with $\gamma_{\text{opt}} = 0.89$ and the step responses are depicted in Fig. 13.11.

After this basic analysis we are in a position to synthesize a robust controller by using the D-K iteration. The D-K iteration gives after 3 steps a maximum value of $\mu_{\max} = 1.367$ for μ (then, no further improvement is achieved). The structured singular value now has to be taken with respect to $\mathbf{M}$ and the given robustness and performance structure (cf. Theorem 13.4.2). Thus robust performance is not achievable for this specification. Since we suppose the uncertainty weights not to be at our disposition, the performance requirements have to be relaxed. Let

$$\omega_1 = 1.2, \; S_\infty = 1.8, \; k_2 = 100$$

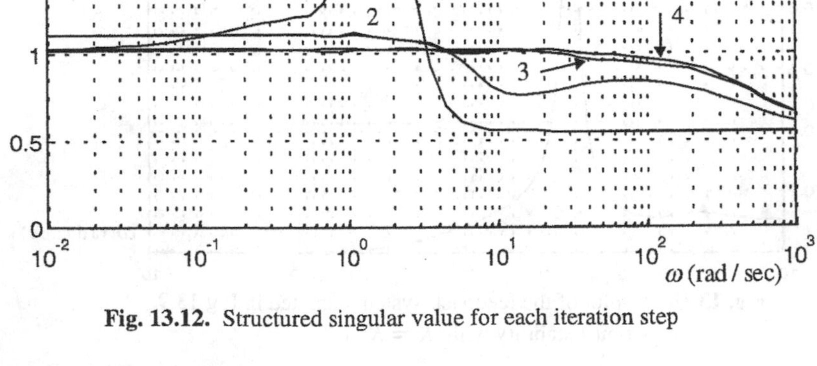

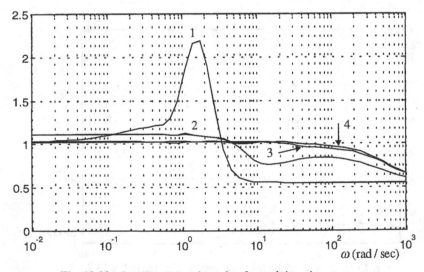

Fig. 13.12. Structured singular value for each iteration step

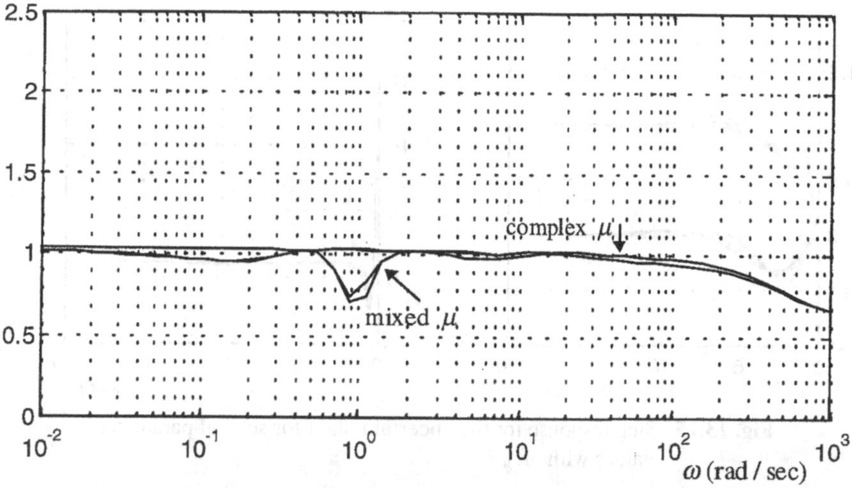

Fig. 13.13. Structured singular value (complex and mixed) for iteration step 4

and leave all other numerical values unchanged. The maximum values for μ are now $\mu_{\max} =$ 2.193, 1.120, 1.034, 1.028. Thus robust performance is now (almost) attained. The dependency of μ on the frequency ω can be seen from Fig. 13.12.

It should be noted that for our system we have a scalar real block, a scalar complex block and a full complex block (which is caused by the performance weights). Thus the structured singular value has to be calculated with respect to a mixed real and complex structure. In Fig. 13.12, the real block is treated as a scalar complex block. This is automatically done by the D-K iteration. It also calculates μ with respect to the real block. Fig. 13.13 shows the complex and mixed μ for the last iteration step. The maximum value is in both cases the same. The resulting controller is denoted by K_μ.

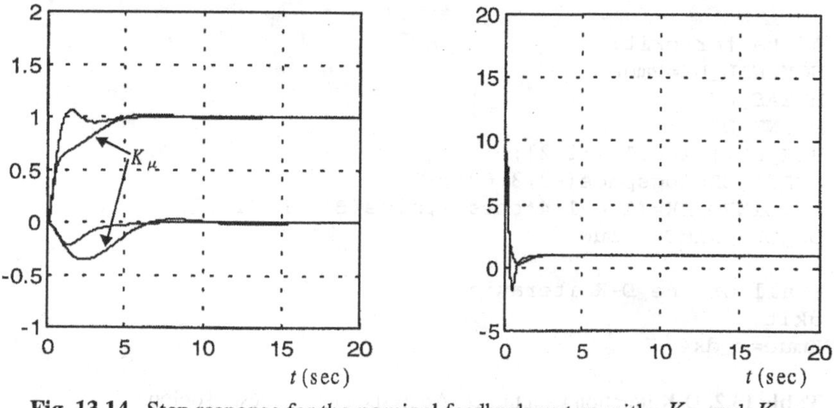

Fig. 13.14. Step response for the nominal feedback system with K_μ and K_{PID}

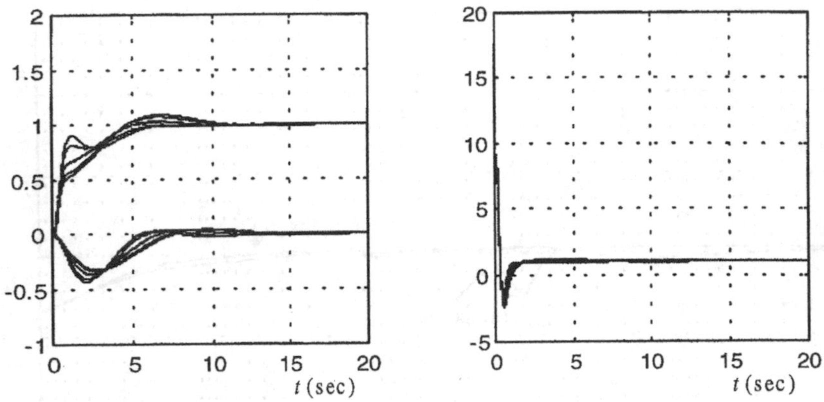

Fig. 13.15. Step response for the uncertain plant for several parameter values with K_μ

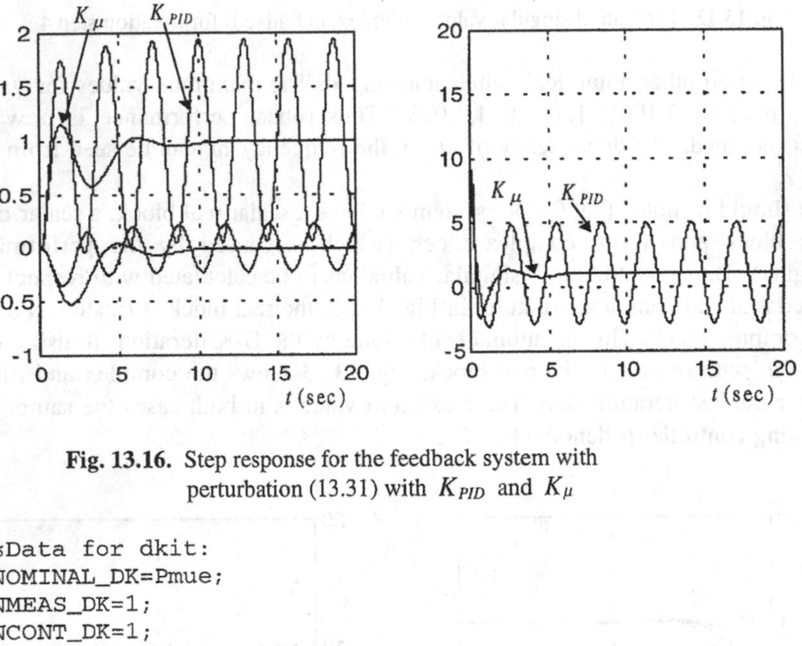

Fig. 13.16. Step response for the feedback system with perturbation (13.31) with K_{PID} and K_μ

```
%Data for dkit:
NOMINAL_DK=Pmue;
NMEAS_DK=1;
NCONT_DK=1;
BLK_DK=[-1 0;1 1;1 2];
OMEGA_DK=logspace(-3,3,60);
AUTOINFO_DK=[1 4 1 4*ones(1,size(BLK_DK,1))];
DK_DEF_NAME='Pmue';

%Call of the D-K iteration:
dkit
Kmue=k_dk4;
```

Table 13.2. D-K iteration with the μ -Analysis and Synthesis Toolbox

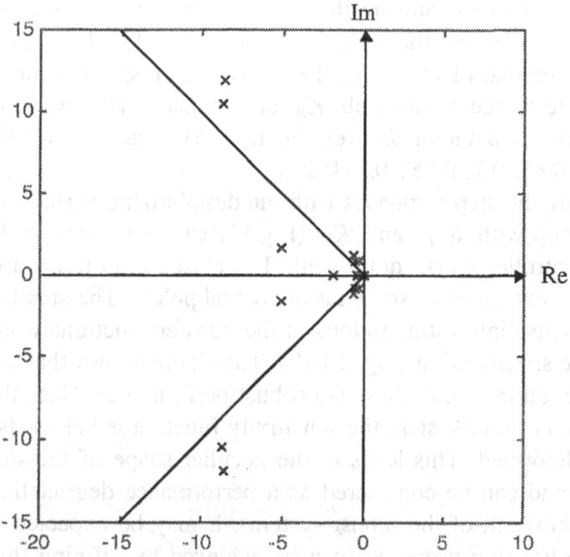

Fig. 13.17. Poles of the feedback system with the perturbation (13.31) and the controller K_μ

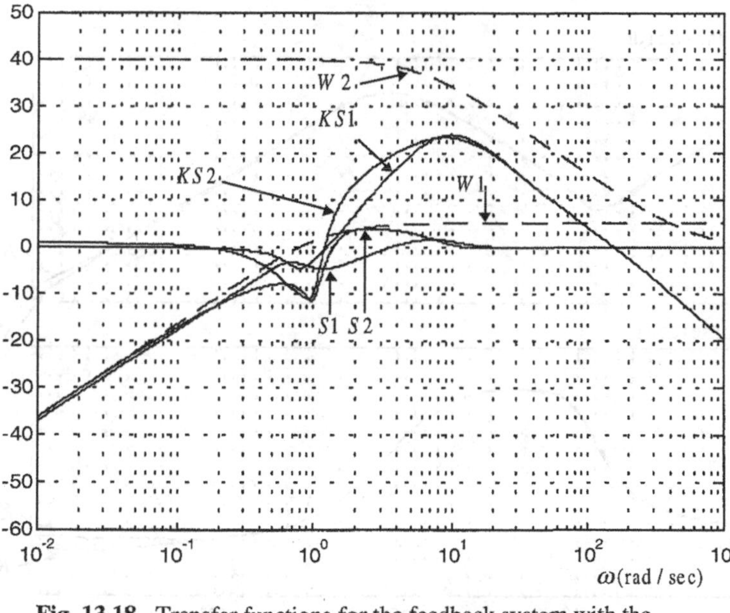

Fig. 13.18. Transfer functions for the feedback system with the perturbation (13.31) and the controller K_μ

$W1$: Weight W_1; $W2$: Weight W_2;
$S1$: Sensitivity function S; $KS1$: transfer function KS (nominal plant);
$S2$: Sensitivity function S; $KS2$: transfer function KS (perturbed plant)

Next we are interested to compare the step responses for the feedback system with the $PIDT_1$ controller and the robust controller. Fig. 13.14 shows the simulation result for the nominal plant. In Fig. 13.15, the step responses for the nominal plant and several perturbed plants with K_μ are depicted. The dynamic perturbation is the multiplicative factor h_2 (cf. Sect. 12.5) with $T_{P2} = 0.035$ and the dampings are $d = 0.85,\ 0.7,\ 0.35\ ,\ 0,\ -0.15$.

We now compare the step responses with the destabilizing perturbation (13.31) for the feedback loop with K_{PID} and K_μ (Fig.13.16). As is seen, in this extreme case the robust controller works quite well. The closed-loop poles are shown in Fig. 13.17 (with the exception of some far-away real poles). The smallest damping is $d = 0.371$. Its is also interesting to look at the transfer functions which are used in the performance specification (Fig. 13.18). The figure shows that the feedback system with perturbation (13.31) has also robust performance. Near the eigenfrequency of the second-order system the sensitivity function is below its bound but looks somewhat deformed. This leads to the peculiar shape of the step response for the command and can be considered as a performance degradation (together with the lower bandwidth of the robust system). It may be expected that an improvement for the tracking properties can be achieved by refining the approach, for example by using a two degrees-of-freedom feedback configuration, but we do not want to investigate this further.

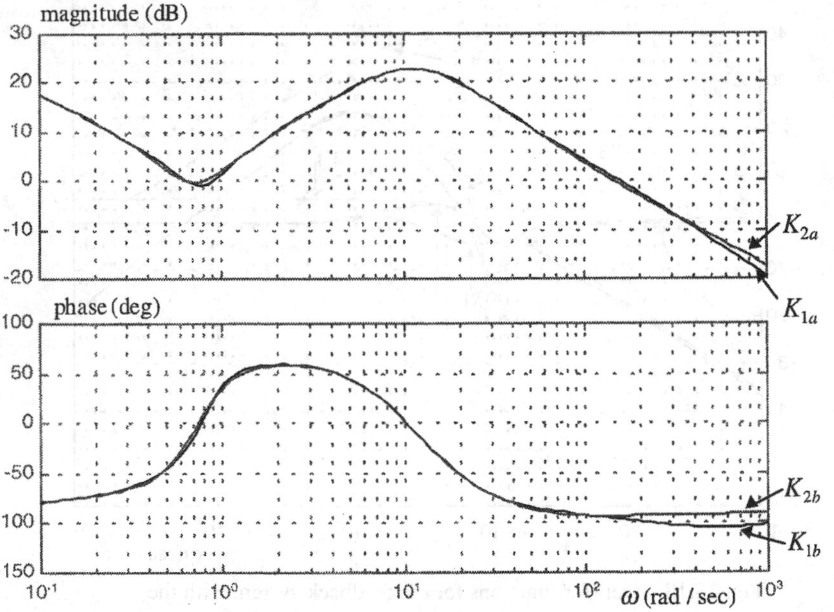

Fig. 13.19. Bode plot of K_μ and $K_{\mu\,\mathrm{red}}$
K_{1a} (K_{2a}) magnitude plot of K_μ ($K_{\mu\,\mathrm{red}}$)
K_{1b} (K_{2b}) phase plot of K_μ ($K_{\mu\,\mathrm{red}}$)

Finally we show the heart of the MATLAB code which is necessary to call the D-K iteration (cf. Table 13.2). The main command is "`dkit`". The procedure is interactive but may alternatively run automatically. Some care has to be taken to provide the iteration with the correct structure of the uncertainty set, which is done by the variable BLK_DK.

The order of K_μ is 21, so that it is desirable to approximate the controller by one with a lower order. We use the method of balanced truncation introduced in Sect. 13.6. The Hankel singular values of K_μ are

$$\sigma_1,\ldots,\sigma_{13} = 298.7091, 7.2942, 6.5097, 0.5909, 0.3506, 0.1926, 0.0612, 0.0568, 0.0555, 0.0417, 0.0197, 0.0181, 0.0072\,;$$

$$\sigma_{21} = 3.9555\times 10^{-7}\,.$$

These numbers suggest to choose a controller approximation with $r=6$. We denote the resulting controller of order 6 by $K_{\mu\,\mathrm{red}}$. Fig. 13.19 shows the Bode plots of K_μ and $K_{\mu\,\mathrm{red}}$. As we can see, the differences in magnitude and phase are only minimal. The same holds for the performance and robustness properties for the system with this controller: they are nearly the same as with the original controller and therefore are not shown here. In this way, we have found a 6th order controller such that the feedback system has robust stability as well as robust performance.

Notes and References

The idea of formulating uncertainty in terms of Δ blocks as described in Chapter 12 appeared in Doyle [32]. The small gain theorem was first shown in Zames [111]. Robust stability is discussed in Doyle, Wall, and Stein [33]. In this paper it is also shown that robust performance can be reduced to robust stability by adding an additional uncertainty block. The structured singular value μ was also introduced in Doyle [32]. A similar quantity is the multivariable stability margin defined by Safonov [78] (it is the reciprocal of μ). The idea of synthesizing μ optimal controllers having robust performance by the D-K iteration appeared in Doyle [34]. Upper and lower bounds for the mixed real/complex structured singular value were derived in Fan, Tits, and Doyle [39] and Young, Newlin, and Doyle [110]. A summary of the research concerning μ can be found in Packard and Doyle [72]. It has to be emphasized that such bounds for the structured singular value are implemented in the MATLAB μ-Analysis and Synthesis Toolbox [7] to calculate μ and to perform the D-K iteration.

Chapter 14

Case Studies for Robust Control

The last chapter of this book is completely devoted to case studies for robust control. The first study continues the controller design for the lateral motion of an aircraft. We introduce different weightings as in Sect. 11.3.5 and obtain in this way a feedback loop with good damping properties without prestabilization. Then a block which models multidimensional input-multiplicative uncertainty is defined and, by applying the D-K iteration, a μ controller is synthesized which improves the robustness properties of the $\mathcal{H}_\infty$ controller as well as that of the LQ controller.

The next example is concerned with the rudder-roll stabilization of ships. As for aircraft, the roll motion of ships is badly damped and this property is particularly unfavorable if waves attack the ship with a frequency that is close to the eigenfrequency of the roll motion. The task is to improve the frequency response of the wave disturbance solely by the rudder and without severely degrading the course-keeping properties of the ship. This is actually achievable in a robust manner but at the price of a higher rudder dynamics in comparison to that in the case of a pure course controller. The uncertainty model comprises input-multiplicative uncertainty and two uncertain critical parameters related to the roll motion, which are only known to lie in a given interval.

The design of a missile autopilot is another challenging control problem, which by now is already classical. For the missile under consideration here, due to its aerodynamical configuration, control of the lateral acceleration is essentially a second-order SISO system, but this has some properties which make the problem difficult. Above all, this is the strong nonlinearity of the missile dynamics. For a full acceleration command there is no valid linear model. Thus the controller must be capable of working for a set of linear models whose eigenvalues lie in a large region of the complex plane. We model the nonlinearity by a suitable parameter uncertainty. Besides this, the limited bandwidth of the actuator and the elastic behavior strongly influence the controller design. As we will see, μ synthesis provides an efficient tool to design a controller which works very well under all these constraints.

Our last case study is concerned with the control of a distillation column, which is again a MIMO problem. As in the previous examples, we make a considerable effort to derive the plant model, which is rather complicated in this case. After some simplifications, there remains a system with a 2×2-state matrix and two inputs and two outputs as the very heart of the problem. This seems to be simple, but is not if high purity distillation is considered because in this case the plant is very ill-conditioned. We start our analysis by designing two different controllers by classical methods. It is shown that they do not lead to satisfactory results since their performance severely degrades, even for small perturbations. This is much better for a μ controller, which is designed with respect to robust performance and actually works as promised by theory.

14.1 Robust Control of Aircraft

14.1.1 $\mathcal{H}_\infty$ Controller for Lateral Control with Good Disturbance Rejection

Our first $\mathcal{H}_\infty$ controller for the lateral motion of an aircraft was based on the mixed-sensitivity approach (Sect. 11.3.5). Due to the pole-zero cancellation inherent in this procedure, the corresponding feedback loop contains the badly damped dutch roll mode, which causes an oscillatory disturbance rejection. To obtain a $\mathcal{H}_\infty$ controller with a better disturbance rejection, we than combined classical prestabilization with the mixed-sensitivity approach, which was now applied to the prestabilized plant. At least from a conceptual point of view, this is not quite satisfactory. Therefore we now present a second approach to obtain a feedback system having only well-damped eigenvalues, which is a pure $\mathcal{H}_\infty$ design and which incorporates additional performance specifications and a model of the ac-

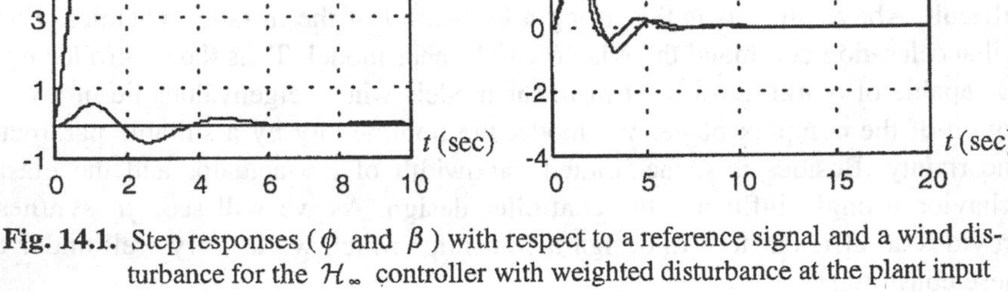

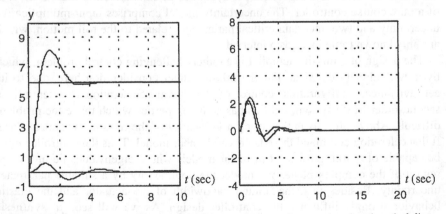

Fig. 14.1. Step responses (ϕ and β) with respect to a reference signal and a wind disturbance for the $\mathcal{H}_\infty$ controller with weighted disturbance at the plant input

tuator. We proceed step by step so that the role of all the weightings becomes clear to the reader. After this, uncertainty is introduced to obtain a robust controller.

The prestabilization is not necessary if a suitable disturbance is introduced in the performance specification of the $\mathcal{H}_\infty$ controller. To be more specific, a fictitious perturbation $\mathbf{d}$ passing a weight $\mathbf{W}_d$ is added to the control. We choose $\mathbf{W}_d = \mathrm{diag}(2 \quad 2)$ and let the weights $\mathbf{W}_1$ and $\mathbf{W}_3$ be exactly as in Sect. 11.3.5.

Figure 14.1 shows step responses for the resulting controller. The step response with respect to the wind disturbance is now rather well damped with a minimum of $d = 0.4$, but the reference signal is tracked only with a large undesired overshoot. Moreover, the sideslip angle does not remain zero during the turn. Thus the controller is not able to guarantee good tracking properties and a good disturbance rejection. The minimum of the $\mathcal{H}_\infty$ optimization is $\gamma = 2.40$, which shows that this specification cannot be exactly fulfilled.

As in comparable classical situations, a possible remedy is given by a *two degrees-of-freedom control configuration*. It has the distinguishing feature that the reference signal enters directly into the controller and not in the form of the control error. Consequently, the controller consists of two blocks, which normally are written as one single block. Figure 14.2 shows the block diagram belonging to this approach. The weights $\mathbf{W}_n$ and $\mathbf{W}_u$ are necessary to fulfill the rank condition. Both are chosen as small constants and their effect on the design is minimal. For this reason, the difference between this and the previous approach is mainly due to the two degrees-of-freedom configuration. In Fig. 14.2, $\mathbf{G}(s)$ denotes the transfer function of the airframe, where $\mathbf{u} = (\zeta \quad \xi)^T$ again is the control and $\mathbf{y} = (\beta \quad \phi)^T$ the variable to be controlled. The reference signal is given by $\mathbf{r} = (\phi_s \quad 0)^T$. The values for the weights are left as before.

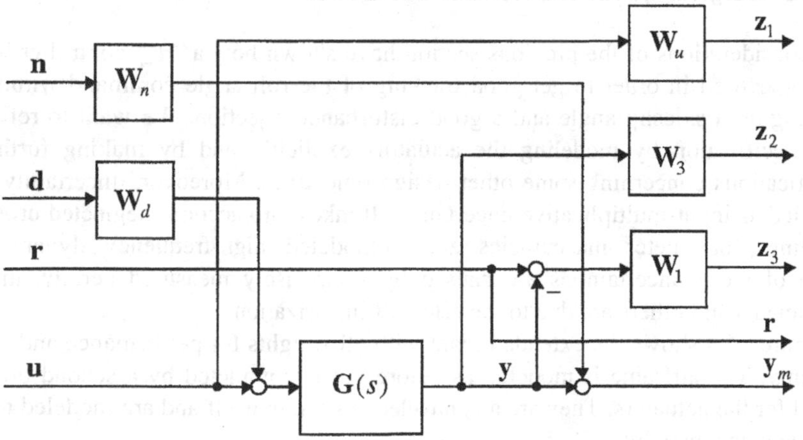

Fig. 14.2. $\mathcal{H}_\infty$ controller design for a two degrees-of-freedom control configuration and a weighted disturbance at the plant input

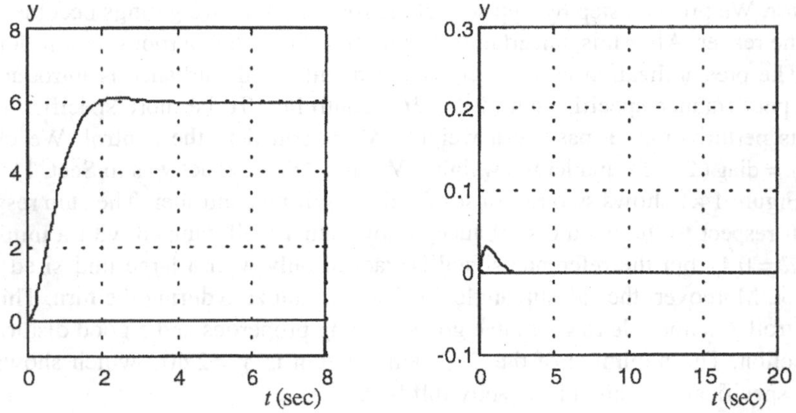

Fig. 14.3. Step responses (ϕ and β) for the lateral motion and weightings according to Fig. 14.2.

Figure 14.3 shows the step responses. The reference signal is tracked exactly as in the mixed-sensitivity approach (cf. Fig. 11.31) and the disturbance rejection is almost perfect. It should be noted that the actuators not have been modeled up to now and it has to be expected that they will lead to a moderate degradation of the performance for the real system. The poles of the closed-loop system have $d = 0.7$ as the minimum value for the damping, and the $\mathcal{H}_\infty$ optimization ends with $\gamma = 1.09$. Hence this controller synthesis can be viewed as successful with respect to our design goals.

14.1.2 Weightings for the Robust Controller

The considerations of the previous section have shown how a $\mathcal{H}_\infty$ controller has to be specified in order to get good tracking of the roll angle command without building up a sideslip angle and a good disturbance rejection. We want to refine this specification by modeling the actuators explicitly and by making further specifications concerning some other design objectives. Moreover, uncertainty is modeled as input-multiplicative uncertainty. It takes into account neglected cross-couplings, parameter inaccuracies and unmodeled high-frequency dynamics. Some of these uncertainties are caused by imprecisely measured aerodynamic coefficients and others are due to the effect of linearization.

Figure 14.4 shows the extended plant with all weights for performance and uncertainty. The airframe is modeled as before but is completed by a second-order model for the actuators. They are a controlled system in itself and are modeled to a sufficient accuracy by

$$G_A(s) = \frac{\omega_A^2}{s^2 + \sqrt{2}\omega_A s + \omega_A^2 s^2} \quad \text{with } \omega_A = 40 s^{-1}.$$

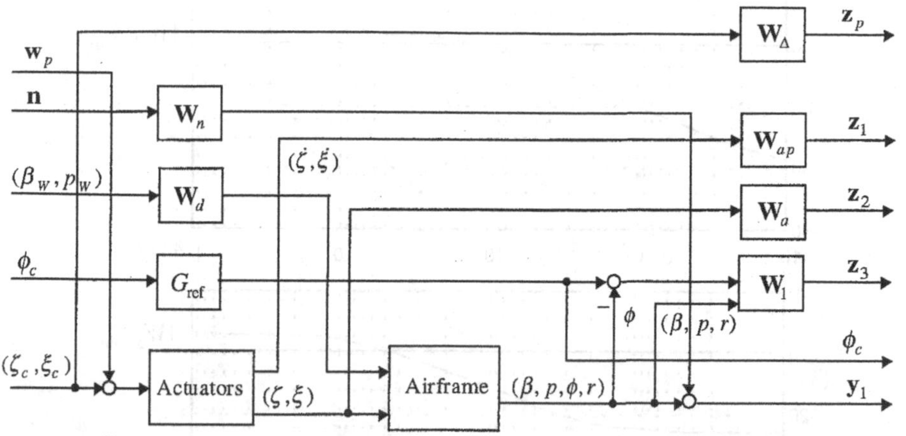

Fig. 14.4. Generalized plant model for the aircraft with all weights for performance and uncertainty

In contrast to the previous example, the controller gets the full state vector for the lateral motion as measurement. This is a realistic assumption for an aircraft since all these quantities are measured by an inertial measurement unit. Possibly it could be necessary to feed back the lateral acceleration instead of the sideslip angle, which would cause no severe problem.

It is important that the performance specification takes the fin angle and the fin rate into account. This will be done by the constant weights

$$\mathbf{W}_a = \mathrm{diag}\,(1/\zeta_m, 1/\xi_m), \quad \mathbf{W}_{ap} = \mathrm{diag}\,(1/\zeta_{pm}, 1/\xi_{pm})$$
$$\text{with} \quad \zeta_m = \xi_m = 20°, \quad \zeta_{pm} = \xi_{pm} = 60°/\sec.$$

The weight

$$\mathbf{W}_1 = \mathrm{diag}\,(W_{11}, W_{11}, W_{12}, W_{12}),$$

$$W_{1j} = \frac{1}{S_{\infty j}}\frac{s+\omega_1}{s+\omega_{2j}} \quad \text{with } S_{\infty 1} = 1.5,\ S_{\infty 2} = 10,\ \omega_1 = 1.5,\ \omega_{21} = 0.01,\ \omega_{22} = 2$$

specifies the performance (cf. Fig. 14.4). Note that $\mathbf{W}_1$ has the control error $\phi_s - \phi$ as input whereas the reference signal ϕ_s enters into the controller.

Moreover, the reference signal has to be tracked in a specific way which is described by the transfer function G_{ref}. In particular, such a specification is made if the dynamic properties of the aircraft have to be improved. Then G_{ref} describes the desired handling qualities of the aircraft. For this aircraft we choose

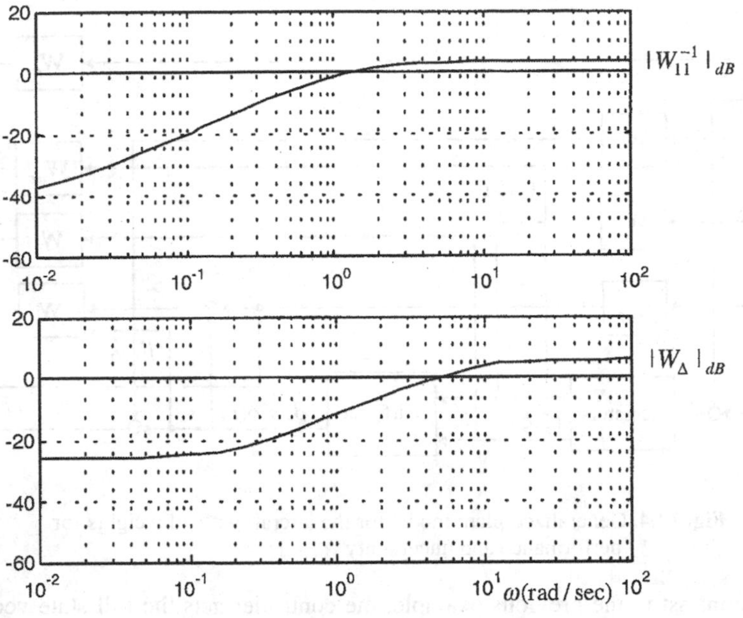

Fig. 14.5. Magnitude plot of W_{11}^{-1} and of W_Δ

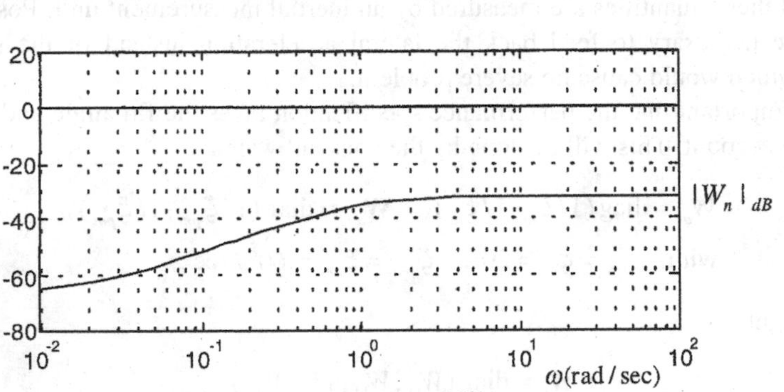

Fig. 14.6. Magnitude plot of W_n

$$G_{\text{ref}}(s) = \frac{\omega_{\text{ref}}^2}{s^2 + \sqrt{2}\,\omega_{\text{ref}}\, s + \omega_{\text{ref}}^2\, s^2} \qquad \text{with } \omega_{\text{ref}} = 1.5\,\text{sec}^{-1}.$$

The weight $\mathbf{W}_d$ for the wind disturbance is chosen as $\mathbf{W}_d = \text{diag}(1, 1)$. Sensor noise is frequency dependent and taken into account by the weight

$$\mathbf{W}_n = \text{diag}(W_n, W_n, W_n, W_n) \quad \text{with} \quad W_n = k_n \frac{\omega_{n2}}{\omega_{n1}} \frac{s + \omega_{n1}}{s + \omega_{n2}},$$

$$k_n = 0.0005, \quad \omega_{n1} = 0.02, \quad \omega_{n2} = 1$$

(cf. Fig. 14.6). As already motivated by the explanation given above, uncertainty is modeled as input-multiplicative uncertainty. The uncertain transfer function Δ is assumed to be a full complex 2×2 block. It is weighted by the diagonal matrix

$$\mathbf{W}_{\Delta} = \mathrm{diag}(W_{\Delta}, W_{\Delta}) \text{ with } W_{\Delta} = W_{\infty} \frac{s + \omega^* k}{s + \omega^* W_{\infty}}, W_{\infty} = 2, k = 0.05, \omega^* = 4 .$$

Thus for low frequencies a value of 5% and for high frequencies a value of up to 200% inaccuracy is allowed. A deviation of 100% is attained approximately at $\omega = 4$ (cf. Fig. 14.5).

14.1.3 D-K Iteration and Simulation Results

Before starting the D-K iteration, we design a $\mathcal{H}_{\infty}$ controller for the weighted plant depicted in Fig. 14.4, where we do not take into consideration the weighting for the uncertainty. The $\mathcal{H}_{\infty}$ norm for the suboptimal controller is $\gamma = 0.356$. Thus nominal performance can be easily achieved for this specification. Figure 14.7 shows the step responses. The roll angle follows the ideal prescribed course (dashed line) almost exactly and the disturbance rejection is even better as for the LQ controller (cf. Fig. 11.29).

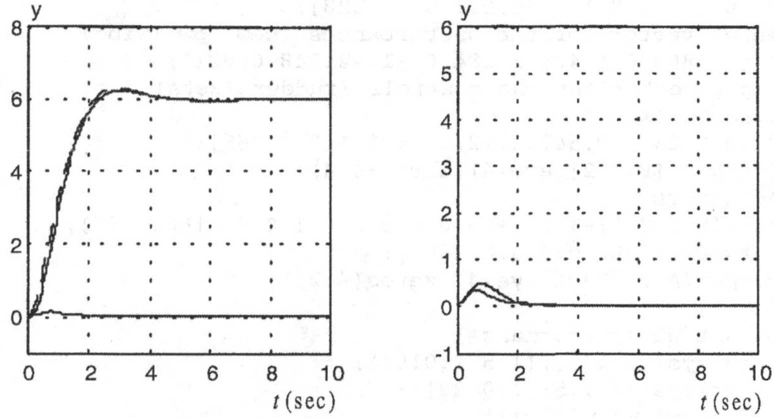

Fig. 14.7. Step responses (ϕ and β) for the lateral motion and weightings according to Fig. 14.4.

A robust controller is obtained by applying the D-K iteration to the extended plant of Fig. 14.4, now of course with the weighting $\mathbf{W}_{\Delta}$ for uncertainty. The necessary MATLAB program is given below in Table 14.1. It is seen that the necessary programming effort even for this rather complex problem is not very high. In the first part of the program, the linear models for the airframe, the actuators and the weights are built. Then the interconnection structure according to Fig. 14.4 is implemented. This can be followed exactly from line to line. The next step is to supply the interactive D-K iteration of the μ - *Analysis and Synthesis Toolbox* with data. A certain care is needed to enter the correct data for the block structure, which represents uncertainty and robustness. The necessary command is

```
BLK_DK=[2 2;7 8];
```

Note that with regard to performance the generalized plant has 7 inputs and 8 outputs. Concerning robustness, the number of inputs and outputs is 2. Since the uncertainty Δ is a full complex 2×2 - block, the parameter pair 2 2 has to be used.

The first "4" in `AUTOINFO_DK` gives the desired number of D-K iterations. Further details with respect to the parameters can be found in the user guide. The iteration is started by the command `dkit`. In each iteration step, the D-K iteration generates a controller which in the 4th step has the name `k_dk4`. The feedback system is built with this controller and can be used for simulation purposes.

```
%Design of a robust controller
%for the lateral motion of an aircraft

%Airframe
%System matrix (state vector: be,p,phi,r)
A=[-0.0869 0 0.039 -1.0;-4.424 -1.184 0 0.335;
    0 1 0 0;2.148 -0.021 0 -0.228];
%Input vector for the disturbances (bew, pw (wind))
B1=[0.0869 0;4.424 1.184;0 -1;-2.148 0.021];
%Input vector for the controls (rudder (zeta),
 aileron (xsi))
B2=[0.0223 0;0.547 2.12;0 0;-1.169 0.065];
AC=pck(A,[B1 B2],eye(4),zeros(4,4));
%Actuators
Aact=[0 1 0 0;-1600 -56 0 0;0 0 0 1;0 0 -1600 -56];
Bact=[0 0;1600 0;0 0;0 1600];
Act=pck(Aact,Bact,eye(4),zeros(4,2));

%Weight W1 (performance)
w1=nd2sys([1 1.5],[1.5 0.015]);
w11=nd2sys([1 1.5],[10 20]);
W1=daug(w1,w1,w11,w11);
%Weights W2, W2p (control)
W2=daug([1/20],[1/20]);
W2p=daug([1/60],[1/60]);
%Weight Wd (disturbance)
Wd=daug([1.0],[1.0]);
%Weight Gref (tracking)
Gref=nd2sys([2.25],[1 2.1 2.25]);
%Weight Wn (noise)
wn=nd2sys([1/0.02 1],[1/1 1],0.0005);
Wn=daug(wn,wn,wn,wn);
%Weight for robustness
```

```
wdel=nd2sys([0.25 0.05],[0.125 1]);
Wdel=daug(wdel,wdel);

%Generalized plant for mue synthesis
systemnames='AC Act Wdel W1 W2 W2p Wd Wn Gref';
inputvar='[wp(2);r;n(4);d{2};uc{2}]';
outputvar='[Wdel;W1;W2;W2p;r;AC+Wn]';
input_to_AC='[Wd;Act(1);Act(3)]';
input_to_Act='[uc+wp]';
input_to_Gref='[r]';
input_to_W1='[AC(3)-Gref;AC(1);AC(2);AC(4)]';
input_to_W2='[uc]';
input_to_W2p='[Act(2);Act(4)]';
input_to_Wd='[d]';
input_to_Wn='[n]';
input_to_Wdel='[uc]';
sysoutname='P';
cleanupsysic='yes';
sysic;

%Data for "dkit"
NOMINAL_DK=P;
NMEAS_DK=5;
NCONT_DK=2;
BLK_DK=[2 2;7 8];
OMEGA_DK=logspace(-3,3,60);
AUTOINFO_DK=[1 4 1 4*ones(1,size(BLK_DK,1))];
DK_DEF_NAME='P';

%D-K iteration
dkit
%Choice of the controller
Kmue=k_dk4;

%Closed feedback loop:
systemnames='[AC Act Kmue]';
inputvar='[r;d(2)]';
outputvar='[AC]';
input_to_AC='[d;Act(1);Act(3)]';
input_to_Act='[Kmue]';
input_to_Kmue='[r;AC]';
sysoutname='Fs';
cleanupsysic='yes';
sysic;
[ar,br,cr,dr]=unpck(Fs);
F=ss(ar,br,cr,dr);
```

Tab. 14.1. MATLAB program for the synthesis of a robust controller for the lateral motion of an aircraft by means of the D-K iteration

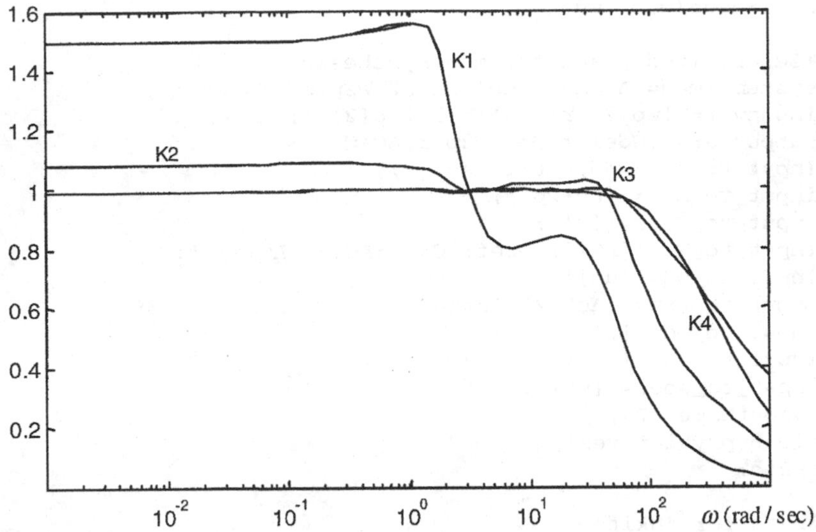

Fig. 14.8. μ plot for each iteration step

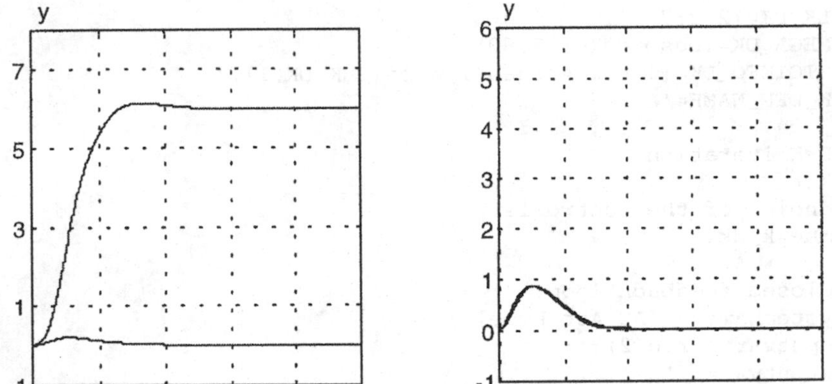

Fig. 14.9. Step responses for the lateral motion with the μ controller

The result of the D-K iteration is depicted in Fig. 14.8. For the optimal μ the values are 1.554, 1.084, 0.995, 0.993, i.e. the iteration attains the optimum essentially after 3 steps. The orders of the controllers are 20, 28, 36, 36. As we see from these results, robust stability and robust performance can be achieved for these weightings.

Figure 14.9 shows the step responses for the robust controller. It is only minimally worse than that for the nominal plant with the $\mathcal{H}_\infty$ controller. The fin angle and the fin rate remain within the prescribed bounds (Fig. 14.10).

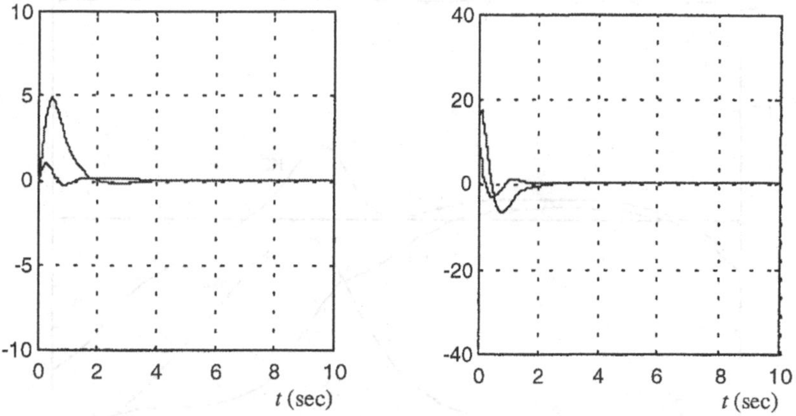

Fig. 14.10. Fin angles (left plots) and fin rates (right plots) belonging to the step responses

14.1.4 Comparison with the LQ Controller

In this section, we want to compare the robust controller with the LQ controller of Sect. 11.3.4. It has the form

$$\begin{pmatrix} \zeta_c \\ \xi_c \end{pmatrix} = (\mathbf{F}_1 \quad \mathbf{F}_2)\,\mathbf{y}_{LQ} \quad \text{with} \quad \mathbf{y}_{LQ} = \begin{pmatrix} \mathbf{x} \\ h_1 \\ h_2 \end{pmatrix}, \quad \mathbf{x} = (\beta \quad p \quad \phi \quad r)^T .$$

Here, $\mathbf{F}_1$ and $\mathbf{F}_2$ are 2×4- and 2×2- matrices, respectively, and h_1, h_2 are the integrals over the control errors. The controller can be written as follows:

$$\begin{pmatrix} \dot{h}_1 \\ \dot{h}_2 \end{pmatrix} = \begin{pmatrix} 0 & 0 \\ 0 & 0 \end{pmatrix} \begin{pmatrix} h_1 \\ h_2 \end{pmatrix} + \begin{pmatrix} 0 & -1 & 0 & 0 & 0 \\ 1 & 0 & 0 & -1 & 0 \end{pmatrix} \begin{pmatrix} \phi_c \\ \mathbf{x} \end{pmatrix}$$

$$\begin{pmatrix} \zeta_c \\ \xi_c \end{pmatrix} = \mathbf{F}_2 \begin{pmatrix} h_1 \\ h_2 \end{pmatrix} + \begin{pmatrix} 0 \vdots & \mathbf{F}_1 \\ 0 \vdots & \end{pmatrix} \begin{pmatrix} \phi_c \\ \mathbf{x} \end{pmatrix}$$

and will be denoted by $\mathbf{K}_{LQ}$. It has exactly the form required in Fig. 14.2 (cf. Sect. 11.3.4).

Next we ask for the performance that can be achieved with this controller. To this end, we have to build up the feedback loop $\mathcal{F}_l(\mathbf{P}, \mathbf{K}_{LQ})$ with the generalized plant $\mathbf{P}$ (which has to be taken without the weight for robustness) and the LQ controller $\mathbf{K}_{LQ}$. Performance of the LQ controller is expressed by the $\mathcal{H}_\infty$ norm of $\mathcal{F}_l(\mathbf{P}, \mathbf{K}_{LQ})$. Figure 14.11 shows the corresponding magnitude plot (curve K1).

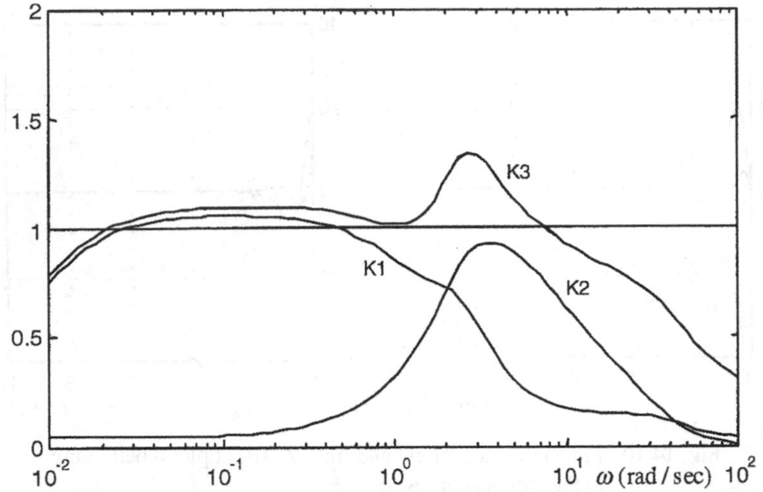

Fig. 14.11. Nominal performance (K1), robust stability (K2) and robust performance (K3) for the LQ controller

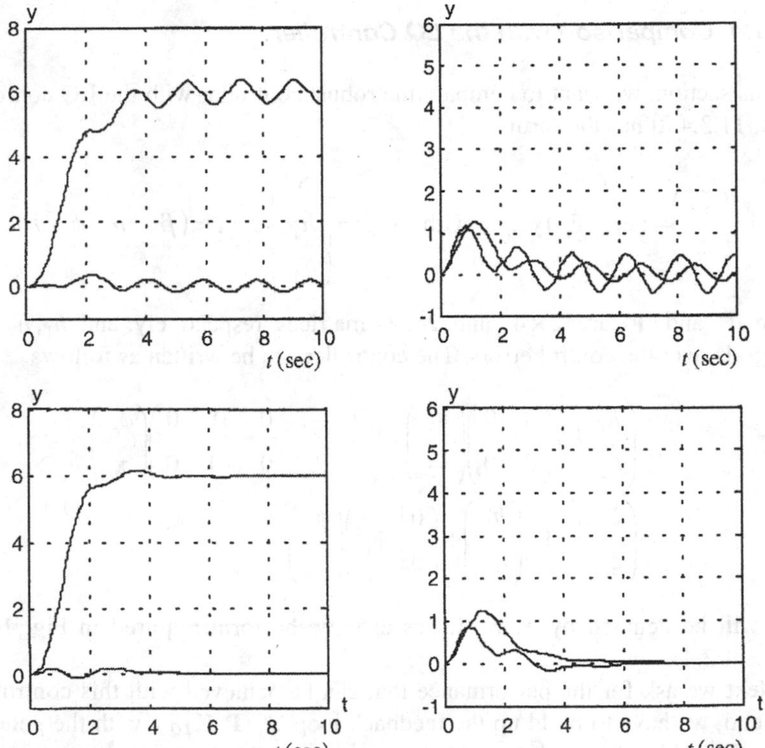

Fig. 14.12. Step response of the feedback system with destabilizing perturbation for the LQ controller (upper plots) and the robust controller (lower plots)

It is seen that nominal performance is not quite achieved by the LQ controller. Probably it is possible to get a somewhat better result by changing the weighting matrices $\mathbf{Q}, \mathbf{R}$, but we do not want to make any effort in this direction, in particular, since there is no obvious systematic way for doing so.

The next question is whether the feedback system with the LQ controller is robustly stable. To get an answer, we have to form the corresponding feedback system with the extended plant $\mathbf{P}_\Delta$, which contains additionally the uncertainty weight:

$$\mathbf{M} = \mathcal{F}_l(\mathbf{P}_\Delta, \mathbf{K}_{LQ}) .$$

Curve K2 in Fig. 14.11 shows the structured singular value of $\mathbf{M}_{11}$ (cf. Theorem 13.4.1). It remains below 1, which means that $\mathbf{K}_{LQ}$ achieves robust stability. Finally, curve K3 depicts the structured singular value of $\mathbf{M}$ (cf. Theorem 13.4.2). Its maximum value clearly exceeds 1. Hence robust performance is not given for the LQ controller.

A destabilizing perturbation for the feedback loop with $\mathbf{K}_{LQ}$ is

$$\Delta(s) = \begin{pmatrix} 0.245\dfrac{s-48.93}{s+48.93} & -0.505\dfrac{(s-48.93)(s-2.719)}{(s+48.93)(s+2.719)} \\ -0.436\dfrac{s-3.594}{s+3.594} & 0.801\dfrac{(s-3.594)(s-2.719)}{(s+3.594)(s+2.719)} \end{pmatrix} .$$

Fig. 14.12 shows the corresponding step responses for the LQ controller and the robust controller. It is seen that the robust controller still works quite well for this perturbation.

14.2 Robust Rudder Roll Stabilization for Ships

14.2.1 Nonlinear and Linear Plant Model

In the next case study, we are concerned with the lateral motion of a ship. Here the main task is to control the course of the ship. Additionally, we require that the controller improves the frequency response of the wave disturbance. As for aircraft, the roll motion of a ship is badly damped. Thus, if waves with a frequency near the eigenfrequency of the roll motion attack the ship, they induce a strong excitement of the roll motion with large roll angles. We try to improve the damping without additional fins, thus solely by the rudder. Since the rudder is not very effective concerning the roll motion and due to the limitations of the fin angle and in particular of the fin rate, this problem is challenging. Damping the roll motion by the rudder is denoted as **rudder roll stabilization (RRS)**.

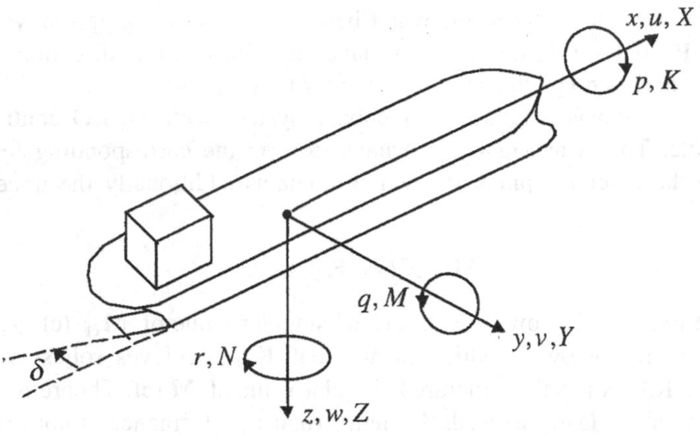

Fig. 14.13. States and controls for the ship dynamics

In principle, the ship dynamics resembles that of aircraft (cf. Sect. 11.3.1), but looking in greater detail at the dynamics, it turns out that there are some big differences. A detailed analysis of ship dynamics can be found in Fossen [44].

For our purposes, it is not necessary to consider the heave and pitch motion of the ship. Thus a model that describes sway, roll and yaw suffices. The nonlinear differential equations are then (cf. Fig. 14.13)

$$(m+m_x)\dot{u}-(m+m_y)vr = X + F_T$$

$$(m+m_y)\dot{v}+(m+m_x)ur+m_y\alpha_y\dot{r}-m_yl_y\dot{p} = Y$$

$$(I_x+J_x)\dot{p}-m_yl_y\dot{v}-m_xl_xur = K - W\,\overline{GM_T}\,\phi$$

$$(I_z+J_z)\dot{r}+m_y\alpha_y\dot{v} = N - x_G Y$$

$$\dot{\phi} = p$$

$$\dot{\psi} = r\cos\phi\ .$$

Here the masses m_x, m_y, the inertia J_x, J_y and the lengths α_y, l_x, l_y take into account the forces and moments of the surrounding water. These are the so-called added masses forces and moments. Furthermore, let F_T be the force exerted by the motor and denote by x_G the distance of the center of gravity to the origin of the coordinate system. The term $W\overline{GM_T}\,\phi$ is the added metacentric restoring moment in roll and is caused by the buoyancy of the ship. Here W denotes the weight of the displaced water and $\overline{GM_T}$ is the so-called transverse metacentric height. Finally, the quantities X, Y, K, N are the hydrodynamic forces and moments. The dependency of the states and controls has the following structure:

$$X = X_1(u) + X_2(v,r,\phi,\delta),\quad Y = Y(v,r,\phi,\delta)$$

$$K = K(v, r, \phi, \delta), \quad N = N(v, r, \phi, \delta).$$

Normally, these functions are approximated by polynomials.

We assume the surge speed to be constant: $u = u_0$. Then the differential equation for u can be skipped and the differential equations for v, p, r reduce to

$$\begin{aligned}
&(m + m_y)\dot{v} + m_y \alpha_y \dot{r} - m_y l_y \dot{p} = -(m + m_x) u_0 r + Y \\
&(I_x + J_x)\dot{p} - m_y l_y \dot{v} = m_x l_x u_0 r + K - W\,\overline{GM_T}\,\phi \\
&(I_z + J_z)\dot{r} + m_y \alpha_y \dot{v} = N - x_G Y.
\end{aligned}$$

Together with the two differential equations for ϕ and ψ they can be linearized in the usual way. This leads to a linearized system of the form

$$\begin{pmatrix} \dot{v} \\ \dot{r} \\ \dot{\psi} \\ \dot{p} \\ \dot{\phi} \end{pmatrix} = \begin{pmatrix} a_{11} & a_{12} & 0 & a_{14} & a_{15} \\ a_{21} & a_{22} & 0 & a_{24} & a_{25} \\ 0 & 1 & 0 & 0 & 0 \\ a_{41} & a_{42} & 0 & a_{44} & a_{45} \\ 0 & 0 & 0 & 1 & 0 \end{pmatrix} \begin{pmatrix} v \\ r \\ \psi \\ p \\ \phi \end{pmatrix} + \begin{pmatrix} b_1 \\ b_2 \\ 0 \\ b_4 \\ 0 \end{pmatrix} \delta .$$

This system can be further simplified if only the most important cross-couplings are taken into account and if, additionally, v is replaced by the sway velocity v' which is caused by the rudder alone. Then, if the resulting elementary linear systems are normalized, the final version of the linear system is (cf. Fossen [44], Sect. 6.6)

$$\begin{pmatrix} \dot{v}' \\ \dot{r} \\ \dot{\psi} \\ \dot{p} \\ \dot{\phi} \end{pmatrix} = \begin{pmatrix} -1/T_2 & 0 & 0 & 0 & 0 \\ K_{vr}/T_1 & -1/T_1 & 0 & 0 & 0 \\ 0 & 1 & 0 & 0 & 0 \\ K_{vp}\omega_0^2 & 0 & 0 & -2d\omega_0 & -\omega_0^2 \\ 0 & 0 & 0 & 1 & 0 \end{pmatrix} \begin{pmatrix} v' \\ r \\ \psi \\ p \\ \phi \end{pmatrix}$$

$$+ \begin{pmatrix} 0 & 0 \\ 1/T_1 & 0 \\ 0 & 0 \\ 0 & \omega_0^2 \\ 0 & 0 \end{pmatrix} \begin{pmatrix} w_\psi \\ w_\varphi \end{pmatrix} + \begin{pmatrix} K_{dv}/T_2 \\ K_{dr}/T_1 \\ 0 \\ K_{dp}\omega_0^2 \\ 0 \end{pmatrix} \delta .$$

Here, w_ψ and w_φ denote the disturbances which are caused by the waves. The transfer functions from rudder to yaw angle and roll angle can be calculated as

$$G_{\psi\delta}(s) = \frac{k_1(1+T_3 s)}{s(1+T_1 s)(1+T_2 s)} \tag{14.1}$$

$$G_{\varphi\delta}(s) = \frac{k_2 \omega_0^2 (1+T_4 s)}{(1+T_2 s)(s^2 + 2d\omega_0 s + \omega_0^2)} \tag{14.2}$$

with the constants

$$k_1 = K_{dv}K_{vr} + K_{dr}, \quad k_2 = K_{dv}K_{vp} + K_{dp},$$
$$T_3 = K_{dr}T_2 / K_1, \quad T_4 = K_{dp}T_2 / K_2 .$$

Their numerical values are also assumed to be given as in Fossen [44], Sect. 6.6 (U denotes the velocity of the ship):

$$K_{dr} = -0.0027U, \; K_{vp} = 0.21U, \; K_{dp} = -0.0014U^2, \; K_{vr} = -0.46, \; K_{dv} = 0.01U,$$
$$T_1 = 13/U, \quad T_2 = 78/U, \quad \omega_0 = 0.63, \quad d = 0.064 + 0.0038U.$$

The choice $U = 7.8\ m/s$ yields

$$k_1 = -0.0569, \quad k_2 = 0.0426,$$
$$T_1 = 1.677, \quad T_2 = 10, \quad T_3 = 3.699, \quad T_4 = -20, \quad d = 0.094 .$$

From the transfer functions and the values of the constants it is seen that the sway dynamics causes no problems. In contrast to this, the roll dynamics is non-minimum phase and has a very small damping. Figure 14.14 shows the corresponding step responses. As we see, a rudder deflection of 30° leads only to a stationary roll angle of 1.3°. Thus the effectiveness of the rudder with respect to the roll motion is very low.

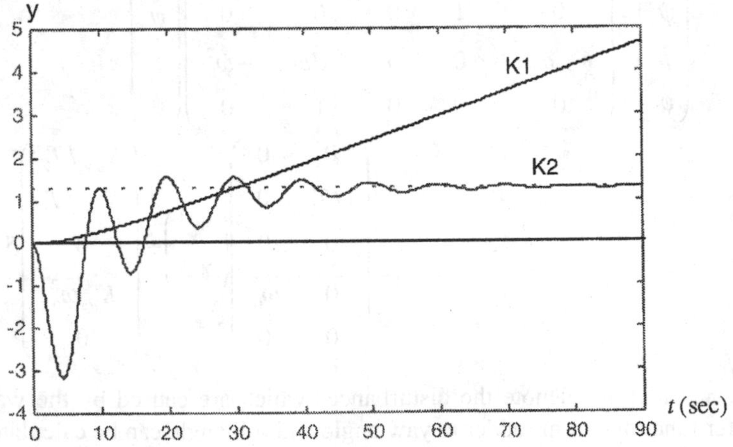

Fig. 14.14. Step responses of the ship: K1: ψ for $\delta = -1°$, K2: φ for $\delta = 30°$

The wave disturbance is modeled by its spectral decomposition, which can be described by the Bretschneider spectrum. The wave motion which results from this spectrum can to a good approximation be obtained by a shaping filter with the transfer function

$$H(s) = \frac{k_W s}{s^2 + 2 d_W \omega_W s + \omega_W^2}.$$

Typical numerical values are given by $d_W = 0.3$ and $0.3 < \omega_W < 1.3$. If the ship travels with velocity U, then ω_W has to be replaced by

$$\omega_e = \omega_W - \frac{\omega_W^2}{g} U \cos \chi .$$

As usual, g denotes the gravitational constant and χ is the angle between the heading and the direction of the wave (for a wave coming from the back, we have $\chi = 0°$). It is very well possible that ω_e coincides with the roll eigenfrequency ω_0 or at least is near to this frequency. Then, due to the low damping of the roll eigenvalue, a wave will cause a strong roll motion of the ship. For this reason it is desirable to obtain a frequency response of the wave disturbance which contains no peak at ω_0.

The actuator is modeled as a controlled system of second order with gain 1 and damping 0.7. It is very important that the controller design takes into account the limitations of the rudder deflection $\vartheta_{\max}$ and the rudder rate $\dot{\vartheta}_{\max}$. A typical value for the maximum deflection is $\vartheta_{\max} = 30°$. The rate $\dot{\vartheta}_{\max}$ lies in the interval from $5°/s$ up to $20°/s$.

14.2.2 Separate Design of the Controllers for Course and Roll Motion

Course controller: classical design. As a first step, we design a conventional PDT_1 controller with transfer function

$$K_{PD}(s) = k_R \frac{1 + T_D s}{1 + T_0 s} \qquad \text{(with } T_0 = 0.2 T_D\text{)}$$

by means of the Nyquist criterion. This puts us in a position to compare the classical and the modern approach.

One simple possibility for a synthesis consists in choosing $T_D = T_2$. Then the controller compensates the slowest part of the plant transfer function (14.1), but simulations show that for such a controller the actuator commands generated by the controller are unrealistically high.

We now try a second approach with a PD controller which makes the feedback system only slightly faster then would be possible with a pure proportional controller and choose $\omega_D = 0.1$ as the gain crossover frequency. Then for a phase

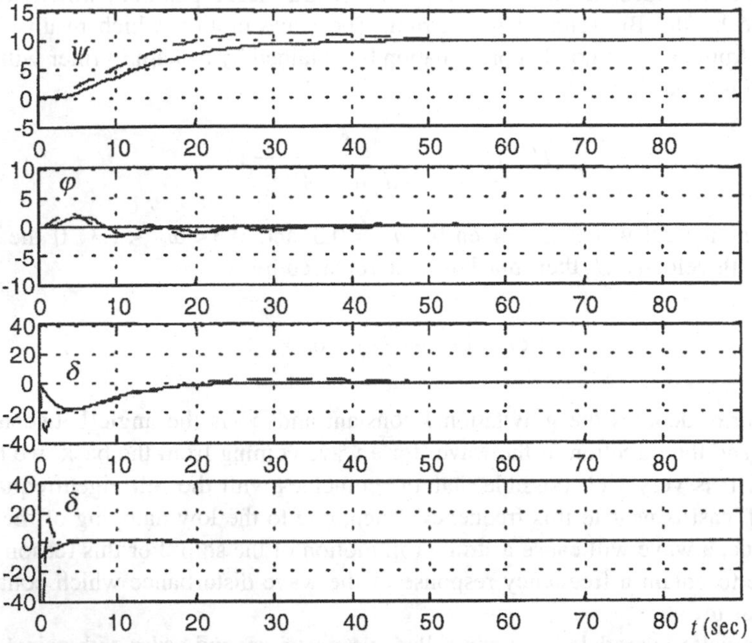

Fig. 14.15. $\psi, \varphi, \delta, \dot{\delta}$ for the course command $\psi_s = 10°$ (solid line: $\mathcal{H}_\infty$ controller; dashed line: PDT_1 controller)

margin of $60°$ the values $k_R = 2.2$, $T_D = 0.8$ result. Figure 14.15 shows the most important dynamic quantities of the feedback system for the course command $\psi_c = 10°$. The course angle and the control angle behave as expected, but the rudder rate becomes unfeasibly large for a short time at the beginning. This causes saturation, but as can shown by simulation, the step response becomes only slightly worse when the actuator limitations are taken into account.

Course controller: $\mathcal{H}_\infty$ design. As we will see now, with a $\mathcal{H}_\infty$ design the limitations of δ and $\dot{\delta}$ can be taken into account in a systematic way by introducing appropriate weights. Figure 14.16 shows the generalized plant with the weights for the course controller.

The weight W_1 for the sensitivity function is chosen dynamically:

$$W_1(s) = \frac{1}{1.5}\frac{s+0.1}{s+0.001}.$$

In contrast to this, the weights $W_\delta, W_{\dot{\delta}}$ for the control are constant

$$W_\delta = 1.33, \quad W_{\dot{\delta}} = 1.4.$$

Since the $\mathcal{H}_\infty$ design doesn't allow plant poles on the imaginary axis, we slightly

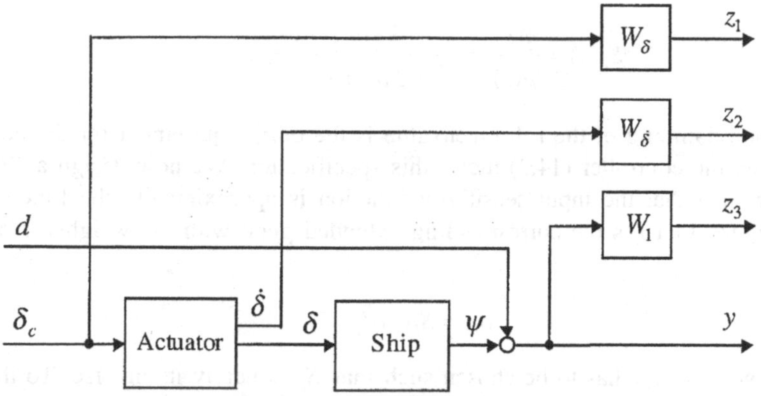

Fig.14.16. Weights for the course controller

perturb the linear factor s and replace it by $s+0.001$. Then the following $\mathcal{H}_\infty$ course controller results:

$$K_\psi(s)=\frac{-378.03\,(s+0.1)\,(s+0.6)\,(s^2+4.2s+9)}{(s+67.3)\,(s+47.25)\,(s+0.357)\,(s^2+0.7755s+0.1584)}\,.$$

The simulation result can also be seen from Fig. 14.15. The rudder rate has become significantly smaller and can be tuned by choosing $W_{\dot\delta}$ adequately.

Roll controller. As we have seen above, it is not possible to track a roll command. The only thing that can be tried is to improve the frequency response for the wave disturbance by feeding back φ. In order to see what can be achieved in the best case, we start our design under the assumption that the rudder can be exclusively used for controlling the roll motion.

Since a simple conventional design using the Nyquist criterion or the root locus method is not obvious, we start directly with the $\mathcal{H}_\infty$ synthesis. Thereby, the choice of the weight for performance is somewhat subtle. For the moment, assume that all three states v', p, φ of the roll dynamics can be measured and that a constant gain controller

$$K_Z=\begin{pmatrix}k_{v'} & k_p & k_\varphi\end{pmatrix} \tag{14.3}$$

has to be used. Such a controller can be designed by pole placement. Choose the closed-loop poles as

$$p_1=-1/T_2, \quad p_{2,3}=-d_i\,\omega_0\pm\sqrt{1-d_i^2}\,\omega_0\,i$$

(e.g. with $d_i=1/\sqrt{2}$). Let $p(s)$ be the characteristic polynomial of the resulting feedback system and denote the denominator of $G_{\varphi\delta}(s)$ by $q(s)$. Then the input sensitivity function is

$$S_i(s) = \frac{q(s)}{p(s)} = \frac{s^2 + 2d\omega_0 s + \omega_0^2}{s^2 + 2d_i\omega_0 s + \omega_0^2}. \tag{14.4}$$

If a good damping of the roll eigenvalue is the only requirement for the controller, then the controller (14.3) meets this specification. We now design a $\mathcal{H}_\infty$ controller such that the input sensitivity function is approximately the function (14.4). Figure 14.16 is the corresponding extended plant with the weights. Note that

$$\hat{\delta}_{c1} = S_i(s)\hat{d}_1.$$

Thus the weight W_{1R} has to be chosen such that S_i is nearly its inverse. To this end, let

$$W_{1R}(s) = \frac{s^2 + 0.891\,s + 0.3969}{s^2 + 0.118\,s + 0.3969}.$$

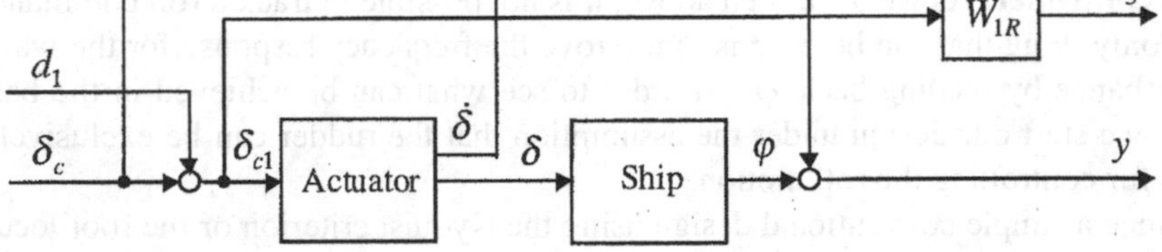

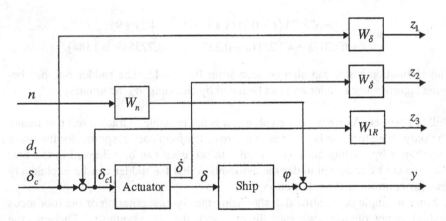

Fig.14.17. Weights for the controller of the roll motion

To satisfy the rank condition, it is necessary to inject small sensor noise (with $W_n = 0.001$). The calculation of the $\mathcal{H}_\infty$ optimal controller ends with $\gamma_{\text{opt}} = 1.22$. Thus the $\mathcal{H}_\infty$ controller is not able to fulfill this specification exactly (where the weights for the control are not active). If a factor α slightly smaller than 1 is added to W_{1R}, a specification is found which can be met by the $\mathcal{H}_\infty$ controller.

The effect of this controller for the roll motion can be seen from the upper plot in Fig. 14.18. The other two plots show that the actuator dynamics is much more demanding than in the case of the course control. If the weightings for δ and $\dot\delta$ are increased, the plots shown in Fig. 14.19 are obtained. As expected, the damping of the roll motion is reduced in this case.

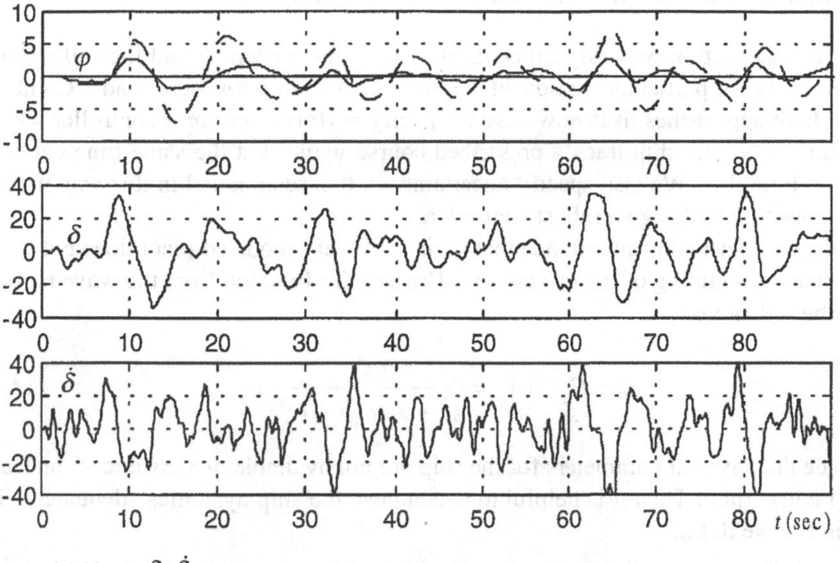

Fig.14.18. $\varphi, \delta, \dot{\delta}$ for the controller of the roll motion (dashed line: no controller)

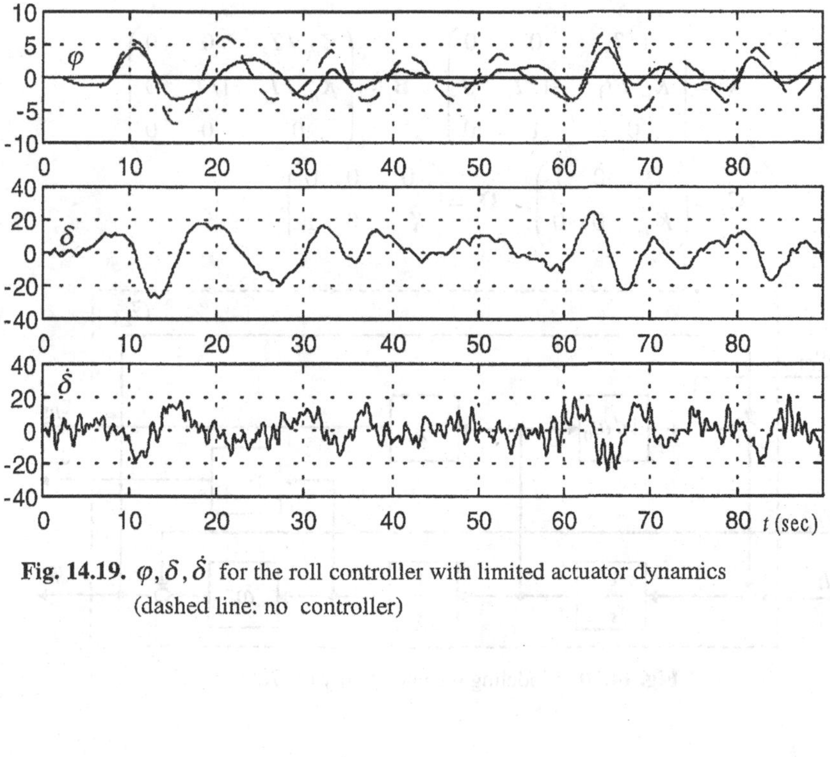

Fig. 14.19. $\varphi, \delta, \dot{\delta}$ for the roll controller with limited actuator dynamics (dashed line: no controller)

14.2.3 Robust Control of Course and Roll Motion

In the last section we designed controllers for course keeping and the roll motion separately. In particular, reasonable choices for the weights were made. Combining both approaches, it is now easy to specify performance for a controller which is able to let the ship track a prescribed course while it at the same time controls the roll motion. We also specify uncertainty in this section and in this way we are in a position to design a robust controller.

It is of particular importance that an inaccurately modeled plant does not deteriorate the damping of the roll motion. The transfer function from the wave motion to the roll angle is

$$\hat{\varphi} = G_2(s)\,\hat{w}_\varphi = \frac{\omega_0^2}{s^2 + 2d\omega_0 s + \omega_0^2}\,\hat{w}_\varphi \,. \tag{14.5}$$

Since the physical parameters for the ship are not available, we assume d and ω_0 to be uncertain. Then it is helpful to decompose the ship dynamics adequately. To this end we define

$$\mathbf{G}_1(s) = \left(\begin{array}{c|c} \mathbf{A}_1 & \mathbf{B}_1 \\ \hline \mathbf{C}_1 & \mathbf{D}_1 \end{array}\right)$$

with the matrices

$$\mathbf{A}_1 = \begin{pmatrix} -1/T_2 & 0 & 0 \\ K_{vr}/T_1 & -1/T_1 & 0 \\ 0 & 1 & 0 \end{pmatrix}, \quad \mathbf{B}_1 = \begin{pmatrix} K_{dv}/T_2 & 0 & 0 \\ K_{dr}/T_1 & 1/T_1 & 0 \\ 0 & 0 & 0 \end{pmatrix}$$

$$\mathbf{C}_1 = \begin{pmatrix} 0 & 0 & 1 \\ K_{dp} & 0 & 0 \end{pmatrix}, \quad \mathbf{D}_1 = \begin{pmatrix} 0 & 0 & 0 \\ K_{dp} & 0 & 1 \end{pmatrix}.$$

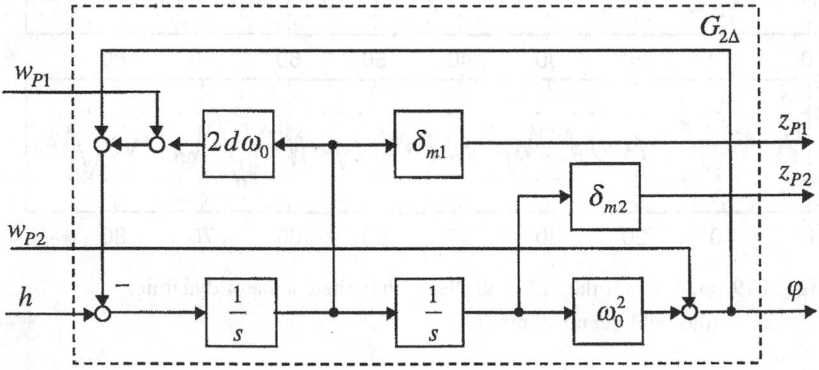

Fig. 14.20. Modeling the uncertainty of G_2

With the transfer function G_2 defined in (14.5) we have

$$\mathbf{G}(s) = \begin{pmatrix} 1 & 0 \\ 0 & G_2(s) \end{pmatrix} \mathbf{G}_1(s),$$

where $\mathbf{G}$ denotes the transfer function of the whole ship with inputs $\delta, w_\psi, w_\varphi$ and outputs ψ, φ. An important property of this decomposition is that the uncertain parameters d, ω_0 do not occur in $\mathbf{G}_1$. Figure 14.20 shows how this parameter uncertainty can be modeled.

We assume $\delta_{m1} = 0.09$ and $\delta_{m2} = 0.2$. This means that a deviation of 30% is allowed for d and ω_0. Further types of uncertainty such as neglected cross-couplings are modeled by input multiplicative uncertainty. It is taken into account by the weight

$$W_\Delta(s) = W_\infty \frac{s + \omega^* k}{s + \omega^* W_\infty} \quad \text{with} \quad k = 0.03,\ \omega^* = 7,\ W_\infty = 1.$$

Figure 14.21 shows the generalized plant with all weights for uncertainty and robustness. In order to obtain robust stability and performance, some of the requirements have to be slightly relaxed. The weight W_1 will be replaced by

$$W_1(s) = \frac{1}{2.5} \frac{s + 0.03}{s + 0.0003}.$$

This relaxes the performance requirement with respect to course control. It is somewhat surprising that better results are obtained when the weight W_{1R} is cho-

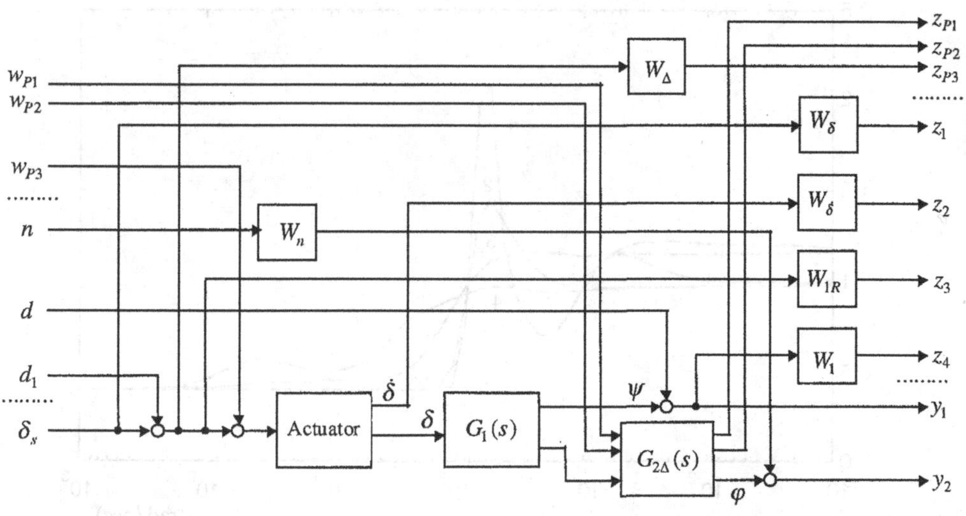

Fig. 14.21. Generalized plant with all weights for performance and robustness

sen to be constant instead of being dynamically as it was for the pure roll control. We put

$$W_{1R}(s) = 0.3 .$$

Finally, let

$$W_\delta = 0.2, \quad W_{\dot{\delta}} = 0.01 .$$

Figure 14.22 shows the result of the D-K iteration. After 4 steps, the value $\mu = 0.97$ is reached. Thus robust stability and robust performance are achievable. The resulting controller $\mathbf{K}_\mu$ has the order 27. The magnitude plot of the transfer function for the wave disturbance (roll motion) is depicted in Fig. 14.23. In contrast to the preceding non-robust version of the pure roll controller of Sect. 14.2.3, the magnitude plot exceeds the 0dB-line slightly in a small region of frequencies. Despite this, the damping of the roll motion is much better than without the roll controller.

The simulation result for the robust controller is shown in Fig. 14.24. It is seen that the roll motion caused by the waves is considerably smaller than without the roll controller. A price that has to be paid for the good damping in combination with robustness is that rather large values for δ and $\dot{\delta}$ occur. In particular, caused by the comparatively high frequency of the waves, the actuator must be able to realize much higher fin rates than in the case of pure course control. This can be relaxed if one is satisfied with a somewhat lower damping. It is also seen from the simulation results that course keeping is not severely degraded by the additional control of the roll motion.

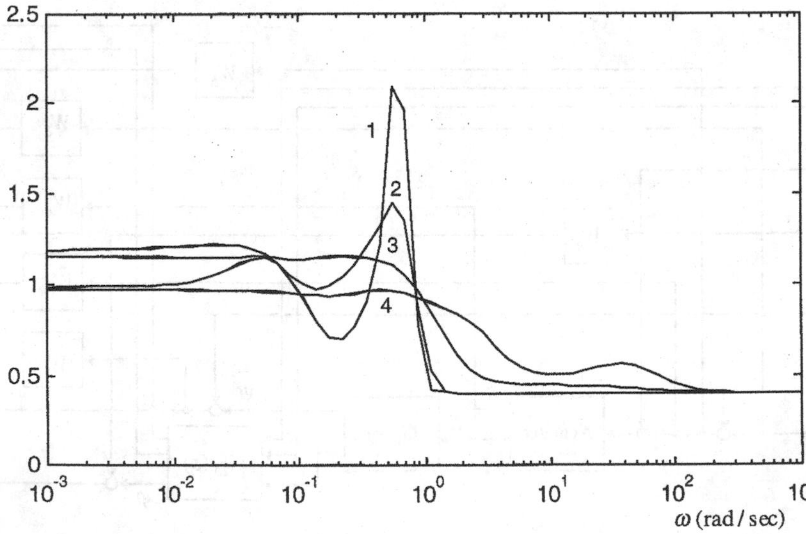

Fig. 14.22. Result of the D-K iteration

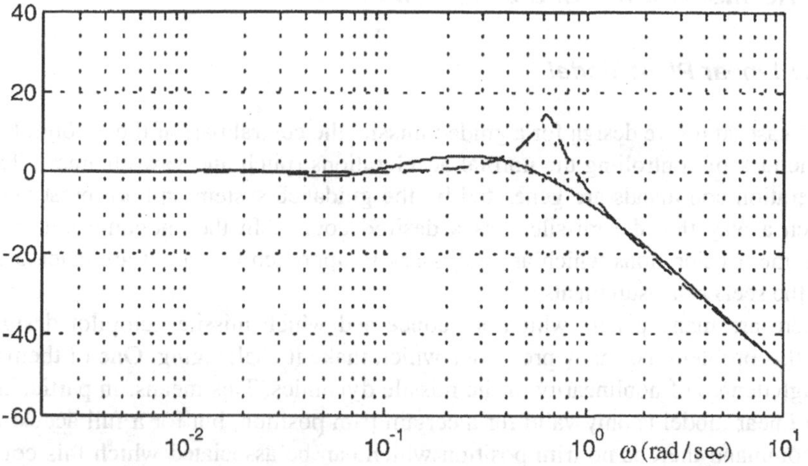

Fig. 14.23. Magnitude plots of the transfer functions for the wave disturbance (roll motion): closed-loop (solid line), uncontrolled ship (dashed line)

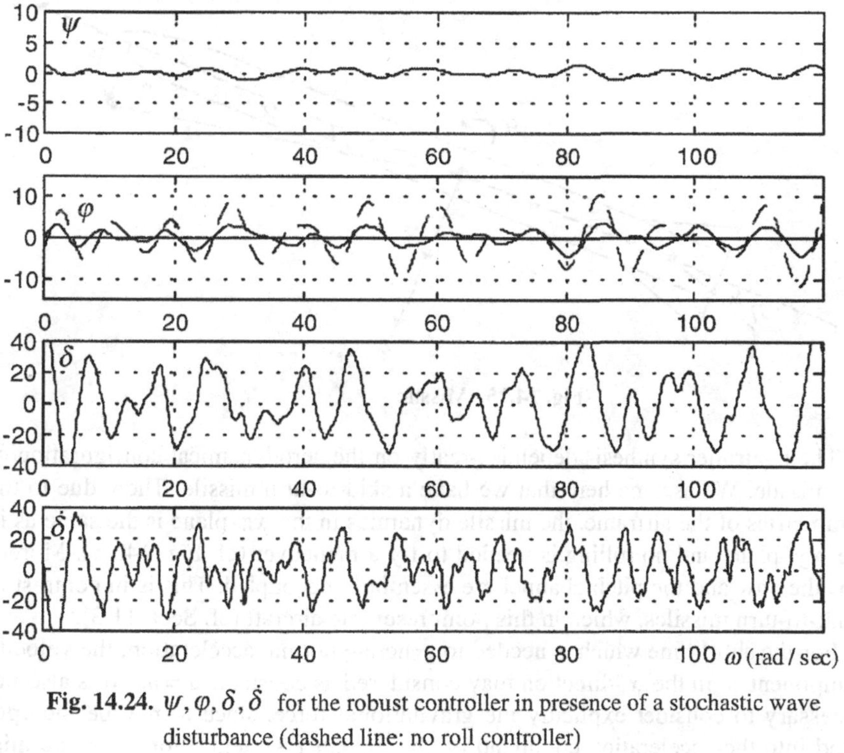

Fig. 14.24. $\psi, \varphi, \delta, \dot{\delta}$ for the robust controller in presence of a stochastic wave disturbance (dashed line: no roll controller)

14.3 Robust Control of a Guided Missile

14.3.1 Linear Plant Model

In this case study we design for a guided missile the central part of the autopilot. It has the task of controlling the normal accelerations (pitch and yaw channel). The acceleration commands are generated by the guidance system and are prescribed in such a way that the missile flies a desired course. In the endgame phase of flight, the accelerations which are necessary to approach the target are calculated from the seeker measurements.

There are many papers which are concerned which missile autopilot design, since this problem has many properties which make it challenging. One of them is the high degree of nonlinearity of the missile dynamics. This means, in particular, that a linear model is only valid for a certain trim position, but for a full acceleration command there is no trim position which can be associated which this command. The linear models describing the dynamical behavior near the trim positions differ considerably in dependence on the stationary acceleration. Later we will show how this problem can be tackled with robust control methods.

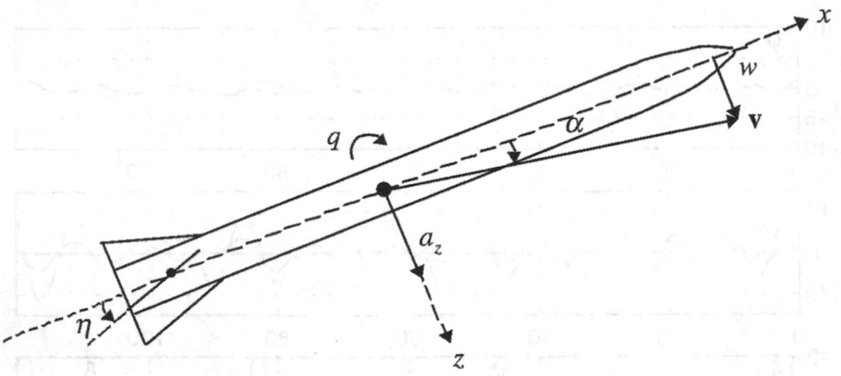

Fig. 14.25. Missile

The controller synthesis depends greatly on the aerodynamical configuration of the missile. We assume here that we have a skid-to-turn missile. Then, due to the symmetries of the airframe, the missile dynamics in the xz-plane is the same as in the xy-plane and no rolling is needed to fly a maneuver (cf. Fig. 14.25). Moreover, the yaw and the pitch channel are essentially decoupled. This is in contrast to bank-to-turn missiles, which in this point resemble aircraft (cf. Sect. 11.3).

For the short time which is needed to generate normal acceleration, the velocity component u in the x-direction may considered as constant: $u = u_0$. It is also not necessary to consider explicitly the gravitational force, since it may be incorporated into the acceleration command by the guidance system. Moreover, the missile is assumed to fly in the xz-plane without rolling. The equations of motion are then obtained by specialization of (11.15), (11.16) as

$$m\dot{w} - mq u_0 = Z(w,q,\eta)$$
$$I_y \dot{q} = M(w,q,\eta).$$

Herein, Z and M are the corresponding components of the aerodynamical force and moment. The variable to be controlled is the normal acceleration of the missile in its center of gravity:

$$a_{zcg} = \frac{1}{m} Z(w,q,\eta).$$

We suppose that Z and M have the following structure (with $\tan\alpha = w/u_0$):

$$Z(w,q,\eta) = Z_1(\alpha) + \gamma_1 \eta$$
$$M(w,q,\eta) = M_1(\alpha) + \gamma_2 q + \gamma_3 \eta.$$

The quantities $\gamma_1, \gamma_2, \gamma_3$ are not functions of α, but, as for Z_1 and M_1, they depend on the operating point of the missile, which in particular is described by its velocity and altitude. We consider here only a single fixed operating point.

The linearized model has now to be calculated for each trim position, i.e. for each stationary angle of attack α_0. Then the remaining trim quantities q_0, η_0 are solutions of the following linear system of equations:

$$m u_0 q_0 + \gamma_1 \eta_0 = -Z_1(\alpha_0)$$
$$\gamma_2 q_0 + \gamma_3 \eta_0 = -M_1(\alpha_0).$$

The linearized system is then given by

$$\begin{pmatrix} \Delta\dot{\alpha} \\ \Delta\dot{q} \end{pmatrix} = \begin{pmatrix} a_{11} & 1 \\ a_{21} & a_{22} \end{pmatrix} \begin{pmatrix} \Delta\alpha \\ \Delta q \end{pmatrix} + \begin{pmatrix} b_1 \\ b_2 \end{pmatrix} \Delta\eta,$$

$$\Delta a_{zcg} = \begin{pmatrix} u_0 a_{11} & 0 \end{pmatrix} \begin{pmatrix} \Delta\alpha \\ \Delta q \end{pmatrix} + u_0 b_1 \Delta\eta$$

with (where $Z_\alpha = \dfrac{\partial Z}{\partial \alpha}$ and so on)

$$a_{11} = \frac{1}{m u_0} Z_\alpha, \quad a_{21} = \frac{1}{I_y} M_\alpha, \quad a_{22} = \frac{1}{I_y} M_q,$$
$$b_1 = \frac{1}{m u_0} Z_\eta, \quad b_2 = \frac{1}{I_y} M_\eta.$$

Instead of discussing the detailed structure of the aerodynamical forces and moments, we simply describe a_{11}, a_{21} by quadratic polynomials:

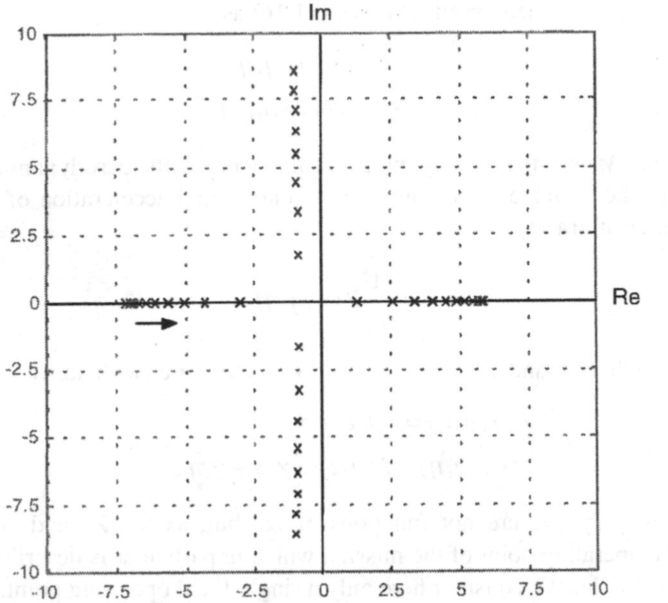

Fig. 14.26. Pole positions in dependency of the angle of attack α

$$a_{i1}(\alpha) = a_{i12}\alpha^2 + a_{i11}\alpha + a_{i10} \quad i = 1, 2 .$$

Although this is a rather crude simplification, the property which causes the greatest difficulty which respect to controller design is still in our model, namely its strong nonlinear dependency of the angle of attack. It should be noted that we are not interested in controlling the normal acceleration in the vicinity of one fixed trim position but in the whole interval of admissible angles of attack. Thus, if a large command is tracked, the region where the linear model is valid will be left.

The transfer function from fin angle η to the normal acceleration a_z has the form

$$G(s) = k\frac{(1+T_1 s)(1-T_2 s)}{as^2 + bs + 1} \quad \text{with } T_1 > 0, T_2 > 0 .$$

All parameters of this transfer function and particularly its poles depend on the Mach number Ma, the altitude h and the angle of attack α, and the possible variations may be very large. The dependency of Ma and h is not problematic, since these quantities vary only slowly with time, but, as explained above, difficulties arise because, even for fixed Ma and h, the parameters strongly depend on α. This is especially the case for slim missiles. We consider here a generic missile of this kind and assume the numerical values to be given as (with $-20° \le \alpha \le 20°$):

$$u_0 = 1100\, m/s,$$

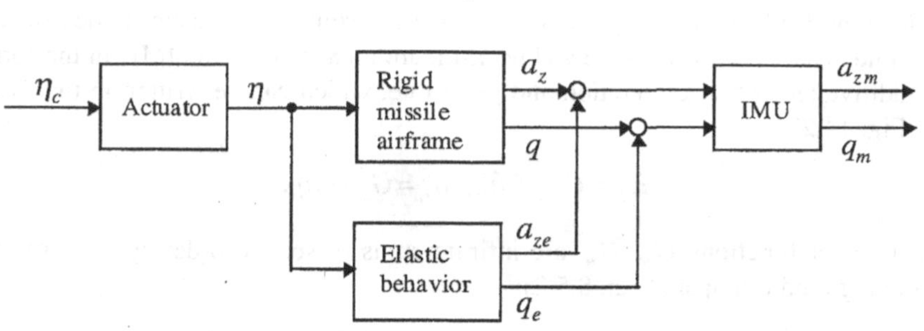

Fig. 14.27. Missile including instrumentation and elastic behavior

$$a_{11}(\alpha) = 0.0005\alpha^2 - 0.0477\alpha - 0.448, \quad b_1 = -0.136,$$
$$a_{21}(\alpha) = -0.335\alpha^2 + 0.299\alpha + 42.09, \quad a_{22} = -0.762, \quad b_2 = -159.1.$$

Figure 14.26 shows the pole positions for some values of α in the admissible interval and an operating point characterized by a great altitude and a high Mach number. It is seen that the missile is unstable for small values of α. If α increases, the missile becomes stable but with some damping, which is typically low. For a different operating point, it is possible that the opposite situation occurs, i.e. stability for small α and instability for large α. The maximum normal acceleration is $35g$.

We assume that the angular velocity q as well as the normal acceleration a_z will be measured by an IMU (inertial measurement unit). The IMU is supposed to be situated at a distance L from the center of gravity from the missile. It measures the acceleration

$$a_z = a_{zcg} + L\dot{q}.$$

Thus we have

$$\Delta a_z = \begin{pmatrix} u_0 a_{11} + La_{21} & La_{22} \end{pmatrix} \begin{pmatrix} \Delta\alpha \\ \Delta q \end{pmatrix} + (u_0 b_1 + Lb_2)\Delta\eta .$$

The sensor dynamics will be neglected. The fin deflection η is generated by an actuator, which itself is a feedback system. We model the transfer function from the commanded fin angle η_c to the actual fin angle η as a linear system of order 2 (with $T_A = 0.005$):

$$G_A(s) = \frac{1}{1 + \sqrt{2}T_A s + T_A^2 s^2}.$$

If the missile has to fly a maneuver with a high lateral acceleration, the fin deflections may cause structural vibrations. In particular, if the missile is long and slender, these vibrations cannot be neglected when the feedback loop is designed, since, in the worst case, they can destabilize the system. More precisely, the fin deflections lead to an aerodynamic force which excites the elastic modes of the airframe. The vibrations at the IMU position are measured by the IMU in the form of additive terms for acceleration and pitch rate, which can be written in the form (cf. Fig. 14.27)

$$\hat{a}_{ze} = G_{ae}(s)\hat{\eta}, \quad \hat{q}_e = G_{qe}(s)\hat{\eta}\,.$$

The transfer functions G_{ae}, G_{qe} are infinite sums of second-order systems of the following kind (compare Sect. 8.5.2):

$$G_{ae\,j}(s) = \frac{k_{ea\,j}s^2}{s^2 + 2d_{e\,j}\omega_{e\,j}s + \omega_{e\,j}^2}, \qquad G_{qe\,j}(s) = \frac{k_{eq\,j}s}{s^2 + 2d_{e\,j}\omega_{e\,j}s + \omega_{e\,j}^2}\,.$$

Here, the gains $k_{ea\,j}, k_{eq\,j}$ tend to 0 and the frequencies $\omega_{e\,j}$ tend to ∞ for $j \to \infty$. The dampings are very small: we assume $d_{e\,j} = 0.01$. The above transfer functions depend on the IMU position. The main part of the vibrational deflection is contributed by the modes with the lowest eigenfrequency, i.e. the first few summands of the series, and only these are incorporated in our missile model.

Instead of modeling the elastic behavior of the missile for controller design in detail, we represent it as additive model uncertainty in the form $\mathbf{W}_e(s)\Delta_e(s)$ with $\|\Delta_e\|_\infty \le 1$. Here, let

$$W_e(s) = k_{E1}\frac{(s+\omega_{E1})^2}{(s+\omega_{E2})(s+\omega_{E3})} \quad \text{with } \omega_{E3} \ge \omega_{E2} \gg \omega_{E1},$$

$$\mathbf{W}_e(s) = \begin{pmatrix} W_e(s) & W_e(s) \end{pmatrix}^T.$$

The modeled dynamics comprises also the actuator. It is necessary that not only the fin deflection η but also the fin rate $\dot{\eta}$ will be weighted, since the latter may limit the bandwidth of the feedback loop. We assume that the feasible maximum values of these quantities are given as follows:

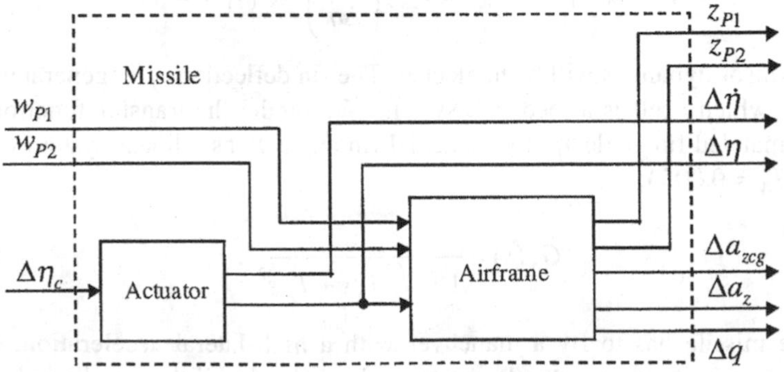

Fig. 14.28. Missile with additional inputs and outputs needed for controller design

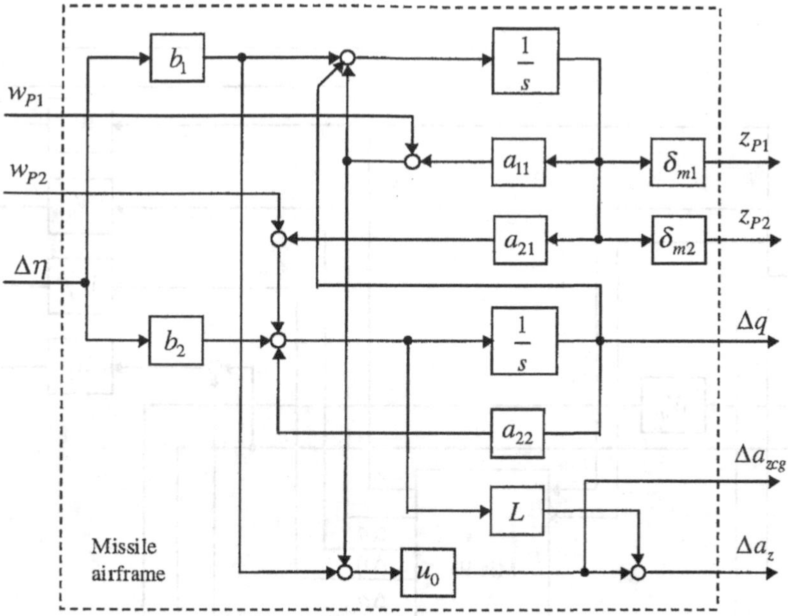

Fig. 14.29. Block diagram of the airframe with uncertainty of the coefficients

$$\eta_{\max} = 20°, \quad \dot{\eta}_{\max} = 400° / \sec .$$

In Fig. 14.28, the missile consisting of airframe and actuator is depicted together with all inputs and outputs which are necessary for controller design.

The dependency of the (rigid) linearized airframe on α will be viewed as the uncertainty of the coefficients a_{11}, a_{21}. This leads to the block diagram in Fig. 14.29 for the uncertain airframe. The values for δ_{m1}, δ_{m2} result from the plots for $a_{11}(\alpha)$, $a_{21}(\alpha)$. One obtains $\delta_{m1} = 0.37, \delta_{m2} = 67$.

14.3.2 Design of a Robust Controller

In a first step one could design the controller by a mixed-sensitivity approach. Then the $\mathcal{H}_\infty$ controller compensates the poles of the missile, but internally these poles are still present in the feedback loop. This undesired property of the feedback loop can be avoided by introducing an input-multiplicative uncertainty. Additionally, in this way, the unmodeled high-frequency dynamics is taken into account. Then the whole model consists of the input-multiplicative uncertainty and the parameter uncertainty introduced in the previous section.

Next we want to discuss the weights for performance. The most important weight is W_1, which shapes the sensitivity function and which has a major influence on the bandwidth of the controlled missile. The weights W_η and $W_{\dot{\eta}}$ are con-

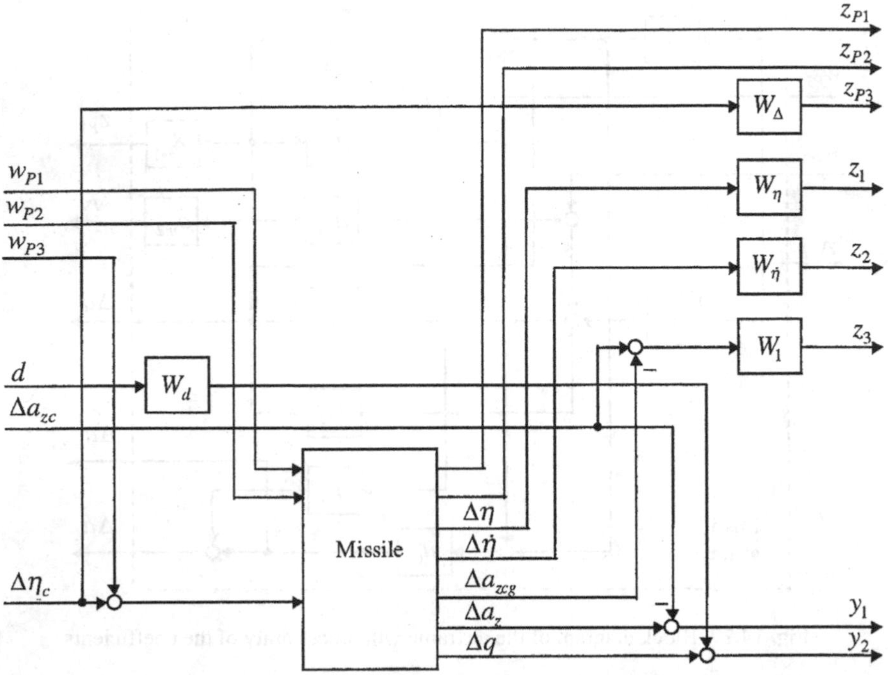

Fig. 14.30. Extended plant with all weights for performance and robustness

stant and limit the fin deflection and the fin rate. The measurement error of the pitch rate q is added in form of a perturbation d. It is needed to fulfill the rank condition imposed by the $\mathcal{H}_\infty$ synthesis via the Riccati approach. The complete extended plant with all necessary weights and incoming and outgoing signals is depicted in Fig. 14.30. The weights for the elastic behavior are not taken into account. We will return to this point in the next section.

The numerical values of W_1 are given by

$$W_1 = \frac{1}{S_\infty}\frac{s+\omega_1}{s+\omega_\varepsilon} \quad \text{with} \quad \omega_1 = 8, \quad \omega_\varepsilon = 0.01, \quad S_\infty = 3.$$

Figure 14.31 shows the magnitude plot of W_1^{-1}. The remaining weights for performance are given by

$$W_\eta = 3, \quad W_{\dot\eta} = 19.1, \quad W_d = 0.005.$$

The inverse of the maximum admissible values of the corresponding signals cannot directly be related to the numerical values chosen in these weights. Some trial and error is necessary to select them reasonably, but then it is possible to adjust the $\mathcal{H}_\infty$ controller exactly to the limitations of the actuator. The robustness weight

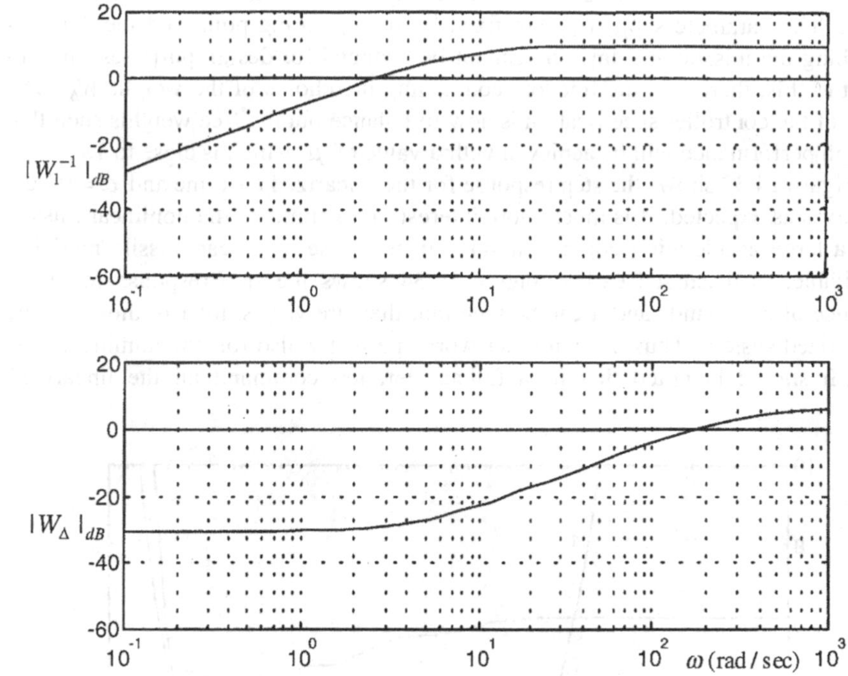

Fig. 14.31. Magnitude plot of W_1^{-1} and of W_Δ

W_Δ is chosen as

$$W_\Delta = W_\infty \frac{s+\omega^* k}{s+\omega^* W_\infty} \quad \text{with} \quad k = 0.03, \quad W_\infty = 2, \quad \omega^* = 150 .$$

The magnitude plot of W_Δ is also depicted in Fig. 14.31.

The extended plant is now completely specified so that we are in a position to start the D-K iteration. Figure 14.32 shows the result for each iteration step. The lower plot is nothing but an enlarged version of the upper one to see the curves for the steps 3,4,5 in greater detail. One observes that the iteration starts with a large value μ, but then quickly converges. After 5 iteration steps, the value $\mu = 1.022$ is obtained. Thus robust stability and robust performance is achievable for this specification. The order of the μ optimal controller is 20.

This result is not obtained in one single step. The performance, which is mainly expressed by a high bandwidth in combination with a good damping of the feedback loop, is mainly influenced by the parameters ω_1 and S_∞. Thus it is of particular interest to get a robust design for a high value of ω_1. These conflict with restrictions of the actuator dynamics, which can be taken into account by the weights W_η and $W_{\dot{\eta}}$. A quick response of the missile is always connected with a high fin rate (if the velocity is high enough, the limitation of the fin deflection is not so restrictive). The weights for uncertainty limit also the achievable performance. The parameters δ_{m1}, δ_{m2} are fixed by the operating point and the data describing the missile airframe and cannot be changed for design purposes. In contrast to this, there is some freedom concerning the choice of the weight W_Δ. The task of the controller specification is now to balance out the free weights such that a high performance will be achieved with a value of μ which is close to 1.

Figure 14.33 shows the step response for the linearized airframe and $\alpha = 0°$ and behaves as expected. It is much more interesting to simulate the nonlinear missile for a large acceleration command, since in this case no linear missile model is available, as discussed earlier. Figure 14.34 shows the step response for a sequence of commands and it can be seen that they are very similar to those for the linearized system. Thus the controller works perfectly also for the nonlinear missile. It should be noted that for a full acceleration command all the linearized

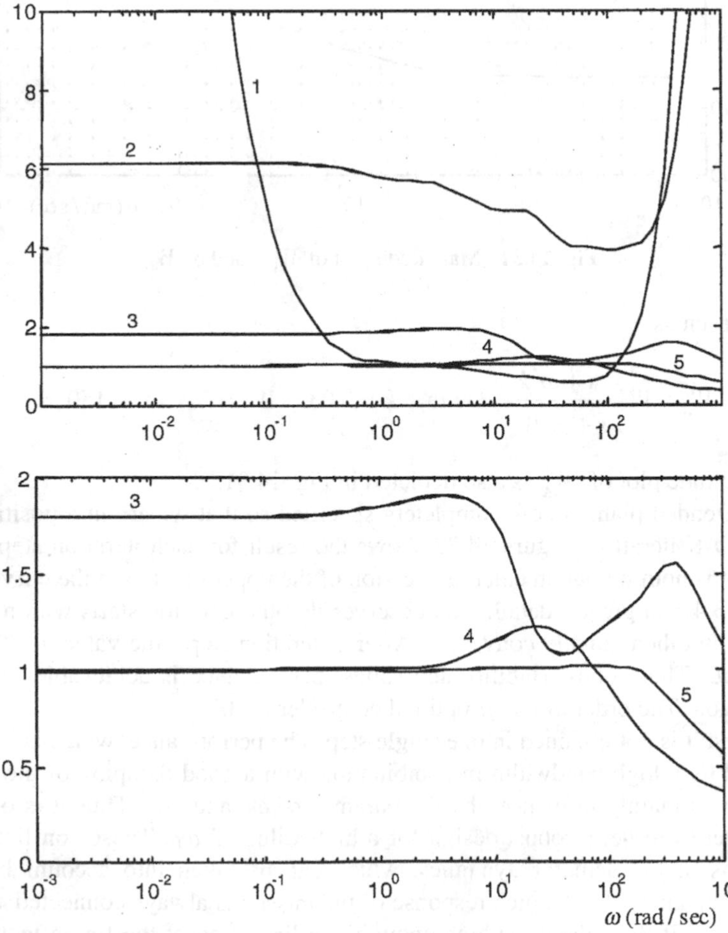

Fig. 14.32. Result of the D-K iteration

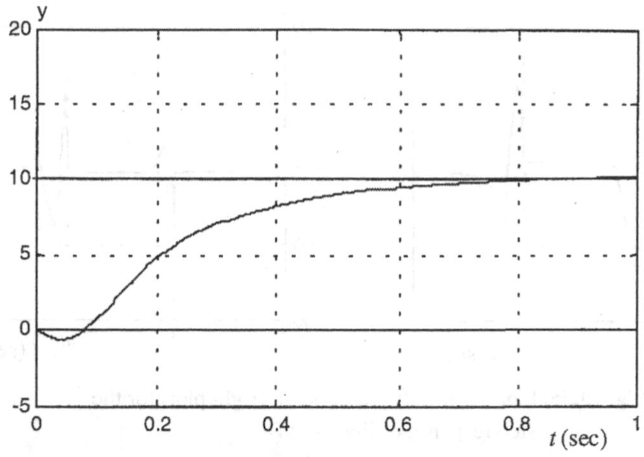

Fig. 14.33. Step response for the linear missile model (for $\alpha = 0°$)

systems which locally represent the missile in the vicinity of a trim position have the very different pole configurations depicted in Fig. 14.26. Figure 14.35 shows that the fin deflection and the fin rate remain within the prescribed bounds.

In our approach for designing the controller, we considered the parameter dependence of a_{11} and a_{21} as uncertainty. This must not necessarily be so since the dependency on the angle of attack is known. Thus another idea would be to proceed as follows: design a controller for each α and adapt it for a large acceleration command with dependency on α. This is a form of gain scheduling, cf. Sect. 10.4. Note that in this approach gain scheduling is done within one control action which is in contrast to the case where the controller is adapted to slowly varying quantities such as altitude or Mach number, where gain scheduling is comparatively harmless.

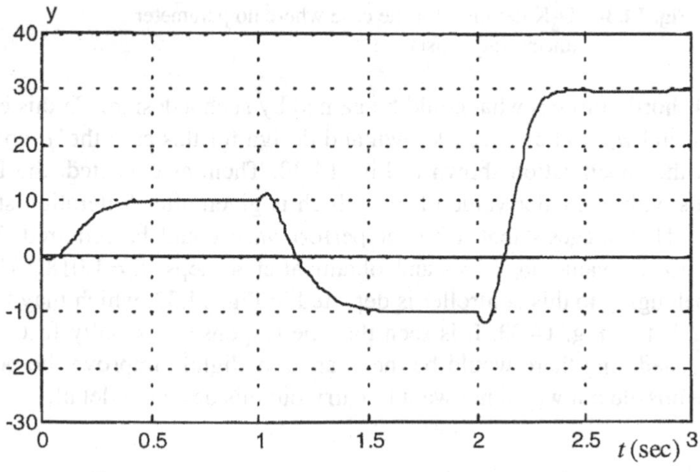

Fig. 14.34. Step responses of the nonlinear missile

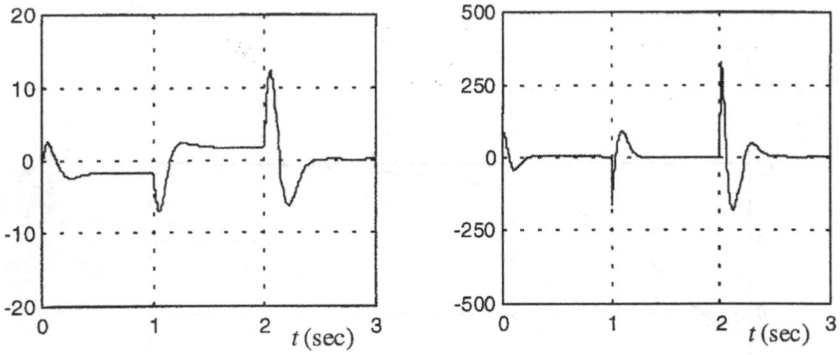

Fig. 14.35. Plots of η (left plot) and $\dot{\eta}$ (right plot) for the step response of Fig. 14.34

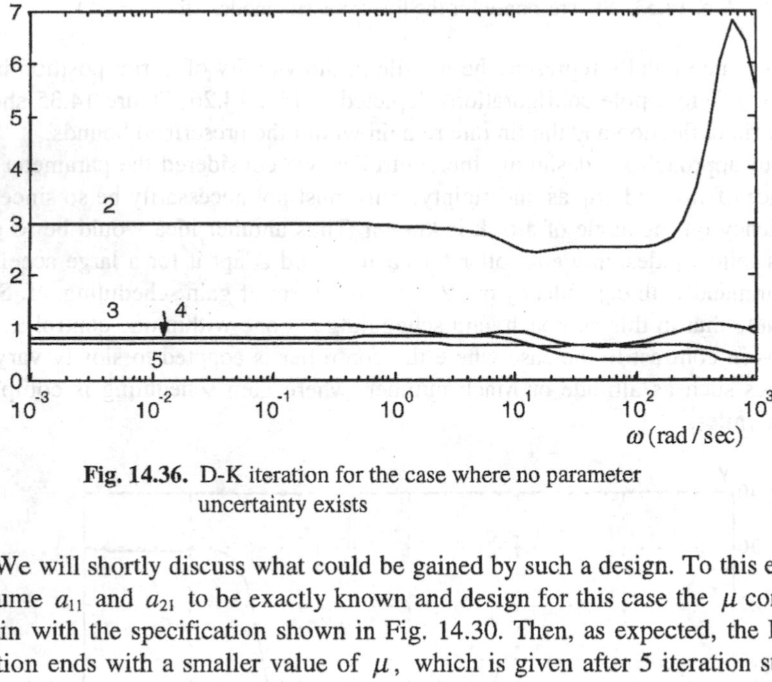

Fig. 14.36. D-K iteration for the case where no parameter uncertainty exists

We will shortly discuss what could be gained by such a design. To this end, we assume a_{11} and a_{21} to be exactly known and design for this case the μ controller again with the specification shown in Fig. 14.30. Then, as expected, the D-K iteration ends with a smaller value of μ, which is given after 5 iteration steps by $\mu = 0.688$. This suggests that a higher performance could be required. We increase ω_1 to the value $\omega_1 = 15.9$ and obtain after 4 steps $\mu = 1.018$. The step response belonging to this controller is depicted in Fig. 14.37, which may be compared with that of Fig. 14.33. It is seen that the response is actually faster. Thus, with gain scheduling there would be the chance to slightly improve the performance of the missile but we do not want to carry out this design in detail.

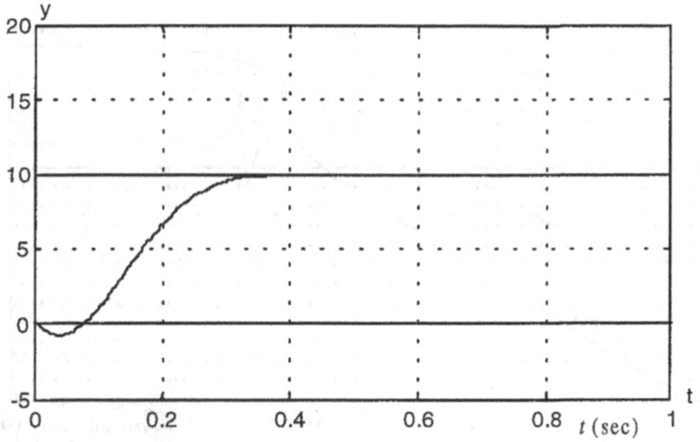

Fig. 14.37. Step response for the linearized missile with a μ controller if no parameter uncertainties are present.

14.3.3 Control of the Elastic Missile

We now want to take into account the elastic behavior of the missile and add, as depicted in Fig. 14.27, to the measured normal acceleration a_z a term a_{ze} and to the measured pitch rate a term q_e. Figure 14.38 shows a typical frequency response of $G_{ae}(s)$ (curve K2), if only the first three bending modes are taken into account. The frequency response of $G_{qe}(s)$ is similar. The parameters of the transfer functions belonging to the lowest eigenfrequency are

$$\omega_{e1} = 150\,s^{-1}, \quad d_{e1} = 0.01, \quad k_{ea1} = 2, \quad k_{eq1} = 200\,.$$

We first simulate the missile with the μ optimal controller of the previous section. The picture shows clearly that the structural vibrations destabilize the feedback loop. In many cases, the situation is not so bad, since the frequency of the lowest bending mode is higher. If ω_{e1} is increased to the value $\omega_{e1} = 300\,s^{-1}$ and if the remaining parameters are left as they are, the performance is much better and the degradation due to elasticity is no longer reflected in the step response. Only the structured singular value becomes somewhat larger.

Our aim is to design a controller, which is robust even for the case $\omega_{e1} = 150\,s^{-1}$. Then we have to incorporate a weight which takes to into account the elastic behavior. This can be achieved by interpreting the effect of elasticity as an additive model uncertainty which is weighted by W_e. In the block diagram of Fig. 14.30, we then have to introduce a fourth output of the form $z_{P4} = \eta$ and also a fourth input w_{P4} which enters $\mathbf{W}_e$. The first output signal of $\mathbf{W}_e$ has to be added to y_1 and the second to y_2 (cf. Fig. 14.30). The data for $\mathbf{W}_e$ are chosen as $k_{E1} = 400$, $\omega_{E1} = 1$, $\omega_{E2} = 100$, $\omega_{E3} = 500$, which leads to the magnitude plot of Fig. 14.38.

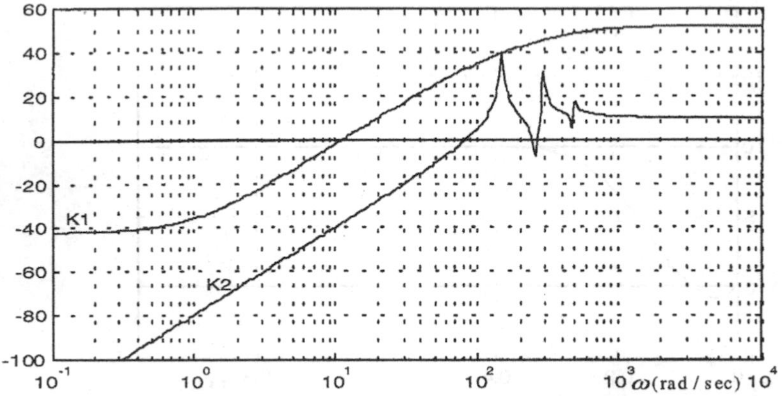

Fig. 14.38. Magnitude plot of the weight W_e (curve K1) and of the transfer function G_{ae} (curve K2)

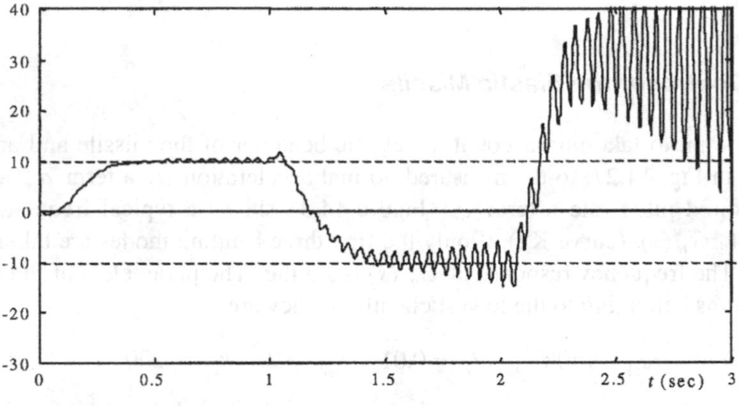

Fig. 14.39. Step response of the elastic missile with the μ optimal controller

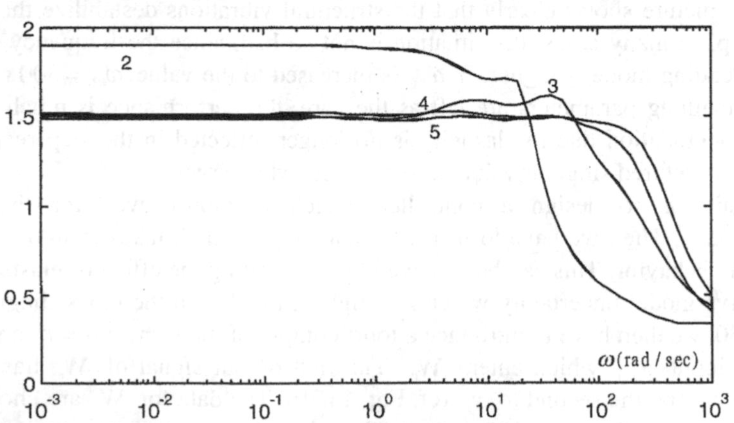

Fig. 14.40. Result of the D-K iteration with the added weight for the elasticity and the other weights left as before

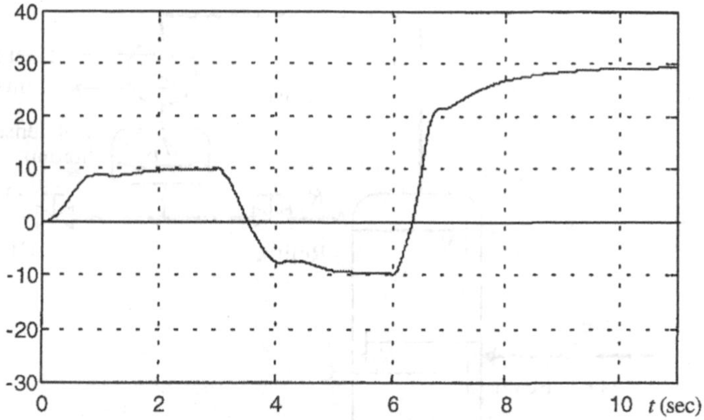

Fig. 14.41. Step response for the elastic missile with the improved robust controller

As can be seen, the uncertainty due to elasticity is within the bounds prescribed by the weight. Concerning performance, we first use the weights, which lead to the design in the last section. The result of the D-K iteration is shown in Fig. 14.40 (without step 1). Then the iteration converges to a value near $\mu = 1.5$, which is, due to the greater degree of uncertainty, much larger as in the previous design. Thus, robust performance is not achievable for this specification.

In order to get again a controller with guaranteed robust stability and performance, the performance requirements have to be relaxed. We try another approach with $\omega_1 = 4$, $S_\infty = 8$. Then the D-K iteration gives after 5 steps $\mu = 1.041$. Figure 14.41 shows the step response again for the sequence of commands $a_{zs} = 10g$, $-10g$, $30g$, but now with a change after every 3s. The response is now much slower, but even in this extreme case the missile can be robustly controlled. In a practical situation, one would probably prefer to make the missile stiffer, but with the μ synthesis it is possible to make a very fine tuned compromise between the conflicting requirements.

14.4 Robust Control of a Distillation Column

14.4.1 Nonlinear Model of the Column

Basic principle of distillation. An interesting MIMO control problem is given by the distillation process. Distillation is used in many chemical processes for separating feed streams for purification of final and intermediate product streams. The problem has attracted much attention in the last years since it serves as an example for an ill-conditioned plant. The basic principle of the distillation process can be explained with the help of Fig. 14.42. There a liquid stream which is a binary mixture (i.e. a mixture consisting of two components) has to be separated. This

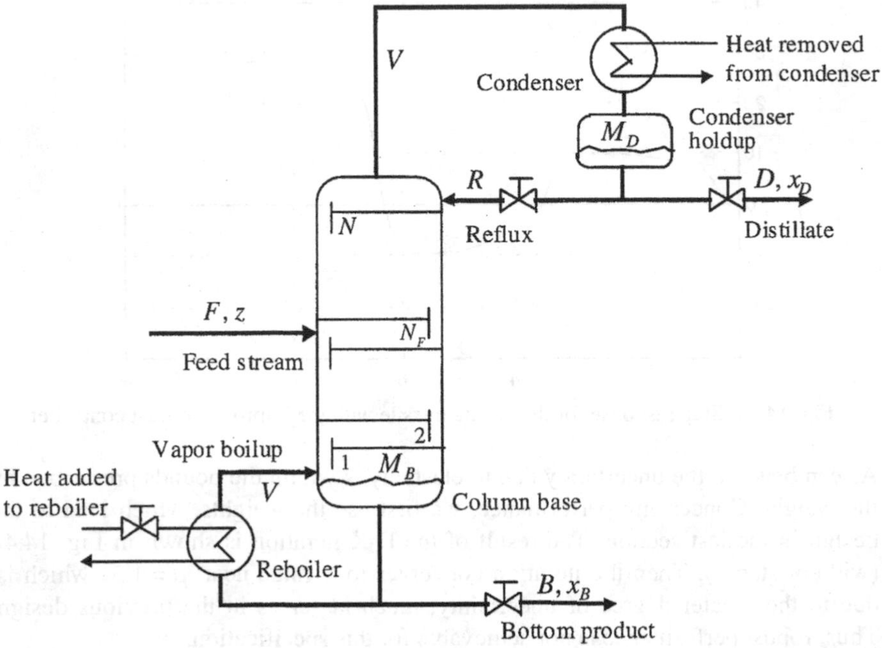

Fig. 14.42. Binary distillation column

is possible due to the different boiling points of the components. The pure component that boils at a lower temperature is denoted as the light component, whereas the component that boils at a higher temperature is denoted as the heavy component.

The feed stream is fed as saturated liquid with flow rate F (mol/min) onto the feed tray with number N_F. The liquid is assumed to be at its bubblepoint. The composition (mole fraction) of the light component is z. Vapor flows from stage to stage up the column, while liquid flows from stage to stage down the column. The overhead vapor is totally condensed in a condenser and flows into a reflux drum, whose holdup of liquid is M_D (moles). The content of the drum is assumed to be perfectly mixed with composition x_D of the light component. A portion of that liquid is returned as reflux, which is pumped back to the top tray (with number N) at rate R. The rest, the overhead distillate product, is removed at rate D by another pump.

At the base of the column, a portion of the liquid bottoms product is removed at a rate B and with composition x_B of the light component. The liquid contains a concentrated amount of the heavy component. The rest is vaporized in the reboiler and returned to the column. The reboiler generates vapor boilup at rate V. The liquids in the reboiler and in the base are also assumed to be perfectly mixed. Their holdup is denoted as M_B (moles).

The main task is to control the top and bottom compositions x_D and x_B. For a

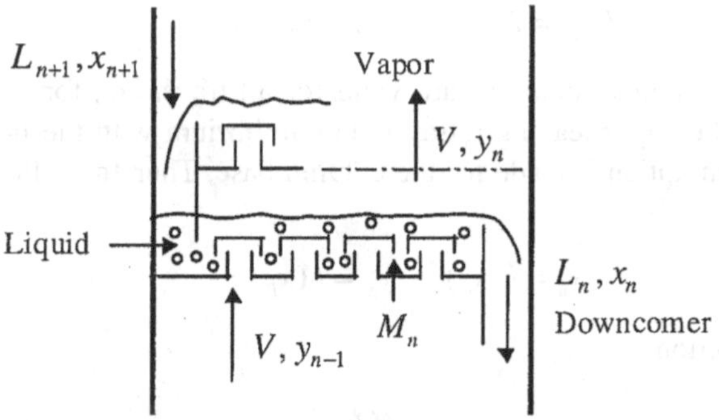

Fig. 14.43. Schematic diagram for a distillation column tray

stable operation of the distillation column it is also necessary to keep the holdups M_B and M_D constant. Thus the controlled variables are x_D, x_B, M_B, M_D. The inputs that can be manipulated are the rates R, V, D, B. They can be adjusted by the four valves depicted in Fig. 14.42. Perturbations occur if the feed stream F or the composition z differ from their nominal values.

Nonlinear mathematical model. We model now step by step the different components of the column.

n th tray $(n = 1, \ldots, N;\ n \neq N_F)$: The liquid from one tray goes over a weir and cascades down to the next tray through a downcomer. Thereby it comes into contact with the vapor from the tray below and we assume that, due to turbulence, vapor and liquid are perfectly mixed. The molar flow rate V of the vapor is assumed to be constant within the column. The light component vapor composition (mole fraction) is denoted by y_n and the light component liquid composition (mole fraction) is denoted by x_n. The vapor leaves the nth tray upwards in direction to the tray with number $n+1$ and the liquid moves downwards in direction to the tray with number $n-1$ (cf. Fig. 14.43). The liquid molar holdup on stage n is M_n and the liquid molar flow rate is L_n.

The dynamics of the nth tray can now be modeled by two continuity equations. The first one is the total continuity :

$$\frac{dM_n}{dt} = L_{n+1} - L_n,$$

and the second one is the continuity of the light component:

$$\frac{d(M_n x_n)}{dt} = L_{n+1} x_{n+1} - L_n x_n + V y_{n-1} - V y_n .$$

This holds for $2 \leq n \leq N-1$. For the bottom tray ($n = 1$) and the top tray ($n = N$)

we define

$$L_{N+1} = R,\ x_{N+1} = x_D,\ y_0 = y_B .$$

Then the above continuity equations are valid for all trays, i.e., for $n = 1,\ldots,N$. We suppose that the vapor leaving a tray is in equilibrium with the liquid on the tray. The same assumption is made for the column base. Then the following equations hold:

$$y_n = h(x_n),\quad y_B = h(x_B),$$

Here, h is the function

$$h(x) = \frac{\alpha x}{1 + (\alpha - 1)x} .$$

The constant α is denoted as relative volatility. A simple functional relationship is also assumed between the liquid rates L_n and the liquid holdups M_n:

$$L_n = f(M_n) .$$

The concrete form of f will be discussed later.

For the remaining components of the column, there are also two equations of continuity. These are given as follows.

Feed tray:

$$\frac{dM_{N_F}}{dt} = L_{N_F+1} - L_{N_F} + F$$

$$\frac{d(M_{N_F} x_{N_F})}{dt} = L_{N_F+1} x_{N_F+1} - L_{N_F} x_{N_F} + V y_{N_F-1} - V y_{N_F} + F z$$

Reboiler and column base:

$$\frac{dM_B}{dt} = L_1 - V - B$$

$$\frac{d(M_B x_B)}{dt} = L_1 x_1 - V y_B - B x_B$$

Condenser and reflux drum:

$$\frac{dM_D}{dt} = V - R - D$$

$$\frac{d(M_D x_D)}{dt} = V y_N - R x_D - D x_D .$$

Thus there are the $2(N+2)$ differential equations for the $2(N+2)$ states

$$M_B, M_1, \ldots, M_N, M_D, \; x_B, x_1, \ldots, x_N, x_D .$$

In these equations, the quantities $L_1, \ldots, L_N$ result from $M_1, \ldots, M_N$ and the quantities $y_B, y_1, \ldots, y_N$ result from $x_B, x_1, \ldots, x_N$. Hence R, V, D, B and F, z remain as free variables. The first four ones are the controls and the last two ones are the disturbances. The variables to be controlled are x_D, x_B, M_D, M_B. The parameters of the plant are the number N of trays, the number N_F of the feed tray, the relative volatility α, and the function $f(x)$.

14.4.2 Nonlinear Model in the Case of Perfect Level Control

Equilibrium point. Let δ_{mn} be the Kronecker symbol. For an equilibrium point, the equations are:

$$L_{10} - V_0 - B_0 = 0$$

$$L_{n+1,0} - L_{n,0} + \delta_{nN_F} F_0 = 0 \qquad \text{for } n = 1, \ldots, N$$

$$V_0 - R_0 - D_0 = 0$$

and

$$L_{10} x_{10} - V_0 y_{B0} - B_0 x_{B0} = 0$$

$$L_{n+1,0} x_{n+1,0} - L_{n0} x_{n0} + V_0 y_{n-1,0} - V_0 y_{n0} + \delta_{nN_F} F_0 z_0 = 0 \qquad \text{for } n = 1, \ldots, N$$

$$V_0 y_{N0} - R_0 x_{D0} - D_0 x_{D0} = 0 .$$

The first set of equations implies

$$B_0 = L_{10} - V_0 ,$$

$$L_{n0} = \begin{cases} R_0 + F_0, & \text{if } n = 1, \ldots, N_F, \\ R_0, & \text{if } n = N_F + 1, \ldots, N+1, \end{cases}$$

$$D_0 = V_0 - R_0 .$$

For given V_0, R_0, F_0, the remaining quantities can be calculated. The second set of equations consists of $N+2$ nonlinear equations for $x_{B0}, x_{10}, \ldots, x_{N0}, x_{D0}$.

Linear differential equations for the liquid holdups. We assume that the functional relationship between L_n and M_n is approximated by a linear function and is given in the form

$$L_n = L_{n0} + \frac{M_n - M_0}{T_L} \qquad n = 1, \ldots, N .$$

Here, T_L is a time constant (typically 3 to 6 s per tray), M_0 is a mean value of the liquid holdups in the trays and the quantities L_{n0} are the stationary liquid rates calculated above. Put with given M_{B0}, M_{D0}

$$\Delta M_n = M_n - M_0, \quad \Delta M_B = M_B - M_{B0}, \quad \Delta M_D = M_D - M_{D0} .$$

Then it is easy to see that (with $\gamma = 1/T_L$)

$$\Delta \dot{M}_B = \gamma \Delta M_1 - \Delta V - \Delta B \tag{14.6}$$

$$\Delta \dot{M}_n = \gamma \Delta M_{n+1} - \gamma \Delta M_n + \delta_{nN_F} \Delta F \qquad n = 1, \ldots, N-1$$

$$\Delta \dot{M}_N = -\gamma \Delta M_N + \Delta R$$

$$\Delta \dot{M}_D = \Delta V - \Delta R - \Delta D \tag{14.7}$$

The differential equations for the liquid holdups are now linear. The quantities describing the linearization are $M_{B0}, M_0, M_{D0}, R_0, V_0, F_0, z_0$.

Level controllers. The stable operation of the column requires that the liquid holdups are controlled. This can be achieved as described now. The holdup ΔM_B is governed by (14.6). We choose ΔB as control and consider ΔM_1 and ΔV as disturbances. This subsystem is a pure integrator with disturbances at the plant input and can be controlled by a PI controller such that for constant disturbances no control error arises. For the holdup ΔM_D which is governed by (14.7), we proceed analogously and choose ΔD as control. Then $\Delta V, \Delta R$ are disturbances and again a PI controller will be used. Hence, with these PI controllers it is possible to keep M_B near the stationary value M_{B0} and M_D near the stationary value M_{D0}. The holdups $M_1, \ldots, M_n$ are not affected by this control action. They depend only on the control ΔR and the disturbance ΔF. If $\Delta F = 0$, then it is easy to see that

$$\Delta \hat{M}_n = \frac{T_L}{(1 + T_L s)^{N+1-n}} \Delta \hat{R} \qquad \text{for } n = 1, \ldots, N .$$

Thus with these two PI controllers the liquid holdups can be stabilized. Normally this feedback loop has a much higher bandwidth than that for the compositions x_B, x_D. For simplicity, we assume therefore that level control perfectly works, i.e.

$$M_B = M_{B0}, \quad M_n = M_0, \quad M_D = M_{D0} .$$

Nonlinear differential equations for the compositions. The differential equations for the compositions can now be written as follows:

$$M_{B0}\dot{x}_B = L_1 x_1 - V h(x_B) - B x_B$$

$$M_0 \dot{x}_n = L_{n+1} x_{n+1} - L_n x_n + V h(x_{n-1}) - V h(x_n) + \delta_{nN_F} F z$$

$$M_{D0}\dot{x}_D = V h(x_N) - R x_D - D x_D \,.$$

The differential equations for the holdups are no longer required and replaced by the algebraic relationships

$$B = L_1 - V \,,$$

$$L_n = \begin{cases} R + F, & \text{if } n = 1, \ldots, N_F \,, \\ R, & \text{if } n = N_F + 1, \ldots, N + 1, \end{cases}$$

$$D = V - R \,.$$

With their help, the following nonlinear system of differential equations for the compositions is obtained:

$$M_{B0}\dot{x}_B = (R + F) x_1 - V h(x_B) - B x_B$$

$$M_0 \dot{x}_n = (R + F) x_{n+1} - (R + F) x_n + V h(x_{n-1}) - V h(x_n) \quad n = 1, \ldots, N_F - 1$$

$$M_0 \dot{x}_{N_F} = R x_{N_F+1} - (R + F) x_{N_F} + V h(x_{N_F-1}) - V h(x_{N_F}) + F z$$

$$M_0 \dot{x}_n = R x_{n+1} - R x_n + V h(x_{n-1}) - V h(x_n) \quad n = N_F + 1, \ldots, N$$

$$M_{D0}\dot{x}_D = V h(x_N) - R x_D - D x_D \,.$$

Herein, the quantities V, R, B, D are the controls. They cannot be independently manipulated because in the case of perfect level control, they obey the equations

$$D + R = V, \quad B + V - R = F \,.$$

If , as described before, B and D are used for level control, the variables V and R remain as controls for the compositions. This gives the so-called LV configuration of the distillation column. There are still other configurations, for example the DV configuration, which arises when the compositions are controlled with D and V. From a control theory point of view, the different configurations may differ considerably. We do not want to investigate this in detail and consider only the LV configuration.

Data of the model column and simulation results. All controller designs and simulations refer to the following model column:

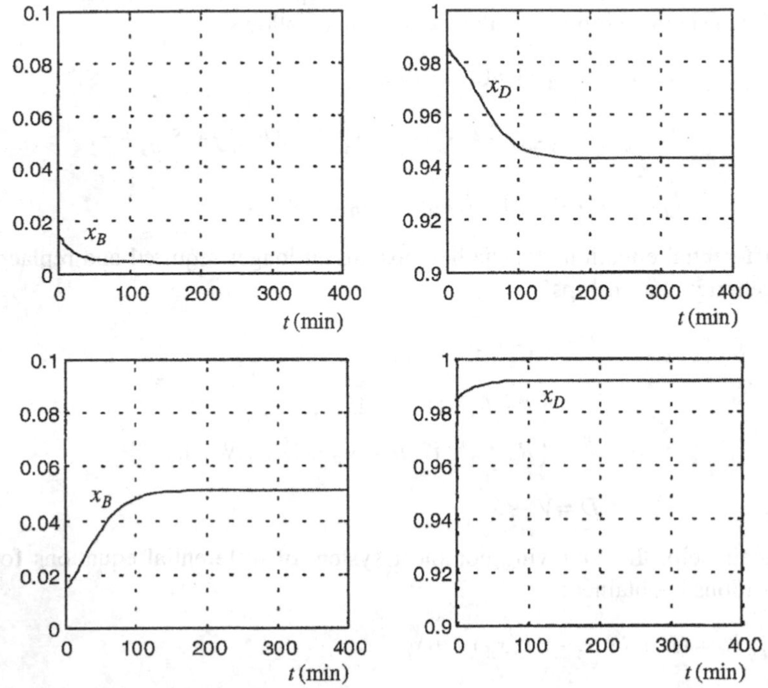

Fig. 14.44. Step response of x_B and x_D if V_0 is increased by 1% (upper plots) and if R_0 is increased by 1% (lower plots)

$$N = 30, \quad N_F = 16, \quad \alpha = 1.6, \quad z_0 = 0.5,$$
$$F_0 = 1\,\text{kmol/min}, \quad D_0 = 0.5F_0, \quad R_0 = 2.305F_0,$$
$$M_{B0} = 2\,\text{kmol}, \quad M_{D0} = 2\,\text{kmol}, \quad M_0 = 0.3\,\text{kmol}.$$

Thereby the feed stream is normed to 1. These data yield $x_{B0} = 0.015$ and $x_{D0} = 0.985$ which means that our column is designed for high-purity distillation.

Figure 14.44 shows the step response for x_B and x_D if the vapor flow rate and the reflux flow rate, respectively, are increased by 1%. The initial values are the corresponding stationary values.

14.4.3 Linear Plant Model and PI Controller

Linearizing the nonlinear system for the compositions in the usual way yields

$$\dot{\mathbf{x}} = \mathbf{A}\,\mathbf{x} + \mathbf{B}_1 \begin{pmatrix} \Delta z \\ \Delta F \end{pmatrix} + \mathbf{B}_2 \begin{pmatrix} \Delta V \\ \Delta R \end{pmatrix}.$$

Here the matrix $\mathbf{A}$ and the vector $\mathbf{x}$ have the form

$$\mathbf{A}=\begin{pmatrix} a_{00} & a_{01} & 0 & \cdots & 0 \\ a_{10} & a_{11} & a_{12} & \ddots & \vdots \\ 0 & \ddots & \ddots & \ddots & 0 \\ \vdots & \ddots & a_{N,N-1} & a_{NN} & a_{N,N+1} \\ 0 & \cdots & 0 & a_{N+1,N} & a_{N+1,N+1} \end{pmatrix}, \qquad \mathbf{x}=\begin{pmatrix} \Delta x_B \\ \Delta x_1 \\ \vdots \\ \Delta x_N \\ \Delta x_D \end{pmatrix}.$$

Hence $\mathbf{A}$ is tridiagonal. The matrices $\mathbf{B}_1, \mathbf{B}_2$ have no particular structure. The formulas for the coefficients that result from the partial derivatives are not given here. The outputs are the quantities Δx_B and Δx_D. The 2×2 transfer matrix of the system is denoted as $\mathbf{G}_0(s)$.

The linearized system has the order $N+2$, which is 32 in our example. Such a high order is typical for high-purity distillation. Some of the eigenvalues of $\mathbf{A}$ are

$$\lambda_1=-0.0148, \quad \lambda_2=-0.1684, \quad \lambda_3=-0.4427, \quad \lambda_4=-1.6651,$$
$$\lambda_5=-2.4980, \quad \lambda_{31}=-44.7266, \quad \lambda_{32}=-48.1834.$$

The ratio of the largest to the smallest eigenvalue is approximately 3250, which means that the system is extremely stiff.

For controller synthesis, it is reasonable to try to approximate the system by one of lower order. This succeeds here in a surprisingly simple way. Figure 14.45 shows the step response of $\mathbf{G}_0(s)$ and a first order approximation of $\mathbf{G}_0(s)$ of the following kind:

$$\mathbf{G}(s)=\frac{1}{1+Ts}\mathbf{G}_0(0)=\frac{1}{1+Ts}\begin{pmatrix} k_{11} & k_{12} \\ k_{21} & k_{22} \end{pmatrix}.$$

Since the step responses of both systems are very similar, controller synthesis will be done for this approximation $\mathbf{G}_0(s)$.

The numerical values of the approximation can be seen from the following formula (where the time constant is in minutes):

$$\mathbf{G}(s)=\frac{1}{1+65s}\begin{pmatrix} -1.1675 & 1.1447 \\ -0.7835 & 0.7952 \end{pmatrix}. \tag{14.8}$$

A minimal state space representation $\mathbf{G}(s)$ is given by

$$\begin{pmatrix} \dot{x}_1 \\ \dot{x}_2 \end{pmatrix}=-\frac{1}{T}\begin{pmatrix} 1 & 0 \\ 0 & 1 \end{pmatrix}\begin{pmatrix} x_1 \\ x_2 \end{pmatrix}+\frac{1}{T}\mathbf{G}_0(0)\begin{pmatrix} \Delta V \\ \Delta R \end{pmatrix}$$
$$y_1=\Delta x_B, \quad y_2=\Delta x_D.$$

The ratio of the largest to the smallest singular value (i.e. the condition number)

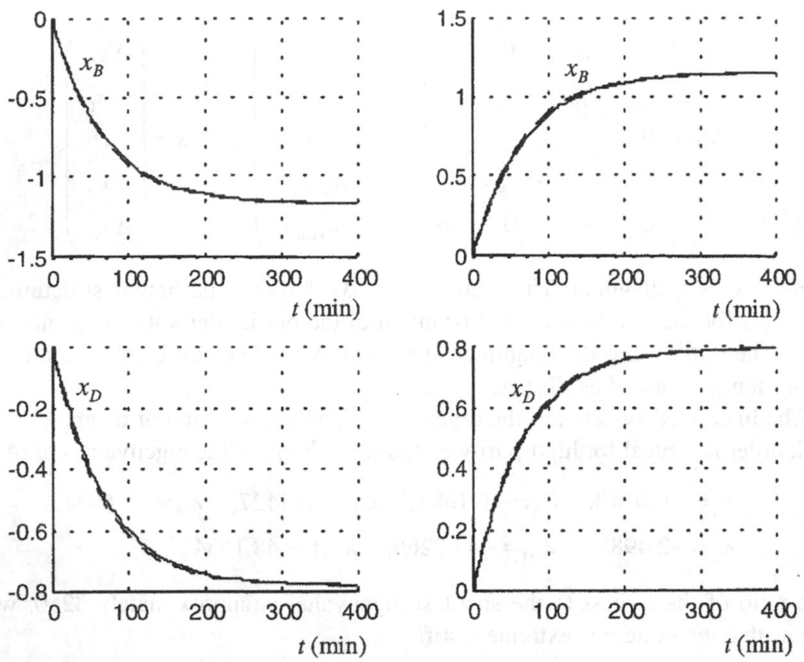

Fig. 14.45. Step response of the linearized system (solid line) and of the first-order approximation (dashed line)

of $\mathbf{G}_0(0)$ is

$$\kappa(\mathbf{G}_0(0)) = 124.3\,.$$

This is a very high value and means that the plant gain depends heavily on the combination of the plant inputs. This fact makes it difficult to design a robust controller for the column, much more than its high order.

Before designing a μ optimal controller, we investigate what is achievable with respect to robustness and performance by means of conventional methods. We suppose that the controls are corrupted by an additive error:

$$\Delta V = (1+\delta_1)\Delta V_c\,, \quad \Delta R = (1+\delta_2)\Delta R_c\,, \quad |\,\delta_i\,| \le 0.2,\; i = 1,2\,.$$

Here $\Delta V_c, \Delta R_c$ are the values demanded by the controller.

1. Two SISO PI controllers. The simplest approach is to ignore the cross-coupling inherent in the 2×2 system and to treat it as two single SISO systems. Then two separate controllers have to be designed for the transfer functions

$$G_{11}(s) = \frac{k_{11}}{1+Ts} \quad \text{and} \quad G_{22}(s) = \frac{k_{22}}{1+Ts}\,.$$

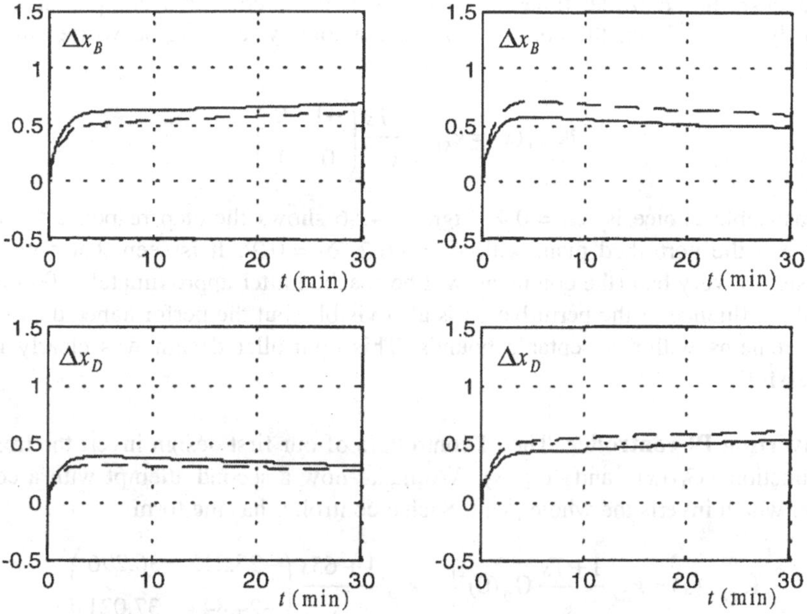

Fig. 14.46. Step response for the controlled column with the two SISO PI controllers (solid line: nominal system; dashed line: perturbed system)

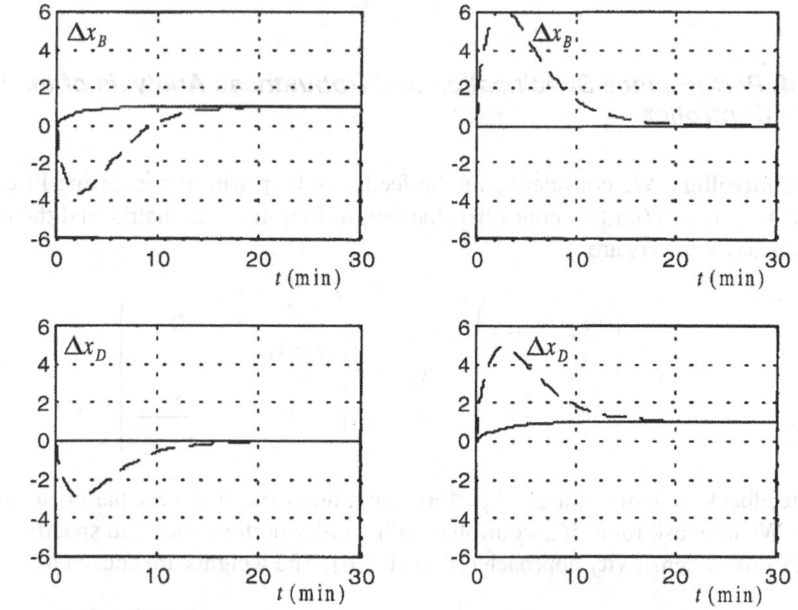

Fig. 14.47. Step response for the controlled column with the inverting PI controller (solid line: nominal system; dashed line: perturbed system

We choose the controller time constant such that the controller compensates the plant dynamics. Then the controller for the whole system can be written in the form

$$\mathbf{K}_{PI\,1}(s) = k_{R1}\frac{1+Ts}{s}\begin{pmatrix} -1 & 0 \\ 0 & 1 \end{pmatrix}.$$

A reasonable choice is $k_{R1} = 0.4$. Figure 14.46 shows the step response for the ideal and the perturbed plant with $\delta_1 = -0.2,\ \delta_2 = 0.2$. It is seen that both responses are very bad (the command will be reached after approximately 300 min) and the influence of the perturbation is also visible, but the performance degradation remains within acceptable bounds. This controller design was clearly not successful.

2. Inverting PI controller. The PI controllers of our first design invert the transfer functions $G_{11}(s)$ and $G_{22}(s)$. We make now a second attempt with a controller which inverts the whole plant. Such a controller has the form

$$\mathbf{K}_{PI\,2}(s) = k_{R2}\frac{1+Ts}{s}\mathbf{G}_0(0)^{-1} = k_{R2}\frac{1+65s}{s}\begin{pmatrix} -25.215 & 36.296 \\ -24.844 & 37.021 \end{pmatrix}.$$

The step responses (tracking) for the nominal plant and $k_{R2} = 0.4$ are now perfect but unacceptable for the perturbed plant (cf. Fig. 14.47).

14.4.4 Performance Specification and Robustness Analysis of the PI Controller

$\mathcal{H}_\infty$ Controller. We consider again the feedback loop with the inverting PI controller $\mathbf{K}_{PI\,2}(s)$. With this controller, the output loop transfer matrix and the output sensitivity matrix are

$$\mathbf{L}_o(s) = \begin{pmatrix} \dfrac{k_{R2}}{s} & 0 \\ 0 & \dfrac{k_{R2}}{s} \end{pmatrix}, \quad \mathbf{S}_o(s) = \begin{pmatrix} \dfrac{s}{s+k_{R2}} & 0 \\ 0 & \dfrac{s}{s+k_{R2}} \end{pmatrix}.$$

This feedback system has a good performance; however, it is very parameter sensitive. We now ask for a $\mathcal{H}_\infty$ controller with similar performance and specify it by a S/KS mixed-sensitivity approach (cf. Sect. 9.6). The weights are chosen as

$$\mathbf{W}_1 = \mathrm{diag}\,(W_1(s), W_2(s)) \quad \text{with } W_1(s) = \frac{s+\omega_1}{s+\omega_\varepsilon}$$

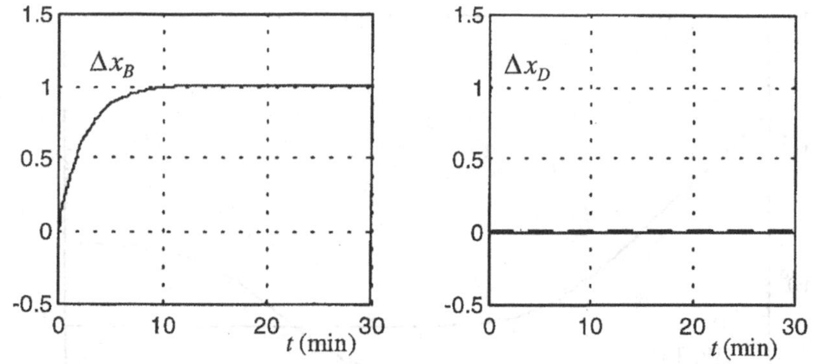

Fig. 14.48. Step response for the feedback loop with the inverting PI controller (solid line) and for the $\mathcal{H}_\infty$ controller (dashed line). Command: x_{Bc}; output signals Δx_B, Δx_D

$$\mathbf{W}_2 = \mathrm{diag}\,(c, c)$$

with the numerical values $\omega_1 = k_{R2} = 0.4$, $\omega_\varepsilon = 10^{-4}$, $c = 10^7$ and the resulting $\mathcal{H}_\infty$ controller is denoted by $\mathbf{K}_\infty$. Figure 14.48 shows that there is no visible difference between the step responses for the controllers $\mathbf{K}_{PI\,2}(s)$ and $\mathbf{K}_\infty$. The first diagonal element of the $\mathcal{H}_\infty$ controller is

$$K_{11\infty} = -\frac{1305406\,(s + 4.265 \cdot 10^4)\,(s + 0.01538)}{(s + 7.848 \cdot 10^4)\,(s + 1082)\,(s + 0.0001)} \approx -10.08\,\frac{1 + 65\,s}{s}\,.$$

This is approximately a PI controller. A similar assertion can be made for the other components of the controller. Thus the above specification for the $\mathcal{H}_\infty$ controller essentially leads to the inverting PI controller $\mathbf{K}_{PI2}$.

Specification of the robust feedback loop. The above specification has to be completed by one for robustness. For this plant it is adequate to model uncertainty as input-multiplicative uncertainty (cf. Fig. 12.2). In this way, the neglected dynamics of the liquid rates at the plant input are taken into account. We select as the weight for robustness

$$\mathbf{W}_\Delta(s) = \mathrm{diag}(W_\Delta(s), W_\Delta(s)) \quad \text{with } W_\Delta(s) = 0.2\frac{s+1}{0.1s+1}\,.$$

Thus for small frequencies $0 \le \omega \le 1$ an uncertainty of 20% with respect to the plant gain is admitted. The maximum value of the uncertainty is 200% and is allowed for frequencies $\omega \ge 10$. For $\omega^* = 5$, the feasible uncertainty is 100%. With respect to performance and robustness, the feedback system is now completely specified. Figure 14.49 shows the magnitude plot of the weights for performance and robustness, but now with a reduced value of ω_1 ($\omega_1 = 0.04$). This

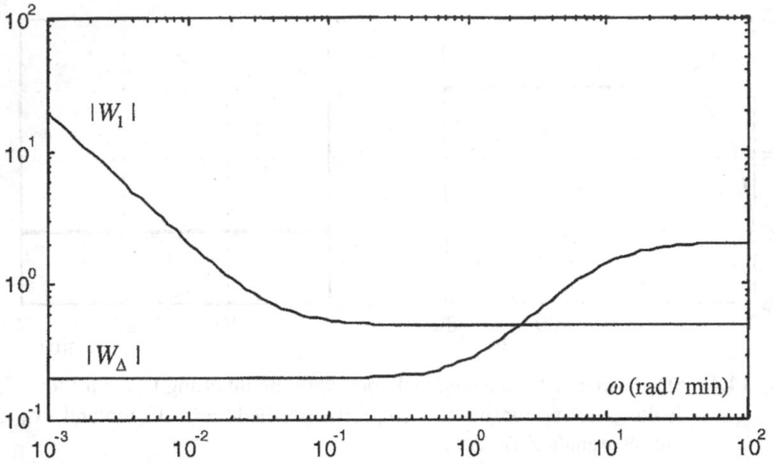

Fig.14.49. Magnitude plot of $W_1(s)$ and $W_\Delta(s)$

value will be used in the next section and there we also give the reason for this reduction. In the following robustness analysis we still use the old weight W_1 with the larger value $\omega_1 = 0.4$.

Robust stability for the inverting PI controller. Let $\mathbf{M}$ be defined as in Sect. 12.4. The maximum value of the norm for a feasible perturbation is by Theorem 13.4.1 the inverse of $\gamma = \sup \mu_\Delta(\mathbf{M}_{11}(i\omega))$. Fig. 14.50 shows $\mu_\Delta(\mathbf{M}_{11}(i\omega))$ for the relevant frequency interval. The maximum is $\gamma = 0.2$; hence the inverse is $1/\gamma = 5$ and the feedback system is robustly stable. In particular, a perturbation, which amounts to 20% of the input signal, does not destabilize the feedback loop as it was also to be expected from the results of the previous section.

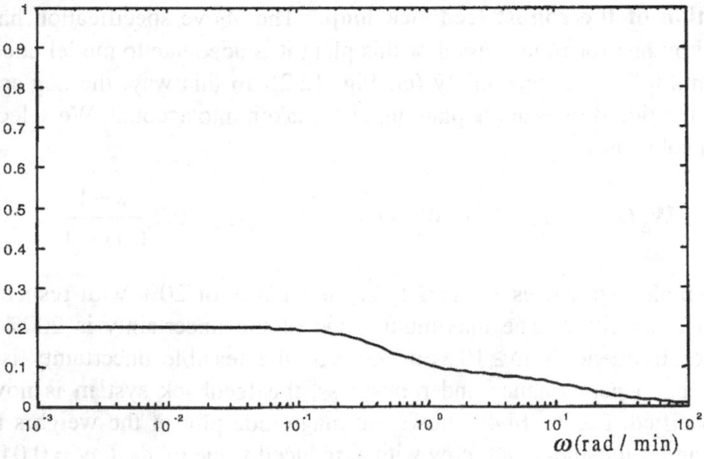

Fig. 14.50. Robust stability for the inverting PI controller

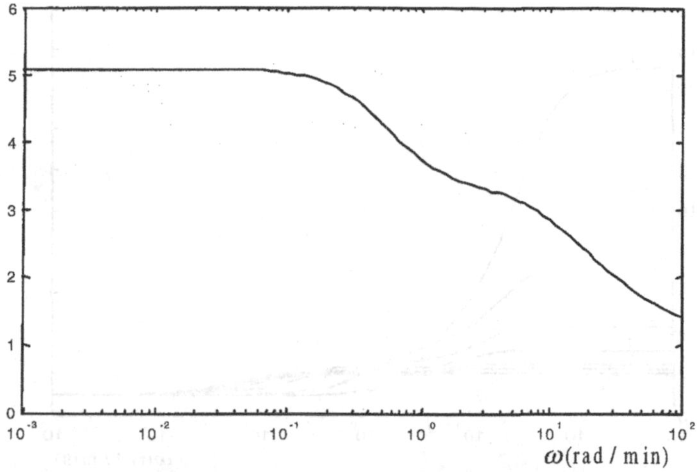

Fig. 14.51. Robust performance for the inverting PI controller

Next we investigate the controlled column with respect to robust performance. To this end, the structured singular value of $\mathbf{M}$ for the set Δ_P has to be calculated (cf. Theorem 13.4.2). In Fig. 14.51, the plot of $\mu_{\Delta_P}(\mathbf{M}(i\omega))$ is depicted. The figure shows that for robust performance only perturbations with a much smaller norm than for robust stability are allowed. The maximum value for the norm of a feasible perturbation is now $1/5.09 = 0.196$. Thus robust performance is not achievable with this controller.

14.4.5 μ Optimal Controller

Controller Synthesis with the D-K iteration. As we have seen, the inverting PI controller and the $\mathcal{H}_\infty$ controller are very sensitive with respect to a disturbance at the plant input. As for the previous case studies, using the D-K iteration it is possible to design a robust controller. All the necessary data have been already given, since we use the extended plant with the weights specified in the previous section. Figure 14.52 shows the μ plots for the first 7 iteration steps. In the last iteration, the maximum of the μ plot is approximately equal to 2, which means that robust performance is not achievable for this specification. Since the robustness property has priority, we relax the performance requirement and replace $W_1(s)$ by

$$W_1(s) = \frac{1}{2}\frac{s + 0.04}{s + \omega_\varepsilon}.$$

This reduces the bandwidth of the feedback loop by the factor 0.1 and the $\mathcal{H}_\infty$ norm of the sensitivity function is now bounded by 2 instead of 1. The result of the D-K iteration is shown in Fig 14.53. Now the value $\mu = 1.042$ is achieved,

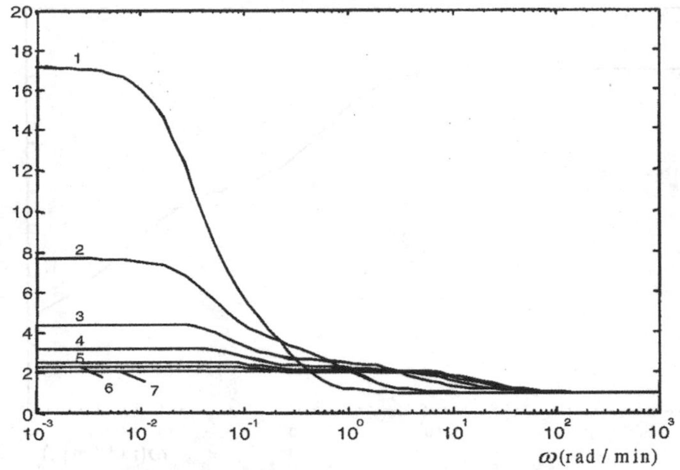

Fig. 14.52. D-K iteration for the weights of Sect. 14.4.4

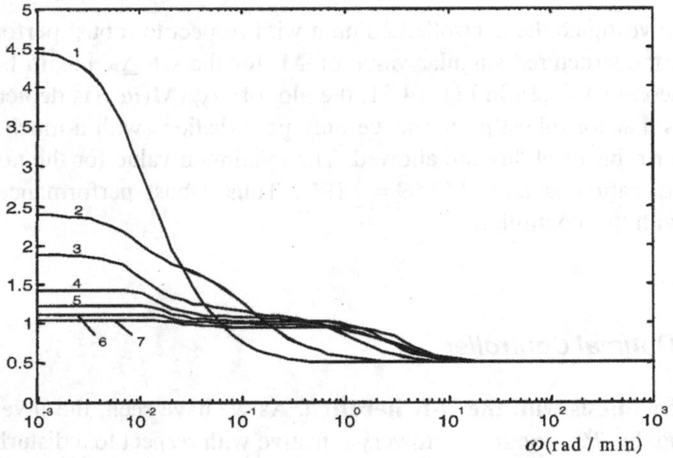

Fig. 14.53. D-K iteration for a relaxed performance requirement

which means that we have robust stability and robust performance. The order of the controller is 22.

Simulation result. Figure 14.54 shows the step response of Δx_B. It first increases very rapidly, but then approaches the command rather slowly. In order to get a more balanced response, it is possible to use a prefilter. This means that the old reference signal r is replaced by a new one R. The connection between these quantities is given by (with $T_r = 7$).

$$\hat{r}(s) = \frac{1}{1+T_r s}\hat{R}(s)\,.$$

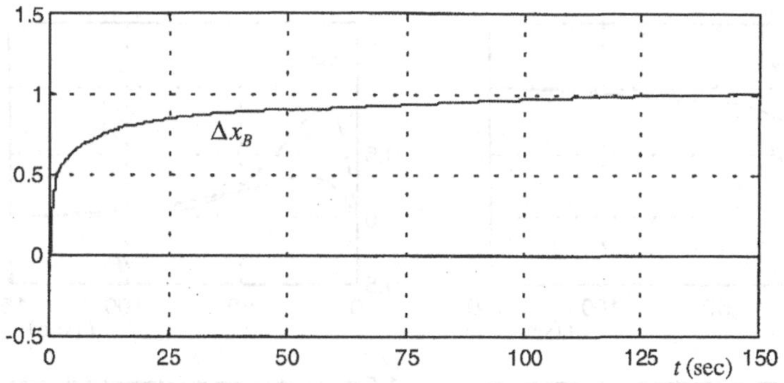

Fig. 14.54. Step response of Δx_B

As can be seen from Fig. 14.54, the prefilter has the desired effect. These pictures can directly be compared with the step responses for the inverting PI controller (but now with $k_{R2} = 0.04$). The μ controller, of course, does not work as perfectly as the inverting PI controller, but the degradation is only moderate. The main difference is that now there is an interaction between vapor rate and reflux. It should be noted that the PI controller works without a prefilter.

It is of particular interest to compare the step response of the μ controller with that of the inverting PI controller for the nominal and the perturbed plant (with the

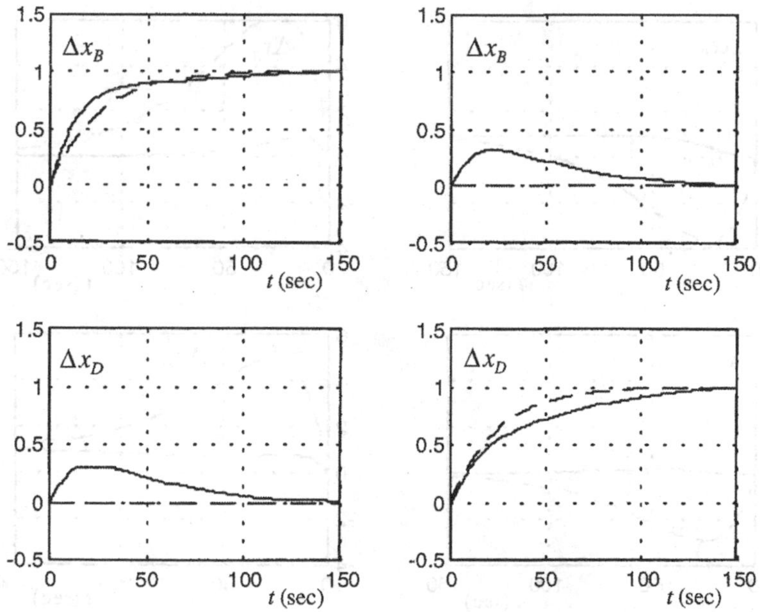

Fig. 14.55. Step response (nominal plant) for the μ controller (solid line) and the inverting PI controller (dashed line).

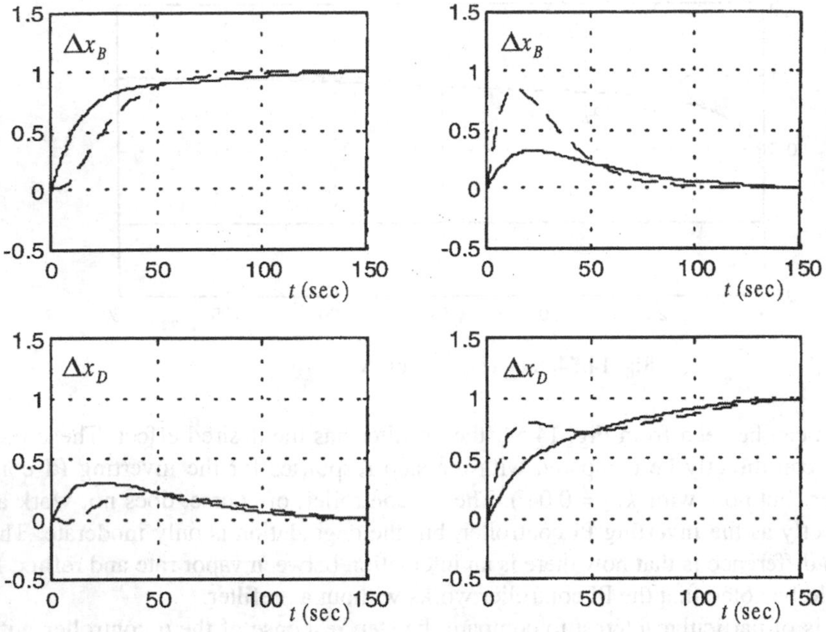

Fig. 14.56. Step responses for the μ controller: nominal plant (solid line) and perturbed plant (dashed line)

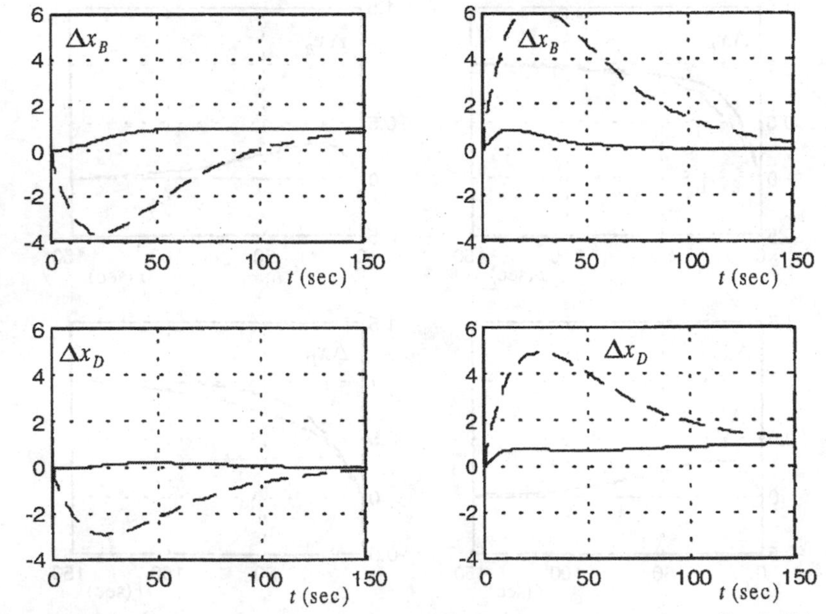

Fig. 14.57. Step responses for the perturbed plant with the μ controller (solid line) and the inverting PI controller (dashed line).

perturbation of the previous section). Figure 14.56 shows the degradation of the step response for the μ controller due to the constant perturbation at the plant input and it is seen that the μ controller still works pretty well for the perturbed plant. This is in contrast to the inverting PI controller, where the effect of the perturbation is disastrous (Fig. 14.57; see also Fig. 14.47).

Notes and References

In the last two decades the methods of robust control have been applied to a large variety of practical problems coming from very different disciplines of the engeneering sciences and branches of industries. In particular, this holds for aircraft and missile technology. There is a great number of papers which report on the successful autopilot design by $\mathcal{H}_\infty$ and μ techniques; see in particular the Journal of Guidance, Control and Dynamics. Some of them are listed in the bibliography. Here we mention only Apkarian and Biannic [5], Fialho et. al. [40], Balas, Lind and Packard [8], and Ferreres and M'Saad [41], but this selection is more or less arbitrary.

An early paper concerning rudder roll stabilization of ships is Van Amerongen, Van der Klugt and Van Nauta Lemke [99]; see also Fossen [44]. An approach using $\mathcal{H}_\infty$ and μ techniques can be found in Yang [108]. The problem of controlling a distillation column has attracted much attention resulting in many papers on this subject. Robust control methods are applied in particular by Skogestad, Morari and Doyle [87] and Hoyle, Hyde, and Limebeer [50].

Appendix A

Mathematical Background

A.1 Linear Algebra

A.1.1 Linear Mappings and Matrices

We first present some basic results from linear algebra that are used in this book. Proofs are only given occasionally.

Let $\mathbb{F}$ be the field $\mathbb{R}$ of real numbers or the field $\mathbb{C}$ of complex numbers and suppose that $\{\mathbf{v}_1,\dots,\mathbf{v}_n\}$ is a basis of $\mathbb{F}^n$ and that $\mathbf{x}\in\mathbb{F}^n$ is an arbitrary vector. Then there exist uniquely determined numbers $\alpha_1,\dots,\alpha_n$ such that

$$\mathbf{x}=\alpha_1\mathbf{v}_1+\dots+\alpha_n\mathbf{v}_n .$$

We define the coordinate vector

$$\mathbf{x}_V=\begin{pmatrix}\alpha_1\\ \vdots\\ \alpha_n\end{pmatrix}.$$

Let $\{\mathbf{u}_1,\dots,\mathbf{u}_n\}$ be a second basis of $\mathbb{F}^n$. Each basis vector $\mathbf{u}_k$ can be represented as

$$\mathbf{u}_k=t_{1k}\mathbf{v}_1+\cdots+t_{nk}\mathbf{v}_n .$$

Define the matrix

$$\mathbf{T}=\begin{pmatrix}t_{11}&\cdots&t_{1n}\\ \vdots&\ddots&\vdots\\ t_{n1}&\cdots&t_{nn}\end{pmatrix}.$$

This matrix is nonsingular and it is easy to see that the following relationship between the coordinate vectors holds:

$$\mathbf{x}_v = \mathbf{T}\mathbf{x}_u .$$

Let $A: \mathbb{F}^n \to \mathbb{F}^m$ be a **linear mapping**, i.e. a mapping such that

$$A(\alpha\mathbf{v} + \beta\mathbf{w}) = \alpha A(\mathbf{v}) + \beta A(\mathbf{w})$$

for every $\mathbf{v}, \mathbf{w} \in \mathbb{F}^n$ and all $\alpha, \beta \in \mathbb{F}^n$. Suppose $\mathbf{x} \in \mathbb{F}^n$ and let $\{\mathbf{w}_1, \ldots, \mathbf{w}_m\}$ be a basis of $\mathbb{F}^m$. Then there are scalars a_{ik} such that

$$A\mathbf{v}_k = a_{1k}\mathbf{w}_1 + \cdots + a_{mk}\mathbf{w}_m .$$

If $\mathbf{x} \in \mathbb{F}^n$ is represented as above, we get

$$A\mathbf{x} = \sum_{i=1}^{m} \left(\sum_{k=1}^{n} a_{ik}\alpha_k \right) \mathbf{w}_i .$$

Defining the $m \times n$-matrix

$$\mathbf{A} = \begin{pmatrix} a_{11} & \cdots & a_{1n} \\ \vdots & \ddots & \vdots \\ a_{m1} & \cdots & a_{mn} \end{pmatrix},$$

one sees that the coordinate vectors $\mathbf{x}_v$ of $\mathbf{x}$ and $\mathbf{y}_w$ of $\mathbf{y} = A\mathbf{x}$ are related by

$$\mathbf{y}_w = \mathbf{A}\mathbf{x}_v .$$

In particular, if $m = n$ and $\mathbf{w}_k = \mathbf{v}_k$ for $k = 1, \ldots, n$, then $\mathbf{y}_v = \mathbf{A}\mathbf{x}_v$. In this sense, it is possible to identify the linear mapping A with the matrix $\mathbf{A}$. This will normally be done in the book.

Let now A be a linear map from $\mathbb{F}^n$ to $\mathbb{F}^n$. We wish to describe A with respect to the second basis $\{\mathbf{u}_1, \ldots, \mathbf{u}_n\}$. From our considerations up to now, the following calculation can be made:

$$\mathbf{y}_u = \mathbf{T}^{-1}\mathbf{y}_v = \mathbf{T}^{-1}\mathbf{A}\mathbf{x}_v = \mathbf{T}^{-1}\mathbf{A}\mathbf{T}\mathbf{x}_u .$$

Thus, with respect to the new basis, the matrix representation of A is $\mathbf{T}^{-1}\mathbf{A}\mathbf{T}$. Such a transformation of $\mathbf{A}$ is denoted as a **similarity transformation.**

The **kernel** of $\mathbf{A}$ is defined by

$$\mathrm{Ker}\,\mathbf{A} = \{\mathbf{x} \in \mathbb{R}^n \mid \mathbf{A}\mathbf{x} = \mathbf{0}\}$$

and the **image** of $\mathbf{A}$ is

$$\mathrm{Im}\,\mathbf{A} = \{\mathbf{A}\mathbf{x} \mid \mathbf{x} \in \mathbb{R}^n\}.$$

If $\mathbf{a}_1, \ldots, \mathbf{a}_n$ are the columns of $\mathbf{A}$, then

$$\mathrm{Im}\,\mathbf{A} = \mathrm{span}\{\mathbf{a}_1, \ldots, \mathbf{a}_n\},$$

where span$\{\mathbf{a}_1,\ldots,\mathbf{a}_n\}$ denotes the set of all linear combinations of $\mathbf{a}_1,\ldots,\mathbf{a}_n$. Kernel and image are linear subspaces. For their dimensions the following relationship holds:

$$\dim(\operatorname{Im}\mathbf{A}) + \dim(\operatorname{Ker}\mathbf{A}) = n\,.$$

The **rank** of a matrix $\mathbf{A}$ is defined by

$$\operatorname{rank}(\mathbf{A}) = \dim(\operatorname{Im}\mathbf{A})\,.$$

The rank equals the maximum number of linearly independent rows or columns. A matrix $\mathbf{A} \in \mathbb{R}^{m\times n}$ is said to have **full row rank** if $m \le n$ and $\operatorname{rank}(\mathbf{A}) = m$. Dually, it is said to have **full column rank** if $n \le m$ and $\operatorname{rank}(\mathbf{A}) = n$.

A1.2 Eigenvalues and Eigenvectors

Let $\mathbf{A} \in \mathbb{C}^{n\times n}$. Then a number $\lambda \in \mathbb{C}$ is denoted as an **eigenvalue** of $\mathbf{A}$ if there is a so-called **eigenvector** $\mathbf{x} \in \mathbb{C}^n$, $\mathbf{x} \ne \mathbf{0}$, such that

$$\mathbf{A}\,\mathbf{x} = \lambda\mathbf{x}\,.$$

Equivalently this means that $\operatorname{Ker}(\lambda\mathbf{I} - \mathbf{A}) \ne \mathbf{0}$ or that $\lambda\mathbf{I} - \mathbf{A}$ is nonsingular. Therefore, since the determinant of a matrix vanishes if and only if the matrix is nonsingular, λ is an eigenvalue of $\mathbf{A}$ if and only if

$$\det(\lambda\mathbf{I} - \mathbf{A}) = 0\,.$$

Regarding λ as a variable, we call the polynomial

$$p_A(\lambda) = \det(\lambda\mathbf{I} - \mathbf{A}) = \lambda^n + a_{n-1}\lambda^{n-1} + \ldots + a_1\lambda + a_0$$

the **characteristic polynomial** of $\mathbf{A}$. It can be decomposed as

$$\det(\lambda\mathbf{I} - \mathbf{A}) = (\lambda - \lambda_1)\cdots(\lambda - \lambda_n)\,.$$

If $\mathbf{A}$ is a real matrix, then complex eigenvalues appear only as conjugate complex pairs. Moreover, a matrix $\mathbf{A}$ is nonsingular if and only if $\lambda = 0$ is an eigenvalue of $\mathbf{A}$.

Suppose that there are n linearly independent eigenvectors $\mathbf{x}_1,\ldots,\mathbf{x}_n$ of $\mathbf{A}$. Then with the matrices

$$\mathbf{D} = \operatorname{diag}(\lambda_1,\ldots,\lambda_n)\,,\quad \mathbf{T} = (\mathbf{x}_1 \;\cdots\; \mathbf{x}_n)$$

the following holds:

$$\mathbf{AT} = (\mathbf{A}\mathbf{x}_1 \;\cdots\; \mathbf{A}\mathbf{x}_n) = (\lambda_1\mathbf{x}_1 \;\cdots\; \lambda_n\mathbf{x}_n) = \mathbf{TD}\,.$$

Thus there is a similarity transformation such that

$$\mathbf{T}^{-1}\mathbf{AT} = \mathbf{D},$$

and in this case the matrix $\mathbf{A}$ is said to be **diagonizable**. A matrix $\mathbf{A}$ is in particular diagonizable if the eigenvalues of $\mathbf{A}$ are pairwise distinct: $\lambda_k \neq \lambda_j$ for $k \neq j$.

There must not necessarily exist a basis of eigenvectors of $\mathbb{C}^n$, but in any case there exists a basis of principal vectors. This is described in Sect 5.2.2. With these principal vectors it possible to derive a decomposition $\mathbf{A} = \mathbf{T}\mathbf{J}\mathbf{T}^{-1}$, where $\mathbf{J}$ is the **Jordan normal form**, but this normal form is not needed in the book.

The maximum modulus of the eigenvalues is called the **spectral radius**, which is denoted by

$$\rho(\mathbf{A}) = \max\{\,|\lambda_i|\,|\,i = 1,\dots,n\}.$$

The following important result holds.

Theorem A.1.1 (Cayley-Hamilton). *Let* $\mathbf{A} \in \mathbb{C}^{n\times n}$. *Then*

$$p_A(\mathbf{A}) = \mathbf{0}.$$

Let $\mathbf{A}$ be a $n \times n$-matrix and let $\mathcal{V}$ be a linear subspace of $\mathbb{C}^n$. The subspace $\mathcal{V}$ is said to be **$\mathbf{A}$-invariant** if $\mathbf{A}\mathbf{x} \in \mathcal{V}$ for every $\mathbf{x} \in \mathcal{V}$. For example, $\operatorname{Im}\mathbf{A}$ and $\operatorname{Ker}\mathbf{A}$ are $\mathbf{A}$-invariant subspaces. An eigenvector of $\mathbf{A}$ spans a further $\mathbf{A}$-invariant subspace. If $\mathcal{V}$ is a nontrivial $\mathbf{A}$-invariant subspace, then there is a λ and a $\mathbf{x} \in \mathcal{V}, \mathbf{x} \neq \mathbf{0}$ such that $\mathbf{A}\mathbf{x} = \lambda\mathbf{x}$.

Let now a linear mapping $A: \mathbb{C}^n \to \mathbb{C}^n$ be given and suppose that $\mathcal{V}$ is an A-invariant subspace with basis $\{\mathbf{v}_1,\dots,\mathbf{v}_l\}$, which is completed by vectors $\{\mathbf{v}_{l+1},\dots,\mathbf{v}_n\}$ to a basis of $\mathbb{C}^n$. Let $\mathbf{A} = (a_{ik})$ be the matrix representing A with respect to this basis. Then for $1 \le k \le l$

$$A\mathbf{v}_k = \sum_{i=1}^{n} a_{ik}\mathbf{v}_i = \sum_{i=1}^{l} a_{ik}\mathbf{v}_i .$$

This means $a_{ik} = 0$ for $k \le l$ and $i > l$, i.e. the matrix $\mathbf{A}$ can be partitioned in the form

$$\mathbf{A} = \begin{pmatrix} \mathbf{A}_{11} & \mathbf{A}_{12} \\ \mathbf{0} & \mathbf{A}_{22} \end{pmatrix}.$$

A.1.3 Self-Adjoint, Unitary and Positive Definite Matrices

A matrix $\mathbf{Q} \in \mathbb{C}^{n\times n}$ is **self-adjoint** or **Hermitian** if

$$\mathbf{Q}^* = \mathbf{Q}.$$

Let $\mathbf{Q}\mathbf{x} = \lambda\mathbf{x}$ with $\mathbf{x} \neq \mathbf{0}$. Then we have

$$\bar{\lambda}\,\mathbf{x}^*\mathbf{x} = (\mathbf{Qx})^*\,\mathbf{x} = \mathbf{x}^*\mathbf{Qx} = \lambda\,\mathbf{x}^*\mathbf{x}\,.$$

Consequently, the eigenvalues of a self-adjoint matrix are always real.

Two vectors $\mathbf{x}, \mathbf{y} \in \mathbb{C}^n$ are **orthogonal** if $\mathbf{y}^*\mathbf{x} = \mathbf{0}$. A set of vectors $\{\mathbf{v}_1, \ldots, \mathbf{v}_r\}$ is **orthonormal** if

$$\mathbf{v}_i^*\mathbf{v}_j = \begin{cases} 1, & \text{if } i = k, \\ 0, & \text{if } i \neq k. \end{cases}$$

Orthonormal vectors are always linearly independent and it is always possible to complete a set of orthonormal vectors with $r < n$ to an orthonormal basis.

A matrix $\mathbf{U} \in \mathbb{C}^{n \times n}$ is called **unitary** if

$$\mathbf{U}^*\mathbf{U} = \mathbf{I}\,.$$

This means that the columns of are orthonormal. Moreover, $\mathbf{U}^{-1} = \mathbf{U}^*$ and therefore $\mathbf{UU}^* = \mathbf{I}$. A key property of unitary matrices is that they leave the norm invariant:

$$\|\,\mathbf{Ux}\,\|^2 = \mathbf{x}^*\mathbf{U}^*\mathbf{Ux} = \|\,\mathbf{x}\,\|^2\,.$$

A real unitary matrix is called **orthogonal.**

Theorem A.1.2. *Let* $\mathbf{A}$ *be a self-adjoint matrix. Then there exist a unitary matrix* $\mathbf{U}$ *and a real diagonal matrix* $\mathbf{D}$ *such that*

$$\mathbf{A} = \mathbf{UDU}^*\,.$$

A self-adjoint matrix $\mathbf{Q}$ with the property

$$\mathbf{x}^*\mathbf{Q}\,\mathbf{x} > 0 \quad \text{for every } \mathbf{x} \neq \mathbf{0}$$

is called **positive definite.** If ">" is relaxed to "≥", the matrix is denoted as **positive semidefinite.** In a similar way, **negative definite** and **negative semidefinite** matrices are defined. The following result is an immediate consequence of Theorem A.1.2.

Corollary A.1.1. *A self-adjoint matrix* $\mathbf{Q}$ *is positive definite (positive semidefinite) if and only if* $\lambda > 0$ *(* $\lambda \geq 0$ *) for every eigenvalue* λ *of* $\mathbf{Q}$.

The largest eigenvalue of a positive semidefinite matrix is denoted by $\lambda_{\max}$. It is possible to define a square root for positive semidefinite matrices. Let $\mathbf{Q} = \mathbf{UDU}^*$. Then with

$$\mathbf{D} = \operatorname{diag}(\lambda_1^{1/2}, \ldots, \lambda_n^{1/2})$$

the matrix

$$\mathbf{Q}^{1/2} = \mathbf{UD}^{1/2}\mathbf{U}^*$$

has the properties

$$\mathbf{Q}^{1/2} \geq \mathbf{0}, \quad \mathbf{Q}^{1/2}\mathbf{Q}^{1/2} = \mathbf{Q}.$$

It is now possible to define an ordering for matrices. We write $\mathbf{Q} > \mathbf{S}$ for two self-adjoint matrices $\mathbf{Q}, \mathbf{S}$ if and only if $\mathbf{Q} - \mathbf{S} > \mathbf{0}$. The relationships "≥", "<" and "≤" are similarly defined. It is easy to see that this ordering has the following properties:

If $\mathbf{Q} > \mathbf{0}$ *and* $\mathbf{A} \in \mathbb{C}^{n \times m}$, *then* $\mathbf{A}^*\mathbf{QA} \geq \mathbf{0}$. *If* $\operatorname{Ker} \mathbf{A} = \mathbf{0}$, *then* $\mathbf{A}^*\mathbf{QA} > \mathbf{0}$.

If $\mathbf{Q}_1 > \mathbf{0}$, $\mathbf{Q}_2 > \mathbf{0}$ *and* $\alpha_1 > 0$, $\alpha_2 \geq 0$, *then* $\alpha_1\mathbf{Q}_1 + \alpha_2\mathbf{Q}_2 > 0$.

Lemma A.1.1. *For a self-adjoint, positive semidefinite* $\mathbf{Q}$ *the following holds:*

(a) $\|\mathbf{Q}\|_2 = \rho(\mathbf{Q})$;

(b) $\rho(\mathbf{Q}^{1/2}) = \sqrt{\rho(\mathbf{Q})}$;

(c) $\rho(\mathbf{Q}) < 1$ *if and only if* $\mathbf{I} - \mathbf{Q} > \mathbf{0}$;

(d) if additionally $\mathbf{S}$ *is a self-adjoint, positive semidefinite matrix, then*

$$\mathbf{Q} > \mathbf{S} \quad \textit{if and only if} \quad \rho(\mathbf{S}\mathbf{Q}^{-1}) < 1.$$

Proof. (a) Let $\mathbf{Q}$ be decomposed as in Theorem A.1.2. Then

$$\mathbf{Q}^*\mathbf{Q} = \mathbf{U}\mathbf{D}^2\mathbf{U}^*$$

and consequently

$$\|\mathbf{Q}\|_2 = \sqrt{\lambda_{\max}(\mathbf{Q}^*\mathbf{Q})} = \lambda_{\max}(\mathbf{Q}) = \rho(\mathbf{Q}).$$

(b) This follows from the definition of $\mathbf{Q}^{1/2}$.
(c) Using (a) and (b), the assertion is true since

$$\begin{aligned} \rho(\mathbf{Q}) < 1 \quad &\Leftrightarrow \quad \sqrt{\rho(\mathbf{Q})} < 1 \\ &\Leftrightarrow \quad \rho(\mathbf{Q}^{1/2}) < 1 \\ &\Leftrightarrow \quad \|\mathbf{Q}^{1/2}\|_2 < 1 \\ &\Leftrightarrow \quad \sup_{\mathbf{x} \neq \mathbf{0}} \frac{\|\mathbf{Q}^{1/2}\mathbf{x}\|_2}{\|\mathbf{x}\|_2} < 1. \end{aligned}$$

(d) We have the following equivalencies:

$$\begin{aligned}\mathbf{Q} > \mathbf{S} \quad &\Leftrightarrow \quad \mathbf{I} - \mathbf{Q}^{-1/2}\mathbf{S}\mathbf{Q}^{-1/2} > \mathbf{0} \\ &\Leftrightarrow \quad \mathbf{I} - \mathbf{Q}^{-1/2}(\mathbf{S}\mathbf{Q}^{-1})\mathbf{Q}^{1/2} > \mathbf{0} \\ &\Leftrightarrow \quad \rho(\mathbf{Q}^{-1/2}(\mathbf{S}\mathbf{Q}^{-1})\mathbf{Q}^{1/2}) < 1 \qquad \text{(because of (c))} \\ &\Leftrightarrow \quad \rho(\mathbf{S}\mathbf{Q}^{-1}) < 1 \quad \text{(since } \mathbf{S}\mathbf{Q}^{-1} \text{ and } \mathbf{Q}^{-1/2}(\mathbf{S}\mathbf{Q}^{-1})\mathbf{Q}^{1/2} \text{ are similar)}.\end{aligned}$$

Theorem A.1.3. *Let* $\mathbf{Q}$ *and* $\mathbf{R}$ *be positive definite matrices. Then there exists a nonsingular matrix* $\mathbf{T}$ *such that*

$$\mathbf{T}\mathbf{Q}\mathbf{T}^T = (\mathbf{T}^{-1})^T\mathbf{R}\mathbf{T}^{-1} = \mathbf{D}$$

where $\mathbf{D}$ *is a diagonal, positive definite matrix.*

Proof. Since $\mathbf{Q}$ and $\mathbf{R}$ are positive definite, the matrix $\mathbf{Q}^{1/2}\mathbf{R}\mathbf{Q}^{1/2}$ is also positive definite. Thus, by Theorem A.1.2 there exist an orthogonal matrix $\mathbf{U}$ and a positive definite diagonal matrix $\mathbf{D}_1$ such that

$$\mathbf{Q}^{1/2}\mathbf{R}\mathbf{Q}^{1/2} = \mathbf{U}\mathbf{D}_1\mathbf{U}^T.$$

With $\mathbf{D} = \mathbf{D}_1^{1/2}$, this implies

$$\mathbf{D}^{-1/2}\mathbf{U}^T\mathbf{Q}^{1/2}\mathbf{R}\mathbf{Q}^{1/2}\mathbf{U}\mathbf{D}^{-1/2} = \mathbf{D}.$$

Now set $\mathbf{T} = \mathbf{D}^{1/2}\mathbf{U}^T\mathbf{Q}^{-1/2}$. Then the last equations yields $(\mathbf{T}^{-1})^T\mathbf{R}\mathbf{T}^{-1} = \mathbf{D}$. The second asserted equation follows from

$$\mathbf{T}\mathbf{Q}\mathbf{T}^T = (\mathbf{D}^{1/2}\mathbf{U}^T\mathbf{Q}^{-1/2})\mathbf{Q}(\mathbf{Q}^{-1/2}\mathbf{U}\mathbf{D}^{1/2}) = \mathbf{D}.$$

With the help of definiteness it is possible to define linear matrix inequalities. Let $\mathcal{W}$ be a linear subspace of $\mathbb{R}^{m\times n}$ and let F be a linear map defined on $\mathcal{W}$ such that $F(\mathbf{X})$ is a self-adjoint matrix for $\mathbf{X} \in \mathcal{W}$. Moreover, suppose that the matrix $\mathbf{Q}$ is also self-adjoint. Then an equation of the form

$$F(\mathbf{X}) < \mathbf{Q}$$

is denoted as a **linear matrix inequality** (LMI). Such an inequality with "$\leq$" instead of "$<$" is also a linear matrix inequality. Let $\mathbf{A} \in \mathbb{R}^{n\times n}$ and suppose that $\mathcal{W}$ is the linear space of all real symmetric $n \times n$-matrices. Then an example of an LMI is given by

$$\mathbf{A}^T\mathbf{X}\mathbf{A} - \mathbf{X} < \mathbf{Q}.$$

An inequality of the form

$$\mathbf{A}^T\mathbf{X}\mathbf{A} + \mathbf{B}\mathbf{Y} + \mathbf{Y}^T\mathbf{B}^T < \mathbf{Q}$$

is also an LMI. In this case, one has to choose $\mathcal{W}$ as the linear space of all matrix pairs of the form $(\mathbf{X},\mathbf{Y})$, where $\mathbf{X}$ is a symmetric and $\mathbf{Y}$ an arbitrary (suitably dimensioned) matrix. Then the linear map F is defined by

$$F(\mathbf{Z}) = \mathbf{A}^T\mathbf{X}\mathbf{A} + \mathbf{B}\mathbf{Y} + \mathbf{Y}^T\mathbf{B}^T \quad \text{with} \quad \mathbf{Z} = (\mathbf{X},\mathbf{Y}).$$

A.1.4 Matrix Inversion and Determinant Formulas

Let $\mathbf{A}$ be a square block matrix

$$\mathbf{A} = \begin{pmatrix} \mathbf{A}_{11} & \mathbf{A}_{12} \\ \mathbf{A}_{21} & \mathbf{A}_{22} \end{pmatrix},$$

where $\mathbf{A}_{11}$ and $\mathbf{A}_{22}$ are also square matrices. If $\mathbf{A}_{11}$ is nonsingular, the following equation holds

$$\begin{pmatrix} \mathbf{I} & \mathbf{0} \\ -\mathbf{A}_{21}\mathbf{A}_{11}^{-1} & \mathbf{I} \end{pmatrix}\begin{pmatrix} \mathbf{A}_{11} & \mathbf{A}_{12} \\ \mathbf{A}_{21} & \mathbf{A}_{22} \end{pmatrix} = \begin{pmatrix} \mathbf{A}_{11} & \mathbf{A}_{12} \\ \mathbf{0} & \mathbf{H} \end{pmatrix},$$

if $\mathbf{H}$ is defined as

$$\mathbf{H} = \mathbf{A}_{22} - \mathbf{A}_{21}\mathbf{A}_{11}^{-1}\mathbf{A}_{12}.$$

This multiplication corresponds to a step in Gaussian elimination for linear systems of equations. Similarly,

$$\begin{pmatrix} \mathbf{A}_{11} & \mathbf{A}_{12} \\ \mathbf{0} & \mathbf{H} \end{pmatrix}\begin{pmatrix} \mathbf{I} & -\mathbf{A}_{11}^{-1}\mathbf{A}_{12} \\ \mathbf{0} & \mathbf{I} \end{pmatrix} = \begin{pmatrix} \mathbf{A}_{11} & \mathbf{0} \\ \mathbf{0} & \mathbf{H} \end{pmatrix}.$$

Since we have for any suitably dimensioned matrix $\mathbf{W}$ the properties

$$\begin{pmatrix} \mathbf{I} & \mathbf{0} \\ \mathbf{W} & \mathbf{I} \end{pmatrix}^{-1} = \begin{pmatrix} \mathbf{I} & \mathbf{0} \\ -\mathbf{W} & \mathbf{I} \end{pmatrix}, \quad \begin{pmatrix} \mathbf{I} & \mathbf{W} \\ \mathbf{0} & \mathbf{I} \end{pmatrix}^{-1} = \begin{pmatrix} \mathbf{I} & -\mathbf{W} \\ \mathbf{0} & \mathbf{I} \end{pmatrix},$$

the following decomposition of $\mathbf{A}$ results:

$$\begin{pmatrix} \mathbf{A}_{11} & \mathbf{A}_{12} \\ \mathbf{A}_{21} & \mathbf{A}_{22} \end{pmatrix} = \begin{pmatrix} \mathbf{I} & \mathbf{0} \\ \mathbf{A}_{21}\mathbf{A}_{11}^{-1} & \mathbf{I} \end{pmatrix}\begin{pmatrix} \mathbf{A}_{11} & \mathbf{0} \\ \mathbf{0} & \mathbf{H} \end{pmatrix}\begin{pmatrix} \mathbf{I} & \mathbf{A}_{11}^{-1}\mathbf{A}_{12} \\ \mathbf{0} & \mathbf{I} \end{pmatrix}.$$

If $\mathbf{A}_{22}$ is nonsingular, put

$$\tilde{\mathbf{H}} = \mathbf{A}_{11} - \mathbf{A}_{12}\mathbf{A}_{22}^{-1}\mathbf{A}_{21}.$$

A similar calculation shows

$$\begin{pmatrix} \mathbf{A}_{11} & \mathbf{A}_{12} \\ \mathbf{A}_{21} & \mathbf{A}_{22} \end{pmatrix} = \begin{pmatrix} \mathbf{I} & \mathbf{A}_{12}\mathbf{A}_{22}^{-1} \\ \mathbf{0} & \mathbf{I} \end{pmatrix} \begin{pmatrix} \tilde{\mathbf{H}} & \mathbf{0} \\ \mathbf{0} & \mathbf{A}_{22} \end{pmatrix} \begin{pmatrix} \mathbf{I} & \mathbf{0} \\ \mathbf{A}_{22}^{-1}\mathbf{A}_{21} & \mathbf{I} \end{pmatrix}.$$

The matrix $\mathbf{H}$ ($\tilde{\mathbf{H}}$) is denoted as the **Schur complement** of $\mathbf{A}_{11}$ ($\mathbf{A}_{22}$) in $\mathbf{A}$.

The decompositions of $\mathbf{A}$ lead immediately to the following determinant formulas:

$$\det \mathbf{A} = \det \mathbf{A}_{11} \det(\mathbf{A}_{22} - \mathbf{A}_{21}\mathbf{A}_{11}^{-1}\mathbf{A}_{12}), \quad \text{if } \mathbf{A}_{11} \text{ is nonsingular,}$$
$$\det \mathbf{A} = \det \mathbf{A}_{22} \det(\mathbf{A}_{11} - \mathbf{A}_{12}\mathbf{A}_{22}^{-1}\mathbf{A}_{21}), \quad \text{if } \mathbf{A}_{22} \text{ is nonsingular.}$$

In particular, for any $\mathbf{B} \in \mathbb{C}^{m \times n}$ and $\mathbf{C} \in \mathbb{C}^{n \times m}$ it follows that

$$\det \begin{pmatrix} \mathbf{I}_m & \mathbf{B} \\ -\mathbf{C} & \mathbf{I}_n \end{pmatrix} = \det(\mathbf{I}_n + \mathbf{C}\mathbf{B}) = \det(\mathbf{I}_m + \mathbf{B}\mathbf{C}).$$

This equation implies for $\mathbf{x}, \mathbf{y} \in \mathbb{C}^n$

$$\det(\mathbf{I}_n + \mathbf{x}\mathbf{y}^*) = 1 + \mathbf{y}^*\mathbf{x}.$$

A.1.5 A Minimum Norm Problem for Matrices

We consider the problem of minimizing

$$\min_{\mathbf{X}} \left\| \begin{pmatrix} \mathbf{X} & \mathbf{B} \\ \mathbf{C} & \mathbf{A} \end{pmatrix} \right\|, \tag{A.1}$$

where $\mathbf{X}, \mathbf{B}, \mathbf{C}$ and $\mathbf{A}$ are matrices of compatible dimensions. Denote this minimum by γ_0.

Theorem A.1.4 (Parrot's Theorem). *The minimum in (A.1) is given by*

$$\gamma_0 = \max\left\{ \left\| (\mathbf{C} \;\; \mathbf{A}) \right\|, \left\| \begin{pmatrix} \mathbf{B} \\ \mathbf{A} \end{pmatrix} \right\| \right\}.$$

A simplified version of the above minimization problem is to find

$$\min_{\mathbf{X}} \left\| \begin{pmatrix} \mathbf{X} \\ \mathbf{A} \end{pmatrix} \right\| \quad \text{or} \quad \min_{\mathbf{X}} \left\| (\mathbf{X} \;\; \mathbf{A}) \right\|.$$

The minimum will also be denoted by γ_0.

Lemma A.1.2. *For $\gamma > \gamma_0$, the following holds:*

(a) $\left\| \begin{pmatrix} \mathbf{X} \\ \mathbf{A} \end{pmatrix} \right\| \leq \gamma$ *if and only if* $\| \mathbf{X}(\gamma^2 \mathbf{I} - \mathbf{A}^T \mathbf{A})^{-1/2} \| \leq 1$;

(b) $\| (\mathbf{X} \quad \mathbf{A}) \| \leq \gamma$ *if and only if* $\| (\gamma^2 \mathbf{I} - \mathbf{A}\mathbf{A}^T)^{-1/2} \mathbf{X} \| \leq 1$.

Note that in this lemma the inverse exists because of

$$\rho(\mathbf{A}^T \mathbf{A}) \leq \| \mathbf{A} \|^2 < \gamma^2 .$$

Lemma A.1.3. *For an arbitrary* $m \times n$ *-matrix* $\mathbf{A}$ *with* $\| \mathbf{A} \| < \gamma$, *the following holds:*

$$\mathbf{A}^T (\gamma^2 \mathbf{I} - \mathbf{A}\mathbf{A}^T)^{1/2} = (\gamma^2 \mathbf{I} - \mathbf{A}^T \mathbf{A})^{1/2} \mathbf{A}^T .$$

Proof. Suppose $m \geq n$ (for $m < n$ the proof works similar). Then, using the singular value decomposition, $\mathbf{A}$ can be decomposed as

$$\mathbf{A} = \mathbf{U} \begin{pmatrix} \boldsymbol{\Sigma} \\ \mathbf{0} \end{pmatrix} \mathbf{V}^T$$

with a positive semidefinite matrix $\boldsymbol{\Sigma}$ and orthogonal matrices $\mathbf{U}, \mathbf{V}$. It is now not difficult to see that

$$\begin{aligned} \mathbf{A}^T (\gamma^2 \mathbf{I} - \mathbf{A}\mathbf{A}^T)^{1/2} &= \mathbf{V}((\gamma^2 \mathbf{I} - \boldsymbol{\Sigma}^2)^{1/2} \boldsymbol{\Sigma} \quad \mathbf{0}) \mathbf{U}^T \\ &= (\gamma^2 \mathbf{I} - \mathbf{A}^T \mathbf{A})^{1/2} \mathbf{A}^T . \end{aligned}$$

Theorem A.1.5. *Suppose* $\gamma \geq \gamma_0$. *Then a matrix* $\mathbf{X}$ *satisfying the inequality*

$$\left\| \begin{pmatrix} \mathbf{X} & \mathbf{B} \\ \mathbf{C} & \mathbf{A} \end{pmatrix} \right\| \leq \gamma$$

is given by $\mathbf{X} = -\mathbf{Y}\mathbf{A}^T \mathbf{Z}$, *where* $\mathbf{Y}, \mathbf{Z}$ *are matrices such that* $\| \mathbf{Y} \| \leq 1$, $\| \mathbf{Z} \| \leq 1$, *which solve the linear equations*

$$\begin{aligned} \mathbf{Y} (\gamma^2 \mathbf{I} - \mathbf{A}^T \mathbf{A})^{1/2} &= \mathbf{B} \\ (\gamma^2 \mathbf{I} - \mathbf{A}\mathbf{A}^T)^{1/2} \mathbf{Z} &= \mathbf{C} . \end{aligned}$$

Corollary A.1.2. *Let* $\gamma > \gamma_0$. *Then the inequality*

$$\left\| \begin{pmatrix} \mathbf{X} & \mathbf{B} \\ \mathbf{C} & \mathbf{A} \end{pmatrix} \right\| < \gamma$$

is satisfied for

$$\mathbf{X} = -\mathbf{B}(\gamma^2\mathbf{I} - \mathbf{A}^T\mathbf{A})^{-1}\mathbf{A}^T\mathbf{C}.$$

Proof. From Lemma A.1.2 we get

$$\|\mathbf{B}(\gamma^2\mathbf{I} - \mathbf{A}^T\mathbf{A})^{-1/2}\| \le 1 \ \text{ and } \ \|(\gamma^2\mathbf{I} - \mathbf{A}\mathbf{A}^T)^{-1/2}\mathbf{C}\| \le 1.$$

Thus a solution $\mathbf{X}$ with the desired property is

$$\begin{aligned}\mathbf{X} &= -\mathbf{B}(\gamma^2\mathbf{I} - \mathbf{A}^T\mathbf{A})^{-1/2}\mathbf{A}^T(\gamma^2\mathbf{I} - \mathbf{A}\mathbf{A}^T)^{-1/2}\mathbf{C} \\ &= -\mathbf{B}(\gamma^2\mathbf{I} - \mathbf{A}^T\mathbf{A})^{-1/2}(\gamma^2\mathbf{I} - \mathbf{A}^T\mathbf{A})^{-1/2}\mathbf{A}^T\mathbf{C} \\ &= -\mathbf{B}(\gamma^2\mathbf{I} - \mathbf{A}^T\mathbf{A})^{-1}\mathbf{A}^T\mathbf{C}.\end{aligned}$$

A.2. Analysis

A.2.1 Banach and Hilbert Spaces

Functional analysis is in this book only used at a very moderate level. It suffices to know some few facts about functional spaces and norms and they will be presented in this section. Again, let $\mathcal{F}$ be the field of real or complex numbers.

Definition. *Let $\mathcal{V}$ be a nonempty set such that for every $u, v \in \mathcal{V}$ a sum $u + v \in \mathcal{V}$ and for every $\alpha \in \mathcal{F}$, $v \in \mathcal{V}$ product αv is defined. Then $\mathcal{V}$ is a* ***linear vector space*** *if for all $u, v, w \in \mathcal{V}$ and all $\alpha, \beta \in \mathcal{F}$ the following properties hold:*

(i) *There is a zero element in $\mathcal{V}$, denoted by 0, such that $u + 0 = u$;*

(ii) *There is an element $-v$, such that $v + (-v) = 0$;*

(iii) $u + v = v + u$;

(iv) $(u + v) + w = u + (v + w)$;

(v) $(\alpha\beta)v = \alpha(\beta v)$;

(vi) $1v = v$;

(vii) $(\alpha + \beta)v = \alpha v + \beta v$;

(viii) $\alpha(u + v) = \alpha u + \alpha v$.

The elements of $\mathcal{V}$ are denoted as ***vectors***.

Definition. *A norm on a vector space is a function* $\|\cdot\|: \mathcal{V} \to [0,\infty)$ *such that for every* $u, v \in \mathcal{V}$, $\alpha \in \mathbb{F}$ *the following properties hold:*

(i) $\|v\| = 0$ *if and only if* $v = 0$;

(ii) $\|\alpha v\| = |\alpha| \|v\|$;

(iii) $\|v + w\| \le \|v\| + \|w\|$.

Sometimes we also write $\|v\|_{\mathcal{V}}$ instead of $\|v\|$. A vector space together with a norm is called a **normed space**. A sequence $v_k \in \mathcal{V}$ is said to converge to an element $v \in \mathcal{V}$ if $\|v_k - v\|$ tends to 0 for $k \to \infty$.

Definition. *A sequence* $v_k \in \mathcal{V}$ *is denoted as a* ***Cauchy sequence****, if for each* $\varepsilon > 0$ *there is a number* n_0 *such that*

$$\|v_m - v_n\| < \varepsilon \quad \textit{for all} \quad m, n \ge n_0 .$$

It is easy to see that each convergent sequence is a Cauchy sequence, but the converse is not necessarily true.

Definition. *A* ***Banach space*** *is a normed space* $\mathcal{V}$ *with the property that every Cauchy sequence has a limit in* $\mathcal{V}$.

Definition. *An* ***inner product*** *is a mapping* $\langle\cdot,\cdot\rangle : \mathcal{V}\times\mathcal{V} \to \mathbb{F}$ *such that for all* $u, v, w \in \mathcal{V}$ *and all* $\alpha, \beta \in \mathbb{F}$ *the following hold:*

(i) $\langle v, v\rangle \ge 0$ *and* $\langle v, v\rangle = 0$ *if and only if* $v = 0$;

(ii) $\langle u, \alpha v + \beta w\rangle = \alpha\langle u, v\rangle + \beta\langle u, w\rangle$;

(iii) $\overline{\langle u, v\rangle} = \langle v, u\rangle$.

We also sometimes write $\langle\cdot,\cdot\rangle_{\mathcal{V}}$ instead of $\langle\cdot,\cdot\rangle$. A vector space with an inner product is called an **inner product space.** It can be verified that

$$\|v\| = \sqrt{\langle v, v\rangle}$$

induces a norm.

Definition. *A* ***Hilbert space*** *is an inner product space which is complete as a normed space under the norm induced by the inner product.*

An important property is the **Cauchy–Schwarz inequality**:

$$|\langle u, v\rangle| \le \|u\| \|v\| .$$

Two vectors $u, v \in \mathcal{V}$ are said to be **orthogonal** if

$$\langle u, v \rangle = 0 .$$

In this case, the following generalization of Pythagoras's theorem holds:

$$\| v + w \|^2 = \| v \|^2 + \| w \|^2 .$$

Let $\mathcal{U}$ be a linear subspace of $\mathcal{V}$. Then the set

$$\mathcal{U}^{\perp} = \{ v \in \mathcal{V} \mid \langle u, v \rangle = 0 \quad \text{for all } u \in \mathcal{U} \}$$

is also a linear subspace, which is the **orthogonal complement** of $\mathcal{U}$. A subspace $\mathcal{U}$ is **closed** if for every convergent sequence in $\mathcal{U}$ the limit vector is also in $\mathcal{U}$. For a closed subspace, the Hilbert space $\mathcal{V}$ can be decomposed in the form

$$\mathcal{V} = \mathcal{U} \oplus \mathcal{U}^{\perp} .$$

This means that any $w \in \mathcal{V}$ has the unique representation $w = u + v$ with $u \in \mathcal{U}$ and $v \in \mathcal{U}^{\perp}$. With this decomposition, it is possible to define a so-called **projection operator** by

$$P : \mathcal{V} \to \mathcal{U}$$

$$P w = v .$$

Function spaces. The most important examples for Banach and Hilbert spaces are function spaces. A few of them, which are of general importance, will be introduced here, whereas some others, which are more specifically related to control theory, are discussed in Chap. 7.

1. By $\mathcal{L}_2^n[0, \infty)$, we denote all measurable functions from $\mathbb{R}$ into $\mathbb{R}^n$ (or $\mathbb{C}^n$) with

$$\int_0^{\infty} \| \mathbf{x}(t) \|_2^2 \, dt < \infty .$$

For this space, a norm is given by

$$\| \mathbf{x} \|_2 = \left(\int_0^{\infty} \| \mathbf{x}(t) \|_2^2 \, dt \right)^{1/2} .$$

The requirement of measurability is needed to guarantee that the above integral is defined at all. We cannot give the precise definition here and refer the reader to any book on integration theory. The equation

$$\langle \mathbf{x}, \mathbf{y} \rangle = \int_0^{\infty} \langle \mathbf{x}(t), \mathbf{y}(t) \rangle \, dt$$

defines a scalar product on this space.

2. Another important function space is $\mathcal{L}_{\infty}^{n}[0,\infty)$. It consists of all functions with values in $\mathbb{R}^n$ (or $\mathbb{C}^n$) such that

$$\|\mathbf{x}\|_{\infty} = \operatorname*{ess\,sup}_{t\in[0,\infty)} \|\mathbf{x}(t)\|_{\infty} < 0$$

is finite. The above equation defines a norm on $\mathcal{L}_{\infty}^{n}[0,\infty)$. Writing "ess sup" instead of "sup" means that the supremum is taken over $[0,\infty)$ with the possible exception of a set of measure zero. In context of control theory applications, nothing goes wrong if the reader unfamiliar with integration theory simply ignores the "ess".

3. For matrix-valued functions $\mathbf{A}(t), \mathbf{B}(t)$, a scalar product can be introduced by

$$\langle \mathbf{A}, \mathbf{B} \rangle = \int_0^\infty \operatorname{trace} \mathbf{A}^*(t)\mathbf{B}(t)\, dt\,.$$

The corresponding norm is

$$\|\mathbf{A}\|_2^2 = \int_0^\infty \|\mathbf{A}(t)\|_F^2\, dt = \int_0^\infty \operatorname{trace}(\mathbf{A}^*(t)\mathbf{A}(t))\, dt\,.$$

The function space consisting of the functions for which this norm is finite will be denoted by $\mathcal{L}_2^{m\times n}[0,\infty)$ or simply by $\mathcal{L}_2[0,\infty)$.

A.2.2 Operators

Definition. Suppose $\mathcal{V}$ *and* $\mathcal{W}$ *are Banach spaces and let* F *be a mapping from* $\mathcal{V}$ *to* $\mathcal{W}$.

(a) F *is called* ***linear*** *if*

$$F(\alpha u + \beta v) = \alpha F(u) + \beta F(v) \qquad \text{for all } u, v \in \mathcal{V} \text{ and } \alpha, \beta \in \mathbb{F}\,.$$

(b) F *is denoted as* ***bounded*** *if there is constant* $c \geq 0$ *such that*

$$\|Fv\|_{\mathcal{W}} \leq c\|v\|_{\mathcal{V}} \quad \text{for all } v \in \mathcal{V}\,.$$

Remark. It is possible to extend the definition of continuity to general mappings from $\mathcal{V}$ to $\mathcal{W}$ and it can be shown that for linear operators continuity is equivalent to boundedness.

The linear space of all linear, bounded operators mapping $\mathcal{V}$ to $\mathcal{W}$ is denoted by $\mathcal{L}(\mathcal{V},\mathcal{W})$. It is an important fact that the norms on $\mathcal{V}$ and $\mathcal{W}$ induce a norm on this space by

$$\| F\|_{\mathcal{L}(\mathcal{V},\mathcal{W})} = \sup_{v\in\mathcal{V}, v\neq 0} \frac{\| Fv\|_{\mathcal{W}}}{\| v\|_{\mathcal{V}}}.$$

Let now a third Banach space $\mathcal{U}$ and a second bounded operator $G:\mathcal{U}\to\mathcal{V}$ be given. Then the composition FG is defined by $(FG)(u) = F(Gu))$ for all $u\in\mathcal{U}$. It is also a bounded operator and its norm can be estimated as follows:

$$\| FG\|_{\mathcal{L}(\mathcal{U},\mathcal{W})} \leq \| F\|_{\mathcal{L}(\mathcal{V},\mathcal{W})} \|G\|_{\mathcal{L}(\mathcal{U},\mathcal{V})}.$$

This property of induced operator norms is denoted as **submultiplicativity**.

Linear operators with range $\mathbb{F}$ are denoted as **functionals**. Let $\mathcal{V}$ be a Hilbert space and $w\in\mathcal{V}$. Then the equation

$$L(v) = \langle v, w\rangle \text{ for every } v\in\mathcal{V} \tag{A.2}$$

defines a bounded linear functional on $\mathcal{V}$. Moreover, every bounded linear functional on $\mathcal{V}$ can be written in this way, as the following theorem states.

Theorem A.2.1 (Riesz Representation Theorem). Let $\mathcal{V}$ *be a Hilbert space and let* L *be a bounded linear functional on* $\mathcal{V}$. *Then there is a unique element* $w\in\mathcal{V}$ *such that (A.2) holds and furthermore,* $\| L\| = \| w\|$.

The following definition makes use of this result.

Definition. *Assume that* $\mathcal{V}$ *and* $\mathcal{W}$ *are Hilbert spaces and let* $F\in\mathcal{L}(\mathcal{V},\mathcal{W})$. *Then the equation*

$$\langle v, F^* w\rangle_{\mathcal{V}} = \langle Fv, w\rangle_{\mathcal{W}} \text{ for all } v\in\mathcal{V},\ w\in\mathcal{W}$$

defines an operator of $\mathcal{L}(\mathcal{W},\mathcal{V})$. *It is called the* ***adjoint operator*** F^* *of* F.

The following lemma contains important properties concerning the norm of the adjoint operator.

Lemma A.2.1. *Suppose* $F\in\mathcal{L}(\mathcal{V},\mathcal{W})$. *Then the following holds:*

(a) $\| F\| = \| F^*\|$;

(b) $\| F\| = \| FF^*\|^{1/2}$.

An $m\times n$-matrix $\mathbf{A}$ can be viewed as a bounded operator from $\mathbb{C}^n$ to $\mathbb{C}^m$. The adjoint of this operator can be identified with the adjoint matrix $\mathbf{A}^*$. Lemma A.2.1 and Lemma A.1.1 imply $\| \mathbf{A}^*\| = \| \mathbf{A}\|$ and

$$\rho(\mathbf{A}^*\mathbf{A}) = \| \mathbf{A}^*\mathbf{A}\| = \| \mathbf{A}\|^2.$$

A.2.3 Complex Analysis

Finally, we discus two results from complex analysis. One of them, namely Lemma A.3.2, will only be needed in Sect. 7.5.

Let $G \subset \mathbb{C}$ be an open set and let $f : G \to \mathbb{C}$ be a complex-valued function on G. The function f is said to be **analytic at a point** $z_0 \in G$ if f is differentiable at each point of some neighborhood of z_0. Then f has continuous derivatives of all orders at z_0. Moreover, a function f, which is analytic at z_0 has a power series representation at z_0. Vice versa, if f has a power series representation at z_0, then it is analytic at z_0. A function f is by definition **analytic in** G if it is analytic at every point of G. A matrix-valued function is said to be analytic in G if each element of the matrix is analytic in G. The function e^{-s} is an example of a function which is analytic in the whole complex plane. A real rational transfer function which is stable is analytic in the open right half-plane (RHP).

An important theorem of the theory of analytic functions is the **maximum principle.**

Theorem A.3.1. *Let f be continuous on a bounded and closed set G and analytic in the interior of G. Then the maximum of $|f(s)|$ will be attained on the boundary ∂G of G, i.e.*

$$\max_{s \in G} |f(s)| = \max_{s \in \partial G} |f(s)|.$$

Lemma A.3.1. *Let F be analytic and of bounded magnitude in the closed RHP. Then for every $s_0 = \sigma_0 + i\omega_0$ with $\sigma_0 > 0$ we have*

$$F(s_0) = \frac{1}{\pi} \int_{-\infty}^{\infty} F(i\omega) \frac{\sigma_0}{\sigma_0^2 + (\omega - \omega_0)^2} d\omega . \tag{A3}$$

Proof. Let C_ρ be a Nyquist contour such that s_0 is encircled by C_ρ. Cauchy's integral formula yields

$$F(s_0) = \frac{1}{2\pi i} \int_{C_\rho} \frac{F(s)}{s - s_0} ds .$$

Since $-\overline{s}_0$ is not encircled by C_ρ, a second application of Cauchy's formula gives

$$0 = \frac{1}{2\pi i} \int_{C_\rho} \frac{F(s)}{s + \overline{s}_0} ds .$$

Subtraction of these two equations leads to

$$F(s_0) = \frac{1}{2\pi i} \int_{C_\rho} F(s) \frac{\overline{s}_0 + s_0}{(s - s_0)(s + \overline{s}_0)} ds .$$

This integral can be written as the sum of two integrals, $F(s_0) = I_1 + I_2$, where

$$I_1 = \frac{1}{\pi} \int_{\rho}^{-\rho} F(i\omega) \frac{\sigma_0}{(i\omega - s_0)(i\omega + \overline{s}_0)} d\omega$$

$$= \frac{1}{\pi} \int_{-\rho}^{\rho} F(i\omega) \frac{\sigma_0}{\sigma_0^2 + (\omega - \omega_0)^2} d\omega$$

$$I_2 = \frac{1}{\pi} \int_{-\pi/2}^{\pi/2} F(\rho e^{i\varphi}) \frac{\sigma_0}{(\rho e^{i\varphi} - s_0)(\rho e^{i\varphi} + \overline{s}_0)} \rho\, e^{i\varphi} d\varphi .$$

The integral I_1 tends to the integral in (A3) for $\rho \to \infty$. Thus it remains to show that I_2 tends to 0 for $\rho \to \infty$.

We have

$$|I_2| \le \frac{\sigma_0}{\pi} \| F \|_\infty \frac{1}{\rho} \int_{-\pi/2}^{\pi/2} \frac{\sigma_0}{|e^{i\varphi} - s_0 \rho^{-1}||e^{i\varphi} + \overline{s}_0 \rho^{-1}|} d\varphi . \tag{A4}$$

Since the integral in (A4) remains bounded for $\rho \to \infty$, the integral I_2 tends to 0 for $\rho \to \infty$.

Notes and References

Standard references on linear algebra are Strang [96] and Horn and Johnson [49]. Concerning functional analysis, the reader can be referred to Bollobas [15] and Young [109]. References for complex analysis are Ahlfors [1] and Churchill et al. [27]. The results in A.1.5 can be found in Zhou, Doyle, and Glover [112], Sect. 2.11. Lemma A.3.1 can be found in Doyle, Francis, and Tannenbaum [36], Sect. 6.2.

Bibliography

[1] Ahlfors, L.V.: *Complex Analysis*, McGraw-Hill, New York, 1966.

[2] Anderson, B. D. O. and Moore, J. B.: *Optimal Control: Linear Quadratic Methods*, Prentice Hall, Englewood Cliffs, New Yersey, 1990.

[3] Apkarian, P., Gahinet, P. and Becker, G.: Self-scheduled $\mathcal{H}_\infty$ control of linear parameter-varying systems: a design example, *Automatica*, 31, pp. 1251–1261, 1995.

[4] Apkarian, P., and Gahinet, P.: A convex characterization of gain-scheduled $\mathcal{H}_\infty$ controllers, *IEEE Trans. Autom. Control*, 40, pp. 853–864, 1995.

[5] Apkarian, P. and Biannic, J.-M.: Self-scheduled $\mathcal{H}_\infty$ control of missile via linear matrix inequalities, *J. Guid. Control Dyn.*, 18, 1995, pp. 532–538, 1998.

[6] Apkarian, P., Biannic, J.-M. and Garrard, W.: Parameter varying control of a high-performance aircraft, *J. Guid. Control Dyn.*, 20, pp. 225–231, 1997.

[7] Balas, G., Doyle, J., Glover, K., Packard, A. and Smith, R.: *μ-Analysis and Synthesis Toolbox*, The Mathworks Inc., 2001.

[8] Balas, G., Lind, R. and Packard, A.: Optimal scaled $\mathcal{H}_\infty$ full information control synthesis with real uncertainty, *J. Guid. Control Dyn.*, 19, pp. 854–862, 1996.

[9] Barker, J. and Balas, G.: Gain-scheduled linear fractional control for active flutter supression, *J. Guid. Control Dyn.*, 22, pp. 507–512, 1999.

[10] Basar, T. and Bernhard, P.: *$\mathcal{H}_\infty$-Optimal Control and Related Mini-Max Design Problems: A Dynamic Game Approach*, Birkhäuser, Boston, 1991.

[11] Bennani, S., Willemsen, D. and Scherer, C.: Robust control of linear parametrically varying systems with bounded rates, *J. Guid. Control Dyn.*, 21, pp. 916–922, 1998.

[12] Bequette, W.: *Process Dynamics, Modelling Analysis and Simulation*, Prentice Hall, Englewood Cliffs, New Yersey , 1998.

[13] Blanke, M. and Christensen, A.: Rudder-roll damping autopilot robustness due to sway-yaw-roll couplings, Proc. *10th Int. Ship Control Systems Sympos. (SCSS'93)*, Ottawa, Canada, A.93–A.119, 1993.

[14] Bode, H. W.: *Network Analysis and Feedback Amplifier Design*, Van Nostrand, Princeton, 1945.

[15] Bollobas, B.: *Linear Analysis*, Cambridge University Press, 1990.

[16] Bryson, A.: *Control of Spacecraft and Aircraft*, Princeton University Press, Princeton, 1994.

[17] Burl, J.: *Linear Optimal Control, $\mathcal{H}_2$ and $\mathcal{H}_\infty$ Methods*, Addison-Wesley, Menlo Park, 1999.

[18] Buschek, H.: Full envelope missile autopilot design using gain scheduled robust control, *J. Guid. Control Dyn.*, 22, pp. 115–122, 1999.

[19] Buschek, H. and Calise, A.: Uncertainty modeling and fixed-order controller design for a hypersonic vehicle model, *J. Guid. Control Dyn.*, 20, pp. 42–48, 1997.

[20] Buschek, H., and Calise, A.: μ controllers: mixed and fixed, *J. Guid. Control Dyn.*, 20, pp. 34–41, 1997.

[21] Carter, L. and Shamma, J.: Gain-scheduled bank-to-turn autopilot design using linear parameter varying transformations, *J. Guid. Control Dyn.*, 19, pp. 1056–1063, 1996.

[22] Chen, B., Lee and T., Venkataramanan, V: *Hard Disk Drive Servo Systems*, Springer-Verlag, London, 2002.

[23] Chen, C.-T: *Linear System Theory and Design*, Oxford University Press, New York, 1999.

[24] Chiang, R. Y. and Safanov, M. G.: *Robust Control Toolbox*, The Mathworks Inc., 1993.

[25] Chilali, M., Gahinet, P. and Apkarian, P.: Robust pole placement in LMI regions, *IEEE Trans. Autom. Control*, 44, pp. 2257–2269, 1999.

[26] Chilali, M., Gahinet., P.: $\mathcal{H}_\infty$ design with pole placement constraints: an LMI approach, *IEEE Trans. Autom. Control*, pp. 358-367, 1996.

[27] Churchill, R.V., Brown, J.W. and Verhey, R.F.: *Complex Variables and Applications*, McGraw-Hill, New York, 1974.

[28] Curtain, R. F. and Zwart, H. J.: *An Introduction to Infinite-Dimensional Linear Systems Theory*, Springer-Verlag, New York, 1995.

[29] Desoer, C. A. and Vidyasagar, M.: *Feedback Systems: Input-Output Properties*, Academic Press, New York, 1975.

[30] Dorf, R. and Bishop, R.: *Modern Control Systems*, Addison-Wesley, Menlo Park, 1998.

[31] Doyle, J. C.: Guaranteed margins for LQG regulators, *IEEE Trans. Autom. Control*, 23, pp. 756–757, 1978.

[32] Doyle, J. C.: Analysis of feedback systems with structured uncertainties, *IEE Proc. D, Control Theory*, 129, pp. 242–250, 1982.

[33] Doyle, J. C., Wall, J. and Stein, G.: Performance and robustness analysis for structured uncertainty, *Proc. 21st IEEE Conf. on Decision and Control*, pp. 629–636, 1982.

[34] Doyle, J. C.: Structured uncertainty in control system design, *Proc. IEEE Conf. on Decision and Control*, Ft. Lauderdale, 1985.

[35] Doyle, J. C., Glover, K., Khargonakar, P. and Francis, B. A.: State-space solutions to standard $\mathcal{H}_2$ and $\mathcal{H}_\infty$ control problems, *IEEE Trans. on Autom. Control*, 34, pp. 831–847, 1989.

[36] Doyle, J. C., Francis, B. and Tannenbaum, A.: *Feedback Control Theory,* Macmillan Publishing Company, New York, 1992.

[37] Dullerud, G., Paganini, F.: *A Course in Robust Control Theory*, Springer-Verlag, New York, 2000.

[38] Duren, P. L.: *Theory of $\mathcal{H}_p$ spaces*, Academic Press, New York, 1970.

[39] Fan, M. K. H., Tits, A. L. and Doyle, J. C.: *Robustness in the presence of mixed parametric uncertainty and unmodeled dyxnamics*, IEEE Trans. Autom. Control, 36(1), pp. 25–38, 1991.

[40] Fialho, I., Balas, G., Packard, A., Renfrow, J. and Mullaney, Ch.: Gain-scheduled lateral control of the F-14 aircraft during powered approach landing, *J. Guid. Control Dyn.*, 23, pp. 450–458, 2000.

[41] Ferreres, G. and M'Saad, M.: Parametric robustness evaluation of a $\mathcal{H}_\infty$ missile autopilot, *J. Guid. Control Dyn.*, 19, pp. 621–627, 1996.

[42] Foias, C., Ozbay, H. and Tannenbaum, A.: *Robust Control of Infinite Dimensional Systems*, Springer-Verlag, New York, 1987.

[43] Föllinger, O.: *Regelungstechnik*, Hüthig Verlag, Heidelberg, 1985

[44] Fossen, T.: *Guidance and Control of Ocean Vehicles*, Wiley, Chichester, 1994.

[45] Francis, B. A.: *A Course in $\mathcal{H}_\infty$ Control Theory*, Springer-Verlag, New York, 1987.

[46] Gahinet, P. and Apkarian, P.: A linear matrix inequality approach to $\mathcal{H}_\infty$ control, *Int. J. Robust Nonlinear Control*, 4, pp. 421–448, 1994.

[47] Gahinet, P., Nemirovski, A., Laub, A. and Chiali M.: *LMI Control Toolbox,* The Mathworks Inc., 1995.

[48] Glover, K., Limebeer, D. J. N., Doyle, J. C., Kasenally, E. .M. and Safanov, M. G.: A characterization of all solutions of the four block general distance problem, *SIAM J. Control Optim.*, 29, pp. 283–324, 1991.

[49] Horn, R.A. and Johnson, C.R.: *Matrix Analysis*, Cambridge University Press, 1991.

[50] Hoyle, D. J., Hyde, R. A. and Limebeer, D. J. N.: *An $\mathcal{H}_\infty$ approach to two degree of freedom design*, Proc. 30th Conf. on Decision and Control, pp. 1581–1585, Brighton, UK, 1991.

[51] Iglesias, P. and Urban, Th.: Loop shaping design for missile autopilots: Controller configuration and weighting filter selection, *J. Guid. Control Dyn.*, 23, pp. 516–525, 2000.

[52] Kailath, T: *Linear Systems*, Prentice Hall, Englewood Cliffs, 1980.

[53] Kalman, R. E.: On the general theory of control systems, in: *Proc. IFAC World Congress*, 1960.

[54] Kalman, R. E.: Mathematical description of linear systems, *SIAM J. Control*, 152-192, 1963.

[55] Kalman, R. E.: A new approach to linear filtering and prediction theory, *ASME Transactions, Series D: Journal of Basic Engineering*, 82, pp. 35–45, 1960.

[56] Kalman, R. E. and Bucy, R. S.: New results in linear filtering and prediction theory, *ASME Transactions, Series D: Journal of Basic Engineering*, 83, pp. 95–108, 1960.

[57] Kautsky, J., Nichols, N. K. and Van Dooren, P.: Robust pole assignment in linear state feedback, *Int. J. Control*, 41, 1129–1155, 1985.

[58] Kwakernaak, H. and Sivan, R.: *Linear Optimal Control Systems*, Wiley-Interscience, New York, 1972.

[59] Lancaster, P. and Rodman, L.: *Algebraic Riccati Equations*, Clarendon Press, 1995.

[60] Lewis, F. and Syrmos, V.: *Optimal Control*, John Wiley, New York, 1995.

[61] Limebeer, D. J. N, Anderson, B. D. O., Khargonekar, P. P. and Green, M.: A game theoretic approach to $\mathcal{H}_\infty$ control for time varying systems, *SIAM J. Control Optim.*, 30, pp. 262–283, 1992.

[62] Limebeer, D. J. N., Kasenally, E. M. and Perkins, J. D.: On the design of robust two degree of freedom controllers, *Automatica*, 29, 157–168, 1993.

[63] Lin, Ch., Cloutier, J. and Evers, J.: High-performance, robust, bank-to-turn missile autopilot design, *J. Guid. Control Dyn.*, 18, pp. 46–53, 1995.

[64] Lions, J. L.: *Optimal Control of Systems Governed by Partial Differential Equations*, Springer-Verlag, Heidelberg, 1971.

[65] Luenberger, D. G.: Observing the state of a linear system, *IEEE Trans. Military Electronics*, 8, pp. 74–80, 1964.

[66] Luyben, W. L.: *Process Modeling, Simulation, and Control for Chemical Engineers*, McGraw-Hill, New York, 1990.

[67] Mackenroth, U.: Convex parabolic boundary control problems with pointwise state constraints, *J. Math. Anal. Appl.*, 87, pp. 256–277, 1982.

[68] Mackenroth, U. and Alt, W.: Convergence of finite element approximations to state constrained convex parabolic boundary control problems, *SIAM J. Control Optim.*, 27, pp. 718–736, 1989.

[69] Malloy, D. and Chang, B: Stabilizing controller design for linear parameter-varying systems using parameter feedback, *J. Guid. Control Dyn.*, 21, pp. 891–898, 1998.

[70] Mangiacasale, L.: *Airplane Control Systems*, Levoretto & Bella, Torino, 1996.

[71] Morari, M. and Zarafiriou, E.: *Robust Process Control*, Prentice Hall, Englewood Cliffs, 1989.

[72] Packard, A. and Doyle, J. C.: The complex structured singular value, *Automatica*, 29, pp. 71–109, 1993.

[73] Packard, A.: Gain scheduling via linear fractional transformations, *Syst. Control Lett.* , 22, pp. 79–92, 1994.

[74] Papoulis, A.: *Probability, Random Variables and Stochastic Processes*, McGraw-Hill, Singapore, 1984.

[75] Popov, V. M.: The solution of a new stability problem for controlled systems, *Autom. Remote Control*, 24, pp. 1-23, 1963.

[76] Ramirez, W.: *Computational Methods for Process Simulation*, Butterworth-Heinemann, Oxford, 1997.

[77] Rudin, W.: *Real and Complex Analysis*, McGraw-Hill, New York, 1987.

[78] Safonov, M. G.: Stability margins of diagonally perturbed multivariable feedback systems, *IEE Proc. D, Control Theory Appl.,* 129, pp. 251–256, 1982.

[79] Safonov, M. G., Limebeer, D. J. N., and Chiang, R. Y.: Simplifying the $\mathcal{H}_\infty$ theory via loop shifting, matriy pencil and descriptor concepts, *Int. J. Control*, 50, pp. 2467-2488, 1990.

[80] Sanchez-Pena, R. and Sznaier, M.: *Robust Control Theory and Application*, Wiley, New York, 1998.

[81] Scherer, C.: $\mathcal{H}_\infty$ optimization without assumptions on finite or infinite zeros, *SIAM J. Control Optim.*, 30, pp. 143–166, 1992.

[82] Scherer, C., Gahinet, P. and Chiali, M.: Multiobjective output-feedback control via LMI-optimization, *IEEE Trans. Autom. Control*, 42, pp. 896–911, 1997.

[83] Schmidt, L.: *Introduction to Aircraft Flight Dynamics*, AIAA Education Series, Reston, 1998.

[84] Schumacher, C. and Khargonegar, P.: Missile autopilot designs using $\mathcal{H}_\infty$-control with gain scheduling and dynamic inversion, *J. Guid. Control Dyn.*, 21, pp. 234–243, 1998.

3, 1998.

[85] Sefton, J. and Glover, K.: Pole/zero cancellations in the general $\mathcal{H}_\infty$-problem with reference to a two-block design, *Syst. Control Lett.*, 14, pp. 295–306, 1990.

[86] Shinskey., F.G.: *Distillation Control*, 2nd ed., MacGraw-Hill, New York, 1984.

[87] Skogestad, S., Morari, M., and Doyle, J. C.: Robust performance of ill-conditioned plants: High-purity distillation, IEEE *Trans. Autom. Control*, 33, pp. 1092–1105, 1988.

[88] Skogestad, S. and Morari, M.: Control configuration selection for distillation columns, *AIChE Journal*, 33, pp. 1620–1635, 1987.

[89] Skogestad, S., Lundström, P. and Jacobsen, E.: Selecting the best distillation control configuration, *AIChE Journal*, 36, pp. 753-764, 1990.

[90] Skogestad, S. and Morari, M.: Variable selection for decentralized control, *Modelling, Identifikation and Control*, 13, pp. 113–125, 1992.

[91] Skogestad, S. and Morari, M.: Understanding the dynamic behavior of distillation columns, *Ind. Eng. Chem. Res.*, 27, pp. 1848–1862, 1988.

[92] Skogestad, S. and Postletwaithe, I.: Multivariable Feedback Control: Analysis and Design, Wiley, Chichester, 1996.

[93] Spillman, M.: Robust longitudinal flight control design using linear parameter-varying feedback, *J. Guid. Control Dyn.*, 23, pp. 101-107, 2000.

[94] Stilwell, D.: State-space interpolation for a gain-scheduled autopilot, *J. Guid. Control Dyn.*, 24, pp. 460–465, 2001.

[95] Stoer, J. and Bulirsch, R.: *Introduction to Numerical Analysis*, Springer-Verlag, New York, 1980.

[96] Strang, G.: *Linear Algebra and its Applications*, Academic Press, New York, 1980.

[97] Trentelman, H., Stoorvogel, A. and Hautus, M.: *Control Theory for Linear Systems*, Springer-Verlag, London, 2001.

[98] Tsourdos, A., Zbikowski, R. and White, B.: Robust autopilot for a quasi-linear parameter-varying missile model, *J. Guid. Control Dyn.*, 24, pp. 287–295, 2001.

[99] Van Amerongen, J., Van der Klugt, P. G. M. and Van Nauta Lemke, H. R.: Rudder Roll Stabilization for Ships, *Automatica*, 26, pp. 679–690, 1990.

[100] Van Amerongen, J. and Van Nautka Lemke, H. R.: Adaptive control aspects of a rudder roll stabilization system, *Proc. of the 10th IFAC World Congress*, München, pp. 215–219, 1987.

[101] Van Keulen: *$\mathcal{H}_\infty$ Control for Distributed Parameter Systems: A State Space Approach*, Birkhäuser, Boston, 1993.

[102] Vidyasagar, M.: Input-output stability of a broad class of linear time-invariant multivariable feedback systems, *SIAM J. Control*, Vol. 10, pp. 203–209, 1972.

[103] Wise, K. and Broy, D.: Agile missile dynamics and control, *J. Guid. Control Dyn.*, 21, pp. 441–449, 1998.

[104] Wolodkin, W., Balas, G. and Garrard, W.: Application of parameter-dependent robust control synthesis to turbofan engines, *J. Guid. Control Dyn.*, 22, pp. 833–838, 1999.

[105] Wonham, W. M.: *Linear Multivariable Control*, Springer-Verlag, New York, 1985.

[106] Wonham, W. M.: On pole assignment in multi-input controllable linear systems, *IEEE Trans. Autom. Control*, 12, pp. 660–665, 1967.

[107] Yang, Ch. and Blanke, M: Rudder-roll damping controller design using μ synthesis, *Proc. 4th IFAC CAMS'98*, Fukuagoka, Japan.

[108] Yang, Ch.: Robust Rudder Roll Damping Control, *Ph. D. Thesis*, Aalborg University, Department of Control Engeneering, 1998.

[109] Young, N.: *An Introduction to Hilbert Space*, Cambridge University Press, 1988.

[110] Young, P. M, Newlin and M., Doyle, J. C.: μ analysis with real parametric uncertainty, *IEEE Proc. 30th Conf. on Decision and Control*, pp. 1251-1235, England, 1991.

[111] Zames, G.: On the input-output stability of time-varying nonlinear feedback systems: Part I: Conditions derived using concepts of loop gain, conicity, and positivity, *IEEE Trans. Autom. Control*, 11, pp. 228–238, 1966.

[112] Zhou, K., Doyle, J. and Glover, K.: *Robust and Optimal Control*, Prentice Hall, Englewood Cliffs, New Yersey, 1995.

[113] *Control Systems Toolbox*, The Mathworks, Inc., 1990.

Notation and Symbols

Symbol	Meaning	Page
$\mathbb{R}$	field of real numbers	–
$\mathbb{C}$	field of complex numbers	–
$\mathbb{C}_-, \overline{\mathbb{C}}_-$	open and closed left half-plane	–
$\mathbb{C}_+, \overline{\mathbb{C}}_+$	open and closed right half-plane	–
$i\mathbb{R}$, I	imaginary axis	–
$\mathrm{Re}\, z$	real part of the complex number z	–
$\mathrm{Im}\, z$	imaginary part of the complex number z	–
$\lvert z \rvert$	absolute value of the complex number z	–
$\arg z$	argument of the complex number z	–
$\mathbf{x}_\infty$	limit of $\mathbf{x}(t)$ for $t \to \infty$	–
$\mathbb{R}^n$	n-dimensional real vector space	–
$\mathbb{C}^n$	n-dimensional complex vector space	–
$\mathbb{R}^{m \times n}$	space of real $m \times n$ -matrices	–
$\mathbb{C}^{m \times n}$	space of complex $m \times n$ -matrices	–
$\mathbf{A}^T$	transpose of the matrix $\mathbf{A}$	–
$\mathbf{A}^*$	complex conjugate transpose of the matrix $\mathbf{A}$	–
A^*	adjoint of an operator A	–
$\mathbf{A}^{-1}$	inverse of the matrix $\mathbf{A}$	–
$\mathbf{I}$	identity matrix	–
$\mathbf{A} \geq \mathbf{0}$	the matrix $\mathbf{A}$ is positive semidefinite	491
$\mathbf{A} > \mathbf{0}$	the matrix $\mathbf{A}$ is positive definite	491
$\mathrm{diag}(a_1, \ldots, a_n)$	an $n \times n$ diagonal matrix with a_i as i th diagonal element	–
$\det(\mathbf{A})$	determinant of $\mathbf{A}$	–
$\mathrm{trace}(\mathbf{A})$	trace of $\mathbf{A}$	–
$\rho(\mathbf{A})$	spectral radius of $\mathbf{A}$	490
$\overline{\sigma}(\mathbf{A})$	largest singular value of $\mathbf{A}$	174
$\underline{\sigma}(\mathbf{A})$	smallest singular value of $\mathbf{A}$	174

$\kappa(\mathbf{A})$	condition number of $\mathbf{A}$	174
$\Vert \mathbf{A} \Vert$	spectral norm of $\mathbf{A}$: $\Vert \mathbf{A} \Vert = \overline{\sigma}(\mathbf{A})$	172
$\mathrm{Im}(\mathbf{A})$	image space of $\mathbf{A}$	488
$\mathrm{Ker}\ \mathbf{A}$	kernel of $\mathbf{A}$	488
$\mathrm{rank}\,(\mathbf{A})$	rank of $\mathbf{A}$	489
$Ric\,(\mathbf{H})$	the stabilizing solution of an algebraic Riccati equation	209
$u_e(t)$	unit step function	24
$\delta(t)$	unit impulse	–
δ_{ij}	Kronecker symbol, $\delta_{ii} = 1$ and $\delta_{ij} = 0$ if $i \neq j$	–
$\langle \cdot, \cdot \rangle$	inner product	498
$\mathcal{V}^{\perp}$	orthogonal complement of the subspace $\mathcal{V}$	499
$\mathbf{C}_{AB}$	controllability matrix	103
$\mathbf{O}_{CA}$	observability matrix	109
$\mathcal{L}_2(-\infty, \infty)$	Lebesque space of square integrable functions on $(-\infty, \infty)$	–
$\mathcal{L}_2[0, \infty)$	subspace of square integrable functions on $[0, \infty)$	499
$\mathcal{L}_2(-\infty, 0]$	subspace of square integrable functions on $(-\infty, 0]$	–
$\mathcal{L}_2(i\mathbb{R})$	Lebesque space of square integrable functions on $i\mathbb{R}$	175
$\mathcal{L}_\infty(i\mathbb{R})$	Lebesque space of essentially bounded functions on $i\mathbb{R}$	176
$\mathcal{H}_2$	Hardy space on $\overline{\mathbb{C}}_+$, subspace of $\mathcal{L}_2(i\mathbb{R})$	177
$\mathcal{H}_\infty$	Hardy space on $\overline{\mathbb{C}}_+$, subspace of $\mathcal{L}_\infty(i\mathbb{R})$	180
$\mathcal{RH}_2$	strictly proper and real rational functions in $\mathcal{H}_2$	179
$\mathcal{RH}_\infty$	proper and real rational functions in $\mathcal{H}_\infty$	180
$\mathcal{F}$	Fourier transformation for functions $\mathbf{f} \in \mathcal{L}_2(-\infty, \infty)$	176
$\mathcal{L}$	Laplace transformation for functions $\mathbf{f} \in \mathcal{L}_2[0, \infty)$	177
$\mathcal{L}_-$	Laplace transformation for functions $\mathbf{f} \in \mathcal{L}_2(-\infty, 0]$	219
$\hat{\mathbf{f}}(t)$	Laplace transform of the (vector valued) function $\mathbf{f}(t)$	–
$\mathbf{G}(s)$	transfer matrix	90
$\mathbf{G}^{\sim}(s)$	shorthand for $\mathbf{G}^T(-s)$	–
$\left(\begin{array}{c\|c} \mathbf{A} & \mathbf{B} \\ \hline \mathbf{C} & \mathbf{D} \end{array}\right)$	shorthand for $\mathbf{C}(s\mathbf{I} - \mathbf{A})^{-1}\mathbf{B} + \mathbf{D}$	90
$\mathcal{F}_l(\mathbf{M}, \mathbf{Q})$	lower linear fractional transformation	220
$\mathcal{F}_u(\mathbf{M}, \mathbf{Q})$	upper linear fractional transformation	381
$\mu_{\mathbf{\Delta}}(\mathbf{M})$	structured singular value with respect to $\mathbf{\Delta}$	398
RHP	right half-plane, $\mathrm{Re}\, s > 0$	–
LHP	left half-plane, $\mathrm{Re}\, s < 0$	–

Index